Electronics Engineering

Electronics Engineering

Sonveer Singh

Alpha Science International Ltd.
Oxford, U.K.

Electronics Engineering
714 pgs. | 616 figs. | 33 tbls.

Sonveer Singh
Department of Electronics Engineering
CMS Government Girls Polytechnic
Meerut

Copyright © 2014

ALPHA SCIENCE INTERNATIONAL LTD.
7200 The Quorum, Oxford Business Park North
Garsington Road, Oxford OX4 2JZ, U.K.

www.alphasci.com

ISBN 978-1-84265-811-6

Printed in India

Dedicated to
My Mother Late Smt. Bhagwan devi
and
My Father Shri B. S. Verma

Preface

It is my great pleasure to present the text book of **Electronics Engineering**. This book introduces students to the theory and practice of basics of electronics.

The study of electronics engineering is essential for students pursuing degree in Mechanical, Electrical, Electrical and Electronics, Electronics and Communication, and Computer Science and Engineering. Now a day's electronics has become the most important and inseparable part of engineering and our social life. More and more people are taking interest in this subject. It is found in broad range of applications within these disciplines. This book is basically written for the average students so the concepts and presentation of the book and the problems is simple.

The book has been compiled to ensure:

- Simple text for better understanding of concepts and basic principles.
- Systematic and stepwise procedure to solve the problems.
- More than 250 solved problems; most of the problems have appeared in UPTU and various university exams.
- More than 200 objective type questions with answers that helps the reader for quick revision.

The author would like to acknowledge the contribution of his colleagues and friends who have contributed many valuable suggestions and their support. I particularly want to thank Prof. D.K. Chaturvedi and Prof. A.K. Saxena, Department of Electrical Engineering, Dayal Bagh Educational Institute (Deemed University) Agra for their suggestions and encouragement which was vital to the completion of this volume.

Sonveer Singh

Contents

Introduction to Electronics 1

1.1 INTRODUCTION

As we know that it is an era of electronics. Everyday, a lot of development is taking place in the field of electronics. Practically we can feel the use of electronics in our daily life because we have a lot of electronic equipments such as TV, Radio, VCR, DVD Player, Mobile Phone, Tape-recorder, Telephone, Fax machine, Computer etc., which have revolutionised our lives.

The word electronics is made from:

$$\underline{Electron} + dynam\underline{ics} = Electronics$$

We can say that electronics is a branch of Science and Engineering which deals with dynamics of electrons in gas, liquids and solids under different operating conditions. In this field, we study the devices based on the flow of electrons and their applications. After the use of semiconductor in electronics, we can say that electronic devices are those in which the current flows due to motion of electrons in semiconductors.

Initially, branch of electronics was considered as an integral part of Electrical Engineering. But due to the several developments in this field, Electronics has got its separate designation in the field of Science and Technology.

1.2 BRIEF HISTORY OF ELECTRONICS

1.2.1 Evolution of Electronics

1883: Thomas Alva Edison discovered that electrons flow from one metal conductor to another through a vacuum. This is known as Edison effect.

1890: The experiment on generation of Electromagnetic waves (E.M. wave) done by Hertz.

1894: The propagation of radio waves discovered by Sir J.C. Bose and Macron given theory of radio-wave propagation.

1897: J.J. Thomson performed experiment on the existence of electron and verified the postulate of existence of electron given by H.A. Lorentz. at the same time the first electron tube was invented by Braun.

1904: John Fleming applied Edison effect inventing a two element electron tube called a diode. It also called as Valve.

1906: Lee De Forest introduced one additional element as a grid in Fleming Valve. This three-element tube is called triode tube. We named it as audion. This was the first amplifier.

1912: First application of radio wave.

1918: Edwin Armstrong invented the "Super-heterodyne receiver" that could select among radio signals or stations and would receive distant signals.

1920: Radio broadcasting grew astronomically.

1930: Black and White T.V. introduced. Amplitude modulation had been used between 1920 to 1935.

1935: Armstrong also invented wide-band frequency modulation.

1950: Color T.V. came into existence. Bell Lab's engineers introduced the cathode ray picture tube and color television.

1.2.2 Evolution of Transistors

1947: John Bardeen and Walter Brattain invented point-contact transistor. Solid state physics group leader William Shockley discovered junction transistor. Shockley, Brattain and Bardeen were jointly awarded the 1956 Nobel Prize in Physics for invention of the transistor.

1948: The point-contact transistor was independently invented by German Physicist Herbert Matare and Heinrich Welker.

1951: The first Germanium transistor produced commercially.

1954: The first Silicon transistor was produced by Texas instruments.

1958: Kilby gave the idea about monolithic.

1960: The first MOS transistor actually built by Kahny and Atalla at Bell Labs.

1.2.3 Evolution of Integrated Circuits

MOSFETS were not originally, better than the junction transistor, but they are much easier to make on integrated circuits or microprocessor, so they became the preferred type of transistor.

1960: Small-Scale Integration (SSI) came into existence. In this, the number of components are less than 100.

1966: Medium-Scale Integration (MSI) came into existence. In this, the number of components are greater than 100 and less than 1000.

1969: Large Scale Integration (LSI) came into existence. In this, the number of components are greater than 1000 and less than 10,000.

1975: Very Large Scale Integration (VLSI) came into existence. In this, the number of components are greater than 10,000.

1980: The chip size was 3×5 cm^2 area and 0.1 mm thick. 10^6 components per chip. 2000 component/mm^2 are used.

1998: Ultra Large Scale Integration (ULSI). 10^3 component per chip. Chip size was 6×6 mm^2 area. More than 10,000 components/mm^2 are used.

1.2.4 Evolution of Computers

1642: Frances Blaise Pascal Invented a machine, called Pascalline, that could add, subtract, and carry between digits.

1674: Germany's Gottfried Wilhelm Lei burtz created a machine that could add, subtract, multiply and divide automatically.

1725: An early form of punch cards were used in textile loom.

1837: Charles Babbage purposed the analytical engine.

1868: Christopher Sholes invented the typewriter in The United States utilizing the QWERTY key board.

1933: Electro-Mechanical Calculator introduced by IBM, USA, size 17 m long and 3 m wide.

1946: Electronic calculator was developed by EC Kest with 18000 vacuum tubes.

1948: General purpose small size Electronic Calculator.

1954: First Generation Computer introduced by IBM and named as IBM 650.

1959: Second generation computer introduced by IBM and named as IBM 7090/7094 series, transistor version.

1965: Third generation computer introduced by IBM and named as IBM 350, IC version.

1970: Computer with semiconductor was introduced by IBM.

1971: First 8" floppy diskelte drive was introduced.

1976: Intel 8085 was introduced in March and 8088 in June.

1978: Entire computer on a single chip. The 5-25" floppy disk became an industry standard.

1979: The Intel 8088 was released on June 1, 1979.

1980: Microcomputer, a general purpose digital processing and control system.

1982: The Intel 80286 was introduced on Feb 1.

1988: The Intel 80386 SX was introduced.

1989: The Intel 486 DX processor with more than 1 million transistor and multitasking capabilities.

1990: Intel releases the 80386 SL processor that uses low power and is found in many portable computer.

1994: Intel released the second generation of Intel Pentium processor on March 7.

1997: Intel Pentium-II was introduced on May 7.

1999: The Intel-Pentium III 500 MHz was released on Feb 26.

2001: Intel introduced Pentium-IV 1.4 GHz processor.

2003: Intel Pentium was in introduced in March.

2006: Intel released the core 2 Duo processor E 6320 April 22 (1-86 G42)

2007: Intel released the core 2 Duo processor E 4400 2.0 GHz.

From the above discussion, we can say that in the last 25 years after the release of microcomputers, a revolution has taken place in the field of computer hardware, size of memory, speed of CPU and clock frequency input/output devices etc.

1.3 APPLICATION OF ELECTRONICS IN DIFFERENT FIELDS

A brief discussion on the application of Electronics in different fields is given below:

(i) Communication: Electronics has a dominant role in communication. Earlier the information was sent from one place to another by telephone or telegraphy. But these days, the information can be received by radio & T.V. at home. There is a lot of development is going on in the field of Information and Technology. We connect with the world while sitting infront of the computer via Internet. Electronics has application in wireless communication such as cellular mobile telephone. Now-a-days the communication technology is improving with the help of optical fibre. Video conferencing, paging, computerized railway reservation, facsimile machine etc., makes our life comfortable.

(ii) Entertainment: Electronics are also used in the field of entertainment. A lot of electronic equipments are available, which help to reduce the tension in our busy life like radio, T.V., Video games, V.C.R., tape recorder, D.V.D./C.D. player etc.

With the help of Internet, we can get any knowledge according to our interest. We can also chat on the Internet. Endless knowledge is available on the Internet about games, Science & Technology, films, share market, employment etc. We can also play games on the computer like chess, car racing, fighting etc.

(iii) Industry: Electronics play an important role in the industry. All machine tools, inspection of machines, robots controlled computer. Different department of industry is upgraded with the computer like Computer Aided Design (CAD), production planning, inventory control, quality control, marketing, sales etc. Different softwares are also used for industry like material requirement planning (MRP), Just In Time Techniques (JIT) etc.

Different work like welding, spray painting, material handling and work in typical environment, chemical industry, blasting material related industry in deep sea, in nuclear reactors operated by the robot.

(iv) Defence: Electronics plays an important role in defence e.g., RADAR With the help of Radio Defection And Ranging equipment, we can get information about enemie's aircraft. The use of electronics in the field of defence was started in World War-II when RADAR was used for the first time. RADAR has a radio transmitter, by which it transmit the radio waves. These waves are reflected from the object and received by the radio receiver. It has many technologies like moving track indication, track while scan, pulse doppler etc.

In our country, the research related to the defence is done by the Defence Research and Development Organization (DRDO).

(v) Medicine Science: Electronic Instrumentation has played a vital role in the field of Medical Science. These instruments are used to observe and treat different diseases.

Electrocardiogram (ECG) is used for testing the patient's heart. X-ray machine is used to take pictures of internal bones and other parts of the body. Computerized Aerial Tomography (CAT) or CT-scan to support X-ray. To study muscle actions in body, CRO is used. CAT or CT scan is used to set 30-image of different parts of body. Endoscopes machine is made with optical fibre and is used to monitor images of different parts of body on a T.V. monitor. This process is called endoscopy.

The treatment of gall bladder; kidney stone and surgery is done by Light Amplification by Stimulated Emission of Radiation (LASER).

(vi) Instrumentation: Electronic instruments are used to measure different physical quantities like temperature, pressures, flow, humidity etc. The example of electronic equipments used for instrumentation are as follows:

Cathode ray oscilloscope (CRO), signal generator, PH meter, electronic counter, A/D or D/A converter, transducer, Q-meter, wave analyzer etc.

(vii) Internet: After the invention of Internet, there has been a revolution in the field of communication. Internet plays a vital role in the field of communication. It has an endless series of knowledge where you can get any type of information within seconds. Basically, we can say that Internet is the mother of human beings and they can demand anything from her mother (Internet). She will fulfill there wishes within seconds. You can shop from any market in the world from your home through Internet. It also provides online education. It is a vast ocean of knowledge.

(viii) Miscellaneous: Weather forecast can be done with the help of satellite. Miscellaneous work can be done with the help of electronic equipments like discovery of space, discovery of new planet, remote sensing, flood control, loss estimation, to find the level of sea etc.

1.4 ELECTRONIC COMPONENTS

Electronic components are used to design an electronic circuit. The electronic components are classified in two categories according to their work.

(i) Passive components

(ii) Active components

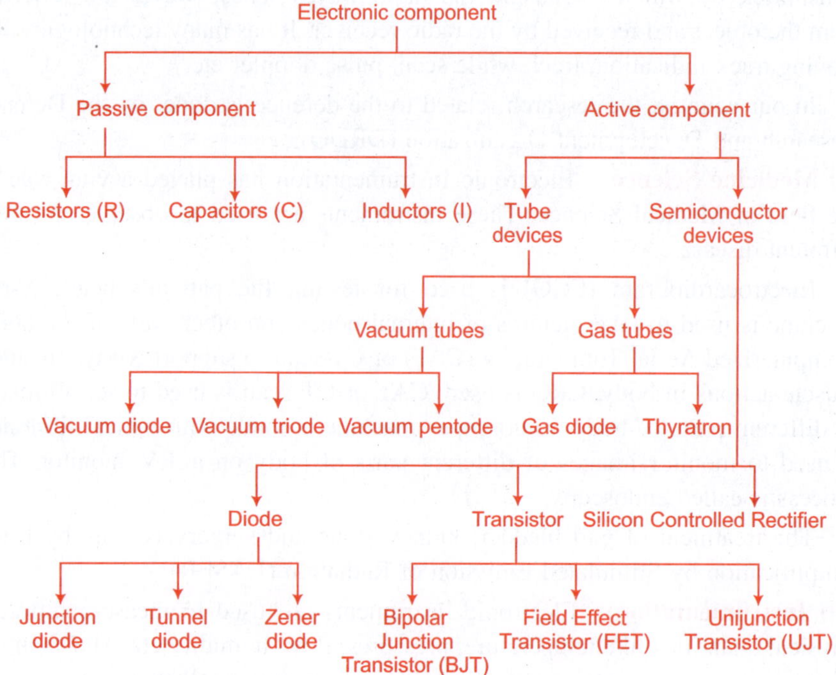

Fig. 1.1 *Classification of electronic components*

1.5 PASSIVE COMPONENTS

The electronic components which are not capable to amplify the electrical signal are called as passive components. The components conduct current in both directions and hence are known as bilateral components or elements. These components are an important part of electronic circuit because without these components, active components are not able to amplify the electrical signal. Passive components are of three types as shown in Fig. 1.1.

1. Resistors

2. Capacitors

3. Inductors.

1.5.1 Resistors

An electronic component which has the properties of resistance is called resistor. Basically it is manufactured with a specified amount of resistance. Resistance can be defined as properties of a substance, which oppsoes a flow of current into the substance. The unit of resistance is ohm and it is represented by greek symbol Ω. The symbol of resistance is shown in Figs. 1.2(a) and (b)

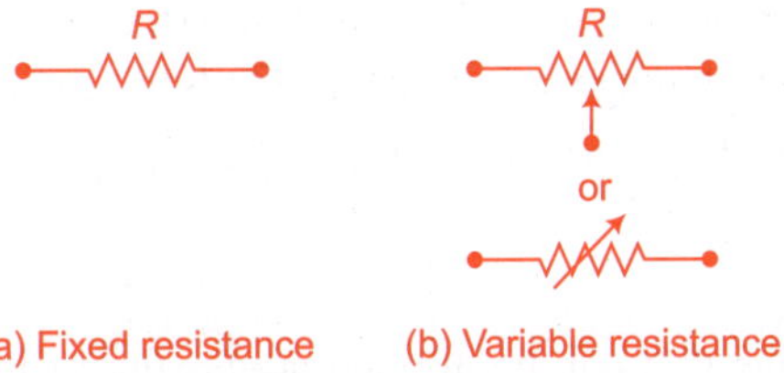

(a) Fixed resistance   (b) Variable resistance

Fig. 1.2 *Symbols of resistance*

Resistivity If l is a length of conductor and A is a cross-sectional area of conductor, then the resistance of a conductor is given by:

$$R = \rho\, \frac{l}{A} \tag{1.1}$$

where, ρ = resistivity of conductor of specific resistance of conductor in Ω-meter.

1.5.1.1 *Important specifications of resistors*

When we have to use any resistance in the electronic circuit or we have to purchase it from the market, we should know the specifications of it. In the same way when we purchase T.V., Refrigerator etc., from the market, then we know all its specifications before the purchase. Following are the different specifications of resistors.

  (i) Resistor value

 (ii) Tolerance

(iii) Power rating

(iv) Voltage rating

 (v) Temperature coefficient of resistor

(i) *Value of resistors*

The value of resistor means how much resistance is created by resistor. The unit of resistance is ohm (Ω). The value of resistance is written on the resistor or obtained from color coding method. The color coding of resistor will be explained in the next section.

(ii) *Tolerance*

Tolerance means, how much per cent deviation will occur in the value of resistance from it specified value. For example, the tolerance of 100 kΩ resistor is ±20% which means the value of resistance will exist between 80 kΩ to 120 kΩ.

Note that the tolerance of resistance should be less for better resistor.

(iii) *Power rating*

Power rating of resistor is the maximum power that can be dissipated by the resistor without any damage of resistor. Carbon composition resistors are available in $\frac{1}{8}$W, $\frac{1}{4}$W, $\frac{1}{2}$W, 1W and 2W of power rating.

(iv) *Voltage rating*

The maximum voltage that can be applied across the resistor (without damage) is called voltage rating. Carbon composition resistors are available 150V, 250V, 350V, 500V of voltage rating. If the voltage rating is not specified by the manufacturer, then it will be calculated by following formula:

$$\text{Voltage rating} = \sqrt{\text{Power rating} \times \text{Value of resistor}}$$

$$\text{Voltage rating} = \sqrt{W \times R}. \tag{1.2}$$

(v) *Temperature coefficient of resistor*

Temperature coefficient of resistor is the rate of change of nominal resistance as the function of temperature. It is expressed as %/°C. In other words, we can say that by the temperature coefficient we know the percent change in the value of resistance with change in temperature. If the temperature coefficient is positive it means the value of resistance will increase with increasing temperature. And if the temperature coefficient is negative it means the value of resistance will decrease with increasing temperature.

For example, the temperature coefficient of any resistor is 0.4%/°C, it means the value of resistance will increase 0.4% of its value after increasing the temperature by 1°C.

1.5.1.2   *Color coding of resistor*

The value of resistor is indicated on the resistor or the value of resistor is also mentioned by color coding. The small size resistor has its value by color coding. The value of small size resistor can be calculated by the color coding method. The strips of different color is available on the resistor. These strips are also called color band as shown in Fig. 1.3. The number of bands in color coding may be three, four, five or six. Generally, four band color coding resistors are mostly used.

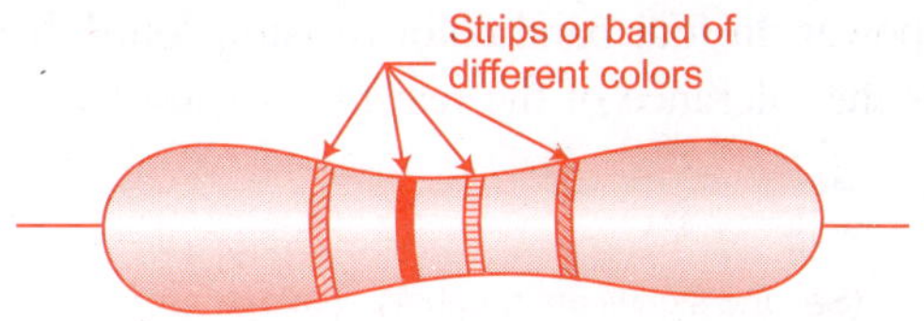

Fig. 1.3   *Resistor with color code*

(i) Four band color coding

Generally, there are four bands or strips made on the resistor. Each color has its numerical value as shown in Table 1.1

Table 1.1

Colour	Digit Numerical Value	Multiplier	Tolerance
Black	0	10^0	–
Brown	1	10^1	–
Red	2	10^2	–
Orange	3	10^3	–
Yellow	4	10^4	–
Green	5	10^5	–
Blue	6	10^6	–
Violet	7	10^7	–
Grey	8	10^8	–
White	9	10^9	–
Gold	–	$10^{-1} = 0.1$	± 5%
Silver	–	$10^{-2} = 0.01$	± 10%
No color	–	–	± 20%

The Table 1.1 can be remembered by the following sentence

B. B. R O Y of Great Britain has Very Good Wife.

0   1 2 3 4   5   6   7   8   9

There are several sentences for color of resistors.

- B.B. Roy of great Britain has very good wife.
- Black Bears run over yellow grass, but vultures glide over water.
- Black Bugs race over yellow grass, beside violent grey water.

In the above sentence, all capital letters indicate the numerical value of color.

In four band color coding, first and second band, represents the significant number of resistance respectively. Third band represents the multiplier in the form of 10 to the power. In four band color resistor, fourth band may be gold or silver. It represents the tolerance of the resistor as shown in Fig. 1.4.

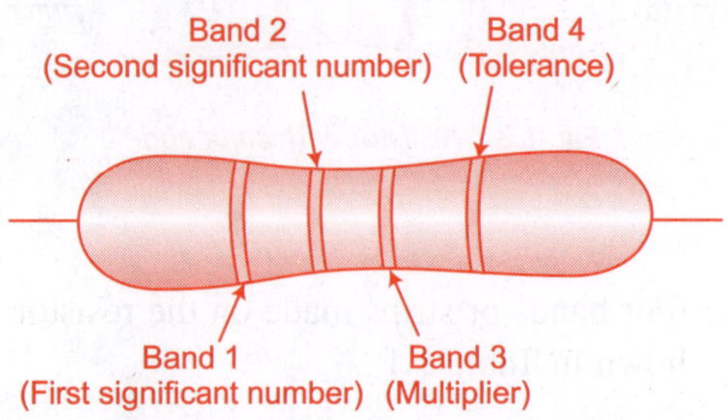

Fig. 1.4   *(a) Four band color coding*

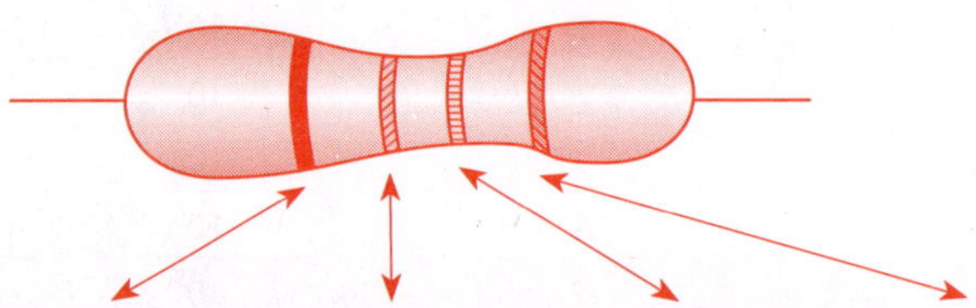

Color	1st band value	2nd band value	Multiplier	Tolerance
Black	0	0	10^0	–
Brown	1	1	10^1	± 1%
Red	2	2	10^2	± 2%
Orange	3	3	10^3	± 3%
Yellow	4	4	10^4	± 4%
Green	5	5	10^5	± 0.5%
Blue	6	6	10^6	± 0.25%
Violet	7	7	10^7	± 0.10%
Grey	8	8	10^8	± 0.05%
White	9	9	10^9	–
Gold	–	–	0.1	± 5%
Silver	–	–	0.01	± 10%
No color	–	–	–	± 20%

Fig. 1.4   *(b) Four band resistor*

Example 1.5.1   Find the value of resistance given in Fig. Ex 1.5.1 by color coding method.

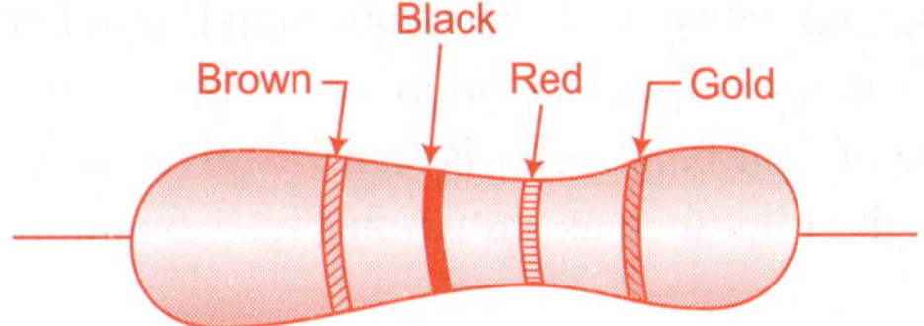

Fig. Ex. 1.5.1

Solution   We keep the tolerance band at the right side and start from the left.

Band	Color		Numerical Value
First band	–	Brown	1
Second band	–	Black	0
Third band	–	Red	$2 \Rightarrow 10^2$
Fourth band	–	Gold	± 5%

The value of resistance

$= 1^{st}$ significant bit second significant bit × Multiplier ± Tolerance

$= 10 \times 10^2 \ \Omega \pm 5\%$

$= 1000 \ \Omega \pm 5\%$

$R = 1 \ k\Omega \pm 5\%$ **(Ans.)**

(ii) Three band color coding

Some resistors have three band only. They do not have fourth band for tolerance. Such resistors have 20% tolerance. Three band have same calculation as four band resistor. For remaining three band as shown in Fig. 1.5.

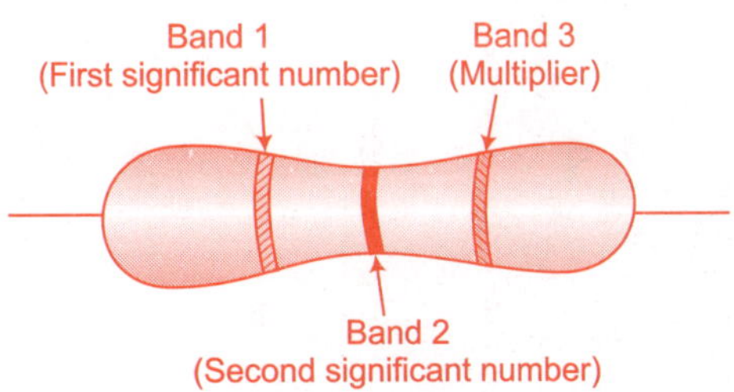

Fig. 1.5   *Three band color coding*

(iii) Five band color coding

Five band color coding is used for better tolerance i.e., the resistors which have tolerance less than equal to ±2% has five bands. In such type of resistor, starting four band indicates the value of resistance and fifth band indicates the tolerance.

First three band indicates the three significant number of resistance and fourth band indicates the multiplier i.e. number of zero after significant numbers. The significant value of four band color is same as given in Table 1.1 but the fifth band color indicates the tolerance as shown in Fig. 1.6 and the value of tolerance is given in Table 1.2.

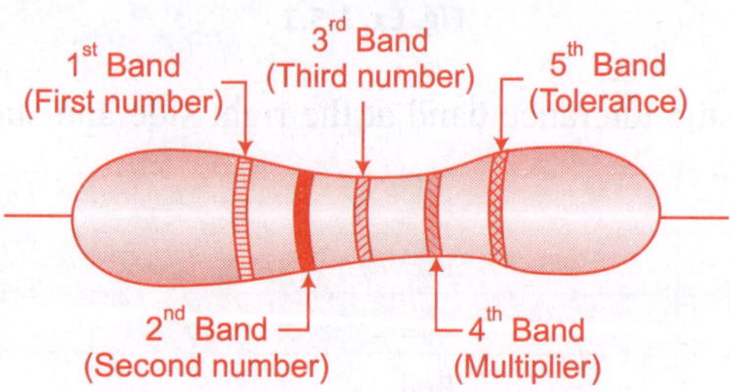

Fig. 1.6 *(a) Five band color coding*

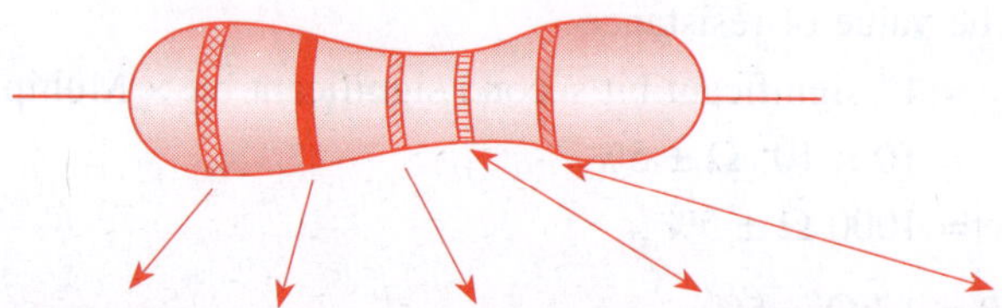

Color	1st band value	2nd band value	3rd band value	Multipler	Tolerance
Black	0	0	0	10^0	–
Brown	1	1	1	10^1	± 1%
Red	2	2	2	10^2	± 2%
Orange	3	3	3	10^3	± 3%
Yellow	4	4	4	10^4	± 4%
Green	5	5	5	10^5	± 0.5%
Blue	6	6	6	10^6	± 0.25%
Violet	7	7	7	10^7	± 0.10%
Grey	8	8	8	10^8	± 0.05%
White	9	9	9	10^9	–
Gold	–	–	-	0.1	± 5%
Silver	–	–	-	0.01	± 10%
No color	–	–	-	–	± 20%

Fig. 1.6 *(b) Five band resistor*

Table 1.2   Tolerance of different colors

Color	Tolerance
Black	–
Brown	± 1%
Red	± 2%
Orange	± 3%
Yellow	± 4%
Green	± 0.5%
Blue	± 0.25%
Violet	± 0.1%
Grey	± 0.05%
White	–
Gold	± 5%
Silver	± 10%
No color	± 20%

Example 1.5.2   Find the value of resistance given in Fig. Ex. 1.5.2 by color coding method.

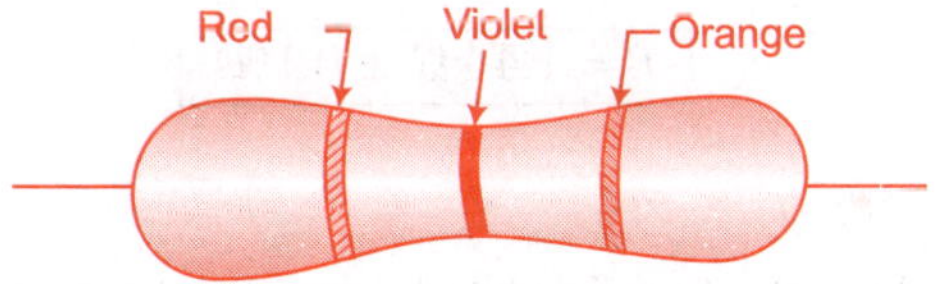

Fig. Ex. 1.5.2

Solution

Band	Color	Numerical Value
First	Red	2
Second	Violet	7
Orange	Orange	$3 \Rightarrow 10^3$

The value *of* resistance = $27 \times 10^3 \ \Omega \pm 20\%$ (no color)

$$= 27000 \ \Omega \pm 20\%$$

$$\boxed{R = 27 \ k\Omega \pm 20\%.}$$ (**Ans.**)

Example 1.5.3   Find the value of resistance given in Fig. Ex. 1.5.3 by color coding method.

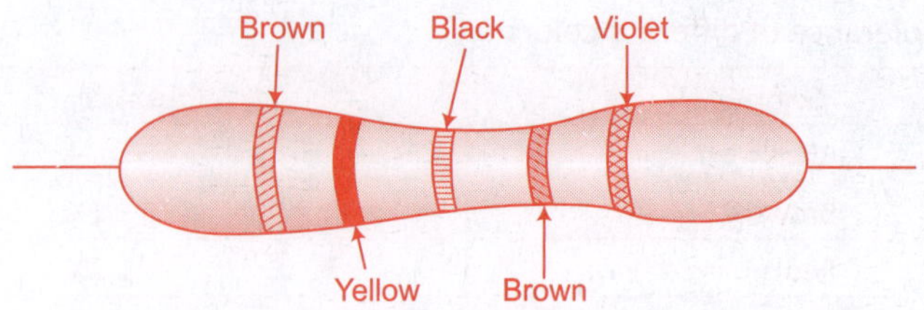

Fig. Ex. 1.5.3

Solution

Band	Color	Numerical Value
First	Brown	1
Second	Yellow	4
Third	Black	0
Fourth	Brown	$1 \Rightarrow 10^1$
Fifth	Violet	0.1%

The value g resistance $= 140 \times 10^1 \ \Omega \pm 0.1\%$

$$= 1400 \ \Omega \pm 0.1\%$$

$$\boxed{R = 1.4 \ k\Omega \pm 0.1\%.}$$ **(Ans.)**

(iv) Six band color coding

In six band color coding, the first five bands, are same as five band color coding. Additional sixth band represents the temperature coefficient of resistance as shown in Fig. 1.7.

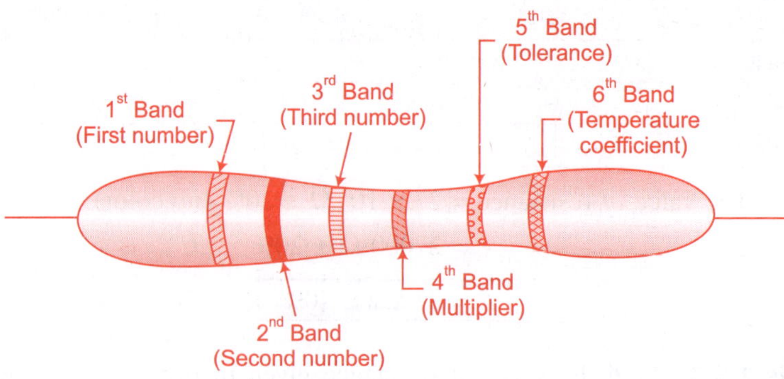

Fig. 1.7   *(a) Six band color coding*

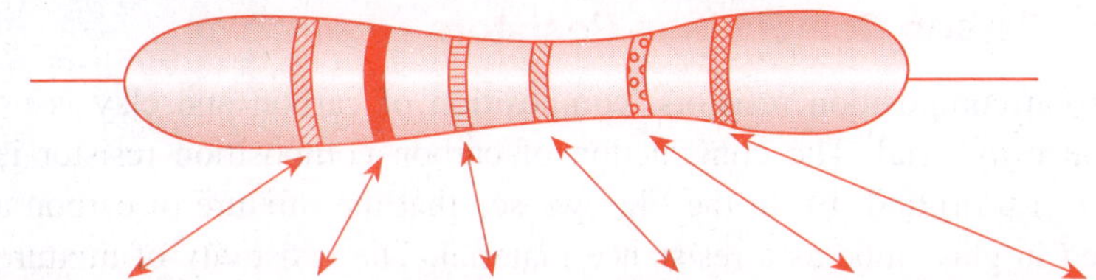

Color	Ist band Value	IInd band Value	IIIrd band Value	Multiplier	Tolerance	Temperature Coeff.
Black	0	0	0	10^0	–	–
Brown	1	1	1	10^1	± 1%	100 ppm/°C
Red	2	2	2	10^2	± 2%	50 ppm/°C
Orange	3	3	3	10^3	± 3%	15 ppm/°C
Yellow	4	4	4	10^4	± 4%	25 ppm/°C
Green	5	5	5	10^5	± 0.5%	
Blue	6	6	6	10^6	± 0.25%	
Violet	7	7	7	10^7	± 0.10%	
Grey	8	8	8	10^8	± 0.05%	
White	9	9	9	10^9	–	
Gold	–	–	–	0.1	± 5%	
Silver	–	–	–	0.01	± 10%	
No band	–	–	–	–	± 20%	

Fig. 1.7   *(b) Six band resistor*

1.6   CLASSIFICATION OF RESISTORS

Resistors are classified as shown in Fig. 1.8

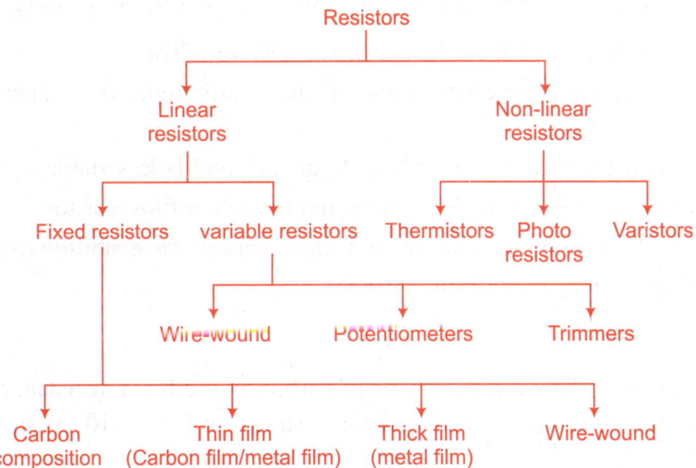

Fig. 1.8   *Classification of resistors*

1.6.1 Carbon Composition Resistors

In carbon composition resistors, composition of carbon and clay is used as a resistance material. The construction of carbon composition resistor is shown in Figs. 1.9 (a) and (b). In the Fig., we see that the mixture of carbon and clay is filled in glass tube as a resistance material. The resistivity of mixture can be increased or decreased by controlling the percentage of carbon. In this way the required value of resistor can be obtained. Then, this glass tube is sealed and copper wires are connected.

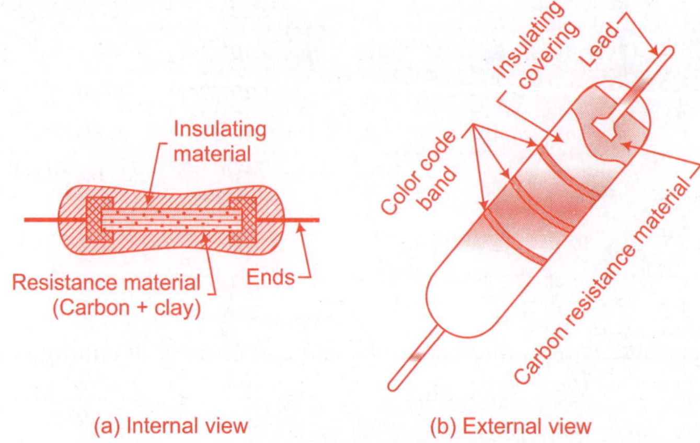

Fig. 1.9 *Construction of carbon composition resistor*

Characteristics of Carbon Composition Resistors

The characteristics of Carbon Composition Resistors are as follows:

  (i) Small in size.

 (ii) These are available in $\frac{1}{8}$W, $\frac{1}{4}$W, $\frac{1}{2}$W, 1W, 2W of power ratings.

(iii) The tolerance of these lies between ±5% to ±20%.

 (iv) The temperature coefficients of these are high. It is approximately ±0.12%/°C.

  (v) Its resistance decreases at high frequency and is less stable.

 (vi) The noise level is high as compared to carbon film resistor.

(vii) Small inherent inductance and capacitances have ability to withstand higher voltage than film resistors.

1.6.2 Carbon Film Resistors

In carbon film resistors, a thin layer of carbon is used as a resistance material. The construction of carbon film resistor is shown in Figs. 1.10 (a) and (b).

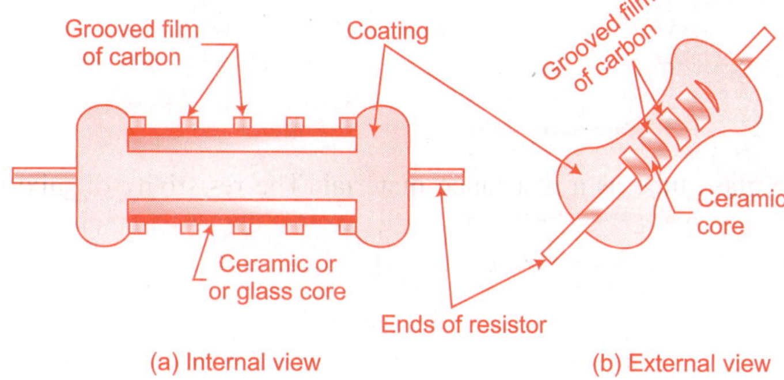

Fig. 1.10 *Carbon film resistors*

In this resistor, a carbon film is deposited on the tube of glass or ceramic (Insulating material). The carbon film can be deposited by pyrolysis of some hydrocarbon like benzene as an insulating core. This process is done at the temperature of 900°C to 1150°C.

Now, derived values are obtained by either training the layer thickness or by cutting helical grooves of suitable pitch along its length. During this process the value of resistance is monitored continuously and process of cutting is continued until the derived value is obtained.

Characteristics of carbon film resistors

The characteristics of carbon composition resistors are as follows :

   (i) The accuracy of resistors can be obtained up to ±1%, so these resistors are called as precision resistor.

  (ii) Carbon film resistors are more stable than the carbon composition resistor.

 (iii) The tolerance of carbon film resistors is better than carbon composition. (Lies between ±1% to ±5%).

  (iv) These are small in size and light in weight.

   (v) These are available in $\frac{1}{8}$W, $\frac{1}{4}$W and $\frac{1}{2}$W of power rating.

  (vi) The temperature coefficient is less than the carbon composition and wire wound (approximate –0.02%/°C to –0.05%/°C).

 (vii) The range of resistance is 1 Ω to 20 kΩ.

(viii) Less noisy as compared to carbon composition.

1.6.3 Metal Film Resistor

In a metal film resistor, a thick film of metal is used as a resistance material. The construction of metal film resistor is shown in Figs. 1.11 (a) and (b).

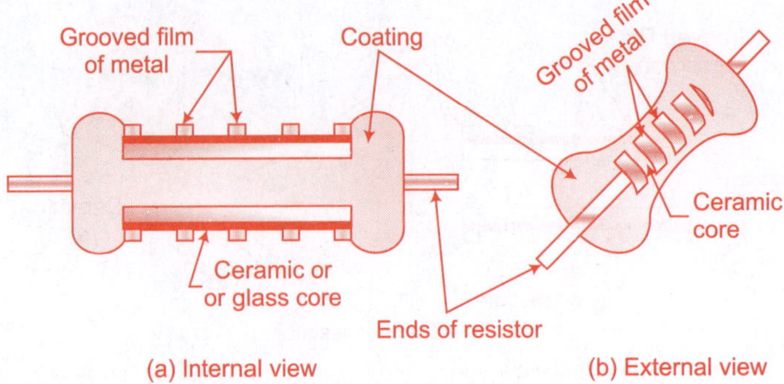

Fig. 1.11 *Metal film resistor*

Metal film resistor is similar in construction as carbon film resistor. Only difference is that we deposit a thin layer of metal or a glass tube in place of carbon layer i.e., grooved film of metal is put in place of grooved films of carbon. The value of resistance is dependent upon the thickness of metal layer. If the thickness of metal layer is less, then the value of resistance is high.

Characteristics of metal film resistors

The characteristics of metal film resistors are as follows:

(i) These are small in size and light in weight.

(ii) These are available in $\frac{1}{8}$ W, $\frac{1}{4}$ W and $\frac{1}{2}$ W of power rating.

(iii) The tolerance of metal film resistors lies between ±1% and ±5%.

(iv) The specificationery carbon film and metal film resistors are approximately equal.

1.6.4 Wire Wound Resistors

A wire is wound on a hollow porcelain core (insulated core) as a resistance material. The wire is made with conducting material like michrome and such resistors are called wire wound resistors.

The construction of wire wound resistor is shown in Fig. 1.12. In this Fig., we can see that a resistance wire, such as michrome, constantan, manganin, is wound into round, hollow porcelain core. The ends of wire are connected to the metal pieces and copper wire leads are welded to these pieces. The wounded wire is coated with glass powder for safety of

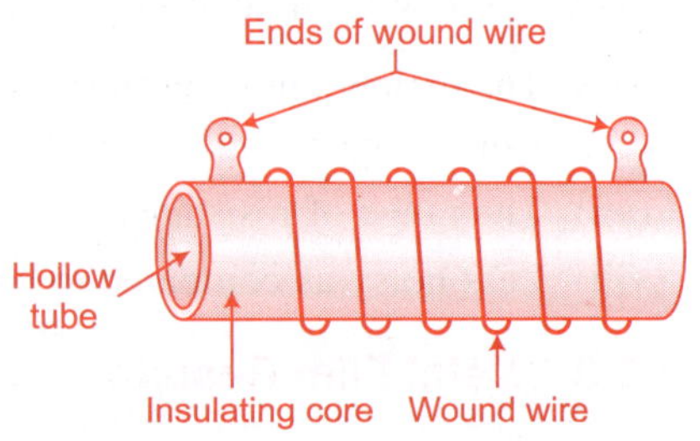

Fig. 1.12 *Wire wound resistor*

wire. This is a special type of coating and is called as vitreous enamel. This coating provides electrical safety to the wire.

Characteristics of wire wound resistors

The characteristics of wire wound resistors are as follows:

   (i) The size of such type of resistors is large.

  (ii) These resistors have tolerance up to ±1%.

 (iii) The temperature coefficient of such resistors is smaller than the carbon composition resistors and greater than the carbon film resistors.

 (iv) These resistors cannot use large frequency (greater than 25 kHz) due to shunting effect. "There is a distributed capacitance existing between the turns of wire. At high frequency, the reactance of these capacitance is low. Due to this, these capacitances are short circuited at high frequency and this effect is called shunting."

  (v) These can manufactured from 1Ω to $100\ k\Omega$.

 (vi) The noise level is very low in such type of resistors.

1.6.5   Variable Resistors

The resistors whose value can be changed in specified range is called as variable resistors. The symbol of variable resistor is shown in Figs. 1.13 (a) and (b).

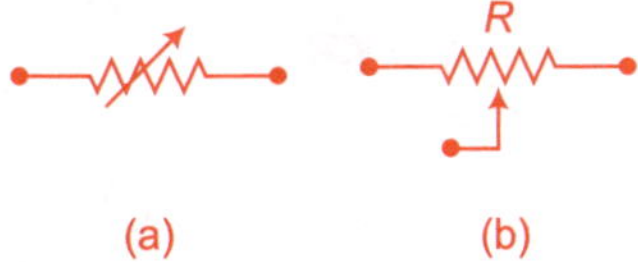

Fig. 1.13   *Symbol of variable resistor*

They have three terminals, out of which, two terminals are fixed and third terminal is connected to the movable contact. This movable contact can be moved along with the length of resistance material. Due to this, the resistance between any fixed terminal and movable terminal can vary. Carbon film, wire wound or, carbon-clay composition, can be used as a resistance material.

(a) Potentiometer or pot

When the movement of movable contact is used frequently and continuously, then the potentiometer is used. The construction of potentiometer is shown in Figs. 1.14 (a) and (b). The resistance wire is wound on a backlighter ceramic. There is a controlling shaft which is used to rotate movable contact on a resistance wire. There is a full resistance between two fixed terminals of potentiometer.

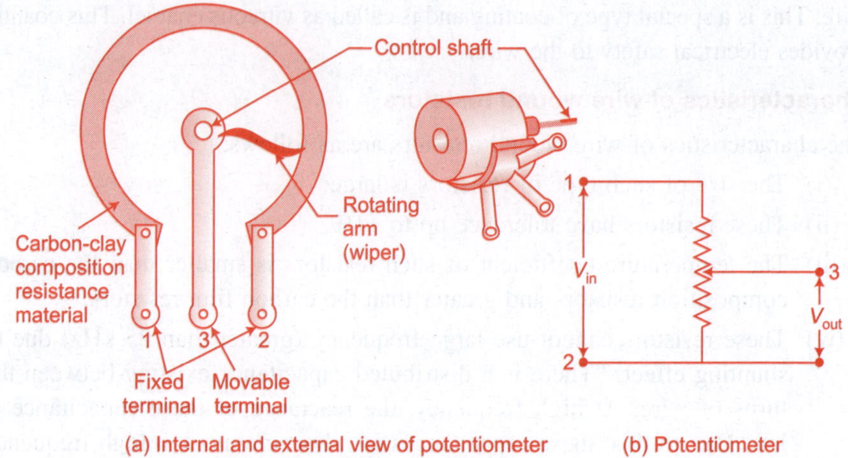

Fig. 1.14 *Potentiometer*

(b) Trimmer or preset

These resistors are used in circuit where resistance varies occasionally. Generally, these are set at the time of testing of the circuit. After that no change is required in a resistor, it is also called as preset. The construction of a resistor is shown in Fig. 1.15.

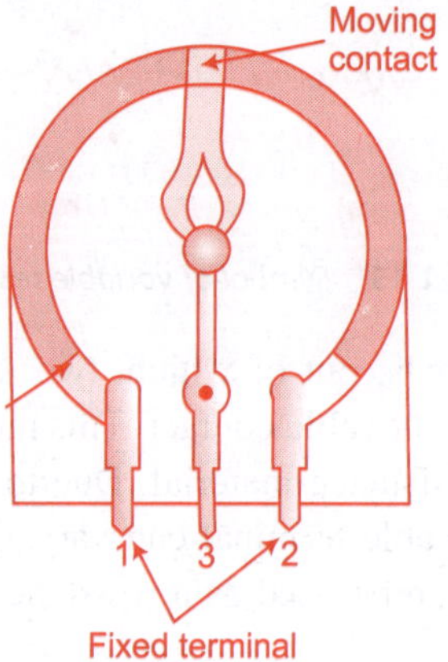

Fig. 1.15 *Trimmer or preset*

(c) Rheostat

Rheostat is used to control the current into the circuit. It is a wire wound resistor with higher wattage. It has two fixed terminals, wherein and one terminal is adjustable and can be wiped all over the length of the wound wire. The value of the resistance varies while changing the length of the wire. The construction of Rheostat is shown in Fig. 1.16.

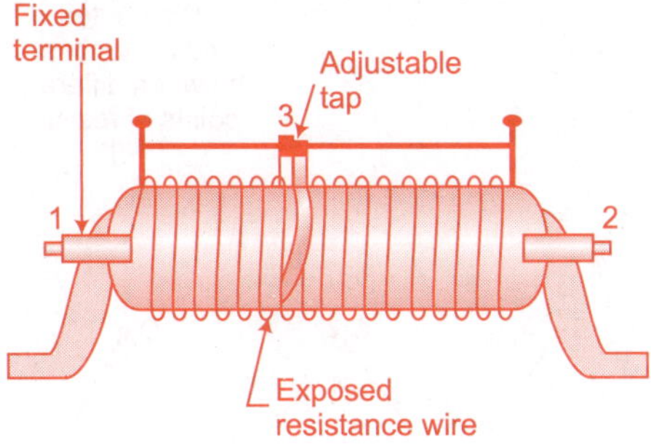

Fig. 1.16   *Rheostat*

Properties of variable resistors

The properties of variable resistors are based on the resistive material used. The power rating, tolerance, noise level, size and temperature coefficient are based on resistance material used in variable resistor.

Different terms used in variable resistors.

Following are the terms used in variable resistors:

   (a) **Total resistance**   It is the resistance between two fixed terminals of the variable resistor.

   (b) **Resolution**   It is the smallest incremental resistance change that is possible in a wire wound potentiometer.

   (c) **Linearity**   It is the deviation of the output versus rotation characteristics from a straight line.

1.7   FREQUENCY DEPENDENCE OF RESISTORS

The property of ideal resistor is that it should be purely resistive at any applied voltage. The applied voltage is DC or AC of any frequency. The value of resistor must be constant. But practically the value of resistors are dependent upon the frequency of applied input due to the capacitive and inductive effects.

(a) Capacitive effect

Let us consider a resistor in which, the current is flows from point L to point M. Hence, voltage drop will occur between point L and M but the complete voltage drop will not be done at a single point where as it occurs from one end of the resistor to another end of resistor at different points as shown is Fig. 1.17. Hence, some potential drop will occur between point A and B, some potential drop will occur between point B and C, similarly between point C and D, D and E.

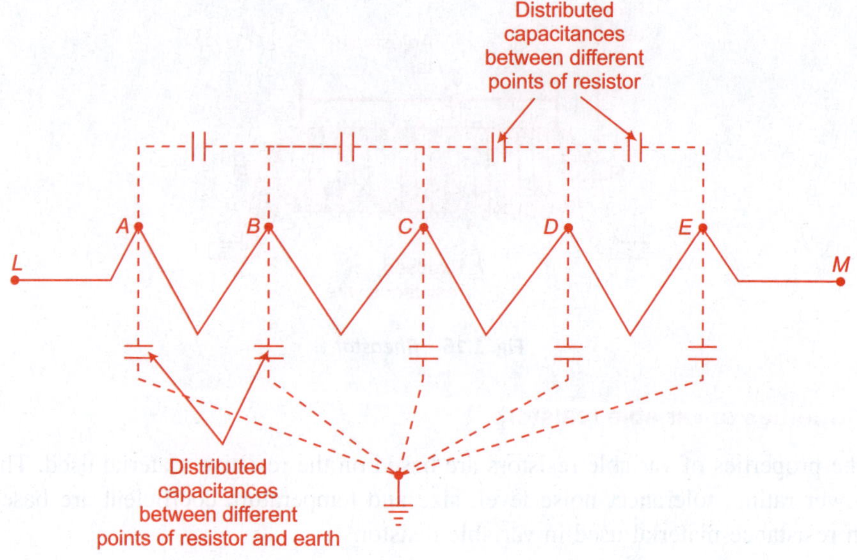

Fig. 1.17   *Distributed capacitances*

It is clear from the Fig. 1.17 that the total voltage drop in a resistor does not occurs at a single point. It equally occurs at different points, of resistor from one end to another end. Hence, there is some potential drop between two points of resistor. This potential drop will similarly occur between two plates of changed capacitor. Hence, the capacitance is created between different point of resistor. And the potential at different points of resistor is not equal to the earth potential. Hence, another type of capacitance is created between different points of resistor and earth. These capacitances are called distributed capacitance of resistor because these are distributed all over the length of the resistor equally. Due to these capacitances, the resistor is treated as a charged capacitor and this effect is called capacitive effect.

As we know that the capacitive reactance of capacitor is given by the following formula:

$$X_c = \frac{1}{2\,\pi f c}. \tag{1.3}$$

From Eqn. (1.3), it is clear that the reactance is high for low frequencies and capacitances will be open circuited approximately and negligible current flow in capacitor. Hence, it will not effect on the value of resistor. But the reactance is low for high frequencies and capacitances will be short circuited, capacitive current will increase and the distribution of current in resistor will be affected.

(b) Inductive effect

In wire wound resistor, the wire is wound in the form of coil on the insulated core. Due to the resistance in the form of coil, an inductive effect will occur. In low value resistors more inductive effect will occur. Hence, we see that inherited capacitance and inductance will occur in practical resistor and these cannot be removed due to these effect. The value of resistor depends on the frequency.

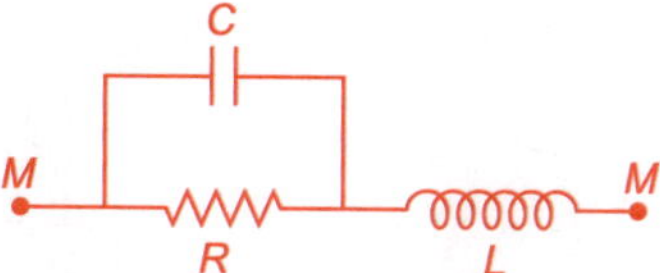

Fig. 1.18   *Dependence of resistor value on frequency*

Let us consider an equivalent circuit of resistor with equivalent capacitance and inductance as shown in Fig. 1.18.

In Fig. 1.18, R is the actual value of resistor while C is an equivalent capacitive effect and L is an equivalent inductive effect.

When R, L and C has the following reaction, then the frequency error will be less in the resistor.

$$R^2 = \frac{L}{C}. \tag{1.4}$$

From Eqn. (1.4), we can state that,

   (i) For low value resistance, L must be high and C must be low.

  (ii) For high value resistance, C must be low.

1.8   COMPARISON OF DIFFERENT RESISTOR

Comparison of different resistors is shown in Table 1.3.

1.9   NON-LINEAR RESISTORS

Some resistors are non-linear. The resistors, which do not follow ohms law are called as non-linear resistors. In such resistors, the current is not proportional to voltage. The value of these resistors, is not constant and depends upon the applied voltage.

1.9.1   Varistors

The resistors whose values depend upon the applied voltage are called Voltage Dependent Resistors or VDR. These resistors are also called as Varistors. The varistors are used to protect the circuit from high energy voltage transients by rapidly changing from high stand by resistance to low conducting resistance.

The symbol of voltage dependent resistor is shown in Fig. 1.19 (a).

Construction

According to the construction, VDR are of two types:

   (i) Silicon carbide VDR

  (ii) Metal oxide VDR and Silicon Carbide VDR are made when a silicon carbide powder is bind by ceramic, while the metal oxide is used in metal oxide VDR.

Fig. 1.19   *(a) Symbol of VDR*

Table 1.3 Comparison between different resistors

S. No.	Type of resistor	Resistive material	Range	Power rating	Tolerance	Cost	Noise	High frequency Response	Temp. coefficients
1.	Carbon composition	Carbon-clay composition	Up to 20 mΩ	Upto 2 W	±5% to ± 20%	Medium	Low	Better	Low
2.	Carbon and metal film	Thin layer of carbon or metal	Up to 20 mΩ	Greater than carbon composition	±1% to ± 5%	Low	High	Normal	Greater than carbon composition
3.	Wire wound	Wounded wire on insulated core	Up to l00 k Ω	Up to 200 W	Approximately ±1%	Costly (high)	Very less	Not good	Medium

The current flow in VDR is given by the following formula:-

$$i = kV^n \tag{1.5}$$

where, v = Voltage across VDR

k = Constant (Amp/volt)

n = 1 to 6 (for Silicon Carbide VDR)

n = 25 and greater (for metal oxide VDR).

The VI characteristic of varistor is shown is Fig. 1.19 (b). Since, in metal oxide VDR, n is large, so metal oxide VDR is core non-linear and there is large change in current by small change in voltage.

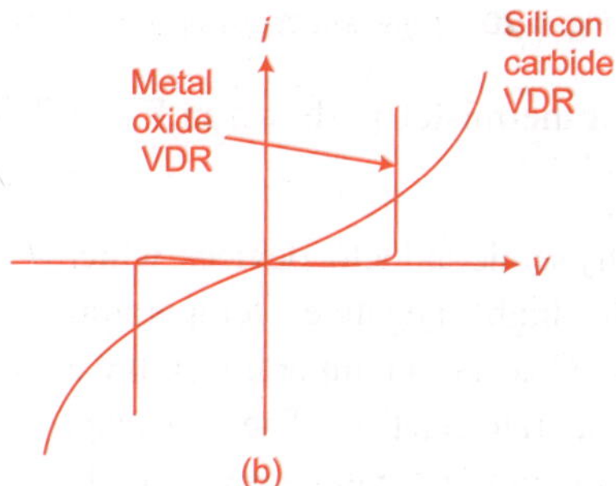

Fig. 1.19   *(b) VI characteristics of VDR*

Application of varistor:

(i) It is used to power supply protection.

(ii) It is also used for transistor protection.

(iii) It provides transient suppression in inductive and transformer switching circuit.

(iv) It is also used for diode and capacitor protection.

1.9.2 Thermistor

In thermal resistors resistance depends upon the temperature. The temperature coefficients of thermistor is very high and these are available with both positive and negative temperature coefficients. Thermistor is used to detect very small changes in temperature. This variation in temperature is reflected in appreciable variable of resistance of the device. Negative temperature coefficients are widely used.

The negative temperature coefficients means the resistance decreases with increasing temperature and positive temperature coefficients means the resistance decreases with increasing temperature.

Construction

Thermistor is made in the form of probes, beads, disc, washers and rods as shown in Fig. 1.20. These are made with mixture of metal oxide like, *Ni*, *Co*, *Fe*, *Cu* etc. The range of thermistors are 0.5 to 75 mΩ

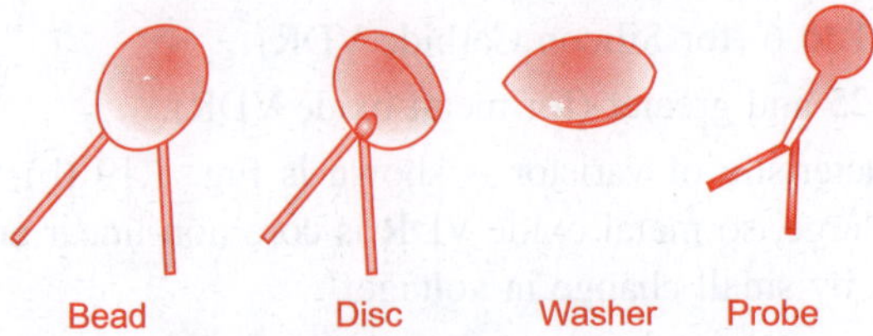

Fig. 1.20 *Different types of thermistor*

The schematic symbol of thermistor is shown in Fig. 1.21.

Characteristics

Thermistors are essentially semiconductor devices, which enables resistors with high negative temperature coefficient (usually 0.04/°C at room temperature). It has non-linear resistance temperate relation. The resistance R of a thermistor at a temperature T can be given by the equation,

Fig. 1.21 Symbol of thermistor

$$R = R_o\, \alpha\, e^{\beta/T} \tag{1.6}$$

where α, β = constant depends upon the type of material.

The characteristics of thermistor is shown in Fig. 1.22. It is drawn between temperature and resistance. In Fig. 1.22, we observe that for a negative temperature coefficient, the resistance decreases with increase in temperature. On the other hand, the resistance increases with increase in temperature for positive temperature coefficient. At the particular temperature, the resistance rise rapidly. The temperature at which the resistance increases rapidly is known as switching point. The typical values of switching point temperature ranges from 30°C to 165°C.

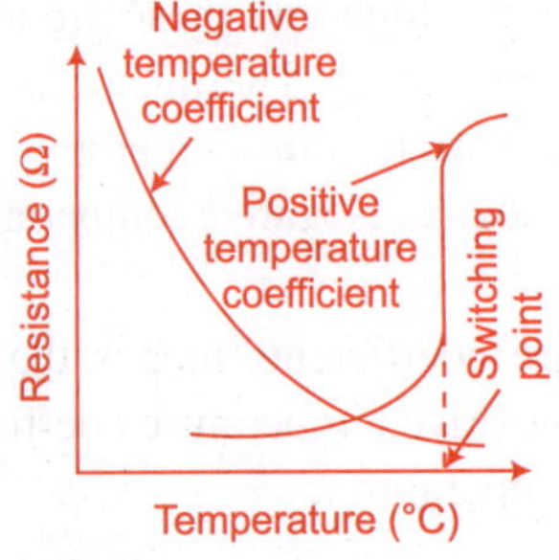

Fig. 1.22 *Thermistor characteristics*

Application of thermistor

These are widely used in industrial, commercial, medical applications as follows:

(i) It is used in time delay and time delay circuits.

(ii) It is used for temperature compensation in electronic circuits.

(iii) It is also used for power measurement at radio frequency.

(iv) Temperature measurement and control.

1.9.3 Photoresistors

These are also known as photoconductive cells. These are made with semiconductor material and resistance of changes when illuminated with light. The examples of photoresistors are cadmium selenide (CdSe), cadmium sulphide (CdS) and lead sulphide (PbS). When light is incident on these materials, the covalent bonds are broken. This creates change carriers, electrons in the conduction band and holes in the valence band. The number of electron-hole pair that breaks is dependent upon the amount of illumination on the surface of the material and it determines the resistance of the material.

The symbol of photoresistor is shown in Fig. 1.23. The resistance of the material varies inversely with the amount of light energy.

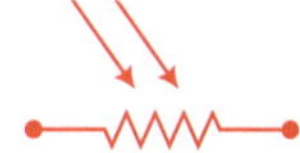

Fig. 1.23   *Symbol of photoresistor*

Characteristics

The characteristics of photoresistor is shown is Fig. 1.24. From this Fig., it can be observed that with increase in illumination, the resistance of the material decreases, due to the increases, in electron-holes pair.

The resistance of photoconductors can be several megaohms in total darkness and less than 100 Ω in well illuminated condition.

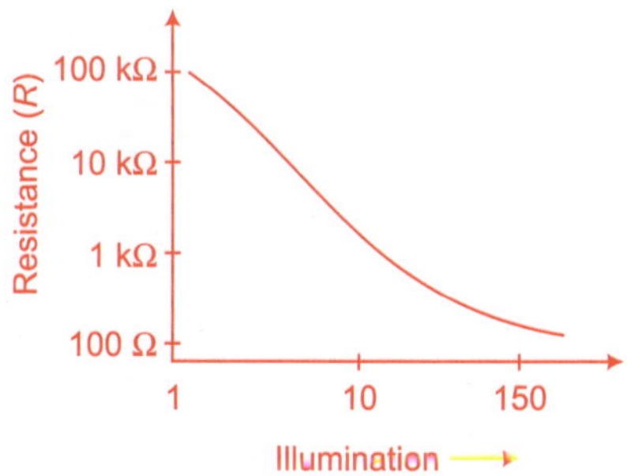

Fig. 1.24   *Characteristic of photoresistor (Cds)*

Applications

The photoresistors are used as follows:

(i) It is used in light activated relay control circuits.

(ii) It can also be used in light meters.

1.10 CAPACITORS

A capacitor is a physical device which is capable of storing energy by virtue of existing voltage. Capacitance is the property of a capacitor that opposes the changes in voltage by means of energy storage in the form of electrostatic energy.

A capacitor consists of two conducting plates, separated by an insulating material, called the dielectric. Capacitors, like resistors can either be fixed or variable. The unit of capacitance is Farad as shown in Fig. 1.25 (a)

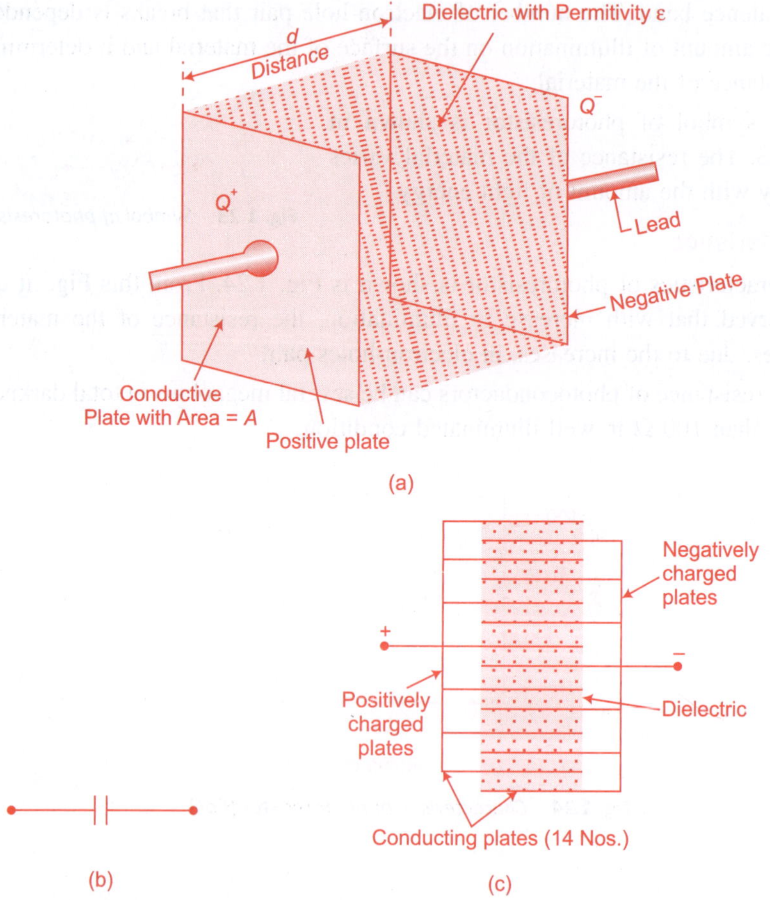

Fig. 1.25 *(a) 2 Plates capacitor (b) Symbol of capacitor (c) 14 plates capacitor*

A capacitor consists of two or more parallel plates (conductive) which are not connected or touched (physically), but are eclectrically separated either by air or by some form of insulating material such as paper, mica, ceramic or plastic, which is called the capacitor dielectric. The conductive metal plates of a capacitor can either be square, rectangular, circular, cylindrical or spherical shape with the general shape, size and construction of a parallel plate capacitor depending on its application and voltage rating. The symbol of capacitor is shown in Fig. 1.25 (b). The capacitor for more than two plates is shown in Fig. 1.25 (c).

From Fig. 1.25 (a), we assume that the area of conductive plate is A (in square meter) and the distance between the plate is d (in meter). The permitivity of dielectric is ε.

The equation for capacitance is given by,

$$C = \frac{\varepsilon A}{d} \text{ Farad } (F) \tag{1.7}$$

where, A = Area of conductive plate (in square meter)

  d = Distance between plate (in meter)

  ε = Permitivity of dielectric (F/M)

  $\varepsilon = \varepsilon_o \, \varepsilon_r$

  ε_o = Permitivity of free space $- 8.854 \times 10^{-12}$ F/M

  ε_r = Relative permitivity

i.e. $$C = \frac{\varepsilon_o \varepsilon_r A}{d} \, F. \tag{1.8}$$

1.10.1 Important Specifications of Capacitors

Like a resistor, when we have to use any capacitor in the electronic circuit or we have to purchase it from the market, we should know the specifications of it.

Following are the different specifications of capacitors:-

  (i) Capacitance Value

  (ii) Tolerance

  (iii) Voltage rating

  (iv) Temperature coefficient

(i) Capacitance value: The value of capacitor means how much capacitance is created by capacitor. The unit of capacitance is Farad (F). The value capacitance is written on the capacitor obtained from color coding method.

(ii) Tolerance: Tolerance of the capacitor is same as the resistor which means how much per cent deviation will occur in the value of capacitance from its specified value. For example, the tolerance of 100 F capacitor is ±5%. That means the value of capacitance will exist between 95 F to 105 F.

(iii) Voltage rating: The maximum voltage that can be applied across the capacitor (without damage of it) is called voltage rating. In other words we can say that voltage rating is the maximum voltage of capacitor at which the dielectric material of capacitor can work safely. Above this voltage, the capacitor will damage.

(iv) Temperature coefficients: The temperature coefficients of capacitor will tell about the value of capacitance changes with the change in temperature. It is represented in %/°C.

1.10.2 Color Coding of Capacitor

The color coding of capacitor is not much popular as the color coding of resistor. The value of most of the resistors is marked on them.

Code for ceramic capacitor

If the value of capacitor is less than 1000 pF, then it will indicate on the capacitor. For example, if the value of ceramic capacitor is 220 pF, then only 220 is marked on the capacitor. If the value of capacitor is greater than and equal to 1000 pF, then the three digit code is used for the value of capacitor.

First number – First significant number.

Second number – Second significant number of capacitor.

Third number – It represents the multiplier.

The capacitor is shown in Fig. 1.26 (b). from this Fig., we observe that,

$$\text{First number} = 1$$
$$\text{Second number} = 0$$
$$\text{Third number} = 3 \Rightarrow 10^3$$

The Value of capacitance is,

$$C = 10 \times 10^3 \text{ pF}$$
$$C = 10000 \text{ pF}$$
$$C = 0.01 \text{ μF.}$$

Fig. 1.26 (a)

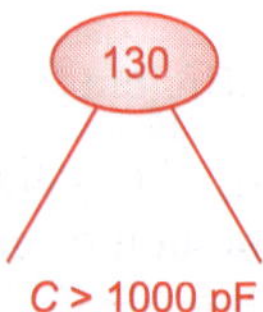

Fig. 1.26 *(b) Coding of ceramic capacitor*

Figure 1.27 shows a different types of ceramic capacitors in which different bands give its value, tolerance temperature coefficient. As shown in Table 1.4.

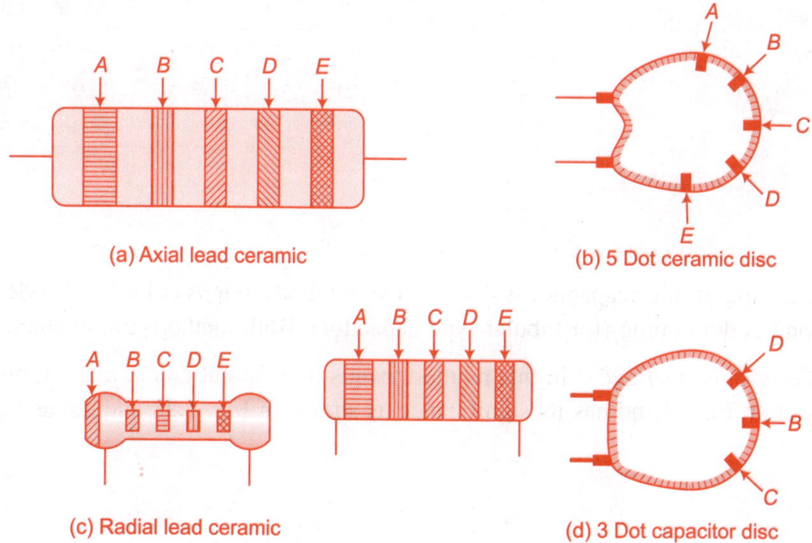

Fig. 1.27 *Different types of ceramic capacitor color coding*

Table 1.4

Band	Indication
A	Temperature coefficients
B	First significant digit
C	Second significant digit
D	Multiplier
E	Tolerance

Table 1.5

Color	1st digit (B)	2nd digit (C)	Multiplier (D)	Tolerance (E)		(A) Temperature Coefficients
				C > 10 pF	C < 10 pF	
Black	0	0	10^0	±20%	±2%	0
Brown	1	1	10^1	± 1%	–	$- 30 \times 10^{-6}$
Red	2	2	10^2	± 2%	–	$- 80 \times 10^{-6}$
Orange	3	3	10^3	–	–	$- 150 \times 10^{-6}$
Yellow	4	4	10^4	–	–	$- 220 \times 10^{-6}$
Green	5	5	–	± 5%	± 0.5%	$- 330 \times 10^{-6}$

Contd...

Contd...

Blue	6	6	–	–	–	-470×10^{-6}
Violet	7	7	–	–	–	-750×10^{-6}
Grey	8	8	0.01	–	± 0.25%	$+30 \times 10^{-6}$
White	9	9	0.1	± 10%	± 1.0%	+ 120 to – 750
Silver	–	–	–	–	–	+ 500 to – 330
Gold	–	–	–	–	–	+ 100

Code for mica capacitor

The coding of mica capacitor is done by two methods, one is color band code and second is dot coding (for tubular type capacitor). Both methods are explained.

1. *Color band method:*   In this method, bands of different colors marked on the capacitor. Each band has its significance as shown in Fig. 1.28 and Table 1.6

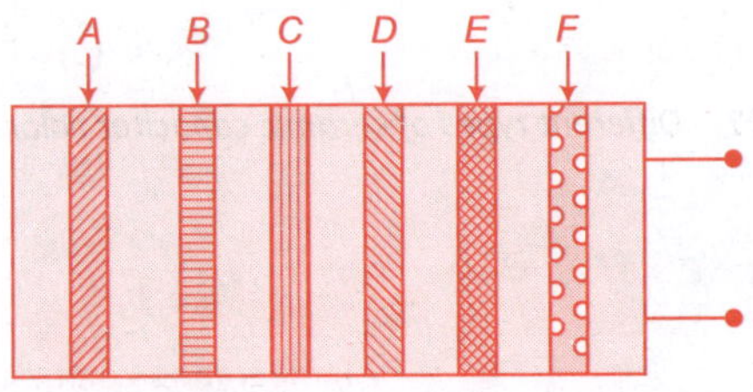

Fig. 1.28   *Color code of mica*

Table 1.6

Band	Indication
A	Temperature coefficient
B	First significant digit
C	Second significant digit
D	Multiplier
E	Tolerance
F	Voltage level

The capacitor color code is shown in Table 1.5 and the voltage levels are shown in Table 1.7.

Table 1.7

Color	Voltage rating of different types of capacitor				
	Dipped tantalum capacitor (V)	Mica capacitor (V)	Polyester/ Polystyrene capacitor (V)	Electrolytic capacitor	
				3 band (V)	4 band (V)
Black	4	100	–	10	10
Brown	6	200	100	–	1.6
Red	10	300	250	35	4
Orange	15	400	–	–	40
Yellow	20	500	400	6	6.3
Green	25	600	–	15	16
Blue	35	700	630	20	–
Violet	50	800	–	–	–
Grey	–	900	–	25	25
White	3	1000	–	3	2.5
Silver	–	–	–	–	–
Gold	–	2000	–	–	–

2. *Dot method:*   In tabular type capacitor, the coding is given as 6 Dot mark on the capacitor, each Dot has its own significance as shown in Figs. 1.29 (a), (b) and (c) below.

The color code table for mica capacitor of tabular capacitor is shown in Table 1.8. The temperature coefficient is same as shown in Table 1.6.

Table 1.8

Color	First digit	Second digit	Multiplier	Tolerance	Voltage rating
Black	0	0	10^0	± 1%	100
Brown	1	1	10^1	± 2%	200
Red	2	2	10^2	± 3%	300
Orange	3	3	10^3	± 4%	400
Yellow	4	4	10^4	± 5%	500
Green	5	5	10^5	± 6%	600
Blue	6	6	10^6	± 7%	700
Violet	7	7	10^7	± 8%	800
Grey	8	8	10^8	± 9%	900
White	9	9	10^9	–	1000
Silver	–	–	0.01	± 20%	–
Gold	–	–	0.1	± 10%	2000
No color	–	–	–	–	–

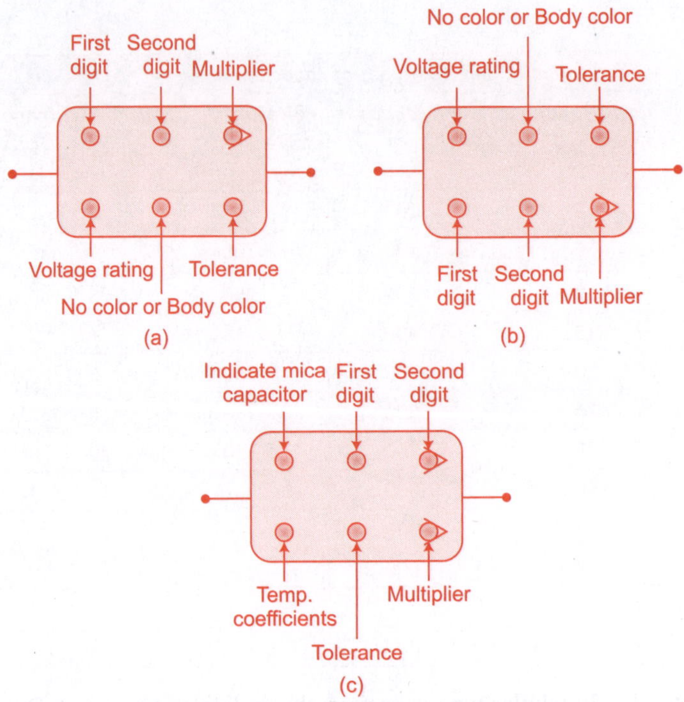

Fig. 1.29 *Dot color coding of mica capacitor*

1.11 DIFFERENT TYPES OF CAPACITORS

Capacitors are classified as shown in Fig. 1.30.

1.11.1 Fixed Capacitors

Fixed capacitors are the ones in which the capacitance value cannot be varied by any reason. The fixed capacitors are of two types, one is electrolytic and second one is electrostatic capacitors.

1.11.1.1 Electrolytic capacitors

The electrolytic capacitor is shown in Fig. 1.31, in which, it has two conducting metal plates having a definite polarity. These conducting plates are separated by dielectric thin metal oxide. The dielectric constant of metal oxide is between 8 and 25. The metal oxide film is deposited on one metal plate and this plate works as a positive electrode or anode. This capacitor is formed by using either a conducting electrolyte as a second electrode or a semiconductor such as manganese dioxide. This capacitor is packed with metal cylinder and the cathode of the capacitor is connected to the metal cylinder. This cylinder is enclosed in a paper tube or cardboard tube and is insulated it from outside.

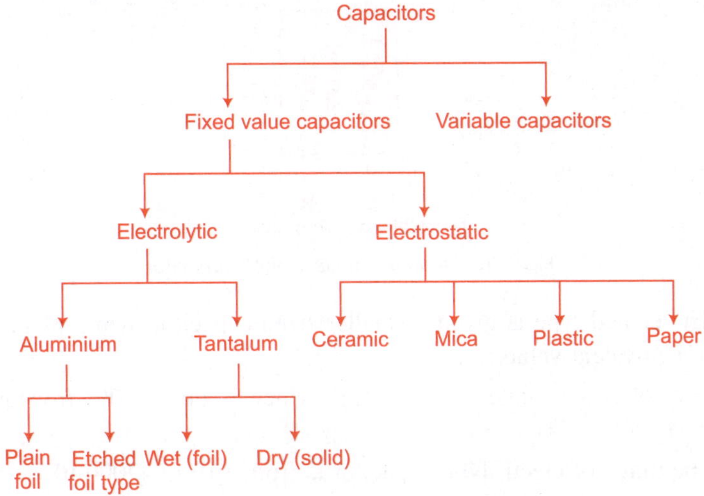

Fig. 1.30 *Classification of capacitor*

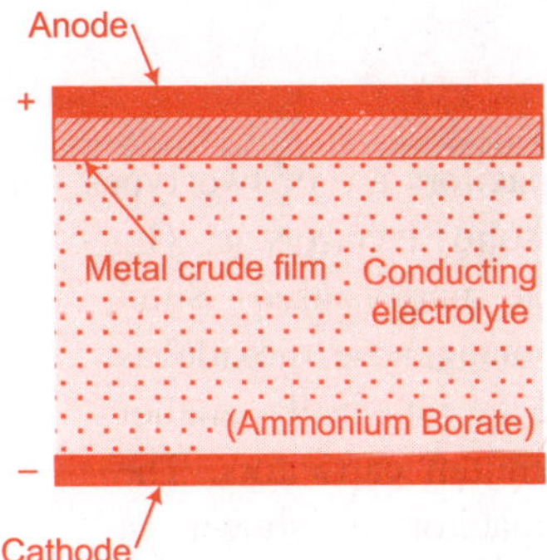

Fig. 1.31 *An electrolytic capacitor*

Range: The value of capacitance ranges from 1 µF to 10,000 µF.

Applications:

  (i) It is used in wide variety of specific applications.

  (ii) It is also used for blocking DC current and passing AC current.

(i) Aluminium electrolytic capacitor

Aluminium electrolytic capacitors are of two types, one is plain foil and another one is etched foil. Due to the thickness of the aluminium oxide film and breakdown voltage, these capacitors have very high capacitance values. The etched foil type differs from the plain foil type in the aluminium oxide. The anode and cathode foils have been chemically etched to increase the surface area and permitivity. The aluminium type capacitor is shown in Fig. 1.32

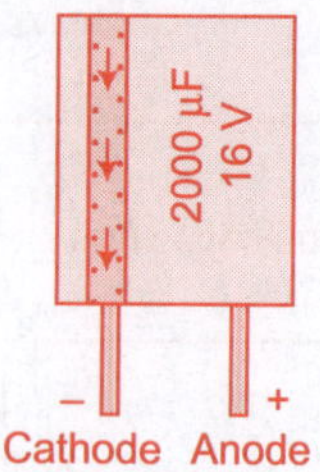

Fig. 1.32 *Aluminium Electrolytic capacitor*

(a) The etched type is given a smaller sized capacitor than a plain foil type of equivalent value.

(b) The tolerance range is quite large which is up to 20% from plain foil type.

(c) The range of electrolytic capacitor is from 1μF to 47000 μF.

(d) Etched foil type is best used in coupling, DC blocking and AC passing circuits while plain foil types are better used as smoothing capacitors in power supplies.

(ii) Tantalum Electrolytic capacitors

Tantalum Electrolytic capacitors are of two types, one is wet (foil) electrolytic and another one is dry (solid) electrolytic. The dry or solid tantalum is mostly used. Solid tantalum capacitors use manganese dioxide as their second terminal and are physically smaller than the equivalent aluminium capacitors. The tantalum electrolytic capacitor is shown in Fig. 1.33.

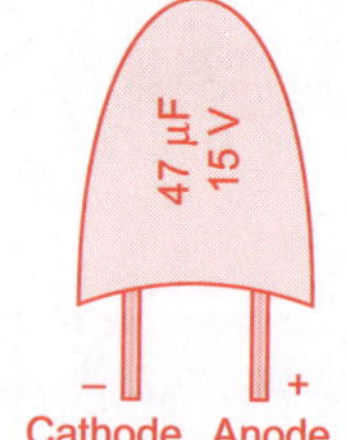

Fig. 1.33 *Tantalum electrolytic capacitor*

(a) The dielectric properties of tantalum oxide is also much better than those of aluminium oxide. Due to this, it has lower leakage current and better capacitance stability.

(b) Due to the capacitance stability, these are used for blocking, by passing, decoupling, filtering and timing application.

(c) Although tantalum capacitors are polarised as aluminium but can tolerate being connected to a reverse voltage much more easily than the aluminium types. They are rated at much lower breaking voltages.

(d) The typical range is from 47 μF to 470 μF.

1.11.1.2 *Electrostatic capacitors*

Electrostatic capacitors are manufactured by two metal conductors i.e., plates which are separated by a dielectric. These capacitors have very low leakage current and high leakage resistance. Following are the different types of electrostatic capacitors:

(i) Ceramic capacitor Ceramic capacitors are those capacitors, which are manufactured by various ceramic materials like barium titanate as a dielectric.

These capacitors are generally manufactured in the shape of tube or disc as shown in Fig. 1.34. For making a ceramic capacitor, copper or silver coating are made on both side of ceramic disc. Due to the coating, the film is form on both sides of the disc and wire leads are attached to each plate. Entire unit is packed into plastic coating.

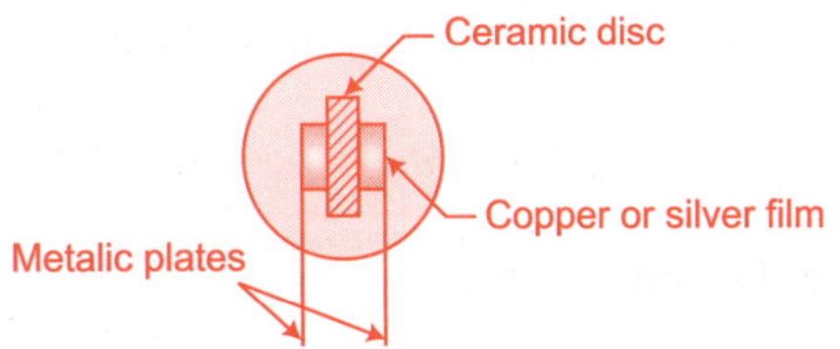

Fig. 1.34 *Ceramic capacitor*

Specifications of ceramic capacitor

    (a) **Range:** It has large capacitance range from 10 pF to 1 μF.

    (b) **Voltage rating:** It has voltage range from 3 V to 6 kV.

    (c) **Tolerance:** It has tolerance from ±10% to ±20%.

    (d) **Merits:**

        1. These capacitors have low leakage current and high leakage resistance approximately 1000 MΩ.

        2. Ceramic capacitors have lower breakdown voltage than that of mica or paper capacitors.

        3. Ceramic capacitors are strong.

        4. These are cheaper and have very good performance up to 200 MHz.

    (e) **Applications:**

        1. These can be used in both AC and DC type of circuits.

        2. These are generally used as coupling capacitor and by pass capacitor.

(ii) Mica capacitor Mica capacitors are those capacitors, which are manufactured by mica as a dielectric.

Mica capacitor is shown in Fig. 1.35. Here, mica is used between the foil of tin or aluminium.

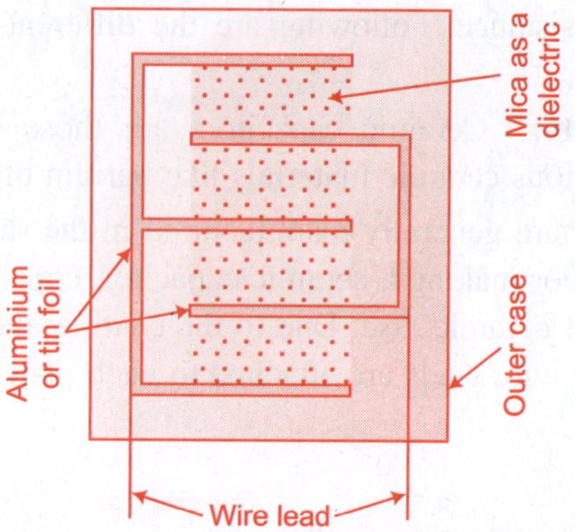

Fig. 1.35 *Mica capacitor*

Specification of mica capacitor

(a) **Range:**   It has capacitance range from 1 pF up to 10,000 pF.

(b) **Voltage rating:**   These are generally available for 500 V but at high frequency, it can be used up to 40 kV.

(c) **Tolerance:**   It has tolerance approximately ± 0.5%.

(d) **Merits:**

    1. This capacitor has less AC current losses.

    2. It has high power factor.

    3. These can be used at high voltage up to 500 V due to high dielectric constant.

    4. It has good electrical stability.

(e) **Applications:**

    1. These are used in high voltage application.

    2. These can also be used in applications where large temperature variations occurs.

    3. These capacitors are widely used in radio and telecommunication applications.

(iii) Plastic or polyester capacitor   Plastic capacitors are those capacitors, which are manufactured by plastic film as dielectric. For this purpose, the plastic used, is polyester, polystyrene, Teflon etc.

Plastic capacitor is shown in Fig. 1.36. In this capacitor, the polyester film is kept between two capacitors, the polyester film is kept between two long and thin metal foils and is rolled in a cylindrical shape. The thickness of polyester is approximately 0.0875 mm.

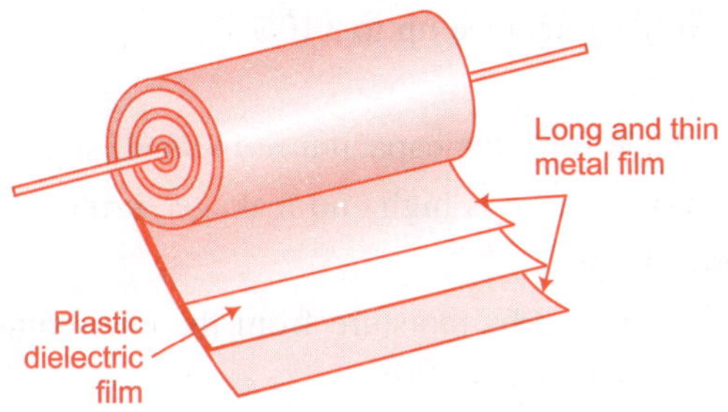

Fig. 1.36   *Plastic capacitor*

Specifications of plastic capacitor

(a) **Range:**   It has capacitance range from 0.0005 µF up to 10 µF.

(b) **Voltage rating:**   It has voltage range up to 2000 V.

(c) **Tolerance:**   It has tolerance up to ±10%.

(d) **Merits:**

   1. It has large insulation resistance than that of paper capacitor.

   2. It has low power factor.

(e) **Applications:**   It is used in digital computers and also in radio frequency and tuned circuits.

(iv) Paper capacitor   Paper capacitors are those capacitors, which are manufactured by impregnated paper as a dielectric.

Paper capacitor is shown in Fig. 1.37. In this capacitor, the paper film is kept between two long and thin metal foils and rolled in the cylindrical shape. The paper is soaked with dielectric wax, oil or plastic. There are of two types, one is hollowax capacitor and second one is metalized paper capacitor.

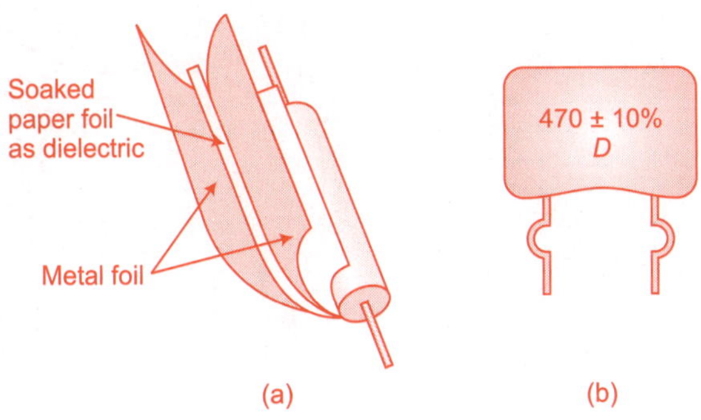

Fig. 1.37   *Paper capacitor*

Specifications of paper capacitor

(a) **Range:**   It has capacitance range from 0.0005 µF up to 10 µF.

(b) **Voltage Rating:**   It has voltage range from 100V up to 1000 Volt.

(c) Tolerance:　　It has tolerance up to ±10%.

(d) Merits:

   (i) It can be made for large capacitance in small size.

   (ii) Its leakage resistance is high and leakage current is low.

   (iii) These are cheaper.

(e) Demerits:　　Paper absorbs moisture from the environment. Due to this, the insulation resistance decreases.

(f) Applications:

   (i) It can equally be used in AC and DC circuits.

   (ii) It can be used for high voltage and high current.

1.11.2 Variable Capacitors

Variable Capacitors are those capacitors in which the capacitance value can be changed by some means. It can be changed by varying the area of the plates and the spacing between them. Variable dielectric capacitors are multi-plate air-spaced types that have a set of fixed plates and are called stator vanes. A set of movable plates are called rotor vanes, which move in between the fixed plates. The position of the moving plates with respect to the fixed plates determines the overall capacitance value. The capacitance is generally at its maximum value when the two sets of plates are fully measured together. The symbol of variable capacitor is shown in Fig. 1.38 and variable capacitor is shown in Fig. 1.39.

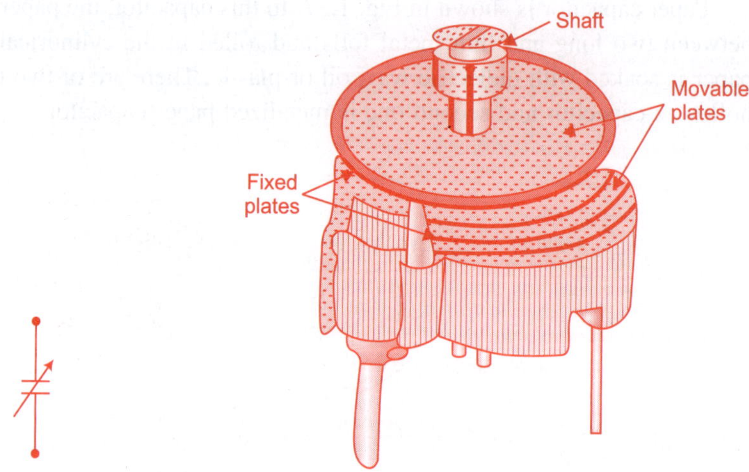

Fig. 1.38　*Variable capacitor symbol*　　　　　　**Fig. 1.39**　*Variable capacitor*

1.11.3 Comparision of Different Capacitors

A comparative study of different capacitors is given in Table 1.9 on the next page.

Table 1.9   Comparative study of different Capacitors

Type of capacitor	Dielectric constant	Range	Voltage Rating	Tolerance	Leakage Current	Polarity	Application
Ceramic	80-1200	10pF to 1µF	3V to 6kV	± 10% to 20%	Low	Unpolarised	Coupling, by pass
Mica	3-8	1pF to 10,000pF	500V, at high frequency µkV	± 0.5%	Low	Unpolarised	Temperature varying application
Plastic	2–0	.0005 µF to 10 µF	2000 V	± 10%	Low	Unpolarised	tuned circuit digital computer
Paper	2-6	.0005 µF to 10 µF	100 V to thousands Volts	± 10%	Low	Unpolarised	AC × DC, high voltage × high current
Electrolytic	Aluminium oxide (7)	1 µF to thousands µF	1 V to 500 V	± 10%	high	Polarised	Filter, coupling circuits

1.11.4 Effect of Temperature and Frequency on Capacitors

The performance of capacitors differs at different frequencies due to the ohmic losses, dielectric losses, resistance of connecting wires and inductive effect of terminals.

If a capacitor is charged with DC voltage, it cannot always store the charge due to the leakage current. The leakage current discharges the capacitor. Similarly, when AC voltage is applied across the capacitor, then it will charge and discharge continuously. If the frequency of applied AC is very high, then there will be a difference in applied AC voltage and stored AC voltage. This phenomenon is called as absorption loss. Hence, the equivalent circuit of capacitor is shown in Fig. 1.40. From this Fig., we say that,

$$C = \text{Capacitance of capacitor}$$
$$r = \text{Load resistance}$$
$$R_a = \text{Absorption loss}$$
$$R_l = \text{Leakage resistance.}$$

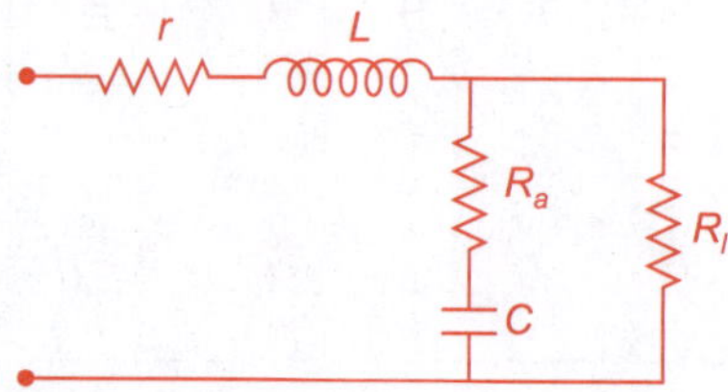

Fig. 1.40 *Equivalent circuit of capacitor*

From the above Fig. 1.40, it is clear that the performance of capacitor changes when the frequency changes and it also depends upon the variation in the temperature and it is represented by its temperature coefficients.

1.11.5 Testing of Capacitors

As we know that the capacitor does not work when it is a open circuit or short circuit. So, for testing a capacitor we connect a ohm meter or multimeter across a capacitor. It is required to keep it in the range of MΩ. When we connect multimeter across the capacitor then the battery of the multimeter is connected across the capacitor and it starts charging the capacitor. The pointer of meter moves towards the zero resistance due to the charging current. The charging current reduces slowly and the pointer of meter moves towards the high resistance as shown in Figs. 1.41 (a) and (b) respectively. When capacitor is charged fully, then only leakage current will flow. Hence, ohmmeter indicates the high resistance.

From the above discussion, we can say that,

(a) If pointer of ohmmeter stays at zero or indicates zero, then the capacitor is short circuit or out of order.

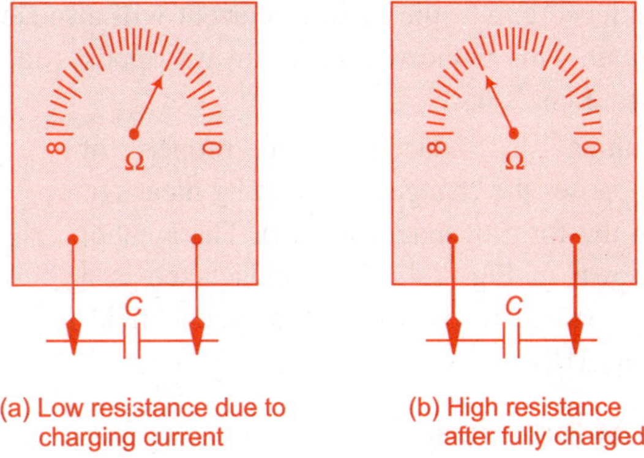

(a) Low resistance due to
charging current

(b) High resistance
after fully charged

Fig. 1.41   *Testing of capacitor*

(b) If pointer of ohmmeter stays at infinite or indicates infinite value, then the capacitor is an open circuit i.e., out of order.

(c) If the pointer of ohmmeter moves towards the low resistance and then slowly moves towards the high resistance, then the capacitor is in working condition.

(d) If pointer of ohmmeter indicates low resistance and then indicates the low resistance than its normal value i.e., the more leakage current is flowing in capacitor and the capacitor is leaky.

1.12   INDUCTORS

An inductor may be defined as an electrical component which is designed with a specific amount of inductance. An inductance is the property of an inductor that opposes the changes in current by means of energy storage in the form of magnetic field.

When a current flows through a wire which is in the form of coil, then the magnetic field is developed surrounding the coil. Due to this magnetic field, the magnetic flux is associated with the coil.

When there is a rate of change of flux due to the change in current flowing through it or by some other means, then it will induce an electronic force (e.m.f.).

$$e \propto -N \frac{d\phi}{dt} \tag{1.9}$$

where, N = Number of turns

$\dfrac{d\phi}{dt}$ = Rate of change of flux

e = e.m.f. (electromotive force).

Due to the induced e.m.f., the induced current will also flow with the main current into the coil. This induced current always opposes the change in main current as given by Lenz's law.

Hence, we can say that inductance is the property of an inductor that opposes the change in current by means of energy storage in the form of magnetic field. The symbol of inductor is shown in Fig. 1.42. This inductance is a measure of energy stored in the form of magnetic field in a coil. The unit of inductance is Henry (H).

Fig. 1.42 *Symbol of inductor*

1.12.1 Types of Inductors

The inductors are classified in following types of cores:-

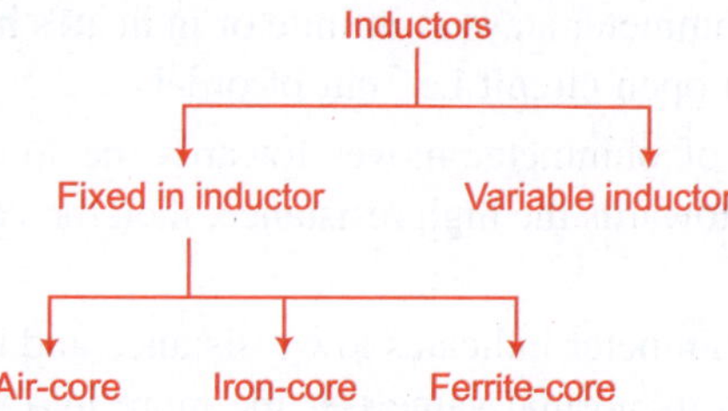

1.12.1.1 *Fixed inductors*

If the inductance of an inductor is fixed, then it is called a fixed inductor i.e, if the core of the inductor is fixed then it is called fixed inductor and is classified as:

   (i) Air-core inductors

  (ii) Iron-core inductors

 (iii) Ferrite-core inductors

(i) Air-core inductors: In air-core inductor, air plays the role of a core where no other core is used. These are made of coils of wire wound on a simple cardboard. The shape and symbol of an iron-core inductor is shown in Fig. 1.43. The air-core inductors have a very low value of inductance (L). Therefore, they are used in radio-frequency applications.

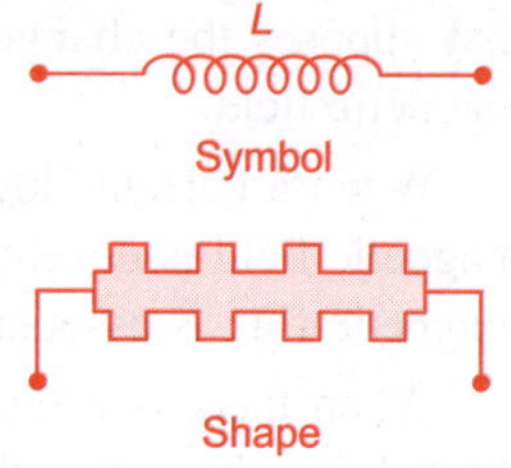

Fig. 1.43 *Air-core inductor*

(ii) Iron-core Inductors: In iron-core inductors, iron acts as a core i.e., These are made of coils of wire wound on a solid iron-core. The iron core of an inductor is laminated to avoid the eddy current loss. The iron-core inductor is shown in Fig. 1.44. These inductors are used in audio-frequency applications.

(iii) Ferrite-core inductors: In ferrite-core inductors, Ferrite play act as a core. These are made of coils of wire wound on a ferrite core. Basically, a ferrite is a

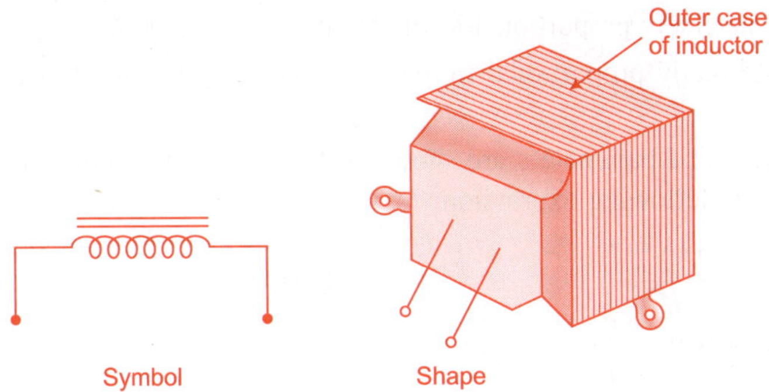

Fig. 1.44 *Iron-core inductor*

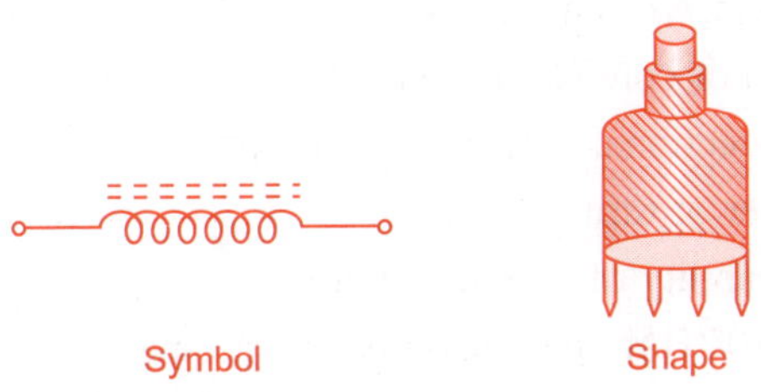

Fig. 1.45 *Ferrite-core inductors*

magnetic material containing five particles of iron, nickel or cobalt embedded in an insulator binder. It has very low eddy current loss. Due to the low eddy current loss such inductors are used at higher frequencies. The ferrite-core inductor is shown in Fig. 1.45.

1.12.1.2 Variable inductors

If the inductance of an inductor is variable, then it is called a variable inductor i.e., if the core of the inductor is adjustable, then it is called a variable inductor. Such type of inductors are widely used in tunning and filter circuits. The variable inductor is shown in Fig. 1.46. Ferrite-core inductors are variable inductors, if the core of these inductors are adjustable.

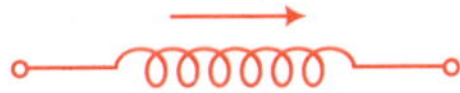

Fig. 1.46 *Variable ferrite-core inductor*

1.12.2 Inductance of a Coil

Inductor is a storage element which can store energy in the form of magnetic field. It can deliver energy but its energy handling capacity is limited. The practical unit of inductance is Henry (H). A practical inductance is called an inductor.

The inductance of a coil or inductor depends on the following parameters:

   (i) It is directly proportional to the permeability (μ) of the magnetic material over which coil is wound.

   (ii) It is directly proportional to the cross-sectional (A) area of the coil.

(iii) It is inversely proportional to the length (l) of the coil.

(iv) It is directly proportional to the sequence of the number of turns of the coil.

These physical parameters are shown in Fig. 1.47. The inductance of a coil is given by the following expression;

$$L = \frac{\mu A N^2}{l} \qquad (1.10)$$

$$\because \quad \mu = \mu_0 \mu_r$$

$$\therefore \quad L = \frac{\mu_0 \mu_r A N^2}{l} \qquad (1.11)$$

where,

Fig. 1.47 *Inductance of a core*

L = Inductance of an inductor

l = Length of core (in meter)

A = Cross-sectional area of core (in m^2)

N = Number of turns of coil

μ = Permeability of magnetic material

μ_0 = $4\pi \times 10^{-7}$ H/m, permeability of free space

μ_v = Relative permeability of core material.

The electronic components which are capable of processing or amplifying an electrical signal are called active components. There are many active components used in electronic circuits. These components are further classified as shown in Fig. 1.1 and will be explained later.

Examples

Examples 1.1 Specify the colour code for the following resistor values.

    (a) 1 kΩ ± 20%   (b) 3.7 kΩ ± 10%

    (c) 100 Ω ± 5%   (d) 470 Ω ± 10%

Solution (a) 1 kΩ ± 20%

i.e.   1000 Ω ± 20%

          1 → Brown

          0 → Black   From Table 1.1

          10^2 → Red

          ± 20% No Color

Hence, the color code for given value is Brown, Black, Red.   (**Ans.**)

(b)   3.7 kΩ ± 10%

i.e., $3700\ \Omega \pm 10\%$

$\left.\begin{array}{l} 3 - \text{Orange} \\ 7 - \text{Violet} \\ 10^2 - \text{Red} \\ \pm\,10\%\ \text{Silver} \end{array}\right\}$ From Table 1.1

Hence, the color code for given value is Orange, Violet, Red, Silver   (**Ans.**)

(c) $100\ \Omega \pm 5\%$

$\left.\begin{array}{l} 1 - \text{Brown} \\ 0 - \text{Black} \\ 10^1 - \text{Brown} \\ \pm\,5\% - \text{Gold} \end{array}\right\}$ From Table 1.1

Hence, the color code for given value is:

Brown, Black, Brown, Gold.   (**Ans.**)

(d) $470\ \Omega \pm 10\%$

$\left.\begin{array}{l} 4 = \text{Yellow} \\ 7 = \text{Violet} \\ 10^1 = \text{Brown} \\ \pm\,10\% = \text{Silver} \end{array}\right\}$ From Table 1.1

Hence, the color code for given value is:

Yellow, Violet, Brown, Silver.   (**Ans.**)

Examples 1.2   A series of colour band of resistor are Grey, Violet, Orange, Gold. Find the value and tolerance of resistor.

Solution   From Table 1.1

1st band Grey – 8

2nd band Violet – 7

3rd band Orange – 10^3

4th band Gold – $\pm 5\%$

The value of resistance $= 87 \times 10^3\ \Omega \pm 5\%$

$$= \boxed{87\ \text{k}\Omega \pm 5\%.}$$ (**Ans.**)

Examples 1.3   Find the capacitance of Parallel plate capacitor having the plate area 0.05m^2 and the distance between plates is 0.08 m. The dielectric between plate is mica, which has dielectric constant of 5.

Solution   Given that,

$A = 0.05\ \text{m}^2$

$d = 0.08\ \text{m}$

We know that,

Permitivity $\varepsilon = \varepsilon_0\,\varepsilon_r$

$$\varepsilon_0 = 8.854 \times 10^{-12} \text{ F/m}$$

For mica $\quad \varepsilon_r = 5$

From Eqn. (1.8), we know that,

$$C = \frac{\varepsilon_0 \varepsilon_r A}{d} = \frac{8.854 \times 10^{-12} \times 5 \times 0.05}{0.08}\, F$$

$$= \frac{2.21 \times 10^{-12}}{0.08}\, F$$

$$= 27.67 \times 10^{-12}\, F$$

$$\boxed{C = 27.67 \text{ pF.}}$$
(**Ans.**)

Examples 1.4 Find the value and tolerance of the resistor which has the following color band.

  (a) Red, Black, Grey, Gold

  (b) Brown, Grey, Yellow, Gold

  (c) Orange, Black, Gold, Silver

Solution   Red, Black, Grey, Gold

From Table 1.1

$$\left.\begin{array}{l} \text{Red} - 2 \\ \text{Black} - 0 \\ \text{Grey} - 10^8 \\ \text{Gold} - \pm 5\% \end{array}\right\} \text{From Table 1.1}$$

Hence, the value of resistance is $20 \times 10^8\ \Omega = 2000\ \text{M}\Omega$ and the tolerance is $\pm 5\%$.
(**Ans.**)

(b) Brown, Grey, Yellow, Gold

$$\left.\begin{array}{l} \text{Brown} - 1 \\ \text{Grey} - 8 \\ \text{Yellow} - 10^4 \\ \text{Gold} - \pm 5\% \end{array}\right\} \text{From Table 1.1}$$

Hence, the value of resistance is $18 \times 10^4\ \Omega = 180\ \text{k}\Omega$

The tolerance is $\pm 5\%$.
(**Ans.**)

(c) Orange, Black, Gold, Silver

$$\left.\begin{array}{l} \text{Orange} - 3 \\ \text{Black} - 0 \\ \text{Gold} - 10^{-1} \\ \text{Silver} - \pm 10\% \end{array}\right\} \text{From Table 1.1}$$

The value and tolerance of the resistance is $30 \times 10^{-1} \, \Omega \pm 10\%$

$$= 3 \, \Omega \pm 10\%. \hspace{3cm} \textbf{(Ans.)}$$

Examples 1.5   There are two wires x and y of same material. x is 25 times longer than y and $\dfrac{1}{4}$ time cross section that of y. If resistance of x is 2 Ω, find the resistance of y.

Solution   Let the length of wire y be $= l_y$ m

So, the length of wire $\quad x = 25 \, ly^m$

Let the Area of wire $\quad y = A_y$ m^2

So, area of wire $\quad x = \dfrac{A_y}{4}$ m^2

Resistance of wire $\quad x = R_x = 2 \, \Omega.$

Since both resistance material is same, so ρ will be same.

i.e., $\hspace{3cm} \rho_x = \rho_y$

we know that,

$$R_x = \rho_x \frac{l_x}{A_x} \hspace{4cm} (1)$$

$$R_y = \rho_y \frac{l_y}{A_y} \hspace{4cm} (2)$$

Now, dividing the Eqns. (2) by (1)

$$\frac{R_y}{R_x} = \frac{\rho_y \dfrac{l_y}{A_y}}{\rho_x \dfrac{l_x}{A_x}} = \frac{\rho_y}{\rho_x} \times \frac{l_y}{A_y} \times \frac{A_x}{l_x}$$

$$\frac{R_y}{2 \, \Omega} = \frac{\rho_y}{\rho_y} \times \frac{ly}{Ay} \times \frac{Ay/4}{25 \, ly}$$

$$R_y = \frac{2}{100} \, \Omega = 0.02 \, \Omega$$

$$\boxed{R_y = 0.02 \, \Omega.} \hspace{3cm} \textbf{(Ans.)}$$

Examples 1.6   Three resistors of 3 k$\Omega \pm 10\%$ 1 W, 2 k$\Omega \pm 5\%$, 2 W 8 kΩ $\pm 5\%$ 3 Ω are connected in series. Find the equivalent resistance, tolerance and power rating.

Solution   Equivalent resistance in series is,

$$R_{eq} = R_1 + R_2 + R_3$$
$$= 3 \text{ k}\Omega + 2 \text{ k}\Omega + 3 \text{ k}\Omega$$
$$= 13 \text{ k}\Omega.$$

- Equivalent tolerance will be the average value of individual tolerances.

  i.e., $\qquad$ Tolerance $= \pm \left(\dfrac{10 + 5 + 5}{3} \right)\%$

  $\qquad\qquad\qquad\quad = \pm \left(\dfrac{20}{3} \right)\% = \pm\, 6.6\%.$

- The power rating of series combination is equal to the minimum voltage of resistors.

  i.e., power rating $= 1$ W.

Examples 1.7 Three resistors of 22 k$\Omega \pm 5\%$ 1 W, 11 k$\Omega \pm 5\%$ 2 W, 33 kΩ $\pm 5\%$ 2 W are connected in parallely. Find the equivalent resistance, tolerance and power rating of combination.

Solution The equivalent resistance in parallel is,

$$\frac{1}{R_{eq}} = \frac{1}{R_1} + \frac{1}{R_2} + \frac{1}{R_3}$$

$$\frac{1}{R_{eq}} = \frac{R_2 R_3 + R_1 R_3 + R_1 R_2}{R_1\, R_2\, R_3}$$

$$R_{eq} = \frac{R_1\, R_2\, R_3}{R_2 R_3 + R_1 R_3 + R_1 R_2}$$

$$= \frac{(11 \times 22 \times 33)}{(11 \times 33) + (22 \times 33) + (22 \times 11)}\ \text{k}\Omega$$

$$= \frac{7986}{363 + 726 + 242}\ \text{k}\Omega$$

$$= \frac{7986}{1331}\ \text{k}\Omega$$

$$\boxed{R_{eq} = 6\ \text{k}\Omega.} \qquad\qquad\qquad \textbf{(Ans.)}$$

- The equivalent tolerance will be the average value of individual tolerance.

  i.e., $\qquad$ Tolerance $= \pm \left(\dfrac{5 + 5 + 5}{3} \right)\%$

  $\qquad\qquad\qquad\quad = \pm\, 5\%.$

- The power rating of parallel combination is equal to the addition of individual voltage of resistors.

  i.e., $\qquad$ power rating $= (1 + 2 + 2)$ W.

$$\boxed{= 5\ \text{watt.}} \qquad\qquad\qquad \textbf{(Ans.)}$$

Examples 1.8   An air-core inductor, whose length is 10 cm, cross sectional area is 1 m^2 and number of turns has an inductance of 100 µH. If this coil is wound on former which has permeability 900, then calculate the inductance of it when the length of magnetic path, cross-sectional area and number of turns are same in both inductors.

Solution   We know that the inductance of inductor is

$$L = \frac{\mu_0 \, \mu_r \, N^2 A}{l} \quad \text{from Eqn. (1.9)}$$

For inductor 1

$$L_1 = \frac{\mu_{01} \, \mu r_1 \, N_1^2 A_1}{l_1} \tag{1}$$

For inductor 2

$$L_2 = \frac{\mu_{02} \, \mu r_2 \, N_2^2 \, A_2 L}{l_2} \tag{2}$$

dividing Eqn. (2) by Eqn. (1), we obtain,

$$\frac{L_2}{L_1} = \frac{\mu_{r2}}{\mu_{r1}} \times \left(\frac{N_2}{N_1}\right)^2 \times \left(\frac{A_2}{A_1}\right) \times \left(\frac{l_1}{l_2}\right)$$

$$\therefore \qquad N_1 = N_2, \; A_1 = A_2 \text{ and } l_1 = l_2$$

$$\frac{\mu_{r2}}{\mu_{r1}} = \frac{900}{1} \qquad\qquad \text{(For air } \mu \simeq 1)$$

Hence,

$$\frac{L_2}{L_1} = \frac{900}{1}$$

$$L_2 = 900 \times L_1$$

$$= 900 \times 100 \; \mu H$$

$$= 90000 \; \mu H$$

$$\boxed{L_2 = 90 \text{ mH.}} \qquad\qquad \textbf{(Ans.)}$$

Examples 1.9   Determine the cross-sectional area of a cable of 1200 m length to transmit 550 amp. so that the total drop in voltage along the cable may not exceed 25 volts. Take $\rho = 1.7 \times 10^{-6}$ Ω/cm^3.

Solution   The resistance of the cable is,

$$R = \frac{\text{Drop in volts (V)}}{\text{current flowing (I)}}$$

$$= \frac{25}{550} = 0.045 \; \Omega$$

Now, we know that,

$$R = \rho \frac{l}{A}$$

$$A = \frac{\rho l}{R}$$

$$A = \frac{1.7 \times 10^{-6} \times 1.2 \times 10^5}{0.045}$$

$$\boxed{A = 4.53 \text{ cm}^2.}$$ (Ans.)

Examples 1.10   A current of 2.1 amperes passed through a wire whose length is 1.5 m and diameter is 2 mm. If the specific resistance or resistivity of the material is 2.52×10^{-6}. What will be the difference of potential between the ends of the wire?

Solution   The resistance of the wire is given by equation,

$$R = \rho \frac{l}{A} = \frac{\rho\, l}{\pi r^2} = \frac{4\, \rho l}{\pi\, D^2}$$

$$= \frac{2.52 \times 10^{-6} \times 1.50 \times 4}{\pi\, r^2 (0.2)^2}$$

$$= \frac{4 \times 2.52 \times 10^{-6} \times 100}{3.14 \times 0.04}$$

$$= \frac{1.51 \times 10^{-3}}{0.126}$$

$$R = 11.98 \times 10^{-3} \ \Omega.$$

According to the ohm's law, the potential difference between the ends of the wire.

$$V = IR$$

$$= 2.5 \times 11.98 \times 10^{-3}$$

$$\boxed{V = 0.03 \text{ volt.}}$$ (Ans.)

Objective Type Questions

1. Which of the following is the passive element?

   (a) Ideal current source   (b) Ideal voltage source

   (c) Capacitance   (d) All of these

   Information for 2 and 3 is given below. Circuit element may be,

   (i) Active   (ii) Passive

   (iii) Unilateral   (iv) Bilateral

2. Which of the following is valid for a capacitor?

   (a) (i) and (iii) only   (b) (ii) and (iv) only

   (c) (ii) and (iii) only   (d) (i) and (iv) only

3. Which of the following is valid for a constant voltage source

   (a) (i) and (iii) only   (b) (ii) and (iii) only

   (c) (i) and (iv) only   (d) (ii) and (iv) only

4. Which of the following is a bilateral element?

   (a) Constant current source   (b) Constant voltage source

   (c) Inductance   (d) FET

5. A $100\ \Omega$ resistor is needed in an electric circuit to carry a current of 0.3A. Which of the following resistors would you select?

   (a) $100\ \Omega$, 1 Watt   (b) $100\ \Omega$, 10 Watt

   (c) $100\ \Omega$, 5 Watt   (d) $100\ \Omega$, 100 Watt

6. Ferrite-cores are useful at,

   (a) Low frequencies   (b) High frequencies

   (c) Very high frequencies   (d) None of these

7. In the circuit shown in Fig. OT 1.1. P is the power dissipated in the resistance R_2, the power dissipated in the circuit will be,

   (a) 4 P

   (b) 5/4 P

   (c) 4/3 P

   (d) 16 P

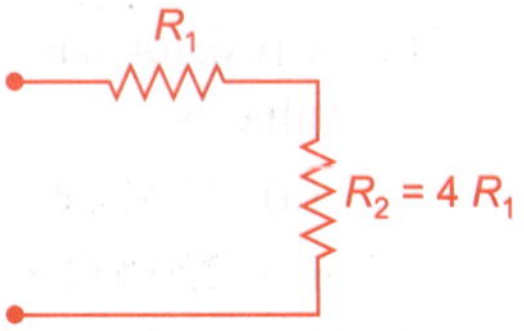

Fig. OT 1.1

8. The inductance of a coil can be increased by:

   (a) Using core material of high relative permeability.

   (b) Increasing core length.

   (c) Reducing number of turns.

   (d) Any of the above

9. An air gap provided in the iron-core of an inductor prevents

   (a) Flux leakage   (b) Core saturation

   (c) Hysteresis loss   (d) Heat generation

10. Which of the following capacitors have the highest capacitance value?

   (a) Mica   (b) Ceramic

   (c) Paper   (d) Electrolytic

11. An inductor stores energy in
    (a) Electrostatic field   (b) Electromagnetic field
    (c) Core   (d) Magnetic field
12. An capacitor stores energy in
    (a) Electrostatic field   (b) Electromagnetic field
    (c) Core   (d) Magnetic field
13. The electronic components which can process the signal are known as
    (a) Passive components   (b) Active components
    (c) Both (a) and (b)   (d) None of these
14. Which statement about the inductance is incorrect?
    (a) The inductive reactance varies directly as the frequency of the applied voltage.
    (b) The inductance of a coil can be increased by adding few more turns to the coil.
    (c) An inductance does not oppose direct current.
    (d) Inductive reactance can be measured by an ohm meter.
15. One of the examples of an active device is
    (a) Transformer   (b) A microphone
    (c) Capacitor   (d) A Transistor
16. A resistor has color code bands Red, Red, Orange and Gold. Its normal value is
    (a) $22 \text{ k}\Omega \pm 5\%$ (b) $22 \text{ k}\Omega \pm 10\%$
    (c) $220 \text{ k}\Omega \pm 5\%$ (d) $220 \text{ k}\Omega \pm 10\%$
17. Which of the following is not a non-linear element?
    (a) Pn junction diode   (b) Electric arc
    (c) Heater coil   (d) Gas diode
18. A resistor has $1 \text{ k}\Omega \pm 5\%$ the color band for the resistor is
    (a) Brown, Brown, Black, Gold
    (b) Brown, Black, Red, Gold
    (c) Brown, Black, Orange, Silver
    (d) None of these
19. Electronics is the branch of science and engineering which deals with
    (a) Static of electrons   (b) Dynamics of electrons
    (c) Both (a) and (b)   (d) None of these
20. The transistor was first invented by
    (a) Armstrong
    (b) J. J. Thomson and John Fleming

(c) John Bardeen and Walter Brattain

(d) None of these

ANSWERS

1 (c)	2 (b)	3 (a)	4 (c)	5 (d)	6 (b)	7 (b)
8 (a)	9 (b)	10 (d)	11 (d)	12 (a)	13 (b)	14 (d)
15 (d)	16 (a)	17 (c)	18 (b)	19 (b)	20 (c)	

Exercise

1.1. Explain electronic components and classify them.

1.2. Define active and passive components.

1.3. Explain the following

    (a) Linear resistor   (b) Non-linear resistor

1.4. Explain the ceramic capacitor.

1.5. Construct and explain carbon film resistor and compare it with metal film and wire wound resistors.

1.6. Explain the following

    (a) Photoresistor   (b) Thermistor   (c) Varistors

1.7. Compare the characteristics of polyester and ceramic capacitors.

1.8. Explain different types of electrolytic capacitors.

1.9. Explain different types of inductors.

1.10. Explain the inductance of a coil.

1.11. What do you understand by color coding of resistors? Explain four band color coding.

1.12. What do you understand by color coding of capacitors?

1.13. Explain different types of fixed inductors.

1.14. Explain different types of fixed resistors.

1.15. What do you mean by variable resistor?

1.16. What do you mean by variable inductors?

1.17. Specify the color code for the following resistors.

    (a) $2.5\ \Omega \pm 10\%$ (b) $3.6\ k\Omega \pm 5\%$

               (Ans.) (a) Red, Green, Gold, Silver

                           (b) Orange, Blue, Red, Gold

1.18. Explain the history of Electronics.

1.19. Explain the evolution of transistor.

1.20. Explain the application of Electronics in different fields.

Voltage and Current Sources 2

2.1 INTRODUCTION

This chapter will provide sufficient knowledge about voltage and current sources. As we know that the operation of any circuit is either electrical or electronic which requires electrical power. This electrical power is provided by different types of electrical power sources. This power can be supplied either in the form of voltage or current. The source which delivers power in the form of a voltage is known as the voltage source and the source which delivers power in the form of a current is known as a Current source.

A source is used to deliver power to the load which is connected as shown in Fig. 2.1.

All the voltage and current sources can be classified into two broad categories:

1. Direct current (DC)
2. Alternating current (AC)

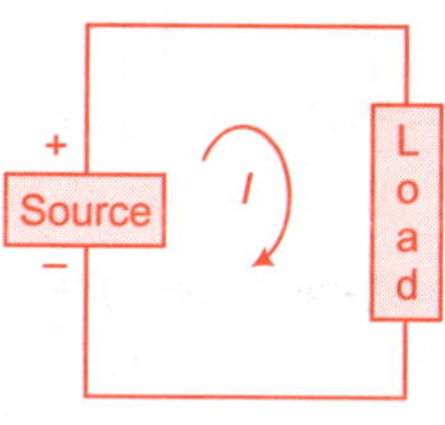

Fig. 2.1

DC Source

When a source supplies direct current to the load it is known as DC source. The examples of DC sources are, DC regulated power supply, DC generator, batteries etc.

AC Source

When a source supplies alternating current to the load it is known as AC source. The example of AC sources are oscillators, alternators and signal generators.

We can say that the batteries, regulated DC power supplies and oscillators are the main sources of energy in the field of Electronics.

Battery

The term battery means one or more electrochemical cells. Basically battery converts chemical energy into electrical energy. A cell may be defined as a fundamental source of energy. It contains a combination of some materials, which produces DC electrical energy from internal chemical reaction.

There are two types of cells: Primary cells or disposable cells, which are designed to be used once and discarded from circuit after use, and secondary cells are rechargable cells, which are designed to be recharged and used multiple times.

Examples of primary cells: Carbon-zinc cells, manganese-alkaline cells, mercury cells etc. Carbon-zinc cell is mostly used as commercial dry cell. Manganese alkaline cell in generally used in T.V., Radios, Tape recorders, Cameras etc.

Examples of secondary cells: Lead-acid cell is widely used in automobile batteries. As we know that a battery is a collection of multiple cells, but in practical usage battery often refers to a single cell. For example, a 1.5 volt AAA battery in a single 1.5 volt, and a 6 volt battery has four 1.5 volt cells in series.

Different types of batteries are shown in Fig. 2.2

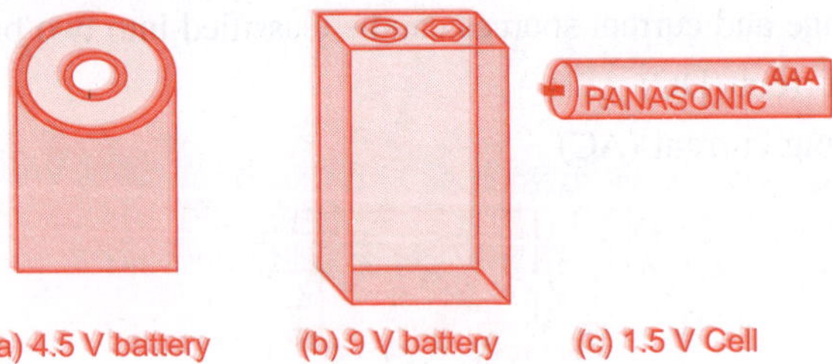

Fig. 2.2 *Different types of batteries*

DC Regulated Power Supply

Generally in all laboratories, DC regulated power supply is available. With the help of DC regulated power supply, the different values of current is measured at different values of voltages in a circuit for obtaining the characteristics of different components. The AC input is given to the supply. It consists rectifier circuit for conversion of AC into DC voltage and filter circuit for removing the AC ripples from DC i.e., the filter circuit smoothes the output of rectifier. The output of filter circuit gives the input to the voltage regulator circuit. It makes the DC output voltage independent from any variation in the rectifier input voltage or load resistance. The block diagram of DC regulated power supply is shown in Fig. 2.3.

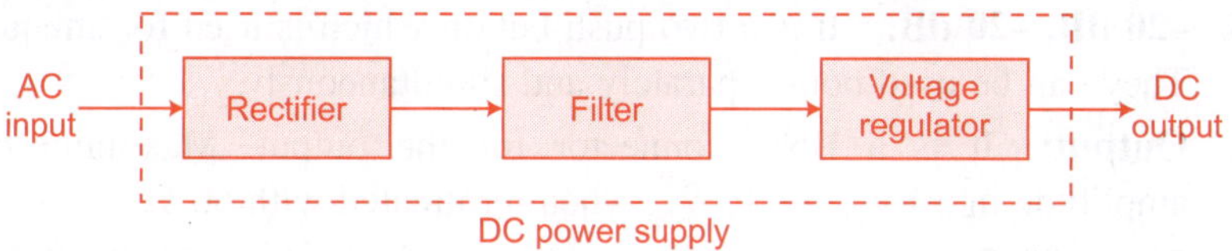

Fig. 2.3 *DC regulated power supply*

Oscillator or Signal Generator

Generally in all laboratories, signal generator or oscillator are also available. Oscillator or signal generator is a device which supplies AC voltages. This voltage is used as a signal to test the operation of various electronic circuits such as filters, amplifiers, rectifiers etc. It produces different types of signal like sine wave, triangular wave, square wave etc. The amplitude and frequency of these signals also can be changed. The front panel of function generator or signal generator is shown in Fig. 2.4.

1. **Power:** Push button for supplying power to instrument.
2. **Digital Display:** It is a 4-digit frequency/amplitude meter and also indicates Hz, kHz, mV and V.
3. **FREQ & AMP:** It is a push button to select the display for frequency or amplitude.
4. **AMP:** It is an adjusting knob, which is used to adjust the amplitude of output from 0 to –20 dB.

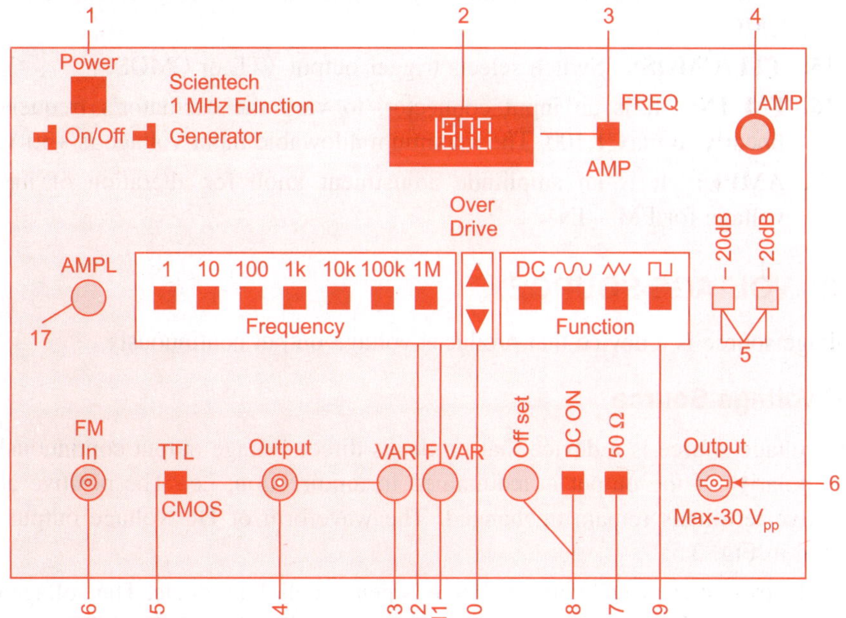

Fig. 2.4 *Front panel control of 1 MHz function generator*

5. **–20 dB, –20 dB:** It is a two push button which is used for attenuation. They can be used both separately and simultaneously.

6. **Output:** It is a BNC connector for the output. Maximum output amplitude in 30 V_{pp}. or 15 V_{pp} when terminated with 50 Ω.

7. **50 Ω/600 Ω:** It is a push button, when pressed selects 600 Ω else 50 Ω in released position.

8. **DC On:** It is an offset voltage adjustment knob. This DC voltage can be superimposed on the output signal. The maximum offset voltage is ± 12.5 V.

9. **Function:** It is a four position push button, used for selecting mode; DC - sin - Triangle - square.

10. **Over drive:** When working in offset voltage mode and the output amplifier is overdriven either in positive or negative direction, the corresponding LED lights up.

11. **FVAR:** It is a frequency adjustment knob, that can adjust frequency from 1 Hz to 1 MHz continuously.

12. **FREQUENCY:** Frequency coarse adjustment from 1 Hz to 1 MHz in 7 decade steps.

13. **VAR:** When trigger output selects in CMOS, output can be set with VAR, to approx. 15 V_{pp}.

14. **Trig output:** This short circuit proof output supplies a square wave signal in synchronys with the output signal. It is a switch selectable TTL/CMOS.

15. **TTL/CMOS:** Switch selects trigger output TTL or CMOS.

16. **FM IN:** It is an input connector, to vary the oscillator's frequency linearly to max 1:100. The maximum allowable input voltage is +30 V.

17. **AMPL:** It is an amplitude adjustment knob for alteration of input voltage for FM – IN.

2.2 VOLTAGE SOURCE

Voltage source is a device that produces voltage output continuously.

DC Voltage Source

DC voltage source is a device that produces direct voltage output continuously. The polarity of the output is maintained in unidirection: i.e., The positive and negative terminals remain unchanged. The waveform of DC voltage output is shown in Fig. 2.5.

The examples of DC voltage source is pencil cell, battery etc. The voltage of pencil is 1.5 V DC and voltage of battery is 6 V DC, 9 V DC, 12 V DC etc.

Where A = Amplitude of DC voltage in volts.

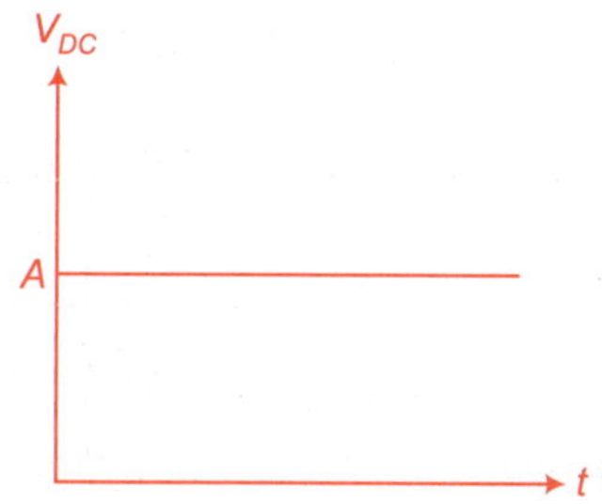

Fig. 2.5   *Waveform of DC voltage source*

AC Voltage Source

AC voltage source is a device that produces alternating voltage output continuously. The polarity of the output of AC voltage source changes periodically. i.e., the positive and negative terminals of the source changes periodically. The waveform of AC voltage output is shown in Fig. 2.6.

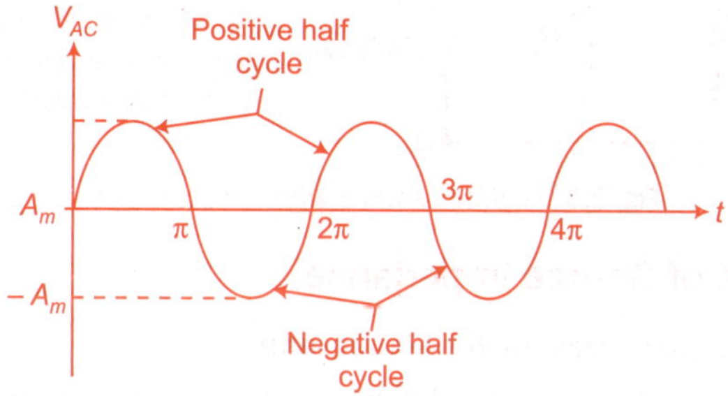

Fig. 2.6   *Waveform of AC voltage source*

Where A_m = Maximum amplitude of AC voltage output in volts.

The example of AC voltage source in the electric supply to the domestic purpose is 220 Vrms and 50 Hz.

Let us connect a load resistance R_L across any voltage source. Now, we will observe the effect on the terminal voltage V_t of voltage source. Let us connect R_L across the voltage source of 1.5 volt and measure the voltage across the terminal of voltage source with the help of multimeter, then you will find that the terminal voltage is less than 1.5 V DC when the load is connected. In other words we can state that the open circuit voltage, which is also called as electromotive source of e.m.f. across voltage source will be greater than the terminal voltage of voltage source, when the load is connected across the voltage source as shown in Fig. 2.7(b).

After connecting the load across the terminals of voltage source, the current will start flowing into the circuit. Due to the flow of current, some voltage drop will occur inside the source. When the value of load resistance is reduced, then the amount of current into the circuit will increase. Due to this, the voltage drop inside the source will also increase. Now, question arises that where this voltage drop

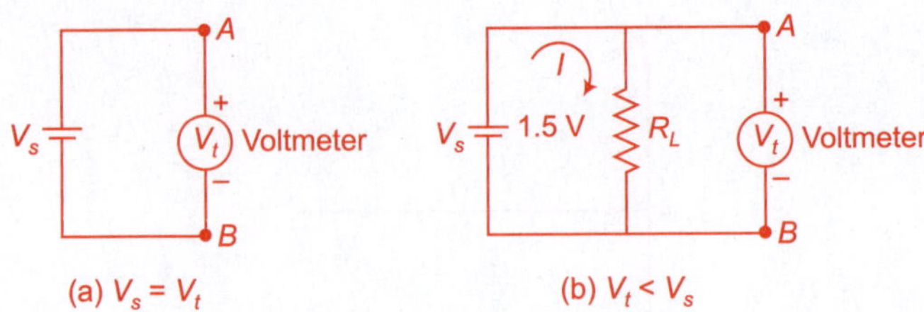

(a) $V_s = V_t$　　(b) $V_t < V_s$

Fig. 2.7 *(a) Voltage source without load (b) Voltage source with load R_L*

is held into the voltage source? Basically, there is an internal impedance existing inside the source. This internal impedance is called as source impedance. If the source is DC, then the internal resistance will exist inside the source. Hence, the Fig. 2.7 (b) can be redrawn and shown in Fig. 2.8.

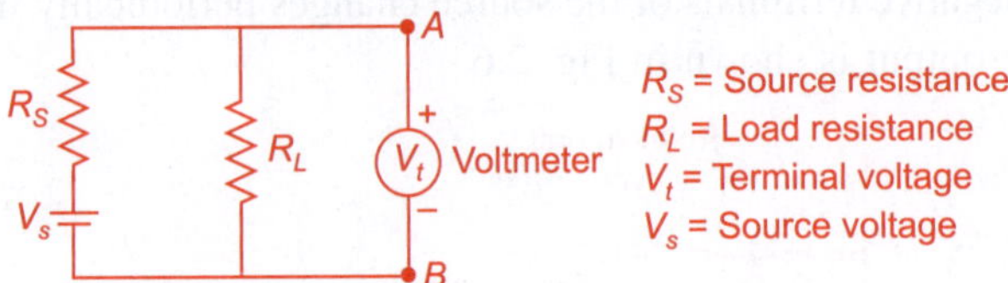

R_S = Source resistance
R_L = Load resistance
V_t = Terminal voltage
V_s = Source voltage

Fig. 2.8 *Voltage source with load resistance*

Main Reasons of Source Impedance

Following are the main reasons for source impedance

  (i) In battery, electrolyte is used between the positive and negative electrode. This electrolyte has some resistance.

  (ii) In AC and DC generator, armature winding is used. This armature winding has some resistance.

  (iii) Due to the resistance of active devices used in oscillator and rectifier circuit like diode transistor.

2.2.1 Concept of Voltage Source (Practical Source)

Source impedance is connected in series with the voltage source. This source impedance is also called as internal impedance of the voltage source. The voltage source AC and DC are shown in Figs. 2.9 (a) and (b) respectively. The terminal voltage V_t is equal to the open circuit voltage V_S when no load is connected.

    Let us consider a DC source. The variable load resistance R_L is connected across the voltage source. Let V_S be the open circuit voltage. Fig. 2.10 shows the concept of voltage source.

where,

$$R_S = \text{Internal resistance of the voltage source}$$
$$R_L = \text{Variable load resistance}$$
$$V_S = \text{Open circuit voltage of the voltage source.}$$

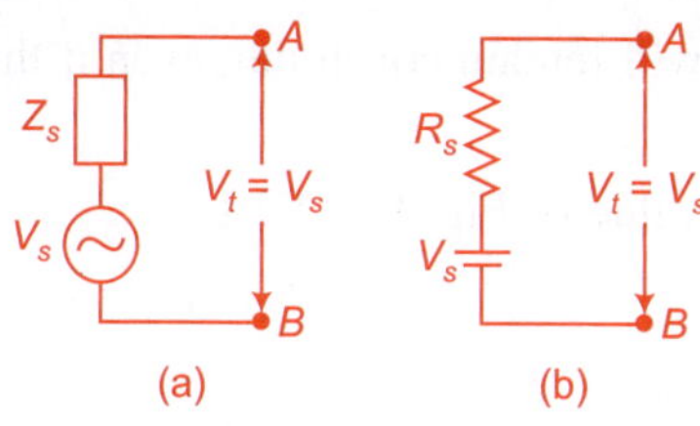

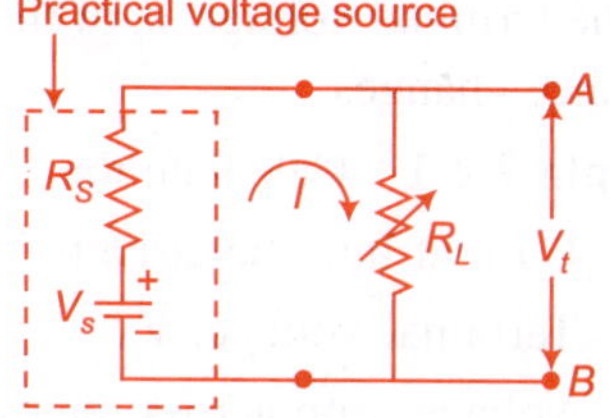

Fig. 2.9 *Symbol of voltage source*   **Fig. 2.10**

From Fig. 2.10, the equivalent resistance,

$$R_{eq} = R_S + R_L \tag{2.1}$$

The current flow into the circuit is I

i.e. $$I = \frac{V_S}{R_{eq}}$$

$$I = \frac{V_S}{R_S} + R_L \tag{2.2}$$

The terminal voltage across the load resistance,

$$V_t = V_L = I \times R_L$$

$$V_t = \frac{V_S}{R_S + R_L} \times R_L$$

$$V_t = \frac{V_S}{\dfrac{R_S}{R_L} + \dfrac{R_L}{R_L}} = \frac{V_S}{1 + \dfrac{R_S}{R_L}}$$

$$V_t = \frac{V_S}{1 + \dfrac{R_S}{R_L}} \cdot \tag{2.3}$$

From Eqn. (2.3), we can state that,

If, $$\frac{R_S}{R_L} \ll 1$$

Then, $V_t = V_S$.

So, we can say that the terminal voltage V_t will be equal to the open circuit voltage V_S, if $R_S \lll R_L$.

The terminal voltage V_t of the source will remain constraint, even if the load resistance changes.

Example 2.2.1 Calculate the following value of Fig. Ex. 2.2.1

  (i) Current into the circuit, I.

  (ii) Terminal voltage, V_t.

  (iii) Voltage drop across internal resistance.

Solution (i) From Eqn. (2.1), the equivalent resistance is,

$$R_{eq} = R_S + R_L$$
$$= 2 + 20 = 22 \ \Omega$$

we know that,

$$I = \frac{V_S}{R_{eq}} = \frac{12V}{22 \ \Omega} = 0.55 \text{ Amp.}$$

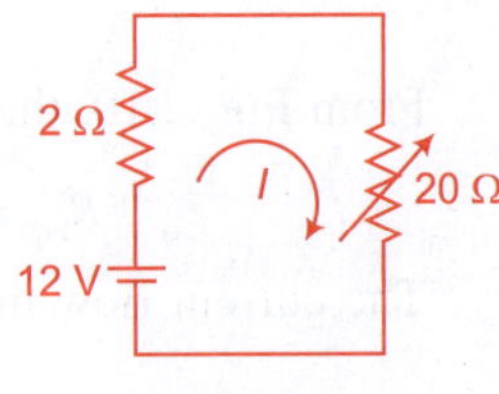

Fig. Ex. 2.2.1

(ii) We know that,

$$V_t = \frac{V_S}{1 + \dfrac{R_S}{R_L}} \quad \text{from Eqn. (2.3).}$$

$$V_t = \frac{12}{1 + \dfrac{2}{20}} = \frac{240}{22} = 10.9 \text{ volt.}$$

(iii) Voltage drop across $R_S = I \times R_S$

$$= 0.55 \times 2$$

$$= 1.1 \text{ volt.}$$

Hence, we see that 1.1 volt. Voltage drop occurs across the source resistance when 0.55 Amp current flows into the circuit. The terminal voltage of the battery is,

$$V_t = V_S - IR_S$$

$$= 12 - 1.1$$

$$\boxed{= 10.9 \text{ volt.}}$$

If the value of load resistance reduces, then the amount of current increases into the circuit. Due to this, the voltage drops and R_S will also increase. Hence, the terminal voltage will reduce.

"A practical voltage source consists of an ideal voltage source in series with the internal impedance of the source as shown in Figs. 2.9 (a) and (b)".

2.2.2 Concept of Current Source (Practical Source)

For practical current source, source impedance is connected parallelly with the current source. The AC and DC current sources are shown in Figs. 2.11(a) and (b) respectively.

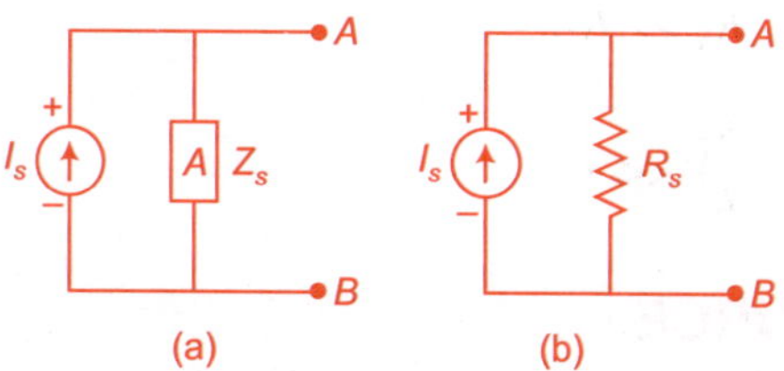

(a)   (b)

Fig. 2.11 *Symbols of current source*

Let us consider a DC current source. The load resistance R_L is connected across the current source. Let I_S be the current of the current source. Shows that

"A practical current source consists of an ideal current source in parallel with an impedance Z_S as shown in Figs. 2.11(a) and (b)."

From Fig. 2.12, the current I_S is given as:

$$I_S = I'_S + I_L$$

or, $$I'_S = I_S - I_L \qquad (2.4)$$

Since, the load resistance R_L is parallel with source resistance R_S, therefore the voltage drop across each of them must be equal,

i.e.

$$I'_S = I_L R_L$$

$\because$ $$I'_S = I_S - I_L$$

$\therefore$ $$(I_S - I_L)\, R_S = I_L R_L$$

$$I_S R_S - I_L R_S = I_L R_L$$

$$I_S R_S = I_L\,(R_S + R_L)$$

$$I_L = \frac{I_S R_S}{R_S + R_L}$$

$$I_L = \frac{I_S}{\dfrac{R_S}{R_S} + \dfrac{R_L}{R_S}} = \frac{I_S}{1 + \dfrac{R_L}{R_S}}$$

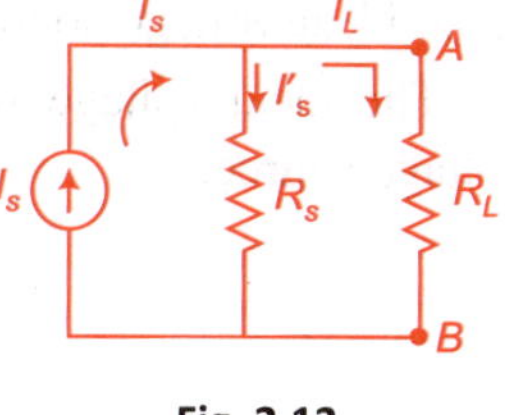

Fig. 2.12

$$I_L = \dfrac{I_S}{1 + \dfrac{R_L}{R_S}}. \tag{2.5}$$

From Eqn. (2.5), we can state that, the load current I_L will remain almost same as the current I_S if $\dfrac{R_L}{R_S} \ll 1$.

If,
$$R_S \gg R_L,$$

Then,
$$\dfrac{R_L}{R_S} \ll 1.$$

2.3 IDEAL SOURCE

In this section, we will study about ideal voltage and ideal current source. In the previous section we have studied about the practical source, in which there is existance of internal impedance of the source.

Ideal Voltage Source

We have seen in earlier section that when the load is connected across the voltage source, then the terminal voltage of the source is reduced due to the voltage drop (IR_S) across the internal impedance of the source.

IF $R_S = 0$ i.e., internal resistance or impedance of the voltage source is zero, then the terminal voltage of the voltage source will be constant and the terminal voltage V_t will be equal to the open circuit voltage (V_S) of the voltage source. Such voltage source is called an ideal voltage source. In other words, we can say that,

"An ideal voltage source is capable of supplying any amount of current at a constant voltage in a voltage source with zero internal impedance, and behaves as an ideal voltage source."

The ideal voltage source is shown in Fig. 2.13, Mathematically, for ideal voltage source,

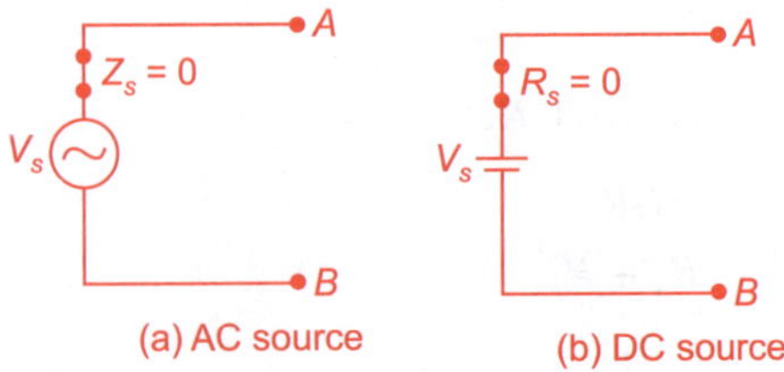

Fig. 2.13 *Ideal voltage source*

$$R_S = 0 \tag{2.6}$$

$$V_t = V_S, \text{ For all values of load impedance.} \tag{2.7}$$

Ideal Current Source

"Ideal current source is a source which is capable of supplying a constant current to a load, even if its load impendence varies. i.e., a current source with infinite internal impedance behaves as an ideal current source."

The ideal current source is shown in Fig. 2.14.

Mathematically, for ideal current source,

$$R_S = \infty \tag{2.8}$$

$$I_L = I_S \tag{2.9}$$

i.e., the load current is equal to the source current at every value of load.

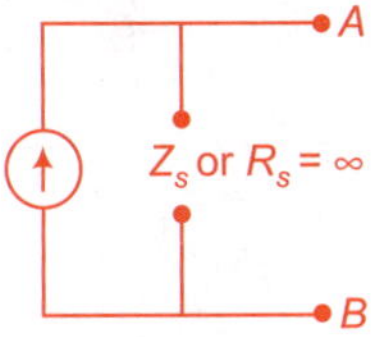

Fig. 2.14   *Ideal current source*

Existance of Ideal Current Source is Impossible, Why?

Practically, the existance of ideal current source is impossible due to the following reasons.

(a) According to the definition of an ideal current source, the current supplied by the current source is constant (I_S) at every value of load. If an open circuit is used in place of load impedance i.e., $Z_L = \infty$, then the current I_S must be supplied by the source. But the current in open circuit is zero practically because there is no current flow in an open circuit as shown in Fig. 2.15.

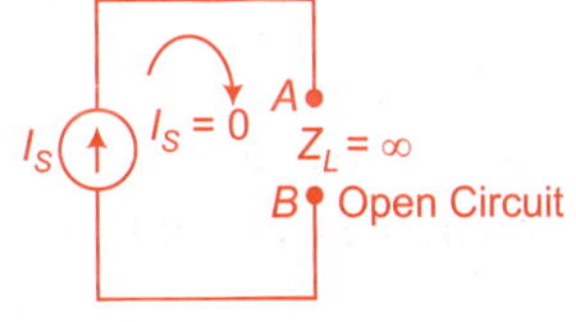

Fig. 2.15

if, $Z_L = \infty$ (Open circuit) $I_S = 0$ (Open circuit current).

(b) If it is possible to supply a constant current (I_S) to the load which has infinite value, then the voltage drop across the load will be infinite because there is no source that can supply infinite power. So, there is no existence of an ideal current source.

i.e.   if $Z_L = \infty$ and $I_S \neq 0$

Hence, $I_S Z_L = \infty$

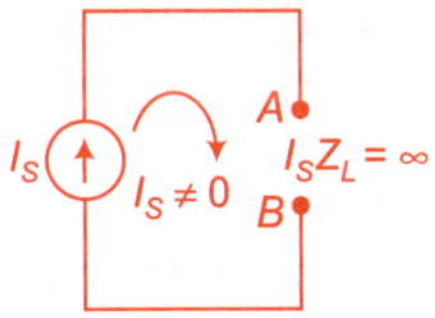

Fig. 2.16

i.e., Voltage drop across infinite load Z_L is also infinite. Hence, the power supply by the source must be infinite.

2.4 V-I CHARACTERISTICS OF PRACTICAL AND IDEAL SOURCES

The V-I Characteristics of ideal, practical voltage and current source is shown in Figs. 2.17 (a) and (b) respectively.

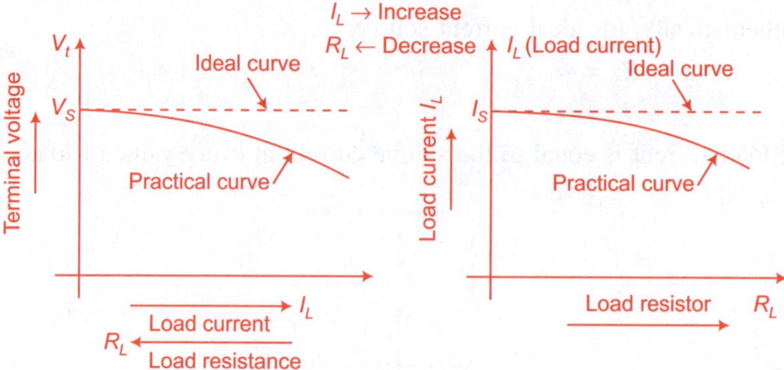

Fig. 2.17 *(a) V-I characteristics of voltage source (b) V-I characteristics of current source*

From Fig. 2.17 (a), we observe that the value of terminal voltage is constant when the load current increases for ideal curve but for pratical curve, the value of terminal voltage reduces with increasing load current.

$$V_t = V_S \text{ for ideal voltage source}$$

$$V_t = V_S - IR_S \text{ for practical voltage source.}$$

From Fig. 2.17 (b), we observe that the value of load current is constant and equal to source current for ideal curve but for practical curve the value of load current reduces with increase in the value of load resistor because,

$$I_L = I_S \text{ for ideal current source}$$

$$I_L = \frac{I_S}{1 + \dfrac{R_L}{R_S}} \text{ for practical current source.}$$

2.5 CONVERSION OF VOLTAGE SOURCE INTO CURRENT SOURCE AND VICE-VERSA

Voltage source can be converted into current source and current source can be converted into voltage source. In other words, we can say that the voltage source can be used as a current source and current source can be used as a voltage source, it depends upon the comparative value of load impedance Z_L and source impedance Z_S as follows:

(i) If the load impedance value is quite large as compared to the internal impedance Z_S of source.

i.e. $Z_L \gg Z_S.$

Then, it will be better to use the source as a voltage source.

(ii) If the load impedance (Z_L) value is quite small as compared to the internal impedance Z_S of source.

i.e. $Z_L \ll Z_S.$

Then, it will be better used as a current source.

The conversion of voltage source into current source and vice versa is very important and useful for the analytical purpose of the circuits, because at some place the analysis of the circuit is easier after conversion of the source.

2.5.1 Conversion of Voltage Source into Current Source

Let us consider a practical voltage source whose voltage is V_S and source impedance is Z_S, as shown in Fig. 2.18 (a). Now, we have to convert this voltage source into an equivalent current source. The current supplied by the required current source is I_S and the source impedance is Z_{S_2} as shown in Fig. 2.18 (b).

If the sources shown in Figs. 2.18 (a) and (b) are equivalent, then it must satisfy the following two conditions:

1. Short circuit current I_{SC} for both the circuits must be equal.

2. Open circuit voltage V_{OC} for both the circuits must be equal.

Condition 1 Short circuit current (I_{SC_1}) for Fig. 2.18 (a) is,

$$I_{SC_1} = \frac{V_S}{Z_{S_1}} \text{ from Fig. 2.19 (a)} \tag{2.10}$$

Similarly, short circuit current (I_{SC_2}) for Fig. 2.18 (b) is,

$$I_{SC_2} = I_S \tag{2.11}$$

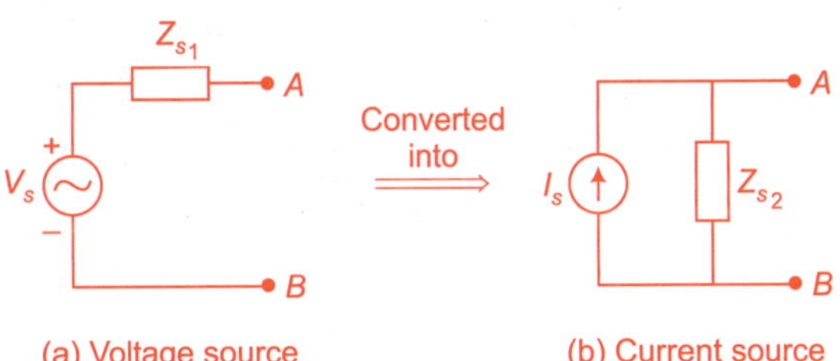

(a) Voltage source   (b) Current source

Fig. 2.18

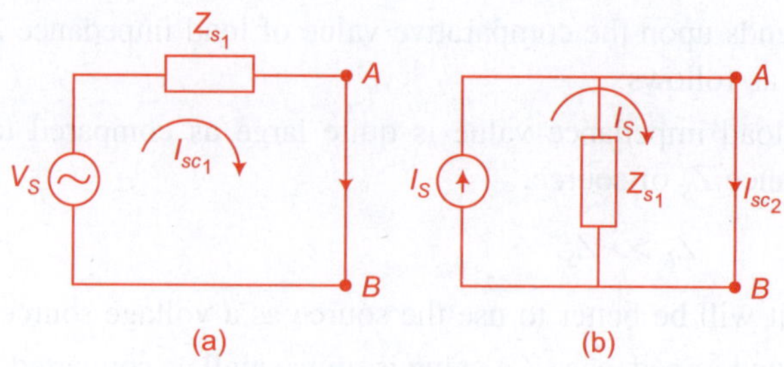

Fig. 2.19

because the entire current will flow through the short circuit path. There is no current flow through Z_{S_2}.

Equating Eqns. (2.10) and (2.11), we obtained:

$$I_{SC_1} = I_{SC_2}$$

$$\frac{V_S}{Z_{S_1}} = I_S \qquad (2.12)$$

Hence, the value of current source would be $\dfrac{V_S}{Z_{S_1}}$.

Condition 2 Open circuit voltage (V_{OC_1}) for Fig. 2.18 (a) is,

$$V_{OC_1} = V_S \text{ from Fig. 2.20 (a)} \qquad (2.13)$$

because the load current is zero.

Now, open circuit voltage (V_{oc_2}) for Fig. 2.18 (b) is,

$$V_{OC_2} = I_S Z_{S_2} \text{ from Fig. 2.20 (b).} \qquad (2.14)$$

Equating Eqns. (2.13) and (2.14), we obtain

$$V_{oc_1} = V_{oc_2}$$

$$V_s = I_s Z_{s_2}$$

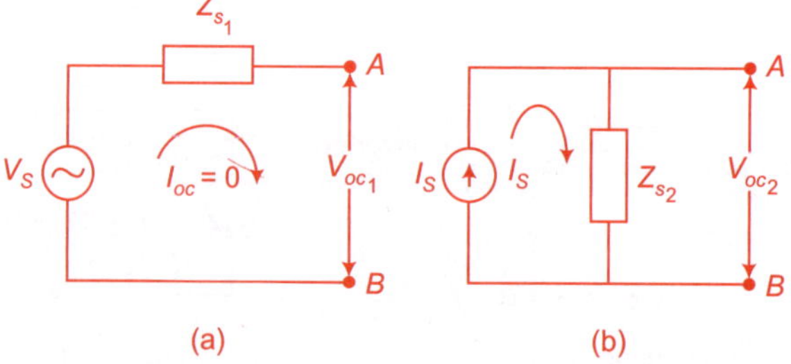

Fig. 2.20

$$\frac{V_s}{I_s} = Z_{s_2} \tag{2.15}$$

From Eqn. (2.12), we know that,

$$\frac{V_s}{I_s} = Z_{s1}$$

Hence, $\qquad\qquad Z_{s_1} = Z_{s_2} \tag{2.16}$

Hence, the equivalent current source of Fig. 2.18 (a) is shown in Fig. 2.21.

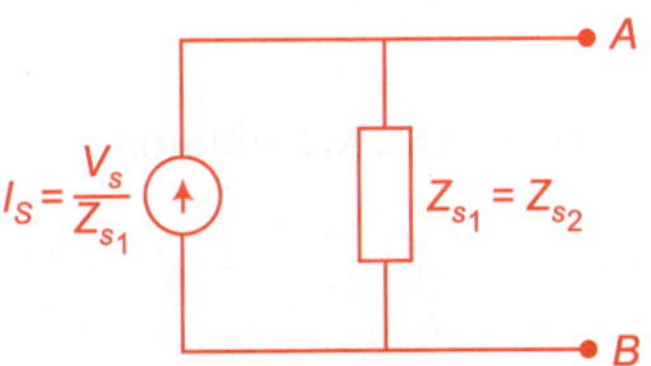

Fig. 2.21 *Equivalent current source after conversion*

2.5.2 Conversion of Current Source into Voltage Source

Let us consider a practical current source whose source current is I_s and source impedance is Z_{s_1} as shown in Fig. 2.22 (a). Now, we have to convert this current source into equivalent voltage source. The voltage of the required voltage source is V_s and the source impedance in Z_{s_2} as shown in Fig. 2.22 (b).

If the sources shown in Figs. 2.22 (a) and (b) are equivalent, then it must satisfy two conditions as explained earlier.

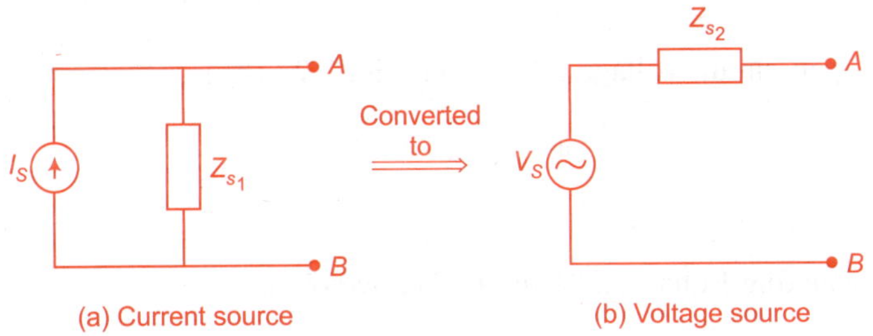

Fig. 2.22

Condition 1 Short circuit current (I_{sc_1}) for Fig. 2.22 (a) is,

$$I_{sc_1} = I_s \tag{2.17}$$

From Fig. 2.23 (a), we see that all the current will flow through the short circuit path. No current flows through Z_{s_1}.

Similarly, short circuit current (I_{sc_2}) for Fig. 2.22 (b) is,

$$I_{sc_2} = \frac{V_s}{Z_{s_2}} \tag{2.18}$$

from Fig. 2.23 (b)

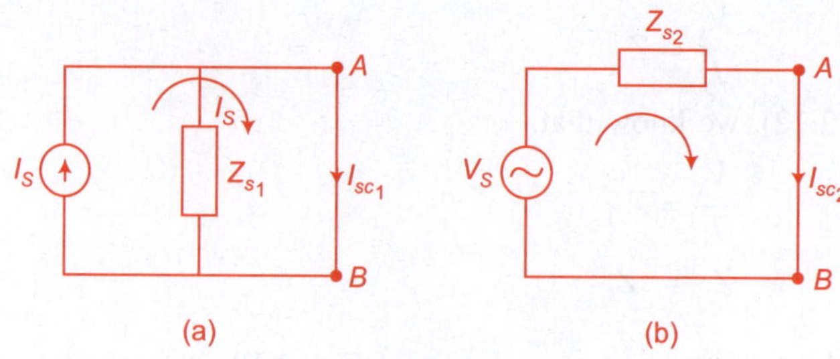

(a)   (b)

Fig. 2.23

Equating Eqns. (2.17) and (2.18), we obtain,

$$I_{sc_1} = I_{sc_2}$$

$$I_s = \frac{V_s}{Z_{s_2}}$$

$$V_s = I_s Z_{s_2} \tag{2.19}$$

Hence, the value of voltage source would be $I_s Z_{s_2}$.

Condition 2 Open circuit voltage (V_{oc_1}) for Fig. 2.22 (a) is,

$$V_{oc_1} = I_s Z_{s_1} \text{ from Fig. 2.24 (a)} \tag{2.20}$$

Now, open circuit voltage (V_{oc_2}) from Fig. 2.22 (b) is

$$V_{oc_2} = V_s \text{ from Fig. 2.24 (a)} \tag{2.21}$$

$\because$

$$I_{oc} = 0.$$

Now, equating Eqns. (2.20) and (2.21), we obtain,

$$V_{oc_1} = V_{oc_2}$$

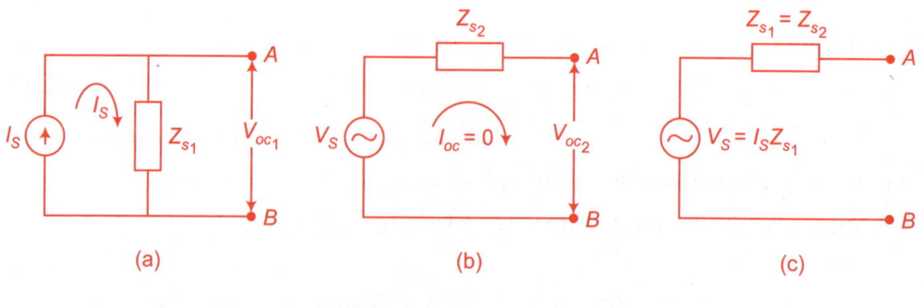

(a)   (b)   (c)

Fig. 2.24 *Equivalent voltage source after conversion*

$$I_s Z_{s_1} = V_s$$

$$\frac{V_s}{I_s} = Z_{s_1} \tag{2.22}$$

from Eqn. (2.19), we know that,

$$\frac{V_s}{I_s} = Z_{s_2}$$

Hence, $\qquad Z_{s_1} = Z_{s_2}.$ $\hfill$ (2.23)

The equivalent voltage source of Fig. 2.22 (a) is shown in Fig. 2.24 (c).

Example 2.5.1 Convert the voltage source into equivalent current source as shown in Fig. Ex. 2.5.1 (a).

$$V_s = 12\ V \quad \text{and } Z_{s_1} = 4\ \Omega.$$

Solution The value of source currents is given by Eqn. (2.12), we obtain

$$I_s = \frac{V_s}{Z_{s_1}} = \frac{12\ V}{4\ \Omega}$$

$$I_s - 3\ A.$$

The internal impedance of current source is given by Eqn. (2.16), we obtain

$$Z_{s_1} = Z_{s_2} = 4\ \Omega.$$

Hence, the equivalent current source is shown in Fig. Ex. 2.5.1 (b)

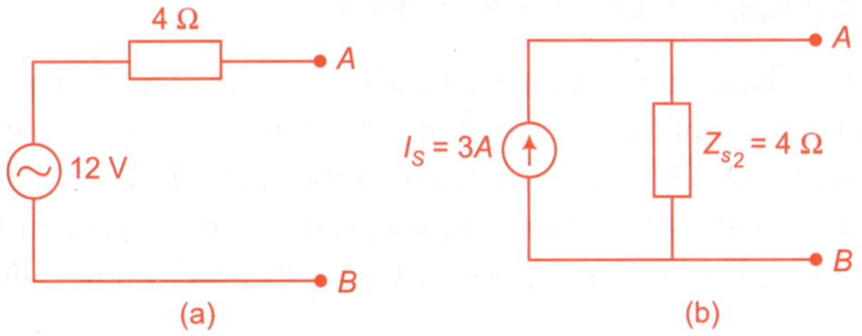

Fig. Ex. 2.5.1

Example 2.5.2 Connect the current source into equivalent voltage source as shown in Fig. Ex. 2.5.2 (a).

Solution The value of voltage source is given by Eqn. (2.21). we obtain,

$$V_s = I_s\ Z_{s_1}$$

$$V_s = 3 \text{ mA} \times 15 \text{ k}\Omega$$

$$V_s = 45 \text{ V}.$$

The internal impedance of voltage source is given by Eqn. (2.23), we obtain,

$$Z_{s_1} = Z_{s_2} = 15 \text{ k}\Omega.$$

Hence, the equivalent voltage source is shown in Fig. Ex. 2.5.2 (b)

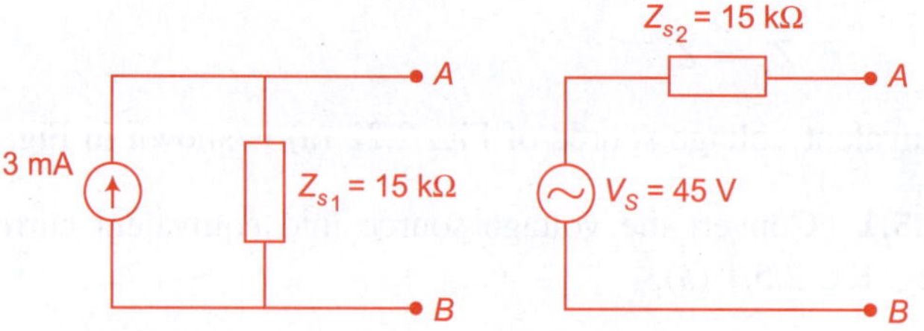

Fig. Ex. 2.5.2 *(a) and (b) Equivalent voltage source*

2.6 APPLICATIONS OF CONSTANT VOLTAGE SOURCES

The internal impedance of constant voltage source is very less as compared to load impedance. Due to this, the output voltage is approximately constant. The constant voltage source have various applications, such as the power is supplied to different electronic circuit used in TV, radio, calculator, tape recorder, computer etc., by DC constant voltage source. Similarly, AC voltage source operates different domestic appliances like bulb, fan, refrigerator, cooler, air conditioner etc.

2.7 GROUNDED POWER SUPPLY

The supply in which one terminal (Positive or Negative) is connected to the metal base or chasis of device, such power supply is called grounded power supply. The terminal which is connected to the metal base or chasis is known as reference terminal. All voltages are measured with respect to this terminal. The grounded power supply is shown in Fig. 2.25 in which negative terminal is grounded.

For example, if we say that the voltage at point A is 9 V, it means that the voltage drop between point A and ground is 9 V. In other words, we can say that the voltage level of point A is 9 V greater than the ground.

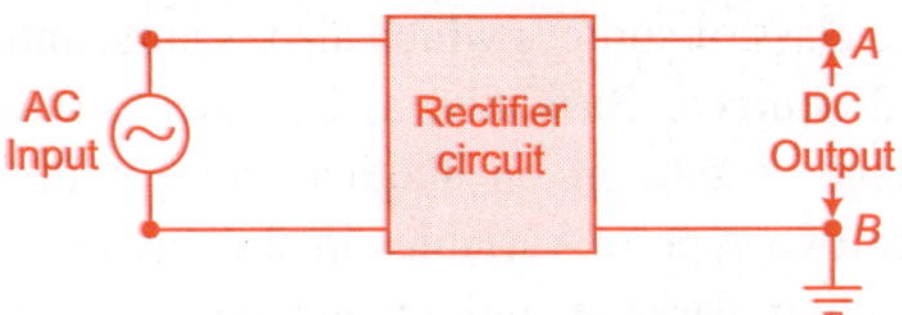

Fig. 2.25 *Grounded power supply*

Generally, the metal base or chasis is called as ground terminal. The point connected to chasis will be of same potential. It makes the electronic circuits easy.

If negative terminal of the supply is grounded, then the rating of the supply is 0–300 V and if the positive terminal is grounded, then the rating of the supply can be up to 0–6 V or 0–9 V.

2.8 FLOATING POWER SUPPLY

The power supply in which any terminal of terminals is not connected to metal base or chasis is called as floating power supply. In this, DC supply is connected across potential divider and the voltages are measured with respect to center point of potential divider i.e., the centre point of potential divider is called as reference terminal. All voltages of the circuit are measured with respect to this terminal as shown in Fig. 2.26

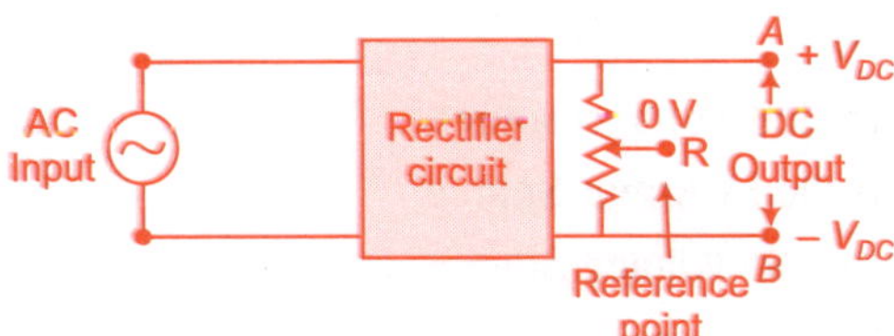

Fig. 2.26 *Floating power supply*

For example, if we say that the voltage at point A is 9 V. This means that the voltage drop between point A and reference point is 9 V. In other words, we can say that the voltage level of point A is 9 V greater than the reference point. If any voltage measured at point B is –9 V, then we will say that the voltage level at point B is 9 V lesser than the reference point.

2.9 DEPENDENT OR CONTROLLED SOURCES

We have discussed different sources in previous sections. These sources can deliver a constant voltage or current which is independent on the load impedance that is connected across it. These sources are called independent sources.

There is one more class of sources which also occurs, and is called dependent sources or controlled sources. Basically, a dependent source is a source that depends on the voltage or current on another element in the circuit. i.e., we can say that there are two type of variables in the circuit which has dependent source, one is controlling variable and second one is controlled variable. All dependent sources are linear elements, if k = constant and output is proportional to the controlling variable. The controlling variable is also called as 'cause' and controlled output is also called as 'effect'.

These dependent or controlled sources consist of four terminals (Two ports), instead of two terminals as in independent sources as shown in Fig. 2.27. Thus, there are two input and two output terminals.

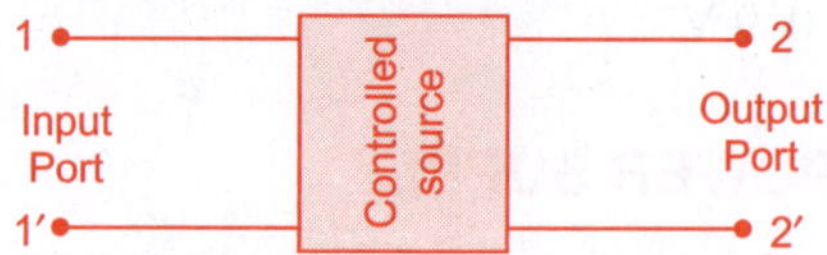

Fig. 2.27 *A dependent or controlled source*

In Fig. 2.27, the Port 1 is an input port and Port 2 is an output port. The output of controlled source is dependent upon the input applied on the controlled source.

Basically, the dependent sources are classified according to the input applied at the source as follows:

1. Voltage Controlled Voltage Source
2. Voltage Controlled Current Source
3. Current Controlled Voltage Source
4. Current Controlled Current Source

2.9.1 Voltage Controlled Voltage Source

Figure 2.28 shows a voltage controlled voltage source. The control voltage V_c is applied at the input terminals and the voltage source of kV_c (effect) volts acts at output terminal as Voltage dependent voltage source because its value is dependent upon the control voltage V_c (cause). The constant k relates the strength of the source with the control voltage and is known as voltage gain (A_v).

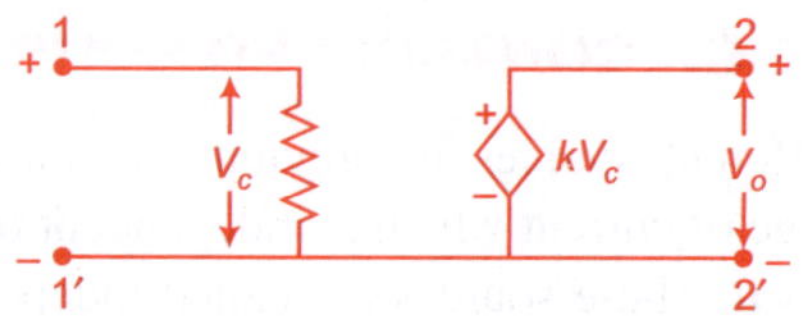

Fig. 2.28 *Voltage controlled voltage source*

2.9.2 Voltage Controlled Current Source

Figure 2.29 shows a voltage controlled current source. The control voltage V_c (cause) is applied at the input terminals and the current source of kV_c (effect). Amp acts at output terminals as a voltage dependent current source because its value is dependent upon the control voltage V_c (cause).

The constant k relates the strength of the source with the control voltage (V_c) and is known as transconductance or mutual conductance (g_m).

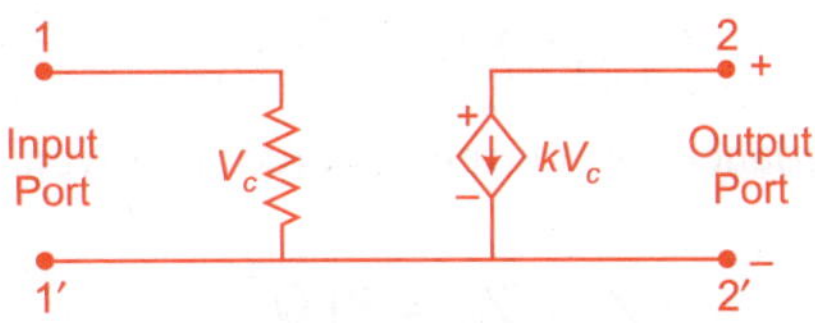

Fig. 2.29   *A voltage controlled current source*

2.9.3 Current Controlled Voltage Source

Figure 2.30 shows a current controlled voltage source. The control current I_c (cause) flows at input terminal and the voltage source of kI_c (effect) Amp acts at output terminal as a current dependent voltage source because its value is dependent upon the control current I_c (cause). The constant k relates the strength of the source with the control current (I_c) and is known as transresistance (r_m).

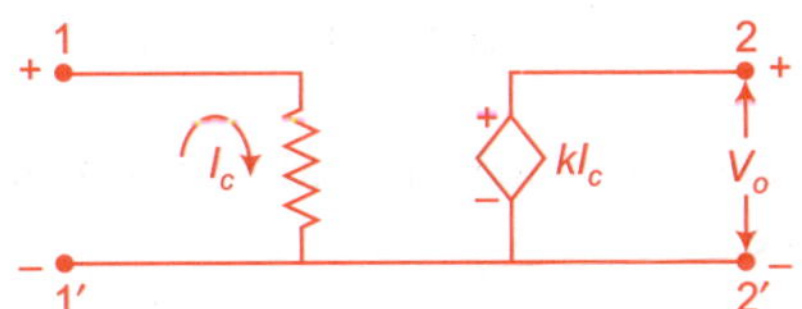

Fig. 2.30   *Current controlled voltage source*

2.9.4 Current Controlled Current Source

Figure 2.31 shows a current controlled current source. The control current I_c (cause) flows at input terminal and the current source of kI_c (effect) Amp acts at output terminal as a current dependent current source because its value is dependent upon the control current I_c (cause). The constant k relates the strength of the source with the control current (I_c) and is known as current gain (A_i).

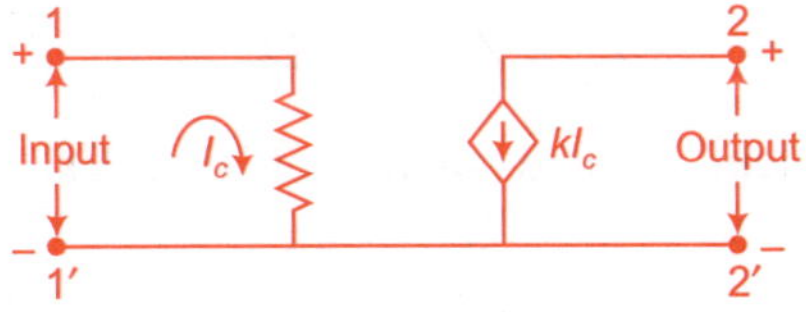

Fig. 2.31   *Current controlled current source*

Examples

Example 2.1 A 240 V source has 30 Ω resistance in series. Find its equivalent current source.

Solution The strength of current source is given by Eqn. (2.12) we obtain,

$$I_S = \frac{V_S}{Z_{S_1}} = \frac{240\ V}{30\ \Omega}$$

$$\boxed{I_S = 8A.}$$

The internal impedance of current source is given by Eqn. (2.16), we obtain,

$$Z_{S_1} = Z_{S_2} = 30\ \Omega.$$

Hence, the equivalent current source of Fig. Ex. 2.1(a) is shown in Fig. Ex. 2.1(b)

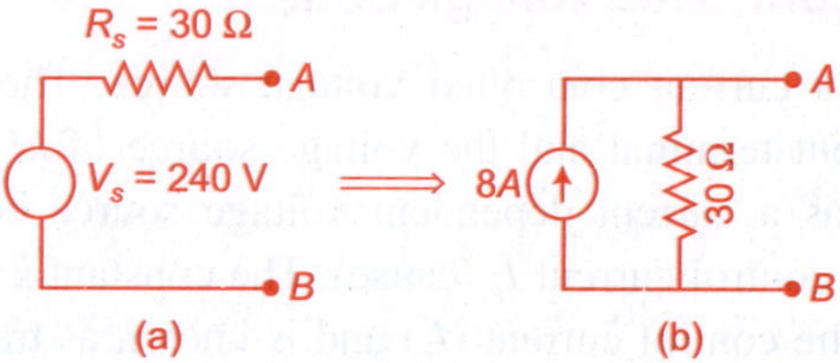

Fig. Ex. 2.1

Example 2.2 A 5 A current source has 20 Ω resistane in parallel. Find its equivalent voltage source.

Solution Given that,

$$I_S = 5\ A, \quad Z_{S_1} = 20\ \Omega$$

The strength of current source is given by Eqn. (2.22), we obtain,

$$V_S = I_S\ Z_{S_1} = 5\ A \times 20\ \Omega$$

$$\boxed{V_S = 100\ \text{volt.}}$$

The internal impedance of voltage source is given by Eqn. (2.23), we obtain,

$$Z_{S_1} = Z_{S_2} = 20\ \Omega.$$

Hence, the equivalent voltage source of Fig. Ex. 2.2 (a) is shown in Fig. Ex. 2.2 (b).

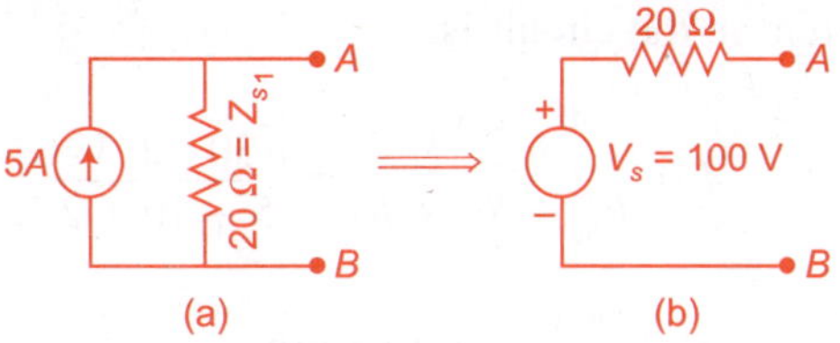

Fig. Ex. 2.2

Example 2.3   Calculate current in 10 Ω resistance by source transformation of Fig. Ex. 2.3 (a).

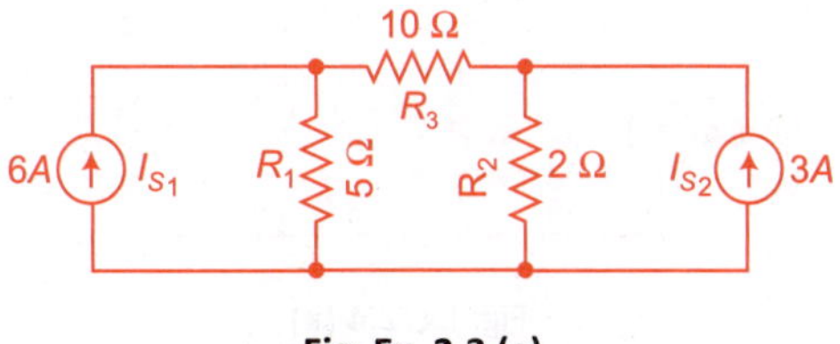

Fig. Ex. 2.3 (a)

Solution   The equivalent voltage source of 6 A, 5 Ω current source is,

$$V_{S_1} = I_{S_1} \times R_1$$

$$V_{S_1} = 6\,A \times 5\,\Omega - 30\,V$$

$$R_{S_1} = 5\,\Omega$$

The equivalent voltage source of 3 A, 2 Ω current source is,

$$V_{S_2} = I_{S_2} \times R_2$$

$$V_{S_2} = 3\,A \times 2\Omega = 6\,V$$

$$V_{S_2} = 6\,V$$

$$R_{S_2} = 2\,\Omega.$$

The equivalent circuit of Fig. Ex. 2.3 (a) is shown in Fig. Ex. 2.3 (b).

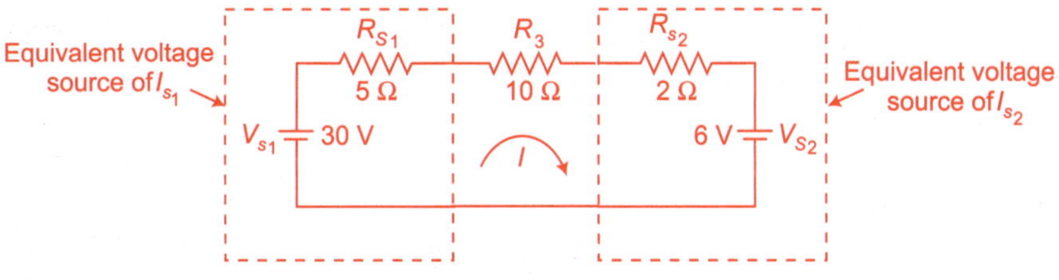

Fig. Ex. 2.3 (b)

Hence, the current in the circuit is,

$$I = \frac{V_{S_1} - V_{S_2}}{R_{S_1} + R_{S_2} + R_3} = \frac{30 - 6\ V}{5 + 10 + 2}$$

$$I = \frac{24\ V}{17\ \Omega} = 1.41\ \text{AMP}$$

$$\boxed{I = 1.41\ \text{A.}}$$

(**Ans.**)

Example 2.4   Calculate the load current for given circuit in Fig. Ex. 2.4 (a) by source transformation.

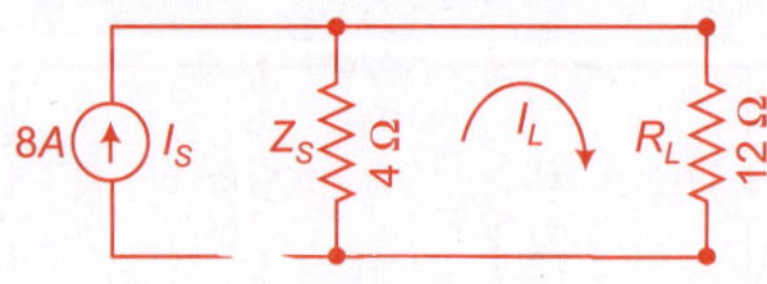

Fig. Ex. 2.4 (a)

Solution   The equivalent voltage source of 8 A, 4 Ω current source is,

$$V_S = I_S \times Z_{S_1}$$
$$V_S = 8\ \text{A} \times 4\ \Omega$$
$$V_S = 32\ V$$

and,
$$R_s = 4\ \Omega.$$

The equivalent voltage source of 8 A, 4 Ω current source is shown in Fig. Ex. 2.4 (b).

The load current is given by,

$$I_L = \frac{V_S}{R_S + R_L} = \frac{32\ V}{4\ \Omega + 12\ \Omega}$$

$$I_L = \frac{32\ V}{16\ \Omega} = 2\ \text{A}$$

$$I_L = 2\ \text{A}$$

(**Ans.**)

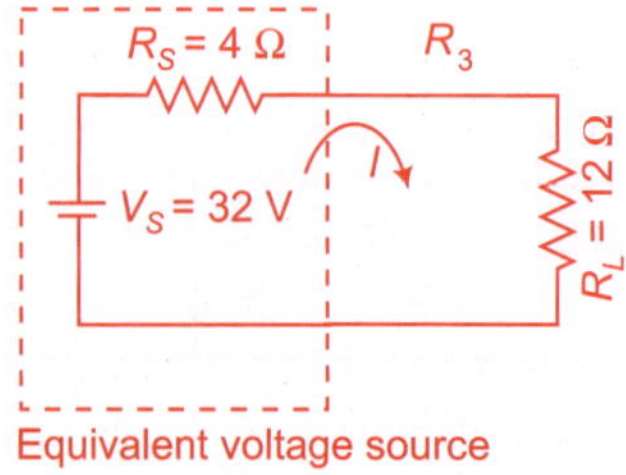

Fig. Ex. 2.4 (b)

Objective Type Questions

1. A source whose internal impedance is 50 Ω. At what value of load resistance, will give the maximum power?
    (a) 100 Ω (b) 25 Ω
    (c) 50 Ω (d) None of these

2. The e.m.f. of the cell is 8 V and the terminal voltage V_t is 6 Volt. If 1 A current is flowing through the cell, what will be the internal impedance of the cell?
    (a) 1 Ω (b) 2 Ω
    (c) 10 Ω (d) 5 Ω

3. .... work as a voltage source.
    (a) Transistor   (b) Tunnel diode
    (c) Zener diode   (d) Photo diode

4. The internal impedance of the ideal current source is
    (a) Infinite   (b) Zero
    (c) Low   (d) High

5. The internal impedance of the ideal voltage source is
    (a) Infinite   (b) Zero
    (c) Low   (d) High

6. A 12 V, 3 Ω voltage source connects .... into current source.
    (a) 12 A, 12 Ω (b) 6 A, 3 Ω
    (c) 4 A, 6 Ω (d) 4 A, 3 Ω

7. The value of terminal voltage of practical voltage source .... with increasing load current.
    (a) Increases   (b) Stable
    (c) Decreases   (d) None of these

8. The example of DC source is
    (a) Oscillator   (b) Alternator
    (c) Signal generator   (d) Battery

9. The equivalent voltage source of 2 mA current and 75 kΩ resistance of practical current source is
    (a) 100 V, 75 kΩ in series   (b) 150 V, 75 kΩ in series
    (a) 150 V, 75 kΩ in parallel   (d) None of these

10. Which one is true in the following?
    (a) The internal impedance of ideal current source is zero
    (b) The internal impedance of ideal voltage source is zero

    (c) Oscillator is a DC source

    (d) None of these

11. Which one is true in the following?

    (a) Primary cell is used for short time of energy

    (b) Cell is an example of rectifier

    (c) The value of primary cell can be changed

    (d) None of these

12. Which one is false in the following?

    (a) Secondary cell can be recharged

    (b) Oscillator is an AC source

    (c) Chemical energy is converted into electrical energy by cell

    (d) The internal impedance of secondary cell is very high

13. Example of an AC source is

    (a) Cell   (b) Rectifier

    (c) Oscillator   (d) DC generator

14. The equivalent current source of 10 V and 10 Ω practical voltage source is

    (a) 1 A, 1 Ω in series   (b) 1 A, 10 Ω in parallel

    (c) 1 A, 10 Ω in series   (d) A, 1 Ω in parallel

15. In Fig. OT 2.1, the value of V is

    (a) 8 V

    (b) 4 V

    (c) 6 V

    (d) 2 V

16. In Fig. OT 2.2, the value of I is

    (a) 30 A   (b) 20 A

    (c) 10 A   (d) 5 A

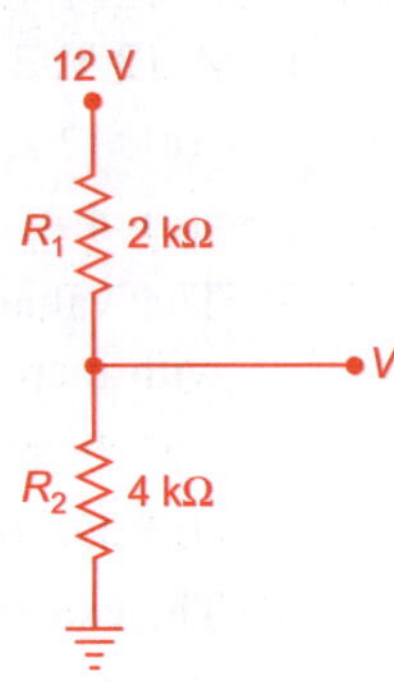

Fig. OT 2.1

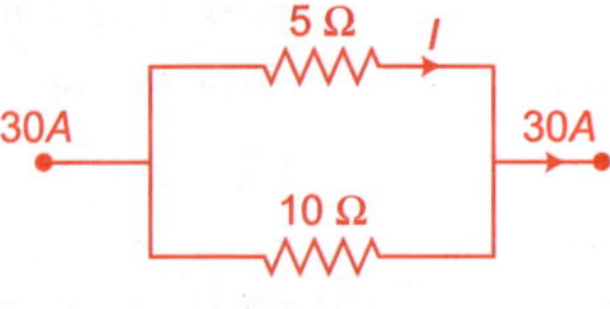

Fig. OT 2.2

17. In Fig. OT 2.3, the value of I is

    (a) 20 A   (b) 30 A

    (c) 5 A   (d) 40 A

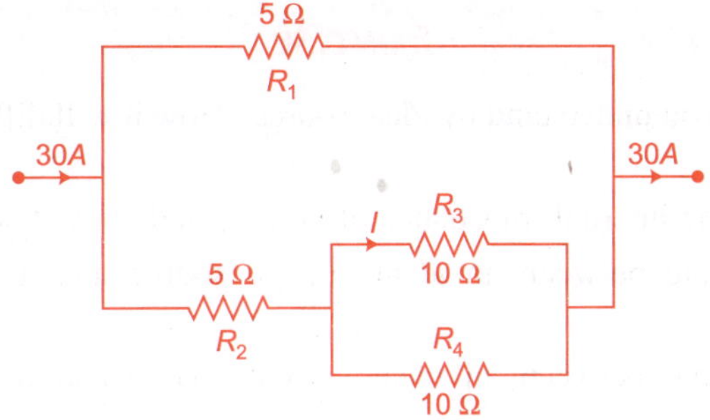

Fig. OT 2.3

18. In Fig. OT 2.3, the voltage drop across R_2 is
    - (a) 50 V
    - (b) 100 V
    - (c) 25 V
    - (d) 75 V

19. In Fig. OT 2.3, the voltage drop across R_1 is
    - (a) 50 V
    - (b) 100 V
    - (c) 20 V
    - (d) 10 V

20. In Fig. OT 2.3, the voltage drop across R_3 is
    - (a) 50 V
    - (b) 100 V
    - (c) 150 V
    - (d) 20 V

21. Which of the following has a high internal resistance?
    - (a) Constant current source
    - (b) Constant voltage source
    - (c) Both (a) and (b)
    - (d) None of these

22. A constant voltage source has
    - (a) High internal resistance
    - (b) Low internal resistance
    - (c) Minimum efficiency
    - (d) Minimum current capacity

23. 20-Zinc carbon cells, each of 1.5 V, are connected in series, the total output voltage will be
    - (a) 1.5 V
    - (b) 40 V
    - (c) 30 V
    - (d) 60 V

24. A DC current source has a short circuit current of 5 mA and an open circuit voltage of 20 V. The internal resistance is
    - (a) 5 kΩ
    - (b) 10 kΩ
    - (c) 4 kΩ
    - (d) 100 kΩ

ANSWERS

1. (c)	2. (b)	3. (c)	4. (a)	5. (b)	6. (d)	7. (c)
8. (d)	9. (b)	10. (b)	11. (c)	12. (d)	13. (c)	14. (b)
15. (a)	16. (b)	17. (c)	18. (a)	19. (b)	20. (a)	21. (a)
22. (b)	23. (c)	24. (c)				

Exercise

2.1. What do you understand by ideal source? How it will differ from practical source?

2.2. Explain the internal impedance of voltage and current source.

2.3. Differentiate between an ideal voltage source and a practical voltage source.

2.4. Differentiate between an ideal current source and a practical current source.

2.5. What is a constant voltage source?

2.6. What do you understand by grounded and floating DC power supply? What is the need of it?

2.7. How can we convert a current source into voltage source?

2.8. Write short notes on the following:

    (a) Voltage source   (b) Current source

    (c) Dependent voltage source.

2.9. Convert 1 mA current and 100 kΩ resistance, current source into equivalent voltage source.

2.10. Name different types of sources which are used for operation of electronic circuits and explain them briefly.

Semiconductor Physics   3

3.1 INTRODUCTION

In this chapter we will tell you about semiconductor. The text given in this chapter will help you to understand different electronic devices like diode, bipolar transistor, field effect transistor etc. So, we can say that this chapter is the heart of electronic. We will study different types of semiconductors and the effect of temperature on them. The basic concept behind using the semiconductor material for manufacturing the electronic devices is that the current in a semiconductor is due to the flow of both negative charges (electrons) and positive charges (holes) whereas the current in a metal is due to the flow of negative charges (electrons) only.

3.2 CLASSIFICATION OF SOLIDS

The solids can be classified in three parts according to conductivity.

3.2.1 Conductor

A conductor is the material which supports the flow of current (or charge) when a voltage of limited magnitude is applied across it. In other words, we can say that the conductor is the material in which number of free electrons are more (approximately 10^{22} per unit volume), due to this, after applying small amount of electric field, the current will flow easily. The conductivity of such metals are more and the resistivity is less. Generally, all the metals and alloys are conductors like silver, aluminium etc.

3.2.2 Insulator

An insulator is a material that resists the flow of current. It has atoms with strongly bonded valence electrons. In other words, we can say that an insulator is a material that does not have free electrons, due to this there is no flow of current, after applying high electric field. So, we can say that the electric resistivity of insulator is high and electric conductivity is very low, e.g., glass, paper, rubber, mica, teflon etc. These materials are used in parts of electrical equipment for insulation, intended to support or separate electrical conductors without passing current through themselves.

A large class of materials, such as rubber and plastics are good enough to insulate electric wiring and cables.

3.2.3 Semiconductor

A semiconductor is a material that has an electrical conductivity between an insulator and a conductor, that is, generally in the range of 10^3 to 10^{-8} siemens/meter like germanium, silicon etc. Semiconductor materials play a vital role in the manufacturing of electronic devices like radio, telephones, computer and many other devices. The beauty of semiconductor material is that an external electrical field may change a semiconductor's resistivity. In other words, we can say that the flow of current in semiconductor material can be controlled by applying external electrical field. In semiconductor materials, current can be carried either by the flow of electrons (negative charges) or by the flow of holes (positive charges). Where as in a conductor, current is carried by the flow of electrons only. Basically, the semiconductor materials are of two types, as explained below:

3.2.3.1 *Intrinsic semiconductor*

A pure semiconductor is often called an intrinsic semiconductor. In other words, we can say that an intrinsic semiconductor is the one which is made of the semiconductor material in its extremely pure form. Basically, intrinsic semiconductors have been carefully refined to reduce the impurity to a very low level and as pure as possible through modern technology like silicon (Si). Germanium (Ge), Selenium (Se), Gallium-Arsenide (GaAs) etc.

Atomic structure of silicon and germanium

As you know that the atom is made of the following small particles

   (i) Proton   (ii) Neutron

 (iii) Electron.

The protons and neutrons are bound with each other in the nucleus where as the electrons move around the nucleus in to the different elliptical orbits.

For Si;   Number of atoms are 14

               Number of neutrons are 14

               Number of atoms = Number of protons

                           = Number of electrons.

According to Fig. 3.1, there are 14 electrons revolving around the nucleus; two in the first orbit, eight in the second orbit and four in the third (outermost) orbit. Thus, there are four valence electrons called tetravalent. The crystal structure of silicon is shown in Fig. 3.2.

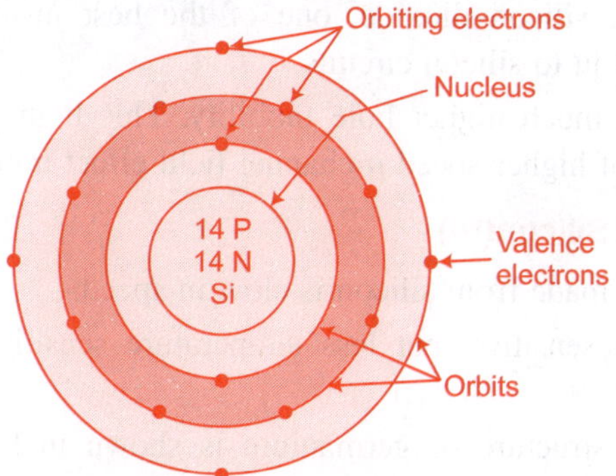

Fig. 3.1 *Atomic structure of Silicon*

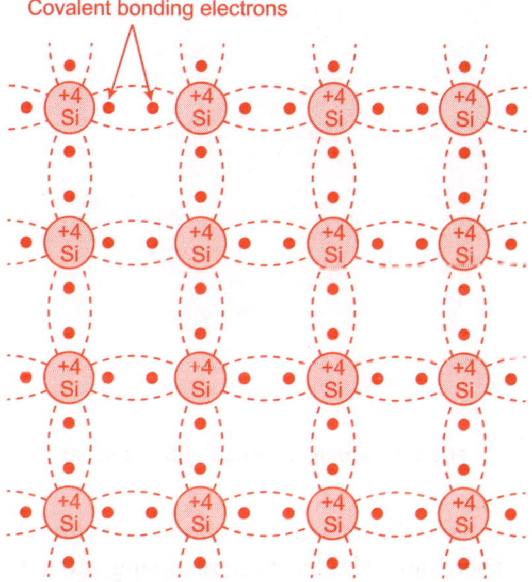

Fig. 3.2 *Crystal structure of silicon*

Silicon has many industrial uses. It is the principal component of most semiconductor devices, most importantly Integrated Circuits (IC's) or microchips. Silicon is widely used in miconductors because it remains a semiconductor germanium since it is native oxide and is easily grown in a furnace. It forms a better semiconductor/dielectric interface than any other material.

Advantages of Silicon (Si)

1. Silicon is abundant and cheap to process. Si is highly abundant in the Earth's crust, in the form of silicate minerals.

2. Existence of silicon dioxide, one of the best insulator can easily be incorporated in to silicon circuits.

3. It possess a much higher hole mobility. This high mobility allows the fabrication of higher-speed *p*-channel field effect transistors.

Disadvantages of Silicon(Si)

1. The devices made from silicon is slow in speed.

2. Temperature sensitive, but less temperature sensitive than germanium. For Ge;

   The atomic structure of germanium is shown in Fig. 3.3. Number of atoms are 32

   Number of neutrons are 41

Number of electrons = Number of protons = Number of atoms.

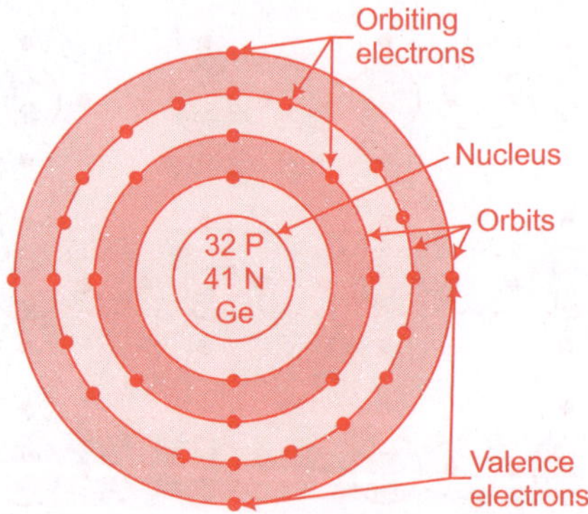

Fig. 3.3 *Atomic structure of Germanium*

According to Fig. 3.3, the nucleus of Germanium atoms contain 32 protons and 41 neutrons. There are 32 electrons revolving around the nucleus; two electrons in the first orbit, eight in the second orbit, eighteen in the third orbit and four in the fourth (outermost) orbit. Thus, germanium like silicon has four valence electrons. Due to this, many properties of germanium and silicon are similar and are called tetravalent. Crystal structure of germanium is similar to the structure of silicon as shown in Fig. 3.2.

Germanium is an important semiconductor material used in transistors (explained in later chapters) and various other devices. It's major end uses are fibre-optic systems and infrared optics. It is also used for polymerization catalyst, in electronics and in solar electric applications.

Advantages of Germanium (Ge)

1. It is easy to find and is available in fairly large quantities.
2. It is easy to refine and obtain very high level of purity.

Disadvantages of Germanium (Ge)

1. It has low level of reliability.
2. More temperature sensitive as compared to Silicon.

Gallium arsenide (GaAs)

Gallium-Arsenide (GaAs) is a compound of the elements Gallium and Arsenic. It is an important semiconductor and is used in the manufacturing of devices such as microwave frequency integrated circuits like monolithic microwave integrated circuits, infrared light-emitting diodes, laser diodes, solar cells etc.

According to Fig. 3.4, there are thirty one electrons revolving around the nucleus of Gallium; two in the first orbit, eight in the second orbit, eighteen in the third orbit and three in the fourth (outermost orbit).

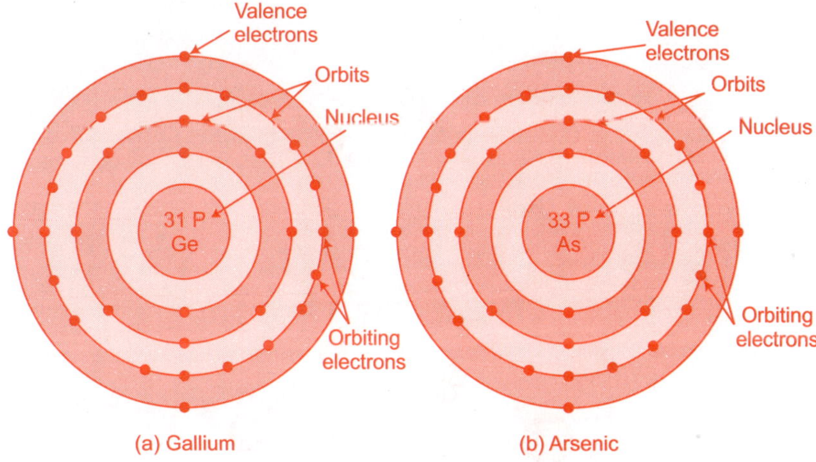

Fig. 3.4 *Atomic structure of Gallium and Arsenic*

Thus, there are three valence electrons. There are 33 electrons revolving around the nucleus of Arsenic; two in the first orbit, eight in the second orbit, eighteen in the third orbit and five in the fourth (outermost) orbit. Thus, there are five valence electrons, so Gallium is called trivalent, and arsenic with five valence electrons is called pentavalent. The crystal structure of GaAs is shown in Fig. 3.5 (a).

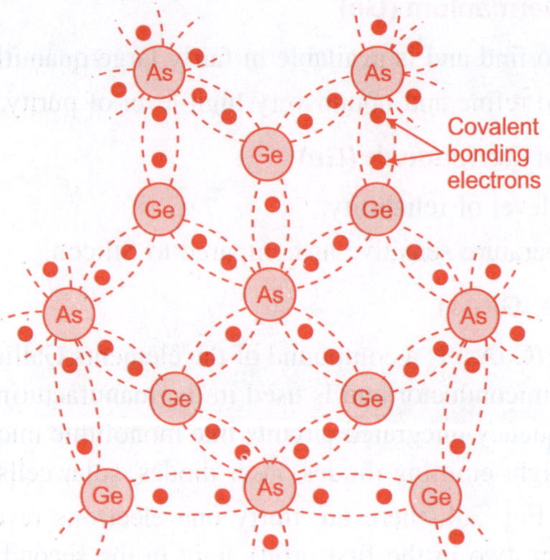

Fig. 3.5 (a) *Crystal structure of GaAs*

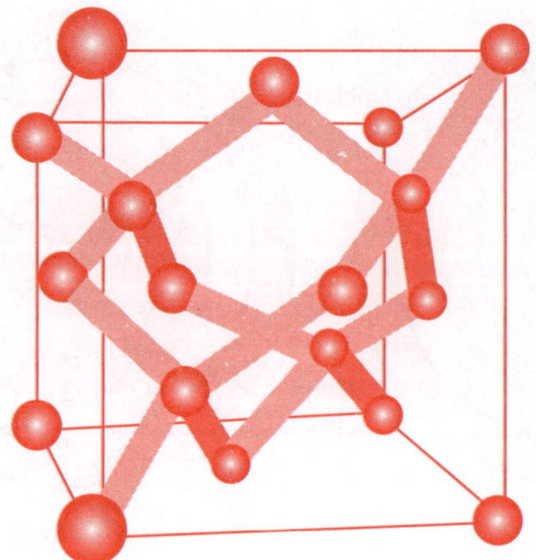

Fig. 3.5 (b) *GaAs until cell 3D balls*

Advantages of GaAs

1. It has higher saturated electrons velocity and higher electron mobility.
2. It is insensitive to heat.
3. Devices made from it, function at higher frequency.
4. Devices made from it, generate less noise when operated at high frequency.
5. They can be operated at higher power level because they have higher breakdown voltages.

6. It has direct bandgap, where silicon has an indirect band gap. Due to this, GaAs emits light efficiently.

Disadvantages of GaAs

1. It is difficult to manufacture high levels of purity as compared to silicon.

2. It is more expensive as compared to silicon.

Application of GaAs

Due to different advantages as discussed above, GaAs is used in mobile phones, satellite communication, microwave, point to point links and some radar systems. It is also used in manufacturing of Gunn diodes for generation of microwaves.

Charge carriers in Intrinsic Semiconductor

We have already studied that there are four valence electrons in Silicon and Germanium. These four electrons make covalent bond with adjacent electrons. There are two electrons in one covalent bond in which both are for different atoms as shown in Fig. 3.2. At absolute zero temperature, all valence electrons are tightly bound with atoms. So, we can say that at absolute zero temperature there is no free electron in semiconductor and it is treated as insulator.

At the room temperature, some bonding electrons get sufficient energy to break covalent bond. When an electron breaks a covalent bond and moves away, a vacancy is created in the broken covalent bond. This vacancy is called a hole. So we can say when a free electron is generated, a hole is created simultaneously. It means, free electrons and holes are always generated in pairs. Therefore, the concentration of free electrons and holes will always be equal in an intrinsic semiconductor. This type of generation of free electron-hole pair is called as thermal generation as shown in Fig. 3.6.

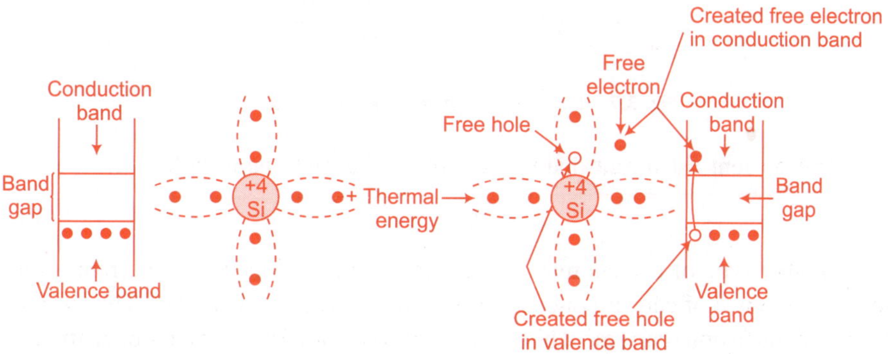

Fig. 3.6   *Generation of electron hole pair due to thermal energy*

Since, an electron is negatively charged, the place of a hole will be left with a net positive charge (equal in magnitude to the charge of electron). Thus, we can say that a positive charge is associated with a hole and the hole is positively charged.

Conduction in Intrinsic Semiconductor

When we connect a battery across a semiconductor as shown in Fig. 3.7, the electron experiences a force towards the positive terminal of the battery and holes experience a force towards negative terminal of battery. The free electrons move towards the positive terminal and the holes toward the negative terminal of the battery. As we know that the electric current flows through the semiconductor in the opposite direction in which the electrons are moving. Although the two charge carriers (electrons and holes) move in opposite direction, the two currents in the same direction, one current is due to the flow of electrons and is represented by I_e and the second flow of drift current is due to the flow of holes and is represented by I_h, they add together. When the flow of carriers is due to an applied voltage, the resultant current is called a drift current.

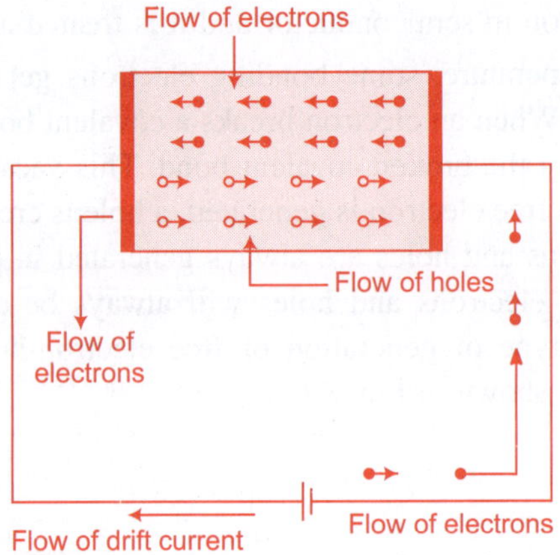

Fig. 3.7 *Current flow in intrinsic semiconductor*

Drift current = Current due to electrons + Current due to holes

$$I = I_e + I_h. \tag{3.1}$$

Another type of current may also exists in semiconductor. This current is due to the difference of carrier concentration from one region to another. This current is called diffusion current. This current flows until the carrier concentration is uniformly distributed in complete crystal structure of semiconductor. The diffusion current is also due to the motion of both holes and electrons.

Semiconductors have negative temperature coefficient of resistance. So, we can say that when we raise the temperature further, more electron hole pairs are generated. That means higher the temperature, higher is the concentration of charge carriers, so the conductivity of an intrinsic semiconductor increase with temperature. In other words, the resistivity decreases.

Table 3.1

Material	Resistivity ρ (Ωm)	Conductivity $\sigma = \dfrac{1}{\rho}$ $(\Omega^{-1}m^{-1})$	Classification
Silver (Ag)	10^{-9}	10^{9}	
Copper (Cu)	10^{-9}	10^{9}	Conductor
Iron (Fe)	10^{-7}	10^{7}	
Germanium (Ge)	0.5	2	Semiconductor
Silicon (Si)	500	0.002	
Backelite	10^{9}	10^{-9}	Insulator
Mica	10^{13}	10^{-13}	

Energy band diagram of metals, insulators and semiconductors

At room temperature, metals such as copper or silver contain large number of free electrons. There is no forbidden-energy gap between the valence and conduction bands. These bands overlap. Due to the overlapping of bands as shown in Fig. 3.8, the conduction band energies are same as valence band energies in the metal and it has large number of free electrons. That is why it works as a good conductor.

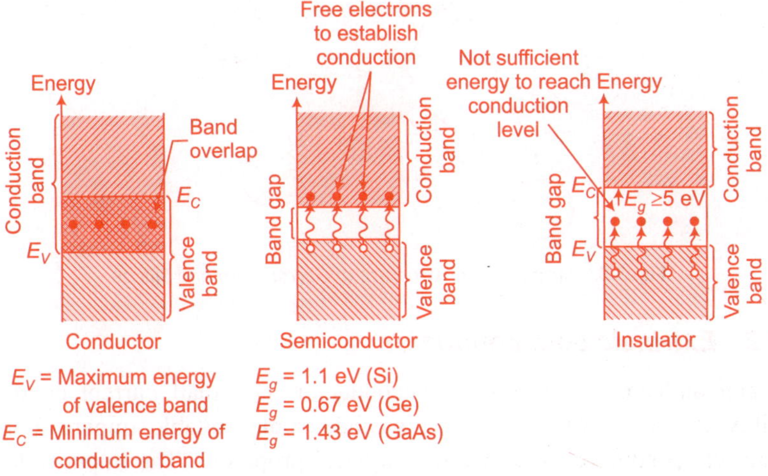

Fig. 3.8   *Band diagram of conductor, semiconductor and insulator*

An insulator material has a very wide forbidden energy gap > 5 eV. Hence, it is not possible for an electron to jump the gap from valence band to conduction band.

In a semiconductor, the forbidden energy gap is not wide. It is of the order of 1 eV (for Ge, $E_G = 0.72$ eV and for Si, $E_G = 1.12$ eV). The energy provided by the heat at room temperature is sufficient to lift electrons from the valence band to the conduction band. Therefore, at room temperature, semiconductors are capable of conducting same electric current as shown in Fig. 3.8.

Why we use germanium and silicon?

There are following reasons to use Germanium and Silicon:

1. We can make these semiconductors pure as compared to other semiconductors. The purity level of such material can be achieved up to $1:10^{10}$ by latest technology. We need purest it semiconductors because in purest semiconductors we add suitable impurities in the ratio of $1:10^6$. Due to these impurities, there is a large change in the conductivity of the semiconductor material. The process of adding suitable impurities in pure semiconductor is called doping. In other words, we can say that the characteristics of semiconductor material can be changed by doping.

2. The second reason to use Ge and Si is that the atom of this material makes a set pattern which is a periodic. A complete pattern is called crystal. The crystal structure of Ge and Si repeats itself i.e., it has a single crystal structure as shown in Fig. 3.9, unlike GaAs as shown in Fig. 3.5 (b). After doping, only electrical properties of this material will change and crystal structure will remain unchanged.

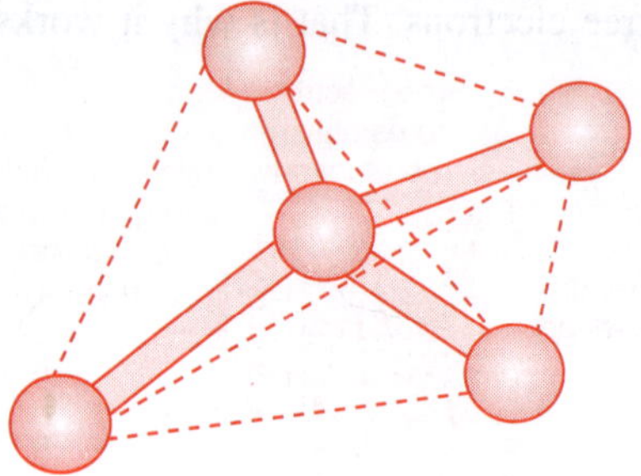

Fig. 3.9 *Single crystal structure of Germanium and Silicon*

3.2.3.2 *Extrinsic semiconductor*

Pure semiconductor or intrinsic semiconductor has small current conduction capability at room temperature. So, we add some suitable impure atoms in intrinsic semiconductor to change its electrical properties. The impurity level is very less (approximately $1:10^6$) and is capable to change its electrical properties completely. The intrinsic semiconductor after adding some suitable impurity is called extrinsic semiconductor. The addition of some suitable impurity in to intrinsic semiconductor is called doping. *In other words we can say that a semiconductor material that has been subjected to doping process is called an extrinsic semiconductor material.*

According to the type of impurity, the extrinsic semiconductors are of following two types:

(i) *n*-type   (ii) *p*-type.

n-type semiconductor

The purpose of adding impurity is to change the electrical properties of semiconductor material. In other words, to increase the number of free electrons in the semiconductor crystal. If a pentavalent impurity (having 5 valence electrons) is added to the semiconductor, a large number of free electrons are produced in the semiconductor. Such type of material having pentavalent impurity is called *n*-type semiconductor material. When a pentavalent or donor impurity is added to semiconductor, the impurity atoms form covalent bonds with four atoms of pure semiconductor material. Due to this, one extra valence electron is produced for each impurity atom added. Each extra electron produced enters the conduction band of pure semiconductor as a free electron, shown in Fig. 3.10. The pentavalent impurity is called the donor type impurities as it donates one electron to the conduction band of a pure semiconductor. *In other words we can say that when a small amount of pentavalent impurity is added to a pure semiconductor, it is known as n-type semiconductor.*

Such pentavalent impurities are Arsenic (As), atomic number is 33 and antimony (Sb), atomic number is 51. The atomic structure of *n*-type material is shown in Fig. 3.10.

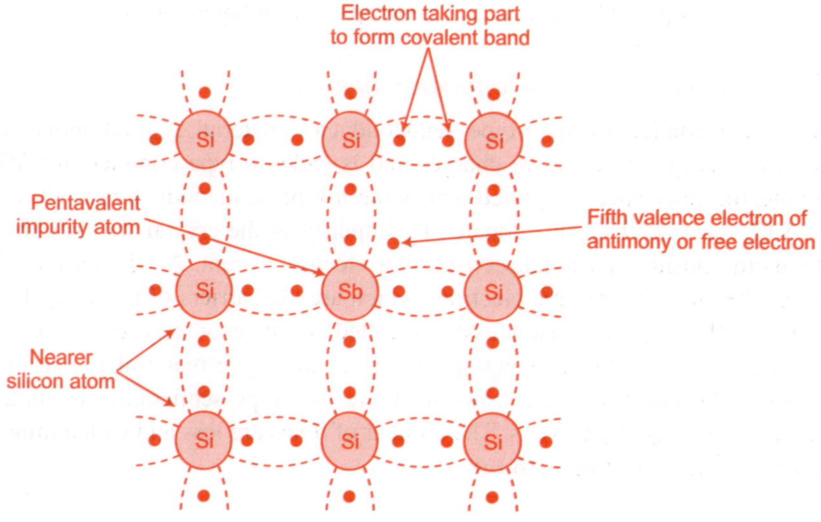

Fig. 3.10   *Antimony impurity in n-type material*

The impurities having five valence electrons is called donor impurities or pentavalent impurities.

Energy band diagram of n-type semiconductor material

The effect of doping process also can be understood with the help of energy band diagram as shown in Fig. 3.11. As we know that the forbidden energy gap for Silicon is 1.1 eV and for Germanium is 0.67 eV.

When a donor or pentavalent impurity is added in pure semiconductor, a new donor level is created in forbidden gap. The free electrons, free from the impurities occupy this donor level. The energy gap between conduction energy level (E_C) and donor energy level (E_D) in Silicon is very less (approximately in Silicon 0.05 eV) and in Germanium 0.01 eV.

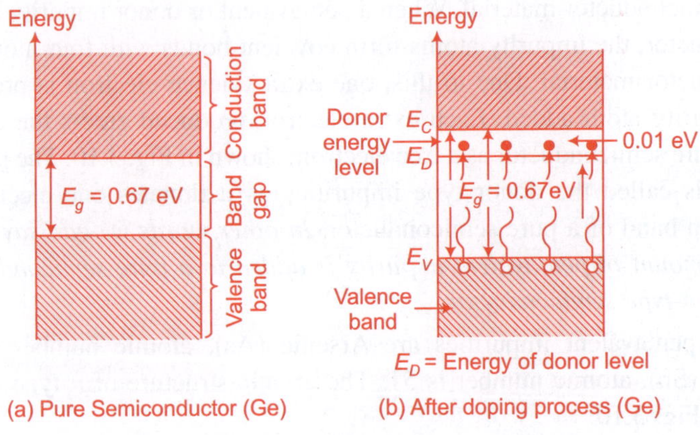

Fig. 3.11 *Energy band diagram after doping (pentavalent)*

Conduction in n-type semiconductor material

The current conduction in *n*-type semiconductor material is predominated by free electrons i.e., by negative charges and is called *n*-type conductivity. When the potential difference is applied across the *n*-type semiconductor, as shown in Fig. 3.12, the free electrons, donated by impurity in the crystal will be directed towards the positive terminal, constituting electric current. As the current flow through the crystal is by free electrons which are the carriers of negative charge. Therefore, this type of conductivity is called negative or *n*-type conductivity. So, we can say that the conduction of current in *n*-type material is due to the electrons. The number of electrons are more in *n*-type semiconductor material and are called majority carriers. The number of holes are less and called minority carriers, as shown in Fig. 3.16 (a).

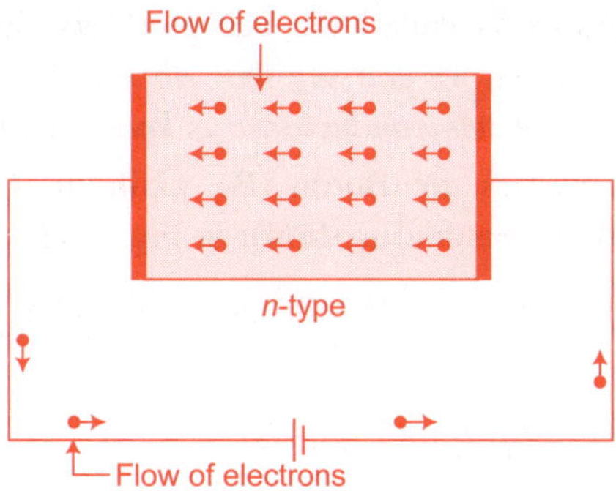

Fig. 3.12   *Current conduction in n-type material*

p-type semiconductor

The purpose of adding impurity is to change the electrical properties of semiconductor material to increase the number of holes in the semiconductor crystal. If a trivalent impurity (having 3 valence electrons) is added to the semiconductor, large number of holes are produced in the semiconductor. Such type of material having trivalent impurity is called *p*-type semiconductor material.

When a trivalent or acceptor impurity is added to semiconductor, the impurity atoms form covalent bonds with three atoms of pure semiconductor material. In the fourth covalent bond, only pure semiconductor atom contributes valence electron while impurity atom has no valence electron to contribute, since all its three valence electrons are already engaged in the covalent bonds with neighbouring pure semiconductor atoms. In other words, fourth bond is incomplete; being short of one electron. This missing electron is called a hole as shown in Fig. 3.13.

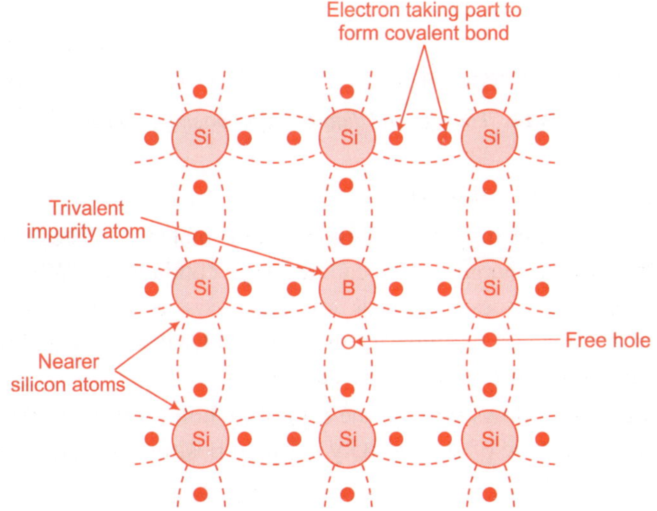

Fig. 3.13   *Boron impurity in p-type material*

The trivalent impurity is called the acceptor type impurity as it accepts one electron. *In other words, we can say that when a small amount of trivalent impurity is added to a pure semiconductor, it is known as p-type semiconductor.*

Such trivalent impurities are Boron (B), Gallium (Ga) and Indium. The atomic structure of *p*-type material is shown in Fig. 3.13.

> The impurities having three valence electrons are called acceptor impurities or trivalent impurities.

Energy band diagram of p-type material

The effect of doping process can also be understood with the help of energy band diagram as shown Fig. 3.14. As we know that the forbidden energy gap in Silicon is 1.1 eV and Germanium is 0.67 eV.

When an acceptor or trivalent impurity is added in pure semiconductor, a new acceptor level is created in forbidden gap. The free holes which are free from the impurity occupy acceptor level. The energy gap between valence band and acceptor band is very less, approximately in Silicon 0.08 eV and in Germanium 0.01 eV.

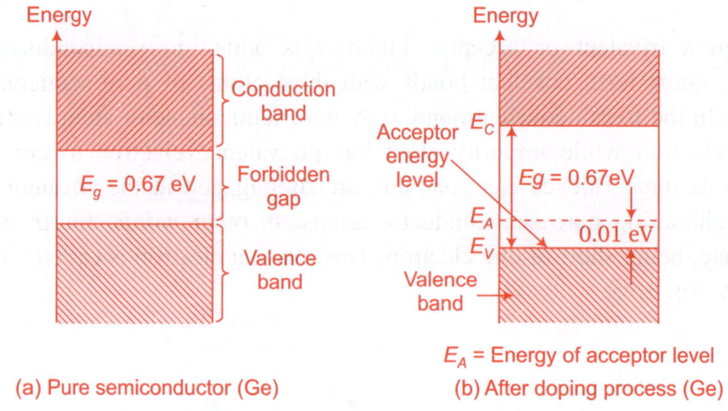

Fig. 3.14 *Energy band diagram after doping (trivalent)*

Conduction in p-type material

The current conduction in *p*-type semiconductor material is predominanted by holes i.e., by positive charges and is called *p*-type conductivity. When a potential difference is applied across the *p*-type semiconductor as shown in Fig. 3.15, the holes, which are donated by impurity in the crystal will be directed towards the negative terminal, constituting electric current.

When the current flows through the crystal by holes which are the carriers of positive charge, this type of conductivity is called positive or *p*-type conductivity.

So, we can say that the conduction of current in *p*-type material is due to the holes. The number of holes are more in *p*-type semiconductor material and are called majority carriers. When the number of electrons are less, they are called minority carriers, as shown in Fig. 3.16 (b).

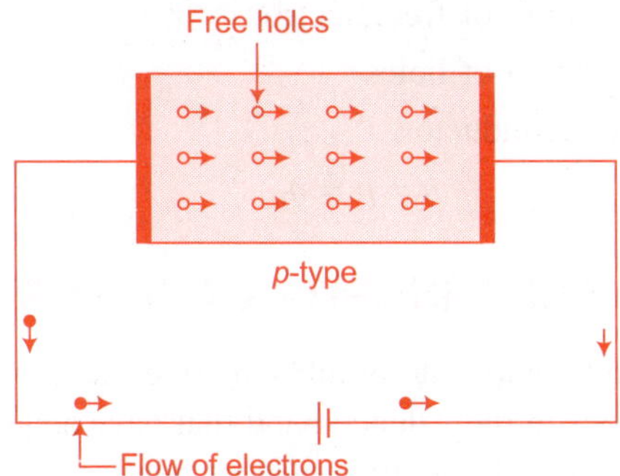

Fig. 3.15 *Current conduction in p-type material*

In *n*-type materials, the majority carriers are electrons and minority carriers are holes whereas in *p*-type semiconductor material the majority carriers are holes and minority carriers are electrons.

Effect of temperature on extrinsic semiconductor

When we add small amount of impurity (pentavalent or trivalent), it produces a large number of charge carriers in extrinsic semiconductor. The conductivity of an extrinsic semiconductor is much more than of an intrinsic semiconductor at room temperature.

If the temperature of an extrinsic semiconductor is increased, the additional thermal energy increases the thermally generated carriers. Due to this, the concentration of minority carriers increases. Eventually, the temperature is reduced, called the critical temperature. (85°C in case of Ge and 200°C in case of Si). Which is the number of covalent bonds, broken are very large, the number of holes become approximately same as the number of electrons. *The extrinsic semiconductor behaves like an intrinsic semiconductor but with higher conductivity.*

As we know that the concentration of free electrons and holes is always equal in an intrinsic semiconductor. When pentavalent impurity is added to an intrinsic semiconductor, the concentration of free electrons is increased (or the concentration of holes is reduced below the intrinsic value). Similarly, addition of trivalent impurity results in the reduction of concentration of free electrons below the intrinsic value. Under thermal equilibrium, the product of concentration of free electrons and concentration of holes is constant and is independent of amount of doping by donor and acceptor impurities. This is known as mass action law. Thus,

$$np = n_i^2 \tag{3.2}$$

where, n_i = Intrinsic concentration

n = Concentration of free electrons

p = Concentration of holes

For an intrinsic semiconductor,

$$n = p = n_i. \tag{3.3}$$

3.3 MAJORITY AND MINORITY CARRIERS

In the intrinsic semiconductor, the number of free electrons in Si or Ge is very less due to few electrons in the valence band that have acquired sufficient energy from heat or light source to break the covalent bond.

In n-type material, the number of free electrons are much more that the number of free holes after doping process because the number of free holes does not change significantly from this intrinsic level and the number of free electrons are much more due to the addition of thermally generated and are added through doping. Hence, in an n-type material electrons are majority carriers and responsible for flow of current.

Similarly in p-type material, the number of free holes are much more than the number of free electrons after doping process because the number of free electrons does not change significantly from this intrinsic level and the number of free holes are much more due to the addition of thermally generated and are added through doping. Hence, in a p-type material holes are the majority carriers and responds for flow of current.

In other words, we can say that, in n-type material the electrons are the majority carriers and holes are minority carriers, as shown in Figs. 3.16 (a) and (b).

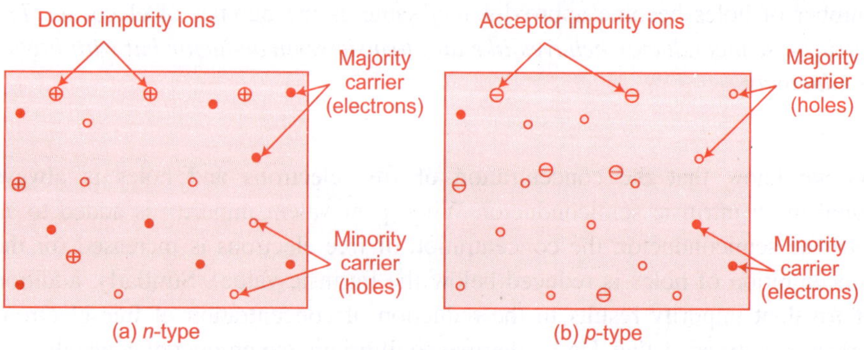

Fig. 3.16 *Majority and minority carrier in p-type and n-type material*

As we can see in Fig. 3.16 (a) the electrons are majority carriers and holes are minority can and it also have has ions similar as in Fig. 3.16 (b) the holes are majority carriers and electrons are minority carriers and it also has acceptor impurities ions. But electrons and holes have free charge and are free to move in crystal. But donor and acceptor ions have fixed charge and are immobile in crystal.

3.4   FERMI LEVEL

Fermi level represents the energy state in the energy band diagram which has a 50% probability of being filled with electrons from the valence band.

The probability of occuring electrons at specific energy level E is represented by Fermi Dirac probability function $P(E)$.

$$P(E) = \frac{1}{1 + e^{(E - E_F)/KT}} \tag{3.4}$$

if,

$$E = E_F$$

then,

$$P(E) = \frac{1}{1 + e^{(E_F - E_F)/KT}}$$

$$P(E) = \frac{1}{1 + 1} = \frac{1}{2}$$

$$P(E) = 0.5. \tag{3.5}$$

So, by Eqn. (3.5) we can say that the probability of finding an electron at Fermi level E_F is 0.5 or 50%.

Fermi level in semiconductor

The Fermi level in intrinsic semiconductor occurs at the mid-level of forbidden energy gap as shown in Fig. 3.17 (a), whereas in n-type semiconductor it occurs near the conduction band as shown in Fig. 3.17 (b) while in p-type semiconductor it occurs near the valence band as shown in Fig. 3.17 (c).

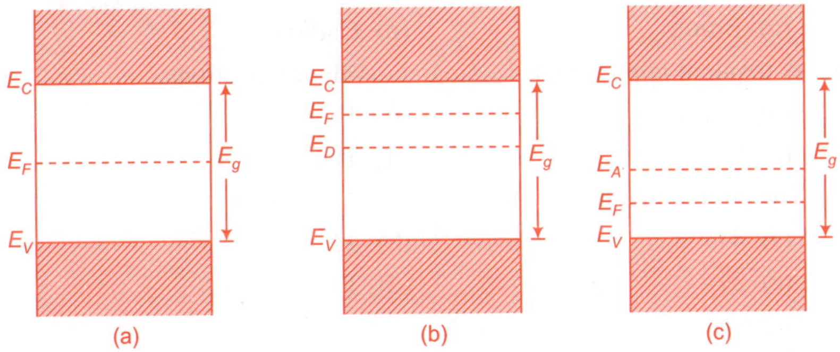

Fig. 3.17 (a) *pure semiconductor* (b) *n-type semiconductor* (c) *p-type semiconductor*

where, E_C = Conduction energy level
E_A = Acceptor energy level
E_V = Valence energy level
E_D = Donor energy level
E_F = Fermi energy level.

Objective Type Questions

1. The flow of current in metal is due to
   (a) Flow of positive charge
   (b) Flow of negative charge
   (c) Flow of both charges
   (d) None of these

2. Solid can be classified in
   (a) Two parts
   (b) Four parts
   (c) Three parts
   (d) Five parts

3. A semiconductor is formed by
   (a) Electrovalent
   (b) Covalent
   (c) Coordinate
   (d) All of these

4. A pentavalent impurity has .... valence electrons.
   (a) 4
   (b) 3
   (c) 5
   (d) 6

5. Rubber is classified as
   (a) Conductor
   (b) Insulator
   (c) Semiconductor
   (d) None of these

6. A pure semiconductor is called as
   (a) Intrinsic
   (b) Extrinsic
   (c) Conductor
   (d) None of these

7. The most commonly used semiconductor is ......
   (a) Silicon
   (b) Germanium
   (c) Boron
   (d) Carbon

8. A trivalent impurity has .... valence electrons.
   (a) Four
   (b) Two
   (c) Five
   (d) Three

9. The atomic number of Ge is
   (a) 41
   (b) 32
   (c) 31
   (d) 30

10. GaAs is a
   (a) Conductor
   (b) Insulator
   (c) Semiconductor
   (d) None of these

11. A semiconductor has .... temperature coefficients of resistance.
    - (a) Zero
    - (b) Negative
    - (c) Positive
    - (d) None of these

12. Device made from GaAs have breakdown voltages
    - (a) Lower
    - (b) Higher
    - (c) Zero
    - (d) Negative

13. The resistivity of pure silicon is about ....
    - (a) $1000 \ \Omega$ cm
    - (b) $6 \times 10^3 \ \Omega$ cm
    - (c) $1.6 \times 10^{-9} \ \Omega$ cm
    - (d) $3 \times 10^6 \ \Omega$ cm

14. Addition of trivalent impurity to a semiconductor creates many ....
    - (a) Free electrons
    - (b) Holes
    - (c) Valence electrons
    - (d) Bound electrons

15. GaAs has an advantage of
    - (a) Direct band gap
    - (b) Indirect band gap
    - (c) No band gap
    - (d) None of these

16. The resistivity of pure Germanium under standard condition is about ....
    - (a) 6 kΩ
    - (h) $60 \ \Omega$ cm
    - (c) 3 MΩ cm
    - (d) $6\mu\Omega$ cm

17. A semiconductor has generally .... valence electrons.
    - (a) Three
    - (b) Two
    - (c) Four
    - (d) Five

18. The devices made from Silicon is
    - (a) High in speed
    - (b) Slow in speed
    - (c) Normal speed
    - (d) None of these

19. A hole in semiconductor is defined as
    - (a) A free proton
    - (b) A free electron
    - (c) An incomplete part of an electron pair bond
    - (d) Free neutron

20. The random motion of electrons and holes due to thermal agitation is called ....
    - (a) Pressure
    - (b) Ionisation
    - (c) Diffusion
    - (d) None of these

21. When a pentavalent impurity is added in a pure semiconductor, such type of semiconductor is called
    (a) p-type
    (b) n-type
    (c) Conductor
    (d) None of these

22. The conductivity of Germanium is .... than Silicon
    (a) Less
    (b) Greater
    (c) Equal
    (d) None of these

23. The forbidden energy gap for Germanium is ....
    (a) 1.12 eV
    (b) 0.72 eV
    (c) 2.02 eV
    (d) None of these

24. At room temperature, an intrinsic semiconductor at approximately is: ....
    (a) A conductor
    (b) An insulator
    (c) A battery
    (d) None of these

25. When small amount of impurity is added in pure semiconductor, this process is called
    (a) Diffusion
    (b) Doping
    (c) Conduction
    (d) All of the these

26. The majority carriers for p-type semiconductors are ....
    (a) Electrons
    (b) Holes
    (c) Impurity ion
    (d) None of these

27. When an acceptor impurity is added to a semiconductor, then it is called ....
    (a) p-type semiconductor
    (b) n-type semiconductor
    (c) Insulator
    (d) All of the these

28. The impurities having five valence electrons are called ....
    (a) Acceptor impurities
    (b) Donor impurities
    (c) Both (a) and (b)
    (d) None of these

29. In a n-type material the ....
    (a) Majority carriers are electrons and minority carriers are holes
    (b) Majority carriers are holes and minority carriers are electrons
    (c) Both (a) and (b)
    (d) None of the above

30. According to mass action law,
    (a) $n^2 p^2 = n_i^4$
    (b) $n^2 p = n_i^2$
    (c) $np = n_i^2$
    (d) None of these

31. The strength of a semiconductor crystal comes from
    - (a) Force between protons
    - (b) Forces between nuclei
    - (c) Covalent bond
    - (d) None of these

32. In a semiconductor, current conduction is due
    - (a) to electrons only
    - (b) to holes only
    - (c) to electrons and holes
    - (d) None of these

33. When a pure semiconductor is heated, its resistance
    - (a) Goes up
    - (b) Goes down
    - (c) Remains the same
    - (d) Cannot say.

ANSWER

1. (b)	2. (c)	3. (b)	4. (c)	5. (b)	6. (a)	7. (a)
8. (d)	9. (b)	10. (c)	11. (b)	12. (b)	13. (b)	14. (b)
15. (a)	16. (b)	17. (c)	18. (b)	19. (c)	20. (c)	21. (b)
22. (b)	23. (b)	24. (b)	25. (b)	26. (b)	27. (a)	28. (b)
29. (a)	30. (c)	31. (c)	32. (c)	33. (b).		

Exercise

3.1. What is a semiconductor? Discuss properties of semiconductors.

3.2. Classify and explain the different types of material.

3.3. Which are the most commonly used semiconductors and why?

3.4. What do you understand by intrinsic and extrinsic semiconductor?

3.5. Draw and explain the crystal structure of GaAs.

3.6. What are the advantages, disadvantages and application of GaAs.

3.7. What do you understand by current conduction in intrinsic semiconductor?

3.8. Draw and explain energy band diagram for semiconductors.

3.9. Explain conduction in *p*-type and *n*-type semiconductor.

3.10. Why is a semiconductor an insulator at normal temperature?

3.11. Discuss the effect of temperature on semiconductors.

Semiconductor Diode and its Applications

4

4.1 *p-n* JUNCTION

When a *p*-type semiconductor is suitably joined to *n*-type semiconductor, the contact surface is called *p-n* junction. In other words, we can say that a *p-n* junction is a junction formed by joining *n*-type and *p*-type semiconductor material together in a very close contact. The term junction refers to the joining line where the two regions of the semiconductor meet. As shown in Fig. 4.1. *p-n* junctions are oftenly created in a single crystal of semiconductor by doping, for example by diffusion of dopants, ion-implantation or by epitaxy. One common method of making *p-n* junction is called alloying.

Fig. 4.1   *p-n junction*

Generally, if *p*-type material just touched with *n*-type material, then the *p-n* junction cannot be formed, because we cannot set single crystal structure. So, for forming *p-n* junction, we need some special techniques as named above.

The special feature of *p-n* junction is that it conducts in only one direction. It has high impedance in another direction.

4.2   FORMATION OF DEPLETION LAYER

The formation of depletion layer in *p-n* junction diode is shown in Fig. 4.1, when external voltage is zero i.e., $(V_s = 0V)$.

4.2.1   *p-n* Junction without External Voltage ($V_S = 0V$)

The Fig. 4.2. shows the *p-n* junction, just after the formation. In this Fig., the left half part is a *p*-type and right half part is *n*-type. *p*-type part contains holes and negatively charged acceptor impurity ion, and *n*-type part contains electrons and positively charged donor impurity ion. Electrons and holes are free charge carriers, but the impurity ions are immobile.

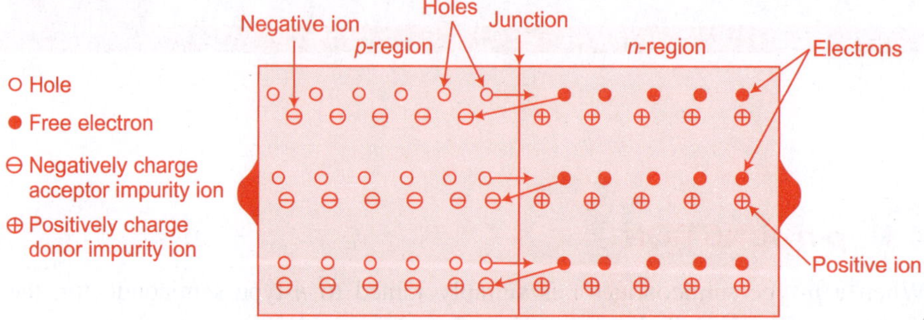

Fig. 4.2 *p-n* junction just after the formation

Just after the formation of *p-n* junction, the following action will take place (when there is no voltage apply):

(i) Holes from the *p*-region diffuse into the *n*-region, combine with the free electrons in the *n*-region and neutralize them. Similarly, electrons from *n*-region diffuse into the *p*-region, combine with the holes in the *p*-region and neutralize them. Remember that the holes from *p*-region move towards *n*-region and electrons from *n*-region move towards *p*-region duo to the difference in concentration of *p* and *n*-region i.e., the concentration of holes in *p*-region is much more than the *n*-region and concentration of electrons in *n*-region is much more than the *p*-region. Due to this diffusion will take place.

(ii) Remember that the diffusion of holes and electrons is due to the thermal agitation and difference in their concentrations in the two regions.

(iii) The diffusion of holes and free electrons across the junction occurs for a very short time. After a few recombinations of electrons and holes in the immediate neighbourhood of the junction, restraining force is set-up automatically. This force is known as barrier. Further diffusion of electrons and holes from one side to the other side is stopped by this barrier.

(iv) Due to the diffusion process, as the free electrons move across the junction from *n*-type to *p*-type, positive donor ions are uncovered (or uncompensated) i.e., they are robbed of free electrons. Hence, a positive charge is built on the *n*-side of junction. At the same time, the free electrons cross the junction and uncover (or uncompensated) the negative acceptor ions by filling in the holes. Due to this, a net negative charge is established on *p*-side of the junction. When a sufficient number of donor and acceptor ions are uncompensated, further diffusion is stopped. It is because now positive charge on *n*-side repels holes to cross from *p*-side to *n*-side and negative charge on *p*-side repels free electrons to cross from *n*-side to *p*-side. Thus, a barrier is set-up against further movement of charge carriers. This is called junction barrier or potential barrier. As shown in Fig. 4.3.

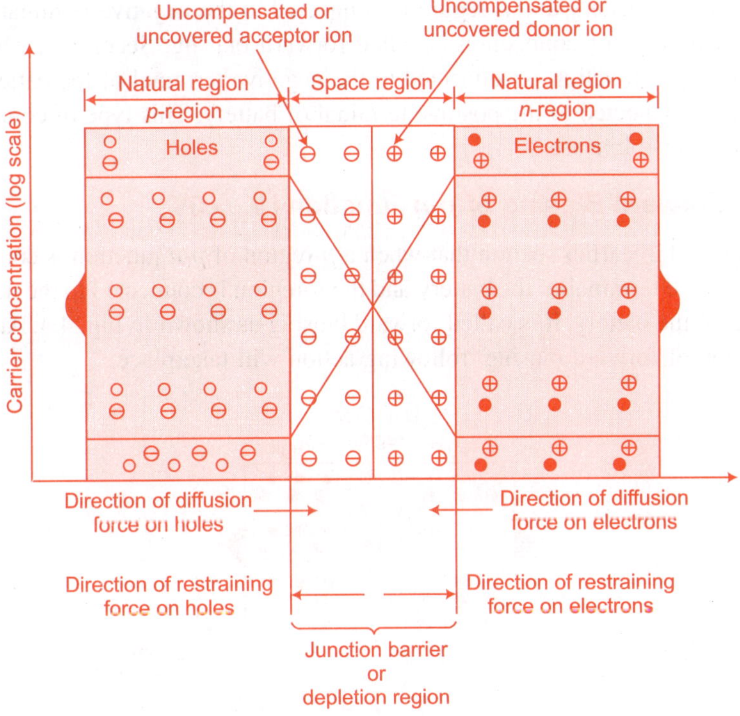

Fig. 4.3   *Creation of restraining force*

(v) The region near the junction, which contains uncompensated or uncovered acceptor and donor ions is called depletion region. In this region, there are no free charge carriers, only fixed or immobile charge. The electric field between acceptor and donor ions is called barrier, so we can say that the potential difference is established across the two ends of the barrier. The value of potential difference depends upon the temperature of material. It is approximately 0.3 V for Germanium and 0.7 V for Silicon.

(vi) The width of depletion region depends upon the doping level of *p*-region and *n*-region. If the doping level is larger, then the width of depletion region will be less and vice versa.

Junction barrier helps to the minority carrier to cross the junction but it prevents the majority carriers to cross the junction.

4.3 BIASING OF *p-n* JUNCTION

When we connect the voltage source across the *p-n* junction, it is called biasing. In this section, we will study the action of *p-n* junction when it is connected by external source. According to the connection of external source, the biasing is of two types. One is when the *p*-region of *p-n* junction is connected to the positive terminal of the battery and *n*-region is connected to the negative terminal of the battery. This type of connection is called forward biasing. Second is, when the *p*-region of *p-n* junction is connected to the negative terminal of the battery and *n*-region is connected to the positive terminal of battery. This type of connection is called reverse biasing.

4.3.1 Forward Biasing of *p-n* Junction ($V_S > 0V$)

We discussed in earlier section that when a *p*-region of *p-n* junction is connected to the positive terminal of the battery and the *n*-region is connected to the negative terminal of the battery, it is called forward biasing as shown in Fig. 4.4. After the connection of forward biasing, following action will take place:

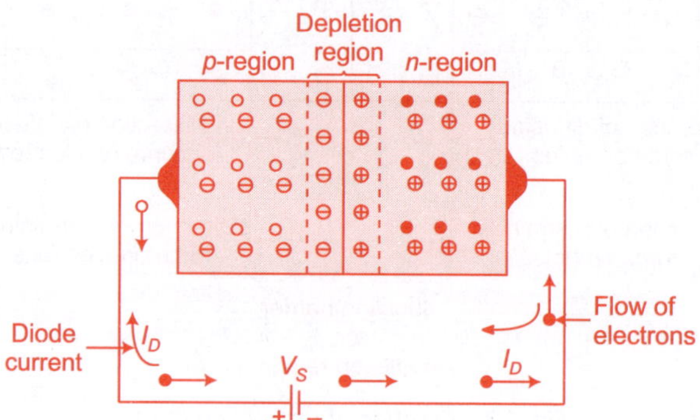

Fig. 4.4 *Forward bias p-n junction*

(i) In forward bias, the applied voltage opposes the barrier in the opposite direction of the barrier.

(ii) In forward bias, the bias voltage repels the majority carriers, holes of *p*-region and electrons of *n*-region towards the junction. If applied voltage (V_S) is greater than the potential barrier voltage (V_B), then the holes of the *p*-region and electrons of *n*-region are capable to cross the junction. Due to this, the positive donor ions and negative acceptor ions of depletion region will get electrons and holes respectively. Due to this, the width of depletion region will reduce.

(iii) If the bias voltage is further increased, then the width of depletion region is further reduced and is very less. Due to this, the large number of majority carriers (electrons and holes) cross the junction and the current flow in the circuit increases rapidly.

(iv) Even though the large number of holes and electrons cross the junction, the number of holes in *p*-region and the number of electrons in *n*-region is constant. To understand this, we can consider one example.

One hole of *p*-region moves in *n*-region and combines with electron. Due to the combination, there is a loss of one hole in *p*-region and loss of one electron in *n*-region. One electron from *p*-region moves towards the positive terminal of the battery after breaking of any covalent bond of *p*-region to compensate the loss of one hole in *p*-region. Similarly, one electron moves towards the *n*-region from negative terminal of battery to compensate the loss of one electron in *n*-region. In this way, the loss of holes and electrons of *p*-region and *n*-region are compensated by the battery.

(v) As we can see in the Fig. 4.5 the holes and electrons are current carriers in *p*-region and *n*-region respectively, but only electrons are current carriers in external circuit.

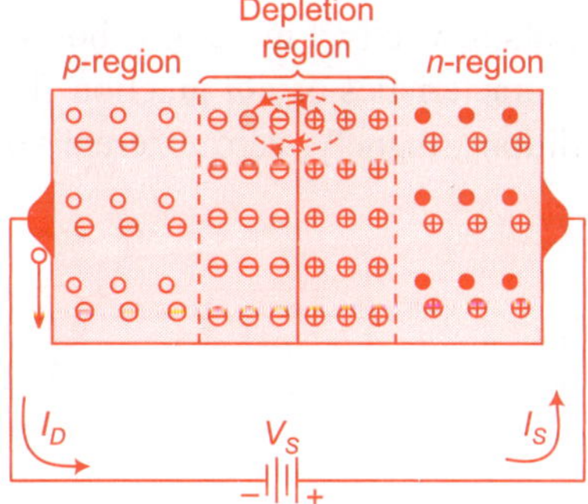

Fig. 4.5   *Reverse bias p-n junction*

4.3.2   Reverse Biasing of *p-n* Junction ($V_S < 0V$)

As we discussed earlier that when a *p*-region of *p-n* junction is connected to the negative terminal of the battery and the *n*-region is connected to the positive terminal of battery it is called reverse biasing as shown in Fig. 4.5. After the connection of forward biasing the following action will take place:

(i) In reverse bias, the applied voltage supports the barrier in the same direction as the barrier.

(ii) In reverse bias, the bias voltage attract the majority carriers, holes of *p*-region and electrons of *n*-region away from the junction. Due to this, the number of uncovered or uncompensated donor and acceptor ions near the junction will increase. So, the width of depletion region also increases. It is not easy for the majority carriers to cross the junction.

(iii) The flow of current in the reverse bias due to the majority carrier is zero but the flow of current due to the minority carrier is small in amount (in µA). We have already told that the barrier prevents the majority carrier to cross the junction but it helps the minority carrier to cross the junction. Since the number of minority carriers are less therefore, the reverse current is small. (For Si its nA and for Ge its µA). This current is called Reverse Saturation Current and is denoted by I_S.

(iv) Minority carriers are generated due to the temperature. So, the reverse saturation current is constant for constant temperature and independent on bias voltage. If the temperature increases, then the current increases rapidly. For Ge and Si, it increases approximately 7% per °C, or we can say that it doubles at 10°C.

4.3.3 Volt-ampere (VI) Characteristics of *p-n* Junction

Volt-ampere or *V-I* characteristics of *p-n* junction is the curve between voltage across the junction and the current in the circuit. Generally, voltage is taken along *x*-axis and current along *y*-axis. Figure 4.6 shows the circuit arrangement for determining the *V-I* characteristics of *p-n* junction. The characteristics can be studied under three conditions, namely; zero external voltage, forward bias and reverse bias.

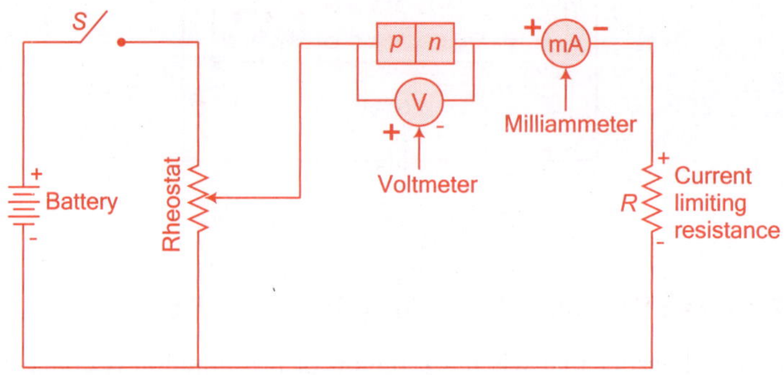

Fig. 4.6 *Circuit diagram for V-I characteristics*

(i) When the external voltage is zero i.e., circuit is open at *S*, the potential barrier at the junction does not permit flow of current. Therefore, the circuit current $I_C = 0$ as indicated by point *O* in Fig. 4.7.

(ii) We discussed earlier that in forward bias, the potential barrier is reduced. At some forward voltage (0.3 V for Ge and 0.7 V for Si), the potential barrier is cancelled and the current starts flowing in the circuit. The current increases with further increase in voltage. Thus, a rising curve *OB* is obtained with forward bias as shown in Fig. 4.7 (a). From the forward characteristics, it is seen that at first region *OA,* the current increases very slowly and the curve is non-linear. It is because the external applied voltage applied is used to overcome the potential barrier. However, once the external voltage exceeds the potential barrier voltage, the *p-n* junction behaves like an ordinary conductor. Therefore, the current rises rapidly with increase in applied voltage, represented by region *AB* on the curve is almost linear.

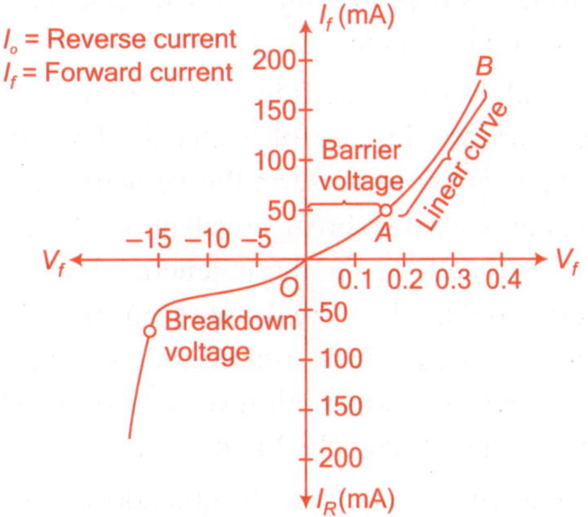

(a) *VI*-characteristic circuit (Forward voltage versus circuit current)

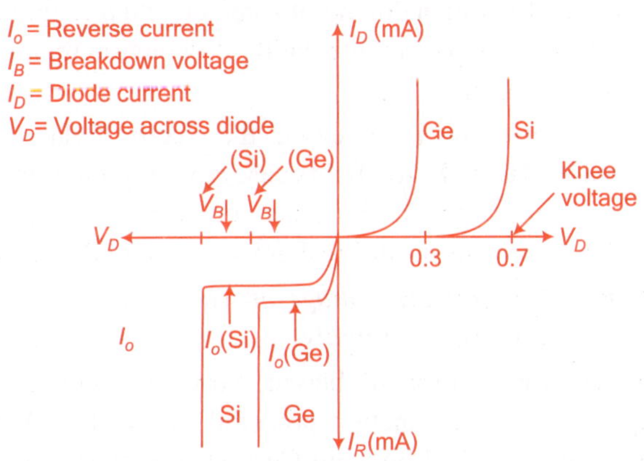

(b) *VI*-characteristic circuit (Voltage across diode versus circuit current)

Fig. 4.7   *VI-characteristics of p-n junction*

(iii) We discussed earlier that in the reverse bias, potential barrier at the junction is increased. Therefore, the junction resistance becomes very high due to no current flow through the circuit. A very small current (in μA) flow in the circuit with reverse bias is shown in Fig. 4.7 and is called reverse saturation current and is represented by I_o. If reverse voltage is increased continuously, the kinetic energy of electrons may become high enough to knock out electrons from the semiconductor atoms. At this stage the breakdown of junction occurs.

When a *p*-type semiconductor is suitably joined to *n*-type semiconductor, the contact surface is called *p-n* junction. *p-n* junction also called crystal or semiconductor diode.

Terms related to VI characteristics

(i) **Breakdown Voltage:**   It is the minimum reverse voltage at which *p-n* junction breakdown with sudden rise in reverse current.

(a) Avalanche breakdown: The flow of current in reverse bias is due to the minority carriers. In case of avalanche breakdown, the increase reverse voltage causes increase in the velocities of minority carriers. These high energy carriers break covalent bonds, and due to this more carriers are generated. Again, these generated carriers are accelerated by the electric field. They break more covalent bonds during their travel. Thus, chain reaction is established creating a large number of carriers. This gives rise to a high reverse current. This mechanism of breakdown is called Avalanche breakdown.

(b) Zener Breakdown: This type of breakdown occurs at very less reverse voltage (approximate 6V). This breakdown occurs in such type of diode, for which doping of *n*-region and *p*-region is high. Due to the high level of doping, the width of depletion region is very less (approximate 10^{-8}m).

When reverse bias is increased, the electric field at the junction also increases. High electric field cause covalent bonds to break. Thus, a large number of carriers are generated. This causes a large current to flow. This mechanism of breakdown is called Zener breakdown.

(ii) **Knee Voltage:**   *It is forward voltage at which the current through the p-n junction starts increasing rapidly.*

When the *p-n* junction is forward biased, very little current flows until the forward voltage exceeds the junction potential barrier (0.3 V for Ge and 0.7 V for Si). This current is called Forward Current. The characteristics follows exponential law as shown in Fig. 4.7 (b). As the forward voltage is increased to the knee of the characteristic, the barrier potential is progressively reduced to zero, allowing more majority charge carriers to flow across the junction. Beyond the *knee of the characteristics, the potential barrier is completely eliminated, forward current increases almost linearly with the increase in forward voltage and the p-n-junction* starts behaving as a resistor. If the forward voltage is increased beyond a certain value, extremely large current will flow and the *p-n* junction may get destroyed due to over heating.

4.4 CURRENT AND VOLTAGE RELATIONSHIP OF *p-n* JUNCTION DIODE

It can be demonstrated through the use of solid-state physics that the general characteristics of a semiconductor diode can be defined by the following equation referred, to as Shockley's equation, for the forward and reverse bias regions:

$$I_D = I_o \, (e^{V_D/nV_T} - 1) \tag{4.1}$$

where, I_D = Diode current

V_D = Forward voltage across diode

I_o = Reverse saturation current

n = Idealist factor ($n = 1$ for Ge and $n = 2$ for Si)

V_T = Thermal voltage

$$V_T = \frac{kT}{q} \tag{4.2}$$

where, k = Boltzmann's constant

$k = 1.38 \times 10^{-23}$ J/k

T = Absolute temperature in kelvin

T = 273 + temperature in °C

q = Magnitude of electronic charge

$q = 1.6 \times 10^{-19}$ C.

Case I In a forward bias, the value of V_D is positive. So, by the Eqn. (4.1).

$$e^{V_D/nV_T} >> 1$$

We can ignore 1, now the Eqn. (4.1) becomes:

$$I_D = I_o \, e^{V_D/nV_T} \quad \text{(in forward bias)} \tag{4.3}$$

So, we can say that, in forward bias, diode current depends exponentially on diode voltage.

Case II In a reverse bias, the value of V_D is negative. So by the Eqn. (4.1).

$$e^{-V_D/nV_T} << 1.$$

We can ignore the term e^{V_D/nV_T}, and the Eqn. 4.1 becomes:

$$I_D = -\, I_o.$$

So we can say that, in reverse bias, diode current will be equal to the reverse saturation current.

Example 4.4.1 At normal temperature, the leakage current of Si-diode is 0.25 μA. If the diode current is 10 mA (in forward bias), then calculate the forward voltage across the diode.

Solution   Given that,

$$T = 273 + 27°C$$

$$T = 300 \text{ K}$$

$$I_0 = 0.25 \text{ μA}$$

$$I_D = 10 \text{ mA}$$

$$V_D = ? \qquad n = 2 \text{ for Si diode}$$

By the Eqn. (4.2), we get,

$$V_T = \frac{kT}{q} = \frac{1.38 \times 10^{-23} \times 300}{1.6 \times 10^{-19}}$$

$$VT = \frac{4.14 \times 10^{-4}}{1.6} = 258.75 \times 10^{-4} \text{ V}$$

$$V_T = 0.0258 \text{ V}$$

$$V_T = 0.026 \text{ V.}$$

Now, by the Eqn. (4.1), we get,

$$I_D = I_0 \left(e^{V_D/nV_T} - 1\right)$$

$$10 \times 10^{-3} = 0.25 \times 10^{-6} \left(e^{V_D/nV_T} - 1\right)$$

$$e^{V_D/nV_T} = \frac{10 \times 10^{-3}}{0.25 \times 10^{-6}} + 1$$

$$e^{V_D/nV_T} = 40 \times 10^3 + 1$$

$$e^{V_D/nV_T} = 40001.$$

Taking the natural log on both sides,

$$\frac{V_D}{nV_T} = \log_e 40001$$

$$V_D = nV_T \log_e 40001$$

$$V_D = 2 \times 0.026 \times 10.6 \quad (\because n = 2 \text{ for Si})$$

$$\boxed{V_D = 0.55 \text{ volt.}} \qquad \text{(Ans.)}$$

4.5 PROPERTIES OF *p-n* JUNCTION

The *p-n* Junction possess some interesting properties which have useful application in modern Electronics. A *p*-doped and *n*-doped semiconductor is relatively conductive. But the junction between them is not conductive. This non-conductive layer, called the depletion zone, occurs because of the electrical charge carriers (electrons and holes) which attract and eliminate each other in a process of recombination. By manipulating this non-conductive layer, *p-n* junctions are commonly used as diodes; circuit elements that allow the flow of electricity in one direction but not in other direction. This property has been explained earlier in terms of forward bias and reverse bias, where the term bias refers to an application of electric voltage to the *p-n* junction.

4.6 LIMITATIONS IN THE OPERATING CONDITIONS OF *p-n* JUNCTION

Each *p-n* junction has limiting value of maximum forward current, maximum power rating and peak inverse voltage. The *p-n* junction will give satisfactory performance, if it is operated within these limiting values. If these values are exceeded, the *p-n* junction may be damaged due to overheat.

(i) **Peak Inverse Voltage (PIV):**   Peak Inverse Voltage of a diode is the maximum reverse voltage that can be applied to the *p-n* junction without damaging the junction.

(ii) **Maximum forward current:**   It is the maximum instantaneous forward current that can flow through *p-n* junction without damaging it,

(iii) **Maximum power rating:**   It is the maximum power that can be dissipated at the junction without damaging it.

4.7 EQUIVALENT CIRCUIT FOR PRACTICAL DIODE

The equivalent circuit for practical diode is shown in Fig. 4.8 (a). In this circuit, the barrier voltage V_B is represented by a battery of 0.7 V for Si and 0.3 V for Ge. The diode resistance is represented by r_f. Notice that the V_B voltage source is not a physical source which means if you connect the voltmeter across it, you will not get any reading. It only denotes that if we apply voltage above this barrier voltage V_B on a practical diode, then diode current will conduct; Ideal diode (D_I) shown in Fig. 4.8 (a) is treated as short circuit in forward bias and open circuit in reverse bias. So, we can say that zero current will flow in reverse bias as shown in Figs. 4.8 (b) and (c).

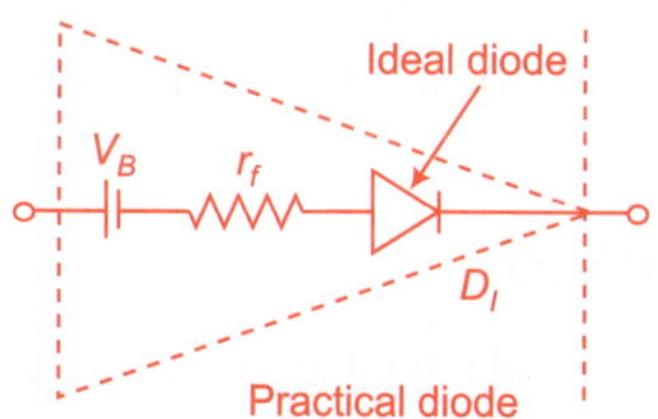

Fig. 4.8   *(a) Equivalent circuit of Practical diode*

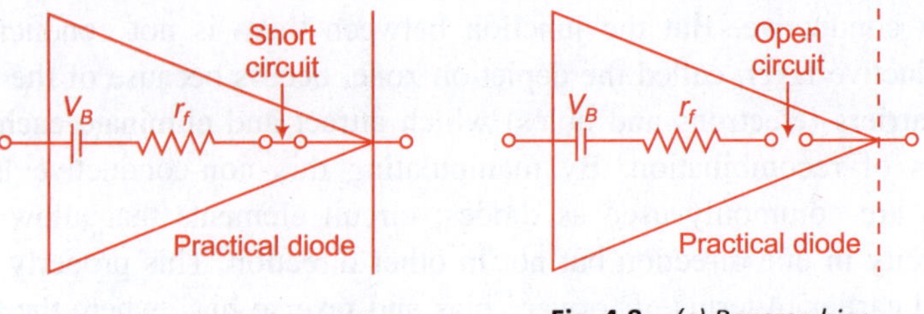

Fig. 4.8 *(b) Forward bias*   **Fig. 4.8** *(c) Reverse bias*

From Fig. 4.8 (b), we can say that when the applied voltage on practical diode is greater than barrier voltage V_B, then the conduction of current will occur. The *VI* characteristics of equivalent circuit is shown in Fig. 4.10. When the applied voltage is much greater than the barrier voltage and the resistance, of the circuit is much greater than the diode resistance, then we can ignore the V_B and r_f. The *VI* characteristics of ideal diode in shown in Fig 4.9.

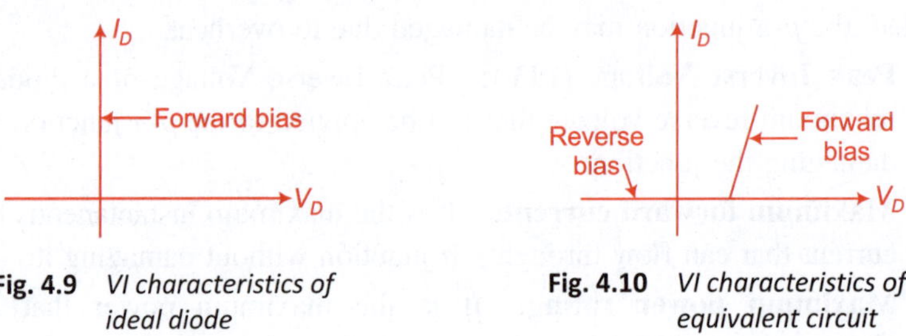

Fig. 4.9 *VI characteristics of ideal diode*   **Fig. 4.10** *VI characteristics of equivalent circuit*

4.8 STATIC AND DYNAMIC RESISTANCE OF DIODE

The resistance of diode is of two types according to the flow of current through it.

4.8.1 Static Resistance

When the steady dc current flows through the diode, then the resistance of the diode is called as static resistance and it is denoted by R_{dc}. The static resistance of the diode is the ratio of the applied voltage on the diode to the current flow through the diode i.e.,

$$R_{dc} = \frac{V_D}{I_D} \tag{4.4}$$

The resistance of diode in forward bias is denoted by R_F and in reverse bias is denoted by R_R.

4.8.2 Dynamic Resistance

When the current in diode is changed of its average value, then the ratio of small change in diode voltage (ΔV_D) to the corresponding small change in diode current (ΔI) is called as dynamic resistance and is denoted by r_d. In other words, we can state that when the AC voltage is applied on the diode with DC voltage, then the resistance of the diode is called dynamic resistance.

$$r_d = \frac{\Delta V_D}{\Delta I_D} \tag{4.5}$$

where, $\Delta V_D = V_{D_2} - V_{D_1}$ small change in diode voltage

$\Delta I_D = I_{D_2} - I_{D_1}$ small change in diode current.

Example 4.8.1   Calculate static resistance from diode characteristics given in Fig. 4.11.

Solution   The voltage and current at point A,

$$V_D = 0.7 \text{ V}, \quad I_D = 10 \text{ mA}.$$

Hence, forward resistance $R_F = \dfrac{V_D}{I_D} = \dfrac{0.7}{10 \times 10^{-3}}$

$$\boxed{R_F = 70\Omega} \text{ at point } A.$$

The voltage and current at point B

$$V_D = 0.6\text{V} \quad \text{and} \quad I_D = 5 \text{ mA}$$

Hence, $R_F = \dfrac{0.6}{5 \times 10^{-3}} = 120\Omega$ at point B

$$\boxed{R_F = 120\Omega.}$$

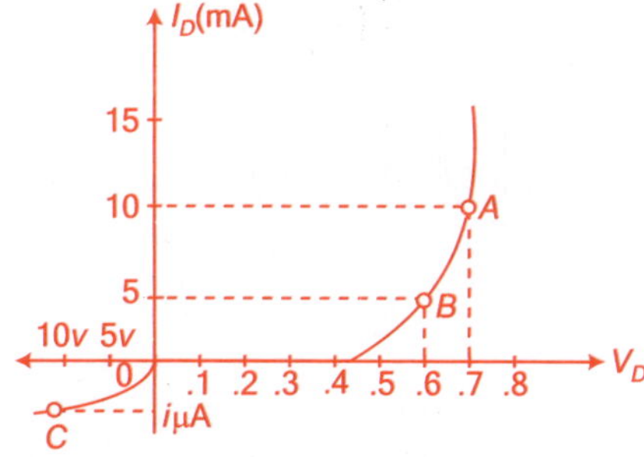

Fig. Ex. 4.8.1   *Diode characteristics*

Now, voltage and current at point C

$$V_D = -1\mu\text{A}$$

and, $$V_D = -10\text{V}$$

Hence,
$$R_R = \frac{V_D}{I_D} = \frac{-10}{-1 \times 10^{-6}} = 10^7 \ \Omega$$

$$\boxed{R_R = 10^7 \ \Omega.}$$ **(Ans.)**

where, R_F = Forward static resistance

R_R = Reverse static resistance.

Since diode is non-linear, therefore the static resistance is different at every point.

Example 4.8.2 Calculate the dynamic resistance from diode characteristic given in Fig. Ex. 4.8.2.

Solution From Fig. Ex. 4.7.2 $V_{D_1} = 0.55$V, $I_{D_1} = 3$ mA

$$V_{D_2} = 0.65\text{V}, \ I_{D_2} = 10 \text{ mA}.$$

Hence, the dynamic resistance,

$$r_d = \frac{\Delta V}{\Delta I_D}$$

$$r_d = \frac{V_{D_2} - V_{D_1}}{I_{D_2} - I_{D_1}}$$

$$= \frac{(0.65 - 0.55) \ V}{(10 - 3) \ mA}$$

$$r_d = \frac{0.1 \ V}{7mA}$$

$$r_d = \frac{100}{7} \ \Omega$$

$$\boxed{r_d = 14.29 \ \Omega.}$$ **(Ans.)**

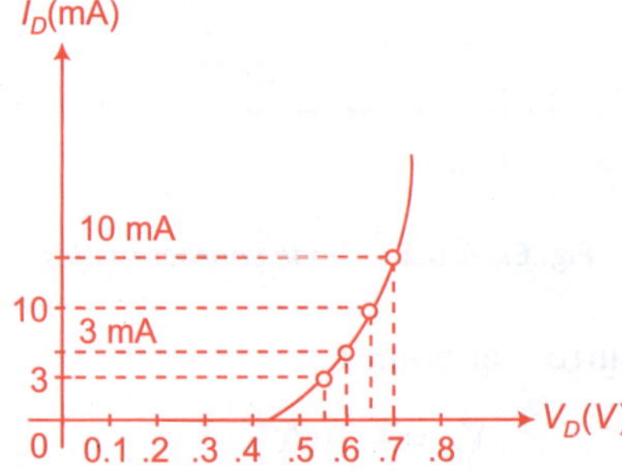

Fig. Ex. 4.8.2   *Diode characteristics*

4.9 JUNCTION CAPACITANCES IN FORWARD AND REVERSE DIRECTION

Basically, there are two types of capacitance that occurs in *p-n* junction. One is in reverse bias of *p-n* junction diode and is called depletion region capacitance or transition capacitance. Second one is in forward bias of *p-n* junction diode and is called diffusion capacitance.

4.9.1 Transition or Depletion Region Capacitance

This capacitance is effective when the diode is in reverse bias mode. As we know that the capacitance of parallel plate is given by the following formula:-

$$C = \frac{\in A}{d} \tag{4.6}$$

where, C = Capacitance

$\in$ = Permitivity

A = Area between plates

d = Distance between plates.

In reverse bias, the depletion region acts as a dielectric medium and *p-n* region acts as the plates of capacitor. Due to this, capacitance will exist at the junction of diode. This capacitance is called depletion region capacitance or transition capacitance. Since the width of depletion region (d) will increase by increasing the reverse voltage due to $\left(C \propto \dfrac{1}{d} \right)$, thus the capacitance will reduce as shown in Fig. 4.11 (a). So we can say that the transition capacitance of the *p-n* junction diode is dependent upon the applied voltage. These properties of diode is used in varactor diode.

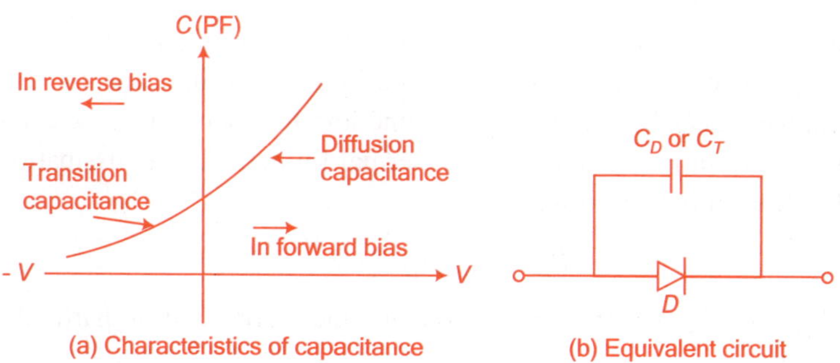

Fig. 4.11   *Capacitance in diode*

This capacitance is effective when the diode is in the forward bias mode. In forward bias, the width of depletion region will reduce. Due to this, the holes from *p*-region move in *n*-region and the electrons from *n*-region move in *p*-region. These charge carriers enter in *p* and *n*-region as minority carried to these minority carriers. The diffusion capacitance occurs due to these minority carriers which diffuse away from the junction and progressively recombine. The density of charge carriers is high near the junction and decays exponentially with distance. Thus, a charge is stored on both sides of the junction with forward bias. The diffusion capacitance is much larger than the transition capacitance.

The amount of stored charge varies with the applied potential for a capacitor.

4.9.3 Diode Ratings or Specifications

The ratings or specifications of specific semiconductor device are necessary, when we want to purchase it from the market for our use. The ratings or specifications of specific semiconductor devices are normally provided by the manufacturer in one of two forms. One is of a very brief description which is frequently provided by the manufacturer. Second is a thorough examination of the characteristics using graphs, artwork, tables and so on. In either case, there are specific pieces of data that must be included for proper utilization of two diode. These include:

Peak inverse voltage (PIV)

Peak inverse voltage of a diode is the maximum reverse voltage that can be applied to the *p-n* junction without damaging the junction. PIV is an important specification of diode which is provided by the manufacturer. For example, we consider that the PIV of diode is 120 V, then we should ensure that the reverse voltage at the diode must not exceed 120 V otherwise diode will damage. Ideally, this figure would be infinite.

Average current

Average current is the maximum average amount of current that can be conducted by a diode in a forward bias and is denoted by $I_{F(AV)}$. It indicates the amount of much heat which can be handled by *p-n* junction. In other words, we can say that the average current is fundamentally a thermal limitation. The dissipation power is equal to the following product,

$$P_{D_{max}} = V_D I_D$$

where, V_D and I_D are the diode voltage and current at a particular point of operation.

And the diode voltage is dependent upon both current and junction temperature. Ideally this figure would be infinite.

Repetitive peak current

Repetitive peak current is the maximum amount of peak current of the diode that can be conducted in forward bias mode, in repetitive pulses. Again, this rating is limited by the diode junction's thermal capacity, which is usually much higher than the average current rating due to thermal inertia.

Reverse recovery time

Reverse recovery time is the amount of time it takes for a diode to "turn off", when the voltage across it alternates from forward bias to reverse bias polarity. The diode must immediately prevent conduction upon polarity reversal. Ideally, this figure would be zero. For a typical rectifier diode, reverse recovery time is in the range of 10s of μs for a "fast switching" diode, it may only be a few nanoseconds. It is denoted by t_{rr}.

Maximum forward voltage

Maximum forward voltage is usually specified at the diode's rated forward current. Ideally this figure would be zero and is denoted by V_F. The diode provides no opposition to forward current. In reality, the forward voltage is described by the "diode equation".

Maximum power dissipation

The maximum power dissipation is the amount of power (in watts) allowed for the diode to dissipate, and is denoted by P_D. It is the multiplication of diode current I_D and voltage drop across diode V_D.

$$P_D = V_D I_D \qquad (4.7)$$

where, P_D = Power dissipated

V_D = Diode voltage drop (0.7V for Si and 0.3 V for Ge)

I_D = Diode current

or, $P_D = I_D^2 R_B \qquad (4.8)$

where, R_B = Bulk resistance.

Reverse saturation current

Reverse saturation current is the amount of current through the diode in reverse bias operation, with the maximum rated inverse voltage applied. Some times referred as leakage current and is denoted by I_R. Ideally this figure would be zero, as an ideal diode would block all current in reverse bias. But practically, it is very small (in μA) as compared to the maximum forward current.

Junction capacitance

Junction capacitance is the typical amount of capacitance intrinsic to the junction, due to the depletion region acting as a dielectric separating the anode and cathode connections. This is usually a very small figure, measured in the range of picofarad (pF) and is denoted by C_j.

Operating junction temperature

Junction temperature is the maximum allowable temperature for the diode junction, usually given in °C. Heat is the "Achilles heat" of semiconductor devices. They must be kept cool to junction properly and give long service life.

Depending on the type of diode being considered, additional data may also be provided, such as frequency range, noise level, switching time, thermal resistance level etc. Most of these parameters vary with temperature or other operating conditions. Therefore, a single figure fails to fully describe any given rating. Therefore, manufacturers provide graphs of component ratings plotted against other variables (such as temperature), so that the user has a better idea of what the device is capable of.

Example 4.9.1 Calculate current I and output voltage V_o in Fig. Ex. 4.9.1

(i) By practical Si diode model in which $r_f = 20\ \Omega$

(ii) When the diode is ideal.

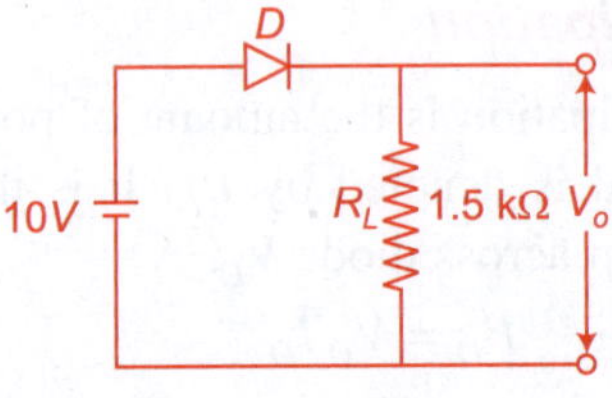

Fig. Ex. 4.9.1

Solution In Fig. Ex. 4.9.1, the diode is in forward bias because the anode of diode is connected to the positive terminal of battery and cathode is connected to the negative terminal of battery.

(i) Since the diode is in forward bias, so replace the diode by its equivalent as shown in Fig. Ex. 4.9.2.

$$I = \frac{V_S - V_B}{(R_L + r_f)}$$

$$I = \frac{(10 - 0.7)\ V}{1.5\ k\Omega + 20\ \Omega}$$

$$\boxed{I = 6.12\text{mA.}}$$ **(Ans.)**

Output voltage V_0

$$V_0 = IR_L$$
$$V_0 = 6.12\text{mA} \times 1.5\ k\Omega$$

$$\boxed{V_0 = 9.18\text{ volt.}}$$ **(Ans.)**

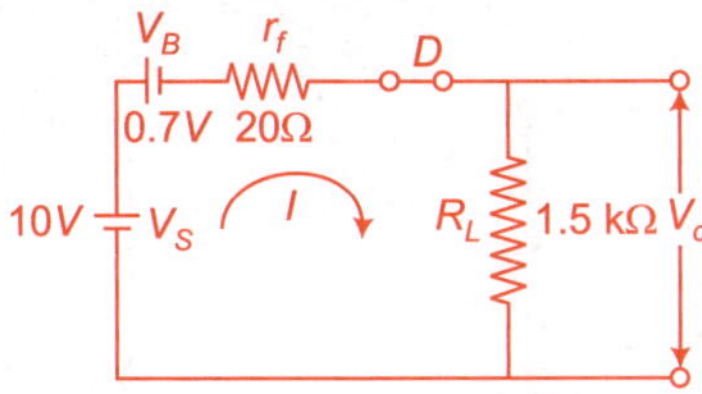

Fig. Ex 4.9.2 *Equivalent model of practical diode*

(ii) When the diode is ideal, then it will treat as a short circuit. Hence, the equivalent model for an ideal diode is shown in Fig. Ex. 4.9.3.

$$I = \frac{V_S}{R_L} = \frac{10 \text{ V}}{1.5 \text{ k}\Omega}$$

$$\boxed{I = 6.7 \text{ mA.}}$$ **(Ans.)**

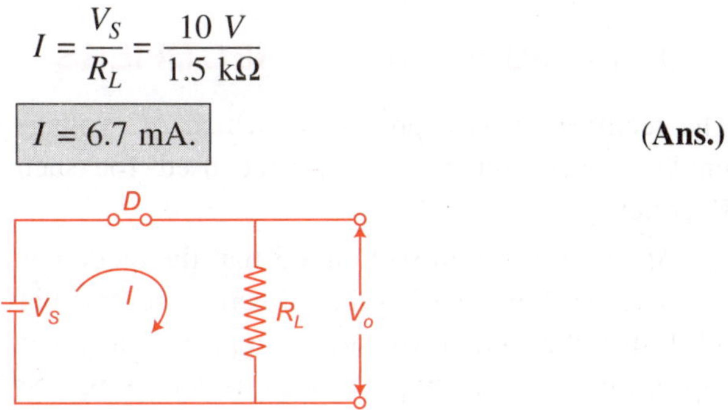

Fig. Ex. 4.9.3 *Equivalent model of ideal diode*

Output voltage,

$$V_o = IR_L$$
$$V_o = 6.7 \text{ mA} \times 1.5 \text{ k}\Omega$$

$$\boxed{V = 10 \text{ volt.}}$$ **(Ans.)**

In above example we observe that the current will flow easily in forward bias. When the diode is ideal, then the output voltage is equal to the input voltage.

Example 4.9.2 In Fig. Ex. 4.9.1, when the terminal of the battery is reversed, Calculate the value of current and output voltage.

Solution In Fig. Ex. 4.9.1, when the terminal of battery is reversed, then the anode of diode is connected to the negative terminal of battery and cathode is connected to the positive terminal of battery. Hence, diode is in reverse bias as shown in Fig. Ex. 4.9.4.

In Fig. Ex. 4.9.5, we see that the diode will be treated as an open circuit because it is in reverse bias. Due to this, no current will flow into the circuit i.e.

$$I = 0\text{mA}$$

Hence, $$V_o = 0\text{V}$$

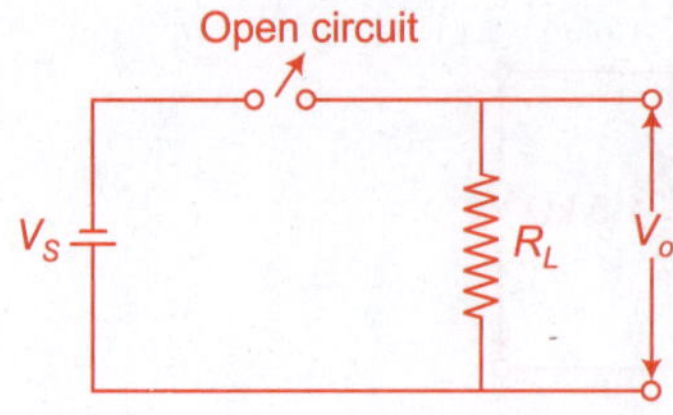

Fig. Ex. 4.9.4   *Equivalent model*

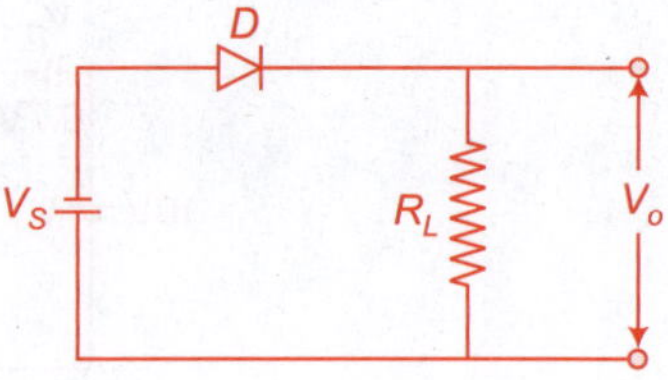

Fig. Ex. 4.9.5   *After changing the terminal of battery*

From above discussion, we can say that there is no current flow in reverse bias so the obtained output is zero.

4.10   *p-n* JUNCTION AS RECTIFIERS

The rectification is a process in which AC voltage is converted in to DC or unidirectional voltage. The circuit used for such conversion is known as Rectifier.

As you can see in section 4.3 that the diode conducts only in one direction (forward bias) and not in other direction (reverse bias). You will see in Example 4.9.1 that the output voltage is approximately equal to the input voltage in forward bias but the output voltage is zero in reverse bias. So, we can state that diode can be used as a rectifier. As you know that the direction of AC voltage is changed regularly i.e., it has two half cycles; one is positive, other is negative as shown is Fig. 4.12.

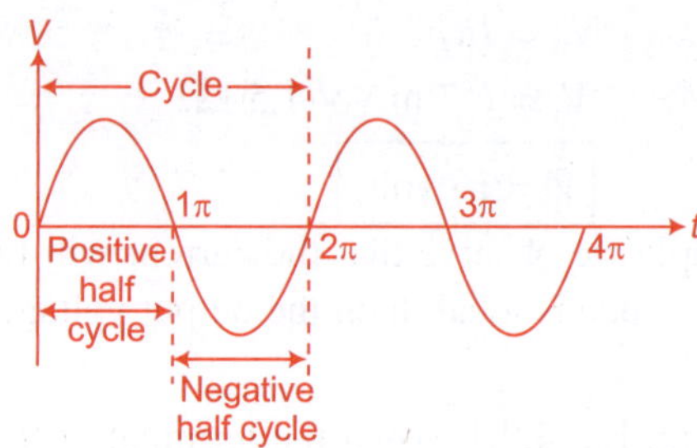

Fig. 4.12   *AC Voltage waveform*

If AC is applied on diode, then it will conduct in half cycle and not in second half cycle i.e., will be in forward bias for one half cycle and in reverse bias for second half cycle. So, the current will flow in one direction only when the diode is in forward bias and current will not flow in other direction when the diode is in reverse bias. These properties of diode is used as a rectifier.

4.10.1   Half Wave Rectifier

The circuit in which unidirectional or DC output is obtained in half cycle of the input is called as half wave rectifier.

The half wave rectifier circuit is shown in Fig. 4.13 (a). In the circuit we can see that the AC voltage is connected at the primary winding (P_1P_2) of the transformer. The anode of diode is connected to S_1 terminal of secondary winding. The cathode of the diode is connected to the load resistance R_L and second terminal of R_L is connected to the S_2 terminal of secondary winding of transformer. Transformer is used to step up and step down the AC input according to requirement.

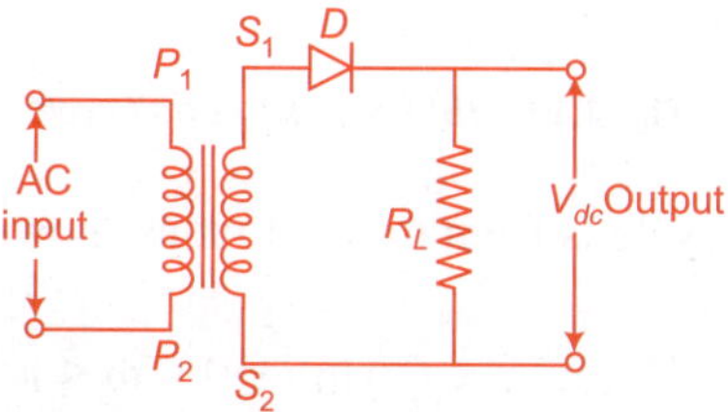

Fig. 4.13   *(a) Half wave rectifier circuit*

Working of half wave rectifier

When the primary of transformer is connected to AC mains supply, then the AC voltage obtained at the secondary $(S_1 \, S_2)$ of the transformer is represented by the equation as follows:

$$V_i = V_m \sin \omega t \tag{4.9}$$

where, V_m = Maximum amplitude of voltage in volts

ω = Angular frequency in rad/sec.

Since the input applied at the rectifier is AC voltage therefore the half ware rectifier will work according to the positive and negative half cycle as follows:

(i)   Positive half cycle

For the positive half cycle, the secondary terminals S_1 is positive with respect to S_2. As we can see in Fig. 4.13 (b), the anode of the diode is connected at the positive terminal cathode and the diode is connected at the negative terminal of secondary. The diode is in forward bias and will conduct, therefore the diode is replaced by their equivalent circuits of forward bias (in Fig. 4.13 (b)) V_B is ignored in analysis. The current will flow in a closed loop $S_1 D R_L S_2 S_1$.

The current in a loop is,

$$i = \frac{V_i}{R_L + r_f}, \quad i = \frac{V_m \sin \omega t}{R_L + r_f} \tag{4.10}$$

Figure 4.13 (b) for positive half cycle,

$$i = \frac{V_m}{R_L + r_f} \sin \omega t = I_m \sin \omega t \tag{4.11}$$

where, $I_m = \dfrac{V_m}{R_L + r_f}$

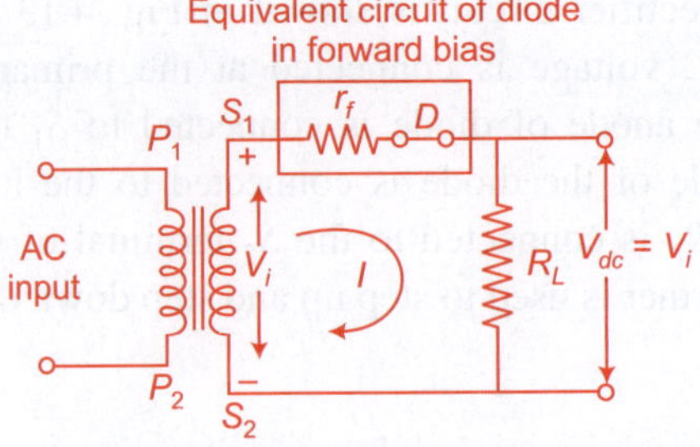

Fig. 4.13 *(b) For positive half cycle*

Since positive half cycle is from $\omega t = 0$ to π as shown is Fig. 4.13, so we can say that,

$$i = I_m \sin \omega t \ \ 0 \le \omega \le \pi \tag{4.12}$$

Since $Vd_c = iR_L$ from Fig. 4.13 (b).

Substituting the value of i from Eqns. (4.10) to (4.12) (b), we get,

$$V_{dc} = \frac{V_m}{R_L + r_f} \sin \omega t . R_L$$

$$V_{dc} = V_m \left(\frac{R_L}{R_L + r_f} \right) \sin \omega t$$

If $R_L >>> r_f$, then $\frac{R_L}{R_L + r_f} \simeq 1$

$$V_{dc} \simeq V_m \sin \omega t = V_i.$$

So, we can say that the output voltage in positive half cycle is approximately equal to the input voltage.

(ii) **Negative half cycle**

For the negative half cycle the secondary terminal S_1 is negative with respect to S_2. As we can see in Fig. 4.14, the anode of the diode is connected at the negative terminal and cathode of the diode is connected at the positive terminal of secondary i.e., the diode is in reverse bias and will not conduct, therefore the diode is replaced by their equivalent circuit of reverse bias in Fig. 4.14. We can see in figure that diode treats as open circuit in reverse bias so the current flow in the circuit is zero.

i.e. $\qquad\qquad\qquad i = 0 \quad \therefore \quad V_{dc} = 0$

$\because$ Negative cycle is from $\omega t = \pi$ to $\omega t = 2\pi$

$$i = 0 \quad \pi \le \omega t \le 2\pi$$

$$V_{dc} = 0 \quad \pi \le \omega t \le 2\pi$$

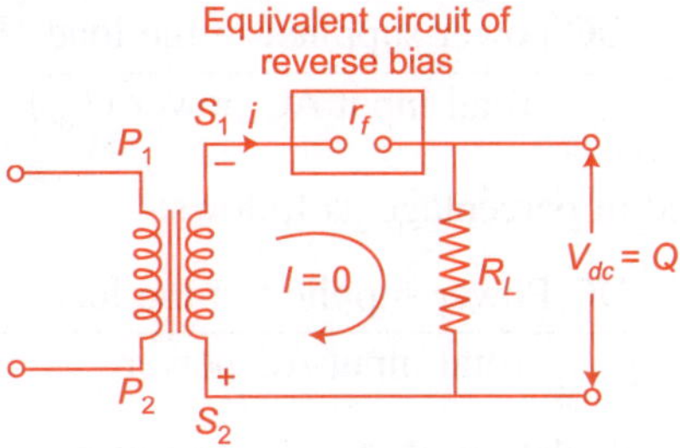

Fig. 4.14 *For negative half cycle*

By the above analysis we can say that the output voltage V_{dc} is approximately equal to the input voltage V_i in positive half cycle while the output voltage is zero in negative half cycle as shown in Fig. 4.15.

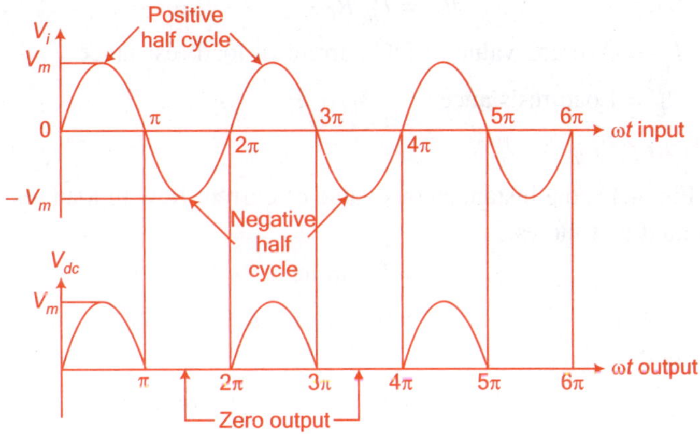

Fig. 4.15 *Input and output wave of half wave rectifier*

PIV of half wave rectifier

As we saw in the previous section, during the negative half cycles of the AC input voltage V_i, the diode is reverse biased and hence it does not conduct. PIV of the rectifier is the maximum reverse voltage across the diode. In half wave rectifier, without filter, the peak inverse voltage across the diode is the maximum transfer voltage V_m across the secondary. Hence, for half wave rectifier,

$$\text{PIV} = V_m. \qquad (4.13)$$

Rectification efficiency of half wave rectifier

As you know that rectifier is a device which is used to convert AC in to DC. The rectification efficiency of the rectifier is defined as the ratio of output DC power to the input AC power and it is denoted by η.

$$\eta = \frac{\text{DC power supplied to the load } (P_{dc})}{\text{Total input AC power } (P_{ac})}$$

usually η is represented in percentage as follows:-

$$\%\eta = \frac{\text{DC Power supplied to the load}}{\text{Total input AC power}} \times 100\%. \qquad (4.14)$$

Hence, before the calculation of η, we need to calculate DC output power and total input AC power first for the calculation of rectification efficiency,

(i) DC output power (P_{dc})

The DC output power in load resistance R_L can be shown by the following formula:

$$P_{dc} = I_{dc}^2 R_L \qquad (4.15)$$

where, I_{dc} = Average value of DC current in load resistance

R_L = Load resistance.

Calculation of I_{dc}

From the Fig. 4.15, the instantaneous value of current flow in load resistance can be represented as follows:

$$\left.\begin{array}{ll} i = I_m \sin \omega t & 0 \leq \omega t \leq \pi \\ = 0 & \pi \leq \omega t \leq 2\pi \end{array}\right\} \qquad (4.16)$$

Let $\omega t = \theta$

then,

$$\left.\begin{array}{ll} i = I_m \sin \theta & 0 \leq \omega t \leq \pi \\ = 0 & \pi \leq \theta \leq 2\pi \end{array}\right\}. \qquad (4.17)$$

As you know that for the calculation of DC of the periodic wave, we integrate it for a complete cycle. Hence, if T is a time period for a wave i, then the DC value for it will be given by the following formula:

$$I_{dc} = \frac{1}{T} \int_0^T i \, d\theta \qquad (4.18)$$

$\because \quad \tau = 2\pi$

$$I_{dc} = \frac{1}{2\pi} \int_0^{2\pi} i \, d\theta \qquad (4.19)$$

Now, dividing the integral in two parts, we get,

$$I_{dc} = \frac{1}{2\pi} \left[\int_0^{2\pi} i \, d\theta + \int_\pi^{2\pi} i \, d\theta \right] \qquad (4.20)$$

Now, substituting the value of i from Eqn. (4.17) into the Eqn. (4.20), we get,

$$I_{dc} = \frac{1}{2\pi} \left[\int_0^{\pi} I_m \sin\theta + \int_{\pi}^{2\pi} i \, d\theta \right]$$

$$I_{dc} = \frac{I_m}{2\pi} \left[\int_0^{\pi} \sin\theta \, d\theta + 0 \right] \qquad \because \quad I_m = \text{constant}$$

$$I_{dc} = I_m \int_0^{\pi} \sin\theta \, d\theta$$

$$I_{dc} = \frac{I_m}{2\pi} \left[-\cos\theta \right]_0^{\pi} \qquad \because \quad \int \sin\theta \, d\theta = -\cos\theta$$

$$I_{dc} = \frac{I_m}{2\pi} \left[(-\cos\pi) - (-\cos 0) \right]$$

$$I_{dc} = \frac{I_m}{2\pi} \left[-(-1) - (-(1)) \right] \qquad \begin{aligned} &\because \quad \cos\pi = -1 \\ &\because \quad \cos 0 = 1 \end{aligned}$$

$$I_{dc} = \frac{I_m}{2\pi} [1+1]$$

$$\boxed{I_{dc} = \frac{2I_m}{2\pi}}$$

$$\boxed{I_{dc} = \frac{I_m}{\pi}} \tag{4.21}$$

Now, substituting the value of I_{dc} in Eqn. (4.15), we get,

$$\boxed{P_{dc} = \left(\frac{I_m}{\pi}\right)^2 R_L.} \tag{4.22}$$

(ii) AC power supplied to the circuit

The AC power supplied to the circuit is equal to the power supplied to both diode and load resistance, so the power loss will occur in both R_L and r_f. Hence, the AC input power is as follows:

$$\boxed{P_{ac} = I_{rms}^2 \, (R_L + r_f)} \tag{4.23}$$

where, I_{rms} = Root means square of load current

      R_L = Load resistance

      r_f = Diode resistance.

Calculation of I_{rms}

The r.m.s. value of load current is represented by the following formula:

$$I_{rms} = \sqrt{\frac{1}{T} \int_0^{T} i^2 \, d\theta} \tag{4.24}$$

$$\because \qquad\qquad T = 2\pi$$

$$\therefore \qquad I_{rms} = \sqrt{\frac{1}{2\pi} \int_0^{2\pi} i^2 \, d\theta}$$

$$I_{rms} = \sqrt{\frac{1}{2\pi} \left[\int_0^{\pi} i^2 \, d\theta + \int_{\pi}^{2\pi} i^2 \, d\theta \right]}$$

Substituting the value of i from Eqn. (4.17), we get,

$$I_{rms} = \sqrt{\frac{1}{2\pi} \left[\int_0^{\pi} (I_m \sin \theta)^2 \, d\theta + \int_{\pi}^{2\pi} 0 \, d\theta \right]}$$

$$I_{rms} = \sqrt{\frac{1}{2\pi} \int_0^{\pi} (I_m^2 \sin^2 \theta \, d\theta)}$$

$$I_{rms} = \sqrt{\frac{I_m^2}{2\pi} \int_0^{\pi} \sin^2 \theta \, d\theta}$$

$$\because \qquad \sin^2 \theta = \frac{1 - \cos 2\theta}{2}.$$

Hence,

$$I_{rms} = \sqrt{\frac{I_m^2}{2\pi} \int_0^{\pi} \left[\frac{1 - \cos 2\theta}{2} \right] d\theta}$$

$$I_{rms} = \sqrt{\frac{I_m^2}{4\pi} \int_0^{\pi} (1 - \cos 2\theta) \, d\theta}$$

$$I_{rms} = \sqrt{\frac{I_m^2}{4\pi} \left[\int_0^{\pi} (1) \, d\theta - \int_0^{\pi} (\cos 2\theta) \, d\theta \right]}$$

$$I_{rms} = \sqrt{\frac{I_m^2}{4\pi} \left\{ [\theta]_0^{\pi} - \left[\frac{\sin 2\theta}{2} \right]_0^{\pi} \right\}} \qquad \therefore \quad \cos 2\theta \, d\theta = \frac{\sin 2\theta}{2}$$

$$I_{rms} = \sqrt{\frac{I_m^2}{4\pi} \left\{ \pi - 0) - \left(\frac{\sin 2\pi}{2} - \frac{\sin 0}{2} \right) \right\}}$$

$$I_{rms} = \sqrt{\frac{I_m^2}{4\pi} [\pi - 0]}$$

$$I_{rms} = \sqrt{\frac{I_m^2}{4\pi} \times \pi}$$

$$I_{rms} = \sqrt{\frac{I_m^2}{4}}$$

$$I_{rms} = \frac{I_m}{2}. \qquad (4.25)$$

Now, substituting the value of I_{rms} in Eqn. (4.23), we get,

$$P_{ac} = \left(\frac{I_m}{2}\right)^2 (R_L + r_f). \qquad (4.26)$$

Calculation of rectification efficiency

The rectification efficiency is given by Eqn. (4.48) as follows:-

$$\eta = \frac{\text{DC output power } (P_{dc})}{\text{Input AC power } (P_{ac})} \times 100\%.$$

Now, substituting the value of P_{dc} and P_{ac} from Eqns. (4.22) and (4.26) respectively, we get,

$$\eta = \frac{(I_{dc})^2 R_L}{I_{rms}^2 (R_L + r_f)} \times 100\%$$

$$\eta = \frac{\left(\dfrac{I_m}{\pi}\right)^2 \times R_L}{\left(\dfrac{I_m}{2}\right)^2 \times (R_L + r_f)} \times 100\%$$

$$\eta = \left(\frac{2}{\pi}\right)^2 \times \frac{R_l}{R_L + r_f} \times 100\%$$

$$\eta = \left(\frac{2}{\pi}\right)^2 \frac{1}{\left(1 + \dfrac{r_f}{R_L}\right)} \times 100\% = \left(\frac{2}{3.14}\right)^2 \frac{1}{\left(1 + \dfrac{r_f}{R_L}\right)} \times 100\%$$

$$\eta = \frac{0.406}{\left(1 + \dfrac{r_f}{R_L}\right)} \times 100\%$$

$$\eta = \frac{40.6\%}{\left(1 + \dfrac{r_f}{R}\right)} \qquad (4.27)$$

If $r_f \ll R_L$ then,

$$\eta = 40.6\%. \qquad (4.28)$$

From Eqn. (4.28), we can state that the rectification efficiency of half wave rectifier is 40.6%.

Ripple factor

In rectifier circuit, the input current is alternating and we get unidirectional current as output, which has a DC component and also has some AC components called ripples. Ripple factor is the measure of AC components or ripples present in the output of a rectifier as shown in Fig. 4.16.

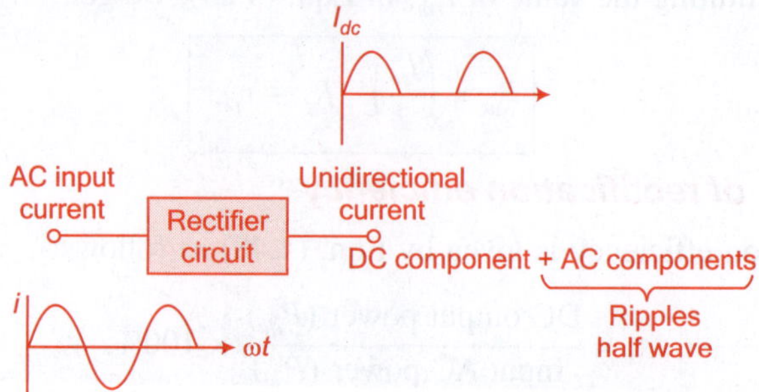

Fig. 4.16   *Block diagram of half wave rectifier*

The ripples factor is denoted by γ.

$$\gamma = \frac{\text{RMS value of AC components}}{\text{Average value of DC components}}$$

$$\gamma = \frac{I'_{rms}}{I_{dc}} = \frac{V'_{rms}}{V_{dc}} \tag{4.29}$$

where, I'_{rms} = *rms* value of AC components of output current

V'_{rms} = *rms* value of AC components of output voltage

I_{dc} = Average value of output DC current

V_{dc} = Average value of output DC voltage.

According to AC circuit theory, the relation between r.m.s. value of total output current (I_{rms}), average DC current (I_{dc}) and *rms* value of AC component (I'_{rms}) is given by,

$$I^2_{rms} = I^2_{dc} + I'^2_{rms}$$

$$I'^2_{rms} = I^2_{rms} - I^2_{dC} \tag{4.30}$$

Dividing the Eqn. (4.30) by I^2_{dc}, we get,

$$\left(\frac{I'_{rms}}{I_{dc}}\right)^2 = \left(\frac{I_{rms}}{I_{dc}}\right)^2 - 1$$

$$\frac{I'_{rms}}{I_{dc}} = \sqrt{\left(\frac{I_{rms}}{I_{dc}}\right)^2 - 1}$$

$\therefore$
$$\gamma = \frac{I'_{rms}}{I_{dc}} \quad \text{from Eqn. (4.29)}$$

Hence,

$$\gamma = \sqrt{\left(\frac{I_{rms}}{I_{dc}}\right)^2 - 1}.$$

(4.31)

For half wave rectifier substituting the value I_{rms} and I_{dc} from Eqns. (4.25) and (4.21) respectively, we get,

$$\gamma = \sqrt{\left(\frac{\dfrac{I_m}{2}}{\dfrac{I_m}{\pi}}\right)^2 - 1} = \sqrt{\left(\frac{\pi}{2}\right)^2 - 1}$$

$$\gamma = \sqrt{(1.57)^2 - 1}$$

$$\boxed{\gamma = 1.21.}$$

(4.32)

Hence, the ripple factor for half wave rectifier is 1.21. So, we can state that the AC components in the output current of half wave rectifier are more.

4.10.2 Full wave Rectifier

The circuit in which unidirectional or DC output is obtained in complete cycle (positive and negative) of the input is called Full wave rectifier.

The full wave rectifier circuit is shown in Fig. 4.17. In the circuit we can see that AC voltage is connected at the primary winding $(P_1 P_2)$ of the transformer. Notice that the transformer used in full wave rectifier is centre tapped. It is called Centre tapped because one connection is taken out from the centre of the secondary winding $(S_1\ S_2)$ of the transformer.

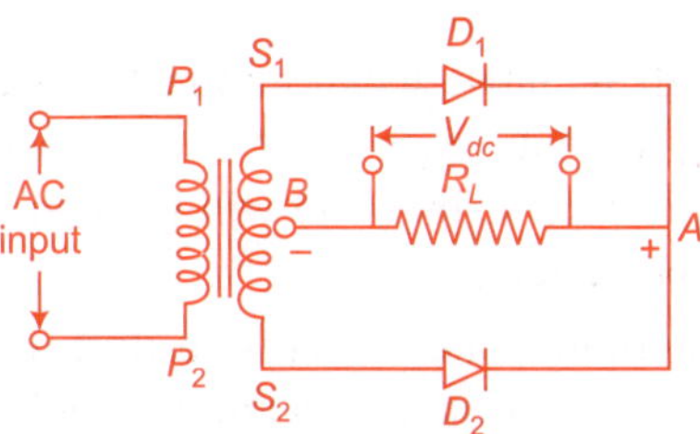

Fig. 4.17 *Full wave rectifier circuit*

Two diodes are used in full wave rectifier circuit. The anodes of two diodes are connected with the secondary winding of the transformer and cathodes of both the diodes are connected with each other. The load resistance R_L is connected between centre tape of transformer and common point (A) of the cathodes of diode.

Working of full wave rectifier

The working of full wave rectifier can be understood as follows:

(i) Positive half cycle

In a positive half cycle of AC input, the S_1 terminal of the transformer secondary is positive with respect to centre tape point B as shown in Fig. 4.17 (a). Due to this, the diode D_2 is in forward bias and will conduct the current easily.

In Fig. 4.17 (a), the diode D_1 is replaced by their forward resistance r_f while the S_2 terminal of the transformer is negative with respect to centre tape point B. Due to this, the diode D_2 is in reverse bias so diode D_2 is replaced by an open circuit in Fig. 4.17 (a).

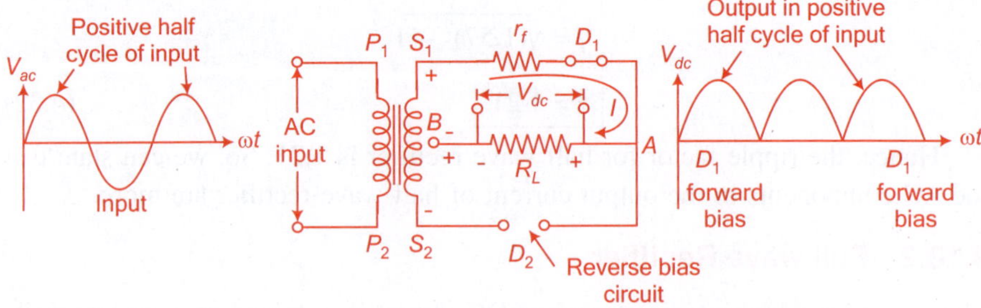

Fig. 4.17 *(a)Full wave rectifier circuit in positive half cycle*

As we can see in Fig. 4.17 (a) that the current flows in closed loop S_1 D_1 A B S_1 because in this path the diode D_1 is in forward bias and there is no current flow in loop $S_2 D_2$ A B S_2 because in this path the diode D_2 is reverse bias. The direction of current in load resistance R_L is from A to B.

(ii) Negative half cycle

In a negative half cycle of AC input, the S_1 terminal of the transformer secondary is negative with respect to centre tape point B as shown in Fig. 4.17 (b). Due to this, the diode D_1 is, in reverse bias and will not conduct the current because diode D_1 is treated as open circuit.

While the S_2 terminal of the transformer is positive with respect to centre tape point B. Due to this, the diode D_2 is in forward bias and will conduct the current easily. In Fig. 4.17 (b), diode D_2 is replaced by their forward resistance rf.

As we see in Fig. 4.17 (b) that the current flows in closed path $S_2 D_2$ A B S_2 because in this path the diode D_2 is in forward bias and there is no current flow in path $S_1 D_1$ A B S_1 because in this path the diode D_1 is in reverse bias. The direction of current in load resistance R_L is from A to B.

The complete input and output waveforms are shown in Fig. 4.17 (c).

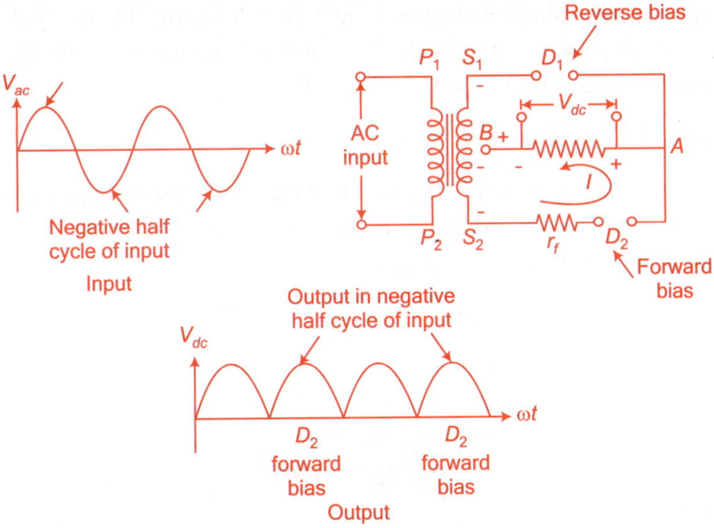

Fig. 4.17   *(b) Full wave rectifier circuit in negative half cycle*

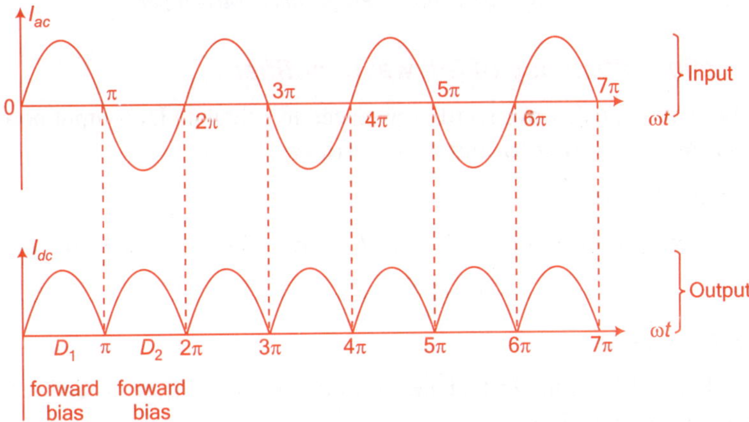

Fig. 4.17   *(c) Input and outpur waveforms of full wave rectifier circuit*

As we can see in Figs. 4.17 (a) and (b) the current flow in load resistance is in same direction in both the half cycles i.e., from A to B. So, we can say that the output of rectifier is unidirectional.

PIV of full wave rectifier

As you know that the peak inverse voltage is the maximum possible reverse voltage across a diode, when it is reverse biased.

In positive half cycle, diode D_1 is in forward bias and diode D_2 is in reverse bias. The maximum voltage across R_L is V_m due to the diode D_1. The forward bias completes the voltage between S_1 and B will supply to the load R_L. The polarity of V_m is shown in Fig. 4.18. The total voltage across diode D_2 is given by Kirchhoff voltage low $V_{D_2} = V_m + V_m = 2V_m$.

Similarly, PIV of diode $D_1 = 2V_m$.

Hence, we can say that the $2V_m$ is a maximum reverse voltage across diode D_2 for positive half cycle and across diode for negative half cycle.

Hence, PIV of full wave rectifier is:

$$\boxed{PIV = 2Vm} \tag{4.33}$$

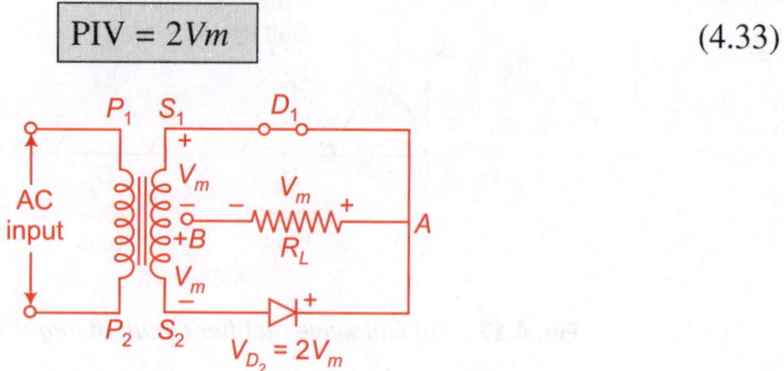

Fig. 4.18 *Calculation of PIV for full wave rectifier*

Rectification efficiency of full wave rectifier

Similarly, as in the half wave rectifier, we need to calculate DC output power and total input AC power first for full wave rectifier,

(i) DC output power

The DC output power in load resistance R_L can be shown by the Eqn. (4.15).

$$P_{dc} = I_{dc}^2 \, R_L \tag{4.34}$$

Calculation of I_{dc}

From the Fig. 4.17 (c), the instantaneous value of current flow in load resistance can be represented as follows:

$$\left. \begin{array}{ll} i = I_m \sin \omega t & 0 \le \omega t \le \pi \\ i = -I_m \sin \omega t & \pi \le \omega t \le 2\pi \end{array} \right\}. \tag{4.35}$$

Let $\omega t = \theta$

then,

$$\left. \begin{array}{ll} i = I_m \sin \theta & 0 \le \theta \le \pi \\ i = -I_m \sin \theta & \pi \le \theta \le 2\pi \end{array} \right\} \tag{4.36}$$

Now, from Eqn. (4.29) (b), we get,

$$I_{dc} = \frac{1}{2\pi} \left[\int_0^\pi i \, d\theta + \int_\pi^{2\pi} i \, d\theta \right] \tag{4.37}$$

Now, substituting the value of i from Eqn. (4.16) (a) into the Eq. (4.37), we get,

$$I_{dc} = \frac{1}{2\pi} \left[\int_0^\pi I_m \sin \theta + \int_\pi^{2\pi} - I_m \sin \theta \; d\theta \right] \qquad \because \quad I_m = \text{constant}$$

$$\therefore \qquad I_{dc} = \frac{1}{2\pi} \left[I_m \int_0^\pi \sin \theta \; d\theta - I_m \int_\pi^{2\pi} \sin \theta \; d\theta \right]$$

$$I_{dc} = \frac{I_m}{2\pi} \left[\int_0^\pi \sin \theta \; d\theta - \int_\pi^{2\pi} \sin \theta \; d\theta \right] \qquad \because \quad \int \sin \theta \; d\theta = - \cos \theta$$

Hence, $\quad I_{dc} = \dfrac{I_m}{2\pi} \left\{ [- \cos \theta]_0^\pi - [- \cos \theta]_\pi^{2\pi} \right\}$

$$I_{dc} = \frac{I_m}{2\pi} \left\{ [(- \cos \pi) - (- \cos 0)] - [(- \cos 2\pi) - (- \cos \pi)] \right\}$$

$$I_{dc} = \frac{I_m}{2\pi} \left\{ [(- (-1) - (-1)] - [(-1) - (- (-1))] \right\}$$

$$I_{dc} = \frac{I_m}{2\pi} \left\{ (1 + 1] - [-1 - 1] \right\}$$

$$I_{dc} = \frac{I_m}{2\pi} [2 + 2]$$

$$I_{dc} = \frac{4 I_m}{2\pi}$$

$$\boxed{I_{dc} = \frac{2 I_m}{\pi}} \tag{4.38}$$

Now, substituting the value of I_{dc} in Eqn. (4.15), we get,

$$P_{dc} = \left(\frac{2 I_m}{\pi} \right)^2 R_L. \tag{4.39}$$

(ii) AC power supplied to the circuit

The AC power supplied to the circuit is equal to the power supplied to the diode and load resistance both, so the power loss will occur at both R_L and r_f.

Hence, the AC input power is as follows:

$$\boxed{P_{ac} = I_{rms}^2 \; (R_L + r_f).} \tag{4.40}$$

Calculation of I_{rms}

The I_{rms} given by the formula as follows:

$$I_{rms} = \sqrt{\frac{1}{T} \int_0^T i^2 \, d\theta}$$

$$I_{rms} = \sqrt{\frac{1}{2\pi} \int_0^{2\pi} i^2 \, d\theta} \qquad \because \ T = 2\pi$$

$$I_{rms} = \sqrt{\frac{1}{2\pi} \left[\int_0^\pi i^2 \, d\theta + \int_\pi^{2\pi} i^2 \, d\theta \right]}$$

Substituting the value of i from Eqn. 4.17 (a), we get,

$$I_{rms} = \sqrt{\frac{1}{2\pi} \left\{ \int_0^\pi I_m^2 \sin^2 \theta \, d\theta + \int_\pi^{2\pi} I_m^2 \sin^2 \theta \, d\theta \right\}}$$

$$= \sqrt{\frac{I_m^2}{2\pi} \left\{ \int_0^\pi \sin^2 \theta \, d\theta + \int_\pi^{2\pi} \sin^2 \theta \, d\theta \right\}}$$

$$= \sqrt{\frac{I_m^2}{2\pi} \left\{ \int_0^\pi \left(\frac{1 - \cos 2\pi}{2} \right) d\theta + \int_\pi^{2\pi} \left(\frac{1 - \cos 2\theta}{2} \right) d\theta \right\}}$$

$$= \sqrt{\frac{I_m^2}{4\pi} \left\{ \int_0^\pi (1 - \cos 2\theta) \, d\theta + \int_\pi^{2\pi} (1 - \cos 2\theta) \, d\theta \right\}}$$

$$= \sqrt{\frac{I_m^2}{4\pi} \left\{ \left[\theta - \frac{\sin 2\theta}{2} \right]_0^\pi + \left[\theta - \frac{\sin 2\theta}{2} \right]_\pi^{2\pi} \right\}}$$

$$= \sqrt{\frac{I_m^2}{4\pi} \{ ((\pi - 0) - (0 - 0)) + 2\pi - \pi) - (0 - 0)) \}}$$

$$I_{rms} = \sqrt{\frac{I_m^2}{4\pi} [\pi + 2\pi - \pi]}$$

$$I_{rms} = \sqrt{\frac{I_m^2}{4\pi} \times 2\pi} = \sqrt{\frac{I_m^2}{2}}$$

$$\boxed{I_{rms} = \frac{I_m}{\sqrt{2}}} \qquad\qquad (4.41)$$

Now, substituting the value of I_{rms} in Eqn. (4.40), we get,

$$P_{ac} = \left(\frac{I_m}{\sqrt{2}}\right)^2 (R_L + r_f). \qquad (4.42)$$

(iii) Calculation of Rectification efficiency

The rectification efficiency is given by Eqn. (4.14) as follows:

$$\eta = \frac{P_{dc}}{P_{ac}} \times 100$$

Now, substituting the value of P_{dc} and P_{ac} from Eqns. (4.39) and (4.42) respectively, we get,

$$\eta = \frac{\left(\dfrac{2I_m}{\pi}\right)^2 R_L}{\left(\dfrac{I_m}{\sqrt{2}}\right)^2 (R_L + r_f)} \times 100\%$$

$$\eta = \frac{\dfrac{4I_m^2}{\pi^2} R_L}{\dfrac{I_m^2}{2} (R_L + r_f)} \times 100 = \frac{8}{\pi^2} \frac{R_L}{(R_L + r_f)} \times 100\%$$

$$\eta = 0.812 \times \frac{R_L}{R_L + r_f} \times 100\%$$

$$\eta = \frac{0.812}{1 + \dfrac{r_f}{R_L}} \times 100\%$$

$$\eta = \frac{81.2}{1 + \dfrac{r_f}{R_L}} \%.$$

If $r_f <<< R_L$ then,

$$\boxed{\eta = 81.2\%.} \qquad (4.43)$$

From Eqn. (4.43), we can say that the maximum rectification efficiency of full wave rectifier is 81.2%.

Ripple factor

The ripple factor given by Eqn. (4.31), is,

$$\gamma = \sqrt{\left(\frac{I_{rms}}{I_{dc}}\right)^2 - 1}$$

$\therefore$

$$I_{rms} = \frac{I_m}{\sqrt{2}} \quad \text{and} \quad I_{dc} = \frac{2I_m}{\pi}$$

Hence,

$$r = \sqrt{\left\{ \frac{\left(\frac{I_m}{\sqrt{2}}\right)^2}{\left(\frac{2I_m}{\pi}\right)^2} \right\} - 1} = \sqrt{\left[\frac{\frac{I_m^2}{2}}{\frac{4I_m^2}{\pi^2}} \right] - 1}$$

$$r = \sqrt{1.23 - 1}$$

$$r = \sqrt{0.23}$$

$$\boxed{\gamma = 0.48.}$$

(4.44)

Hence, the ripple factor for full wave rectifier is less than the ripple factor of half wave rectifier.

4.10.3 Full wave Bridge Rectifier

Full wave bridge rectifier is a circuit in which bridge is made with the help of four diodes and it provides DC output for both half cycle of AC input, therefore, it is called as full wave bridge rectifier. The bridge rectifier circuit is shown in Fig. 4.19 in which four diodes D_1, D_2, D_3, D_4 are used for making a bridge *ABCD*. The two opposite ends of bridge (*A* and *C*) are connected to the secondary of transformer. The load resistance R_L is connected between other two opposite ends *(B* and *D)*.

(i) In positive half cycle

The positive half cycle of AC input is applied at the primary terminal of transformer (P_1P_2). In that condition, the S_1 of secondary is positive with respect to the S_2 of secondary. Due to this, the diode D_1 and D_3 are in forward bias and will conduct the current while the diode D_2 and D_4 are in reverse bias and will not conduct the current as shown in Fig. 4.19 (a). Hence, the current will flow in closed path S_1A_1, BDC S_2 S_1. The direction of current in load is from *B* to *D*.

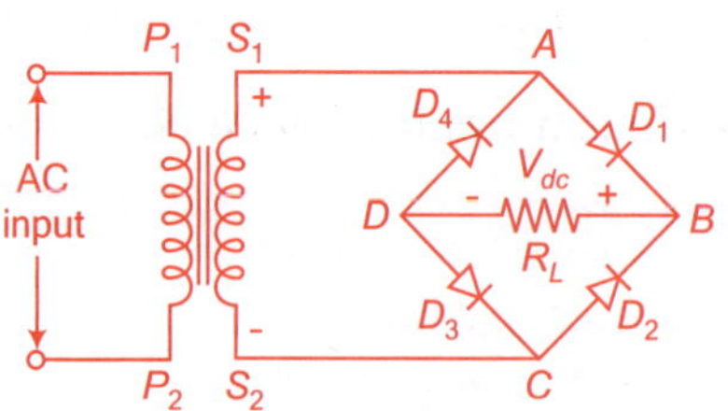

Fig. 4.19 *Full wave bridge rectifier circuit*

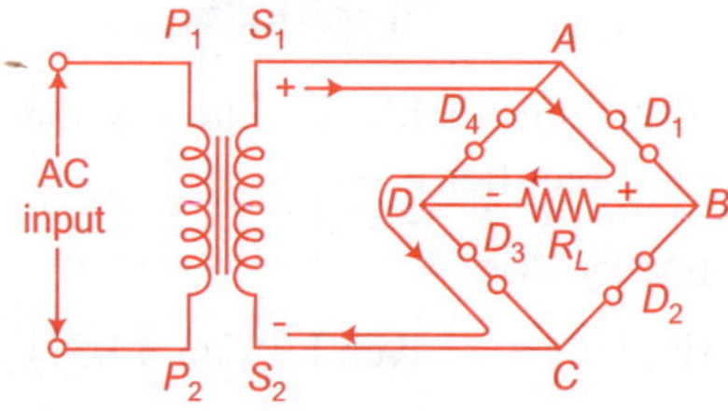

Fig. 4.19 *(a) In positive half cycle*

(ii) In negative half cycle

The negative half cycle of AC input is applied at the primary terminal $(P_1 P_2)$ of the transformer. In that condition, the S_1 of secondary is negative with respect to the S_2 of secondary. Due to this, the diode D_2 and D_4 are in forward bias and will conduct the current while the diode D_1 and D_3 are reverse bias and will not conduct the current as shown in Fig. 4.19 (b). Hence, the current will flow in closed path S_2 *CBDA* $S_1 S_2$. The direction of current in load is from B to D.

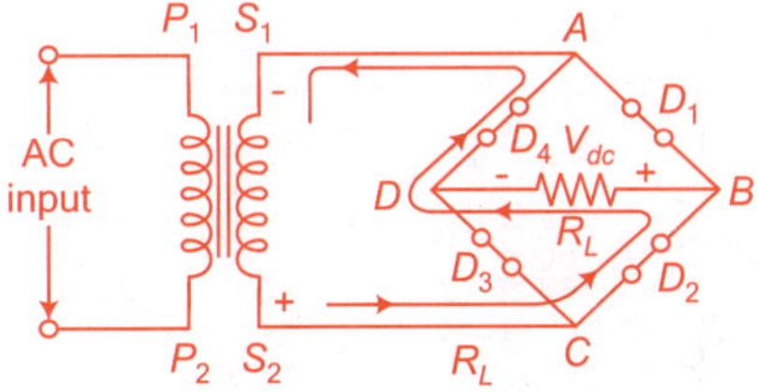

Fig. 4.19 *(b) In negative half cycle*

As we can see in both half cycles of input, the current in load resistance is unidirectional, as shown in Fig 4.20.

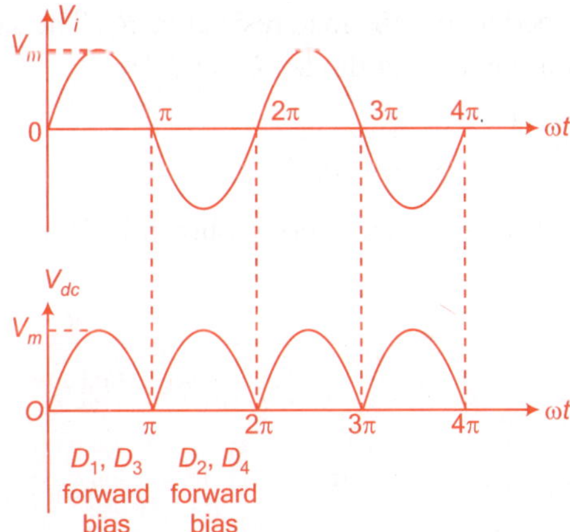

Fig. 4.20 *Input and output waveform for full wave bridge rectifier*

Advantages of bridge rectifier

(i) There is no need of centre tape transformer.

(ii) The secondary winding of transformer is used completely, so the output of bridge rectifier is double than the two diode full wave rectifier.

(iii) PIV is half as compared to two diode full wave rectifier.

Disadvantages of bridge rectifiers

(i) It needs four diodes. So it is costly.

(ii) Voltage regulation is not good.

PIV of diode in bridge rectifier

In Fig. 4.21, bridge rectifier is shown in positive half cycle. When the secondary voltage attains its positive peak value V_m, diode D_1 and D_3 are conducting whereas diodes D_2 and D_4 are non-conducting being reverse biased.

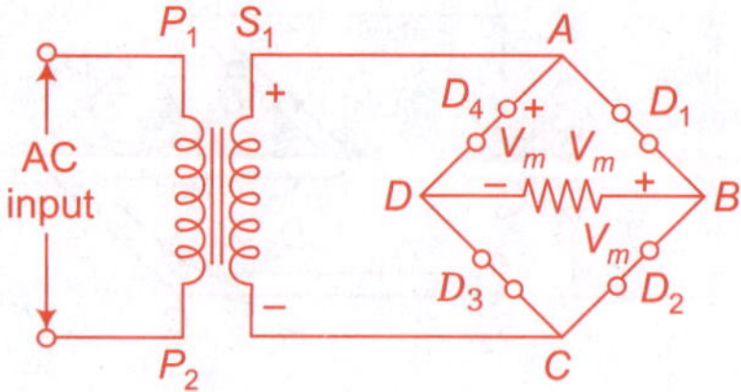

Fig. 4.21 *Calculation of PIV of diode for bridge rectifier*

The conducting diodes D_1 and D_3 have almost zero resistance (zero voltage drop across them). Point B has the same potential as point A. Similarly, point D has the same potential as point C. The entire voltage of transformer secondary winding, V_m is developed across the load resistance R_L. The same voltage V_m acts across each of the non-conducting diodes D_2 and D_4.

Thus, for bridge rectifier,

$$\boxed{PIV = V_m.}$$
(4.45)

The comparison of different rectifiers is shown in Table 4.1.

Table 4.1

S. No.	Parameter	Half wave rectifier	Full wave rectifier	
			Centre tapped	Bridge
1.	Number of diodes	1	2	4
2.	Peak inverse voltage	V_m	$2V_m$	V_m
3.	Maximum load current	$\dfrac{V_m}{R_L + r_f}$	$\dfrac{V_m}{R_L + r_f}$	$\dfrac{V_m}{R_L + 2r_f}$
4.	I_{rms}	$\dfrac{I_m}{2}$	$\dfrac{I_m}{\sqrt{2}}$	$\dfrac{I_m}{\sqrt{2}}$
5.	I_{dc}	$\dfrac{I_m}{\pi}$	$\dfrac{2I_m}{\pi}$	$\dfrac{2I_m}{\pi}$
6.	Ripple factor γ	1.21	0.482	0.482
7.	Maximum efficiency η	40.6%	81.2%	81.2%
8.	Minimum ripple frequency	f_i	$2f_i$	$2f_i$

4.11 FILTERS

You studied in previous section about the rectifier circuit. As we know that rectifier is a device which is used to convert AC signal into DC output. But we have seen that the output of rectifier is not pure DC. It is a pulsating output due to the ripples in output i.e., output of rectifier contains DC components and AC components (ripples) both. Due to this the output of rectifier is unidirectional but pulsating, so we need a filter circuit, as the output of rectifier. So, we can say that the output of rectifier is given input to the filter circuit and it removes the ripples from the input and provides steady DC output.

"In other words we can say that a filter is a device that converts pulsating output of a rectifier into a steady DC level i.e., it removes ripples from the output of rectifier".

Basically, we can say that filter are those devices that separate two things i.e., it separate AC ripples (those have high frequency), from DC components (those have zero frequency). Generally inductor or capacitor is used in filters because their reactance depends on the frequency.

As you know that,

The reactance of capacitor $X_C = \dfrac{1}{2\pi fc}$

$$X_C = \frac{1}{2\pi fc} \tag{4.46}$$

For DC, $\qquad\qquad f = 0$

then, $\qquad\qquad X_C = \infty$ (Infinite)

For high frequency,

$$X_C = \text{Low}$$

From the above equation, we can say that capacitor blocks the DC (open circuit for DC) because at DC its reactance is infinite and it passes the AC (short circuit for AC) because at AC, its reactance is very less.

Similarly,

The reactance of inductor,

$$X_L = 2\pi fL \tag{4.47}$$

For DC, $\qquad\qquad f = 0$

then, $\qquad\qquad X_L = 0$ (zero)

For high frequency,

$$X_L = \text{high.}$$

From the above equation, we can say that inductor blocks the AC (open circuit for AC) because at AC its reactance is high and it passes the DC (short circuit for DC) because at DC, its reactance is very less or zero.

Different types of filters

Filters are classified as follows:

   (i) Shunt capacitor filter

  (ii) Series inductor filter

 (iii) LC filter.

4.11.1 Shunt Capacitor Filter

Figure 4.22 shows a circuit for shunt capacitor filter in which large value of capacitor is connected in parallel with load and the ripples get bypassed through the capacitor C. Only DC components flow through the load.

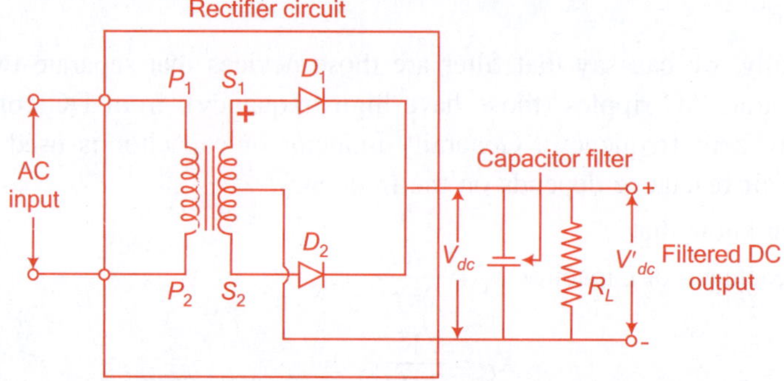

Fig. 4.22   *Shunt capacitor filter with rectifier circuit*

Working

As we have discussed earlier that the reactance of capacitor is infinite for DC so it blocks the DC and its reactance is less for AC (high frequency), so it passes the AC.

Hence, AC ripples come out from the rectifier circuit through the capacitor and go back to the secondary of the transformer. Capacitor blocks the DC components which comes as output from rectifier circuit. So, it flow through the load resistance. Pure DC is obtained at the load.

When the rectifier voltage is increasing i.e., in forward condition, then capacitor filter is charged to V_m and when the rectifier voltage tries to fall, the charged capacitor immediately tries to send the current back to the rectifier (i.e., discharged). In this process, the rectifier diodes are reverse biased. Thus, the capacitor discharges through load. The capacitor continues to discharge until the source voltage is greater than the capacitor voltage. Again diode starts conducting and capacitor is charged to peak value V_m. During this time the rectifier supplies the charging current I_c and load current I_L. During forward bias, shunt capacitor is charged and during reverse bias, shunt capacitor is discharged. The capacitor not only removes the AC component but also improves the output voltage. The magnitude of the output voltage depends upon the time constant R_LC. When the time constant is large, there is slight decay in the output voltage and hence low ripples are obtained. For this, it is necessary to use the capacitor of large value.

The output waveform of rectifier circuit and filter circuit is shown in Figs. 4.23 (a) and (b) respectively.

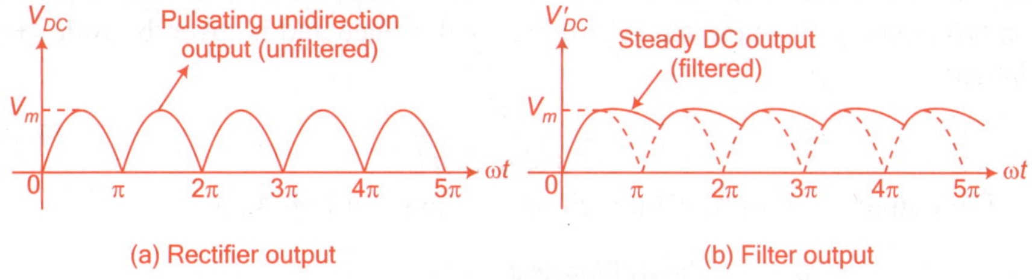

Fig. 4.23 *Output waveform*

If R_LC is greater than the time period of input wave, then the capacitor will discharge slowly. Due to this, the output voltage is constant, because the AC ripples will reduce and are given by following relation:

$$\gamma = \frac{1}{4\sqrt{3}\,f\,R_LC}. \tag{4.48}$$

4.11.2 Series Inductor Filter

Figure 4.24 shows a circuit for series inductor filter in which suitable value of inductor is connected in series with load and it blocks the AC ripples and pass through the DC components only.

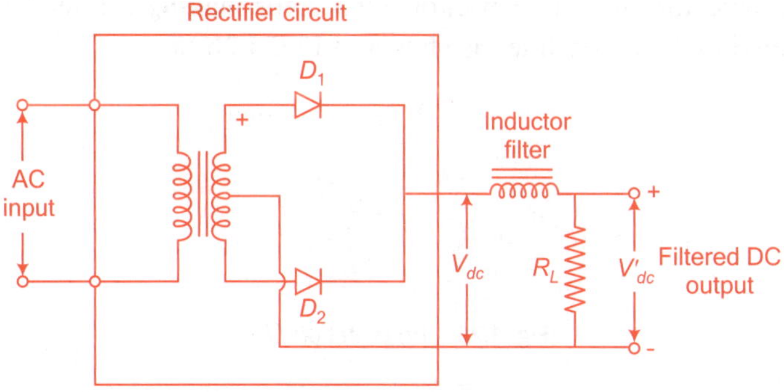

Fig. 4.24 Series inductor filter with rectifier circuit

Working

The reactance of inductor is zero for DC because $X_L = 2\pi fL$. Hence, it passes the DC current and it also flows through the load R_L. The reactance of inductor is high for high frequency ripples so it blocks the ripples. Hence, only DC current will flow through the load resistance R_L.

As we know that inductor always opposes the charges in current flowing into it. When the output current rises above the average value, the magnetic energy is stored in the inductor. When the output current falls below the average value, then it releases the stored energy to the circuit. Hence, the output current is approximately constant and AC ripples will reduce and is given by following relation:

$$\gamma = \frac{R_L}{3\sqrt{2} \times 2\pi f L}.$$

(4.49)

The output waveform of filter circuit is shown in Fig. 4.25.

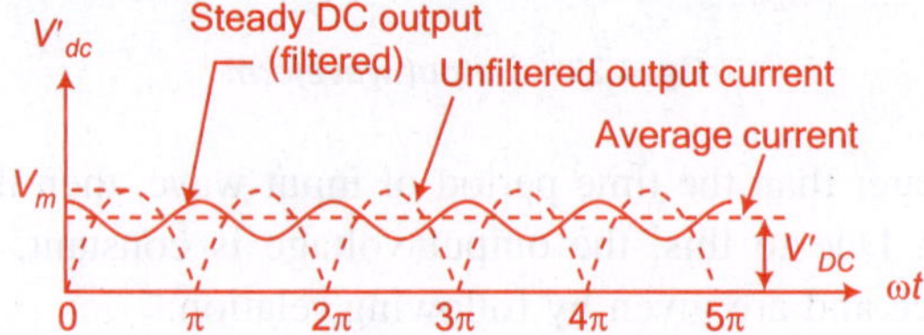

Fig. 4.25 *Filter output waveform*

4.11.3 LC Filter

When both inductor and capacitor are used for making a filter, such type of filter is called as LC filter.

L-section filter

When one inductor and one capacitor is used for making a filter circuit, such filter is called as *L*-section filter as shown in Fig. 4.26 (a).

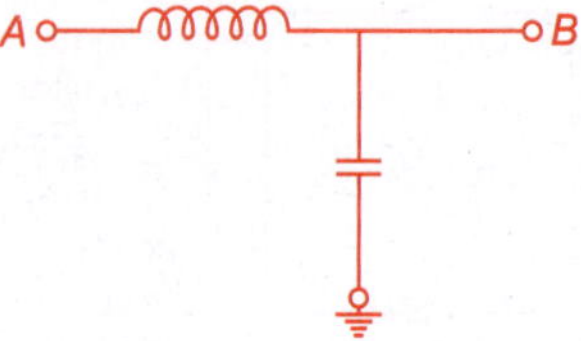

Fig. 4.26 *(a) L-section filter*

π *and T-section filter*

When inductor and capacitor are connected as the T and π shape, such type of filters are called as T and π-section filter. In T-section filter, two inductors and one capacitor are used where as in π-section filter two capacitors and one inductor are used as shown in Figs. 4.26 (b) and (c).

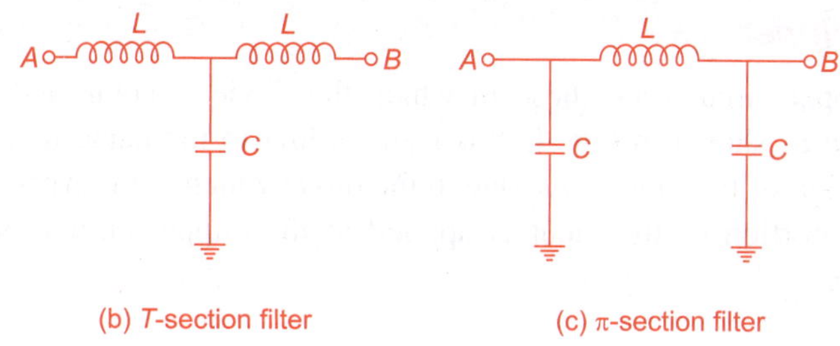

(b) *T-section filter*   (c) *π-section filter*

Fig. 4.26 *LC filters*

The detailed explanation of LC filters are out of scope of this book.

4.12 CLIPPINGS CIRCUITS

A circuit which 'dip' off or remove a portion of the input waveform is known as a clipper. It clips the input signal without distorting the remaining part of the input signal. Generally, the clipper circuits are designed with the help of diode. The half wave rectifier of section 4.10.1 is an example of the simplest form of diode clipper.

According to the orientation of the diode, there are two general categories of clippers; series (negative) and parallel (positive).

Negative clipper

Negative clipper circuits are those in which the diode is connected in series with the load as shown in Fig. 4.27 (b). According to the name it 'clip' off the negative portion of the input wave due to the diode, which is in reverse bias when the negative portion of the input is applied at the clipper circuits as shown in Fig. 4.27 (c). The different types of input signals are shown in Fig. 4.27 (a) that can be applied to a clipper because that are no boundaries on the type of input signals.

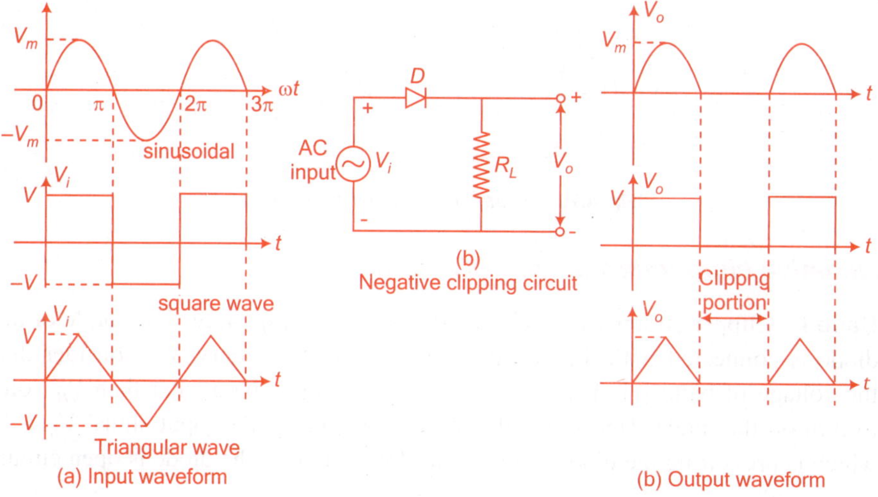

(a) Input waveform   (b) Output waveform

Fig. 4.27

Positive clipper

Positive clipper circuits are those in which the diode is connected in parallel with the load as shown in Fig. 4.28 (b). According to the name it 'clip' off the positive portion of the input wave due to the diode which is in reverse bias when the positive portion of the input is applied at the clipper circuit as shown in Fig. 4.28 (c).

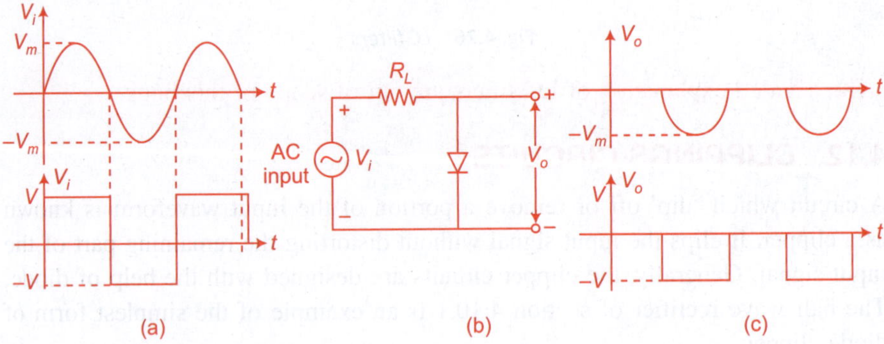

Fig. 4.28   (a) *Input wave;* (b) *Positive clipper circuit;* (c) *Output wave*

4.12.1   Clipper Circuits with DC Supply

The clipper circuits which we have studied earlier are used to clip the complete portion (positive or negative) of the input wave and the peak value of output wave which is equal to the input wave as shown in Figs. 4.27 and 4.28. But in this section, we will see the clipper circuits with DC supply and the peak value of the output which is not same with the peak value of input. It will differ according to the connection of battery.

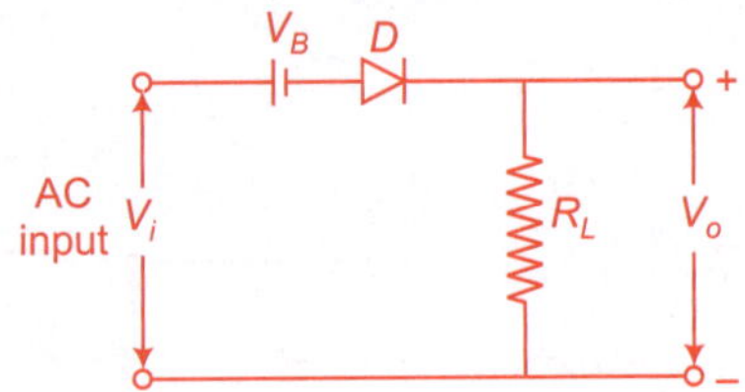

Fig. 4.29   *Negative clipper with DC supply*

Negative clipper circuit with DC supply

Case I   Clipper circuit with DC supply is shown in Fig. 4.29. The anode of the diode is connected with the negative terminal of the battery. We consider that the voltage of battery is V_B. The input voltage V_i must be greater than V_B volts to turn on the diode. Hence, for the negative region of the input signal $V_B > V_i$ which is pressuring the diode in to the turn off state (i.e. the diode is open circuit as shown in Fig. 4.29 (b). So, the output is zero for negative region. When the positive region of the input signal is applied and $V_B < V_i$, then the diode is in forward bias and treated as short circuit as shown in Fig. 4.30 (b).

Now, applying KVL in loop $AB\ CDA$ in Fig. 4.30 (b), we will get,

$$V_i - V_B - V_o = 0 \tag{4.50}$$

Hence,
$$\boxed{V_0 = V_i - V_B.} \tag{4.51}$$

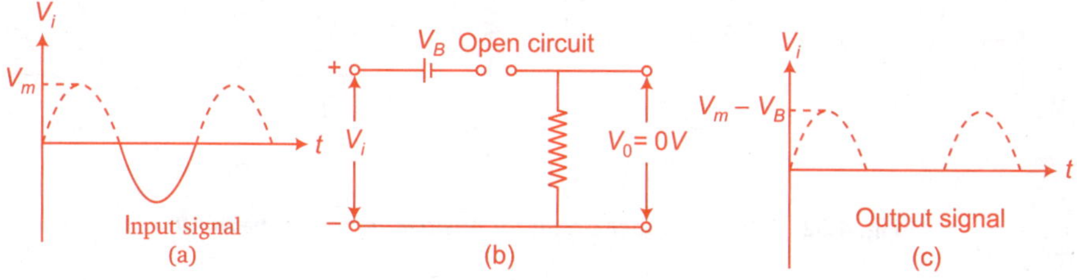

Fig 4.29 *(a) Input signal (b) The clipper circuit for the negative region of input (c) The output for negative region is shown by dark line*

From the Eqn. (4.51), we observe that the output is reduced by V_B from input signal, as shown in Fig. 4.30.

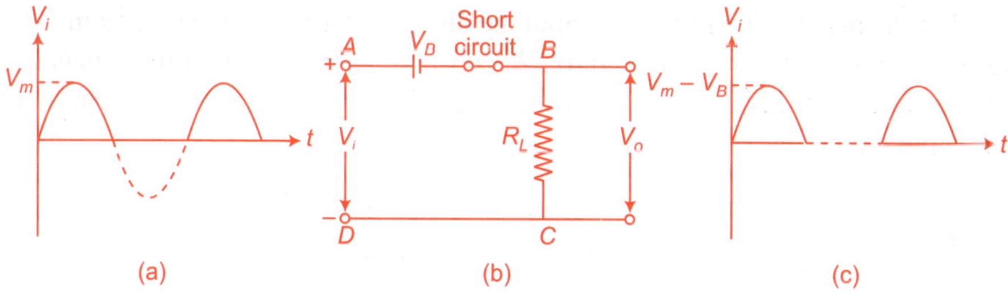

Fig. 4.30 *(a) Input signal (b) The clipper circuit for the positive region of input (c) Output for the positive region is shown by the dark line*

Hence, the final input and output waveform for the negative clipper is shown in Figs. 4.31 (a) and (b) respectively.

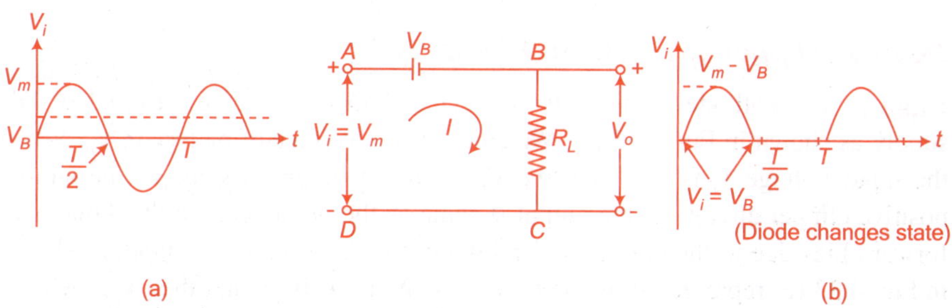

Fig. 4.31 *(a) Input signal for negative clipper; (b) Output signal for negative clipper*

Case II The anode of the diode is connected with the positive terminal of the battery. In this case, V_B will support the input voltage for the positive region. The diode is in forward bias as shown in Fig. 4.32 (a) applying *KVL* in loop *ABCDA* we get,

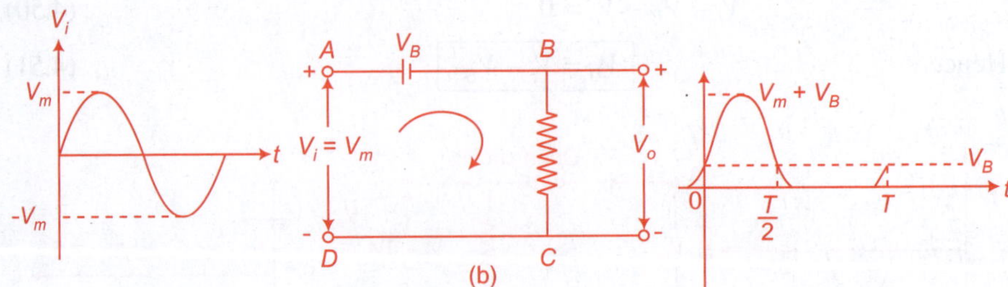

Fig. 4.32 *(a) The clipper circuit for positive region of the input*

$$V_i + V_B - V_o = 0 \qquad (4.52)$$

$$\boxed{V_o = V_i + V_B.} \qquad (4.53)$$

When we see the output waveform of the clipper circuit of Fig. 4.32 (a), we observe that when $V_i = 0$, then the output signal has $V_0 = V_B$ and the peak of the signal will increase by V_B.

For the negative region of the input the diode will turn off in the same manner as in case I, so the output is zero. We observe in output waveform negative clipper circuit in Fig. 4.32 (b) that the output is not zero for negative value of the input signal until $-V_i = V_B$.

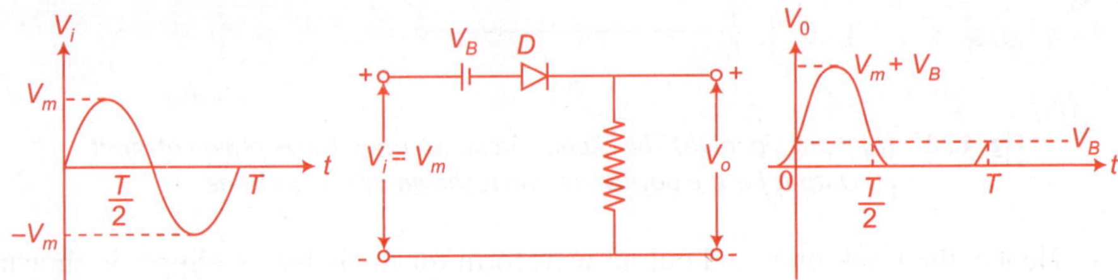

Fig. 4.32 *(b) Negative clipper circuit with DC supply*

Positive clipper circuit with DC supply

Case I The cathode of diode is connected with the positive terminal of DC supply as shown in Fig. 4.33 (a). We consider the voltage of battery in V_B. When the input voltage V_i is less than the V_B, then the output is same as the simple positive clipper circuit i.e. the output is same as the input because the diode is in forward bias due to the voltage V_B and the output is as similar to input as shown in Fig. 4.33 (c) represented by the curve *OA*. When V_i is greater then V_B, then the diode conduct the constant current of V_B voltage as in curve *AB* and when V_i is decreasing below V_B, then it gives the output as similar as the input as shown in curve BC.

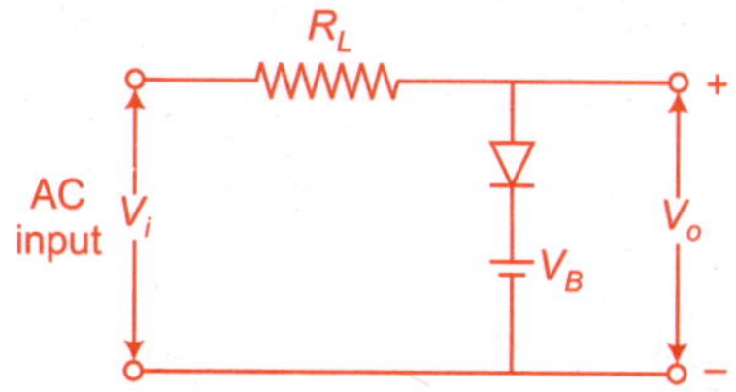

Fig. 4.33 *(a) Positive clipper with DC supply*

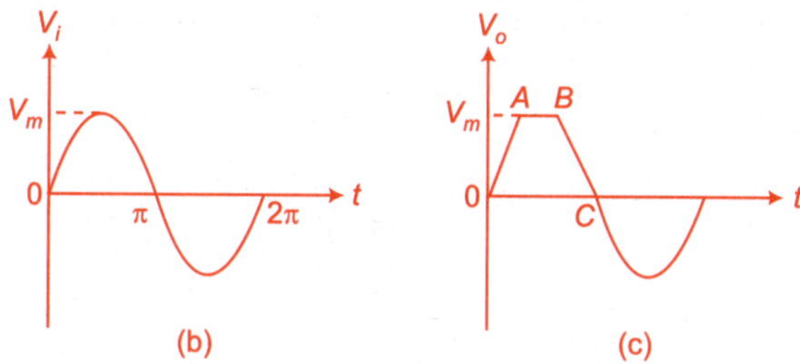

Fig. 4.33 *(b) Input to the clipper circuit (c) Output of the clipper circuit*

Case II When the cathode of the diode is connected with the negative terminal of the battery as shown in Fig. 4.34 (b). The voltage of the battery is V_B.

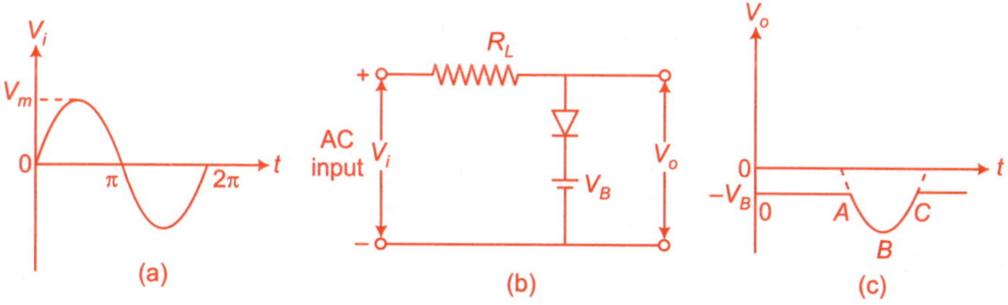

Fig. 4.34 *(a) Input (b) Clipper circuit (c) Output*

When the input voltage V_i is greater than the diode which is conducted only due to the battery $-V_B$, it gives constant current equal to $-V_B$ for positive half cycle of the input as shown in curve OA. When V_i is negative and greater than V_B, then it gives the output as input shown in curve ABC.

Output waveform for variety of clipping circuits

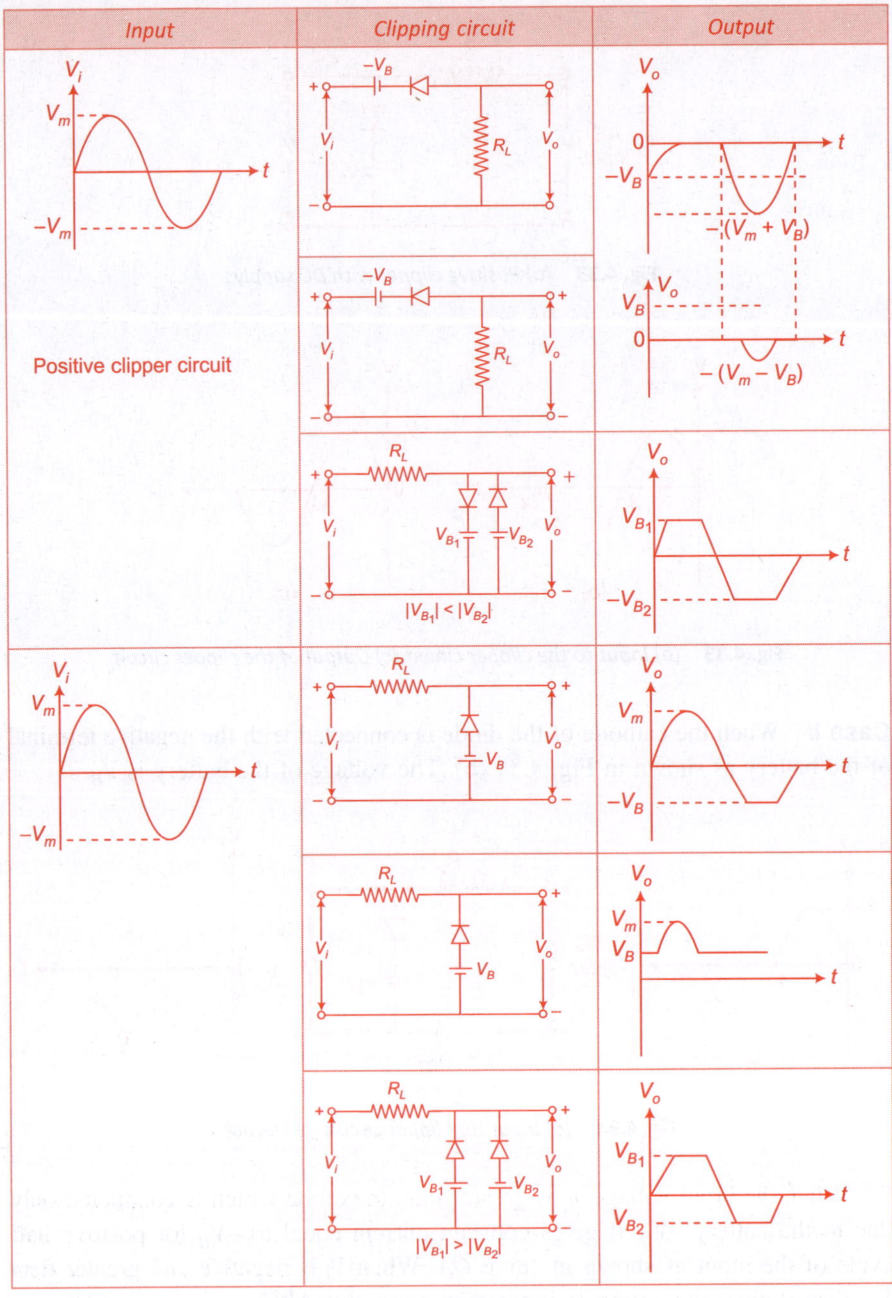

Example 4.12.1 Determine V_0 for the network of Fig. Ex. 4.12.1

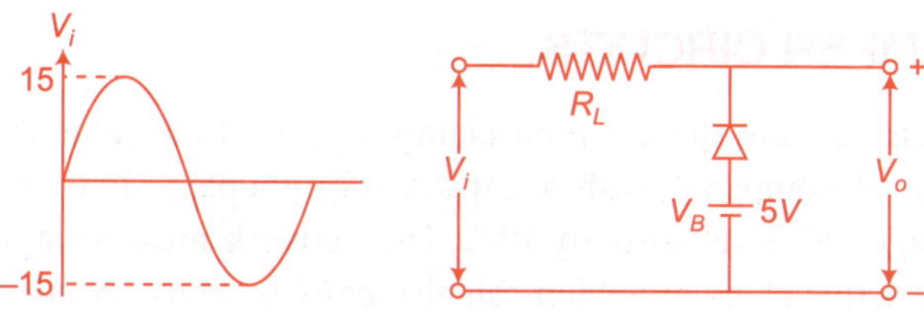

Fig. Ex. 4.12.1

Solution According to the polarity of the DC supply V_B and the direction of the diode, the diode will be in 'on' state for the negative region of the input signal. For this region, the network will be shown in Fig. Ex. 4.12.2. Applying *KVL* at the output side we get,

$$V_o - V_B = 0$$
$$V_o = V_B$$
$$\boxed{V_o = 5 \text{ volt.}}$$

Fig. Ex. 4.12.2

The transition state can be determined from Fig. Ex. 4.12.3, where the condition $i_d = 0A$ at $V_d = 0V$ has been imposed.

$$\text{Transition} = V_B$$
$$= 5V.$$

Since the DC supply V_B is obviously "pressuring" the diode to conduct in forward bias, the input voltage must be greater than 5V for the diode to be in the 'off state. i.e., Any input voltage less than 5V will result in a short circuited diode. For the open circuit state, the network will appear as shown in Fig. Ex. 4.12.4,. where $V_D = V_i$.

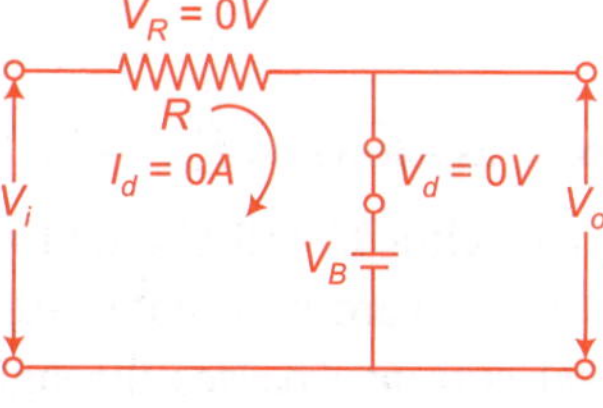

Fig. Ex. 4.12.3

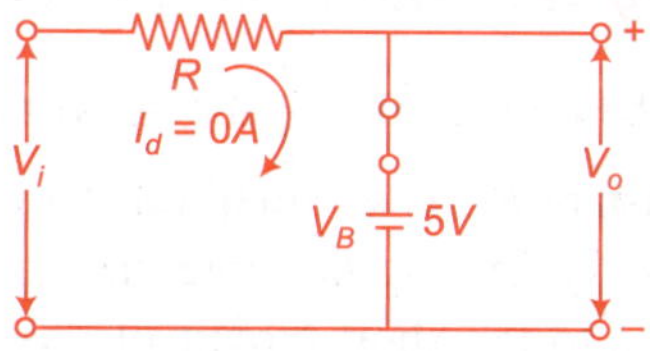

Fig. Ex. 4.12.4

Now, sketch the output V_0 as shown in Fig. Ex. 4.12.5.

4.13 CLAMPER CIRCUITS

The clamper circuits are those which clamp a peak of an input waveform to a specific DC level compared with a capacitively coupled signal which swings about its average DC level (usually 0V). The network must have a capacitor, a diode, and a resistive element, which can also employ an independent DC supply to introduce an addition shift.

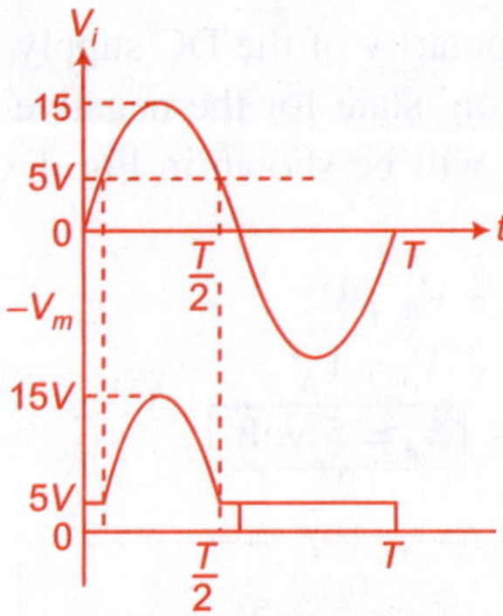

Fig. Ex. 4.12.5

4.13.1 Positive Clamper

The circuit shown in Fig. 4.35 shows a clamp circuit which clamps the input signal to the zero level if the diode is ideal. The resistor R_L is a load resistor or a parallel combination of the load resistor.

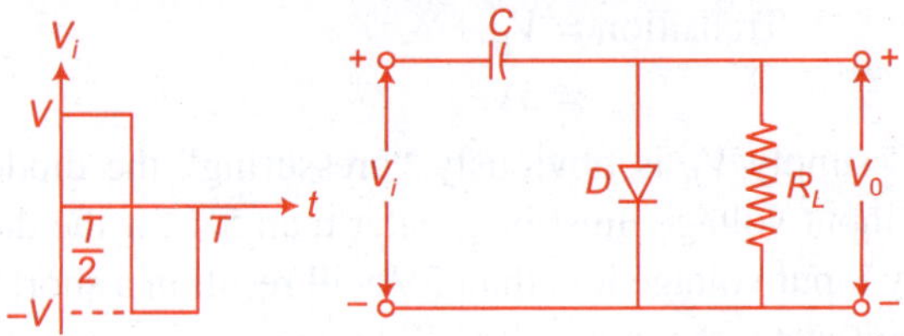

Fig. 4.35 *Positive clamper circuit with input signal*

Working of positive clamper circuit

During the interval 0 to $\dfrac{T}{2}$ the network will appear as shown in Fig. 4.36 (a). The time constant RC is so small due to which the capacitor C will charge to V volts very quickly. During this interval, the diode is in forward bias state i.e., in 'on' state. Due to the short circuit of diode, the total current i passes through diode where there is no current flow through R_L, so the output (V_0) is zero in this interval.

i.e. $\qquad V_o = 0$ volt when $0 \leq t \leq \dfrac{T}{2}$

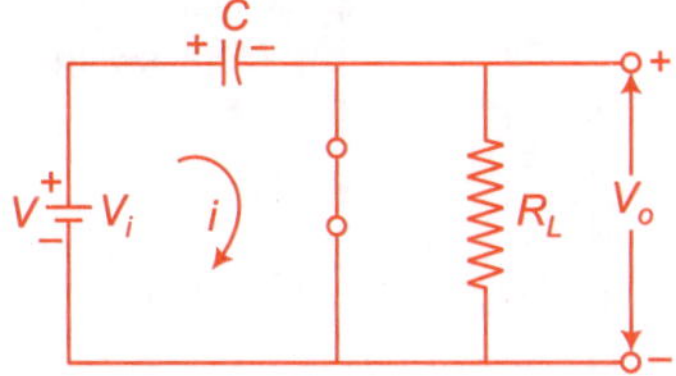

Fig. 4.36   (a) *Clamper circuit during interval* $0 \rightarrow \dfrac{T}{2}$

when the input switches to the $-V$ state, the network will appear as shown in Fig. 4.36 (b).

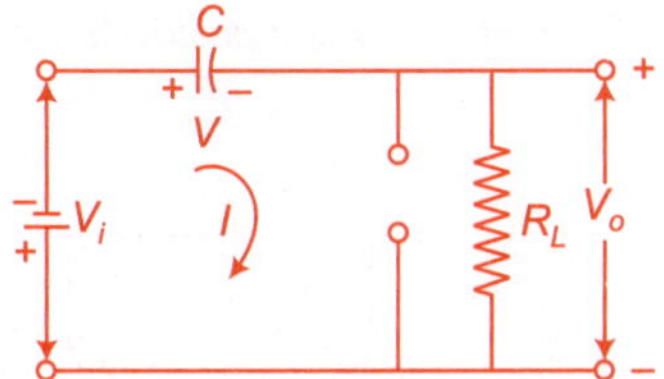

Fig. 4.36   (b) *Clamper circuit during interval* $\dfrac{T}{2} \rightarrow T$

In this interval, the diode is in reverse bias and treated as an open circuit. Now the R_L is back in the network, so the time constant determined by the RC product is sufficiently large to establish a discharge period which is greater than the period $\dfrac{T}{2} \rightarrow T$. So, it can be assumed that the capacitor holds on to all its charge and, voltage during this period.

Now, applying KVL in loop $V_i\ C\ R_L\ V_i$, we get,

$$- V - V - V_0 = 0$$

$$\boxed{V_0 = -2\ V \text{ volt.}}$$

The resulting output waveform appears in Fig. 4.37 with the input signal. The output signal is clamped to 0 level, during the interval $0 \rightarrow \dfrac{T}{2}$ and it maintains the same total swing (2 V) as the input. For a clamper circuit, we can state that the total swing of the output signal is equal to the total swing of the input signal.

4.13.2   Negative Clamper

The circuit shown in Fig. 4.38 shows a negative clamper which clamps the negative region of the input signal to the zero level. In this type of circuit, the anode of the diode is connected to the negative terminal of the battery as shown in Fig. 4.38.

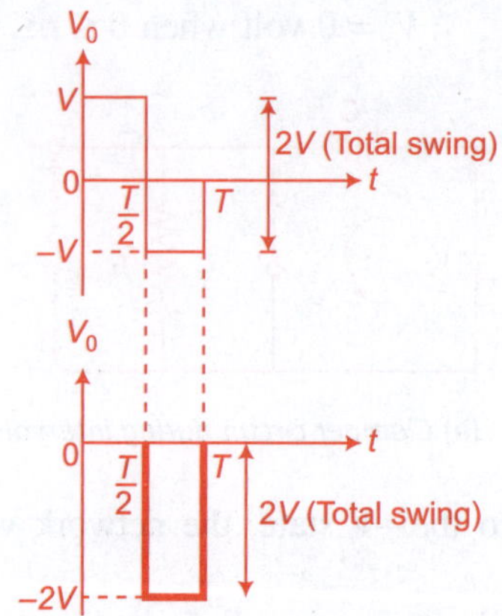

Fig. 4.37 *Sketch of V_0 for positive clamper*

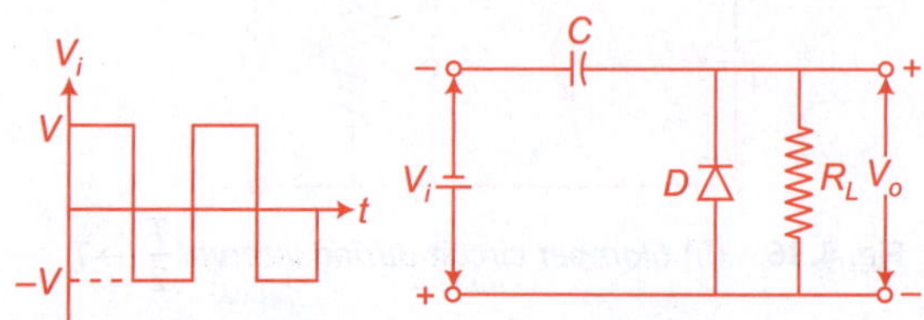

Fig. 4.38 *Negative clamper with input signal*

Working of negative clamper circuit

The working of negative clamper circuit is similar to the positive clamper circuit with only difference that during the positive interval $0 \to \dfrac{T}{2}$ the diode is in 'off' state and the total current will pass through the load R_L which will produce the output $V_0 = 2V$ similar to the positive clamper.

During the interval $\dfrac{T}{2} \to T$, the diode is in 'on' state. So, all the current will pass through the diode which will produce zero output during this interval.

The resultant output waveforms appear in Fig. 4.39 with the input signal.

The output signal is clamped to 0 level during the $\dfrac{T}{2} \to T$ and it maintains the same total swing 2V as the input.

Start the analysis of clamping circuits by considering that point of the input signal which will forward bias the diode.

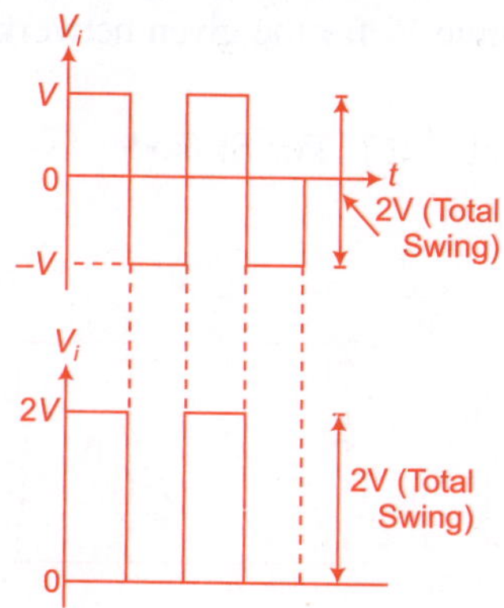

Fig. 4.39 *Sketch of V_0 for negative clamper*

Output waveform for a variety of clamping circuits

Input	Clipping circuit	Output
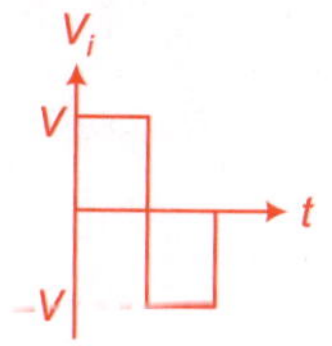	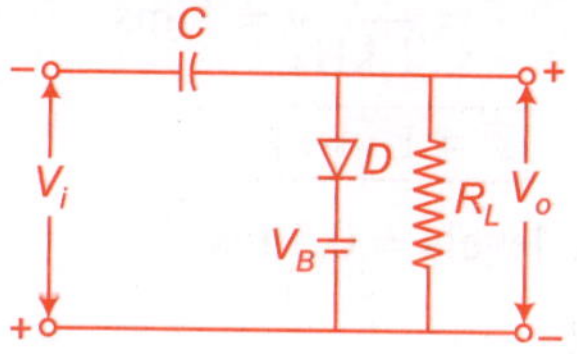	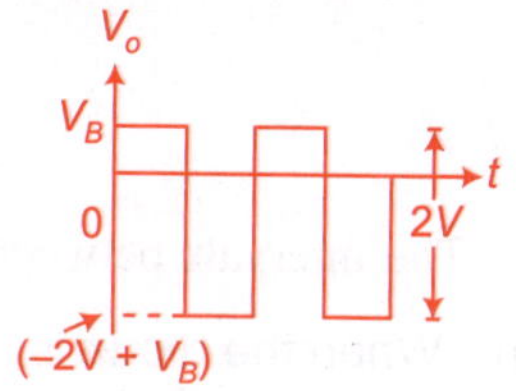
	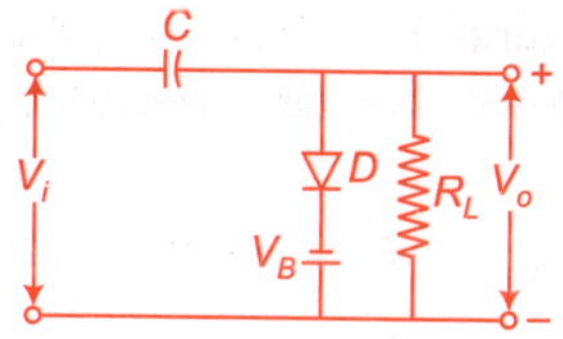	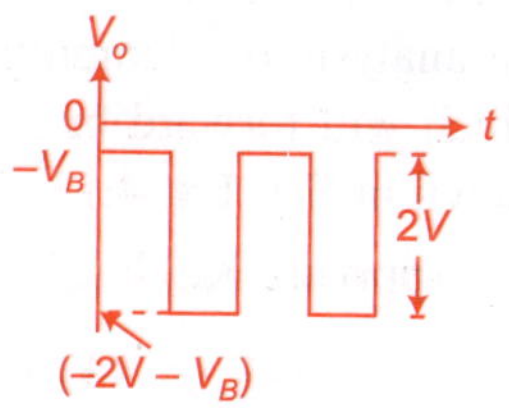
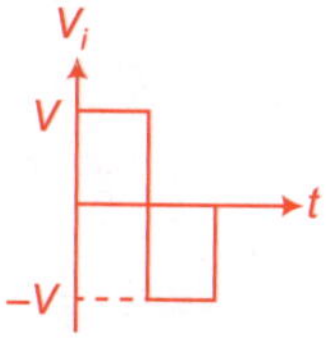	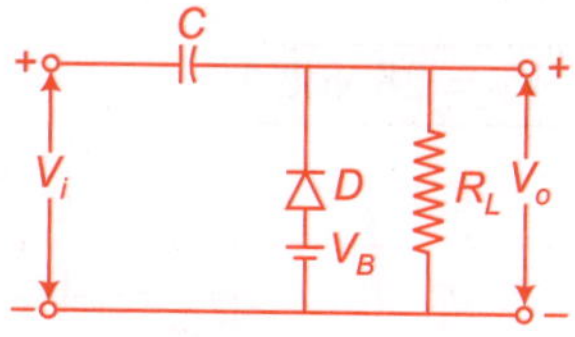	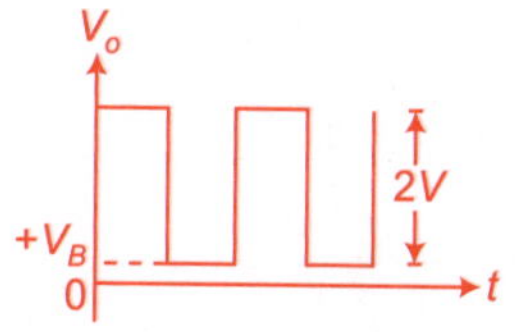
	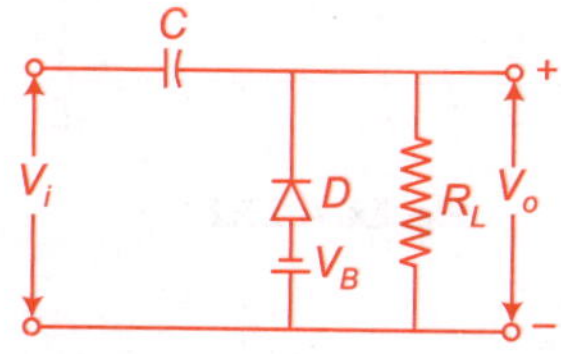	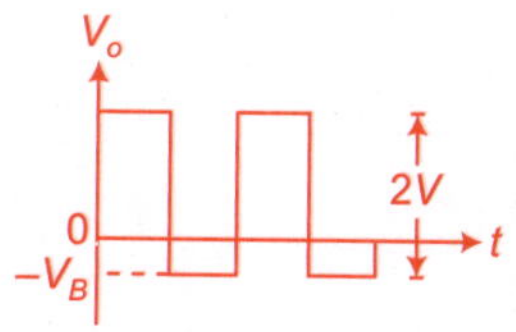

Example 4.13.1 Determine V_0 for the given network of Fig. Ex. 4.13.1 for the given input,

   (a) When diode is ideal   (b) For Si diode

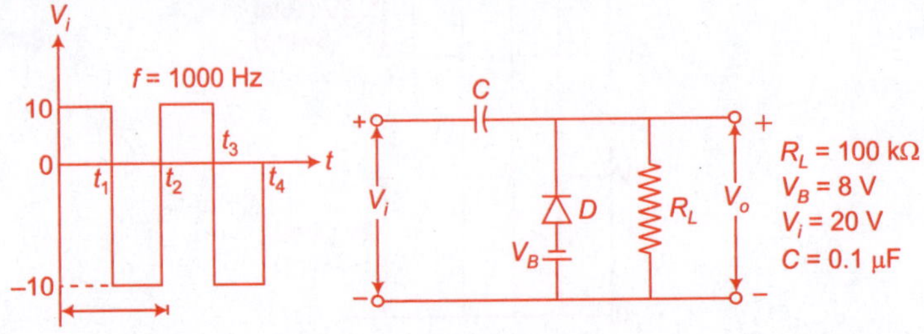

Fig. Ex. 4.13.1

Solution Given that,

$$f = 1 \text{ kHz}$$

So, the time period $T = \dfrac{1}{f}$ sec.

$$T = \frac{1}{1 \text{ kHz}} = 1 \text{ ms}$$

$$\boxed{T = 1 \text{ ms.}}$$

The intervals between the levels = 0.5 ms.

(a) When the diode is ideal

The analysis will start with the period $t_1 - t_2$ of the input signal because we start the analysis of clamping networks by considering that part of the input signal which will forward bias the diode. For this interval, the network will appear as shown in Fig. Ex. 4.13. 2.

Applying Kirchhoff's voltage law around the input loop (*ABEFA*) will give,

$$-V_i + V_C - 8V = 0$$

$$-20 + V_C - 8V = 0$$

$$\boxed{V_C = 28 \text{ volt.}}$$

Fig. Ex. 4.13.2

The capacitor will charge up to 28 V because **"during this period the diode is in 'on' state. The capacitor will charge up instantaneously to a voltage level determined by the network (V_C)".**

During this interval the output voltage $V_0 = 8$V.

i.e.
$$\boxed{V_0 = 8 \text{ volt.}}$$

For the interval $t_2 - t_3$, the network will appear as shown in Fig. Ex. 4.13.3. Applying the Kirchhoff's voltage law around the outside loop of the network will give:

$$V_i + V_C - V_0 = 0$$
$$10V + 28V - V_0 = 0$$
$$\boxed{V_0 = 38 \text{ volt.}}$$

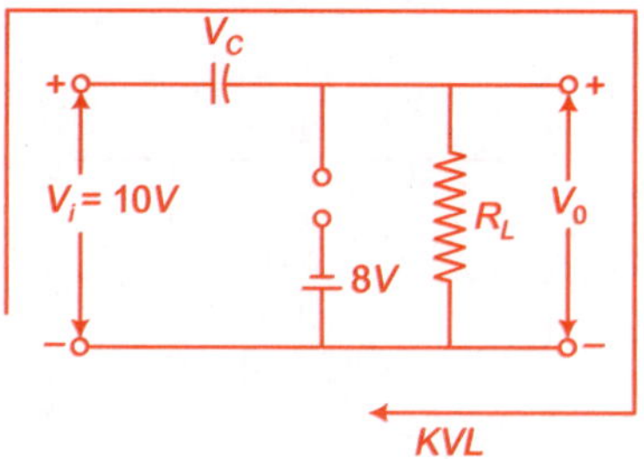

Fig. Ex. 4.13.3

The time constant of the discharging network of Fig. Ex. 4.13.3 is determined by the product RC and has the magnitude,

$$T = RC = (100 \text{ k}\Omega)\ (0.1\ \mu F)$$
$$= 0.01 \text{ sec}$$
$$= 10 \text{ ms.}$$

The total discharge time is therefore,

$$5T = 5(10 \text{ ms})$$
$$= 50 \text{ ms.}$$

Since the interval $t_2 \rightarrow t_3$ will be only 0.5 ms and the discharging time constant is 50 ms i.e., the capacitor will hold its voltage during the discharge period between the pulses of the input signal. The output wave is shown in Fig. Ex. 4.13.4.

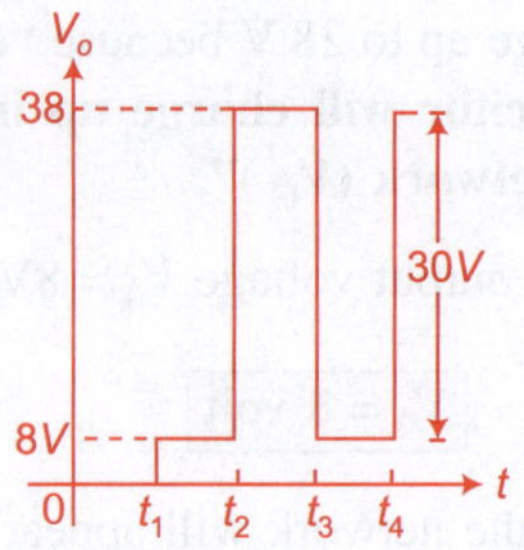

Fig. Ex. 4.13.4 *Output waveform*

(b) For Si diode

As we know that for Si diode the voltage across diode is 0.7V.

For the short circuit state i.e., for interval $t_1 \rightarrow t_2$ the network will appear as shown in Fig. Ex. 4.13.5. Applying Kirchhoff's voltage law in the output side (*BCDEB*).

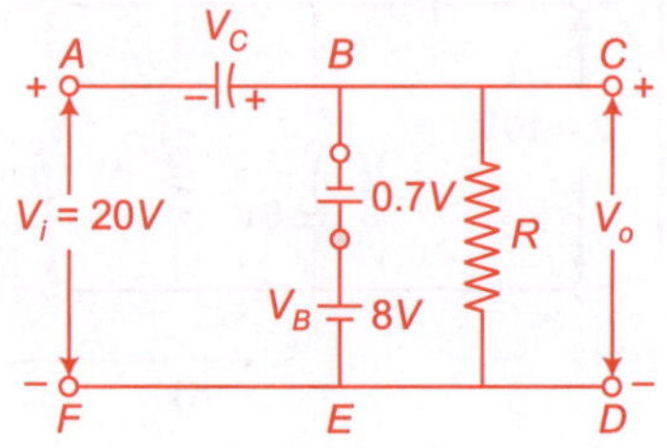

Fig. Ex. 4.13.5

$$8V - 0.7V - V_0 = 0$$

$$\boxed{V_0 = 7.3 \text{ volt.}}$$

For the input side, applying Kirchhoff's voltage law in loop (*ABEFA*), we get,

$$-20V + V_c + 0.7V - 8V = 0$$

$$-20V + V_c + 0.7V - 8V = 0$$

$$V_c = 20V - 0.7V + 8V$$

$$\boxed{V_c = 27.3 \text{ volt.}}$$

For a period $t_2 \rightarrow t_3$ the network will appear as shown in Fig. Ex. 4.13.6. Applying Kirchhoff's voltage law, around the output side of the loop (*ABCDEFA*).

$$10V + 27.3V - V_0 = 0$$

$$\boxed{V_0 = 37.3 \text{ volt.}}$$

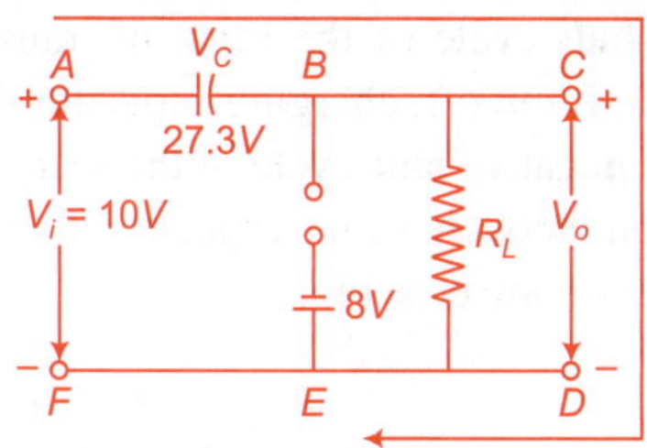

Fig. Ex. 4.13.6

The resultant output appears in Fig. Ex. 4.13.7, verifying the statement that the input and output swings are the same.

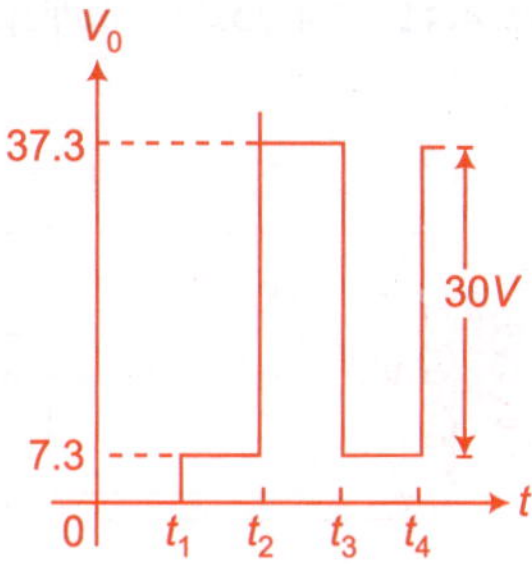

Fig. Ex. 4.13.7

4.14 VOLTAGE MULTIPLIERS

A voltage multiplier is a specialized rectifier circuit producing an output which is theoretically an integer time i.e., two, three, four or more times the AC peak input. Thus, by the voltage multiplier, it is possible to get 200 V DC from a 100 V peak AC source, using a multiplier which is also called as doubter, 400 V DC from a quadruples.

Voltage doubler (Half wave)

The half wave voltage double network is shown in Fig. 4.40.

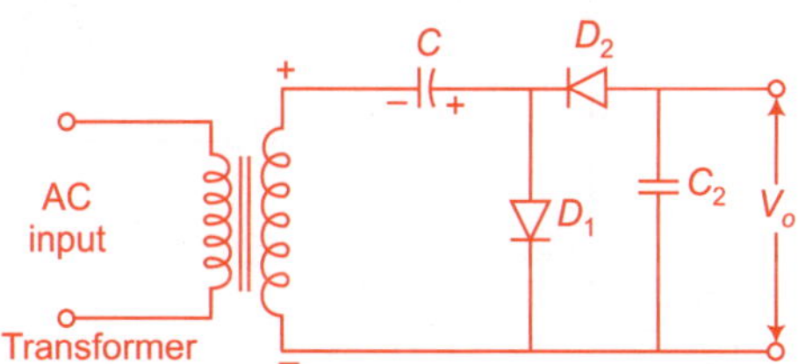

Fig. 4.40 *Half wave voltage doubler*

During the positive half cycle of the input of transformer, secondary diode D_1 conducts and diode D_2 is cut off, charging capacitor C_1 up to the peak rectified voltage (V_m). During the negative half cycle of the secondary voltage, diode D_1 is cut off and diode D_2 conducts charging capacitor C_2. We can sum the voltage around the outside loop as shown in Fig. 4.42

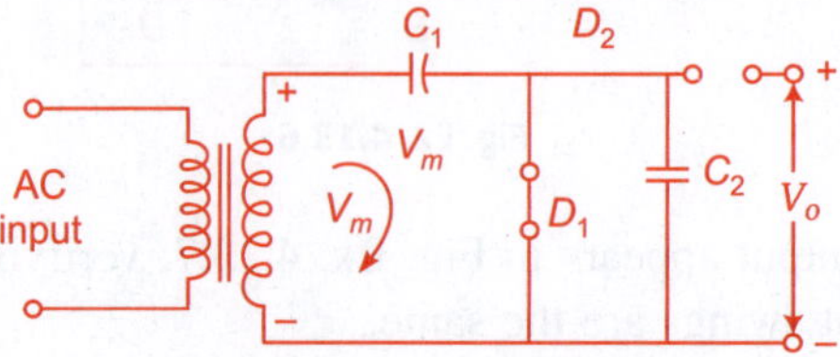

Fig. 4.41 *For positive half cycle*

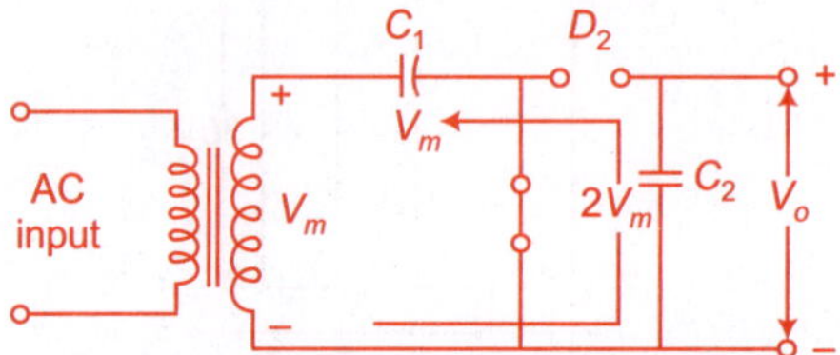

Fig. 4.42 *For negative half cycle*

When next positive half cycle occurs, diode D_2 is non-conducting and capacitor C_2 will discharge through the load. If the load is not connected across capacitor C_2, both capacitor C_1 and C_2 stay charged up to V_m and $2V_m$ respectively.

Full wave voltage doubler

The full wave voltage doubler network is shown in Fig. 4.43 (a). The positive half cycle of transformer secondary voltage diode D_1 conducts charging capacitor C_1 to a peak voltage V_m. Diode D_2 is non-conducting as shown in Fig. 4.43 (b).

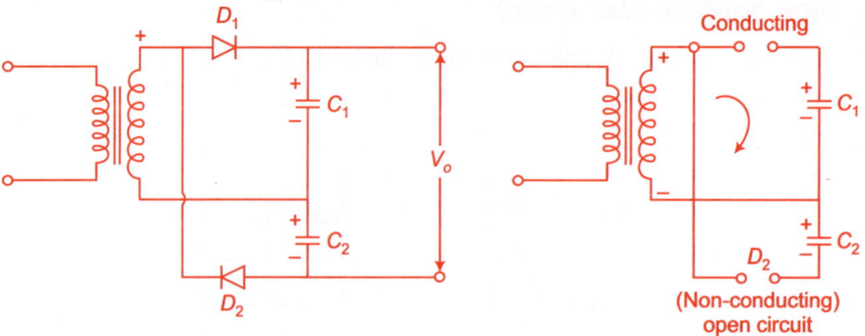

Fig. 4.43 *(a) Full wave voltage doubler*   **Fig. 4.43** *(b) Positive half cycle*

During the negative half cycle, diode D_2 conducts charging capacitor C_2 while diode D_1 is non-conducting, as shown in Fig. 4.43 (c). If no load current is drawn from the circuit, the voltage across capacitors C_1 and C_2 is $2V_m$. If load current is drawn from the circuit, the voltage across capacitor C_1 and C_2 is same as that across a capacitor fed by full wave rectifier circuit. One difference is that the effective capacitance is that of C_1 and C_2 in series, which is less than the capacitance of either C_1 or C_2 alone. Due to this, the filtering action is poorer than the simple capacitor filter.

So, we can state that half wave or full wave voltage doubler circuits provide twice the peak voltage of the transformer secondary, requiring no centre tapped transformer and only $2V_m$ PIV rating for the diode is required.

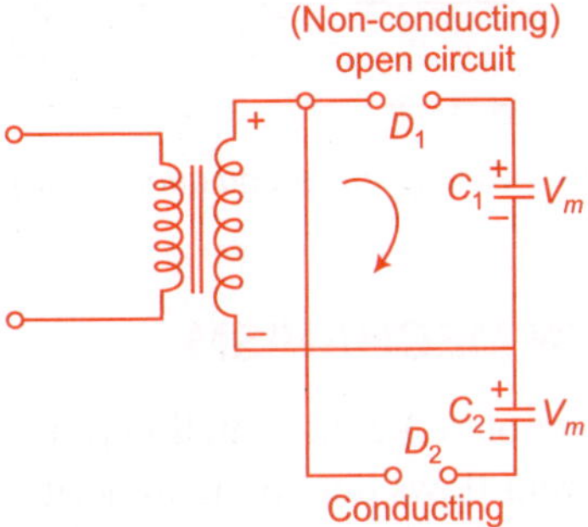

Fig. 4.43 *(c) Negative half cycle*

Voltage trippler and quadrupler

The network for voltage tripler and quadrupler is shown in Fig. 4.44. Basically it is an extension of the half wave voltage doubler which develops three and four times the peak input voltage.

During the positive half cycle, capacitor C_1 charges through diode D_1 to a peak voltage V_m. Capacitor C_2 charges to $2V_m$ developed by the sum of the voltage across capacitor C_1 and the transformer, during the negative half cycle of the transformer secondary voltage. Hence, the voltage across capacitor C_1 and C_3 is $3V_m$. So, it is operated as voltage triplet.

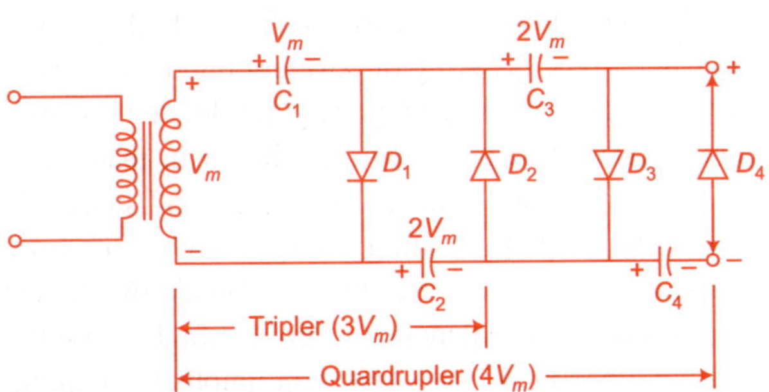

Fig. 4.44 *Voltage tripler and quadrupler*

During the positive half cycle, diode D_3 is conducted and the voltage across capacitor C_2 charges capacitor C_3 to the same peak voltage $2V_m$. And during negative half cycle, diode D_2 and D_4 conducts with capacitor C_3, charging C_4 to $2V_m$. Hence, the voltage across C_2 and C_4 is $4V_m$. So, it is operated as voltage quadrupler. Network shown in Fig. 4.44 is redrawn and shown in Fig. 4.45.

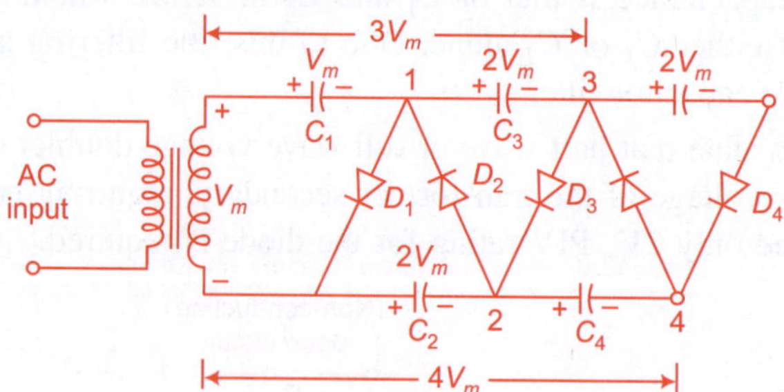

Fig. 4.45 *Voltage tripler and quadrupler five times*

4.15 BREAKDOWN MECHANISM

As we have studied in section 4.3.2, the small current flows in reverse biased of the diode. This small current flows due to the minority carriers. Further when we increase the reverse voltage, the reverse current increases rapidly. The process of increasing the reverse current rapidly by increasing the reverse voltage is called Breakdown. The voltage at which this phenomenon occurs is called breakdown voltage. At Breakdown Voltage, the crystal structure of the diode will break. Generally, we avoid this phenomenon. When we remove the excess reverse voltage, it gets normal crystal structure.

Junction breakdown is of following two types:

  (i) Avalanche breakdown   (ii) Zener breakdown.

4.15.1 Avalanche Breakdown

As we have discussed in section 4.3.2, the current flow in diode is due to the minority carriers in reverse bias. If we increase the reverse voltage, there is an increase in the velocities of minority carriers. These high energy carriers break the covalent bond, due to the breaking of covalent bond, generated at ion of more carriers. Again, these generated carriers are accelerated by the electric field. They break more covalent bonds during their travel in material. That generates more electron-hole pairs. Due to this, continuous collisions of carriers break the covalent bond continuously, creating a large number of carriers. This gives rise to high reverse current. This mechanism of breakdown is called avalanche breakdown. It is also called as avalanche multiplication because the free electron-hole pair which are generated, increases continuously in multiplication. The voltage at which this breakdown occurs is called avalanche breakdown voltage. The avalanche breakdown voltage increases with increasing the temperature because the amplitude of vibrations of atoms of crystal increases. Due to this, the possibility of collisions of electrons with atoms of crystal increase. So, there is a loss of energy. To complete this loss of energy, we need more reverse voltage to start the avalanche mechanism.

4.15.2   Zener Breakdown

This type of breakdown occurs at very less reverse voltage approximately 5 V. The avalanche mechanism cannot occur at this voltage. This type of breakdown occurs in a type of diode whose doping level of p-region and n-region is very high. Due to the high level of doping, the width of depletion region is small (proximal 10^{-8}m). Due to this, the electric field at junction is high because,

$$E \propto \frac{1}{d} \tag{4.54}$$

where, E = Electric field at junction

d = Width of depletion region.

The high electric field, causes covalent bonds to break. Thus, a large number of carriers are generated. This causes large current to flow. This mechanism of breakdown is called Zener breakdown.

4.16   ZENER DIODE

Zener diode is a type of p-n junction diode which is specially designed to operate in breakdown region. As we have studied in section 4.15.2, with increasing the reverse voltage, the reverse current increase rapidly at specific reverse voltage. This voltage is called as Zener voltage and is denoted by V_Z. The diode voltage is constant after zener voltage. The breakdown voltage of diode can be changed from 2V to 100 V by changing the doping level of p and n-region.

The symbol of zener diode is shown in Fig. 4.46. The zener model for 'on' state is shown in Fig. 4.47 (a) for $V = V_Z$.

The zener model for 'OFF' state is shown in Fig. 4.47 (b) for voltage less than V_Z but greater than 0V ($V_Z > V > 0$), with the polarity indicated.

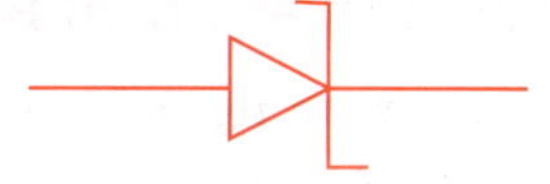

Fig. 4.46   *Symbol of zener diode*

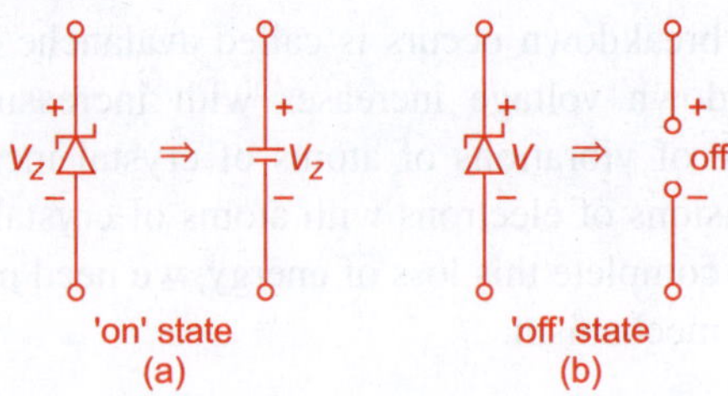

Fig. 4.47 *Zener diode equivalent for 'on' and 'off' state*

VI Characteristics of zener diode

The V-I characteristics of zener diode is shown in Fig. 4.48. From the characteristics we observe that,

 (i) VI characteristics of zener diode like an ordinary diode, only differs that its curve is very sharp.

 (ii) Zener diode can be operated in forward and reverse bias like the normal diode, but it is specially designed to operate in breakdown region. (Reverse biased).

 (iii) The breakdown voltage of zener diode in operating region is approximately constant.

 (iv) The zener breakdown voltage (V_z) and zener test current (I_{ZT}) are the important specifications of zener diode and are provided in the data sheet.

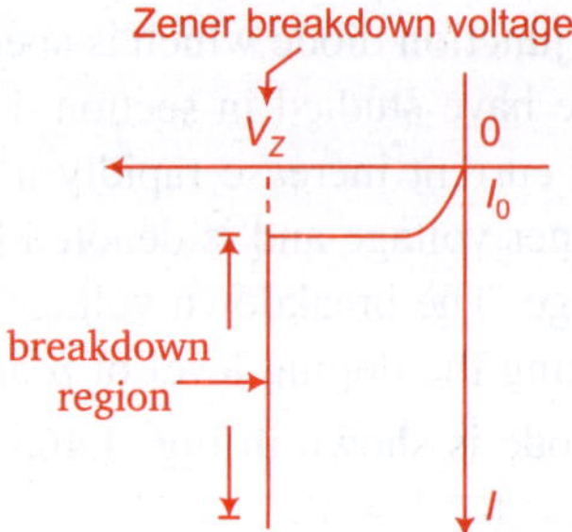

Fig. 4.48 *Characteristics*

4.16.1 Zener Diode Application as Shunt Regulator

A zener diode can be used as a voltage regulator to provide a constant output voltage from a source whose voltage (input voltage) may vary over sufficient range. The output voltage of voltage regulator is also constant when the load current changes.

So, we can say that a zener diode can provide a constant output voltage to the load even though the source voltage may vary. An ordinary diode can damage in breakdown region but zener diode is specially designed to operate in breakdown region. The circuit arrangement for voltage regulator is shown in Fig. 4.49. The zener diode of zener voltage V_Z is connected in reverse bias across the load R_L across which the constant output is desired. In Fig. 4.49, the I_Z and I_L = Current through zener diode and load resistance respectively.

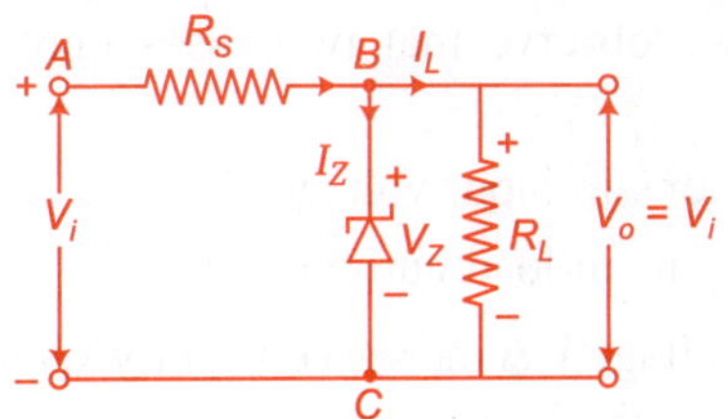

Fig. 4.49   *Zener diode as voltage regulator*

where, R_S = Series resistance

V_i = Input DC voltage

V_0 = Output voltage of regulator

R_L = Load resistance

When the circuit is properly designed, the load voltage V_o remains essentially constant equal to V_Z, even though the input voltage V_i and load resistance R_L may vary over a wide range.

Let us assume that V_i is the input DC voltage whose variations are to be regulated with the help of zener diode. The zener diode is connected across V_i in reverse bias as shown in Fig. 4.49. The zener diode conducts and relatively large current flows through the series resistance R_s, when potential difference across the diode is greater than V_Z. The load resistance R_L is connected in parallel with the diode.

Applying Kirchhoff's current law at node B, we get,

$$I_S = I_Z + I_L. \tag{4.55}$$

Under all conditions,

$$V_0 = V_Z. \tag{4.56}$$

Now, apply Kirchhoff's voltage law in loop *ABCA*.

$$V_i = I_S R_S + V_Z \tag{4.57}$$

$$I_S = \frac{V_i - V_Z}{R_S} \tag{4.58}$$

Now, substituting the value of I_S from Eqn. (4.55) to equal Eqn. (4.57), we get,

$$V_i = (I_Z + I_L) R_S + V_Z$$

$$\boxed{V_{out} = V_Z = V_i - (I_Z + I_L) R_S.} \tag{4.59}$$

From Eqn. (4.33), we observe that two cases may cause the variation in output voltage:

(i) If there are variations in input voltage (V_i)

(ii) If there are variations in load current (I_L).

Case I Suppose input voltage V_i is increased slightly keeping resistance R_L fixed. Since, the zener diode is in the breakdown region, the zener diode is equivalent to battery V_Z. Due to the increase in I_i, the total current I_S will increase from Eqn. (4.58). This increase in total current will be absorbed by zener diode with out affecting load current I_L. The increase in input voltage V_i will be dropped across R_s. Hence, output voltage V_o remains constant irrespective of the increase in the input voltage.

On the other hand, if input voltage V_i slightly decreases, the diode takes smaller current and voltage drop across R_S is reduced. Thus, keeping output voltage V_o constant.

Hence, when input voltage V_i changes, the voltage across the fixed resistance changes to keep $V_o = V_Z$ constant.

Case II Suppose input voltage is constant but the load resistance R_L decreases. This will cause an increase in load current. When load current I_Z is increased, diode current I_z decrease by Eqn. (4.50). Due to this, the total current I_S and $I_S R_S$ drop is constant, so the output voltage V_o remains constant.

If the load resistance R_L increases, this will cause a decrease in load current. When load current I_L is decreased, diode current I_L will increase in order to keep I_S and $I_S R_S$ drop constant. So, the output voltage V_o will remain constant.

4.16.2 Methods to Solve Zener Diode Circuits

The analysis of zener diode circuits is quite similar to that applied to the analysis of an ordinary diode. There are two steps to solve the circuits; first step is to determine the state of zener diode i.e., whether the zener diode is in 'on' state or 'off state. Second step is, substitute the appropriate equivalent circuit of zener diode and solve for the desired unknowns. There are three conditions which can be enlisted in the circuits:

(i) Input voltage V_i and load resistance R_L are fixed.

(ii) Input voltage V_i is fixed and load resistance R_L is variable.

(iii) Load resistance R_L is fixed and input voltage V_i is variable.

4.16.1.1 Input voltage V_i and load resistance R_L are fixed

This type of circuit is the simplest one of zener diode networks as shown in Fig. 4.50 (a). In this circuit, the input voltage V_i and load resistance R_L are fixed.

Step I Determine the state of zener diode by removing it from the network and calculate the voltage across the resulting open circuit as shown in Fig. (4.50. (b)).

Applying voltage dividing rule,

$$V_o = \frac{R_L}{R_S + R_L}\, V_i$$

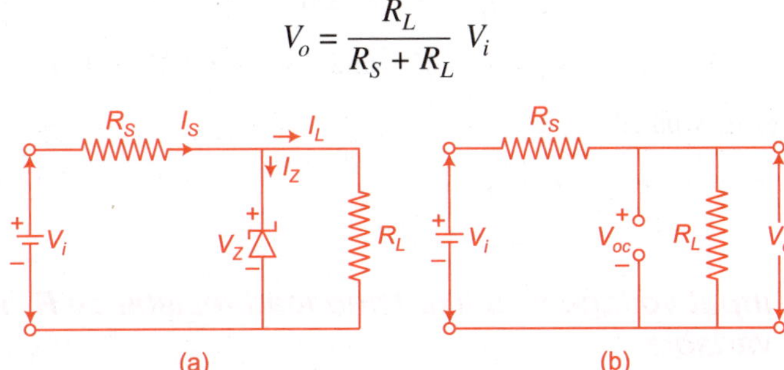

(a)   (b)

Fig. 4.50   (a) *Zener regulator*, (b) *Determining the state of zener diode*

$\therefore$
$$V_0 = V_{oc}$$

i.e.
$$\boxed{V_{oc} = \frac{R_L}{R_S + R_L}\, V_i} \tag{4.60}$$

where, V_{oc} = Voltage across the resulting open circuit.

Step II   If $V_Z \le V_{oc}$, the zener diode is in 'on' state and its equivalent circuit can be substituted as shown in Fig. 4.51 (a).

If $V_z > V_{oc}$, the zener diode is in 'off' state as shown in Fig. 4.51 (b).

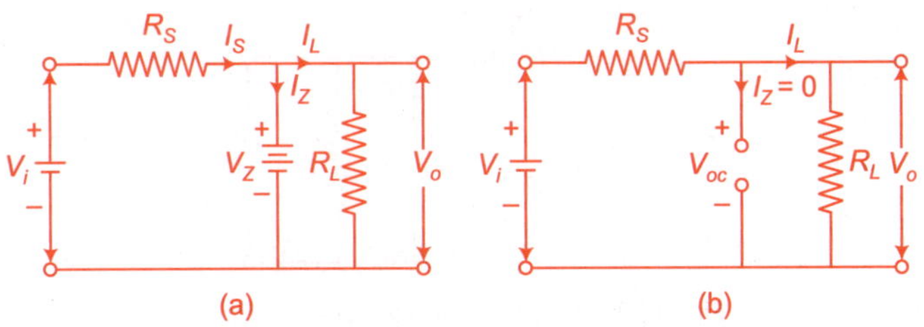

(a)   (b)

Fig. 4.51   (a) *Substitute the zener equivalent for 'on' state;*

(b) *Substitute the zener equivalent for 'off' state*

(i) From Fig. 4.51 (a) 'on' state,

$$V_o = V_Z \tag{4.61}$$

Applying Kirchhoff's current law, that is,

$$I_S = I_Z + I_L$$

and, $$\boxed{I_Z = I_S - I_L}$$ (4.62)

where, $$I_L = \frac{V_o}{R_L} \quad \text{and} \quad I_S = \frac{V_S}{R} = \frac{V_i - V_o}{R}$$

Power dissipated in zener diode, $P_Z = V_Z I_Z$.

(ii) From Fig. 4.51 (b), 'off' state,

$$I_S = I_L \quad \because \quad I_Z = 0$$

$$V_S = V_i - V_o \quad \text{and} \quad V_{oc} = V_o \ (V_{oc} < V_Z)$$

Power dissipated

$$P_z \, V_{oc} \, I_Z = V_{oc} \, (0)$$

$$= 0.$$

4.16.1.2 Input voltage V_i is fixed and load resistance R_L is variable

This type of circuit is shown in Fig. 4.52. In this circuit, the input voltage V_i is fixed and load resistance R_L is variable. Here, the load resistance R_L changes. Hence, the load current also changes.

Step 1 There is a specific range of resistor values which will ensure that the zener diode is in 'on' state. When the load resistance is too small, it will result in a voltage across load (V_L) resistor less than V_Z and the zener diode will be in 'off state.

Step 2 When the load resistance R_L is minimum i.e., $R_{L\,min}$ and load current will be maximum i.e., $I_{L\,max}$ of Fig. 4.50, it will turn 'on' the zener diode.

We know that the Eqn. (4.60),

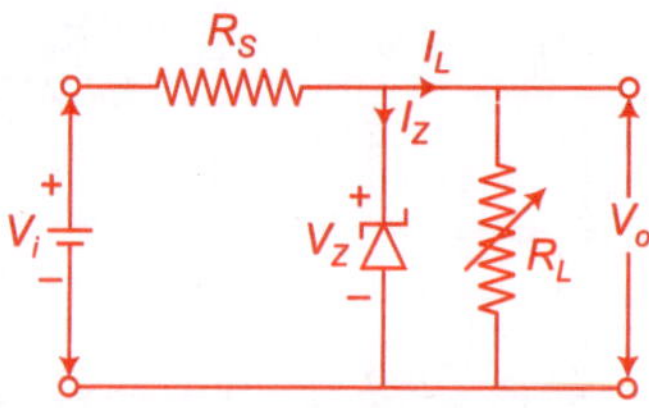

Fig. 4.52 *Fixed Vi and variable RL circuit*

$$V_o = V_Z = \frac{R_L}{R_S + R_L} \, V_i$$

for R_L, we have,

$$V_i R_L = V_Z \, (R_S + R_L)$$

$$V_i R_L = V_Z R_S + V_Z R_L$$

$$(V_i - V_Z)\, R_L = V_Z R_S$$

$$\boxed{R_L = \frac{V_Z R_S}{V_i - V_Z}.}$$
(4.63)

Substituting $R_L = R_{L\,min}$ in Eqn. (4.36), we get,

$$R_{L\,min} = \frac{V_Z R_S}{V_i - V_Z}$$
(4.64)

Any load resistance value greater than R_L obtained from Eqn. (4.64) will ensure that the zener diode is in 'on' state and the diode can be replaced by its V_Z source equivalent.

The conditions defined by Eqn. (4.64) establishes the minimum R_L but it specifies the minimum load current.

$$\boxed{I_{L\,max} = \frac{V_L}{R_L} = \frac{V_Z}{R_{L\,min}}}$$
(4.65)

Once the diode is in 'on' state, the voltage across R_S remains fixed at,

$$V_S = V_i - V_Z$$
(4.66)

and I_S remains fixed at,

$$I_s = \frac{V_S}{R_s}$$
(4.67)

When the load resistance R_L i.e., maximum i.c., $R_{L\,max}$ and load current will be minimum i.e., $I_{L\,min}$

The zener current,

$$I_z = I_s - I_L$$
(4.68)

when the zener is in 'on' state, I_s remains fixed. This means that when I_L is maximum i.e., $I_{L\,max}$, I_Z will be minimum. On the other hand, when I_L is minimum $I_{L\,min}$, I_Z is maximum. If the maximum current that a zener can carry safely is $I_{Z\,max}$ then,

and,

$$\left. \begin{aligned} I_{L\,min} &= I_S - I_{Z\,max} \\[2mm] R_{L\,max} &= \frac{V_o}{I_{L\,min}} = \frac{V_Z}{I_{L\,min}} \end{aligned} \right\}.$$
(4.69)

and,

4.16.1.3   *Load resistance R_L is fixed and input voltage V_i is variable*

This type of circuit is shown in Fig. 4.53. In this circuit, the input voltage V_i varies and load resistance R_L is constant. Note that there is a definite range of V_i

values that will ensure that zener diode is in 'on' state. When the applied input voltage is minimum i.e., $V_{i\,min}$, it will turn the zener diode 'on' state

From Eqn. (4.34), we have,

$$V_o = V_z = \frac{R_L}{R_s + R_L}\,V_i$$

$$V_z\,(R_s + R_L) = V_i R_L$$

$$V_i = \frac{V_Z\,(R_S + R_L)}{R_L}$$

if,

$$V_i = V_{i\,min}$$

$\therefore$ then,

$$V_{i\,min} = \frac{V_Z\,(R_S + R_L)}{R_L} \tag{4.70}$$

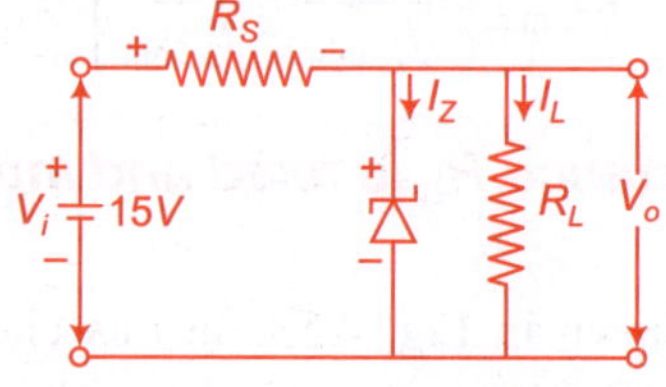

Fig. 4.53 *Fixed R_L and variable V_i circuit*

The maximum value of V_i is limited by the maximum zener current $I_{Z\,max}$.

Since,

$$I_{Z\,max} = I_S - I_L$$

$$I_{S\,max} = I_{S\,max} + I_L \tag{4.71}$$

Since I_L is fixed at $\dfrac{V_Z}{R_L}$ and $I_{Z\,max}$ is the maximum value of I_Z, the maximum value V_i is defined by,

$$V_{i\,max} = V_{Z\,max} + V_Z$$

$$V_{i\,max} = I_{S\,max}\,R_S + V_Z \tag{4.72}$$

Example 4.16.1 Determine V_o, V_Z, I_Z, and P_Z for the given zener diode network of Fig. Ex. 4.16.1 for given data $R_s = 1\ k\Omega$

Fig. Ex. 4.16.1 Zener diode regulator

$$R_L = 1.5 \text{ k}\Omega$$

$$V_Z = 10 \text{ V}$$

$$V_i = 15 \text{ V}$$

$$P_{Z\,max} = 30 \text{ mW.}$$

Step 1 We will determine the state of zener diode by removing it from the network and calculating the voltage V_{oc} as shown in Fig. Ex. 4.16.1(a)

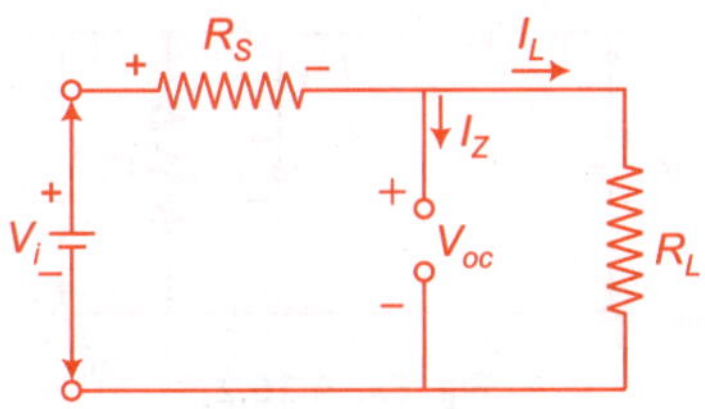

Fig. Ex. 4.16.1 (a)

The open circuit voltage is given from the Eqn. (4.60), we get,

$$V_{oc} = \frac{R_L}{R_L + R_s} \, V_i$$

$$= \frac{1.5 \text{ k}\Omega \,(15 \text{ V})}{1.5 \text{ k}\Omega + 1 \text{ k}\Omega} = \frac{225 \times 10^3}{25 \times 10^3} \text{ V}$$

$$\boxed{V_{oc} = 9 \text{ volt.}}$$

Since $V_{oc} = 9$ V is less than $V_z = 10$V i.e., $V_{oc} < V_Z$, the diode is in 'off' state. Substituting the open circuit equivalent will result in same network as in Fig. Ex. 4.16.2. where we find that,

$$V_o = V_{oc} = 9\text{V}$$

$$\boxed{V_o = 9 \text{ volt.}} \qquad \text{(Ans.)}$$

$$V_S = V_i - V_o$$

$$= (15 - 9) \text{ V} = 06 \text{ V}$$

i.e. $\qquad \boxed{V_S = 6 \text{ volt.}} \qquad \text{(Ans.)}$

$$I_Z = 0A \qquad \text{(Ans.)}$$

and, $\qquad P_Z = V_Z I_Z = V_2 (0A) = 0\text{W}$

$$\boxed{P_Z = 0\text{W.}} \qquad \text{(Ans.)}$$

Example 4.16.2 Repeat the Fig. Ex. 4.16.1 when $R_L = 4.0$ kΩ

Solution Applying Eqn. (4.60) will now result in,

$$V_{oc} = \frac{R_L}{R_S + R_L} \, V_i$$

$$V_{oc} = \frac{4\ k\Omega\ (15V)}{1\ k\Omega + 4\ k\Omega} = \frac{60}{5}\ V$$

$$V_{oc} = 12\ V.$$

Since $V_{oc} = 12$ V is greater than $V_Z = 10$ V, the diode is in 'on' state and the network of Fig. Ex. 4.16.2 will give output

Applying Eqn. (4.60) yields

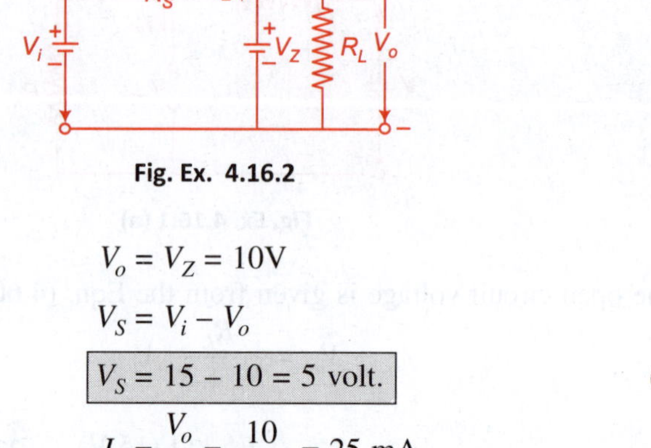

Fig. Ex. 4.16.2

$$V_o = V_Z = 10V$$

and,
$$V_S = V_i - V_o$$

$$\boxed{V_S = 15 - 10 = 5\ \text{volt.}}\qquad \textbf{(Ans.)}$$

with,
$$I_Z = \frac{V_o}{R_L} = \frac{10}{4k\Omega} = 25\ \text{mA}$$

$$\boxed{I_L = 25\ \text{mA.}}\qquad \textbf{(Ans.)}$$

and,
$$I_S = \frac{V}{R} = \frac{5V}{1k\Omega} = 5\text{mA}$$

$$\boxed{I_S = 5\ \text{mA.}}$$

so that,
$$I_Z = I_S = I_L$$

$$I_Z = 5\ \text{mA} - 25\ \text{mA}$$

$$\boxed{I_Z = 25\ \text{mA.}}$$

The power dissipated,

$$P_Z = V_Z I_Z = (10\ V)\ (2.5\ \text{mA})$$

$$\boxed{P_Z = 25\ \text{mW.}}\qquad \textbf{(Ans.)}$$

Example 4.16.3 Determine the range of R_L and I_L that will result in V_0 being maintained at 10V for the network given in Fig. Ex. 4.16.3. Also determine the maximum voltage rating of the diode, for the data given as follows:

$$V_Z = 10V$$

$$R_S = 1k\Omega$$

$$V_i = 60V$$

$$V_o = 10V$$

$$I_{Z\,max} = 30 \text{ mA.}$$

Fig. Ex. 4.16.3   *Voltage regulator for Ex. 4.16.2*

Solution   The value of R_L that will turn on the zener diode is given by Eqn. (4.64).

$$R_{L\,min} = \frac{V_Z R_S}{V_i - V_Z} = \frac{(10V)\,(1\text{ k}\Omega}{60V - 10V}$$

$$R_{L\,min} = \frac{10,000}{50} = 200\ \Omega. \qquad \textbf{(Ans.)}$$

The voltage across R_S is given by Eqn. (4.39), we get,

$$V_S = V_i - V_Z$$

$$V_S = 60 - 10V = 50V$$

$$V_S = 50V$$

and,
$$I_S = \frac{V_S}{R_S} = \frac{50V}{1\text{ k}\Omega} = 50 \text{ mA.}$$

The minimum level of I_L is given by the Eqn. (4.42), we get,

$$I_{L\,min} = I_S - I_{Z\,max}$$

$$I_{L\,min} = 50 \text{ mA} - 30 \text{ mA}$$

$$I_{L\,min} = 20 \text{ mA.} \qquad \textbf{(Ans.)}$$

The maximum value of R_L is given by Eqn. (4.42)

$$R_{L\,max} = \frac{V_o}{I_{L\,min}} = \frac{10V}{20 \text{ mA}}$$

$$R_{L\,max} = \frac{10,000}{20} = 500\ \Omega$$

$$R_{L\,max} = 500\ \Omega. \qquad \textbf{(Ans.)}$$

The maximum power,

$$P_{max} = V_Z I_{Z\,max}$$

$$P_{max} = (10\text{ V})(30\text{mA})$$

$$\boxed{P_{max} = 300\text{mW.}}$$ **(Ans.)**

Example 4.16.4 Determine the range of values of V_i that will maintain the zener diode of network of Fig. Ex. 4.16.4 in 'on' state for the given data.

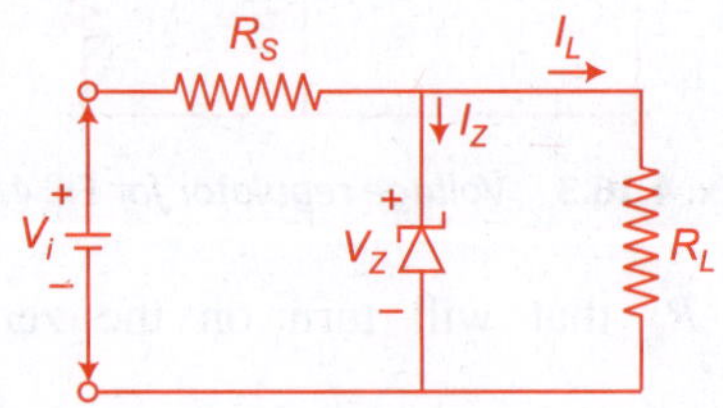

Fig. Ex. 4.16.4 *Regular circuit for Ex. 4.16.3*

$$R_S = 220\ \Omega$$

$$V_Z = 15\text{V}$$

$$I_{Z\,max} = 50\text{ mA}$$

$$R_L = 1.2\text{ k}\Omega.$$

Solution The minimum input voltage is given by Eqn. (4.70), we get,

$$V_{i\,min} = \frac{V_Z(R_L + R_S)}{R_L}$$

$$= \frac{(15\text{V})(1200\ \Omega + 220\ \Omega)}{1200\ \Omega}$$

$$\boxed{V_{i\,min} = 17.75\text{ volt.}}$$ **(Ans.)**

and,

$$I_L = \frac{V_o}{R_L} = \frac{V_Z}{R_L} = \frac{15\text{V}}{1.2\text{k}\Omega}$$

$$I_L = 12.5\text{ mA.}$$

The maximum current through resistor R_s is given by Eqn. (4.71).

$$I_{S\,max} = I_{Z\,max} + I_L$$

$$I_{S\,max} = 50\text{ mA} + 12.5\text{ mA}$$

$$\boxed{I_{S\,max} = 62.5\text{mA.}}$$ **(Ans.)**

The maximum value of input voltage is given by Eqn. (4.72) we get,

$$V_{i\,max} = I_{S\,max} R_S + V_Z$$

$$V_{i\,max} = (62.5\ \text{mA})\ (0.22\text{k}\Omega) + 15\text{V}$$

$$V_{i\,max} = 13.75 + 15Vt$$

$$V_{i\,max} = 28.75\ \text{V}$$

$$\boxed{V_{i\,max} = 28.75\ \text{volt.}}$$ (Ans.)

Examples

Numerical problems based on rectifier circuits.

Example 4.1 Forward resistance (r_f) of a Si diode is 25Ω. It is used for half wave rectifier. The input voltage of the transformer is 220 V. The turns ratio of the transformer is 12 : 1. The load resistance is 1 kΩ. Determine the following:

- (a) DC load current
- (b) Average value of DC load current
- (c) DC output power
- (d) Input AC power
- (e) % efficiency
- (f) Ripple factor
- (g) PIV of diode

Solution   The RMS value of input AC = 220 V

The RMS value of the secondary of the transformer,

$$V_2 = \frac{N_2}{N_1}\,V_1 \qquad\qquad \because\ \ \frac{N_2}{N_1} = \frac{V_2}{V_1}$$

$$V_2 = 220 \times \frac{1}{12}$$

$$\boxed{V_2 = 18.33\ \text{mt.}}$$

Maximum voltage at the secondary of the transformer,

$$V_m = V_{rms}\ \sqrt{2}$$

$$V_m = 18.33 \times \sqrt{2}$$

$$\boxed{V_m = 25.85\ \text{volt.}}$$

Peak value of output DC current,

$$\text{(a)}\ \ I_m = \frac{V_m}{R_L + r_f} = \frac{25.85}{1000 + 25}$$

$$I_m = 0.03\text{A}$$

$$\boxed{I_m = 30\ \text{mA.}}$$ (Ans.)

(b) Average value of output DC current, we know from Eqn. (4.21)

$$I_{dc} = \frac{I_m}{\pi} = \frac{30 \times 10^{-3}}{314}$$

$$I_{dc} = 0.0096 \text{ Amp}$$

$$\boxed{I_{dc} = 9.6 \text{ mA.}}$$

(Ans.)

(c) Output DC power, we know from Eqn. (4.15)

$$P_{dc} = I_d^2 R_L$$

$$P_{dc} = (9.6 \times 10^{-3})^2 \times 1 \text{ k}\Omega$$

$$P_{dc} = 0.09 \text{ watt.}$$

(d) RMS value of load current, we know from Eqn. (4.25)

$$I_{rms} = \frac{I_m}{2} = \frac{30}{2} = 15 \text{ mA}$$

$$I_{rms} = 15 \text{ mA}$$

Now, input AC power, we know from Eqn. (4.23)

$$P_{ac} = I_{rms}^2 \ (R_L + r_f)$$

$$P_{ac} = (15 \times 10^{-3})^2 \times (1 \text{ k}\Omega + 25 \ \Omega)$$

$$P_{ac} = (15 \times 10^{-3})^2 \times 1025 \ \Omega$$

$$\boxed{P_{ac} = 0.226 \text{ watt.}}$$

(Ans.)

(e) Percentage efficiency, we know from Eqn. (4.14)

$$\%\eta = \frac{P_{dc}}{P_{ac}} \times 100 = \frac{0.09W}{0.226W} \times 100$$

$$\boxed{\eta = 39.8\%}$$

(Ans.)

(f) Ripple factor, we know from Eqn. (4.31)

$$r = \sqrt{\left(\frac{I_{rms}}{I_{dc}}\right)^2 - 1}$$

$$\sqrt{\left(\frac{\frac{I_m}{2}}{\frac{I_m}{\pi}}\right)^2 - 1}$$

$$\sqrt{\left(\frac{\pi}{2}\right)^2 - 1} = \sqrt{(1.57)^2 - 1}$$

$$r = \sqrt{2.46 - 1}$$

$$r = \sqrt{1.46} = 1.21.$$

(g) PIV of diode, we know from Eqn. (4.13)

$$PIV = V_m$$

$$\boxed{PIV = 25.85 \text{ volt.}}$$ (Ans.)

Example 4.2 For a full wave rectifier, the date is given as $R_L = 1 \text{ k}\Omega$, $r_f = 95 \ \Omega$ and the voltage applied at each diode $250 \sin 150 \ \pi t$. Determine

(a) Maximum current   (b) DC current

(c) RMS value of current   (d) Ripple factor

(e) Output DC power   (f) Input AC factor

(g) Efficiency of full wave rectifier   (h) PIV.

Solution The applied voltage is,

$$V_i = 250 \sin 150 \ \pi t$$

comparing the above equation with $V_m \sin 2\pi \ ft$, we get,

(a) $V_m = 250$ V.

The maximum value of the current,

$$I_m = \frac{V_m}{R_L + r_f}$$

$$I_m = \frac{250}{1000 \ \Omega + 95 \ \Omega}$$

$$I_m = \frac{250 \text{ V}}{1095 \ \Omega} = 0.228 \text{ Amp}$$

$$\boxed{I_m = 228 \text{ mA.}}$$ (Ans.)

(b) DC current, we know from Eqn. (4.38)

$$I_{dc} = \frac{2I_m}{\pi}$$

$$I_{dc} = \frac{2 \times 22 \times 10^{-3}}{3.14} = \frac{0.456}{3.14}$$

$$I_{dc} = 0.1452 \text{ A}$$

$$\boxed{I_{dc} = 145.2 \text{ mA.}}$$ (Ans.)

(c) RMS value of current, we know from Eqn. (4.41)

$$I_{\text{rms}} = \frac{I_m}{\sqrt{2}}$$

$$I_{rms} = \frac{0.228 \text{ A}}{\sqrt{2}} = 0.1617 \text{ Amp}$$

$$\boxed{I_{rms} = 162 \text{ mA.}}$$ **(Ans.)**

(d) Output DC power, we know from Eqn. (4.15)

$$P_{dc} = I_{dc}^2 R_L$$

$$P_{dc} = (145.2 \times 10^{-3})^2 \times 1000$$

$$P_{dc} = 21.08 \text{ watt.}$$ **(Ans.)**

(e) Input AC power, we know from Eqn. (4.42)

$$P_{ac} = \left(\frac{I_m}{\sqrt{2}}\right)^2 (R_L + r_f) = I_{rms}^2 (R_L + r_f)$$

$$P_{ac} = (162 \times 10^{-3})^2 (10000 + 95 \ \Omega)$$

$$\boxed{P_{ac} = 28.73 \text{ watt.}}$$ **(Ans.)**

(f) Efficiency of rectifier, we know from Eqn. (4.14)

$$\% \ \eta = \frac{P_{dc}}{P_{ac}} \times 100$$

$$\% \ \eta = \frac{21.08}{28.73} \times 100$$

$$\boxed{\% \ \eta = 73.7\%.}$$ **(Ans.)**

(g) Ripple factor, we know from Eqn. (4.31)

$$r = \sqrt{\left(\frac{I_{rms}}{I_{dc}}\right)^2 - 1} = \sqrt{\left(\frac{162}{145.2}\right)^2 - 1}$$

$$r = \sqrt{1.21 - 1} = \sqrt{0.21}$$

$$\boxed{r = 0.46.}$$ **(Ans.)**

(h) Peak inverse voltage of diode, we know from Eqn. (4.33)

$$PIV = 2V_m$$

$$PIV = 2 \times 250$$

$$\boxed{PIV = 500 \text{ volt.}}$$ **(Ans.)**

Example 4.3 The applied input AC power to half wave rectifier is 120 watts. The DC output power obtained is 50 watts. Determine the rectification efficiency and why the approximate 50% power is lost?

Solution We know that the rectification efficiency from Eqn. (4.14)

$$\% \ \eta = \frac{P_{dc}}{P_{ac}} \times 100$$

Given that $P_{dc} = 50$ watts

$$P_{ac} = 120 \text{ watts}$$

Now,

$$\% \eta = \frac{50}{120} \times 100$$

$$\boxed{\eta = 41.67\%.}$$ **(Ans.)**

Approximate 60% power is lost in the rectifier circuit because it is half wave rectifier and 120 watt AC power is contained as 60 watt in positive half cycle and other 60 watts in negative half cycle. The 60 watts in negative half cycle does not supply to the load. i.e., only 60 watts power is converted into 50 watts.

Example 4.4 Determine the output DC voltage and the peak inverse voltage for a given circuit of half wave rectifier in Fig. Ex. 4.4.1. Assume the diode is ideal.

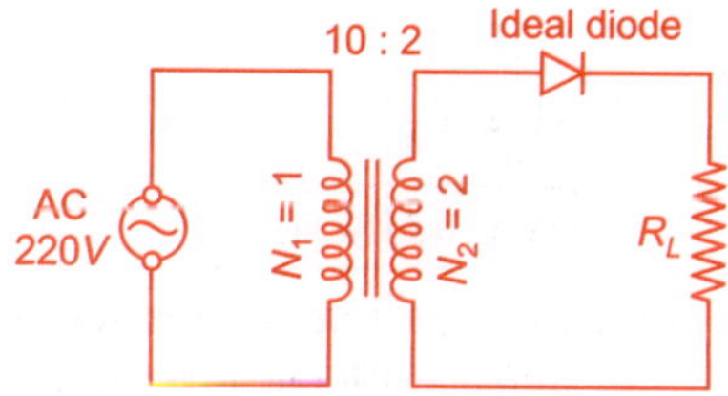

Fig. Ex. 4.4.1 *Half wave rectifier circuit*

Solution Given that RMS primary voltage,

$$V_{rms_1} = 220 \text{ volt.}$$

Number of turns for primary $N_1 = 10$ and number of turns for secondary $N_2 = 2$. The maximum primary voltage is given by equation,

$$V_{m_1} = \sqrt{2} \times V_{rms_1}$$

$$V_{m_1} = \sqrt{2} \times 200 \text{ V}$$

$$V_{m_1} = 310.2 \text{ volt.}$$

We know the equation,

$$\frac{V_{m_2}}{V_{m_1}} = \frac{N_2}{N_1}$$

$$V_{m_2} = V_{m_1} \times \frac{N_2}{N_1} = 310.2V \times \frac{2}{10}$$

$$V_{m_2} = 62.04 \text{ volt.}$$

The output DC voltage,

$$V_{dc} = I_{dc}R_L \qquad \because I_{dc} = \frac{I_m}{\pi}$$

Now,

$$V_{dc} = \frac{I_m}{\pi} \cdot R_L$$

$$V_{dc} = \frac{V_{m_2}}{\pi} \qquad \because I_m R_L = V_{m_2}$$

$$V_{dc} = \frac{62.04}{3.14} = 19.76 \text{ volt}$$

$$V_{dc} = 19.76 \text{ volt.} \qquad \textbf{(Ans.)}$$

During the negative half cycle of AC supply, the diode conducts no current due to the reverse bias. Therefore, the maximum secondary voltage appears across the diode,

$$PIV = V_{m_2}$$

$$PIV = 62.04 \text{ volt.} \qquad \textbf{(Ans.)}$$

Example 4.6   A crystal diode having internal resistance $r_f = 15\ \Omega$ is used for half wave rectification. If the applied voltage $V = 40 \sin \omega t$ and load resistance $R_L = 500\ \Omega$. Determine:

   (a)   Maximum current I_m, DC current I_{dc} and RMS value of current I_{rms}

   (b)   AC input power and DC output power

   (c)   DC output voltage

   (d)   Rectification efficiency.

Solution   Given that,

$$r_f = 15\ \Omega, \quad R_L = 500\ \Omega$$
$$V_m = 40 \text{ V.}$$

   (a)   The maximum current,

$$I_m = \frac{V_m}{R_L + r_f}$$

$$I_m = \frac{40}{500 + 15} = 0.0777 \text{ A}$$

$$I_m = 77.7 \text{ mA.} \qquad \textbf{(Ans.)}$$

Now,
$$I_{dc} = \frac{I_m}{\pi} = \frac{77.7}{3.14}$$

$$\boxed{I_{dc} = 24.75 \text{ mA.}}$$ **(Ans.)**

$$I_{rms} = \frac{Im}{2} = \frac{7.77}{2}$$

$$\boxed{I_{rms} = 38.9 \text{ mA.}}$$ **(Ans.)**

(b) AC input power,

$$P_{ac} = I_{rms}^2 (R_L + r_f)$$

$$P_{ac} = (38.9 \times 10^{-3})^2 (500 + 15)\ \Omega$$

$$\boxed{P_{ac} = 0.778 \text{ watt.}}$$ **(Ans.)**

DC output power,

$$P_{dc} = I_{dc}^2\ R_L$$

$$P_{dc} = (24.75 \times 10^{-3}) \times 500\ \Omega$$

$$\boxed{P_{dc} = 0.305 \text{ volt.}}$$ **(Ans.)**

(c) DC output voltage,

$$V_{dc} = I_{dc}R_L$$

$$V_{dc} = (24.75 \times 10^{-3})\ A \times 500\ \Omega$$

$$\boxed{V_{dc} = 12 \text{ volt.}}$$ **(Ans.)**

(d) Rectification efficiency,

$$\eta = \frac{P_{dc}}{P_{ac}} \times 100$$

$$\eta = \frac{0.305}{0.778} \times 100$$

$$\boxed{\eta = 39.2\%.}$$ **(Ans.)**

Example 4.7 In a bridge rectifier circuit shown in Fig. Ex. 4.7.1, the diodes are assumed to be ideal.

Determine (a) DC output voltage (b) *PIV* (c) Output frequency. Assume the turns ratio is 10 : 2

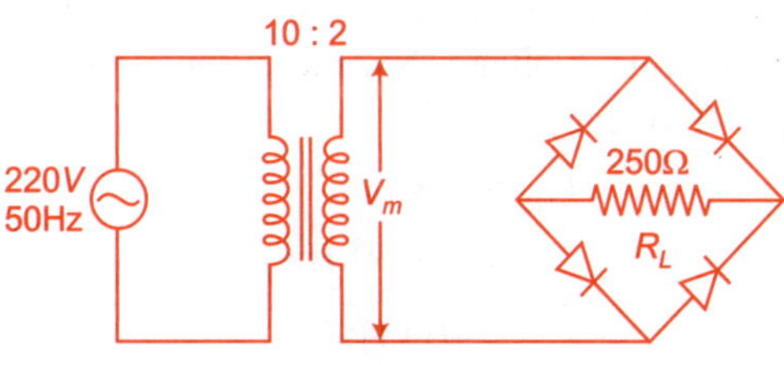

Fig. Ex. 4.7.1

Solution RMS value of primary voltage,

$$V_{1(rms)} = 220 \text{ V}$$

RMS value of secondary voltage,

$$V_{2(rms)} = \frac{N_2}{N_1} \times V_1$$

$$= \frac{2}{10} \times 220$$

$$V_{2(rms)} = 44 \text{ volt.}$$

Maximum voltage across secondary is,

$$V_m = V_{rms} \sqrt{2}$$

$$V_m = 44\sqrt{2} = 62.04 \text{ volt}$$

$$V_m = 62.04 \text{ volt.}$$

DC output current,

$$I_{dc} = \frac{2I_m}{\pi} = \frac{2V_m}{\pi R_L} \qquad \because \quad I_m = \frac{V_m}{R_L}$$

$$I_{dc} = \frac{2 \times 62.04}{3.14 \times 250} = \frac{124.08}{785}$$

$$I_{dc} = 0.16 \text{ A.}$$

(a) The DC output voltage,

$$V_{dc} = I_{dc} R_L$$

$$V_{dc} = 0.16 \times 250$$

$$\boxed{V_{dc} = 40 \text{ volt.}}$$ (**Ans.**)

(b) The peak inverse voltage is equal to the maximum secondary voltage,

$$PIV = V_m$$

$$\boxed{PIV = 62.04 \text{ volt.}}$$ (**Ans.**)

(c) In a full wave rectification, there are two output pulses for each complete cycle of the input AC voltage. So, the output frequency is twice that of the AC supply frequency i.e.

$$f_{out} = 2 \times f_{in}$$

$$= 2 \times 50 \text{ Hz}$$

$$\boxed{f_{out} = 100 \text{ Hz.}}$$ (**Ans.**)

Example 4.8　Figures. Ex. 4.8.1 and 4.8.1 (a) shows the centre-tap and bridge type circuit having the same load resistance and transformer turns ratio. The primary of each circuit is connected to 220V, 50Hz supply. Determine as the DC output voltage PIV for each case for the same output. Assume the diodes to be ideal.

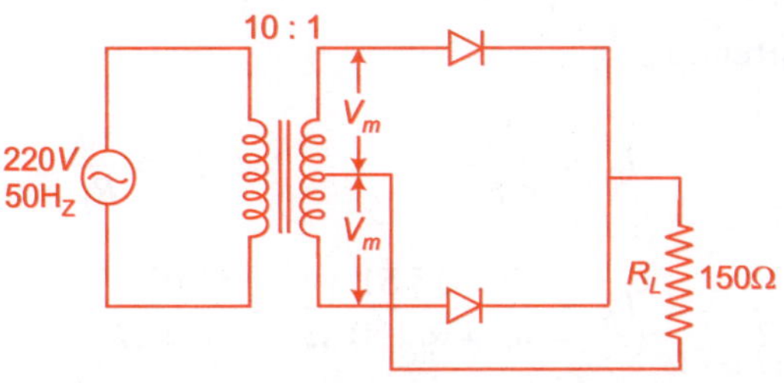

Fig. Ex. 4.8.1

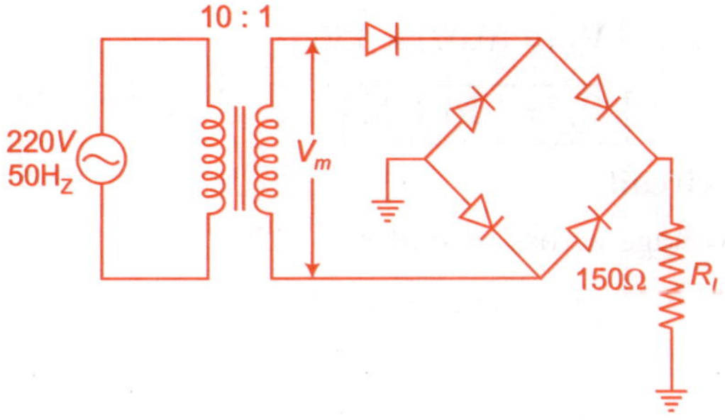

Fig. Ex. 4.8.1 (a)

Solution　(a) For centre-tap circuit RMS value of primary voltage,

$$V_{rms_1} = 220\text{V}$$

RMS value of secondary voltage is given by the equation,

$$V_{rms_2} = \frac{N_2}{N_1} \times V_{rms_1}$$

$$V_{rms_2} = \frac{1}{10} \times 220$$

$$V_{rms_2} = 22 \text{ volt.}$$

Maximum voltage across secondary,

$$V_{m_2} = V_{rms_2} \times \sqrt{2}$$
$$V_{m_2} = 22 \times \sqrt{2}$$
$$V_{m_2} = 31.02 \text{ volt.}$$

Maximum voltage appearing across half secondary winding is,

$$V_m = \frac{V_{m_2}}{2} = \frac{31.02}{2}$$

$$V_m = 15.51 \text{ volt}$$

Average current, $I_{dc} = \dfrac{2I_m}{\pi}$

$$I_{dc} = \frac{2V_m}{\pi\, R_L} \qquad\qquad \because\quad I_m = \frac{V_m}{R_L}$$

$$I_{dc} = \frac{2 \times 15.51 \text{ V}}{3.14 \times 150\ \Omega} = \frac{31.02 \text{ V}}{471\ \Omega}$$

$$I_{dc} = 0.07 \text{ Amp}$$

Now, DC output Voltage, $V_{dc} = I_{dc}R_L$

$$V_{dc} = (0.07) \times 150$$

$$\boxed{V_{dc} = 10.5 \text{ volt.}}$$ 　　　　　　　　　**(Ans.)**

For bridge circuit

Maximum voltage across secondary,

$$V_m = 31.02 \text{ volt}$$

DC output Voltage, $V_{dc} = I_{dc}\, R_L$ 　　　　$\because\ I_{dc} = \dfrac{2I_m}{\pi}$

$$\therefore\quad V_{dc} = \frac{2I_m}{\pi} \times R_L \qquad \because\quad I_m = \frac{V_m}{R_L}$$

$$\therefore\quad V_{dc} = \frac{2V_m}{\pi\, R_L} \times R_L = \frac{2Vm}{\pi}$$

$$V_{dc} = \frac{2 \times 31.02}{3.14} = \frac{62.04}{3.14}$$

$$\boxed{V_{dc} = 19.75 \text{ volt.}}$$ 　　　　　　　　　**(Ans.)**

The above calculation, shows that for the same secondary voltage, the DC output voltage of bridge rectifier circuit is twice that of centre tape circuit.

(b) The DC output voltage of the two circuits will be same, if V_m is same. For this to happen, the turns ratio of the transformer should be as follows.

For centre tap,

$$V_2 = \frac{N_2}{N_1} V_1 = \frac{1}{10} \times 220$$

$$V_2 = 22 \text{ volt.}$$

$$V_2 = 22 \text{ volt (RMS value)}$$

Maximum voltage across secondary,

$$V_m = V_2 \sqrt{2} = 22\sqrt{2}$$

$$V_m = 31.02 \text{ volt.}$$

Maximum voltage appearing across half secondary winding,

$$V_m = \frac{31.02}{2} = 15.51 \text{ volt.}$$

i.e. $\quad N_1 : N_2 = 10:1.$

For bridge circuit,

$$V_2 = \frac{N_2}{2N_1} V_1 = \frac{1}{20} \times 220$$

$$V_2 = 11 \text{ volt (RMS value)}$$

$$V_m = V_{rms} \sqrt{2} = 11\sqrt{2} = 15.51 \text{ volt}$$

i.e. $\quad N_1 : N_2 = 20:1.$

For Centre tap circuit,

RMS value of secondary voltage,

$$V_2 = \frac{N_2}{N_1} V_1$$

$$V_2 = \frac{1}{10} \times 220 = 22 \text{ volt.}$$

Maximum voltage across secondary,

$$V_m = V_{rms} \sqrt{2}$$

$$V_m = 22 \sqrt{2} = 31.02 \text{ volt.}$$

Maximum voltage across half secondary winding is,

$$V_m = \frac{31.02}{2} = 15.51 \text{ volt.}$$

Now,

$$PIV = 2V_m = 2 \times 15.51$$

$$\boxed{PIV = 31.02 \text{ volt.}}$$

(Ans.)

For Bridge circuit

RMS value of secondary,

$$V_Z = \frac{N_2}{N_1} \times V_1$$

$$V_Z = \frac{1}{20} \times 220$$

$$V_Z = 11 \text{ volt.}$$

Maximum voltage across secondary,

$$V_m = V_{rms}\, \sqrt{2} = 11\, \sqrt{2}$$

$$V_m = 15.51$$

$$PIV = V_m = 15.51 \text{ volt.}$$

$$\boxed{PIV = 15.51 \text{ volt.}}$$ **(Ans.)**

From the above calculation, we observe that for the same DC output voltage, PIV of bridge circuit is half of centre tap circuit. This is a distinct advantage of bridge circuit.

Example 4.9 A full wave rectifier with centre-tapped transformation supplies DC current of 100mA to load resistance of $R_L - 20\,\Omega$. The secondary resistance of transformer is $1\,\Omega$. Each diode has a forward resistance of $0.5\,\Omega$. Determine the following: **(UPTU 2007-08)**

(a) RMS value of the signal voltage across each half of the secondary.

(b) DC power supplied to the load

(c) PIV rating for each diode

(d) AC power input to the rectifier

(e) Conversion efficiency.

Solution   Given that,

$$I_{dc} = 100 \text{ mA}, \quad R_L = 20\,\Omega, \quad R_S = 1\,\Omega$$

$$R_f = 0.5\,\Omega$$

(a) The RMS value of the signal voltage across half of secondary

$$I_{rms} = 1.11\, I_{dc}$$

$$I_{rms} = 1.11 \times 100 \text{ mA} = 111 \text{ mA}$$

$$\boxed{I_{rms} = 111\text{mA.}}$$ **(Ans.)**

We know that,

$$V_{rms} = I_{rms} R_T$$

$$\because \qquad R_T = R_L + R_f + \frac{R_S}{2}$$

$$R_T = 20 + 1.5 + 0.5 = 21\,\Omega$$

Now, $\qquad V_{rms} = I_{rms} \times R_T$

$$V_{rms} = 111 \times 10^{-3} \times 21 = 231 \text{ mA.}$$

$$\boxed{V_{rms} = 2331 \text{ mA.}} \qquad \textbf{(Ans.)}$$

(b) DC power supplied to the load

$$P_{dc} = I_{dc}^2 \, R_L$$

$$P_{dc} = (100 \times 10^{-3})^2 \times 20 \ \Omega$$

$$\boxed{P_{dc} = 0.2 \text{ watt.}} \qquad \textbf{(Ans.)}$$

(c) PIV rating for each diode

$$PIV = 2V_m$$

$$PIV = 2\sqrt{2}\ V_{rms}$$

$$PIV = 2\sqrt{2} \times 2331 \text{ mV}$$

$$PIV = 2 \times 1.414 \times 2331 \text{ mV}$$

$$PIV = 6592.06 \text{ mV}$$

$$\boxed{PIV = 6.59 \text{ volt.}} \qquad \textbf{(Ans.)}$$

(d) AC power input to rectifier

$$P_{ac} = I_{rms}^2 \, (R_f + R_L)$$

$$P_{ac} = (111 \times 10^{-3})^2 \, (20 + 0.5)$$

$$P_{ac} = 0.01232 \times 20.5$$

$$\boxed{P_{ac} = 0.252 \text{ watt.}} \qquad \textbf{(Ans.)}$$

(e) Rectification efficiency

$$\eta = \frac{P_{dc}}{P_{ac}} \times 100 = \frac{0.2}{0.252} \times 100$$

$$\boxed{\eta = 79.37\%.} \qquad \textbf{(Ans.)}$$

Problems based on diode and V-I relationship.

Example 4.10. A forward bias Si diode whose reverse saturation current is 15µA. And diode current 1.2A is flowing through the diode at 27°C. Calculate forward diode voltage.

Solution   Given that $I_0 = 15 \Omega A$

$$I_D = 1.2 \text{ A}$$

and,

$$T = 273 + 27°C = 300 \text{ K}$$

We know that, from Eqn. (4.14), we get,

$$V_T = \frac{KT}{q} = \frac{1.38 \times 10^{-23} \times 300}{1.6 \times 10^{-19}}$$

$$\boxed{V_T = 0.026 \text{ volt.}} \hspace{3cm} \textbf{(Ans.)}$$

Now, by the Eqn. (4.1), we get,

$$I_D = I_0 \left(e^{V_D/nV_T} - 1 \right)$$

$$e^{V_D/nV_T} - 1 = \frac{I_D}{I_0}$$

$$e^{V_D/nV_T} = \frac{I_D}{I_0} + 1$$

$$e^{V_D/nV_T} = \frac{1.2}{15 \times 10^{-6}} + 1$$

$$e^{V_D/nV_T} = 100000 + 1 = 100001$$

Taking the natural log on both sides,

$$\frac{V_D}{nV_T} = \log_e 100001$$

$$V_D = 2V_T \log_e 100001$$

$$n = 2 \text{ for Si diode}$$

$$V_D = 2 \times 0.026 \times 11.5$$

$$\boxed{V_D = 0.59.} \hspace{3cm} \textbf{(Ans.)}$$

Example 4.11   The leakage current of *p-n* junction (Ge) diode at normal temperature is 0.18 µA. If the forward voltage applied on this diode is 0.15 volt, then calculate the value of current flowing in diode.

Solution   Given that $I_0 = 0.18$ µA

$$V_D = 0.12 \text{ volt}, \quad \text{for Ge; } n = 1$$

and $V_T = 0.26$ volt at normal temperature,

$$I_D = ?$$

By the Eqn. (4.1), we get,

$$I_D = I_0 \left(e^{V_D/nV_T} - 1 \right)$$

$$= 0.18 \times 10^{-6} \times \left(e^{\frac{0.12}{1 \times 0.026}} - 1\right)$$

$$= 0.18 \times 10^{-6} \times (101.02 - 1)$$

$$\boxed{I_D = 18 \ \mu A.}$$ (Ans.)

Example 4.12 Determine V_0, I_1, I_{D_1} and I_{D_2} for parallel diode configuration in Fig. Ex. 4.12.1. **(UPTU 2008-09)**

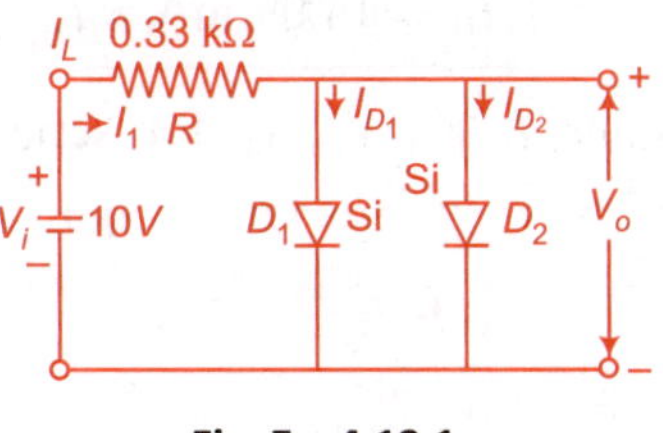

Fig. Ex. 4.12.1

Solution   Redraw the circuit of Fig. Ex. 4.12.1.

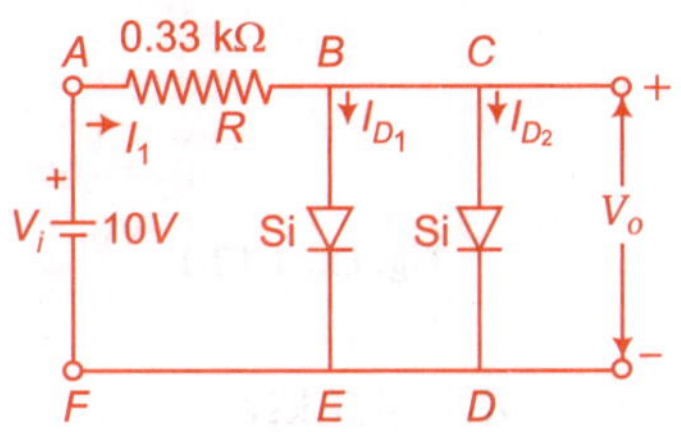

Fig. Ex. 4.12.1 (a)

Applying Kirchhoff's voltage law in loop $ABEFA$,

$$V_i - I_1 R_1 - V_{BE} = 0.$$

Due to 10V supply, both diodes are conducting and the voltage drop across each diode is 0.7 volt (for Si diode), i.e., $V_{BE} = 0.7$ volt.

Now,

$$I_1 R_1 = V_i - V_{BE}$$

$$I_1 = \frac{V_i - V_{BE}}{R_1} = \frac{10 - 0.7}{0.33 \times 10^3}$$

$$I_1 = \frac{9.3 \times 10^{-3}}{0.33} = 28.18 \times 10^{-3} \ A$$

$$\boxed{I_1 = 28.18 \ mA.}$$ (Ans.)

The output voltage $V_0 =$ voltage drop across diode D_2

$$\boxed{V_0 = 0.7 \ volt.}$$ (Ans.)

Applying Kirchhoff's current law at node *B,* we get,

$$I_1 = I_{D_1} + I_{D_2}$$

$\because \quad I_{D_1} = I_{D_2}$ because both diodes are same (Si)

Hence,
$$I_1 = 2I_{D_1}$$

$$I_{D_1} = \frac{I_1}{2} = \frac{28.18 \text{ mA}}{2}$$

$$\boxed{I_{D_1} = 14.09 \text{ mA} = I_{D_2}.} \qquad \text{(Ans.)}$$

Example 4.13   Determine I, V_1, V_2, V_o for series DC configuration of the following Fig. Ex. 4.13.1.   **(UPTU 2006-07)**

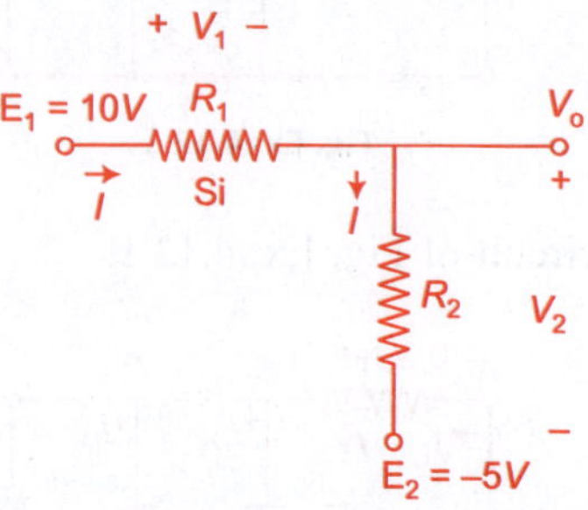

Fig. Ex. 4.13.1

Given that,
$$R_1 = 4.7 \text{ k}\Omega$$

$$R_2 = 2.2 \text{ k}\Omega$$

$$E_1 = 10\text{V}, \quad E_2 = -5 \text{ volt.}$$

Solution   Redraw Fig. Ex. 4.13.1 as shown in Fig. Ex. 4.13.1 (a)

Applying Kirchhoff's voltage law in loop $E_1A\ E_2E_1$.

$$E_1 - IR_1 - V_{DE} - R_2I - E_2 = 0$$

$$10\text{V} - I_1(4.7\ K) - 0.7 - (2.2\ K)I + 5\text{V} = 0$$

$$15 - 0.7 = I_1(4.7 + 2.2)\ K$$

$$14.3 = I(6.9\ K)$$

$$I_0 = \frac{14.3 \times 10^{-3}}{6.9}$$

$$\boxed{I = 2.07 \text{ mA.}} \qquad \text{(Ans.)}$$

Fig. Ex. 4.13.1 (a)

Now,
$$V_1 = IR_1$$

$$= 2.07 \text{ mA} \times 4.7 \text{ k}\Omega$$

$$\boxed{V_1 = 9.73 \text{ volt.}}$$ **(Ans.)**

and,

$$V_2 = IR_2$$

$$V_2 = 2.07 \text{ mA} \times 2.2 \text{ k}\Omega$$

$$\boxed{V_2 = 4.55 \text{ volt.}}$$ **(Ans.)**

Now, applying KVL at the output side loop AV_0E_2A

$$V_0 - IR_2 - E_2 = 0$$

$$V_0 = IR_2 + E_2$$

$$V_0 = 2.07 \text{ mA} \times 2.2 \text{ k}\Omega - 5V$$

$$\boxed{V_0 = -4.55 \text{ volt.}}$$ **(Ans.)**

Example 4.14 Calculate V_0 and I_D for the following Fig. Ex. 4.14.1. Data is given as follows:

$$I = 10 \text{ mA}$$

$$R_1 = 2.2 \text{ k}\Omega$$

$$R_2 = 1.2 \text{ k}\Omega.$$

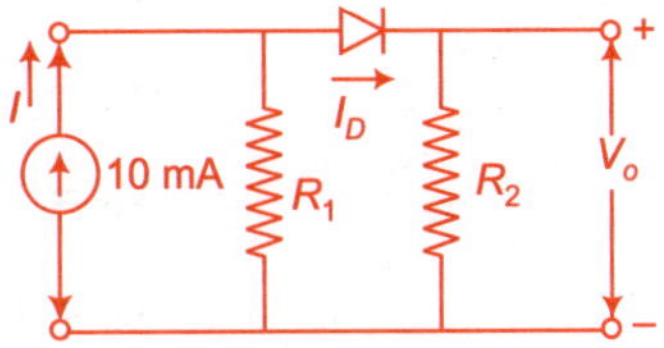

Fig. Ex. 4.14.1

Solution The network shown in Fig. Ex. 4.14.1, is redrawn by source transformation where the current source is converted into voltage source as shown in Fig. Ex. 4.14.1(a) **(UFTU 2006-07)**

Applying Kirchhoff's voltage law in loop ABCD,

$$V_i - I_D R_1 - V_{DE} - I_D R_2 = 0$$

$$\because \qquad V_i = IR_1 = 10 \text{ mA} \times 2.2 \text{ k}\Omega$$

$$V_i = 22 \text{ volt.}$$

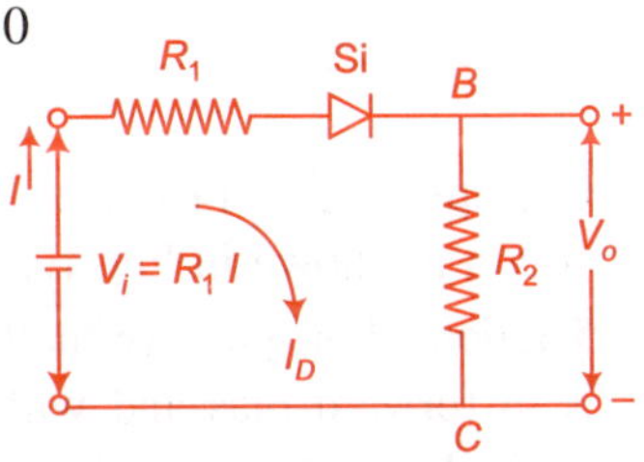

Now, $22 - I_D(22 \text{ k}\Omega) - 0.7 - I_D(1.2 \text{ k}\Omega) = 0$

$$22 - 0.7 = I_D(22 + 1.2) \text{ k}\Omega$$

$$I_D(3.4) \text{ k}\Omega = 21.3 \text{ V}$$

$$I_D = \frac{21.3 \text{ V}}{3.4 \text{ k}\Omega} = 6.26 \times 10^{-3}$$

$$\boxed{I_D = 6.26 \text{ mA.}}$$ **(Ans.)**

Fig. Ex. 4.14.1 (a)

The output voltage V_0 is given as,

$$V_0 = I_D R_2$$

$$V_0 = 6.26 \times 10^{-3} A \times 1.2 \times 10^3 \ \Omega$$

$$\boxed{V_0 = 7.512 \text{ volt.}}$$ **(Ans.)**

Example 4.15 Find the voltage V_0 and the current flowing in circuit shown in Fig. Ex. 4.15.1.

Solution   Given that,

$$V_S = 15 \text{ V}$$

$$R = 7 \text{ k}\Omega$$

Diodes are made with Si material.

So, voltage drop across each diodes is 0.7V.

Now, the redrawn circuit of Fig. Ex. 4.15.1 and is shown in Fig. Ex. 4.15.1 (a)

$$V_{D_1} = V_{D_2} = 0.7 \text{ volt.}$$

Applying Kirchhoff's voltage law in a loop ABCDA.

$$V_S - V_{D_1} - V_{D_2} - V_0 = 0$$

$$V_0 = V_S - V_{D_1} - V_{D_2}$$

$$V_0 = 15 - 0.7 - 0.7$$

$$\boxed{V_0 = 13.6 \text{ volt.}}$$ (Ans.)

Current flow into the circuit is,

$$I = \frac{V_0}{R} = \frac{13.6 \text{ V}}{7 \text{ k}\Omega}$$

$$I = 1.9 \times 10^{-3} \text{ Amp}$$

$$\boxed{I = 1.9 \text{ mA.}}$$ (Ans.)

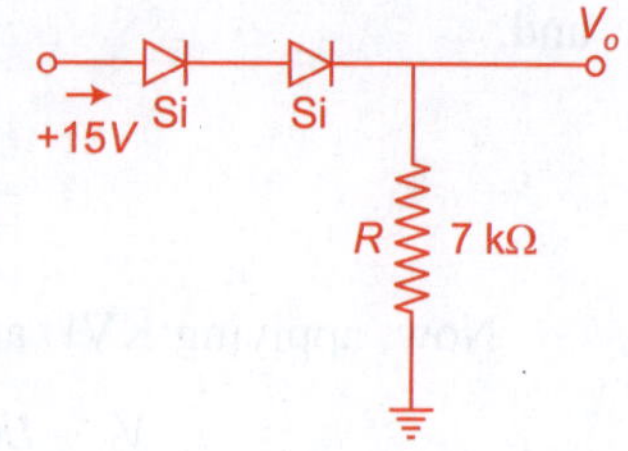

Fig. Ex. 4.15.1

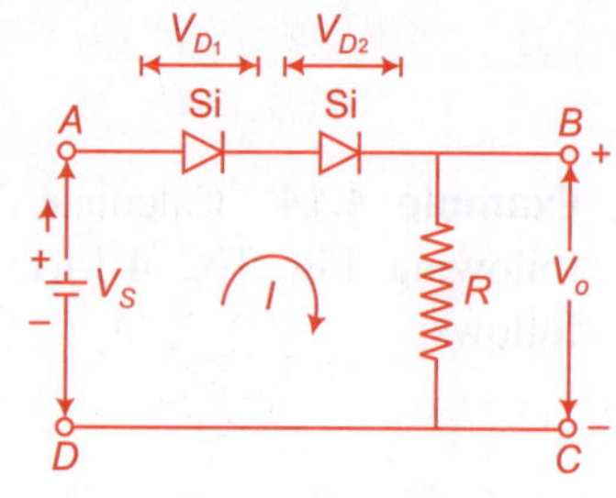

Fig. Ex. 4.15.1 (a)

Example 4.16 Find the voltage drop across 2 kΩ resistance in the following circuit as shown in Fig. Ex. 4.16.1. Assume diode to be ideal. **(UPTU 2001-02)**

Solution   For positive half cycle, diode is in forward bias and will conduct. The equivalent resistance of circuit,

$$R_{eq} = R_1 + R_2 \parallel R_3$$

$$R_{eq} = R_1 + \frac{R_2 R_3}{R_2 + R_3}$$

$$R_{eq} = 1 + \frac{1 \times 2}{1 + 2} = 1 + \frac{2}{3}$$

$$R_{eq} = \frac{5}{3} \text{ k}\Omega$$

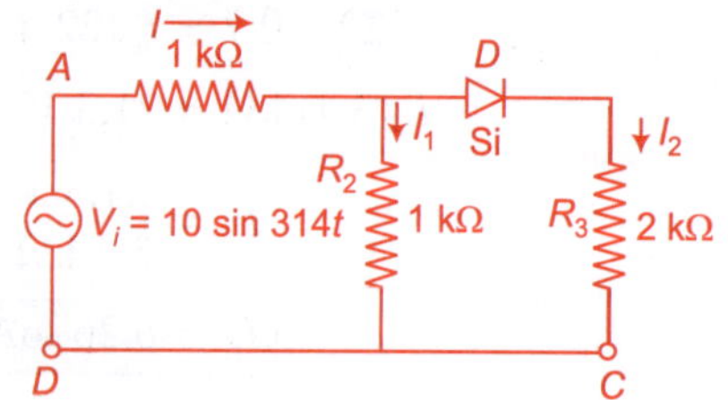

Fig. Ex. 4.16.1

Current through the circuit,

$$I = \frac{V_i}{R_{eq}} = \frac{10 \sin 314t}{\dfrac{5}{3} \times 10^3}$$

$$I = 6 \sin 314t \text{ mA}.$$

Current through resistance R_3 is,

$$I_2 = \frac{R_2}{R_2 + R_3} \times I$$

$$I_2 = \frac{1}{3} \times 6 \sin 314t \text{ mA}$$

$$I_2 = 2 \sin 314t \text{ mA}.$$

Now, the voltage across resistance A_3 is,

$$V_{R_3} = I_2 R_3$$

$$V_{R_3} = 2 \sin 314t \text{ mA} \times 2 \text{ k}\Omega$$

$$V_{R_3} = 4 \sin 314t \text{ volt}$$

So, the maximum voltage is,

$$\boxed{V_{R_3} = 4 \text{ volt.}}$$ (**Ans.**)

Example 4.17   Fig. Ex. 4.17.1 shows the circuit of series diode configuration. Find the value of V_f, V_R and I_f.

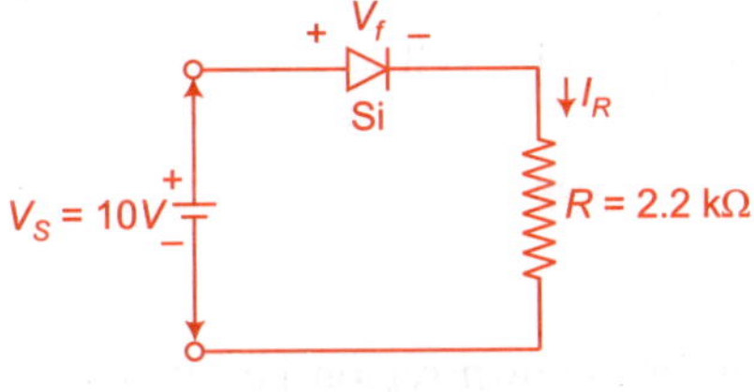

Fig. Ex. 4.17.1

Solution   The diode is in forward bias. Here, the diode is Si diode. So, the voltage drops across diode

$$\boxed{V_f = 0.7 \text{ volt.}}$$ (**Ans.**)

Applying the Kirchhoff's voltage low across the loop,

$$V_S - V_f - V_R = 0$$

$$V_R = V_S - V_f = 10 - 0.7$$

$$\boxed{V_R = 9.3 \text{ volt.}}$$ (**Ans.**)

From the circuit,

$$I_f = I_R = \frac{V_R}{R} = \frac{9.3V}{22 \times 10^3 \; \Omega}$$

$$\boxed{I_f = 4.23 \text{ mA.}}$$

(Ans.)

Problems based on zener diode

Example 4.18 Calculate the value of I_L, I_Z, I_S and V_0 for a given circuit in Fig. Ex. 4.18.1.

Solution Given that,

$$V_i = 25 \text{ V}$$

$$R_S = 1.2 \text{ k}\Omega$$

$$R_L = 1.6 \text{ k}\Omega$$

$$V_Z = 10 \text{ volt.}$$

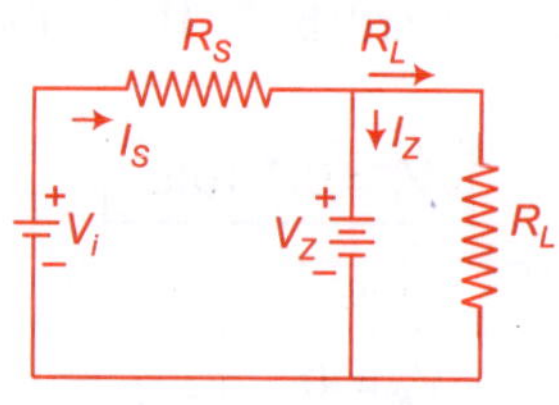

Fig. Ex. 4.18.1

Let us check whether the state of zener diode is 'on' or 'off'. Voltage across zener diode after removing it from the circuit is V_{oc} and is shown in Fig. Ex. 4.18.1 (a)

By voltage dividing rule,

$$V_{oc} = \frac{R_L}{R_S + R_L} V_i$$

$$V_{oc} = \frac{1.6}{1.6 + 1.2} \times 25 = \frac{1.6}{2.8} \times 25$$

$$V_{oc} = \frac{40}{2.8} = 14.3 \text{ volt}$$

Fig. Ex. 4.18.1 (a)

$$V_{oc} = 14.3 \text{ volt}$$

$$\therefore \qquad V_{oc} > V_Z$$

Hence, diode is in breakdown region i.e., it is 'on' state and its equivalent circuit can be established as shown in Fig. Ex 4.18.1 (b)

Fig. Ex 4.18.1 (b)

From Fig. Ex 4.18.1 (b)

$$V_0 = V_z = 10 \text{ volt.}$$

Load current,
$$I_L = \frac{V_0}{R_L} = \frac{V_Z}{R_L} = \frac{10 \text{ mA}}{1.6 \text{ k}\Omega}$$

$$\boxed{I_Z = 6.3 \text{ mA.}} \qquad \textbf{(Ans.)}$$

Current flowing through R_S, i.e., I_S

$$I_S = \frac{V_i - V_Z}{R_S} = \frac{25 - 10}{1.2 \text{ k}\Omega}$$

$$I_S = \frac{15V}{1.2 \text{ k}\Omega} = 12.5 \text{ mA}$$

$$\boxed{I_S = 12.5 \text{ mA.}} \qquad \textbf{(Ans.)}$$

Now, the zener current,

$$I_Z = I_S - I_L$$

$$I_Z = (12.5 - 6.3) \text{ mA}$$

$$\boxed{I_Z = 6.2 \text{ mA.}} \qquad \textbf{(Ans.)}$$

Example 4.19   Calculate the value of I_L, I_Z, I_S and V_0 for the given circuit in Fig. Ex. 4.19.1.

Solution   Given that,
$$V_i = 20V$$
$$V_Z = 10 \text{ volt}$$
$$R_S = 1 \text{ k}\Omega \quad \text{and} \quad R_L = 1.5 \ \Omega.$$

Fig. Ex. 4.19.1

Let us check the state of zener diode after removing it from the circuit and calculate the open circuit voltage V_{oc} as shown in Fig. Ex. 4.19.1 (a)

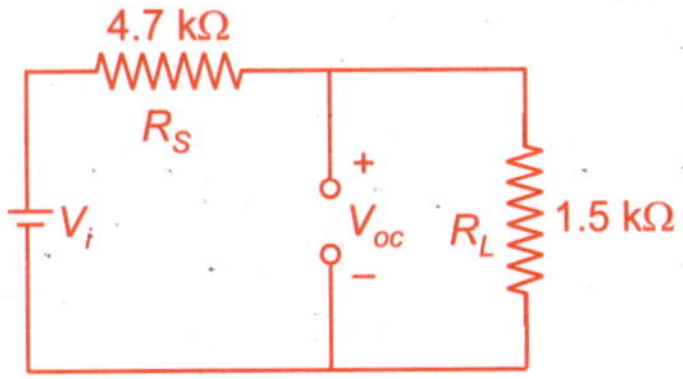

Fig. Ex. 4.19.1 (a)

By voltage dividing rule,

$$V_{oc} = \frac{R_L}{R_S + R_L} \times V_i$$

$$V_{oc} = \frac{1.5}{1.5 + 4.7} \times 20$$

$$V_{oc} = \frac{30}{6.2} = 12 \text{ volt}$$

$$V_{oc} = 4.8 \text{ volt.}$$

$\because$ $V_{oc} < V_Z$, the zener diode is in 'off' state and will be an open circuit.

Hence, $\boxed{V_0 = V_Z = 4.8 \text{ volt.}}$ **(Ans.)**

$$I_L = \frac{V_0}{R_L} = \frac{4.8V}{1.5 \text{ k}\Omega}$$

$\boxed{I_L = 3.2 \text{ mA.}}$ **(Ans.)**

$\boxed{I_Z = 3.2 \text{ mA.}}$ $\because$ zener is in 'off' state **(Ans.)**

$$I_S = I_L + I_Z$$

$$I_S = 3.2 \text{ mA} + 0 = 3.2 \text{ mA}$$

$\boxed{I_S = 3.2 \text{ mA.}}$ **(Ans.)**

Example 4.20   For the circuit shown in Fig. Ex. 4.20.1. Determine V_0, I_L, I_Z and I_S with $R_S = 220$ kΩ and $R_L = 50$ kΩ, $V_Z = 10$V.

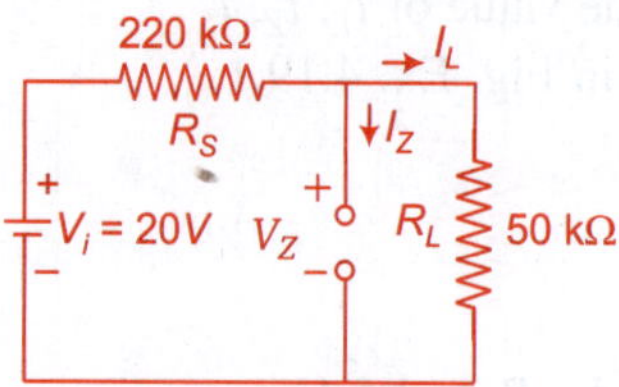

Fig. Ex. 4.20.1

Solution   Let us check the state of zener diode. Establish an equivalent circuit as shown in Fig. Ex. 4.20.1 (a)

By voltage dividing rule,

$$V_{oc} = \frac{R_L}{R_L + R_S} \times V_i$$

$$V_{oc} = \frac{50}{50 + 220} \, 20$$

$$V_{oc} = \frac{1000}{270} = 3.7 \text{ volt.}$$

Fig. Ex. 4.20.1 (a)

Since, $V_{oc} < V_z$, so the zener diode is in 'off' state.

Hence, $$V_0 = V_{oc} = 3.7 \text{ volt.}$$

$$V_S = V_i - V_o = 20 - 3.7$$

$\boxed{V_S = 16.30.}$ **(Ans.)**

The zener current,

$$\boxed{I_Z = 0.}$$ **(Ans.)**

The load current,

$$I_L = \frac{V_0}{R_L} = \frac{3.7}{50\ k\Omega}$$

$$\boxed{I_L = 74.1\ \mu A.}$$ **(Ans.)**

$$I_S = I_L + I_Z$$

$$I_S = (74.1 + 0)\ \mu A$$

$$\boxed{I_S = 74.1\ \mu A.}$$ **(Ans.)**

Example 4.21   Calculate the values of R_S to maintain V_L at 12V for varying I_L, from 0 to 200 mA. Also find V_Z and $P_{Z\,(max)}$.   **(UPTU 2006-07)**

Solution   Given that,

$$V_Z = 12V$$

$$V_i = 16V$$

$$I_{Z\,(max)} = 200\ mA.$$

Redrawing the circuit of Fig. Ex. 4.21.1 and shown in Fig. Ex. 4.21.1 (a)

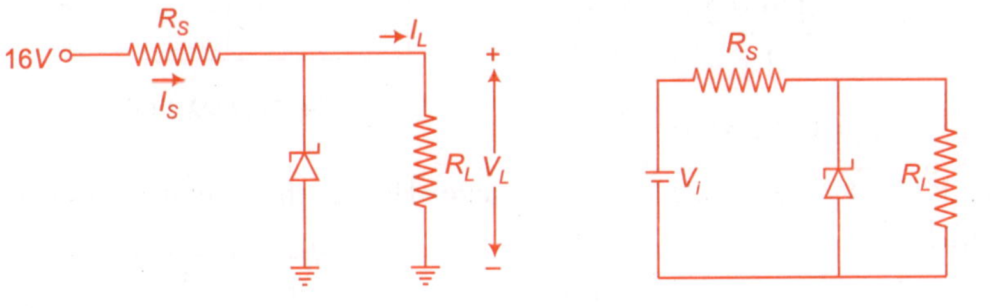

Fig. Ex. 4.21.1   **Fig. Ex. 4.21.1 (a)**

We have,

$$V_Z = 12\ volt$$

Voltage drop across R_S,

$$V_S = V_i - V_L$$

$$V_S = 16 - 12\ = 4\ volt.$$

Now,

$$R_S = \frac{V_S}{I_{Z\,(max)}} = \frac{4V}{200\ mA} = 20\ \Omega$$

$$\boxed{R_S = 20\ \Omega.}$$ **(Ans.)**

$$R_{L\,(min)} = \frac{V_L}{I_{Z\,(max)}} = \frac{12A}{200\ mA} = 60\ \Omega$$

$$R_{L\,(max)} = \infty \text{ at } I_L = 0$$

$$P_{L\,(max)} = V_Z \times I_{Z\,(max)} \text{ (at } I_L = 0)$$

$$\boxed{P_{Z\,(max)} = 24\ watt.}$$
(Ans.)

Example 4.22 Determine V_L, I_L, I_Z and I_S for the circuit shown in Fig. Ex. 4.22.1.
(UPTU 2007-08)

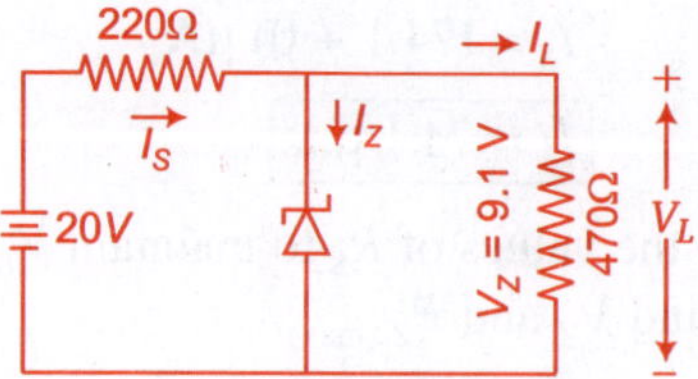

Fig. Ex. 4.22.1

Solution   Let us check the state of zener diode by removing it from the circuit as shown in Fig. Ex. 4.22.1(a)

$$V_{oc} = \frac{R_L}{R_S + R_L}\, V_i$$

$$V_{oc} = \frac{470 \times 20}{470 + 220} = \frac{9400}{690}$$

$$V_{oc} = 13.62 \text{ volt.}$$

Fig. Ex. 4.22.1 (a)

Since, $V_{oc} > V_2$, the diode is in 'on' state. Hence, the circuit is shown in Fig. Ex. 4.22.1 (b)

$$\because \qquad V_Z = 9.1$$

Hence, $V_L = V_Z = 9.1$ volt
Voltage across resistance R_S is,

$$V_S = V_i - V_L$$

$$V_S = 20 - 9.1 = 11.9 \text{ volt}$$

$$\boxed{V_S = 11.9 \text{ volt.}}$$
(Ans.)

Fig. Ex. 4.22.1 (b)

The load current I_L,

$$I_L = \frac{V_L}{R_L} = \frac{9.1\,V}{470\ \Omega}$$

$$\boxed{I_L = 0.019 \text{ A.}}$$
(Ans.)

And, $$I_S = \frac{V_S}{R_S} = \frac{11.9V}{220\Omega}$$

$$\boxed{I_S = 0.054 \text{ A.}}$$ (**Ans.**)

Now, $$I_Z = I_S - I_L$$

$$I_Z = 0.054 - 0.019$$

$$\boxed{I_Z = 0.035 \text{ A.}}$$ (**Ans.**)

Example 4.23 A 20V, 500 mW, zener diode is used as 20V regulated power supply.

(a) if V_i is 30V, then calculate series resistance R_S

(b) if $R_1 = 1$ kΩ, then what will be the zener current I_Z.

Solution Given that, zener voltage $V_Z = 20$ V

maximum power rating $P_{Z(\text{max})} = 500$ mW

(a) Maximum current rating $$I_{Z(\text{max})} = \frac{P_{Z\,(\text{max})}}{V_Z}$$

$$I_{Z\,(\text{max})} = \frac{500 \text{ mW}}{20 \text{ V}}$$

$$I_{Z\,(\text{max})} = 25 \text{ mA}$$

Series resistance, $$R_S = \frac{V_i - V_0}{I_{Z\,(\text{max})}}$$

$\because$ $$V_o = V_Z = 20\text{V}$$

and, $$V_i = 30\text{V}$$

Hence, $$R_S = \frac{(30 - 20)\text{V}}{25 \text{ mA}} = 0.4 \ \Omega$$

$$\boxed{R_S = 0.4 \text{ k}\Omega.}$$ (**Ans.**)

(b) If $R_L = 1$ kΩ

$$I_L = \frac{V_0}{R_L} = \frac{20\text{V}}{1 \text{ k}\Omega}$$

$$= 20 \text{ mA.}$$

$$I_S = \frac{V_i - V_o}{R_S} = \frac{(30 - 20)\text{V}}{0.4 \text{ k}\Omega} = 25 \text{ mA}$$

$$I_Z = I_S - I_L$$

$$I_Z = 25 - 20 = 5 \text{ mA}$$

$$\boxed{I_Z = 5 \text{ mA.}}$$ (**Ans.**)

Example 4.24 For a given circuit, calculate safe value of R_S.

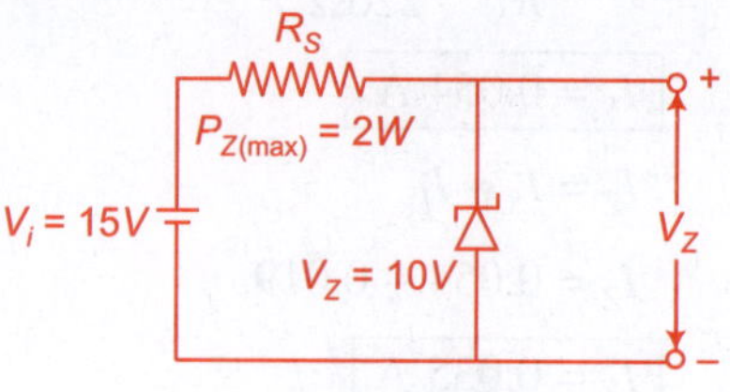

Fig. Ex. 4.24

Solution Given that,

$$P_{Z\,(max)} = 1W$$

$$I_{Z\,(max)} = \frac{P_{z\,(max)}}{V_z}$$

$$I_{Z\,(max)} = \frac{2W}{10V} = 0.2A$$

$$I_{Z\,(max)} = 0.2\ A$$

The value of series resistance, R_S

$$R_S = \frac{V_i - V_o}{I_{Z\,(max)}} = \frac{(15V - 10V)}{0.2A}$$

$$R_S = \frac{5V}{0.2A} = 25\ \Omega$$

$$\boxed{R_S = 25\ \Omega.}$$
(**Ans.**)

Example 4.25 For the circuit shown in Fig. Ex. 4.25.1, find the maximum and minimum values of zener diode current.

Solution Given that,

$$R_S = 10\ k\Omega$$

$$R_L = 15\ k\Omega$$

$$V_i = 60\ \text{to}\ 100V$$

$$\boxed{V_Z = 40\ \text{volt.}}$$

Fig. Ex. 4.25.1

Maximum zener current

When the input voltage is maximum, i.e., 100 V, then zener will conduct maximum current.

Under such conditions,

Voltage across 10 kΩ resistance i.e.,

$$V_S = V_i - V_Z = 100V - 40V$$

$$V_S = 60 \text{ volt.}$$

Current though 10 kΩ resistance i.e.,

$$I_S = \frac{V_S}{R_S} = \frac{60V}{10 \text{ k}\Omega}$$

$$I_S = 6mA$$

and load current,
$$I_L = \frac{V_Z}{R_L} = \frac{40V}{15 \text{ k}\Omega}$$

$$I_L = 2.66 \text{ mA.}$$

We know that from circuit,

$$I_S = I_Z + I_L$$

$$I_S \cdot I_Z = I_S - I_L = 6 \text{ mA} - 2.266 \text{ mA}$$

$$\boxed{I_Z = 3.34 \text{ mA.}} \qquad \textbf{(Ans.)}$$

Minimum zener current

When the input voltage is minimum i.e., 60V, then zener will conduct minimum current. Voltage across l0 kΩ resistance V_S.

$$V_S = V_i - V_Z$$

$$V_S = 60 - 40 = 20 \text{ V}$$

$$\boxed{V_S = 20 \text{ volt.}} \qquad \textbf{(Ans.)}$$

Current through 10 k resistor, I_S

$$I_S = \frac{V_S}{R_S} = \frac{20V}{10k\Omega}$$

$$I_S = 2mA$$

Load current,
$$I_L = \frac{V_Z}{R_L} = \frac{20V}{15k\Omega}$$

$$I_L = 1.33 \text{ mA}$$

$\therefore$ Zener current
$$I_Z = I_S - I_L$$

$$= (2 - 1.33) \text{ mA}$$

$$\boxed{I_Z = 0.67mA.} \qquad \textbf{(Ans.)}$$

Example 4.26   A 7.5V zener is used in the circuit shown in Fig. Ex. 4.26.1 and the load current varies from 10 to 100 mA. Find the value of series resistance.

R_S to maintain a voltage of 7.5V across the load. The input voltage is constant at 15V and the minimum zener current is 10 mA.

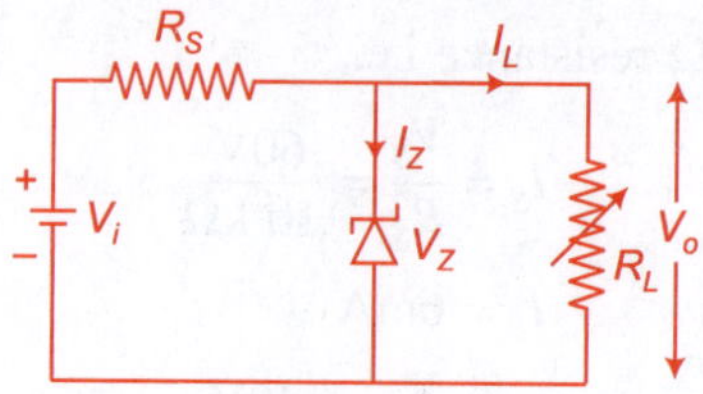

Fig. Ex. 4.26.1

Solution Given that,

$$V_i = 15V, \quad V_z = 7.5 \text{ V}$$

$$R_S = \frac{V_i - V_o}{I_Z + I_L} = \frac{V_S}{I_S}.$$

The voltage across R_S remains constant at $15 - 7.5 = 7.5$V, as the load current changes from 10 to 100 mA. The minimum zener current will occur when the load current is maximum.

$$R_S = \frac{V_i - V_o}{I_{Z\,(max)} + I_{Z\,(min)}} = \frac{(15 - 7.5)V}{(10 + 100)\text{mA}}$$

$$R_S = \frac{7.5 \text{ V}}{110\text{mA}} = 68.2 \text{ }\Omega$$

$$\boxed{R_S = 68.2\Omega.}$$ **(Ans.)**

If $R_S = 68.2\Omega$ is inserted in the circuit, the output voltage will remain constant over the regulating range. As the load current I_Z decreases, the zener current I_Z will increase to such a value that $I_Z + I_L = 110$ mA.

Example 4.27 A 20V zener diode is used to regulate the voltage across a variable load resistor. The input voltage varies between 25V and 45V and the load current varies between 20 and 120 mA. The minimum zener current is 25 mA. Calculate the value of series resistance.

Solution The zener will conduct minimum current (25 mA), when input voltage is minimum (i.e., 25V).

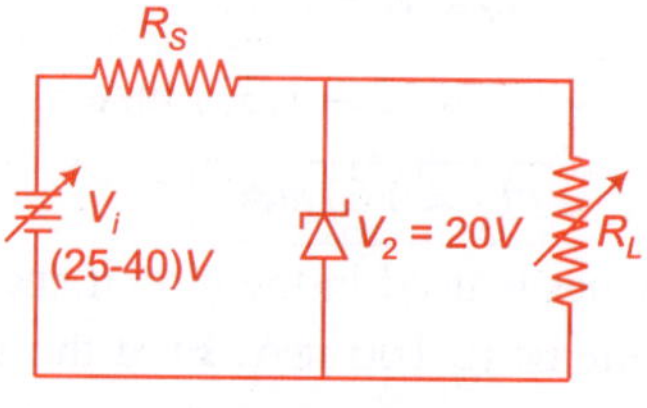

Fig. Ex. 4.27.1

$$\therefore \quad R_S = \frac{V_i - V_o}{I_{Z\,(min)} + I_{L\,(max)}} = \frac{(25 - 20)\ V}{(25 + 120)mA}$$

$$R_S = \frac{5V}{145mA}$$

$$\boxed{R_S = 34.5\ \Omega.}$$

(Ans.)

Example 4.28   What values of series resistance is required when two 10 watts, 12V 600 mA zener diodes are connected in series to obtain 24V regulated output from a 50V DC power source?

Solution   Given that,

Voltage rating of each zener $V_Z = 12$ V

Current rating of each zener $I_Z = 600$ mA

Input voltage $V_i = 50$ V.

Fig. Ex. 4.28.1 shows the desired circuit.

Here, we are considering no load conditions because at this condition zener carry maximum current.

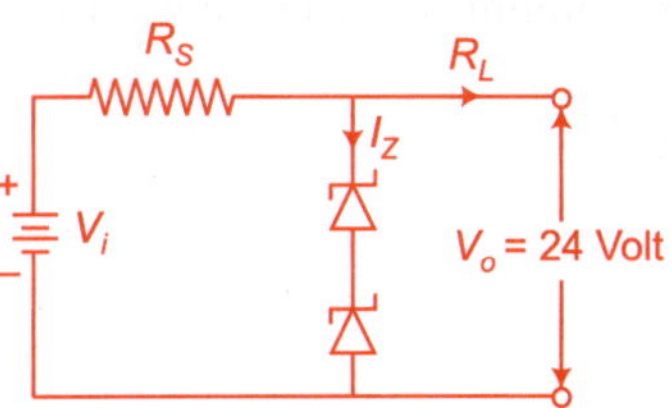

Fig. Ex. 4.28.1

Regulated output voltage, V_o

$$V_o = 12V + 12V = 24V.$$

Voltage across series resistance R_S,

$$V_S = V_i - V_o$$

$$= 50 - 24 = 26 \text{ volt.}$$

$$\therefore \quad R_S = \frac{V_S}{I_S} = \frac{26V}{600mA}$$

$$R_S = 0.0433 \times 10^3\ \Omega$$

$$\boxed{R_S = 43.3\ \Omega.}$$

(Ans.)

Example 4.29   In the circuit shown in Fig. Ex. 4.29.1, the voltage across the load is to be maintained at 10V as load current varies from 0 to 150 mA.

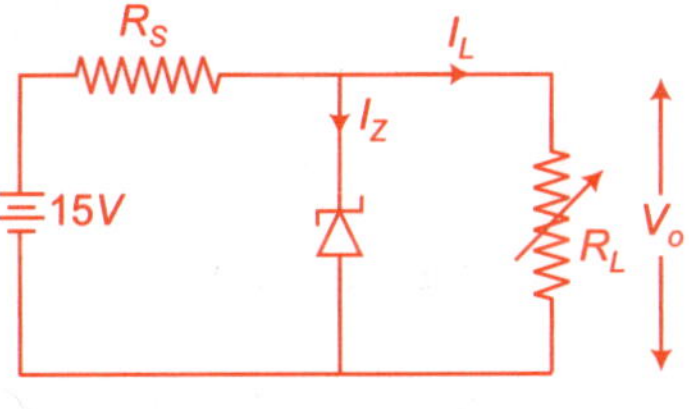

Fig. Ex. 4.29.1

Design the voltage regulator. Also, find the maximum voltage rating of zener diode.

Solution Given that,

$$V_Z = 10V.$$

The voltage across series resistance, R_S

$$V_S = V_i - V_0$$

$$V_S = 15 - 10 = 5 \text{ volt.}$$

V_S remains constant as the load current changes from 0 to 150 mA. The minimum zener current will occur when the load current is maximum.

$\therefore$

$$R_S = \frac{V_i - V_0}{I_{Z(min)} + I_{Z(max)}}$$

$$= \frac{(15 - 10)V}{(0 + 150)mA} = \frac{5V}{15mA}$$

$$\boxed{R_S = 33.3\Omega.}$$ (**Ans.**)

Maximum power rating of zener is,

$$P_{Z\,(max)} = V_Z I_{Z\,(max)}$$

$$= (10C)\,(150mA)$$

$$P_Z = 1.5W$$

$$\boxed{P_{Z\,(max)} = 1.5 \text{ watt.}}$$ (**Ans.**)

Numerical problems based on approximated model of diode

Example 4.30 Determine the diode current I_D and load voltage across the load for a given circuit shown in Fig. Ex. 4.30.1.

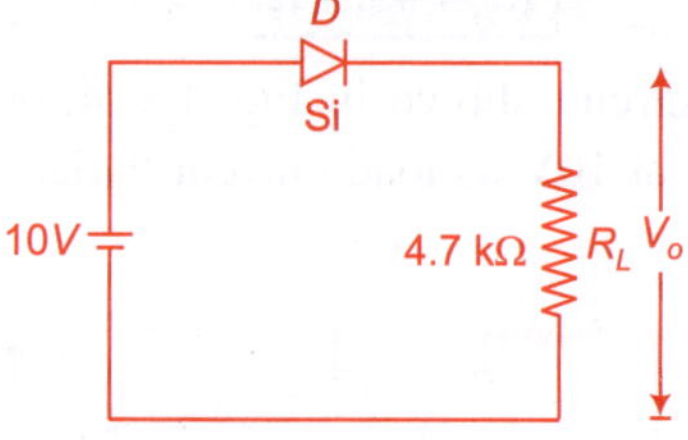

Fig. Ex. 4.30.1

Solution Given that,

$$V_i = 10V$$

The equivalent circuit is shown in Fig. Ex. 4.30.2.

Voltage drop across diode is V_B

$$V_B = 0.7 \text{ for Si diode.}$$

The diode current I_D is obtained by applying KVL in loop,

$$V_i - V_B - I_D R_L = 0$$

$$10 - 0.7 - I_D \, 4.7 \text{ k}\Omega = 0$$

$$I_D = \frac{(10 - 0.7)\text{V}}{4.7 \text{ k}\Omega} = \frac{9.3\text{V}}{4.7 \text{ k}\Omega}$$

$$\boxed{I_D = 1.98\text{mA.}}$$ 　(**Ans.**)

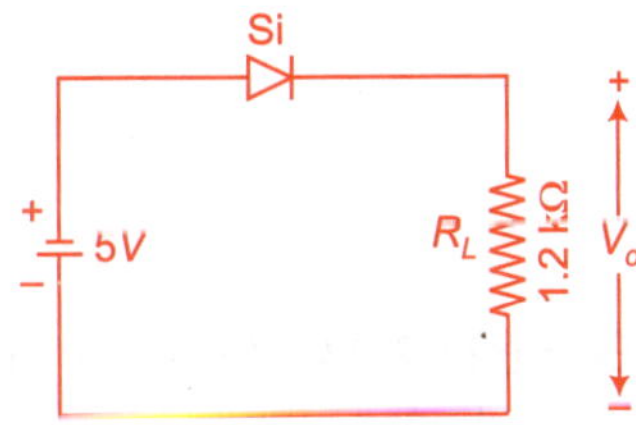

Fig. Ex. 4.30.2

Now, the voltage drop across the load,

$$V_o = R_L I_D$$

$$V_o = 4.7 \text{ k}\Omega \times 1.98\text{mA}$$

$$\boxed{V_o = 9.3 \text{ volt.}}$$ 　(**Ans.**)

Example 4.31   Determine the diode current and output voltage for the circuit shown in Fig. Ex. 4.31.1.

Solution   Given that,

$$R_L = 1.2 \text{ k}\Omega$$

$$R_L = 1.2 \text{ k}\Omega$$

$$V_i = -5 \text{ volt.}$$

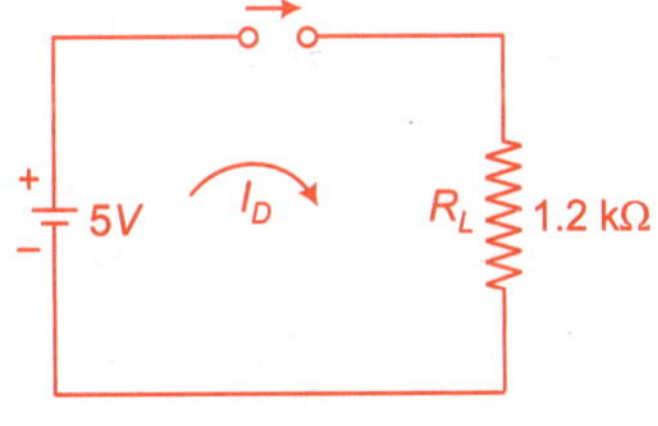

Fig. Ex. 4.31.1

Since the diode is in reverse bias, therefore the diode is replaced by their equivalent model and diode is open circuited as shown in Fig. Ex. 4.31.2.

Hence, the load current,

$$\boxed{I_D = 0.}$$ 　(**Ans.**)

And the voltage across the load,

$$\boxed{V_o = 0 \text{ volt.}}$$ 　(**Ans.**)

Fig. Ex. 4.31.2

Example 4.32   Determine the diode current and output voltage for the given circuit shown in Fig. Ex. 4.32.1.

Solution   Given that,

$$R_L = 2.2 \text{ k}\Omega$$

$$V_i = 10 \text{ volt.}$$

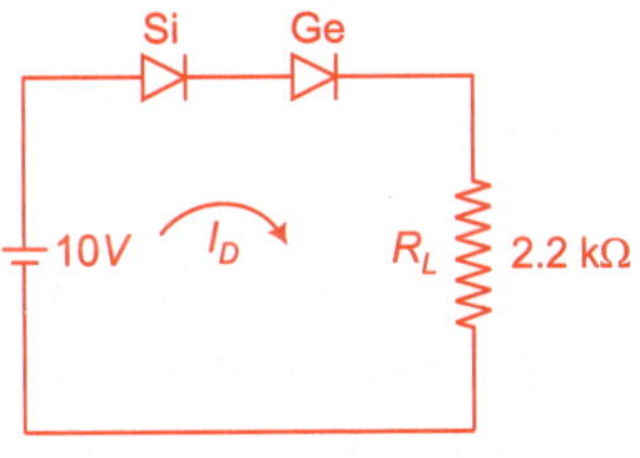

Fig. Ex. 4.32.1

Now, replacing the diode by its equivalent model shown in Fig. Ex. 4.32.2.

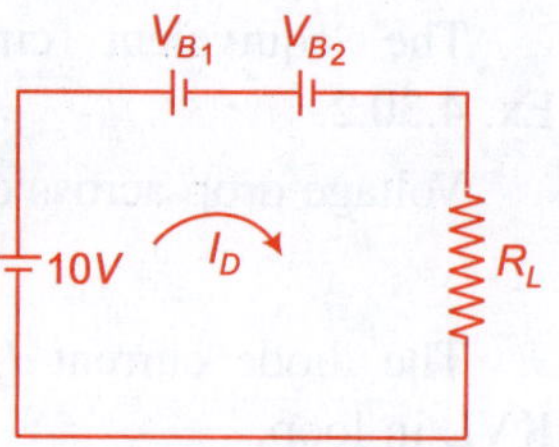

Fig. Ex. 4.32.2

$$V_{B_1} = 0.7 \text{ V}$$

$$V_{B_2} = 0.3\text{V}$$

$$V_i = 10\text{V} \times R_L = 2.2\text{k}\Omega.$$

Applying kirchhoff's voltage law,

$$V_i - V_{B_1} - V_{B_2} - I_D R_L = 0$$

$$10 - 0.7 - 0.3 - I_D\,(2.2\ \text{k}\Omega) = 0$$

$$2.2\ \text{k}\Omega\ I_D = 10 - 1$$

$$I_D = \frac{9}{2.2\ \text{k}\Omega}$$

$$\boxed{I_D = 4.1\text{mA}.}$$ (Ans.)

The output voltage V_o is given by,

$$V_o = I_D R_L$$

$$= 4.1 \times 2.2\ \text{k}\Omega$$

$$\boxed{V_o = 9.01\ \text{volt}.}$$ (Ans.)

Example 4.33   Sketch V_o for the circuit shown in Fig. Ex. 4.33.1. D_1 and D_2 are S_i diodes.

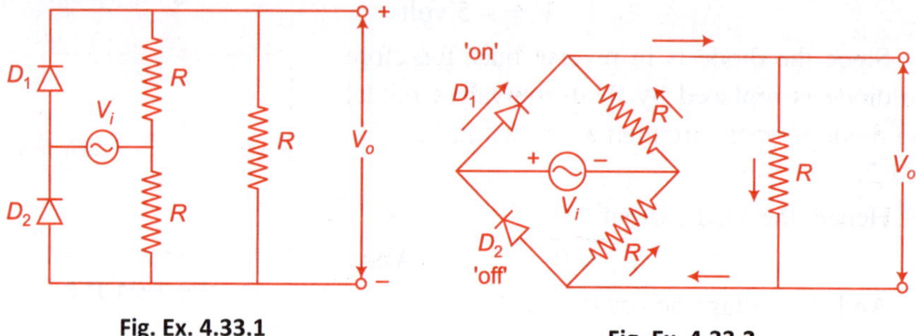

Fig. Ex. 4.33.1

Fig. Ex. 4.33.2

Solution   Given that,

$$V_i = 100 \sin 314\,t$$

The circuit is redrawn and shown in Fig. Ex. 4.33.2.

For '+ve' half cycle of input

Diode D_1 will be in forward bias and will conduct the current after 0.7V, since it is a Si diode.

The diode D_2 will be in reverse bias and will not conduct.

For '–ve' half circle of input

Diode D_2 will be in forward bias and will conduct the current after 0.7V, since it also is a Si diode.

The diode D_1 will be in reverse bias and will not conduct.

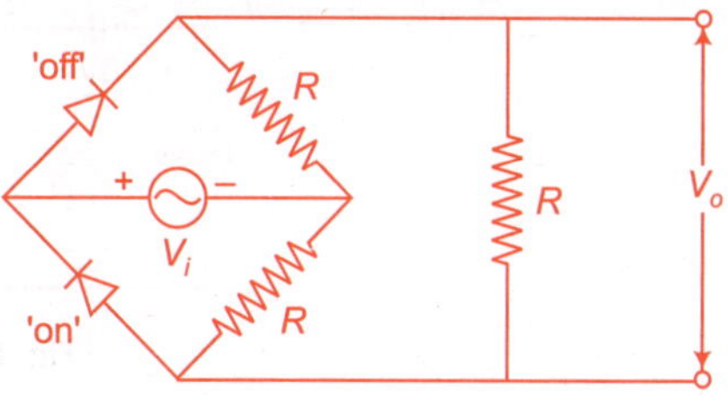

Fig. Ex. 4.33.3

The sketch of output waveform is shown in Fig. Ex. 4.33.4.

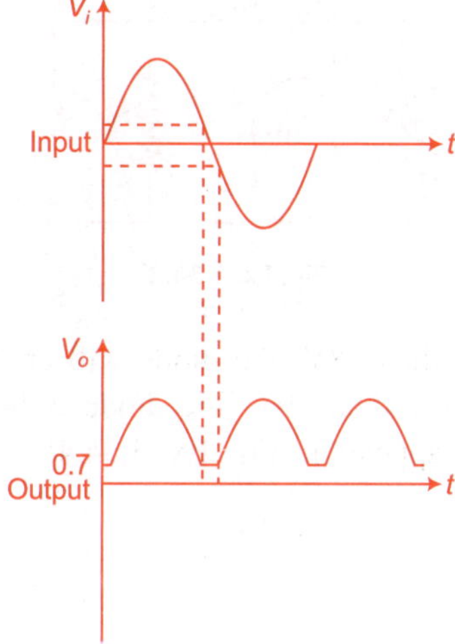

Fig. Ex. 4.33.4

Problems based on clippers and clampers

Example 4.34 Determine the output waveform for the network of Fig. Ex. 4.34.1.

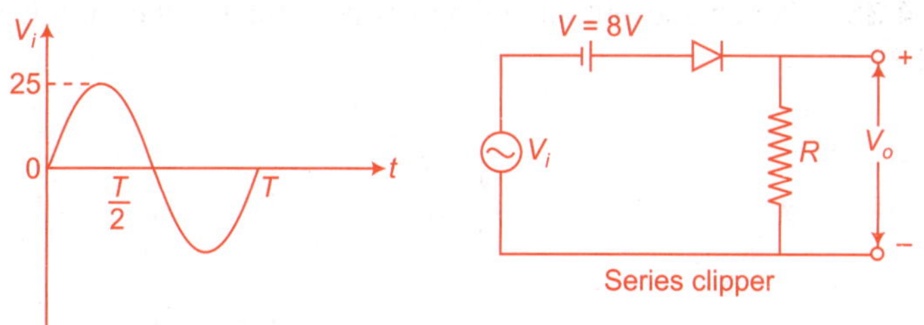

Fig. Ex. 4.34.1

Solution For a positive half cycle, the diode will be in 'on' state. The network will appear as shown in Fig. Ex. 4.34.2. and $V_o = V_i + 8V$.

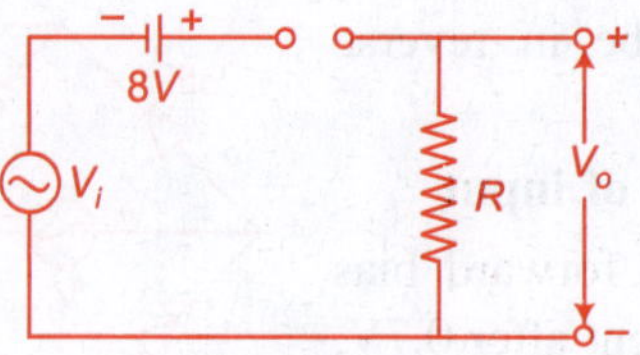

Fig. Ex. 4.34.2

Now substituting $I_d = 0$ at $V_d = 0$ for the transition level, we obtain the network of Fig. Ex. 4.34.3 and $V_i = -8V$.

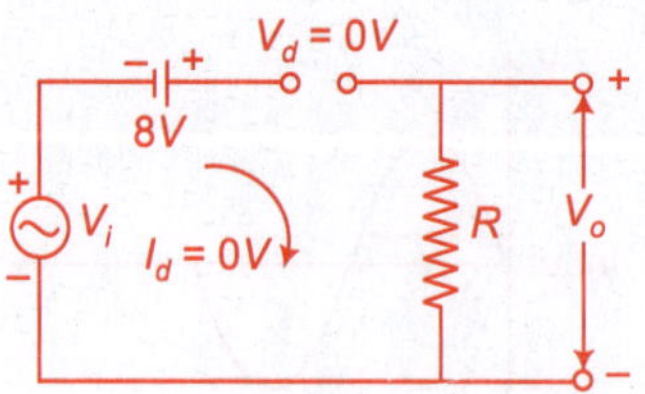

Fig. Ex. 4.34.3

For V_i more negative than $-8V$, the diode will enter in open circular state, while voltage more positive than $-9V$, the diode is in short circuit state. The input and output voltages appears in Fig. Ex. 4.34.4.

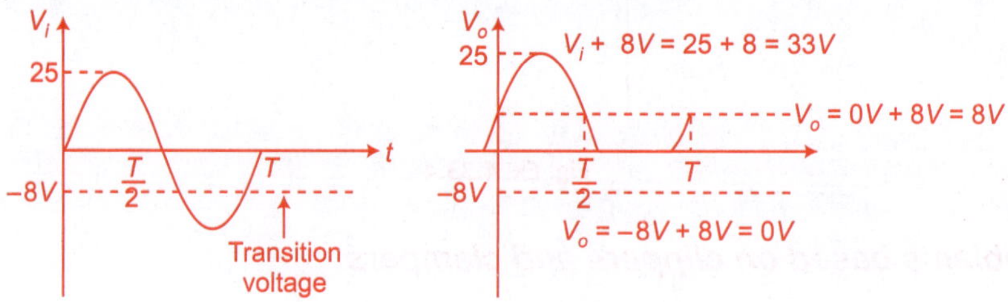

Fig. Ex. 4.34.4

Example 4.35 Sketch the output voltage waveform for the circuit shown in Fig. Ex. 4.35.1

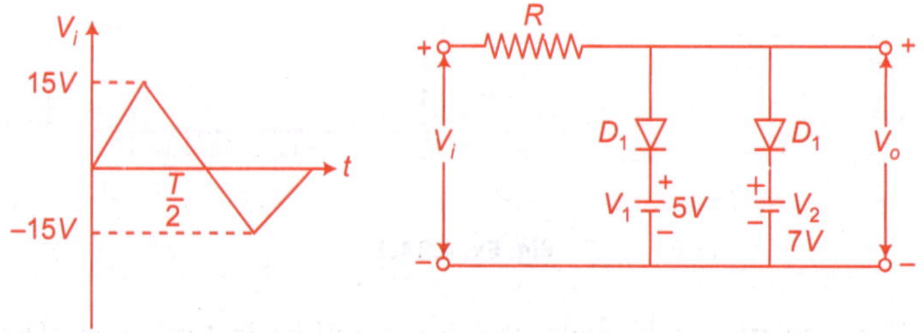

Fig. Ex. 4.35.1

Solution   **(i) During positive half cycle**

When $V_i < V_1$, D_1 and D_2 are 'off', then $V_0 = V_i$

When $Vi > V_1$, D_1 is 'on' and diode D_1 is 'off', then the output voltage

$$V_0 - 0.7V + 5V = 5.7 \text{ V}$$

$$\boxed{V_0 = -7.7 \text{ volt.}}$$ **(Ans.)**

(ii) For negative half cycle

When $V_i < V_1$, D_1 and D_2 are 'off, then $V_0 = V_i$

When $Vi > V_1$, D_2 is 'on' and diode D_1 is 'off', then the output voltage $V_0 - 0.7V$

$$\boxed{V_0 = -7.7 \text{ volt.}}$$ **(Ans.)**

The sketch of the output is shown in Fig. Ex. 4.35.2.

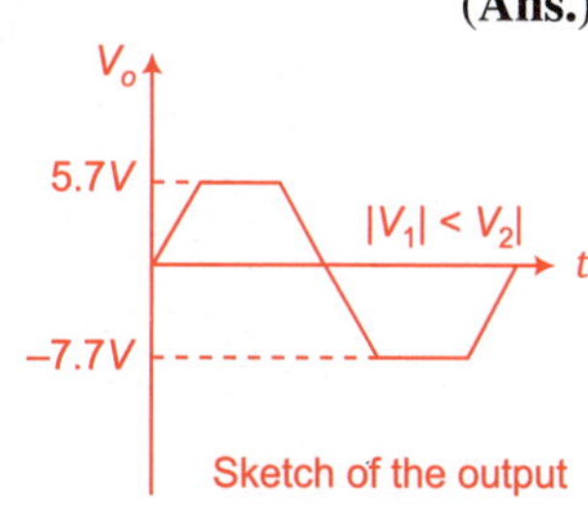

Fig. Ex. 4.35.2

Example 4.36   For the circuit shown in Fig. Ex. 4.36.1. Sketch i_R and V_0. (Diode D_1 and D_2 are Si diode).

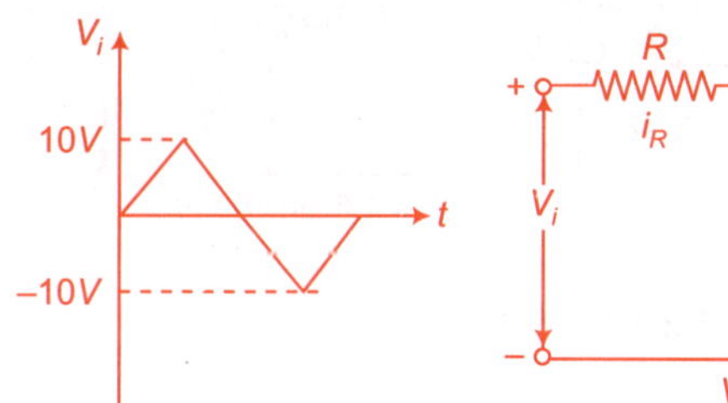

Fig. Ex. 4.36.1

Solution   Both diodes D_1 and D_2 are silicon diodes.

Given that, $V_i = 5.3$ V and $V_2 = 7.3$ volt.

Now, from table 4.1.

$$V_a = V_1 + 0.7 = (5.3 + 0.7) \ V = 6.0 \text{ V}$$

$$V_b = V_2 + 0.7 = -7.3 + 0.7 = -6.6 \ V.$$

For +ve half cycle

When $V_i < V_a$, the diode D_2 is reverse bias and diode is in forward bias.

$\therefore$ $V_0 = V_i$

when $V_i > V_a$, the diode D_1 and D_2 are in forward bias

$\therefore$ $V_0 = V_a.$

For negative half cycle

When $V_i < V_b$, the diode D_1 and D_2 are in reverse bias, then $V_0 = V_i$.

When $V_i > V_b$, then it is in forward bias and D_2 is in reverse bias.

The sketch of output is shown in Fig. Ex. 4.35.2.

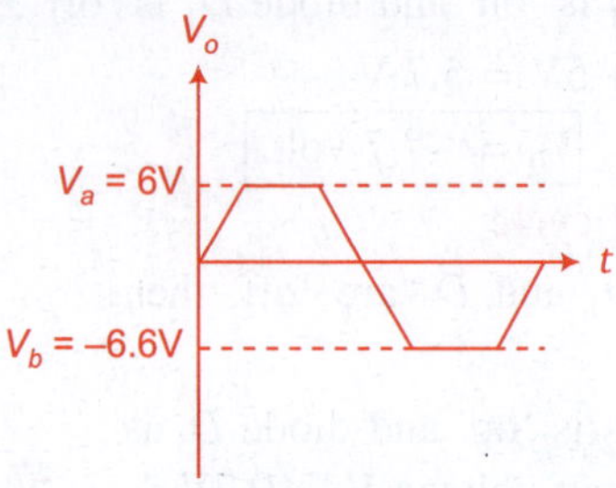

Fig. Ex. 4.36.2

Example 4.37 Sketch the output for the given input of Fig. Ex. 4.37.1.

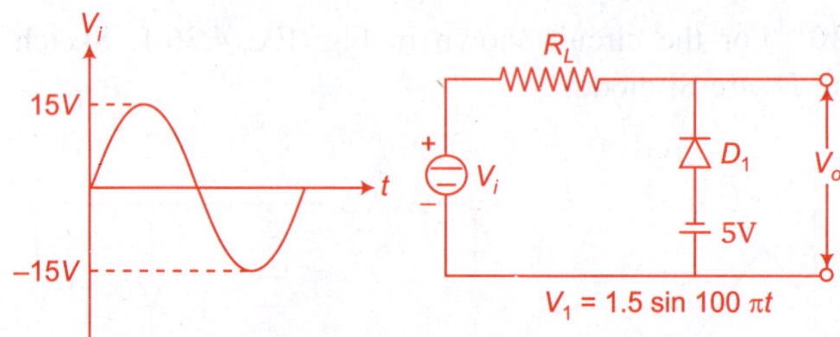

Fig. Ex 4.37.1

Solution Given that,

$$V_B = -5 \text{ V and } V_m = 15 \text{volt.}$$

Assume that the diode is ideal for positive half cycle

For positive half cycle

For positive half cycle, the diode D is in reverse bias so that $V_0 = V_i$.

For negative half cycle

When $V_i < V_B$, then the diode is 'off' and will not conduct. So, the output is.

$$V_0 = V_i.$$

When $V_i > V_B$, then the diode is 'off' and will conduct. So, the output $V_0 = -V_B$.

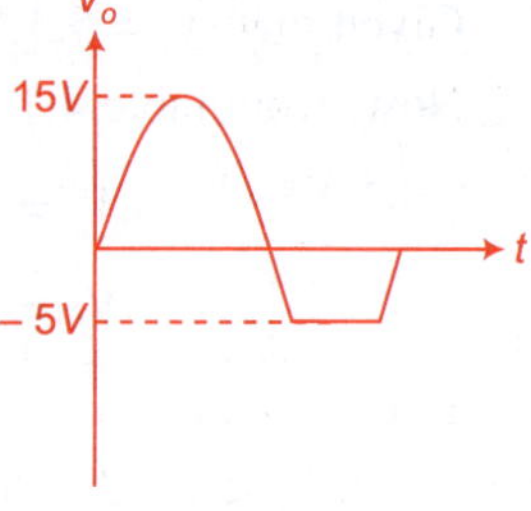

Fig. Ex. 4.37.2

Objective Type Questions

1. A forward bias *p-n* junction has a resistance of the ....
    (a) Order of Ω (b) Order of $k\Omega$
    (c) Order of $M\Omega$ (d) None of these

2. When the *p-n* junction is forward bias, then
   (a) +ve terminal to *n* and –ve terminal to *p*
   (b) +ve terminal to *p* and '–ve' terminal to *n*
   (c) +ve terminal to *p* and *n*
   (d) None of these
3. The carrier voltage at a *p-n* junction for Si is,
   (a) 3.2 volt (b) 0.7 volt
   (c) 0.3 volt (d) Zero
4. A reverse bias *p-n* junction has resistance of the ....
   (a) Order of MΩ (b) Order of kΩ
   (c) Order of Ω (d) None of these
5. A *p-n* junction act as a .... switch.
   (a) Unidirectional (b) Bidirectional
   (c) Controlled (d) None of these
6. The barrier voltage at a *p-n* junction for Ge is about ....
   (a) 0.6 V (b) 0.3 V
   (c) 1.06 V (d) 1.07 V
7. When small amount of impurity is added in pure semiconductor, this process is called
   (a) Diffusion (b) Doping
   (c) Conduction (d) All of these
8. The majority carriers for *p*-type semiconductors are ....
   (a) Electrons (b) Holes
   (c) Impurity ions (d) None of these
9. When an acceptor impurity is added to a semiconductor, then it is called
   ....
   (a) *p*-type semiconductor (b) *n*-type semiconductor
   (c) Insulator (d) All of these
10. The impurities having five valence electrons is called ....
    (a) Acceptor impurities (b) Donor impurities
    (c) Both (a) and (b) (d) None of these
11. According to mass action law
    (a) $n^2 p^2 = n_i^4$ (b) $n^2 p^2 = n_i^2$
    (c) $np = n_i^2$ (d) None of these
12. Junction barrier helps the ...., to cross the junction.
    (a) Majority carriers (b) Minority carriers
    (c) Both (a) and (b) (d) None of these

13. .... is the minimum reverse voltage of *p-n* junction breakdown with sudden rise is reverse current.
    (a) Breakdown voltage
    (b) Knee voltage
    (c) PIV
    (d) None of these

14. Zener breakdown occurs in such type of diode for which doping of *n*-region and *p*-region is ....
    (a) Low
    (b) Medium
    (c) High
    (d) None of these

15. Thermal voltage is given by the equation
    (a) $V_T = \dfrac{T}{Kq}$
    (b) $V_T = \dfrac{Kq}{T}$
    (c) $V_T = \dfrac{kT}{q}$
    (d) None of these

16. At normal temperature, the thermal voltage for Si diode is
    (a) 0.02 V
    (b) 0.026 V
    (c) 0.32 V
    (d) 10.05 V

17. In the reverse bias, the diode current will be .... to the reverse saturation current.
    (a) Greater
    (b) Smaller
    (c) Equal
    (d) None of these

18. Typical value of forbidden energy gap in Ge is
    (a) 1 eV
    (b) 1.4 eV
    (c) 10 eV
    (d) 0.67 eV

19. A *p-n* junction is a transition
    (a) between two semiconductor materials having different electrical properties.
    (b) between two regions having different properties in *n*-type semiconductor material.
    (c) between two semiconductor materials having identical electrical properties
    (d) None of these

20. *P*-side of semiconductor diode is applied on a potential of 2.5 V where as *N*-side is applied on a potential of –5.V the diode shall
    (a) Conduct
    (b) Not conduct
    (c) Conduct partially
    (d) Breakdown

21. Cut-in voltage of a Si diode is
    (a) 0.1 V
    (b) 0.2 V
    (c) 0.6 V
    (d) None of these

22. In the forward bias conditions, the potential of *P*-side with respect to *N*-side is
    - (a) Positive
    - (b) Negative
    - (c) Either positive or negative
    - (d) None of these

23. In a *n*-type material, the electron is called the .... and the hole is ....
    - (a) Majority carrier, minority carrier
    - (b) Minority carrier, Majority carrier
    - (c) Both (a) and (b)
    - (d) None of these

24. In the reverse bias region, the reverse saturation current of a Si diode doubles for every ...., rise in temperature.
    - (a) 5°C
    - (b) 10°C
    - (c) 15°C
    - (d) None of these

25. A zener diode has a *sharp* breakdown voltage at reverse voltage. The above statement is (a) True (b) False.

26. A forward bias *p-n* junction has a resistance of the ....
    - (a) Order of Ω
    - (b) Order of kΩ
    - (c) Order of MΩ
    - (d) None of these

27. When the *p-n* junction is forward bias, then
    - (a) +ive terminal to *n* and −ve terminal to *p*
    - (b) +ve terminal to *p* and −ve terminal to *n*
    - (c) +ve terminal to *p* and +ve terminal to *n*
    - (d) None of these

28. The barrier voltage at a *p-n* junction for silicon is
    - (a) 3.2 volt
    - (b) 0.7 volt
    - (c) 0.3 volt
    - (d) Zero

29. A reverse bias *p-n* junction has resistance of the ....
    - (a) Order of MΩ
    - (b) Order of kΩ
    - (c) Order of Ω
    - (d) None of these

30. A *p-n* junction acts as a .... switch
    - (a) Unidirectional
    - (b) Bidirectional
    - (c) Controlled
    - (d) None of these

31. The barrier voltage at a *p-n* junction for Germanium is about ....
    - (a) 6V
    - (b) 0.3 V
    - (c) 0.6 V
    - (d) 0.7 V

32. Junction barrier helps the .... to cross the junction
    - (a) Majority carriers
    - (b) Minority carriers
    - (c) Both (a) and (b)
    - (d) None of these

33. .... is the minimum reverse voltage at which *p-n* junction breakdown with sudden rise in reverse current
    - (a) Breakdown voltage
    - (b) Knee voltage
    - (c) PIV
    - (d) None of these

34. Zener breakdown occurs in such type of diode for which doping of *n*-region and *p*-region is ....
    - (a) Low
    - (b) Medium
    - (c) High
    - (d) None of these

35. Thermal voltage is given by equation
    - (a) $V_T = \dfrac{T}{kq}$
    - (b) $V_T = \dfrac{kq}{T}$
    - (c) $V_T = \dfrac{kT}{q}$
    - (d) None of there

36. At normal temperature, the thermal voltage for silicon diode is
    - (a) 0.02 V
    - (b) 0.026 V
    - (c) 0.32 V
    - (d) 0.05 V

37. In the reverse bias, the diode current will be .... to the reverse saturation current.
    - (a) Greater
    - (b) Smaller
    - (c) Equal
    - (d) None of these

ANSWER

1. (a)	2. (b)	3. (b)	4. (a)	5. (a)	6. (b)	7. (b)
8. (b)	9. (a)	10. (b)	11. (c)	12. (b)	13. (a)	14. (c)
15. (c)	16. (b)	17. (c)	18. (d)	19. (b)	20. (a)	21. (c)
22. (a)	23. (a)	24. (b)	25. (b)	26. (a)	27. (b)	28. (b)
29. (b)	30. (a)	31. (b)	32. (b)	33. (a)	34. (c)	35. (c)
36. (b)	37. (c)					

Exercise

4.1. What is a *p-n* junction? Explain the formation of potential barrier in a *p-n* junction.

4.2. Draw and explain *V-I* characteristics of *p-n* junction diode.

4.3. Write short notes on:

    (a) Knee voltage   (b) Breakdown voltage

    (c) PIV

4.4. Explain the biasing of p-n junction diode.

4.5. Write short notes on:

    (a) Avalanche breakdown

    (b) Zener breakdown

    (c) Limitation in the operating conditions of p-n junction.

4.6. Derive the expression for conductivity and mobility for intrinsic semiconductor.

4.7. Define Static and Dynamic Forward Resistance for a p-n junction diode.

4.8. Explain the transition capacitance and diffusion capacitance of a p-n junction.

4.9. Derive the diode current equation.

4.10. A Germanium diode carries a current of 1 mA at room temperature, when a forward bias of 0.15V is applied. Estimate the reverse saturation current at room temperature. **(UPTU 2005-06) (Ans. I_S = 3.036 µA)**

4.11. Explain the behaviour of p-n junction at no bias, reverse bias and forward bias. Sketch the *VI* characteristic of p-n junction diode.**(UPTU 2006-07)**

4.12. What do you understand by average current, repetitive peak current, non-repetitive current, peak-inverse voltage and reverse saturation current?

(UPTU 2006-07)

4.13. What are the differences between diffusion and transition capacitance? How will you represent the capacitive effect of a practical diode on an ideal diode? **(UPTU 2006-07)**

4.14. Explain the working of a voltage doubler with neat diagram.

(UPTU 2006-07)

4.15. Define

    (a) Donor and acceptor impurities

    (b) Mobility and conductivity. **(UPTU 2007-08)**

4.16. Explain

    (a) How does the reverse saturation current of a p-n diode vary with temperature?

    (b) How does the diode voltage (at constant current) vary with temperature? **(UPTU 2007-08)**

4.17. Sketch the circuit for a full wave rectifier using two diodes only. Derive the expression for

    (a) DC current   (b) DC load voltage

    (c) DC diode voltage   (d) RMS current   **(UPTU 2007-08)**

4.18. Explain clipping and clamping circuits in detail.

4.19. Describe the difference between majority and minority carriers.

4.20. Define the reverse recovery time of a diode.

4.21. Draw the circuit diagram to show two methods of producing a negative output voltage from half wave rectifiers. Explain the circuit operations in brief. **(UPTU 2008-09)**

4.22. Explain the working of half wave and full wave bridge rectifier. What are the advantages of full wave rectifier? **(UPTU 2008-09)**

4.23. What is a filter? Explain different types of filters.

4.24. What is a ripple factor and load regulation?

4.25. What is a voltage multiplier?

4.26. What is a *p-n* junction? Explain the formation of potential barrier in a *p-n* junction.

4.27. Draw and explain the *VI* characteristics of *p-n* junction diode.

4.28. Write short notes on:

    (a) Knee voltage   (b) Breakdown voltage

    (c) Peak inverse voltage

4.29. Explain the biasing of *p-n* junction diode.

4.30. Write short notes on:

    (a) Avalanche breakdown   (b) Zener breakdown

    (c) Limitation in the operating conditions of *p-n* junction

Bipolar Junction Transistor   5

5.1 INTRODUCTION

The name transistor comes from the combination of the words **Transfer Resistor,** means transfer the resistance or impedance from the input to the output. So, we can say that transistor is a device that is used to amplify the weak signal, or transfer the impedance from input to output. During early days (1994-1947), the vacuum tube was the main electronic device, which were used in most of the electronic circuits. The invention of transistor was revolution in the field of Electronics which was invented by John Bareen and Walter H. Brattain of Bell Telephone Laboratories, USA in December 23, 1947. Now-a-days, transistor has become the heart of most Electronic Applications. **Bipolar transistors are the current amplifying or current regulating devices that control the amount of current flowing through them in proportion to the amount of raising current applied to their base terminal.**

Followings are the advantages of transistor over vacuum tubes.

(i)   Transistors are small in size and light in weight whereas the size of vaccum tube is large and bulky.

(ii) There is no need of heating element in the transistors due to which, transistors work quickly.

(iii) The power consumption of transistors are less due to which, the efficiency is high. Whereas in vaccum tubes, the power consumption is high due to the need of heating element. So, the efficiency is less.

(iv) Transistors operate at low voltage.

(v)   The durability of transistor is large and long lasting.

(vi) There is no damage after falling down and getting shock.

Though transistor is 60 years old, yet it is fast replacing vacuum tubes due to its applications as discussed earlier. In this chapter, we shall focus our attention on the various aspects of transistors and their increasing applications in the fast developing electronics industry.

5.2 CONSTRUCTION OF TRANSISTOR

Transistor is made by Silicon or Geramium. According to the construction, the transistor are of two types, when a *p*-type material is sandwiched between two *n*-type materials such type of transistor is called *n-p-n* transistor as shown in Fig. 5.1 (a) and second type is, when *n*-type material is called *p-n-p* transistor as shown in Fig. 5.1 (b).

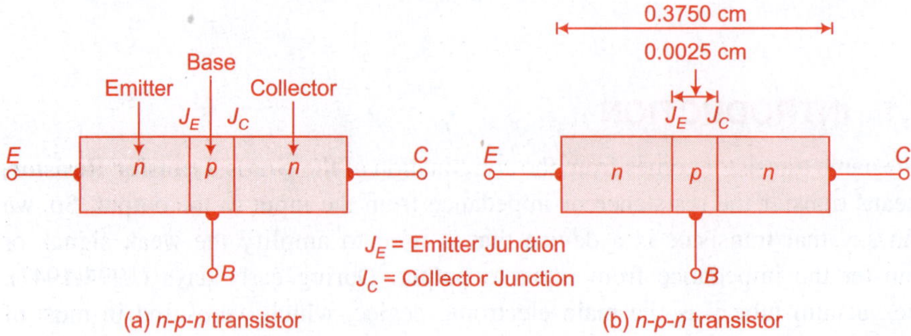

Fig. 5.1 *Types of Transistors*

As we see in Figs 5.1 (a) and (b), we can say that the transistor is a three terminal device. The terminals as named as 'Emitter', 'Base' and 'Collector'. It has two *p-n* junctions called as Emitter junction and Collector junction and denoted by J_E and J_C respectively. In other words, we can say that when two *p-n* junction diodes are connected back to back as shown in Figs. 5.2 (a) and (b), it form a transistor. In this way, transistor has three layer named as Emitter. Base and Collector. Sandwiched layer is called base and the side of it is called Emitter. The larger side is called Collector as shown in Figs. 5.1 (a) and (b).

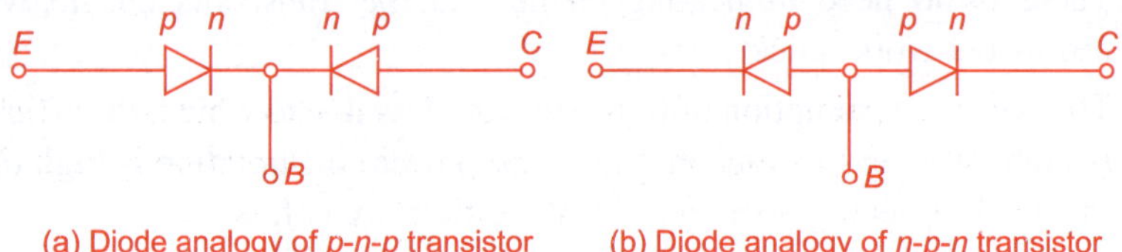

Fig. 5.2 *Diode analog of transistors*

Emitter: Emitter is highly closed. We call it as emit, because it emits majority carriers (electrons or holes).

Base: Base is lightly closed. It is a thin layer approximately 10^{-6}m. It receives the majority carriers from emitter and transistor and majority carriers from the collector.

Collector: The doping level of collector is less than the doping level of emitter and greater than the doping level of base. It collects the majority carriers, so it is called collector. The width of the collector is greater than both emitter and base.

5.3 BIASING OF TRANSISTOR

When there is no DC voltage in a transistor i.e., it is unbiased, then there is no current flow in the transistor because there is a potential barrier on both the junctions which prevents the flow of current, like a diode.

Basically, the biasing means to connect the DC supply across the junction to reduce the effect of barrier and working of transistors. As we have discussed earlier, that biasing is of two types, one is forward biasing and second one is reverse biasing for *p-n* junction because in case of transistor there are two *p-n* junction. So, we need two DC supply. We connect any biasing across two junction but if we use the transistor as an amplifier then the emitter junction is forward bias and collector junction is reverse bias as shown in Fig. 5.3. The resistance of emitter junction (forward biased) is very small as compared to collector junction (reversed biased). Therefore, forward bias applied to the emitter junction is generally very small whereas reverse bias on the collector junction is much higher.

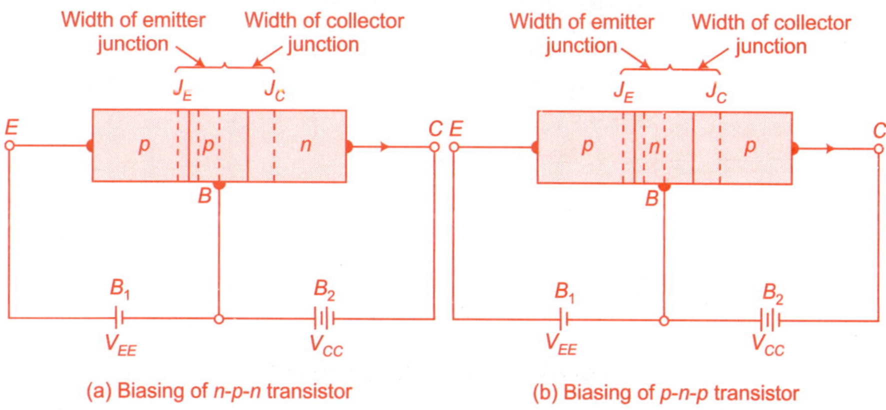

(a) Biasing of *n-p-n* transistor   (b) Biasing of *p-n-p* transistor

Fig. 5.3

The transistor has two *p-n* junctions i.e., it is like two diodes (connected back to back) as shown in Figs. 5.2 (a) and (b). The junction between emitter and base may be called emitter base diode or simply the emitter diode. The junction between the base and collector may be called collector-base diode or simply collector diode.

As we can see in Figs. 5.3 (a) and (b) the junction J_E is forward bias and the width of depletion region at J_E is reduced and the junction J_C is reversed biased and the width of depletion region at J_C is increased i.e., the resistance of forward bias junction is less whereas the resistance of reverse bias junction is high. When we use transistor as an amplifier, then we apply weak signal at base-emitter circuit and we get amplified output at collector base circuit.

5.4 WORKING OF *p-n-p* TRANSISTOR

(i) As we can see in Fig. 5.4, the emitter junction J_E is forward bias and the collector junction J_C reverse biased. Due to the forward bias, the width of the depletion region at J_E is reduced and due to the reverse bias, the width of depletion region at J_C will increase.

(ii) Under the forward bias condition of the emitter-base junction, the large number of majority carriers (Holes) cross the junction enters the base suppose n_H holes/see crosses the junctions J_E. At the same time, a very small number of electrons (suppose n_e electrons/sec) flow from the base to the emitter. Due to this, the current flow in the opposite direction of electrons and in the same direction of holes, due to minority carriers (electrons) and majority carriers (holes) as shown in Fig. 5.4 (a).

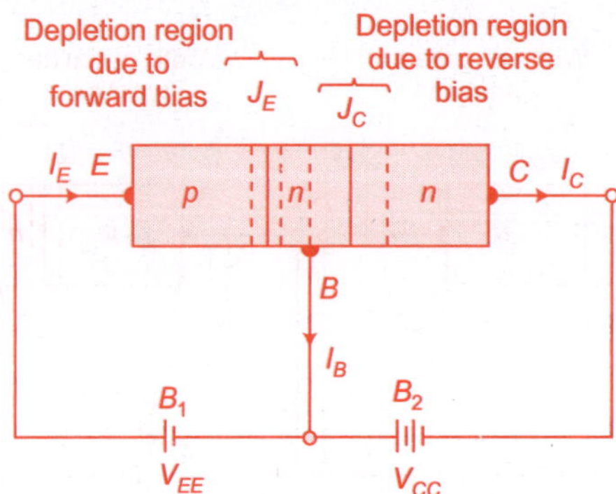

Fig. 5.4 *Biasing of p-n-p transistor*

These electrons (n_e) flowing from base to emitter reaches at the emitter and recombine with an equal number of holes in the emitter.

where, n_h = Number of holes i.e., (majority carrier) (coming from emitter to base),

n_e = Number of electrons i.e., (minority carrier) (coming from base to emitter).

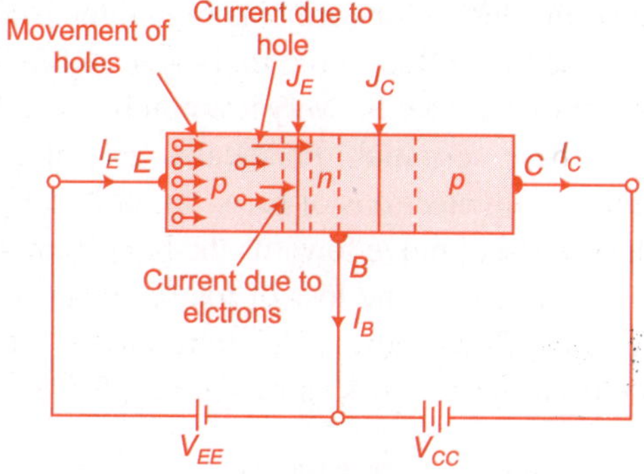

Fig. 5.4   *(a) Current conduction p-n-p transistor*

(iii) The loss of total number of holes $(n_h + n_e)$ in the emitter is made by the flow of an equal number of electrons from the emitter to the positive terminal of battery V_{EE}. These electrons are released from the covalent bonds of the crystal atoms in the emitter and equal number of holes are created. The flow of n_h holes/sec from the emitter to the base and n_e electrons/sec from the base to the emitter gives rise to the emitter current I_E (towards the emitter) as shown in Fig. 5.4 (b).

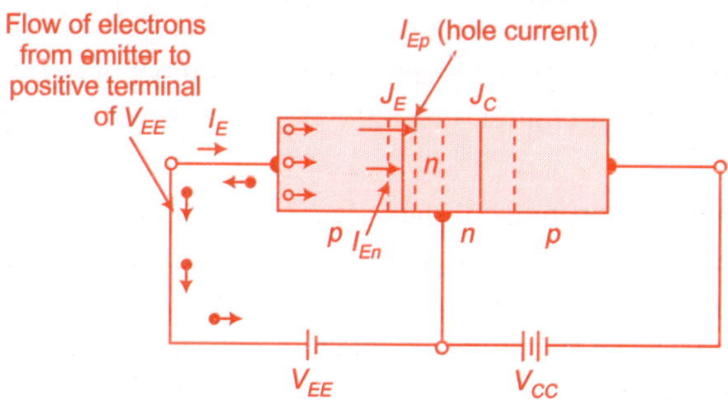

Fig. 5.4   *(b) Flow of emitter current in p-n-p transistor*

The emitter current $\boxed{I_E = I_{E_p} + I_{E_n}}$ (5.1)

where,

I_{E_p} = Emitter current due to majority carriers (holes)

I_{E_n} = Emitter current due to minority carriers (electrons).

Since emitter is highly closed as compared to base so $I_{E_p} >>> I_{E_n}$, therefore we ignore I_{E_n}.

Now, $\boxed{I_E = I_{E_p}}$ (5.2)

So, we can say that the emitter current I_E flows mainly due to flow of holes as shown in Fig. 5.4 (b).

(iv) In *p-n-p* transistor, due to forward bias of emitter junction, large number of holes reach the base. Due to the thin layer and light doping level, the conductivities of base is less. So, very less number of holes (let n_h holes/sec) move towards base terminal. Practically 2% of holes coming from emitter, combine with electrons of base. So, the total number of electrons [$(n_e + n_h)$ electrons/sec] move towards the base from negative terminal of battery V_{EE} to compensate the loss of total number of electrons $(n_e + h_h)$ per sec. Due to the flow of these electrons, base current I_B flows into the circuit (away from the base) as shown in Fig. 5.5 (c).

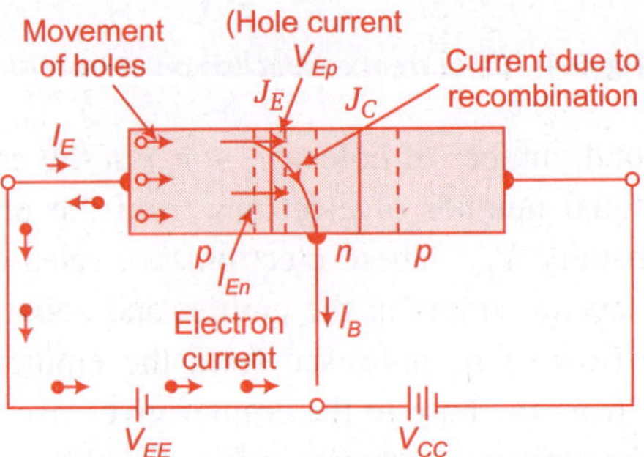

Fig. 5.4   *(c) Flow of base current*

I_B is small in amount (In μA) because large number of holes (approximately 98%) come from emitter, cross the base and move towards the collector. junction,

(v) In *p-n-p* transistor, most of the holes come from the emitter which crosses the base and move towards the collector junction because these holes are minority carriers for *n*-type base. So, they easily cross the collector junction because the collector junction barriers support minority carrier to cross the junction. And these holes reach the collector. After that, same number of electrons move towards the collector from the negative terminal of battery V_{cc} and neutralise the holes. Which are excessive in collector i.e., coming from the base to the collector.

These holes on reaching the collector are neutralised by an equal number of electrons flowing from the negative terminal of battery V_{CC} into the collector. At the same time, an equal number of electrons i.e., $(n_H - n_h)$ flow from the negative terminal of V_{EE} and reach the positive terminal of V_{cc} as shown in Fig. 5.5 (d). So, we can see that the collector current I_{C_p} (away from the collector) flows in collector due to the holes that reach the collector from the emitter. There is another component of collector current flow due to the minority carriers which are generated due to the thermal vibrations in collector junction; called reverse saturation current or leakage current and is represented by I_{CBO}.

So, the collector current $I_C = I_{Cp} + I_{CBO}$ where I_{Cp} = collector current, due to holes coming from the emitter.

$$I_{CB_0} = \text{Leakage current}$$

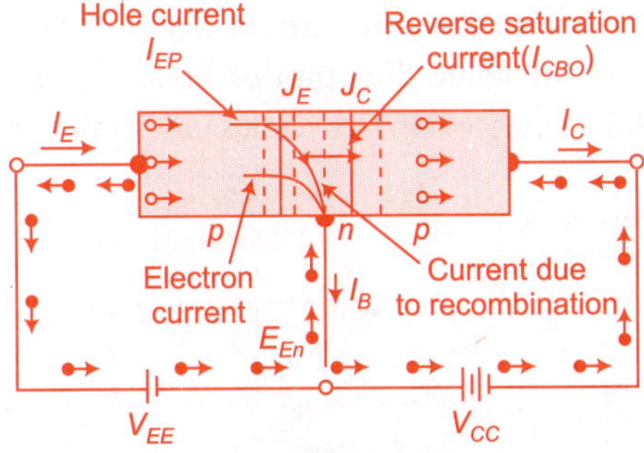

Fig. 5.4   *(d) Different components of current in p-n-p transistor*

It may be noted that current conduction within *p-n-p* transistor is by holes. However, in the external connecting wires, the current is conducted by electrons only as shown in Fig. 5.4 (d).

The emitter current I_E, base current I_B and collector current I_C can be represented by the following Eqn. for internal circuit.

$$I_E = I_B + I_C. \tag{5.3}$$

5.5   WORKING OF *n-p-n* TRANSISTOR

(i)   As we can see in Fig. 5.5, the emitter junction J_E is forward bias and the collector junction J_C is reverse biased. Due to the forward bias, the width of depletion region of J_E is reduced and due to the reverse bias, the width of depletion region of J_C will increase.

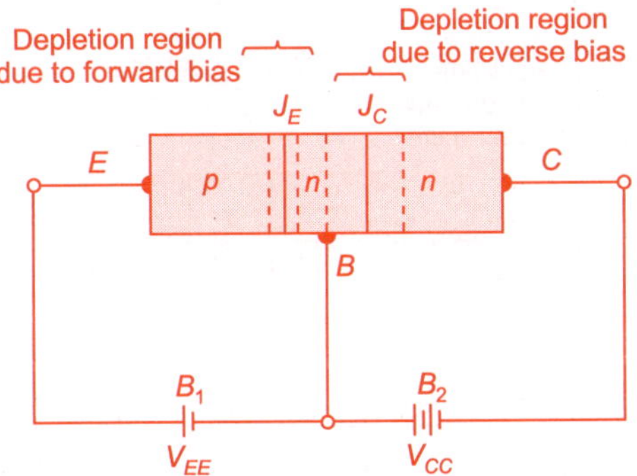

Fig. 5.5   *Biasing of n-p-n transistor*

(ii) Under the forward bias condition of the emitter junction, the large number of majority carriers (electrons) cross the junction and enter in the base suppose n_E electrons/sec crosses the junction J_E. At the same time, very small number of holes (suppose n_h holes/sec) flow from the base to the emitter. Due to this, the current flows in the opposite direction of electrons and in the same direction of holes due to the majority carriers (electrons) and minority carriers (holes) as shown in Fig. 5.5 (a).

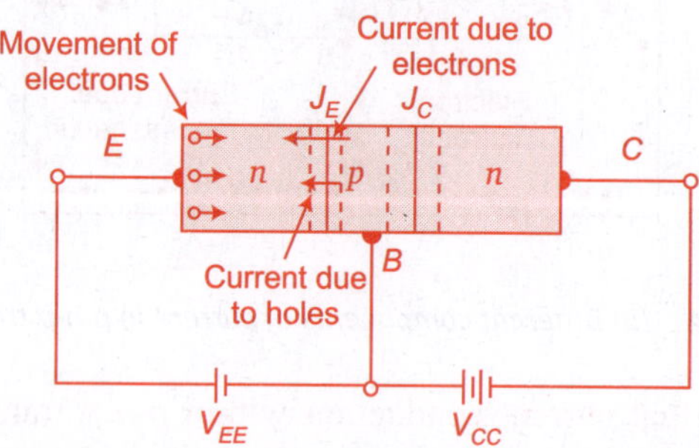

Fig. 5.5   *(a) Current conduction in n-p-n transistor*

These holes n_h flow from base to emitter reaching the emitter and recombine as equal number of electrons in the emitter.

where, n_E = Number of electrons i.e., majority carrier (coming from emitter to base)

n_h = Number of holes i.e., majority carrier (moving from base to emitter).

(iii) The loss of total number of electrons $(n_E + n_h)$ in the emitter is made up by the flow of an equal number of electrons from the negative terminal of battery V_{EE} to the emitter. The flow of n_E electrons/sec from the emitter to the base and n_h holes less from the base to the emitter rise to the emitter current I_E (away from the emitter) as shown in Fig. 5.5 (b).

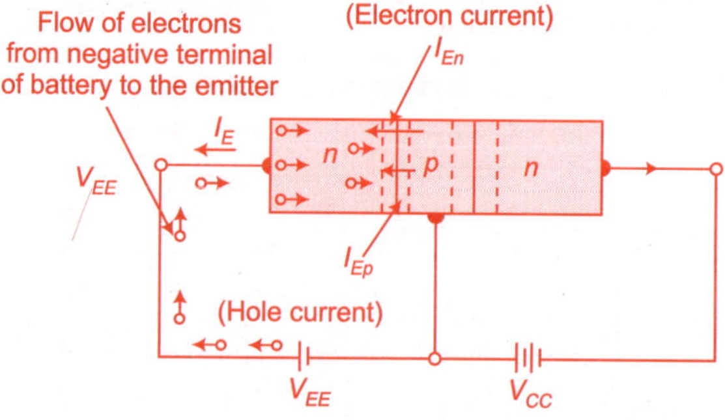

Fig. 5.5   *(b) Flow of emitter current in n-p-n transistor*

The emitter current,

$$I_E = I_{E_n} + I_{E_p}.$$ (5.4)

where, I_{E_n} = Emitter current due to majority carrier (electrons).

Since emitter is highly doped as compared to base so $I_{E_n} >>> I_{E_p}$, therefore we ignore I_{E_p}

Now,

$$I_E \simeq I_{E_p}.$$ (5.5)

So, we can say that the emitter current I_E flows mainly due to the flow of holes as shown in Fig. 5.5 (b).

(iv) In *n-p-n* transistor, due to forward bias of emitter junction a large number of electrons reach the base. Due to the thin layer and light doping level, the conductivity of base is less. So, very small number of holes (let n_h holes less) move towards the base terminal. Practically 2% of electrons coming from emitter, combine with the holes of base. So, the total number of electrons [$(n_h + n_e)$ electrons/sec] move towards the positive terminal of the battery V_{EE} from base to compensate the loss of total number of holes $(n_e + n_h)$ holes/sec. Due to the flow of these holes, base current I_B flows into the circuit (towards the base) as shown is Fig. 5.5 (c).

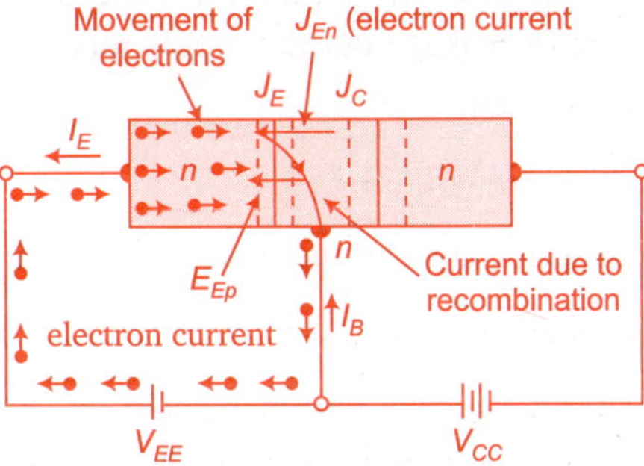

Fig. 5.5   *(c) Flow of base current*

I_B is small in amount (in µA) because large number of electrons (approximately 98%) coming from emitter cross the base and move towards the collector junction.

(v) In *n-p-n* transistor, most of the electrons coming from the emitter cross the base and move towards the collector junction because these electrons are minority carriers for *p*-type base. So, they easily cross the collector junction. As you know that collector junction barriers support the minority carriers to cross the junction. These electrons reach the collector. The same number of electrons move away from the collector towards the positive terminal of battery V_{CC}. These electrons are excess in collector i.e., coming from the base to the collector. These electrons on reaching the collector move towards the positive terminal of the battery V_{CC}. At the same time an equal number of holes $(n_E - n_e)$ flow from the negative terminal of battery V_{CC} and reach the positive terminal of battery V_{BE}, as shown in Fig. 5.5 (d).

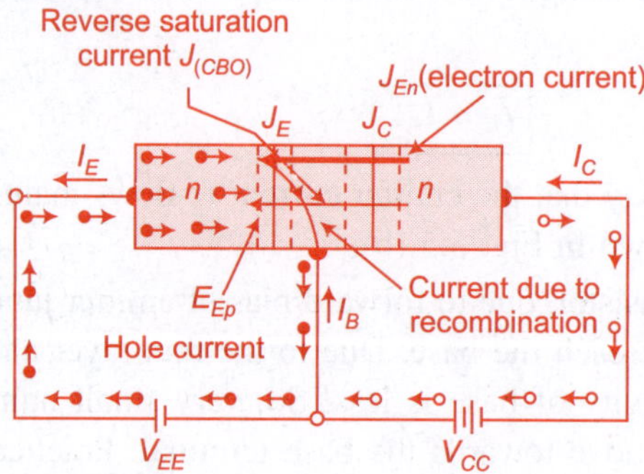

Fig. 5.5   *(d) Different components of current in n-p-n transistor*

So, we can see that the collector current I_{Cn} (towards the collector) flow in collector due to the electrons reach in the collector. There is another component of collector current flow due to the minority carriers generated due to the thermal vibration in collector junction called reverse saturation current or leakage current and is represented by I_{CBO}.

So, the collector current,

$$I_C = I_{Cn} + I_{CBO} \tag{5.6}$$

where, I_{Cn} = Collector current due to electrons coming from the emitter.

I_{CBO} = Leakage current.

The emitter current I_E, base current I_B and collector current I_C can be represented by the following Eqn. as shown in Eqn. (5.3)

$$I_E = I_B + I_C.$$

5.6   SOME IMPORTANT TERMS RELATED TO THE FLOW OF CURRENT IN A TRANSISTOR

(i) **Emitter Injection Efficiency**   A large number of majority carriers cross the emitter junction and reach into base, but the majority carriers of emitter are much greater than the majority carriers of base. So, we can say that the emitter current at the emitter junction J_E is due to the majority carriers of emitter and total emitter current. It is represented by γ.

Now,
$$\gamma = \frac{\text{Emitter current at the emitter junction } J_E \text{ due to the majority carriers of emitter}}{\text{Total emitter current}}$$

For *n-p-n* transistor – (the majority carriers are electrons)

$$\gamma = \frac{I_{En}}{I_{Ep} + I_{En}}$$

By the Eqn. (5.1), we have,

$$\boxed{\gamma = \frac{I_{En}}{I_E}} \tag{5.7}$$

where, γ = Emitter Injection Efficiency

I_{En} = Emitter current at J_E, due to electrons

I_E = Total emitter current.

For *p-n-p* transistor (the majority carriers are holes)

$$\gamma = \frac{I_{E_p}}{I_{E_p} + I_{En}}$$

By the Eqn. (5.3), we have,

$$\boxed{\gamma = \frac{I_{E_p}}{I_E}} \tag{5.8}$$

where, I_{E_p} = Emitter current at J_E due to holes

I_E = Total emitter current.

Practically, the value of γ is 0.98 or 0.99 i.e., close to unity. High emitter injection efficiency is achieved by heavier doping of the emitter than the base.

(ii) **Base transport factor:** We know about the percentage of majority carriers, which enter into emitter junction J_E, cross the base and reach the collector junction. It is denoted by β'. In other words, we can state that base transient factor is the ratio of collector current due to the majority carrier, that cross the base and reach collector junction and emitter current due to the majority carriers which cross the emitter junction J_E.

Now, $\beta' = \dfrac{\text{Collector current due to the majority carriers at collector junction } J_C \text{ and coming from emitter junction}}{\text{Emitter current due to the majority carriers which cross emitter junction } J_E}$

For *n-p-n* transistor,

$$\boxed{\beta' = \frac{I_{Cn}}{I_{En}}} \tag{5.9}$$

where, β' = Base transport factor

I_{Cp} = Collector current due to holes at J_C coming from J_E

I_{E_p} = Emitter current due to holes which cross emitter junction J_E

$$\boxed{\beta' = \frac{I_{Cp}}{I_{Ep}}} \tag{5.10}$$

where, β' = Base transport factor

I_{Cp} = Collector current due to holes at J_C coming from J_E

I_{E_p} = Emitter current due to holes which cross emitter junction J_E.

To achieve I_E close to I_E, we make the base transport factor β' high by making the base width smaller than a diffusion length by which we can reduce the loss of electrons due to recombination and by the low level doping of the base. Generally, the value of β' is approximately 0.995.

(iii) **Current amplification factor:**   Current amplification factor is the ratio of collector current due to the carriers coming from emitter and total emitter current. It is denoted by α_{dc}.

$$\alpha_{dc} = \frac{\text{Collector current due to the carriers coming from emitter}}{\text{Total emitter current}}$$

We know that collector current has two components.

(i) Part of emitter current that reaches the collector

(ii) Collector base saturation current or leakage current I_{CBO}.

By the Eqns. (5.5) and (5.6), we have,

$$I_{Cn} = I_{Cp} = I_C I_{CBO} \tag{5.11}$$

Now, $\alpha_{dc} = \dfrac{I_{Cn}}{I_E}$ (for *n-p-n*) $= \dfrac{I_{Cp}}{I_E}$ (for *p-n-p*)

By Eqn. (5.11) we have,

$$\alpha_{dc} = \frac{I_C - I_{CBO}}{I_E}$$

$$\alpha_{dc} I_E = I_C - I_{CBO}$$

$$\boxed{I_C = \alpha_{dc} I_E + I_{CBO}.} \tag{5.12}$$

If I_{CBO} is negligible then

$$I_C = \alpha_{dc} I_E$$
$$\alpha_{dc} = \frac{I_C}{I_E}.$$

(5.13)

So, we can state from Eqn. (5.13) that the ratio of I_C and I_E is called current amplification factor α_{dc}, if leakage current I_{CBO} is negligible.

For AC situation, α_{ac} is defined as the ratio of change in collector current (ΔI_C) to the change in emitter current (ΔI_E) which is known as common base, short circuit, current-amplification factor.

$$\alpha_{ac} = \frac{\Delta I_C}{\Delta I_E}\bigg|_{V_{CB} = \text{constant}}$$

(iv) **Relation between α_{dc}, β' and γ.** By the Eqn. (5.13), we have,

$$\alpha_{dc} = \frac{I_{Cn}}{I_E}.$$

(5.14)

From the Eqn. (5.9), substitute the value of I_{Cn} in Eqn. (5.14), we have,

$$\alpha_{dc} = \frac{\beta' I_{En}}{I_e}$$

(5.15)

Now, substitute the value of I_E from Eqn. (5.7) in Eqn. (5.15), we have

$$\alpha_{dc} = \frac{\beta' I_{En}}{\dfrac{I_{En}}{\gamma}} = \frac{\beta' \gamma I_{En}}{I_{En}}$$

$$\boxed{\alpha_{dc} = \beta' \gamma.}$$

(5.16)

5.7 EFFECT OF TEMPERATURE ON LEAKAGE CURRENT

As you know that the leakage current exists in reverse bias collector junction J_C, due to the flow of minority carriers. Since these minority carriers generate the thermal vibrations, therefore the leakage current depends upon temperature. Practically this current increases 7% per °C, twice increasing the temperature, at 10°C.

5.8 TRANSISTOR CONFIGURATIONS

Generally transistors are used as amplifiers. We give input to the amplifier and it gives amplified output, so we need two terminals for input and two terminals for output. Total four terminals are needed in amplifier as shown in Fig. 5.6, but transistor has three

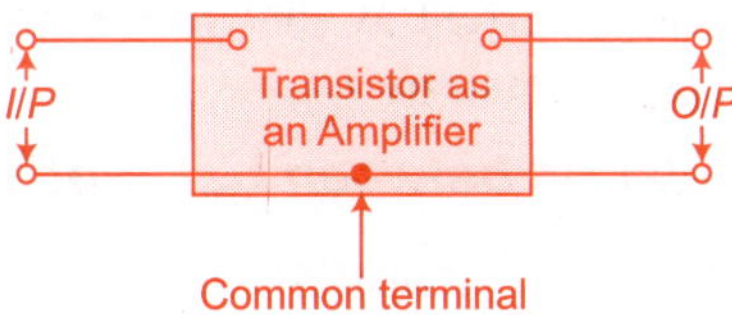

Fig. 5.6 *Transistor as an amplifier*

terminals. So, it is necessary to take one terminal as common in input and output side so we get four terminals effectively.

On the basis of common terminal of the transistor, it has three configurations.

(i) Common emitter configuration

(ii) Common base configuration

(iii) Common collector configuration

The common emitter configuration in shown in Figs. 5.7 (a) and (b) for *n-p-n* and *p-n-p* transistor respectively. This type of configuration is used widely. In this type of configuration, emitter is common reference to input and output. Input is given between base and emitter terminal and output is taken between emitter and collector. So, it is called Common emitter configuration.

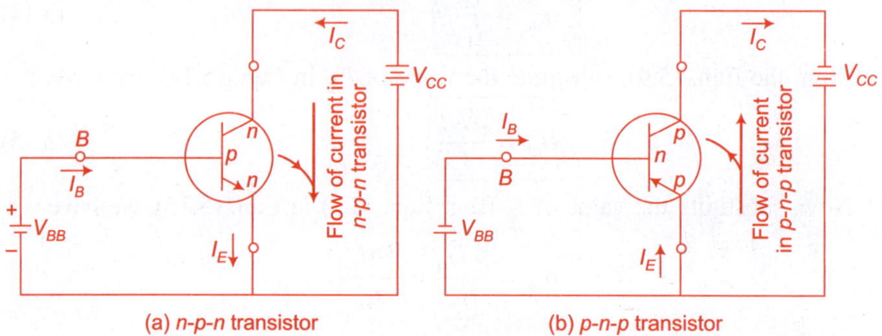

Fig. 5.7 *Common emitter configuration*

We are using common emitter configuration, but the current relation developed earlier for the common base as in Eqn. (5.6) is still applicable.

i.e.
$$I_E = I_C + I_B.$$

Collector-emitter leakage current

In common emitter configuration, if base is open circuit and collector junction is reverse biased, then the current flow between collector and emitter is called leakage current as shown in Fig. 5.8 and is represented by I_{CEO}.

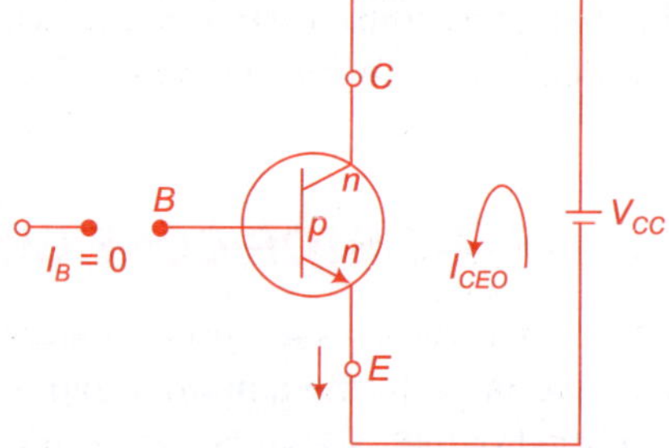

Fig. 5.8 *Collector-emitter leakage current*

Relation between I_C, I_B and I_{CEO}

When the transistor is connected in common emitter configuration and emitter junction in forward bias and collector junction in reverse biased as shown in Fig. 5.8, then the collector current has two components.

1. The part of emitter current that reaches up to collector because some part (approximately 2%) of majority carriers coming from emitter move towards base terminal and recombine with holes (for *n-p-n)* or electrons (for *p-n-p).*

2. Leakage current or emitter-collector current I_{CEO}.

Now,

The part of emitter current that reaches up to collector $= I_C - I_{CEO}$

So, the base current amplification factor

$$\beta_{dc} = \frac{\text{Part of emitter current that reaches up to collector}}{\text{Base current}}$$

$$\beta_{dc} = \frac{I_C - I_{CEO}}{I_B}$$

$$\beta_{dc} I_B = I_C - I_{CEO} \tag{5.17}$$

$$I_C = \beta_{dc} I_B + I_{CEO}$$

if $I_{CEO} <<< \beta_{dc} I_B$, we can ignore it, then

$$\boxed{\beta_{dc} = \frac{I_C}{I_B}.} \tag{5.18}$$

Relation between α_{dc} *and* β_{dc}

We know from the Eqn. (5.6) that,

$$I_E = I_C + I_B \tag{5.19}$$

dividing Eqn. (5.19) by I_C, we get,

$$\frac{I_E}{I_C} = 1 + \frac{I_B}{I_C}. \tag{5.20}$$

From Eqns. (5.18) and (5.13), we know that,

$$\beta_{dc} = \frac{I_C}{I_B} \Rightarrow \frac{1}{\beta_{dc}} = \frac{I_B}{I_C}$$

and,

$$\alpha_{dc} = \frac{I_C}{I_E} \Rightarrow \frac{1}{\alpha_{dc}} = \frac{I_E}{I_C}$$

Putting the above value in Eqn. (5.20) we get,

$$\frac{1}{\alpha_{dc}} = 1 + \frac{1}{\beta_{dc}}$$

$$\frac{1}{\alpha_{dc}} = \frac{\beta_{dc} + 1}{\beta_{dc}}$$

$$\beta_{dc} = \alpha_{dc}\beta_{dc} + \alpha_{dc}$$

$$\beta_{dc} - \alpha_{dc}\beta_{dc} = \alpha_{dc}$$

$$\beta_{de}(1 - \alpha_{dc}) = \alpha_{dc}$$

$$\boxed{\beta_{dc} = \frac{\alpha_{dc}}{1 - \alpha_{dc}}.} \qquad (5.21)$$

From the Eqn. (5.21), it is clear that when 'a' approaches unity, 'β' approaches infinity. In other words, the current gain in common emitter configuration is very high. Due to this reason, the circuit used is about 90 to 95% of all transistor applications.

In most transistors, approximately 2% of emitter current flows as the base current, therefore the value of β is generally greater than 50. For practical devices, the level of β typically ranges from 50 to over 400.

Situation, β_{ac} is defined as the ratio of change in collector current (ΔI_C) to the change in base current (ΔI_B) is known as Base Current Amplification factor or forward current amplification factor.

$$\beta_{ac} = \left.\frac{\Delta I_C}{\Delta I_B}\right|_{V_{CE} = \text{constant}}. \qquad (5.22)$$

5.8.2   Common Base Configuration

The common base configuration is shown in Figs. 5.9 (a) and (b) for *n-p-n* and *p-n-p* transistor respectively. In this type of configuration, base is common or reference to input and output. Input is given between base and emitter terminal

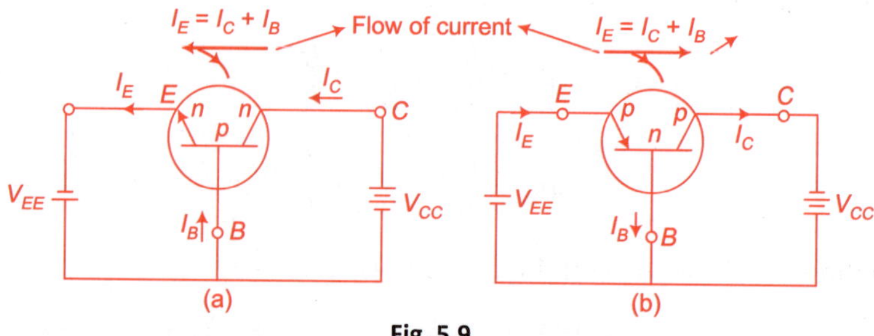

Fig. 5.9

and output is taken between base and collector terminal. So, it is called common base configuration. The current relation as by Eqn. (5.6).

$$I_E = I_C + I_B.$$

Collector to base leakage current

In common base configuration, if emitter is open circuit and collector junction is reverse biased, then the current flow between collector and base is called leakage current as shown in Fig. 5.9 (c). It is represented by I_{CBO}. This current is due to the flow of minority carriers.

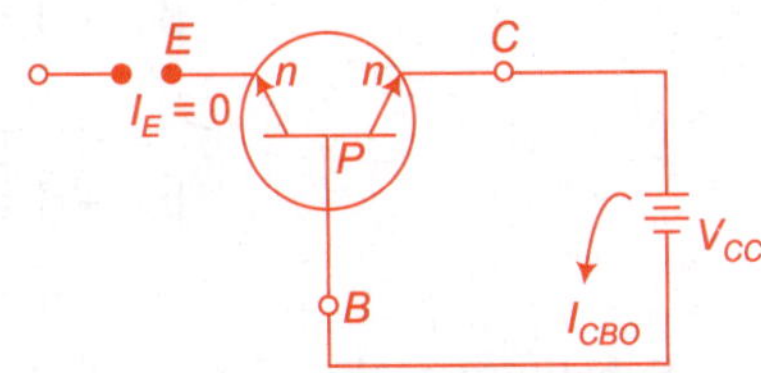

Fig. 5.9   *(c) Collector base leakage current*

Relation between I_C, I_B and I_{CBO}

When the transistor is connected in common base configuration and the emitter junction is forward bias and collector junction is reverse biased as shown in Fig. 5.9, then the collector current has two components.

(i) The part of emitter current that reaches up to collector because some part (approximately 2%) of majority carriers coming from emitter move towards base terminal and recombine with holes (for *n-p-n*) or electrons (for *p-n-p*).

(ii) Leakage current or collector to base current I_{CBO}.

We know that,

$$I_E = I_B + I_C \qquad \text{by Eqn. (5.6)} \qquad (5.23)$$

and,
$$I_C = \alpha_{dc} I_E + I_{CBO} \qquad \text{by Eqn. (5.12)} \qquad (5.24)$$

Put the value of I_E from Eqn. (5.23) to Eqn. (5.24).

$$I_C = \alpha_{dc} (I_B + I_C) + I_{CBO}$$

$$I_C = \alpha_{dc} I_B + \alpha_{dc} I_C + I_{CBO}$$

$$I_C - \alpha_{dc} I_C = \alpha_{dc} I_B + I_{CBO}$$

$$(1 - \alpha_{dc}) I_C = \alpha_{dc} I_B + I_{CBO}$$

$$I_C = \frac{\alpha_{dc}}{1 - \alpha_{dc}} I_B + \frac{1}{1 - \alpha_{dc}} I_{CBO} \qquad (5.25)$$

We know that by Eqn. (5.21)

$$\boxed{\beta_{dc} = \frac{\alpha_{dc}}{1 - \alpha_{dc}}.} \qquad (5.26)$$

Adding 1 on both sides of Eqn. (5.26),

$$\beta_{dc} + 1 = \frac{\alpha_{dc}}{1 - \alpha_{dc}} + 1$$

$$\beta_{dc} + 1 = \frac{\alpha_{dc} + 1 - \alpha_{dc}}{1 - \alpha_{dc}}$$

$$\beta_{dc} + 1 = \frac{1}{1 - \alpha_{dc}}. \tag{5.27}$$

Substituting the value of $\dfrac{1}{1 - \alpha_{dc}}$ and $\dfrac{\alpha_{dc}}{1 - \alpha_{dc}}$ in Eqn. (5.25) from Eqns. (5.26) and (5.27), we get,

$$I_C = \beta_{dc}I_B + (\beta_{dc} + 1)I_{CBO}. \tag{5.28}$$

Relation between I_{CBO} and I_{CEO}

We know that from Eqns. (5.28) and (5.17)

$$I_C = \beta_{dc}I_B + (\beta_{dc} + 1)I_{CBO} \tag{5.29}$$

$$I_C = \beta_{dc}I_B + I_{CEO} \tag{5.30}$$

Comparing Eqns. (5.29) and (5.30), we get,

$$I_{CEO} = (\beta_{dc} + 1)\, I_{CBO}. \tag{5.31}$$

5.8.3 Common Collector Configuration

The common collector configuration is shown in Figs. 5.10 (a) and (b) for *n-p-n* and *p-n-p* transistor respectively. In this type of configuration, collector is common and reference to input and output. Input is given between base and collector terminal and output is taken between emitter and collector. So, it is called common collector configuration.

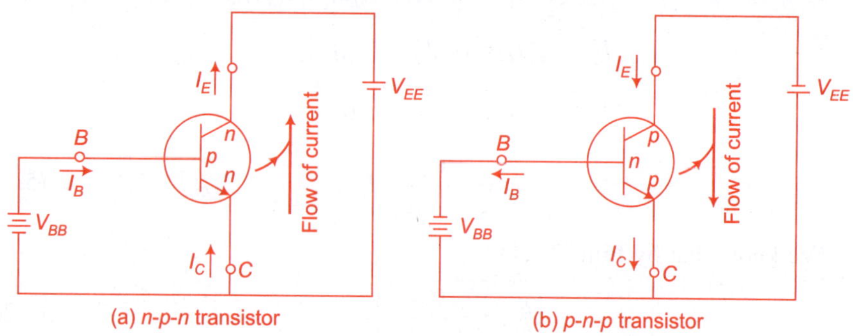

Fig. 5.10 *Common collector configuration*

Expression for emitter current in terms of base current and leakage current is common collector configuration

Current Eqn. of transistor by Eqn. (5.14)

$$I_C = \alpha_{dc}I_E + I_{CBO} \tag{5.32}$$

and,
$$I_E = I_B + I_C \qquad \text{by Eqn. (5.7)} \tag{5.33}$$

Substituting the value of I_C from Eqn. (5.33) to Eqn. (5.32)

$$I_E - I_B = \alpha_{dc}I_E + I_{CBO}$$

$$I_E - \alpha_{dc}I_E = I_B + I_{CBO}$$

$$I_E\,(1 - \alpha_{dc}) = I_B + I_{CBO}$$

$$I_E = \frac{I_B}{1 - \alpha_{dc}} + \frac{I_{CBO}}{1 - \alpha_{dc}}$$

$\because$ From Eqn. (5.27), we know that,

$$\frac{1}{1 - \alpha_{dc}} = \beta_{dc} + 1$$

Now,
$$\boxed{I_E = (\beta_{dc} + 1)I_B + (\beta_{dc} + 1)\,I_{CBO}.} \tag{5.31a}$$

5.9 STANDARD SIGN CONVENTION FOR TRANSISTOR CURRENTS

The current flowing towards (inside) the transistor are considered to be positive and the current flowing away (outside) from the transistor is considered to be negative according to Standard Sign Convention.

For example, in *n-p-n* transistor, I_E flows outside the transistor but I_B and I_C flows inside transistor, so I_E is considered negative while I_C and I_B are considered positive. Where as in *p-n-p* transistor I_E is positive and I_C and I_B are negative.

Example 5.8.1   In a common base configuration, the emitter current is 1.5 mA. If the emitter circuit is open, the collector current is 60 μA. Find the total collector current. Given that $\alpha_{dc} = 0.95$.

Solution   Given that $I_E = 1.5$ mA

$$I_{CBO} = 60\mu A \text{ and } \alpha_{dc} = 0.95$$

We know that by Eqn. (5.14)

$$I_C = \alpha_{dc}I_E + I_{CBO} = [(0.95 \times 1.5) + (60 \times 10^{-3})]\ \text{mA}$$

$$I_C = [1.425 + 0.060]\ \text{mA}$$

$$\boxed{I_C = 1.49\ \text{mA.}} \qquad \textbf{(Ans.)}$$

Example 5.9.2   In a common emitter configuration, $I_B = 0.03$ mA, $I_C = 0.95$ mA. Calculate the value of I_E.

Solution   We know that by Eqn. (5.6)

$$I_E = I_C + I_B$$

$$I_E = 0.95 + 0.03$$

$$\boxed{I_E = 0.98 \text{ mA.}}$$ **(Ans.)**

Example 5.9.3   In a common base configuration, $I_C = 0.96$ mA and $I_B = 0.04$ mA. Find the value of α_{dc}.

Solution   We know that,

$$I_E = I_C + I_B \text{ by Eqn. (5.6)}$$

$$I_E = 0.96 + 0.04$$

$$\boxed{I_E = 1 \text{ mA.}}$$ **(Ans.)**

Again we know by the Eqn. (5.13)

$$\alpha_{dc} = \frac{I_C}{I_E} = \frac{0.96}{1}$$

$$\boxed{\alpha_{dc} = 0.96.}$$ **(Ans.)**

Example 5.9.4   In a common base connection, $\alpha_{dc} = 0.95$ the voltage drop across 2.5 kΩ resistance which is connected in the collector is 2.5 V Find the base current.

Solution   Given that, $\alpha_{dc} = 0.95$

$$V_R = 2.5 \text{ V}, \quad R = 2.5 \text{ k}\Omega$$

$$I_C = \frac{V_R}{R} = \frac{2.5 \text{ V}}{2.5 \text{ k}\Omega}$$

$$\boxed{I_C = 1 \text{ mA.}}$$ **(Ans.)**

We know that $\alpha_{dc} = \dfrac{I_C}{I_E}$ by Eqn. (5.13)

$$I_E = 1.05 \text{ mA}$$

By the relation of Eqn. (5.6)

$$I_E = I_B + I_C$$

$$I_B = I_E - I_C$$

$$I_B = 1.05 - 1$$

$$\boxed{I_B = 0.05 \text{ mA.}}$$ **(Ans.)**

Example 5.9.5 For a transistor β_{dc} is 50 and $I_B = 25\mu A$. Calculate emitter current.

Solution Given that,

$$\beta_{dc} = 50, \ I_B = 25\mu A$$

We know that,

$$\beta_{dc} = \frac{I_C}{I_B} \text{ by Eqn. (5.18)}$$

$$I_C = \beta_{dc}I_B = 50 \times 25 \ \mu A$$

$$I_C = 1.25 \text{ mA.}$$

Now using the Eqn. (5.6)

$$I_E = I_C + I_B$$

$$I_E = 1.25 + 0.025$$

$$I_E = 1.275 \text{ mA}$$

i.e.,
$$\boxed{I_E = 1.28 \text{ mA.}}$$
(Ans.)

Example 5.9.6 In a common emitter configuration, β is 49. The base current is 200 μA and emitter current is 10 mA as shown in Fig. Ex. 5.7.6. Determine the collector current using α and β.

Solution By the Eqn. (5.26), we know that,

$$\beta_{dc} = \frac{a_{dc}}{1 - a_{dc}}$$

$$\beta_{dc} - \beta_{dc}\alpha_{dc} = \alpha_{dc}$$

$$\beta_{dc} = \alpha_{dc} + \beta_{dc} \ \alpha_{dc}$$

$$\beta_{dc} = \alpha_{dc} (\beta_{dc} + 1)$$

$$\alpha_{dc} = \frac{\beta_{dc}}{1 - \beta_{dc}} = \frac{49}{1 + 49} = \frac{49}{50}$$

$$\boxed{\alpha_{dc} = 0.98.}$$
(Ans.)

The value of I_C by Eqns. (5.13) and (5.18)

$$\alpha_{dc} = \frac{I_C}{I_E} \Rightarrow I_C = \alpha_{dc}I_E$$

$$I_C = 0.98 \times 10 = 9.8 \text{ mA}$$

$$\boxed{I_C = 9.8 \text{ mA.}} \quad \text{(Ans.)}$$

$$\beta_{dc} = \frac{I_C}{I_B} \Rightarrow I_C = \beta_{dc}I_B$$

Fig. Ex. 5.9.6

$$I_C = (49 \times 200 \times 10^{-6}) \text{ A}$$

$$\boxed{I_C = 9.8 \text{ mA.}}$$ **(Ans.)**

Example 5.9.7 Find the value of β_{dc} if $\alpha_{dc} = 0.95$.

Solution We know that by Eqn. (5.26)

$$\beta_{dc} = \frac{\alpha_{dc}}{1 - \alpha_{dc}}$$

$$= \frac{0.95}{1 - 0.95} = \frac{0.95}{0.05}$$

$$\boxed{\beta_{dc} = 19.}$$ **(Ans.)**

Example 5.9.8 In a common emitter configuration β_{dc} is given 50, base current is 0.05 mA and the leakage current is 20 µA. Calculate the collector current I_C.

Solution Given that,

$$\beta_{dc} = 50$$

$$I_B = 0.05 \text{ mA}$$

$$I_{CEO} = 20 \mu A$$

$$I_C = ?$$

We know by the Eqn. (5.17)

$$I_C = \beta_{dc} I_B + I_{CBO}$$

$$= (50 \times 0.05 + 20 \times 10^{-3}) \text{ mA}$$

$$I_C = (2.5 + 0.02) \text{ mA}$$

$$\boxed{I_C = 2.52 \text{ mA.}}$$ **(Ans.)**

5.10 TRANSISTOR CHARACTERISTICS

In this section, we will study the following characteristics of transistor for each configuration.

(i) **Input characteristics:** Input characteristic of a transistor is a curve between input voltage and input current at different constant values of output voltage. It is for driving point or input parameters.

) **Output characteristics:** Output characteristic of a transistor is a curve between output voltage and output current at different constant values of input current. It is for output parameters.

5.10.1 Characteristics of Common Base Configuration

The common base configuration is shown in Fig. 5.11. For the characteristics of transistor, we should know the input and output parameters for the transistor as we have discussed earlier. The input and output parameters are as follows:

   (i) Input voltage V_{BE} (ii) Input current I_E

 (iii) Output voltage V_{CB} (iv) Output current I_C

Input characteristics

These characteristics are drawn between input voltage V_{BE} and input I_E, at constant values of output voltage V_{CB}.

The circuit diagram for the input characteristics is shown in Fig. 5.12 (b). The input characteristics of *CB* mode are shown in Fig. 5.13. When we study the input characteristics of common base transistor, then we can state the following important points related to this.

Fig. 5.11 *Common base configuration*

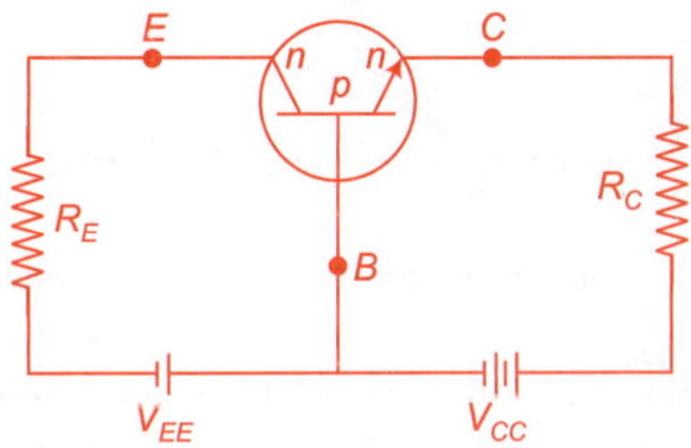

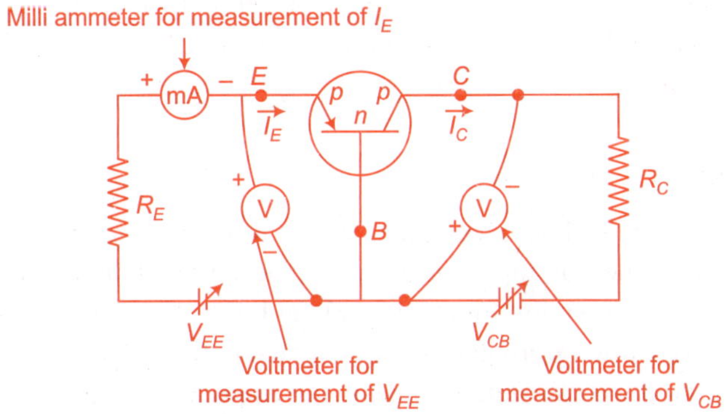

Fig. 5.12 *Circuit diagram for input characteristic of CB configuration*

   (i) The input characteristics of common base configuration is similar to the forward characteristics of diode because the emitter junction is forward bias.

(ii) If we increase the V_{CB}. i.e., the reverse voltage of collector base junction is increased, due to this, the input characteristics curve shift inwards i.e., towards emitter current axis as shown in Fig. 5.13. This effect is called Early effect. The early effect occurs due to the increase in reverse voltage V_{CB}, and due to the increase in V_{CB}, the emitter current will increase rapidly because the majority carriers (holes in *p-n-p* and electron in *n-p-n*) which attract towards collector are increased.

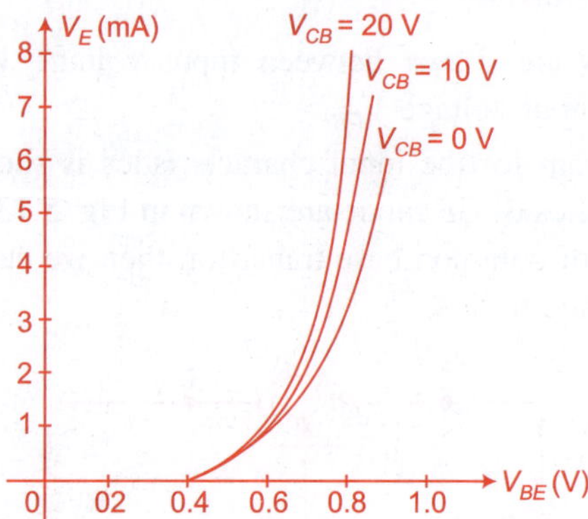

Fig. 5.13 *Input characteristics of transistor in CB mode*

When the reverse voltage at collector junction is increased, due to this, the width of depletion region is increased and the width of base will reduce. The reduction in base width due to the increase in collector base voltage is called base width modulation. Due to the base width modulation, the recombination process in base will reduce, so the emitter current will increase. This effect is called Early effect.

Output characteristics

This characteristic is drawn between output voltage V_{CB} and output current I_C, at constant values of input current I_E. The circuit diagram for the output characteristics is shown in Fig. 5.14 (a) and the characteristic is shown in Fig. 5.14 (b). When we study the output characteristic of *CB* configuration as in Fig. 5.14 (b), then we can state the following important related points:

(i) If the collector junction is reverse biased, then the collector current is constant at constant emitter current, i.e., I_C independent on V_{CB}. This is due to the most of the majority carriers coming from emitter which reach the collector, so that the collector current is not zero, still $V_{CB} = 0$. (Remember it that increasing in V_{CB} the I_C will also slightly increase due to early effect which is negligible).

(ii) I_C is always less than I_E because there is a loss of carriers in base.

(iii) When $I_E = 0$, then $I_C \neq 0$ due to the reverse bias of collector junction the leakage current I_{CBO} flow.

Three regions of output characteristics: The output characteristics of transistor shows three regions.

(i) **Active region** The region of output characteristics at which the collector junction is reverse biased and emitter junction is forward bias is called Active region. In common base configuration, the active region is in right side of $V_{CB} = 0$ and above from $I_E = 0$ as shown in Fig. 5.15 (b) (unshaded area).

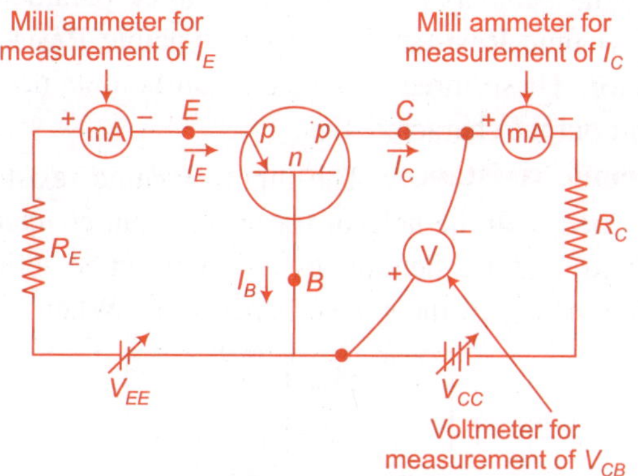

(a) Circuit diagram for output characteristics in common base mode

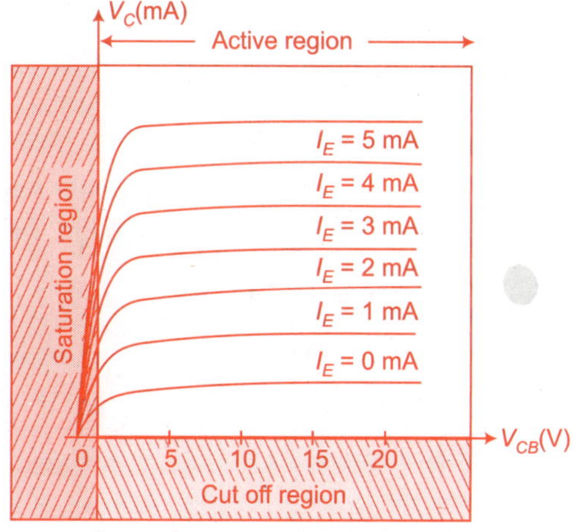

(b) Output characteristics of transistor in common base mode

Fig. 5.14

(ii) **Saturation region** The region of output characteristic at which the emitter junction and collector junction both are forward bias, is called saturation region. In Fig. 5.14 (b), the region to the left of $V_{CB} = 0$ is saturation region (shaded area).

(iii) **Cut off region**   The region of output characteristic at which the emitter junction and collector junction both are reverse bias, is called cut off region. In Fig. 5.14 (b), the region below $I_E = 0$ is cut off region (shaded area).

5.10.2   Determination of Transistor Parameters Using Common Base Characteristics

For understanding the behaviour of transistors, three parameters are important named as Input Dynamic Resistance, Output Dynamic Resistance and Current Amplification Factor. These three parameters can be obtained with the help of transistor input and output characteristics.

(i) **Input dynamic resistance**   The input dynamic resistance of transistor can be obtained with the help of transistor input characteristics as shown in Fig. 5.15 (a) and it is denoted by r_i which can be defined as the ratio of small change in V_{BE} to the small change in I_E. When V_{CB} is constant,

$$r_i = \left. \frac{\Delta V_{BE}}{\Delta I_E} \right|_{V_{CB} = \text{constant}}$$ **(Ans.)**

From Fig. 5.15 (a), we know that,

$$\Delta V_{BE} = (0.75 - 0.7)\text{V}$$
$$= 0.05 \text{ V}$$

and,
$$\Delta I_E = (5 - 2) \text{ mA}$$
$$\Delta I_E = 3 \text{ mA}$$

Hence,
$$r_i = \frac{\Delta V_{BE}}{\Delta I_E} = \frac{0.05}{3 \times 10^{-3} \text{ A}} = \frac{50}{3}$$

$$r_i = 16.7 \ \Omega.$$

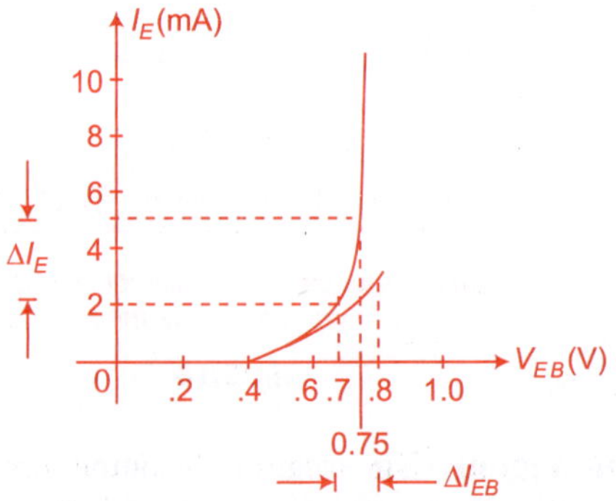

Fig. 5.15   *(a) To obtain r_1 by input characteristics*

So, we can say that the input impedance of transistor in *CB* mode is very low basically in ohms (approximately 10 Ω to 100 Ω).

(ii) **Output dynamic resistance** The output dynamic resistance of transistor can be obtained with the help of transistor output characteristics as shown in Fig. 5.15 (b). It is denoted by r_0 and can be defined as the ratio of small change in V_{CB} to the small change in I_C at constant value of I_E.

$$\gamma_0 = \frac{\Delta V_{CB}}{\Delta I_C}\bigg|_{I_E = \text{constant}}$$

$\because$ small change in I_C due to the change in V_{CB} as shown in Fig. 5.16 (b) So, the output dynamic resistance is very high in mΩ.

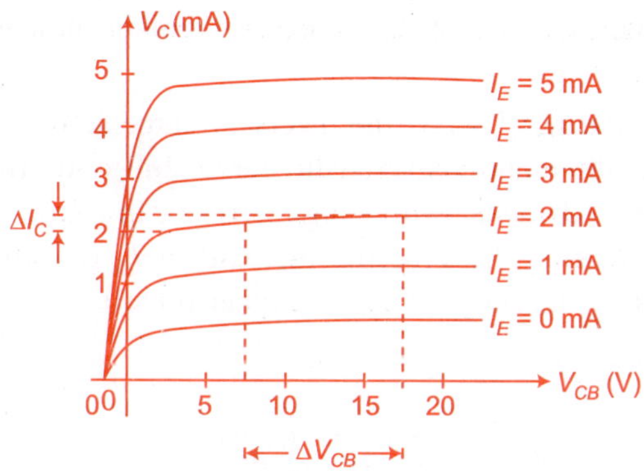

Fig. 5.15 *(b) To obtain γ_0 by output characteristics*

(iii) **Current amplification factor** Current amplification factor is defined as the ratio of small change in I_C to the small change in I_E and it is denoted by α or α_{ac}.

$$\alpha = \frac{\Delta I_C}{\Delta I_E}.$$

(5.34)

Practically, the value of α is less than 1 and it is approximately 0.98.

5.10.3 Characteristics of Common Emitter Configuration

The common emitter configuration is shown in Fig. 5.16 (a), As we have discussed earlier the characteristics of transistor we should know the input and output parameters for the transistor in common emitter mode. The input and output parameters are as following:

    (i) Input voltage V_{BE}

    (ii) Input current I_B

    (iii) Output voltage V_{CE}

    (iv) Output current I_C

Input characteristics

This characteristic is drawn between input voltage V_{BE} and input current I_B, at constant values of output voltage V_{CB}.

The circuit diagram for the input characteristics is shown in Fig. 5.16 (b). The input characteristic of *CE* mode is shown in Fig. 5.16 (c). When we study the input characteristics of *CE* mode transistor, then we can state the following related points:

    (i) The input characteristic of common base configuration is similar to the forward characteristics of diode because emitter junction is forward bias at $V_{CE} = 0$.

    (ii) If we increase the V_{CE} i.e., the reverse voltage across collector junction, due to this, the input characteristic curve shift outwards i.e., away from base current axis as shown in Fig. 5.16 (c). In other words, we can say that by increasing the V_{CE}, the increase in I_B is slow and this can be understood by the early effect as explained earlier.

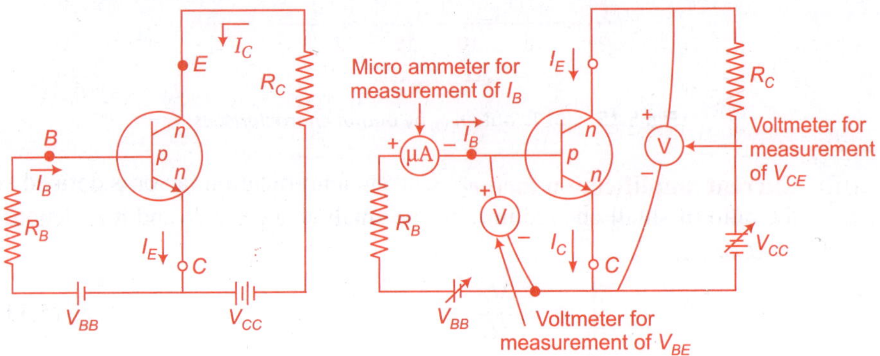

(a) Common emitter circuit   (b) Circuit diagram for input characteristics of *CE* mode

Fig. 5.16 (a) to (b)

Output characteristics

This characteristic is drawn between output voltage V_{CE} and output current I_C at constant values of input current I_B. The circuit diagram for the output characteristics is shown in Fig. 5.16 (d) and the characteristics are shown in Fig. 5.16 (e).

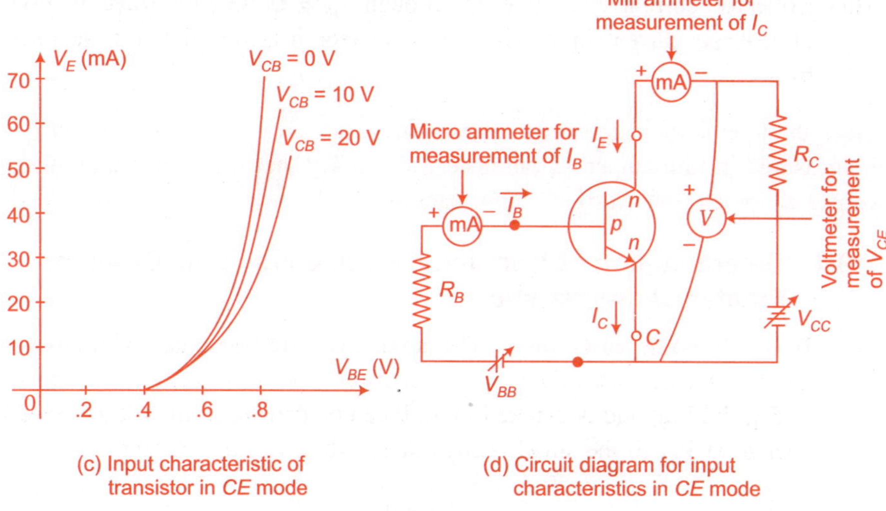

(c) Input characteristic of transistor in *CE* mode

(d) Circuit diagram for input characteristics in *CE* mode

Fig. 5.16 (c) to (d)

When we study the output characteristics of *CE* configuration in Fig. 5.17 (e), then we can state the following important points:

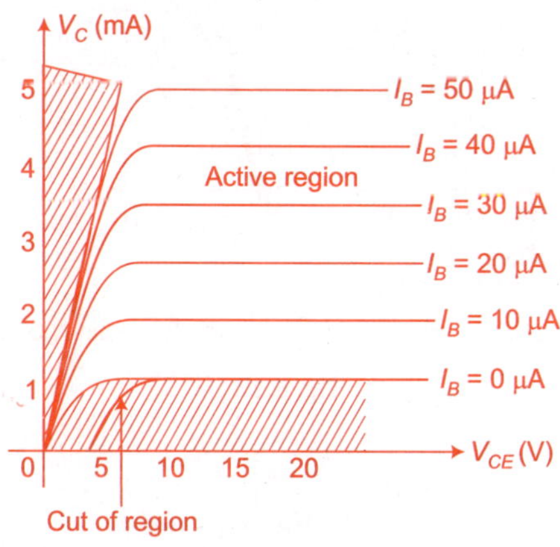

(e) Output characteristics of transistor in *CE* mode

Fig. 5.16 (e)

(i) At very small value of V_{CE}, the I_C will increase rapidly and then remain constant. Since emitter junction is forward bias, therefore at low value of V_{EC}, a large number of majority carriers reach up to collector from emitter. Remember that there is a small increase in I_C due to change in V_{CE} but it is negligible.

(ii) If V_{CE} is reduced up to few tenths of volts, then I_C will reduce rapidly because V_{CE} is less than V_{BE}, due to this, collector junction is forward bias.

(iii) Collector current $I_C \neq 0$ even through $I_B = 0$ because there is flow of leakage current in reverse bias collector junction and it is denoted by I_{CEO}.

Since the input current I_B is measured in micro amperes and output current I_C is measured in milliamperes, therefore, we can say that the common emitter configuration exhibits current amplification.

5.10.4   Determination of Transistor Parameters using Common Emitter Characteristics

(i) **Input dynamic resistance**   The input dynamic resistance of transistor can be obtained with the help of transistor input characteristic as shown in Fig. 5.17 (a) and is denoted by r_i. It can be defined as the ratio of small change in V_{BE} to the small change in I_B, when V_{CE} is constant.

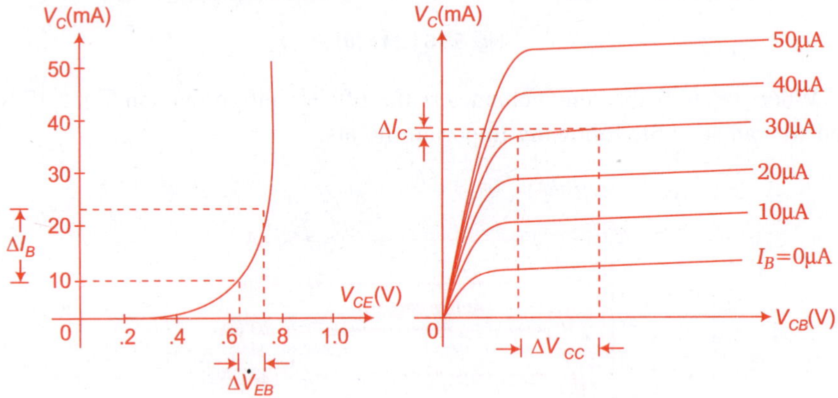

(a) To obtain r_1 input characteristic   (b) To obtain r_1 output characteristic

Fig. 5.17

$$r_i = \left.\frac{\Delta V_{BE}}{\Delta I_B}\right|_{V_{CE}\,=\,\text{constant}} \qquad (5.35)$$

The input resistance of transistor in CE mode is larger (in kΩ) than the input resistance of transistor in CB mode.

(ii) **Output dynamic resistance**   The output dynamic resistance of transistor can be obtained with the help of transistor output characteristics as shown in Fig. 5.17 (b). It is denoted by r_0 and can be defined as the ratio of small change in V_{CE} to the small change in I_C at constant value of I_B.

$$r_0 = \left.\frac{\Delta V_{CB}}{\Delta I_C}\right|_{I_B\,=\,\text{constant}} \qquad (5.36)$$

For transistor BC 108, $I_C = 2$ mA and $V_{CE} = 5$V, then r_0 is approximately 33 kΩ.

(iii) **Base current amplification factor** Current amplification factor is defined as the ratio of small change in I_C to the small change in I_B and it is denoted by β or β_{ac}.

$$\beta = \left.\frac{\Delta I_C}{\Delta I_B}\right|_{V_{CE} = \text{constant}} \tag{5.37}$$

Practically the value of β is approximately 50 to 100.

5.10.5 Characteristics of Common Collector Configuration or Emitter Follower

The circuit diagram for common collector configuration is shown in Fig. 5.18 (a). This circuit is redrawn in Fig. 5.18 (b) in which the emitter is put down and collector is put above conventionally.

In Fig. 5.18 (b), the collector terminal is also common in input and output when we perform AC analysis of this circuit. In case of AC analysis, DC supply is considered as short circuit. Due to this, collector and base both are grounded. AC analysis of transistor is out of scope of this book.

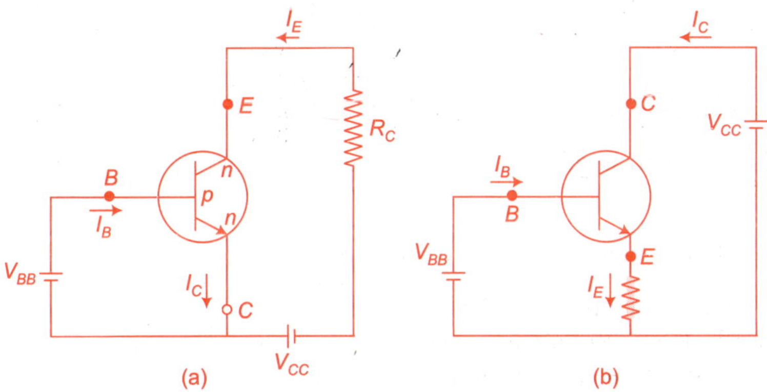

Fig. 5.18 *Common collector configuration*

If we compare this circuit with common emitter mode transistor circuit, then know that in *CE* mode, output is taken from collector terminal where as in *CC* mode, the output is taken from emitter terminal. That is why it is also called Emitter follower.

The working of *CC* mode transistor is equivalent to the *CE* mode so that characteristics of common collector mode is similar to the characteristics of common emitter mode.

The important merit of *CC* mode is that it has high input impedance and low output impedance. Due to this merit, it is used for impedance matching and buffer stage circuits.

Current amplification factor

Current amplification factor is defined as the ratio of small change in I_E to small change in I_B and is denoted by,

$$\gamma = \frac{\Delta I_E}{\Delta I_B} \tag{5.38}$$

∵ We know that, $\qquad \Delta I_B = \Delta I_E - \Delta I_C$

Substituting the above Eqn. in Eqn. (5.36), we get,

$$\gamma = \frac{\Delta I_E}{\Delta I_E - \Delta I_C} = \frac{\dfrac{\Delta I_E}{\Delta I_C}}{\dfrac{\Delta I_E}{\Delta I_C} - 1} \qquad \because \quad \frac{\Delta I_E}{\Delta I_C} = \frac{1}{\alpha}$$

Now, $$\gamma = \frac{\dfrac{1}{\alpha}}{\dfrac{1}{\alpha} - 1} = \frac{\dfrac{1}{\alpha}}{\dfrac{1 - \alpha}{\alpha}} = \frac{1}{1 - \alpha}$$

$$\boxed{\gamma = \frac{1}{1 - \alpha} = \beta + 1.} \tag{5.39}$$

5.11 TRANSISTOR AS AN AMPLIFIER

Transistor is used in amplifier circuits because it raises the strength of a weak signal.

5.11.1 The Common Base Amplifier Circuit

The common base amplifier circuit is shown in Fig. 5.19, in which the weak signal is applied between emitter base junction and output is taken across the load R_C connected in the collector circuit. In order to achieve faithful amplification, the input circuit should always remain forward biased. To do so, DC

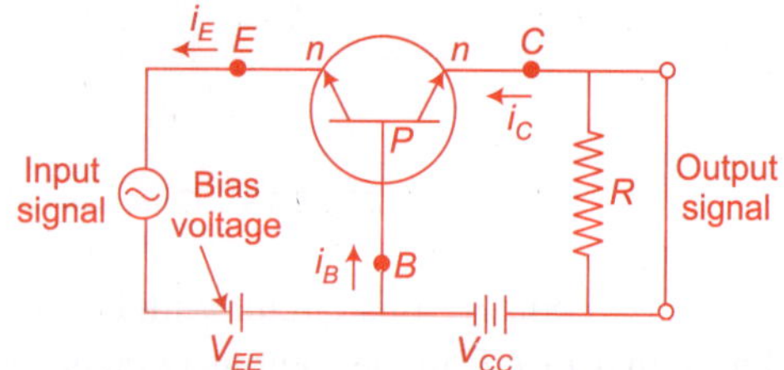

Fig. 5.19   *CB amplifier circuit*

voltage V_{EE} is applied in the input circuit in addition to the signal shown. This DC voltage is known as bias voltage and its magnitude is such that it always keeps the input circuit forward biased, regardless the polarity of the signal.

Since the input circuit has low resistance, a small change in signal voltage causes an appreciable change in emitter current I_E. Due to the change in I_E, I_C will change due to transistor action. The output resistance R_c is high. So, when the collector current flows, it produces a large voltage across it. Thus, a weak signal applied at input circuit appears in the amplified form at the output.

This type of arrangement is not very common due to its unusually high voltage gain characteristics.

Common base voltage gain,

$$A_V = \frac{I_C \times R_C}{I_e \times R_{in}}$$

$$A_V = \alpha \, \frac{R_L}{R_{in}} = \frac{V_{out}}{V_{in}}. \tag{5.40}$$

The common base circuit is generally used in single stage amplifier circuits such as microphone pre amplifier or *RF* radio amplifier due to its high frequency response.

Example 5.11.1   A common base transistor amplifier has an input resistance of 25Ω and output resistance of 100 kΩ. The collector load is 1.5 kΩ. If a signal of 400 mV is applied between emitter and base, find the voltage gain. Assume α_{ac} as nearly one.

Solution   Given that,

$$R_{in} = 25 \ \Omega \quad \text{and} \quad R_C = 1.5 \ \text{k}\Omega$$

$$V_{in} = 400 \ \text{mV}, \ \alpha_{ac} = 1$$

Input current,

$$I_E = \frac{V_{in}}{R_{in}} = \frac{400\text{mV}}{25 \ \Omega}$$

$$I_E = 16 \ \text{mA}$$

Since,

$$\alpha_{ac} = 1$$

so,

$$I_E = I_C = 16 \ \text{mA}$$

Output voltage,

$$V_{out} = I_C R_C$$

$$= 16 \ \text{mA} \times 1.5 \ \text{kW}$$

$$V_{out} = 26 \ \text{V}$$

$\therefore$ Voltage gain,

$$A_v = \frac{V_{out}}{V_{in}} = \frac{24 \ \text{V}}{400 \ \text{V}}$$

$$\boxed{A_V = 60.} \hspace{4cm} \textbf{(Ans.)}$$

Merits and demerits of CB amplifier circuit

Merits

(i) Change in transistor parameters does not effect the common base amplifier where as it effects the CE amplifier comparatively.

(ii) Common base amplifier can operate at high frequency as compared to common emitter amplifiers.

Demerits

(i) Input impedance is low and output impedance is high for common base amplifier. So, it cannot be used for multistage amplification.

(ii) The current gain of common base amplifier is less than one.

5.11.2 The Common Emitter Amplifier Circuit

The common emitter amplifier circuit is shown in Fig. 5.20, in which the weak signal is applied between base-emitter junction and output is taken across the load R_C connected in the collector circuit.

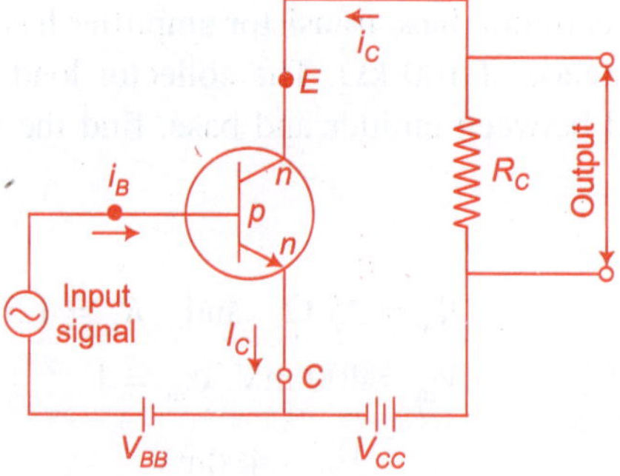

Fig. 5.20 *Common emitter amplifier circuit*

During the positive half cycle of the input signal, the forward bias across emitter base junction is increased. Due to this, the collector current will also increase. The increased collector current produces a large voltage drop across the collector load resistance R_C. However, during the negative half cycle of the input signal, the forward bias cross emitter junction is decreased. Due to this, collector current will decrease. This results in the decreased output voltage in opposite direction. Hence, an amplified output is obtained across the load R_C.

Merits and demerits of common emitter amplifier circuit

Merits

(i) The voltage gain A_v and current gain A_i is high for common emitter amplifier.

(ii) The power gain of common emitter amplifier is greater than common base and common collector.

(iii) There is not much difference between input and output impedances. So, it can easily be used in cascading or multistage.

Demerits

The frequency response curve is not better as compared to common base and common collector amplifier.

5.11.3 The Common Collector Amplifier Circuit

The common collector amplifier circuit is shown in Fig. 5.21, in which weak signal is applied between base and collector junction and output is taken across the load R_E connected between emitter and collector.

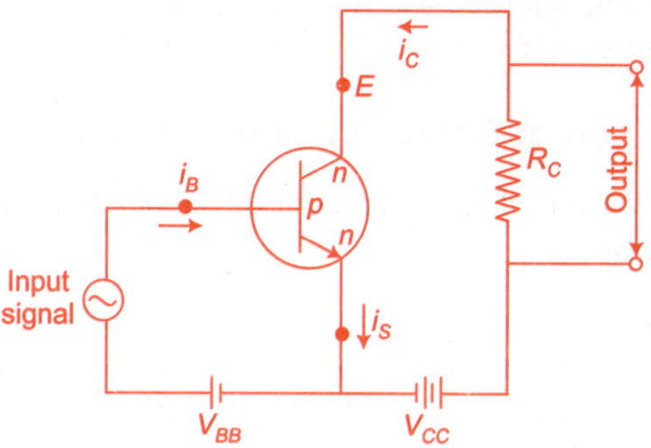

Fig. 5.21

The common emitter configuration has a current gain equal to the β values of the transistor itself.

In the common collector configuration, the load resistance R_E is situated in series with the emitter. So, its current is equal to that of the emitter current. As the $I_E = I_C + I_B$, the load resistance in this type of amplifier configuration also has both collector current and input current of the base flowing through it. Then, the current gain of the circuit is given as,

$$I_E = I_C + I_B$$

$$A_i = \frac{I_E}{I_B} = \frac{I_C + I_B}{I_B}$$

$$A_i = \frac{I_C}{I_B} + 1$$

$$\boxed{A_i = \beta + 1.}$$

(5.41)

Merits and Demerits of CC Amplifier Circuit

Merits

(i) The current gain A_i of common collector amplifier circuit is high.

(ii) The input impendance is high and output impedance is low. So, it is used for impedance matching.

Demerits

The voltage gain of common collector amplifier circuit is less than one.

> The common collector circuit has very high input impedance and very low output impedance. Due to this reason, the voltage gain of it is less than one. Therefore, this circuit is rarely used for amplification. However, due to relatively high input impedance and low output impedance, this circuit is primarily used for impedance matching i.e., for driving a low impedance load for a high impedance source.

5.11.4   Comparison of Transistor Connections

The comparison of various characteristics of different connections is given in the following tabular form:

Table 5.1

S.No.	Characteristics	Common base	Common emitter	Common collector
1.	Input impedance	Low (about 100Ω)	Medium (about 750 Ω)	Very high
2.	Output Impedance	Very high (about 450 Ω)	High (about 50Ω)	Low (about 50Ω)
3.	Phase angle	0°	180°	0°
4.	Voltage gain	High (about 150Ω)	Very high (about 500 Ω)	Low less than 1
5.	Current gain	Low (less than 1)	Very high	High
6.	Power gain	Low	Very high	Medium

According to, the above table, we can say that common emitter amplifier circuit is commonly used for amplification because it has better voltage current and power gain.

5.11.5   Standard Notation for Transistor Voltages and Currents

The standard notation for voltages and current is shown in Table 5.2.

Table 5.2

S. No.	Characteristics		DC value	Instantaneous AC value	r.m.s. value	Instantaneous total value
1.	Emitter current		I_E	i_e	I_e	i_E
2.	Collector current		I_C	i_c	I_b	i_C
3.	Base current		I_B	i_b	I_b	i_B
4.	CE mode	Base-emitter Voltage (input)	V_{BE}	v_{be}	V_{be}	v_{BE}
5.		Collector emitter Voltage (input)	V_{CE}	v_{ce}	V_{ce}	v_{CE}
6.	CB mode	Emitter-base Voltage (input)	V_{EB}	v_{eb}	V_{eb}	v_{EB}
7		Collector base Voltage (output)	V_{CB}	v_{cb}	V_{cb}	v_{CB}
8.	CC mode	Base-collector Voltage (input)	V_{BC}	v_{bc}	V_{bc}	v_{BC}
9.		Emitter-collector Voltage (output)	V_{EC}	v_{ec}	V_{ec}	v_{EC}

Why common emitter mode is used in amplifiers?

In almost all applications, we use a number of amplifier stages, connected in series. The signal to be amplified is fed to the input of the first stage. The output of the first stage is connected to the input of second stage. The output of second stage is fed to the input of third stage, and so on. Finally, the output appears across the load connected to the output of the final stage as shown in Fig. 5.22. Such a connection of amplifier stages is known as Cascade amplifier. A good amplifier is the one which has high input impedance and low output impedance.

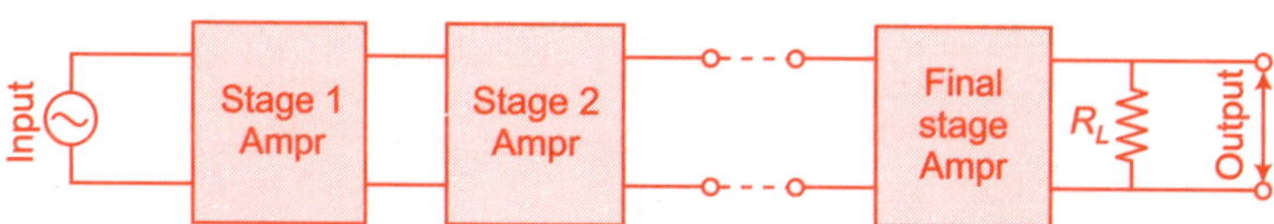

Fig. 5.22   *Multistage amplifier*

A transistor in common base mode has a very low input impedance (about 100 Ω) and very high output impedance (about 1 MΩ). It is just the reverse of an amplifier. That this why the common base mode is not popular. The current, voltage and power gain of common emitter is greater than those of common base.

Common collector mode would have been the best. Its input resistance is high and output resistance is low. However, it is less than one. Therefore, we use common collector amplifier only in such applications, where the requirement of high input resistance is important, such as impedance matching.

Thus, we can say that common emitter configuration is best suited for most of the amplifier circuits.

5.12 TRANSISTOR AS A SWITCH

As we have discussed earlier that a transistor can be used both as an amplifier and a switch. In this section, we will study the transistor as a switch. The common emitter circuit was at in the previous section. In this section the transistor is operated as a switch. It needs to be turned either fully "OFF" (cut off) or fully "ON" (saturated), to control the 'on' and 'off' states of the light-bulb or any other device in the collector branch of the network. An ideal transistor switch would have an infinite resistance when turned 'OFF' resulting in zero current flow, and zero resistance when turned 'ON', resulting in maximum load current. In practical transistor, when turned 'OFF', small leakage current flows through the transistor and when fully 'ON' the device has a low resistance value causing a small saturation voltage V_{CE} across it. In both the 'cut off and 'saturation' regions, the power dissipated by the transistor is at its minimum.

The circuit is shown in Fig. 5.23 in which transistor is used as a switch, which controls the relay operation. When the switch S is in "ON" position, to make the base current flow, the base input terminal must be made more positive than the emitter by increasing it above 0.7 volts, needed for a silicon device. By varying the base-emitter voltage V_{be}, the base current is changed which in turn controls the amount of collector current I_C. When the collector current is maximum, the transistor is said to be saturated i.e., the transistor is said to be 'cut off i.e., the transistor is fully "OFF".

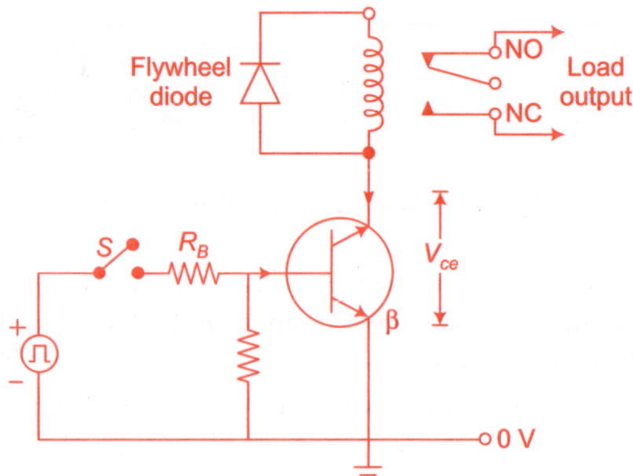

Fig. 5.23 *Transistor as a switch*

Example 5.12.1 For a common emitter transistor circuit, the values are $\beta = 150$, $I_C = 3.5$ mA and $I_B = 15$ mA. Find the value of the base resistor R_B required to switch the load 'ON' when the input terminal voltage exceeds 5.0 V.

Solution Given that,

$$\beta = 150, \qquad I_C = 3.5 \text{ mA}$$

and, $$I_B = 15 \ \mu A, \qquad\qquad V_{in} = 5 \text{ volt}$$

We have, $$R_B = \frac{V_{in} - V_{BE}}{I_B}$$

$$V_{BE} = 0.7 \text{ V for Si device}$$

$$R_B = \frac{5 - 0.7}{15 \times 10^{-6}}$$

$$\boxed{R_B = 286 \text{ k}\Omega.} \qquad\qquad \textbf{(Ans.)}$$

Example 5.12.2 Using the same values given in Example 5.12.1, find the minimum base current required to turn the transistor fully 'ON' for a load that requires 225 mA of current.

Solution Given that,

$$\boxed{I_C = 225 \text{ mA.}}$$

$$\beta = 150$$

$$I_B = \frac{I_C}{\beta} = \frac{225}{150} \text{ mA}$$

$$\boxed{I_B = 1.5 \text{ mA}} \qquad\qquad \textbf{(Ans.)}$$

As we see in Fig. 5.24 the flywheel diode is used for the protection and it eliminates the severe change in current level. When the transistor is in 'ON' state then the diode is reverse based; it is like an open circuit and does not affect the working of transistor.

However, when the transistor turns 'OFF' the voltage across the coil will reverse and diode will forward bias, and it is in 'ON' state. The current though the inductor established during the 'ON' state of the transistor can then continue to flow through the diode, eliminating the severe change in current level. The rating of the diode must have matched to the current through the inductor and the transistor when in the 'ON' state as shown in Figs. 5.24 (a) and (b).

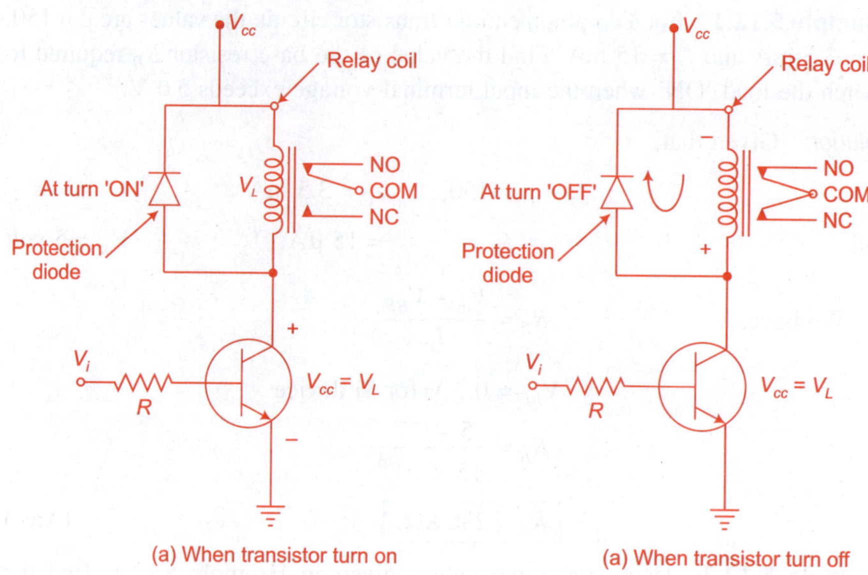

(a) When transistor turn on   (a) When transistor turn off

Fig. 5.24

5.12.1 Digital Logic Transistor Switch

Transistor switches are used for a wide vanity of applications, such as interfacing high voltage on large current devices like motors, relays on large current devices like motors, relays or lamps to low voltage digital logic IC's or gates like AND Gates, OR Gates. In this circuit, the output from a digital logic gate is +5 V but the device to be controlled may require a 12 or 24 volts supply. If DC motor connected to the circuit as a load, then it may need to have its speed controlled using a series of pulses and transistor switches will allow us to do this faster and more easily then with conventional mechanical switches. The circuit for digital logic transistors switch is shown in Fig. 5.25.

In the circuit A and B are the input of the digital gate. R_L is the current limiting resistor. The base resistor R_B is required to limit the output current of the logic gate. When a LED is first turned, on, its resistance is quite low, even through the resistance will increase rapidly the longer the LED is on.

This can cause a momentary high level of collector current I_C, which could damage the LED. An additional small resistor R_L in series with the LED is used to ensure a limit on the initial singe in current when the LED is first turned on.

5.12.2 Darlington Transistors

Sometimes the DC current gain of the bipolar transistor is too low, to directly switch the loads current or voltage, so multiple switching transistors are used. As shown in Fig. 5.26 (a). Here are small input transistor is used to switch 'ON'

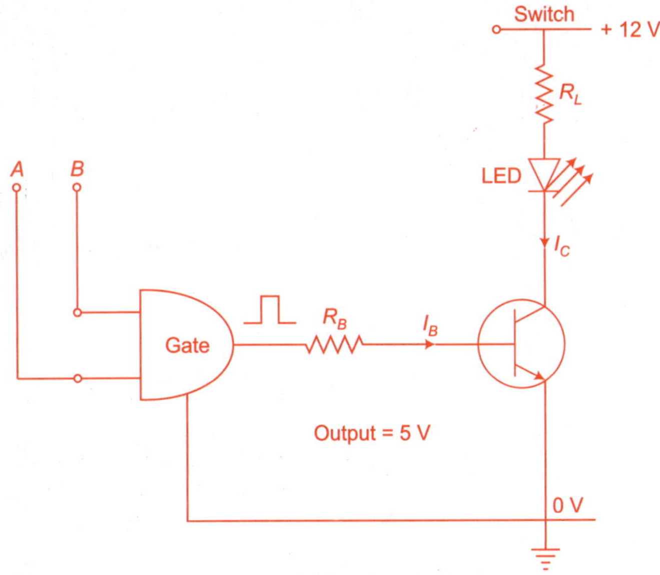

Fig. 5.25   *Digital logic transistor switch*

or 'OFF' a much longer current handling output transistor. To increase the signal gain, the two transistors are connected in a 'complementary gain compounding configuration'. and is called Darlington configuration, where the amplification factor is the product of the two individual transistors.

In other words we can state that the main feature of the Darlington connection is that the composite transistor acts as a single unit as shown in Fig. 5.26 (b) with the current gains that is the product of the current gains of the individual transistors.

If two transistors, having current gains of β_1, and β_2, the Darlington connection provides a current gain β_D.

$$\boxed{\beta_D = \beta_1 \beta_2.}$$
(5.42)

DC Biasing of Darlington Circuit

The circuit diagram for darlington circuit with DC bias is shown in Fig. 5.27. The base current from the circuit can be calculated.

$$I_B = \frac{V_{CC} - V_{BE}}{R_B + \beta_D R_E}$$
(5.43)

and
$$I_E = (\beta_D + 1)\, I_B \qquad \because \beta_D >>> 1$$

$$I_E \approx \beta_D I_B$$
(5.44)

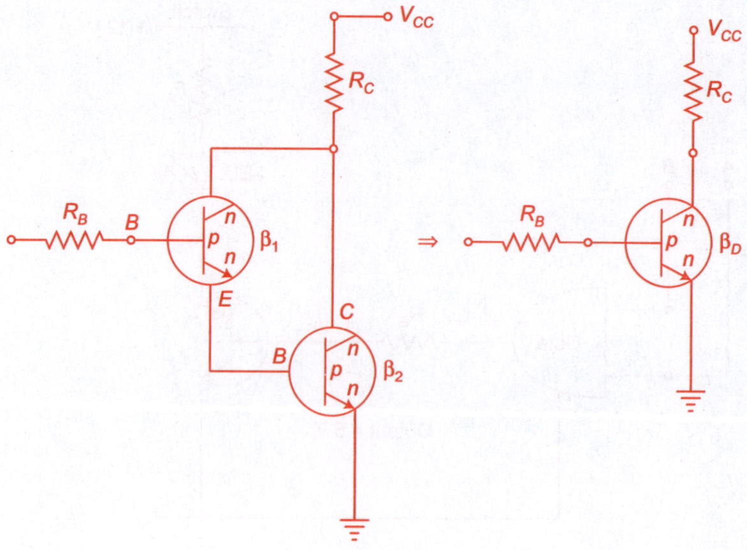

(a) Darlington circuit   (a) Darlington equivalent

Fig. 5.26

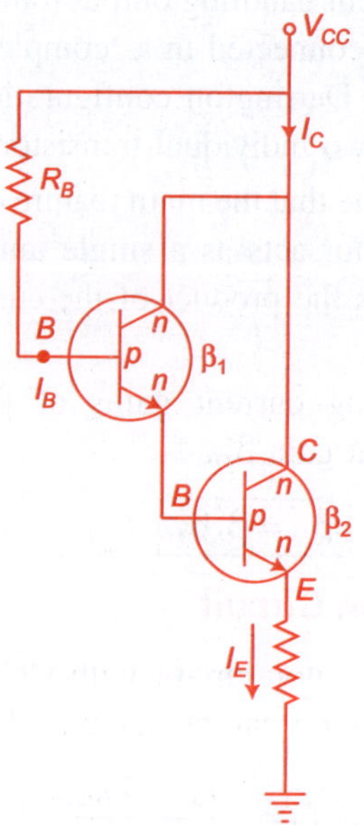

Fig. 5.27   *Darlington circuit*

The dc voltage are

$$V_E = I_E R_E \tag{5.45}$$

$$V_B = V_E + V_{BE}. \tag{5.46}$$

Example 5.12.3 Calculate the DC bias voltage and currents for the circuit of Fig. Ex. 5.12.3.

The data given as $R_B = 3.2$ mΩ, $R_E = 400$ mΩ, $\beta_1 = 90$ and $\beta_2 = 95$ and V_{BE} is 15 volt, $V_{CC} = 15$ V.

Solution Solution that

$$R_B = 3.2 \text{ mΩ}$$

$$V_{CC} = +15 \text{ V}$$

$$\beta_1 = 90 \quad \beta_2 = 95$$

From Eqn. (5.43) the base current is

$$I_B = \frac{V_{CC} - V_{BE}}{R_B + \beta_D R_E}$$

$$\beta_D = \beta_1 \beta_2$$

$$\beta_D = 90 \times 95 = 8550$$

Fig. Ex 5.12.3

Now,

$$I_B = \frac{15 \text{ V} - 1.5 \text{ V}}{3.2 \times 10^6 \text{ Ω} + 8550(400) \text{ Ω}}$$

$$= \frac{13.5 \text{ V}}{(3.2 \times 10^6 + 3.42 \times 10^6) \text{ Ω}}$$

$$= \frac{13.5 \text{ V}}{6.62 \times 10^6 \text{ Ω}}$$

$$\boxed{I_B = 2.04 \text{ μA}} \qquad \textbf{(Ans.)}$$

From Eqn. (5.44), the emitter current

$$I_E = \beta_D I_B$$

$$I_E = 8550 (204 \times 10^{-6}) \text{ A}$$

$$\boxed{I_E = 17.44 \text{ mA}} \qquad \textbf{(Ans.)}$$

The dc voltages from Eqns. (5.45) and (5.46) are

$$V_E = I_E R_E$$

$$= (17.44 \text{ mA}) (400 \text{ Ω})$$

$$\boxed{V_E = 6.98 \text{ volt.}} \qquad \textbf{(Ans.)}$$

$$V_B = V_E + V_{BE}$$

$$V_B = 6.99\ V + 1.5\ V$$

$$\boxed{V_B = 8.48\ \text{volt.}}$$

(Ans.)

5.12.3 Use of Transistor as a Switch for Logic Gates

As we already discussed that the transistor can be used as a switch, in this section we will use the transistor for making a logic gates. The output characteristic of transistor is shown in Fig. 5.28 in which two points S and C on a load line represent the 'saturation' point i.e., the transistor is in fully 'ON' mode, and the 'cut-off point i.e., the transistor is in fully 'OFF' mode.

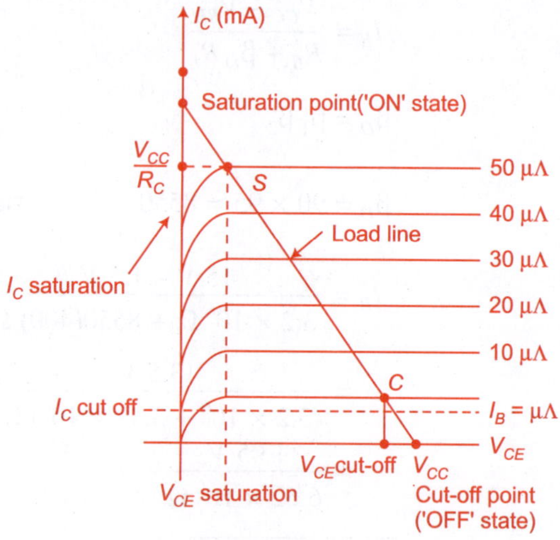

Fig. 5.28 *Darlington bias circuit*

Impedance at collector to emitter terminal i.e., output impedance is low at saturation (In 'ON' state) and high at cut-off (In 'OFF' state). In other words we can say that at saturation, current I_C is high and V_{CE} is low and a cut-off, the collector current I_C is low or near to zero and V_{CE} is high.

OR Gate

The gate make by transistor is shown in Fig. 5.29

Base terminal to turn the transistor ON and if the input is low (0) that means 0 volt is applied at the base terminal ensuring that transistor if OFF.

There are two inputs A and B on the transistors if the input is high (1) that means high voltage at the verification of truth table for OR Gate.

   (i) $A = 0$ i.e., transistor Q_1 is in cut-off (OFF)

        $B = 0$ i.e., transistor Q_2 is in cut-off (OFF)

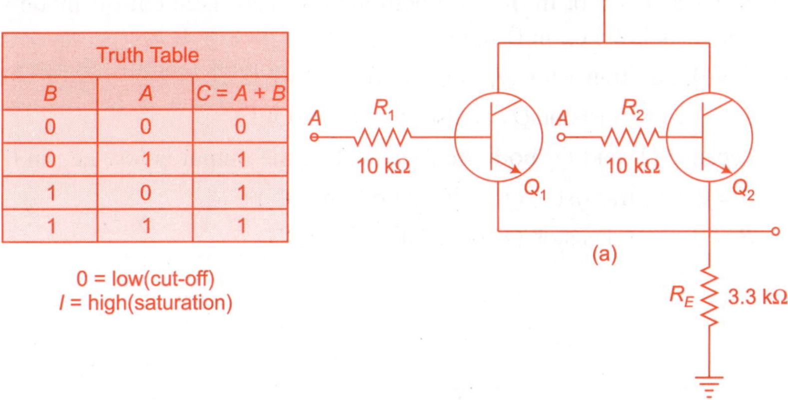

Fig. 5.29 *OR gate*

i.e.,

The impedance between the collector and the emitter of each transistor is approximately open circuit. Due to this the zero current through each transmitter and though the R_E resistor so the output is zero.

(ii) $A = 1$ i.e., transistor Q_1 is in saturation (ON)

  $B = 0$ i.e., transistor Q_2 is in cut-off (OFF)

  i.e. Due to the positive voltage at the base of Q_1 and 0 volt at the base of Q_2, then the short circuit equivalent between the collector and emitter for transistor Q_1 and can be applied, and the voltage at the output is 5 V (high) or 1 state.

(iii) $A = 0$ i.e., transistor Q_1 is in cut-off (OFF)

  $B = 1$ i.e., transistor Q_2 is in saturation (OFF)

  i.e. Due to the positive voltage at base of Q_2 and 0 V at the base of Q_1, then the short circuit equivalent between collector and emitter for transistor Q_2 can be applied, and the voltage is 5 V (high) or 1 state.

(iv) $A = 1$ i.e., transistor Q_1 is in saturation (ON)

  $B = 1$ i.e., transistor Q_2 is in saturation (ON)

  i.e. Due to positive voltage at the base of Q_1 and Q_2 is applied, they will both ensure that the output voltage is 5 V (high) or 1 state.

AND Gate

The AND Gate make by transistors is shown in Fig. 5.30

(i) $A = 0$ i.e., transistor Q_1 is in cut-off (OFF)

  $B = 0$ i.e., transistor Q_2 is in cut-off (OFF)

  So the output is low i.e., in 0 state

(ii) $A = 1$ and $B = 0$, In the case both the transistor is in cut-off mode so the output is low i.e., in 0 state.

(iii) $A = 0$, i.e., transistor Q_1 is in cut-off (OFF)

    $B = 1$ i.e. transistor Q_2 is also in cut-off (OFF)

    because Q_1 and Q_2 both are in cut-off so the output is low i.e., in 0 state

(iv) $A = 1$ i.e., transistor Q_1 is in saturation (ON)

    $B = 0$ i.e., transistor Q_2 is in saturation (ON)

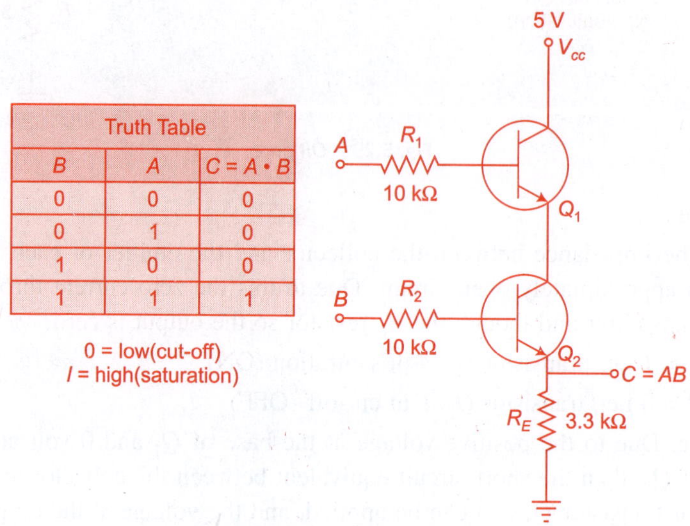

Fig. 5.30 *Verification of truth table for AND gate*

The output is high only if both inputs have a turn on voltage applied. If both are in the 'ON' state, a short circuit equivalent can be used for the connection between the collector and the emitter of each transistor, provide direct both from the V_{CC} to the output.

5.12.4 Some Important Points Related to the Transistor as a Switch

- Transistor as a switch can be used to switch and control, lamps, relays or even motors.
- When we use bipolar transistors as switches they must be fully 'OFF' or fully 'ON'.
- Transistor that are fully "ON" are said to in their saturation region.
- Transistor that are fully "OFF" are said to be in their cut-off region.
- In a transistor switch, a small base current control a much larger collector current.

- When using transistors as a switch inductive relay loads a "Flywheel" diode is required.
- When large currents or voltage need to be controlled, Darlington transistors are used.

Examples

Example 5.1 In a common emitter transistor, the change in base current is 0.3 mA then what will be the change in collector current IC. If $\alpha = 0.98$.

Solution Given that

$$\Delta I_B = 0.3 \text{ mA and } \alpha = 0.98$$

We know that

$$\beta = \frac{\alpha}{1 - \alpha}$$

$$\beta = \frac{0.98}{1 - 0.98} = 49$$

$$\beta = 49$$

and

$$\beta = \frac{\Delta I_C}{\Delta I_B}$$

$$\Delta I_C = \beta \, \Delta I_B$$

$$\Delta I_C = 49 \times 0.3 = 14.7 \text{ mA}$$

$$\boxed{\Delta I_C = 14.7 \text{mA.}}$$

(Ans.)

Example 5.2 In a transistor, $I_{CBO} = 15 \ \mu A$ and β is 50, if the base current is 0.1 mA then calculate the collector current.

Solution Given that

$$I_{CBO} = 15 \ \mu A, \qquad \beta = 50$$

$$I_B = 0.1 \text{ mA}$$

We know that

$$I_C = \beta I_B + (\beta + 1) \, I_{CBO}$$

$$I_C = (50 \times 0.1 \times 10^{-3}) + (50 + 1) \, 15 \times 10^{-6}$$

$$I_C = (5 + 0.765) \times 10^{-3}$$

$$\boxed{I_C = 5.767 \text{ mA.}}$$

(Ans.)

Example 5.3 In a transistor circuit, collector load is 5 kΩ where as quiescent current is 2 mA.

(i) What is the operating point if $V_{CC} = 12\ V$?

(ii) What will be the operating point $R_C = 4\ k\Omega$?

Solution Given that

$$V_{CC} = 12\ V, \quad I_C = 2\ mA$$

(i) When

$$R_C = 4\ k\Omega$$

$$V_{CE} = V_{CC} - I_C R_C$$

$$= [12 - (2\ mA \times 5\ k\Omega)]$$

$$V_{CE} = 12 - 10 = 2\ V$$

$$V_{CE} = 2\ V$$

$\therefore$ The operating point is 2 V, 2 mA.

(ii) When collector load $R_C = 4$

$$V_{CE} = V_{CC} - I_C R_C$$

$$V_{CE} = [12 - (2\ mA \times 4\ k\Omega)]$$

$$V_{CE} = 12 - 8 = 4\ V$$

$$V_{CE} = 4\ V$$

$\therefore$ Operating point is 4 V, 2 mA.

Examples 5.4 In a common emitter transistor, $\beta = 100$ $I_{CBO} = 6\ \mu A$ and $I_C = 15\ mA$. Calculate I_B, I_E, α and I_{CEO}.

Solution Given that

$$\beta = 100, \quad I_{CEO} = 6\ \mu A$$

$$I_C = 1.5\ mA.$$

We know that

$$\alpha = \frac{\beta}{\beta + 1} = \frac{100}{100 + 1} = \frac{100}{101} = 0.99$$

$$\boxed{\alpha = 0.99.}$$

We know that

$$I_C = \beta I_B + (\beta + 1)\ I_{CEO}$$

$$\beta I_B = I_C - (\beta + 1)\ I_{CEO}$$

$$I_B = \frac{I_C - (\beta + 1)\ I_{CEO}}{\beta}$$

$$I_B = \frac{[1.5 - (101 \times 6 \times 10^{-3})]}{100} \text{ mA}$$

$$I_B = \frac{0.894}{100} \text{ mA}$$

$$I_B = 8.94 \times 10^{-3} \text{ mA}$$

$$\boxed{I_B = 8.94 \ \mu\text{A.}} \qquad\qquad \textbf{(Ans.)}$$

$$I_E = I_C + I_B$$

$$= (1.5 + 0.00894) \text{ mA}$$

$$\boxed{I_E = 1.5089 \text{ mA.}} \qquad\qquad \textbf{(Ans.)}$$

Now,
$$I_{CEO} = (\beta + 1)\, I_{CEO}$$

$$I_{CEO} = (100 + 1) \times 6 \ \mu\text{A} \qquad\qquad \textbf{(Ans.)}$$

Example 5.5 For a transistor, $\beta = 49$ and voltage drop across 2 kΩ which is connected in the collector circuit is 2 volt. Find the base current for common emitter connection shown in Fig. Ex. 5.5.

Solution   Given that

Voltage drop across 2 kΩ = 2 V

$$\beta = 49$$

The collector current

$$I_C = \frac{\text{Voltage across } R_C}{R_C}$$

$$I_C = \frac{2\ V}{2\ \text{k}\Omega}$$

$$\boxed{I_C = 1 \text{ mA.}} \qquad\qquad \textbf{(Ans.)}$$

Now,
$$I = \frac{I_C}{\beta} = \frac{1 \text{ mA}}{49}$$

$$\boxed{I_B = 0.02 \text{ mA.}} \qquad\qquad \textbf{(Ans.)}$$

Fig. Ex. 5.5

Example 5.6   In a Ge transistor the leakage current is 15 μA and β is 50.

   (i)   if base current is 0.02 mA then calculate collector current.

   (ii) if temperature increase 40°C then what will be the value of collector current (assume D is constant).

Solution Given that

$$I_{CBO} = 15 \ \mu A, \ \beta = 50$$

(i)
$$I_B = 0.02 \text{ mA}$$

$$I_C = \beta I_B + (\beta + 1) \ I_{CBO}$$

$$I_C = [50 \times 0.02 \text{ mA} + (50 + 1) \times 15 \times 10^{-3}] \text{ mA}$$

$$I_C = [1 + (51 \times 15 \times 10^{-3})] \text{ mA}$$

$$\boxed{I_C = 1.765 \text{ mA.}} \qquad \textbf{(Ans.)}$$

(ii) We know that increasing of temperature 10°C the leakage current will be twice so 40°C increase in temperature, the leakage current will be 2^4 times.

$$I'_{CBO} = (15 \times 10^{-3} \times 2^4) \text{ mA}$$

$$I'_{CBO} = 0.240 \text{ mA.}$$

Now the collector current

$$I'_C = \beta l_B + (\beta + 1) \ I'_{CBO}$$

$$I'_C = [(50 \times 0.002) + (51 \times 0.240)] \text{ mA}$$

$$I'_C = (1 + 1224) \text{ mA}$$

$$\boxed{I'_C = 1324 \text{ mA.}} \qquad \textbf{(Ans.)}$$

Example 5.7 A transistor is connected in common emitter configuration in which collector supply V_{CC} is 10 V and the voltage drop across resistance R_C is 0.6 V. The value of $R_C = 850 \ \Omega$ if $\alpha = 0.99$, determine

(i) Collector emitter voltage   (ii) Base current.

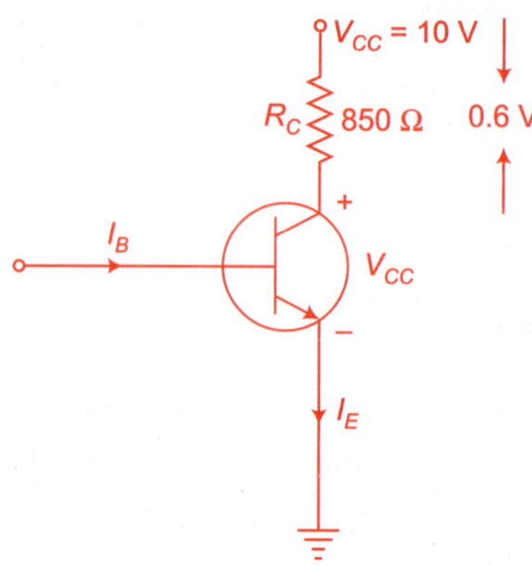

Fig. Ex. 5.7

Solution (i) Collector emitter voltage

$$V_{CE} = V_{CC} - I_C R_C$$
$$V_{CE} = 10\ V - 0.6\ V$$

$$\boxed{V_{CE} = 9.4\ V.}$$ **(Ans.)**

(ii) The voltage drop across $R_C = 850\ \Omega$ is $0.6\ V$

$\therefore$
$$I_C = \frac{0.6\ V}{850\ \Omega} = 0.706\ mA$$

$$\boxed{I_C = 0.706\ mA.}$$ **(Ans.)**

Now,
$$\beta = \frac{\alpha}{1 - \alpha} = \frac{0.99}{1 - 0.99}$$

$$\beta = \frac{0.99}{0.01} = 99$$

$$\beta = 99$$

$\therefore$ Base current
$$I_B = \frac{I_C}{\beta} = \frac{0.706}{99}$$

$$\boxed{I_B = 7.13\ \mu A.}$$ **(Ans.)**

Example 5.8 Using the characteristics of Fig. Ex. 5.8.1, determine

(a) The resulting collector current, if $I_E = 5\ mA$ and $V_{CB} = 15V$.

(b) The resulting I_C, if I_E remains at 5 mA but V_{CB} is reduced to 10 V.

(c) Using the characteristics of Figs. Ex. 5.8.1 and Ex. 5.8.2 determine, V_{BE} if $I_C = 6\ mA$ and $V_{CB} = 20\ V$.

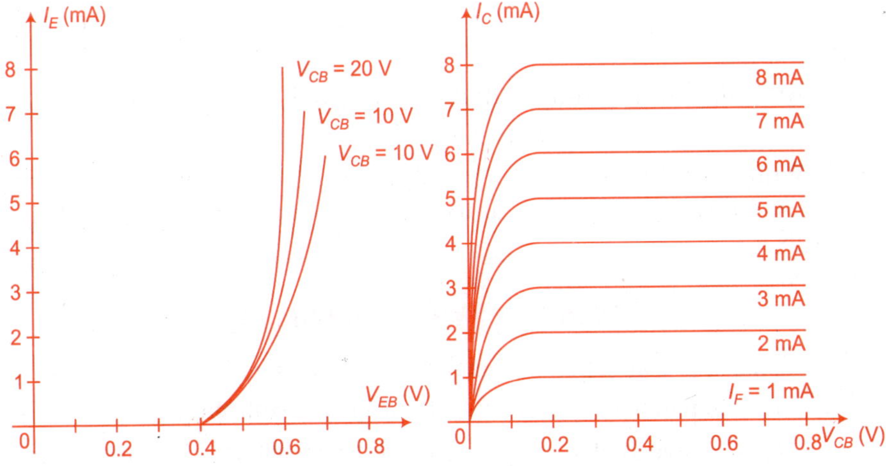

Fig. Ex. 5.8.1 *Input characteristics*   **Fig. Ex. 5.8.2** *Output characteristics*

Solution

(a) The characteristics clearly indicate that I_C = 4.6 mA at I_E = 5 mA and V_{CB} = 15 V.

(b) The effect of changing in V_{CB} is negligible and I_C continue to be 4.6 mA,

(c) From Fig. Ex. 5.8.2 $I_E \cong I_C$ = 6 mA. In Fig. Ex. 5.8.1 the resulting level of V_{BE} is about 0.72 V.

Example 5.9   Using the characteristics of Fig. Ex. 5.9.1, determine

(a) Collector current at I_B = 20 µA and V_{CE} = 10 V.

(b) Collector current at V_{BE} = 0.7 V and V_{CE} = 15 V

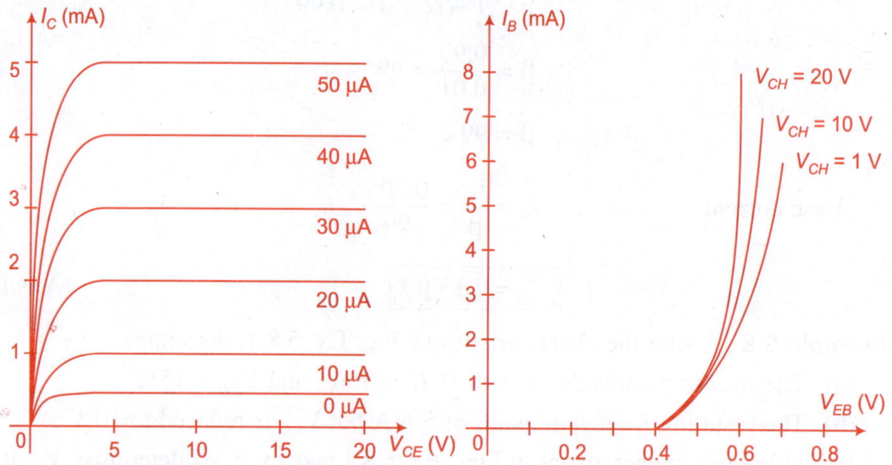

Fig. Ex. 5.9.1 (a)   *Output characteristics*   **Fig. Ex. 5.9.2 (a)**   *Input characteristics*

Solution

(a) At the intersection of I_B = 20 µA and V_{CB} = 10 V,
$$I_C = 23 \text{ mA}.$$

(b) Wring Fig. Ex 5.11.1 (6), I_B = 20 µA at V_{BH} = 0.7 V and from Fig. Ex. 5.9.1(a). I_C = 3.5 mA at the intersection of I_B = 20 µA and V_{CH} = 15 V.

Objective Type Questions

1. .... configuration is most used in amplifier.

   (a) Common base   (b) Common collector

   (c) Common emitter   (c) None of these

2. The value of α is
    - (a) Greater than 1
    - (b) Less than 1
    - (c) Greater than 50
    - (d) 1

3. .... is lightly doped and heavily doped respectively.
    - (a) Base and emitter
    - (b) Emitter and base
    - (c) Collector and base
    - (d) None of these

4. If α is 0.95 then β is
    - (a) 19
    - (b) 0.01
    - (c) 99
    - (d) 49

5. In Amplifier, the biasing of emitter and collector junction is keep .... respectively
    - (a) Reverse bias and forward bias
    - (b) Forward bias and reverse bias
    - (c) Forward bias and forward bias
    - (d) Reverse bias and reverse bias

6. .... is constant to draw input characteristics
    - (a) Input current
    - (b) Input voltage
    - (c) Output voltage
    - (d) Output currents

7. The flow of current in *n-p-n* transistor is due to
    - (a) Electrons
    - (b) Holes
    - (c) Impurity ions
    - (d) Electrons and holes

8. The voltage gain for emitter follower is
    - (a) Less than one
    - (b) Zero
    - (c) Greater than one
    - (d) None of these

9. In the common emitter configuration, $\alpha = 0.98$ if $\Delta I_B = 0.2$ mA then ΔI_C is
    - (a) 10 mA
    - (b) 9.8 mA
    - (c) 10 µA
    - (d) 9.8 µA

10. BJT is .... device
    - (a) Correct controlled, bipolar
    - (b) Current controlled, unipolar
    - (c) Voltage controlled, unipolar
    - (d) None of these

11. Match the following.
    - (a) Active region
    - (b) Cut-off region
    - (c) Saturation region
    - (d) Inverse mode

    1. J_E is forward bias. J_C is reverse bias
    2. J_E, J_C both are forward bias
    3. J_E, J_C both are reverse bias
    4. J_E, is reverse bias, J_E is forward bias

	ABCD		ABCD		ABCD		ABCD
(a)	4 3 2 1	(b)	4 2 3 1	(c)	1 2 3 4	(d)	1 3 2 4

12. The use of emitter follower is
    - (a) Voltage gain
    - (b) In oscillator
    - (c) In rectifier
    - (d) For impedance matching

13. For amplifier, transistor is biased in
    - (a) Cut off region
    - (b) Active region
    - (c) Saturation region
    - (d) Inverse mode

14. Which half cycle clip, when the Q point of an amplifier shift towards Saturation 'and' cut off region respectively
    - (a) Negative, positive
    - (b) Positive, negative
    - (c) Positive, positive
    - (d) Negative, negative

15. In a transistor, the number of junction and terminals...... are respectively
    - (a) 3, 2
    - (b) 2, 3
    - (c) 3, 3
    - (d) 2, 2

16. The value of α is
    - (a) $\dfrac{\beta + 1}{1 - \beta}$
    - (b) $\dfrac{1 - \beta}{1 + \beta}$
    - (c) $\dfrac{\beta}{1 + \beta}$
    - (d) $\dfrac{\beta + 1}{\beta}$

17. Thermal runway is due to
    - (a) Large voltage
    - (b) Large doping
    - (c) Large temperature
    - (d) None of these

18. $I_C = \alpha L_E + \ldots\ldots\ldots\ldots$
    - (a) I_{CEO}
    - (b) I_B
    - (c) I_{CBO}
    - (d) βV_β

19. In a transistor ....
    - (a) $I_C = I_E + I_B$
    - (b) $I_E = I_C + I_B$
    - (c) $I_E = I_C + I_B$
    - (d) $I_B - I_C - I_B$

20. The element that has the biggest size in a transistor is ....
    - (a) Base
    - (b) Emitter
    - (c) Collector
    - (d) Collector base junction

21. The input impedance of a transistor connected in .... configuration is the highest
    - (a) Common emitter
    - (b) Emitter
    - (c) Common base
    - (c) None of these

22. The output impedance of a transistor connected in .... configuration is the highest
    - (a) Common base
    - (b) Common emitter
    - (c) Common collector
    - (d) None of these

23. As the temperature of a transistor goes up, the base emitter resistance ....
    - (a) Increase
    - (b) Decrease
    - (c) Remains the same
    - (d) None of these

24. The phase difference between the input and output voltages of transistor connected in CE mode is
    - (a) $0°$
    - (b) $180°$
    - (c) $90°$
    - (d) $270°$

25. A transistor has ....
    - (a) One p-n junction
    - (b) Two p-n junction
    - (c) Three p-n junction
    - (d) Four p-n junction

26. The phase different between the input and output voltage in *CB* mode is
    - (a) $180°$
    - (b) $90°$
    - (c) $270°$
    - (d) $0°$

27. In a transistor, if $I_{CBO} = 10$ μA and $\beta = 50$ then I_{CEO} will be
    - (a) 510 μA
    - (b) 10 μA
    - (c) 500 μA
    - (d) 60 μA

28. The leakage current I_{CBO} and I_{CEO} of .... mode respectively
    - (a) CE and CC
    - (b) CB and CC
    - (c) CB and CE
    - (d) CE and CB

29. In transistor, $\beta = 49$, $I_{CBO} = I$ μA, $I_B = 10$ μA then the value of I_C is
    - (a) 460 μA
    - (b) 490 μA
    - (c) 450 μA
    - (d) 540 μA

30. Emitter follower is
    - (a) Common collector configuration
    - (b) Common base configuration
    - (c) Common emitter configuration
    - (d) None of these

31. $I_{CEO} =$
    - (a) $(1 + \alpha) I_{CBO}$
    - (b) $\dfrac{I_{CBO}}{I - \alpha}$
    - (c) $\dfrac{I_{CBO}}{I + \alpha}$
    - (d) $(I - \alpha) I_{CBO}$

32. The value of β for a transistor is generally ....
    - (a) 1
    - (b) Less then one
    - (c) Above 500
    - (d) Between 20 and 500

33. The current I_B is
    - (a) Electron current
    - (b) Hole current
    - (c) Donor ion current
    - (d) Acceptor ion current

34. The input impedance of transistor is
    - (a) High
    - (b) Low
    - (c) Very high
    - (d) Almost Zero

35. The output impedance of a transistor is ....
    - (a) High
    - (b) Low
    - (c) Zero
    - (d) Very low

36. The phase difference between the input and output voltages of transistor connected in common collector arrangement is ....
    - (a) 180°
    - (b) 0°
    - (c) 90°
    - (d) 270°

37. $I_C = \beta I_B +$ ....
    - (a) I_{CBO}
    - (b) I_{CEO}
    - (c) I_C
    - (d) αI_E

38. $I_C = \dfrac{\alpha}{1 - \alpha} I_B + \dfrac{A}{1 - \alpha}$, where A is
    - (a) I_{CBO}
    - (b) I_{CEO}
    - (c) I_C
    - (d) I_E

39. A transistor is connected in CB mode. If it is now connected in CE mode with same bias voltages, the value of I_E, I_B and I_C will ....
    - (a) Increase
    - (b) Decrease
    - (c) Remain the same
    - (d) None of these

40. $I_{CEO} =$ .... I_{CBO}
    - (a) β
    - (b) $1 + \alpha$
    - (c) $\beta + 1$
    - (d) None of these

41. The most commonly used semiconductor in the manufacture of a transistor is
    - (a) Ge
    - (b) Si
    - (c) C
    - (d) None of these

42. BC 147 transistor indicates that it is made of ....
    - (a) Ge
    - (b) C
    - (c) Si
    - (d) None of these

43. In a *p-n-p* transistor, the current carriers are
    - (a) Holes
    - (b) Electrons
    - (c) Acceptor ions
    - (d) Donor ions
44. The number of depletion layers in a transistor is ....
    - (a) One
    - (b) Two
    - (c) Three
    - (d) Four
45. In a transistor, the base current is about ....of emitter current
    - (a) 20%
    - (b) 2%
    - (c) 30%
    - (d) 50%
46. In a *n-p-n* transistor, ...are minority carriers
    - (a) Free electrons
    - (b) Holes
    - (c) Donor ions
    - (d) Acceptor ions
47. Some of the majority carriers from the emitter ....
    - (a) Recombine in the base
    - (b) Recombine in emitter
    - (c) Pass through base to the collector
    - (d) None of these
48. The collector of a transistor is .... doped
    - (a) Heavily
    - (h) Lightly
    - (c) Moderately
    - (d) None of these
49. The voltage gain of a transistor connected in .... configuration is the height
    - (a) CB
    - (b) CE
    - (c) CE
    - (d) None of these

—— ANSWERS ——

1. (c)	2. (b)	3. (a)	4. (a)	5. (b)	6. (c)	7. (d)
8. (a)	9. (b)	10. (a)	11. (d)	12. (d)	13. (b)	14. (a)
15. (b)	16. (c)	17. (c)	18. (c)	19. (b)	20. (c)	21. (b)
22. (a)	23. (b)	24. (b)	25. (b)	26. (d)	27. (a)	28. (c)
29. (b)	30. (a)	31. (b)	32. (d)	33. (a)	34. (b)	35. (a)
36. (b)	37. (b)	38. (a)	39. (c)	40. (c)	41. (b)	42. (c)
43. (a)	44. (b)	45. (b)	46. (b)	47. (a)	48. (c)	49. (b)

EXERCISE

5.1. What is a transistor? Why it is so called?

5.2. Explain the construction and working of *p-n-p* transistor.

5.3. Explain the different configuration of transistor and what is the need of those?

5.4. Draw and explain the input and output character of common emitter configuration.

5.5. In a transistor circuit $I_E = 2$ mA and $I_C = 1.5$ mA. What is the value of I_B?   (**Ans.** 0.5 ma)

5.6. Explain the CE and CC configuration of BJT.   (**UPTU 2008-09**)

5.7. Explain why in the active operation, the base current I_B is much smaller than I_C or I_E. What is the relation among the three currents?
                                                       (**UPTU 2008-09**)

5.8. Define α and β with respect to BJT and derive the relationship between than.   (**UPTU 2008-09**)

5.9. Establish the following relations

$$\text{(a) } I_C = \frac{\alpha}{1-\alpha} I_B + \frac{1}{1-\alpha} I_{CBO}$$

$$\text{(b) } \gamma = \frac{1}{1-\alpha}$$

$$\text{(c) } I_E = (\beta + 1) I_B + (\beta + 1) I_{CBO}.$$

5.10. Explain the input and output characteristics of common base configuration.

5.11. In a transistor, $I_B = 65$ mA, $I_B = 25$ mA and $\beta = 450$. Find the value of α and determine the value of I_C.   (**Ans.** 0.99, 29.25)

5.12. Draw the circuit of transistor in CE configuration sketch the output characteristics indicate the active, saturation and cut-off region and explain each region in detail.   (**UPTU 2007-08**)

5.13. Explain the transistor as a switch.

5.14. Write short notes on BIT.

5.15. What is the leakage current I_{CBO} and how it is depend on temperature?

5.16. Explain why a transistor action cannot be achieved by connecting two back-to-back diode. In a transistor explain, why emitter region is heavily doped, based width in small and collector area is large. (**UPTU 2006-07**)

5.17. How much the two transistor junction be biased for proper transistor amplifier operation?

5.18. Which of the transistor current is always the largest? Which is always the smallest? Which two current are relatively closed manufactured?
                                                       (**UPTU 2004-05**)

5.19. Explain why an ordinary junction transistor is called bipolar?

5.20. Sketch a family of CB output characteristics for a transistor, Indicate the active, solution and cut-off region.   (**UPTU 2003-07**)

DC Biasing of BJTs and Stabilization of Operating Point — 6

6.1 INTRODUCTION

DC biasing is required for the transistor as an amplifier. It requires the knowledge of both DC and AC response of the system. If AC signal is applied at the transistor without DC biasing, then the amplification of AC signal is not possible. As we have studied earlier that the transistor can be used as an amplifier i.e., it is a magical device that can increase the level of the applied AC input without the assistance of an external energy source. The improved output AC power level is the result of the transfer of energy from the applied DC supplies. As we have studied earlier that the emitter junction J_E is forward bias and collector junction J_C is reverse bias, when the transistor acts as an amplifier. Due to this, the operating point of transistor (will be discussed in this chapter) comes into the active region of the transistor. So, we can say that the analysis or design of any electronic amplifier has two components: first is DC part and second is AC part. Fortunately, the superposition theories are applicable and the investigation of the DC conditions can be totally separated from the AC response.

The DC operation of a transistor is controlled by a number of factors, including the range of Q-points on the device characteristics.

There are some important basic relationships for a transistor which has recurring use as follows:

$$V_{BE} = 0.7 \text{ V (for } Si) \tag{6.1}$$

$$I_E = (\beta + 1)\, I_B = I_C \tag{6.2}$$

and,
$$I_C = \beta I_B. \tag{6.3}$$

What is the need of transistor biasing? Why operating point must lie in the centre of active region? What are the different biasing circuits? Is the operating point changed with the parameter variations? There are a lot of questions and answer must be known to the electronics engineering students. Generally, you will get the answer of these questions are in this chapter.

6.2 TRANSISTOR LOAD LINE ANALYSIS

In the transistor circuit analysis, it is important to examine the variation in collector current I_C with change in collector emitter voltage V_{CE}. There are two methods to determine the collector current I_C, one is by output characteristics in which we can obtain the collector current I_C at any desired value of V_{CE}. Second method is by load line analysis. This method is easy to determine the collector current I_C at different values of collector emitter voltage V_{CE}. So, we can say that load line analysis method is frequently used in the analysis of transistor applications.

DC Load Line

Considering a common emitter *n-p-n* transistor circuit as shown in Fig. 6.1(a), there is no signal applied on the circuit i.e., we can say that we are considering only DC conditions. The output characteristics of this circuit are shown in Fig. 6.1(b).

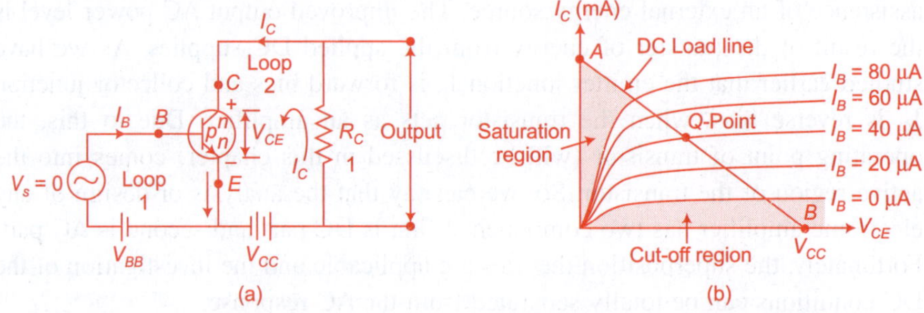

Fig. 6.1 *(a) Common emitter transistor amplifier circuit; (b) DC load line on output characteristics of transistor*

By the output characteristics, we can say that the most important factor is the effect of V_{CE} upon the collector current I_C. V_{CE} is greater than 1 volts. We can see that I_C is largely unaffected by changes in V_{CE}. It is entirely controlled by the base current I_B. When this happens, we can say that the output circuit represents a "constant current source". It can also be seen from the common emitter circuit of Fig. 6.1(a), that the emitter current I_E is the sum of the collector current I_C and the base current I_B, added together so we can also say that "$I_E = I_C + I_B$" for the CE configuration.

Now, applying KVL in output side of Fig. 6.1(a) i.e., in loop (2) we get the equation,

$$V_{CC} - I_C R_C - V_{CE} = 0$$
$$V_{CE} = V_{CC} - I_C R_C$$
$$I_C R_C = V_{CE} - V_{CC}$$

$$I_C = \frac{(V_{CE} - V_{CC})}{R_C}$$

$$I_C = \frac{V_{CC} - V_{CE}}{R_C} \tag{6.4}$$

where,

I_C = Collector current

V_{CC} = Collector supply voltage (bias voltage at collector junction)

R_C = Load resistance

V_{CE} = Collector emitter voltage.

A load line can be drawn directly on the graph of output characteristics above as shown in Fig. 6.1(b). From the point of "saturation", when $V_{CE} = 0$ to the point of "cut-off", when $I_C = 0$, giving us two point on the characteristics. When we join these two points, we will get a straight line and an operating point or Q-point of the transistor. These two points are calculated as:

From Eqn. (6.4)

(i) if $V_{CE} = 0$ (point of saturation)

$$I_C = \frac{V_{CC}}{R_C} - 0$$

$$I_C = \frac{V_{CC}}{R_C}$$

i.e. indicates the point $A \left[0, \dfrac{V_{CC}}{R_C} \right]$ lies on DC load line.

(ii) if $I_C = 0$ (point of cut-off)

$$0 = \frac{V_{CC}}{R_C} - \frac{V_{CC}}{R_C}$$

i.e. $V_{CC} = V_{CE}$ i.e., indicates the point B $(V_{CC}, 0)$ lies on the DC load line.

6.2.1 Operating Point

Biasing of a transistor is needed to restrict the operation of a transistor in the active region. The biasing circuit provides a DC current and voltage so that they correspond to a point in the linear region of the output characteristics. This point is known as Q-point (Quiescent point) or operating point for the transistor. This point is generally mid way in the linear characteristics of transistor to provide linear amplification i.e., output which is proportional to the input.

When the zero signal is applied at the transistor input, in that condition the values of I_C and V_{CE} give one point on the load line. This point is known as operating point or Q-point as shown in Fig. 6.2. In other words, for transistor amplifiers the resulting DC current and voltage establishing an operating point on the output characteristics which defines the region that will be employed for amplification of the applied signal. This operating point is also called as *Q*-point or Quiescent point which means quiet, still, inactive.

In the absence of signal, the base current is 10 µA, then I_C and V_{CE} conditions in the circuit must be represented by some point on $I_B = 10$ µA characteristics. This point is called as *Q*-point or operating point on DC load line *AB*. We will try to locate the operating point at the centre of the active region of output characteristics.

When the operating point is fixed at the centre of the active region because at the centre of active region the output characteristics are most linear and uniformly spaced, variation in base current (input signal) will produce proportional changes in output (collector current). It is necessary that

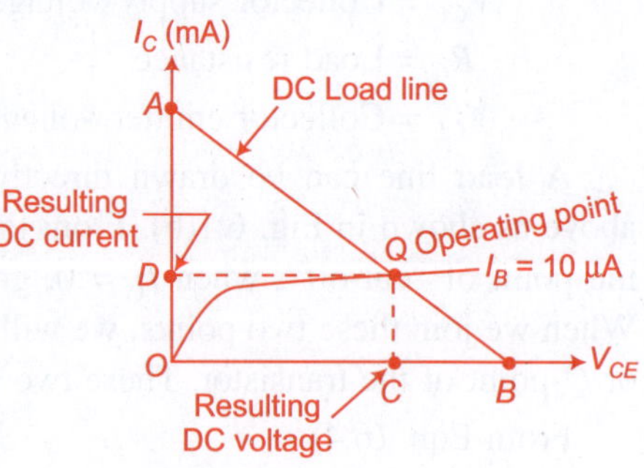

Fig. 6.2 *Operating point*

once the *Q*-point is fixed, its position should not change. If it shifts to a position near the saturation region or to a position near the cut-off region of the output characteristics, the amplified output will be distorted as shown in Figs. 6.3 (a), (b), (c) and (d).

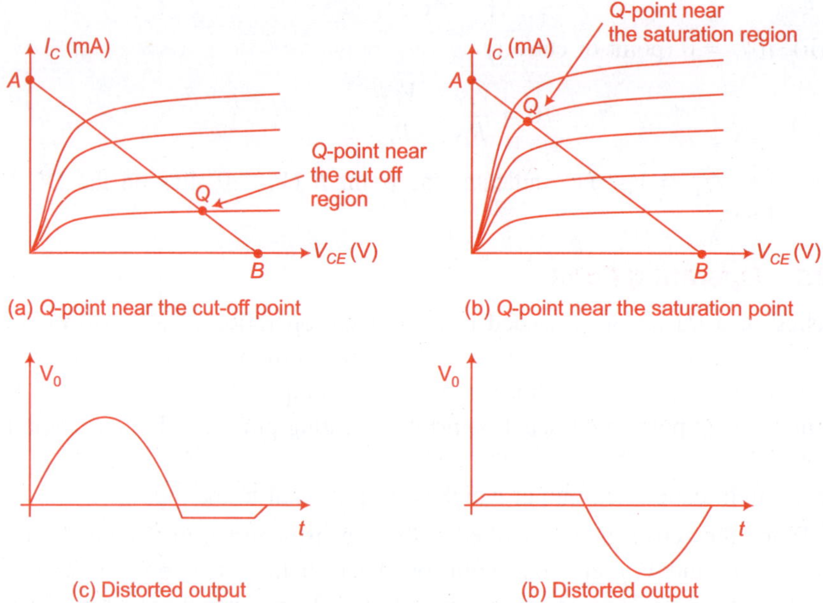

Fig. 6.3

Example 6.2.1   For the circuit shown in Fig. Ex. 6.2.1 (a) draw the DC load line

Solution   Given that,

$$R_C = 2.4 \text{ k}\Omega$$

$$V_{CC} = 12 \text{ V.}$$

The collector emitter voltage is given by,

$$V_{CE} = V_{CC} - I_C R_C \quad \text{by Eqn. (6.1)}$$

$$I_C = 0$$

$$V_{CE} = V_{CC} = 12 \text{ volt.}$$

This indicates point B on the voltage axis.

When $V_{CE} = 0$

$$I_C = \frac{V_{CC}}{R_C} = \frac{12 \text{ V}}{2.4 \text{ k}\Omega}$$

$$I_C = 5 \text{ mA.}$$

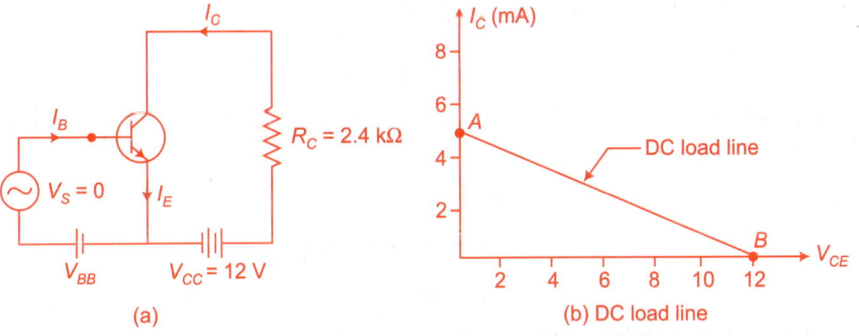

Fig. Ex. 6.2.1   *(a) Circuit for Ex. 6.2.1; (b) DC load line*

Thus, locate point A on the collector current axis. By joining these two points A and B, we will get DC load line as shown in Fig. Ex. 6.2.1(b).

Example 6.2.2   In the circuit diagram of Fig. Ex. 6.2.2 (a) if $V_{CC} = 13$ V and $R_C = 5$ kΩ draw the DC load line. What will be the Q-point, if zero signal I_B is 20 µA and $\beta = 50$?

Solution   Given that,

$$V_{CC} = 13 \text{ V,} \quad I_B = 20 \text{ µA}$$

$$R_C = 5 \text{ k}\Omega, \quad \beta = 50.$$

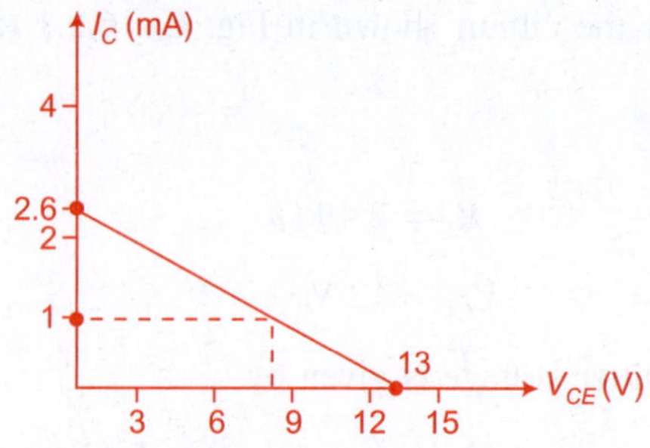

Fig. Ex. 6.2.2 *Q-point on DC load line*

The collector emitter voltage V_{CC} is given by Eqn. (6.1)

$$V_{CE} = V_{CC} - I_C R_C$$

$$I_C = 0$$

$$V_{CE} = V_{CC} = 13 \text{ volt}$$

$$V_{CE} = 13 \text{ V} \quad \text{(locate point } A)$$

When $V_{CE} = 0$

$$V_{CC} = I_C R_C$$

$$I_C = \frac{V_{CC}}{R_C} = \frac{13 \text{ V}}{5 \text{ k}\Omega}$$

$$I_C = 1.6 \text{ mA} \quad \text{(locate point } B).$$

By joining point A and B, load line is constructed as shown in Fig. Ex. 6.2.2. Zero signal collector current

$$I_C = \beta I_B$$

$$= 50 \times 0.02 \text{ mA}$$

$$= 1 \text{ mA}.$$

Zero signal collector emitter voltage

$$V_{CE} = V_{CC} - I_C R_C$$

$$= 13 - (1 \text{ mA} \times 5 \text{ k}\Omega)$$

$$= 13 - 5$$

$$V_{CE} = 8 \text{ V}$$

$\therefore$ Operating point is 8 V and 1 mA.

From the above discussion, we can say that the output characteristics of a transistor is more linear at the centre of the active region and uniformly spaced. Hence, there is sufficient space is available at both sides of operating point as shown in Fig. 6.4. Due to this, there are proper and proportional changes with occur in output signal when input signal is changed. So, the input signal is amplified properly and is equal in shape as the input. Hence, the output is not distorted. For the BJT to be biased in its linear or active operating region, the following must be true:

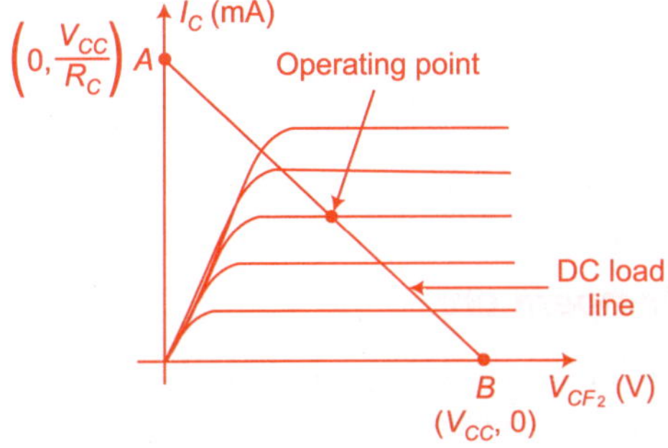

Fig. 6.4 *Correct selection of Q-point (at the centre of active region)*

1. The base-emitter junction J_E must be forward bias (*p*-region voltage more positive) with a resulting forward bias voltage of about 0.6 to 0.7 V.

2. The base collector junction J_C must be reverse biased (*n*-region more positive) with the reverse bias voltage being any value within the maximum limits of the device.

6.2.2 Factor which Cause Shift of Operating Point

Let us assume that the DC voltage is applied at the transistor for biasing of it. Due to the DC voltage, DC current will flow in the transistor. So, I_B, I_C and V_{CB} are fixed. Hence, the operating point is fixed at the particular place of output characteristics. Now, we have to discuss the factors which cause shift of operating point at the output characteristics which are as follows:

(i) Variation of Parameter

The position of operating point or Q-point is changed on the load line with the change in parameters of transistor like α and β i.e., the operating point shifts from one place to another.

As we know that the value of α and β is decided at the time of manufacturing or provided by the manufacturer. Then you will think how the change in parameters like α and β is possible? For the answer of this question, we will consider that, in any circuit we are using the transistor BC147. If this transistor is faulty, then we will replace this transistor by the same number, where as both the transistors are of same type. But there can be small difference in α. As we know that due to small change in α, there is large change in β.

For example,

$$\alpha = 0.99$$

then,

$$\beta = \frac{\alpha}{1 - \alpha} = \frac{0.99}{1 - 0.99} = 99$$

if α is charge and occurs

$$\alpha = 0.98$$

then,

$$\beta = \frac{\alpha}{1 - \alpha} = \frac{0.99}{1 - 0.98} = 49.$$

By the above calculation, we can observe that if the is change in α is 1% it will cause change in β by 50%.

(ii) Variation of Temperature

Shift in operating point due to change in temperature is the main reason. As we know that the leakage current (I_{CEO} & I_{CBO}) or reverse saturation current of transistor is dependent on the temperature. Generally, the leakage current is doubled, when the temperature increase is 10°C. The collector current is dependent upon the leakage current as shown below,

$$I_C = \beta I_B + (\beta + 1)I_{CBO}.$$

Hence, the collector current is also increased with the increase in leakage current. Due to this, the operating point is shifted.

Thermal Runway: Transistor can be seriously affected by the temperature rise. Two of the most temperature sensitive quantities are V_{BE} and I_C.

As we know that an increase in I_{CBO} will cause an increase in I_C and further an increase of I_C will cause rise in the temperature of the junction. Due to this temperature increase, I_{CBO} will again increase and this process will go on till the junction is finally damaged. This process of self-destruction of the transistor is called Thermal Runaway.

What is the need of bias stabilization?

As we have studied in the earlier section that the biasing of transistor is necessary to make the transistor work. It is not sufficient that the operating point of transistor is set at the centre of the active region when the emitter junction is forward bias and collector junction is reverse bias. It is also necessary to fix the operating point at the place where it is set. As we saw in the previous section that the operating point of transistor shifts with the parameter variations. Due to the shifting of operating point, the performance of transistor is bad. Due to the change in parameters of transistor, the operating point shift towards the unuseful region which is called cut-off or saturation region. The output of the transistor is distorted. So, we need a bias circuit which is capable to fix the operating point with parameter and temperature variations.

"So, we can say that the stabilization is the process by which the effect of temperature and variation of transistor parameter has no impact on the operating pointer Q-point".

6.3 BIAS CIRCUITS

Before studying the bias circuits, we should know its properties i.e., the expectation from bias circuit. A good bias circuit is capable of:

    (i) To set the operating point at the centre of active region.

    (ii) The operating point must fix the change in parameter of transistor i.e., α and β.

   (iii) The operating point must fix the change in temperature.

So, the bias circuits are designed to fulfill the above requirements.

The stability of operating point is measured by the stability factor of the bias circuit.

"Stability factor S is defined as the ratio of change in the collector current (δI_C) to the change in the collector base leakage current (δI_{CBO})"

i.e.
$$S = \frac{\delta I_C}{\delta I_{CBO}}\bigg|_{\beta \ \& \ V_{BE} = \text{constant}} \tag{6.5}$$

where,
$$S \geq 1.$$

If the stability factor S of bias circuit is less than the effect of change in leakage current on the collector current is less then, its thermal stability is good. In other words, we can say that the operating point is more stable when the stability factor is less or minimum.

Mathematical Expression for Stability Factor

As we know the equation for collector current is,

$$I_C = \beta I_B + (1 + \beta)\, I_{CBO} \tag{6.6}$$

where, I_C = Collector current

             I_B = Base current

         I_{CEO} = Reverse saturation current

           β = Common-emitter current gain.

Differentiating the Eqn. (6.6) with respect to I_C we get,

$$1 = \frac{d}{dI_C}\,(\beta.I_B) + \frac{d}{dI_C}\,[(1 + \beta)\, I_{CBO}]$$

Let us assume β is constant,

$$1 = \beta\,\frac{dI_B}{dI_C} + (1 + \beta)\,\frac{dI_{CBO}}{dI_C}$$

$$\because \qquad \frac{dI_C}{dI_{CBO}} = S$$

from Eqn. (6.5). Now, we get,

$$1 = \beta \, \frac{dI_B}{dI_C} + (1 + \beta) \cdot \frac{1}{S}$$

$$S = S \, \beta \, \frac{dI_B}{dI_C} + 1 + \beta$$

$$S - S \, \beta \, \frac{dI_B}{dI_C} = 1 + \beta$$

$$S \left[1 - \beta \, \frac{dI_B}{dI_C} \right] = 1 + \beta$$

$$S = \frac{1 + \beta}{1 - \beta \left(\dfrac{dI_B}{dI_C} \right)} \qquad (6.7)$$

The Eqn. (6.6) is a general equation for any type of biasing circuit.

Different Biasing Circuits

We have studied about biasing circuit. In this section, we will study about different biasing circuits as follows:

    (a) Fixed bias circuit

    (b) Collector to base bias circuit

    (c) Emitter bias or self bias circuit

    (d) Potential divider bias circuit

6.3.1.1 Fixed Bias Circuit

The fixed bias circuit is shown in Fig. 6.5 (a). It provides a relatively straight forward and simple introduction to transistor DC bias analysis. As we can see in Fig. 6.5 (a), the battery of V_{CC} volts provides bias voltage to the base and collector. The bias voltage provides to base through resistance R_B and bias voltage provides to the collector through resistance R_C. The biasing of collector and base also can be done by two separate DC supply. But single DC supply is more economic and suitable for biasing. The network employs a *n-p-n* transistor, the equations and calculations apply equally well to a *p-n-p* transistor merely by changing all current directions and voltage polarities. The voltages are defined by the standard double-subscript notation. For the DC analysis, the network can be isolated from the indicated AC levels by replacing the capacitor with an open circuit equivalent.

Analysis of the Circuit

The DC analysis of the bias circuit is divided in two parts, one is input side and another one is output side as shown below,

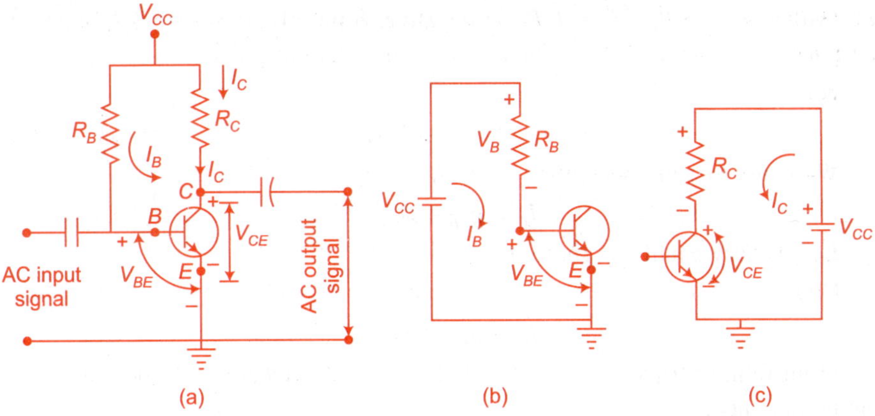

Fig. 6.5 *(a) Fixed bias circuit; (b) Input side loop; (c) Output side loop*

(i) Analysis of Input or Base-Emitter Side

The input side of the circuit of Fig. 6.5 (a) is shown in Fig. 6.5 (b).

Applying the Kirchhoff's voltage low at the input side, we obtain,

$$V_{CC} - I_B R_B - V_{BE} = 0.$$

Note The current I_B carried from the positive terminal of V_{CC} so V_{CC} is positive and the current I_B enter at the positive terminal of the resistance R_B. So, the voltage across R_B is V_B and is negative. Similarly, the current I_B is enters the positive terminal of the voltage V_{BE}, So, it is also negative.

$$I_B = \frac{V_B}{R_B} = \frac{V_{CC} - V_{BE}}{R_B} \tag{6.8}$$

From Eqn. (6.8), we can state that I_B is the base current that flows through R_B and by Ohm's law the current is the voltage across R_B which divided by the resistance R_B. The voltage across R_B is,

$$V_B = V_{CC} - V_{BE}$$

where, V_{CC} = Supply voltage

V_{BE} = Voltage drop across base to emitter junction.

So, by the Eqn. (6.8) we can say that the selection of base resistor, R_B, sets the level of base current for the operating point because V_{CC} and V_{BE} are constant.

(ii) Analysis of Output Side or Collector Emitter Side

The output side of the circuit of Fig. 6.5 (a) is shown in Fig. 6.5 (c).

Applying Kirchhoff's voltage low at the output side, we obtain,

$$V_{CC} - V_{CE} - I_C R_C = 0.$$

Note The current I_C comes out from the positive terminal of battery V_{CC}, so the voltage V_{CC} is positive. And current I_C enters at the positive terminal of R_C, So, the voltage across R_C, $V_C = I_C R_C$ is negative. Similarly, the current I_C enters at the positive terminal of the voltage V_{CE}, so it is also negative.

Now,

$$V_{CE} = V_{CC} - I_C R_C. \qquad (6.9)$$

We know that the collector current I_C is,

$$I_C = \beta I_B + I_{CEO}$$

I_{CEO} is very small so ignore it.

Now,

$$I_C = \beta I_B$$

From Eqn. (6.9), we obtain V_{CE}, I_C by which the location of operating point can be obtained.

Stability Factor for Fixed Bias Circuit

From Eqn. (6.7), we know that the stability factor is,

$$S = \frac{1 + \beta}{1 - \left(\beta \dfrac{dI_B}{dI_C}\right)}.$$

We know that fixed bias circuit I_B is independent of I_C by Eqn. (6.8).

Hence,

$$\frac{dI_B}{dI_C} = 0$$

so,

$$S = \frac{\beta + 1}{1 - \beta(0)} = \beta + 1$$

$$S = \beta + 1. \qquad (6.10)$$

From Eqn. (6.10), we state that the value of S is large due to change in I_{CBO}. Hence, the stability is poor in fixed bias circuit.

6.3.1.2 *Merits and Demerits of Fixed Bias Circuit*

Merits Fixed bias circuit has following merits:

(i) The value of base current I_B can be easily set by the appropriate selection of base resistance R_B.

(ii) Fixed bias circuit is simple in construction and calculations are easy.

(iii) By the proper selection of R_B, the I_B and I_C can be set easily and operating point can be located at the centre of active region easily.

Demerits Fixed bias circuit has following limitations:

(i) When the transistor is not working, then we replace the transistor. Due to this, there is small change in α of transistor, so there is large change in β. As we have discussed earlier since the collector current is dependent upon the β, therefore it will also change. Due to this, the operating point is shifted.

(ii) Due to the change in temperature, the leakage current will increase and I_C will also increase depending on the leakage current. So, the operating point will shift. There is an increase in power loss at the junction due to increase in I_C. So, the temperature will increase, Hence, there is a repetition of process. So, there is a possibility of thermal runway. There is no protection available from thermal runway.

By summarizing the merits and demerits of fixed bias circuit, we can state that the operating point is easily located at the centre of active region in fixed bias circuit but the stability of the operating point is poor due to change in parameters of the transistors. Therefore, fixed bias circuits are not used.

Example 6.3.1.1 Determine the following for the fixed bias circuit of Fig. Ex. 6.3.1.

(a) location of operating point i.e., I_B, I_C and V_{CE} assume that $V_{BE} = 0.7$ volt.

(b) V_B and V_C.

(c) V_E and V_{BC}.

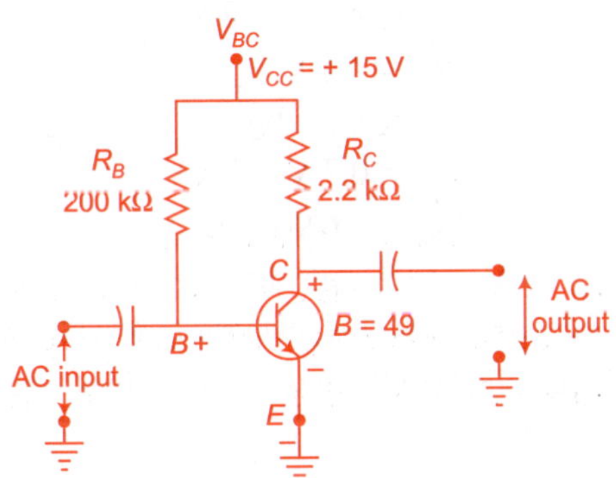

Fig. Ex. 6.3.1 *DC fixed bias circuit for Ex. 6.3.1*

Solution   (a) From Eqn. (6.8)

$$I_B = \frac{V_{CC} - V_{BE}}{R_B}$$

$$I_B = \frac{(15 - 0.7) \text{ V}}{200 \text{ k}\Omega}$$

$$I_B = 71.5 \ \mu\text{A}.$$

We know that,

$$I_C = \beta \, I_B = 49 \times 71.5$$

$$I_C = 3.5 \text{ mA}.$$

From Eqn. (6.9), we obtain,

$$V_{CE} = V_{CC} - I_C R_C$$
$$V_{CE} = 15 \text{ V} - (3.5 \text{ mA}) (2.2 \text{ k}\Omega)$$
$$V_{CE} = (15 \text{ V} - 7.7 \text{ V})$$
$$V_{CE} = 7.3 \text{ volt.}$$

Hence, the operating point is located at,

$$I_C = 3.5 \text{ mA} \quad \text{and} \quad V_{CE} = 7.3 \text{ volt.}$$

(b) Because emitter is grounded. So,

$$V_E = 0$$
$$\therefore \quad V_{BE} = V_B - V_E$$
$$V_B = V_{BE} + V_E = 0.7 \text{ V} + 0 \text{ V}$$
$$V_B = 0.7 \text{ V.}$$

Now,

$$\therefore \quad V_{CE} = V_C - V_E$$
$$V_{CB} = V_{CE} + V_E$$
$$= 7.3 \text{ V} + 0 \text{ V}$$
$$V_C = 7.3 \text{ V}$$

and,

$$V_{BE} = V_B - V_C$$
$$V_{BC} = 0.7 \text{ V} - 7.3 \text{ V}$$
$$V_{BC} = -6.96 \text{ volt.}$$

The negative sign of V_{BC} indicates that the base voltage with respect to collector is negative i.e., base collector junction is reverse biased.

Example 6.3.1.2 Figure Ex. 6.3.2 shows a fixed bias circuit. Determine with $\beta = 50$.

(a) The operating point, Q

(b) The stability factor, S.

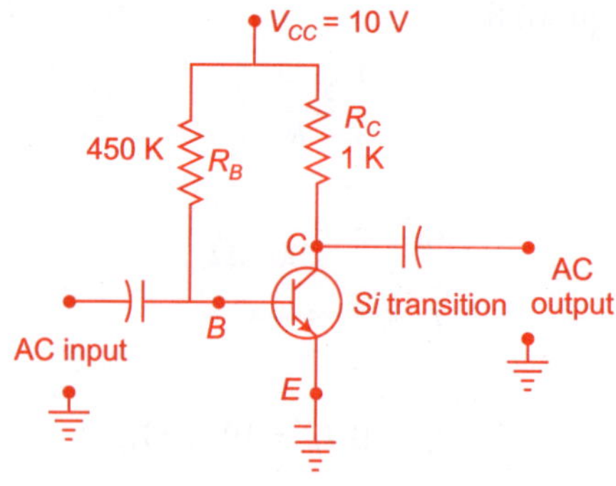

Fig. Ex. 6.3.2 *Fixed bias circuit*

Solution   For a DC analysis, we consider only DC circuit i.e., we remove AC part of the circuit from input and output side. So, the circuit will be similar as shown in Fig. Ex. 6.3.3.

(a) Given that,

$$V_{CC} = 10 \text{ V}$$

Base resistance, $\qquad R_B = 450 \text{ k}\Omega$

Collector resistance, $\qquad R_C = 1 \text{ k}\Omega.$

Since, transistor is silicon transistor therefore the voltage drop across base-emitter junction is 0.7 V

i.e. $\qquad\qquad V_{BE} = 0.7 \text{ V}.$

Now, applying KVL at the input side, the circuit will be shown in Fig. Ex. 6.3.3 (a)

$$V_{CC} - I_B R_B - V_{BE} = 0$$
$$V_{CC} = I_B R_B + V_{BE}$$

$$I_B = \frac{V_{CC} - V_{BE}}{R_B}$$

$$I_B = \frac{10 - 0.7}{450 \text{ k}\Omega} = \frac{9.3 \text{ V}}{450 \text{ k}\Omega}$$

$$I_B = \frac{9.3 \times 10^{-5}}{4.5} = 2.06 \times 10^{-5} \text{ A}$$

$$I_B = 20 \text{ }\mu\text{A}.$$

Fig. Ex. 6.3.3

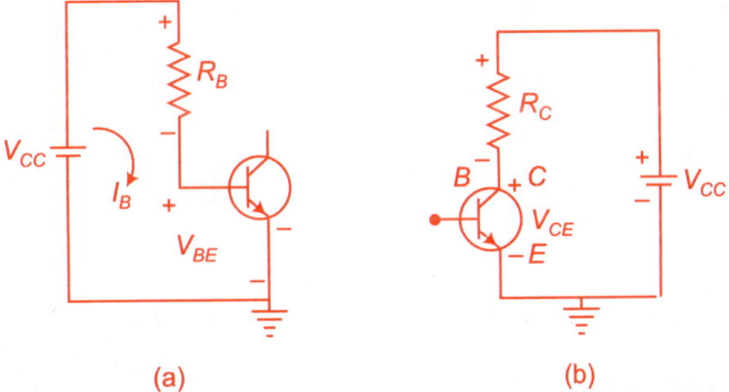

Fig. Ex. 6.3.3   *(a) KVC at input side; (b) KVC at output side*

The collector current is given by,

$$I_C = \beta I_B = 50 \times 20 \text{ }\mu\text{A}$$
$$I_C = 1000 \text{ }\mu\text{A} = 1 \text{ mA}$$
$$I_C = 1 \text{ mA}.$$

Applying KVL at the output side the circuit will be shown in Fig. Ex. 6.3.3 (b)

$$V_{CE} - V_{CC} - I_C R_C = 0$$
$$V_{CE} = V_{CC} - I_C R_C$$
$$V_{CE} = 10 - (1 \times 10^{-3} \times 1 \times 10^3)$$
$$V_{CE} = 10 - 1 = 9 \text{ V}$$

Thus, operating point $Q = (V_{CE}, I_C) = (9 \text{ V}, 1 \text{ mA})$.

(b) For fixed bias circuit the stability factor is given by the Eqn. (6.7)

i.e.
$$S = \beta + 1$$
$$S = 50 + 1 = 51$$
$$S = 51.$$

6.3.2.1 *Collector to Base Bias Circuit*

The collector to base bias circuit is shown in Fig. 6.6 (a). In this circuit, the bias voltage is given to the base from collector terminal in place of V_{CC}. The bias voltage is provided to the base through resistance R_B. This type of biasing is also called as Grounded Emitter Circuit with feedback resistor. The resistance R_B is used as feedback resistor. For the DC analysis, the network can be isolated from the indicated AC levels by replacing the capacitor with an open circuit equivalent.

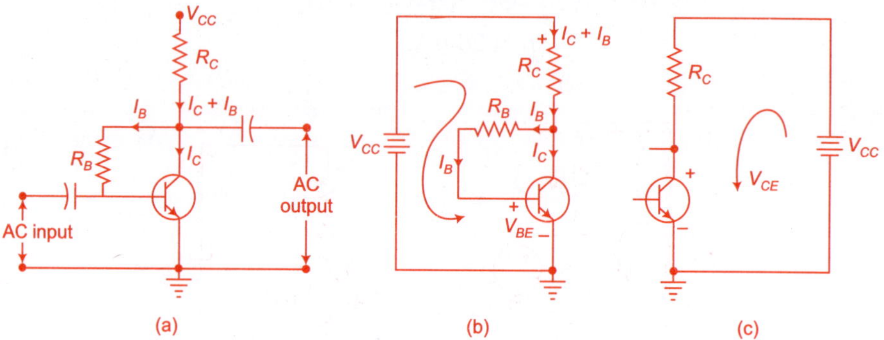

Fig. 6.6 *(a) Collector to base bias circuit; (b) Input side loop; (c) Output side loop*

Analysis of the Circuit

Again the DC analysis of the circuit is divided in two parts, one is input side and another one is output side as shown below.

(i) DC analysis of input side or base-emitter side

The input side or base-emitter side of Fig. 6.6 (a) is shown in Fig. 6.6 (b).

Applying the Kirchhoff's voltage law (KVL) at input side, we obtain,

$$V_{CC} - (I_C + I_B) R_C - I_B R_B - V_{BE} = 0 \tag{6.11}$$

Note The current $(I_B + I_C)$ comes from the positive terminal of V_{CC}, so V_{CC} is positive and the current $(I_B + I_C)$ enters the positive terminal of the resistance R_C, so the voltage across R_C is V_C and is negative. Similarly, the current I_B enters at the positive terminal of the voltage V_B (across R_B) and V_{BE}, so it is also negative.

Now, putting $I_C = \beta I_B$ in Eqn. (6.11) we obtain,

$$V_{CC} - (\beta I_B + I_B)\, R_C - I_B R_B - V_{BE} = 0$$

$$V_{CC} - I_B\, (\beta + 1)\, R_C - I_B R_B - V_{BE} = 0$$

$$V_{CC} - V_{BE} = I_B\, (\beta + 1)\, R_C + I_B R_B$$

$$V_{CC} - V_{BE} = I_B\, [(\beta + 1)\, R_C + R_B]$$

$$I_B = \frac{V_{CC} - V_{BE}}{R_B + (\beta + 1)\, R_C}. \tag{6.12}$$

So, by the Eqn. (6.12), we can say that the selection of base resistance and collector resistance, sets the level as base current for the operating point because V_{CC} and V_{BE} are constant.

(ii) Analysis of output side or collector emitter side

The output side or collector emitter side of Fig. 6.6 (a) is shown in Fig. 6.6 (c).

Applying the Kirchhoff's voltage law (KVL) at output side, we obtain,

$$V_{CC} - (I_C + I_B)\, R_C - V_{CE} = 0$$

$$V_{CE} = V_{CC} - (I_C + I_B)\, R_C. \tag{6.13}$$

From Eqn. (6.13), we can state that the location of operating point can be calculated by the calculation of the value of I_C and V_{CE}.

Effect of temperature rise or change in value β on location of Q-point.

From Eqn. (6.11)

$$V_{CC} - (I_C + I_B)\, R_C - I_B R_B - V_{BE} = 0$$

$$I_B R_B = V_{CC} - (I_C + I_B)\, R_C - V_{BE}$$

$$I_B = \frac{V_{CC} - (I_C + I_B)\, R_C - V_{BE}}{R_B} \tag{6.14}$$

Putting the value of V_{CE} from Eqn. (6.13) in Eqn. (6.14), we obtain,

$$\because \qquad V_{CE} = V_{CC} - (I_B + I_C)\, R_C$$

$$\therefore \qquad I_B = \frac{V_{CE} - V_{BE}}{R_B}. \tag{6.15}$$

* When temperature is increased, the leakage current I_{CEO} increases, the collector current will also increase due to the equation $I_C = \beta I_B + I_{CEO}$.

$$I_C = \beta I_B + I_{CEO}$$

| Temperature $\uparrow$ | $\rightarrow$ | $I_{CE} \uparrow$ | $\rightarrow$ | $I_C \uparrow$ |

* When β is increased due to the replacement of transistor, the collector current I_C will increase due to the equattion $I_C = \beta I_B$. Due to the increase in I_C, the collector-emitter voltage V_{CE} is reduced by Eqn. (6.13). The base current I_B is also reduced by Eqn. (6.15). Due to the reduction in base current I_B, the collector current I_C will be reduced.

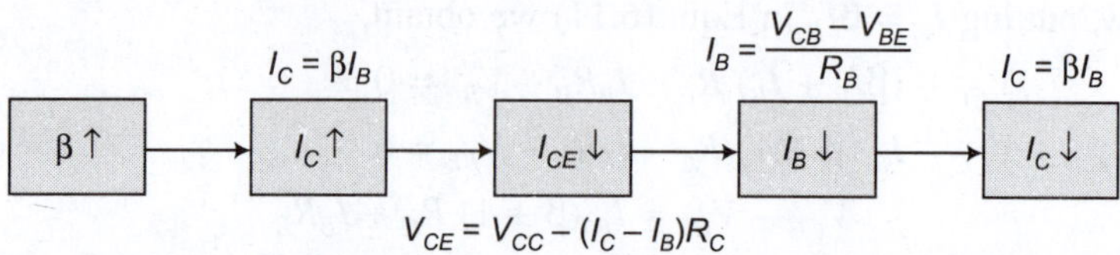

Hence, from the above discussion, we can state that the collector current I_C tries to increase. But this circuit has provision to reduce the I_C i.e., I_C cannot increase rapidly. So, this circuit increases the stability of the operating point.

So, we can say that increase in temperature or value of β, the collector current I_C will not increase as rapidly as increase in fixed bias circuit.

6.3.2.2   Merits and Demerits of Collector in Base Bias Circuit

Merits

(i) The circuit is easy i.e., the analysis of the circuit is simple.

(ii) The stability of the circuit is more than the stability of the fixed bias circuit when there is change in temperature and β.

Demerits

The disadvantage of this circuit is that as R_B is connected to the collector. Hence, the amplified AC signal present at the collector, will also be feedback to the input circuit. The AC signal at output is out of phase with input signal. Due to the feedback of AC signal, the voltage gain of amplifier is reduced. So, this circuit is not used widely.

Stability Factor For Collector to Base Bias Circuit

Stability factor is given by the expression,

$$S = \frac{1 + \beta}{1 - \beta \dfrac{dI_B}{dI_C}}. \qquad (6.16)$$

Applying KVL to the emitter-base portion of collector to base bias circuit of Fig. 6.6 (b)

$$V_{CC} - (I_C + I_B)\,R_C - I_B\,R_B - V_{BE} = 0$$

$$V_{CC} = (I_C + I_B)\,R_C + I_B\,R_B + V_{BE}$$

$$\because \qquad I_C = \beta I_B$$

$$\therefore \qquad V_{CC} = (I_C + I_B)\,R_C + I_B\,R_B + V_{BE}$$

$$V_{CC} = (R_C + R_B)\, I_B + V_{BE} + I_C\, R_C$$

$$I_B = \frac{V_{CC} - V_{BE} - I_C\, R_C}{R_B + R_C}. \tag{6.17}$$

Differentiating the Eqn. (6.17) with respect to I_C, we obtain,

$$\frac{dI_B}{dI_C} = \frac{0 - 0 - 1\,.\,R_C}{R_B + R_C} = -\,\frac{R_C}{R_B + R_C}$$

$$\frac{dI_B}{dI_C} = -\,\frac{R_C}{R_B + R_C} \tag{6.18}$$

Putting the value of $\dfrac{dI_B}{dI_C}$ in Eqn. (6.16), we have,

$$S = \frac{1 + \beta}{1 + \beta\left(\dfrac{R_C}{R_B + R_C}\right)}.$$

From the above expression, it is clear that stability factor of collector-base circuit is smaller than that of fixed bias circuit. Therefore, this biasing circuit has greater stability than fixed bias circuit.

Example 6.3.2.1 Find I_C and V_{CB} from given Fig. Ex. 6.3.2.1 given that $V_{BE} = 0.7$ V

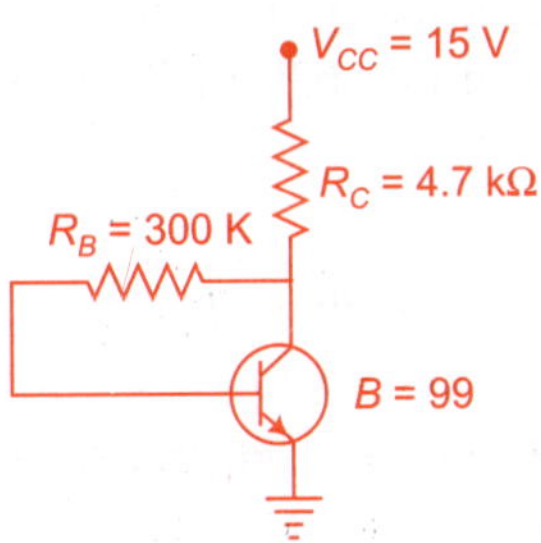

Fig. Ex. 6.3.2.1

Solution Given that,

$$V_{CC} = 15\ \text{V}$$
$$R_B = 300\ \text{k}\Omega \quad \text{and} \quad R_C = 4.7\ \text{k}\Omega$$
$$V_{BE} = 0.7\ \text{V} \quad \text{and} \quad \beta = 99.$$

From Eqn. (6.12)

$$I_B = \frac{V_{CC} - V_{BE}}{R_B + (\beta + 1)\, R_C}$$

$$I_B = \frac{15 - 0.7}{300\ \text{k}\Omega + (99 + 1)\, 4.7\ \text{k}\Omega}$$

$$= \frac{14.3}{300\ \text{k}\Omega + 470\ \text{k}\Omega}$$

$$I_B = \frac{14.3}{770 \text{ k}\Omega} = 0.0186 \text{ mA}$$

$$I_B = 18.6 \text{ μA}$$

Now,

$$I_C = \beta I_B$$
$$I_C = 99 \times 18.6 \text{ μA}$$
$$I_C = 1.84 \text{ mA}$$

From Eqn. (6.13)

$$V_{CE} = V_{CC} - (I_C + I_B) R_C$$
$$V_{CE} = 15 - (1.84 + 0.0186) \text{ mA} \times 4.7 \text{ k}\Omega$$
$$V_{CE} = 15 - 8.74$$
$$V_{CE} = 6.26 \text{ volt.}$$

Hence, the location of Q-point is

$$\boxed{I_C = 1.84 \text{ mA, } V_{CE} = 6.26 \text{ volt.}}$$

6.3.3.1 Emitter Bias or Self Bias Circuit

Fixed bias circuit is shown in Fig. 6.5 (a). If emitter resistance R_E is connected between emitter terminal and ground, then this combination of circuit is called emitter bias circuit as shown in Fig. 6.7 (a). Three resistance R_B, R_C and R_E and one DC supply is used.

Analysis of Circuit

The DC analysis of the circuit is divided in two parts, one is input side and another one is output side as shown below.

(i) DC analysis of input side or base-emitter bias

The input side or base-emitter side of Fig. 6.7 (a) is shown in Fig. 6.7 (b).

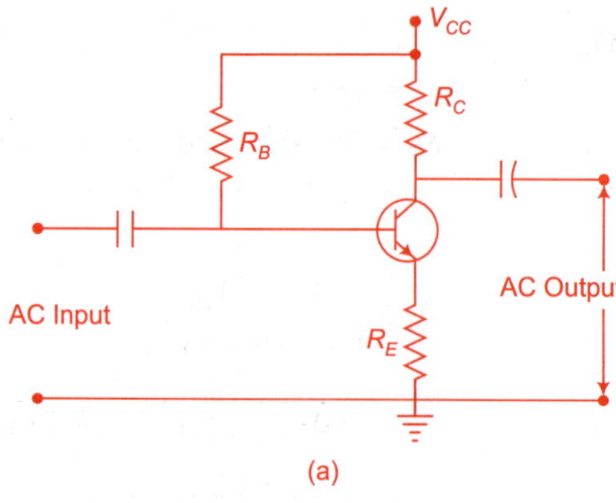

Fig. 6.7 *(a) Emitter bias or self bias circuit*

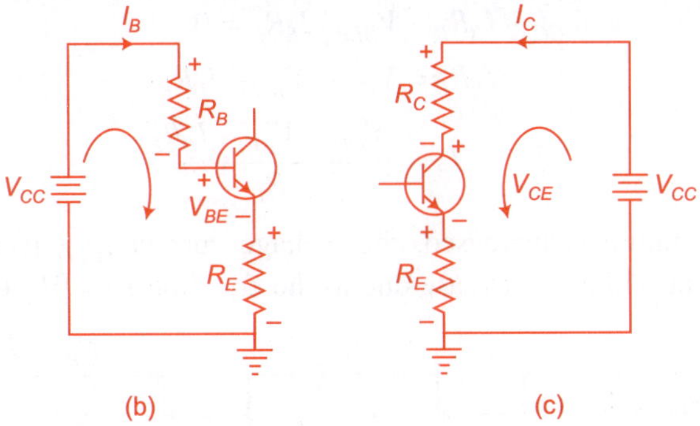

Fig. 6.7 *(b) Input side loop; (c) Output side loop*

Applying Kirchhoff's voltage law (KVL) at input side loop, we obtain,

$$V_{CC} - I_B R_B - V_{BE} - I_E R_E = 0 \tag{6.19}$$

$$\because \qquad I_E = (\beta + 1)\, I_B$$

Hence,

$$V_{CC} - I_B R_B - V_{BE} - (\beta + 1)\, I_B R_E = 0$$

$$I_B R_B + (\beta + 1)\, I_B R_E = V_{CC} - V_{BE}$$

$$I_B\, [R_B + (\beta + 1)\, R_E] = V_{CC} - V_{BE}$$

$$I_B = \frac{V_{CC} - V_{BE}}{R_B + (\beta + 1)\, R_E}. \tag{6.20}$$

(ii) Analysis of Output Side or Collector Emitter Side

The output side or collector emitter side of Fig. 6.7 (a) is shown in Fig. 6.7 (c).

Applying Kirchhoff's voltage law in output loop, we obtain,

$$V_{CC} - I_C R_C - V_{CE} - I_E R_E = 0$$

$$V_{CE} = V_{CC} - I_C R_C - I_E R_E$$

$$\therefore \qquad I_E = I_C$$

$$V_{CE} = V_{CC} - I_C R_C - I_C R_E$$

$$V_{CE} = V_{CC} - I_C\,(R_C + R_E) \tag{6.21}$$

The collector current is given by equation,

$$I_C = \beta I_B + I_{CEO}$$

$\because I_{CEO}$ is negligible.

Hence, $\qquad\qquad I_C = \beta I_B$

So, we can say that by calculation of I_C and V_{CE}, the operating point can be defined.

Effect of temperature rise or change in value of β on location of Q-point in emitter bias circuit.

From Eqn. (6.19),

$$V_{CC} - I_B R_B - V_{BE} - I_E R_E = 0$$

$$I_B R_B = V_{CC} - V_{BE} - I_E R_E$$

$$I_B = \frac{V_{CC} - V_{BE} - I_E R_E}{R_B}. \tag{6.22}$$

* When temperature is increased, the leakage current I_{CEO} increases and the collector current will also increase due to the equation $I_C = \beta I_B + I_{CEO}$.

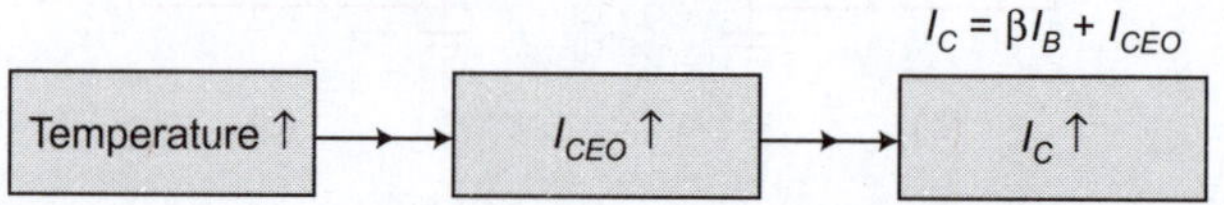

* When β is increased due to the replacement of transistor, the collector current I_C will increase. The emitter current I_E will also increase with increase in I_C due to the equation $I_C = I_E$. So, the voltage drop across V_E across R_E i.e., $I_E R_E$ will increase. According to the Eqn. (6.22) the base current I_B will reduce. Due to the reduction in I_B, the collector current I_C will also reduce. Hence, this circuit has a provision to reduce the collector current if it increases due to change in temperature and the value of β. So, this circuit will provide stability of the operating point.

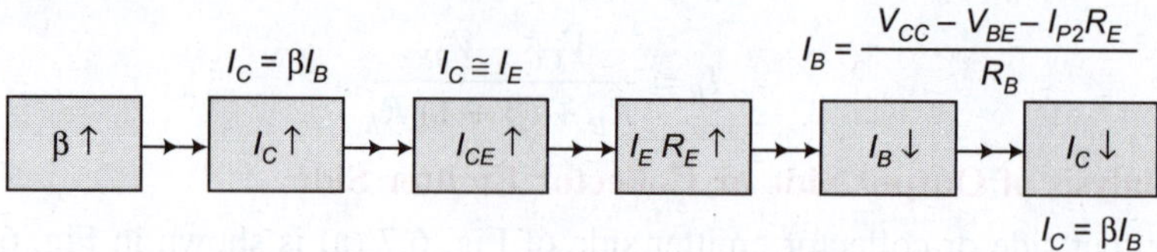

6.3.3.2 Merits and Demerits of Emitter Bias Circuit

Merits

   (i) The analysis of the circuit is easy.

  (ii) The stability of the circuit is more than the stability of the fixed bias circuit, when there is change in β or increment in temperature.

Demerits

This circuit is used rarely because

According to Eqn. (6.20)

$$I_B = \frac{V_{CC} - V_{BE}}{R_B + (\beta + 1) R_E}$$

$$\because \quad I_C = \beta I_B$$

Hence,

$$I_C = \frac{\beta(V_{CC} - V_{BE})}{R_B + (\beta + 1) R_E}$$

$$I_C = \frac{V_{CC} - V_{BE}}{\dfrac{R_B}{\beta} + \dfrac{(\beta + 1)}{\beta} R_E}$$

$$\because \qquad \beta >>> 1$$

Hence, $\qquad \dfrac{(\beta + 1)}{\beta} = 1.$

Now,

$$I_C = \frac{V_{CC} - V_{BE}}{\dfrac{R_B}{\beta} + R_E}. \tag{6.23}$$

In Eqn. (6.23) the value of I_C is dependent upon the value of β. When $R_E >> \dfrac{R_B}{\beta}$ i.e., we increase R_E at high value so the value of $\dfrac{R_B}{\beta}$ is very less. Therefore, we can ignore it at this value of R_E wherein I_C is independent of β.

If we keep R_E at maximum value, then we have to use battery at high voltage for Q-point because there is large voltage drop across R_E.

If we keep R_B at low value, then we need to use separate battery V_{BB} at base to maintain the value of I_B at low. These two problems occur in this type of circuit which cannot be eliminated simultaneously.

Stability Factor of Emitter Bias or Self Bias

To obtain the stability factor, applying KVL to the input side of Fig. 6.7(b), we obtain,

$$V_{CC} - I_B R_B - V_{BE} - I_E R_E = 0$$

$$\therefore \qquad I_E = I_C + I_B$$

$$\therefore \qquad I_B R_B + V_{BE} + (I_C + I_B) R_E = V_{CC}$$

$$I_B [R_B + R_E] = V_{CC} - V_{BE} - I_E R_E$$

$$\therefore \qquad I_B = \frac{V_{CC} - V_{BE} - I_E R_E}{R_B + R_E}. \tag{6.24}$$

Differentiating the above Eqn. (6.24) with respect to I_C

$$\frac{dI_B}{dI_C} = \frac{0 - 0 - 1 \cdot R_E}{R_B + R_E} = -\frac{R_E}{R_B + R_E}$$

Putting the value of dI_B/dI_C in the expression of S,

$$S = \frac{1 + \beta}{1 - \beta \left(\dfrac{dI_B}{dI_C}\right)}$$

$$S = \frac{1 + \beta}{1 - \beta \left(-\dfrac{R_E}{R_B + R_E} \right)}$$

$$= \frac{(1 + \beta)\,(R_B + R_E)}{(R_B + R_E) + \beta\,R_E}$$

$$S = \frac{R_E}{\left(1 + \dfrac{R_B}{R_E} \right)(1 + \beta)/R_E\left(1 + \dfrac{R_B}{R_E} + \beta \right)}$$

$$S = \frac{(1 + \beta)\left(1 + \dfrac{R_B}{R_E} \right)}{1 + \beta + \dfrac{R_B}{R_E}}. \tag{6.25}$$

Example 6.3.3.1   Calculate the following values for given circuit, if ($V_{BE} = 0.7$ V)

(a) I_B (b) V_{CE} (c) I_C (d) V_E
(e) V_C (f) I_E (g) V_B (h) V_{BC}

Solution   According to the Eqn. (6.20), we obtain,

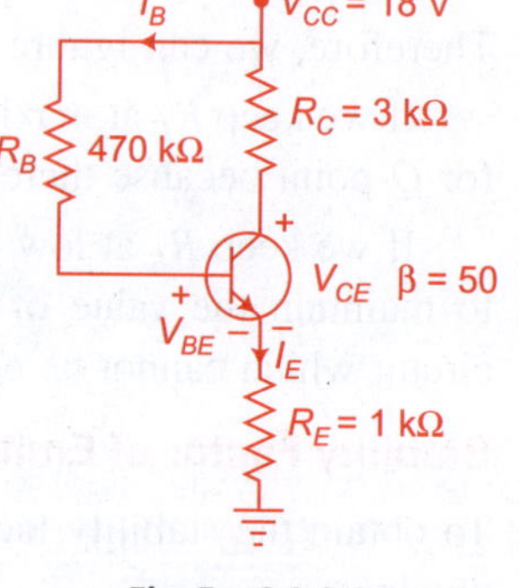

Fig. Ex. 6.3.3.1

(a)
$$I_B = \frac{V_{CC} - V_{BE}}{R_B + (\beta + 1)\,R_E}$$

$$I_B = \frac{18\text{ V} - 0.7\text{ V}}{470\text{ k}\Omega + (51 \times 1\text{ k}\Omega)}$$

$$I_B = \frac{17.3\text{ V}}{521\text{ k}\Omega} = 0.0332\text{ mA}$$

$$I_B = 33.2\ \mu\text{A}.$$

(b) According to Eqn. (6.21), we obtain,

$$V_{CE} = V_{CC} - I_C\,(R_C + R_E)$$
$$V_{CE} = 18 - 1.69\text{ mA }(3\text{ k}\Omega + 1\text{ k}\Omega)$$
$$V_{CE} = 18 - 1.69\text{ mA }(4\text{ k}\Omega)$$
$$V_{CE} = 18 - 6.76\text{ V}$$
$$V_{CE} = 11.24\text{ volt.}$$

(c) We know that,

$$I_C = \beta I_B$$
$$I_C = 51 \times 33.2$$
$$I_C = 1.69\text{ mA}$$
$$I_C = 1.69\text{ mA}$$

(d) We know that,

$$V_E = I_E R_B$$
$$V_E = 1.72 \text{ mA} \times 1 \text{ k}\Omega$$
$$V_E = 1.72 \text{ volt.}$$

(e) $\because$

$$V_{CE} = V_C - V_E$$
$$V_C = V_{CE} + V_E$$
$$= 11.24 + 1.72$$
$$V_C = 12.96 \text{ volt.}$$

(f) $\because$

$$I_E = I_B + I_C$$
$$I_E = 33.2 \text{ μA} + 1.69 \text{ mA}$$
$$I_E = 1.72 \text{ mA.}$$

(g) We know that,

$$V_{BE} = V_B - V_E$$
$$V_B = V_{BE} + V_E$$
$$= 0.7 \text{ V} + 1.72 \text{ V}$$
$$V_B = 2.42 \text{ volt.}$$

(h) We know that,

$$V_{BC} = V_B - V_C$$
$$= 2.42 \text{ V} - 12.96 \text{ V}$$
$$\boxed{V_{BC} = -10.54 \text{ volt.}}$$

6.3.4.1 Potential Divider Bias Circuit

Now, whatsoever bias circuit you will study, it is very important and widely used. Most of the circuits use potential divider bias circuit, so give it maximum attention.

The potential divider bias circuit is shown in Fig. 6.8 (a). In this type of circuit, three resistance R_1, R_2 and R_E are used. R_C is a collector resistance. R_1

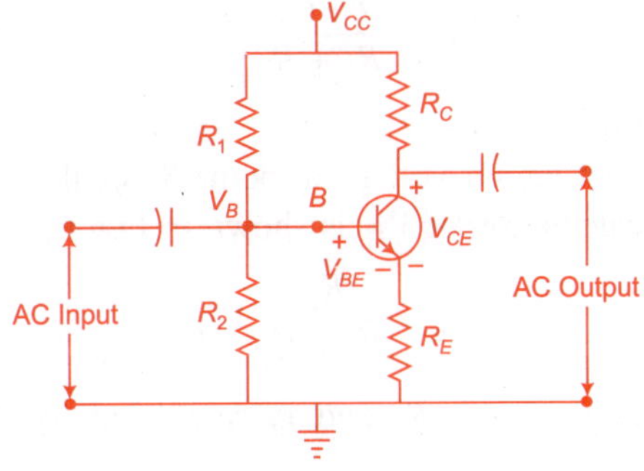

Fig. 6.8 *(a) Potential divider bias circuit*

and R_2 are used in potential divider circuit. Due to the potential divider bias circuit, the base voltage V_B is stable.

Analysis of Circuit

Analysis of DC circuit is divided in two parts, one is input side and another one is output side. When we analyse the circuit the capacitor is considered as open circuit for DC analysis as shown in Fig. 6.8 (b). The analysis of the circuit can be done with the help of Thevenin Theorem. For getting the remaining equivalent, the left hand side of point B circuit is replaced by it's Thevenin equivalent circuit.

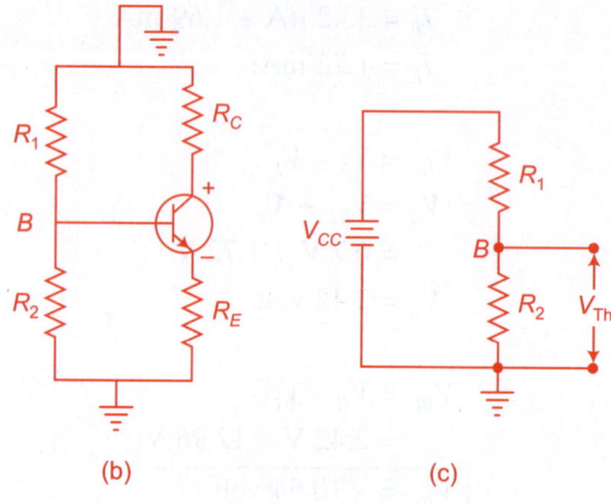

Fig. 6.8 *(b) Circuit for calculation of R_{Th}; (c) Circuit for calculation of V_{Th}*

If V_{CC} is short circuited in Fig. 6.8 (a), then the circuit will be shown in Fig. 6.8 (b) in which one end of R_1 and R_2 will join at point B and the other end will be grounded. Hence, R_1 and R_2 will be in parallel.

$$R_{Th} = R_1 \parallel R_2$$

$$R_{Th} = \frac{R_1\, R_2}{R_1 + R_2}. \tag{6.26}$$

(ii) Calculation of V_{Th}

In Fig. 6.8 (a), the open circuit voltage at point B is called as Thevenin voltage V_{Th}. The circuit for calculation of V_{Th} is shown in Fig. 6.8 (c).

$$V_{Th} = \frac{R_2}{R_1 + R_2}\, V_{CC}. \tag{6.27}$$

Note V_{Th} is the voltage across R_2 and is calculated by the voltage dividing rule.

Put the value of V_{Th} and R_{Th} in Fig. 6.8 (a). The resultant circuit is shown in Fig. 6.8 (d) in which potential divider circuit is replaced by a Thevenin equivalent circuit.

DC Analysis of input side

Applying Kirchhoff's voltage law (KVL) at the input side in Fig. 6.8 (d), we obtain,

$$V_{Th} - I_B R_{Th} - V_{BE} - I_E R_E = 0 \qquad (6.28)$$

$$\because \qquad I_E = (\beta + 1)\, I_B$$

Putting the value of I_E in Eqn. (6.28), we get,

$$V_{Th} - I_B R_{Th} - V_{BE} - (\beta + 1)\, I_B R_E = 0$$

$$V_{Th} - V_{BE} = I_B R_{Th} + (\beta + 1)\, I_B R_E$$

$$I_B = \frac{V_{Th} - V_{BE}}{R_{Th} + (\beta + 1)\, R_E} \cdot \qquad (6.29)$$

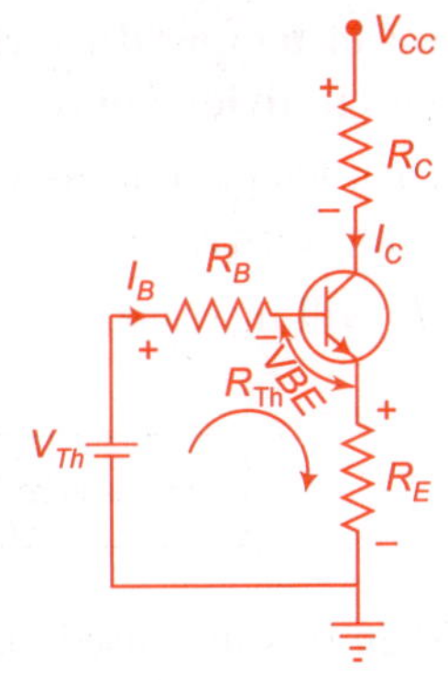

Fig. 6.8 *(d) Simplified potential divider circuit*

DC Analysis of Output Side

The output side of Fig. 6.8 (d) is shown in Fig. 6.8 (e).

Applying Kirchhoff's voltage law (KVL) at output side,

$$V_{CC} - I_C R_C - V_{CE} - I_E R_E = 0$$

$$V_{CE} = V_{CC} - I_C R_C - I_E R_E \qquad (6.30)$$

$$\because \qquad I_E = I_C.$$

Putting the value of I_E in Eqn. (6.30) we obtain,

$$V_{CE} = V_{CC} - I_C R_C - I_C R_E$$

$$V_{CE} = V_{CC} - I_C (R_C + R_E) \qquad (6.31)$$

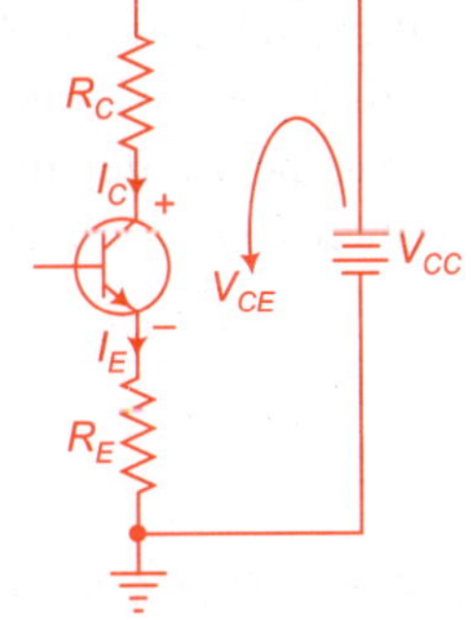

Fig. 6.8 *(e) Output side circuit*

The collector current I_C is given by the equation,

$$I_C = \beta I_B + I_{CEO}$$

I_{CEO} is very less so ignore it,

$$I_C = \beta I_B.$$

Hence, the calculation of I_C and V_{CE}, the location of operating point can be calculated.

Effect of temperature rise or change in value of β on location of *Q*-point in potential divider bias circuit.

* When temperature is increased, the leakage current I_{CEO} will increase due to the collector current.

I_C will increase.

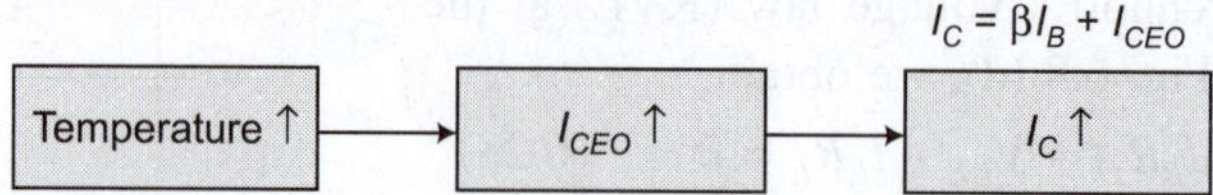

* When β is increased due to the replacement of transistor, the collector current I_C will increase, then I_E will increase due to the equation $I_C = I_E$, then the voltage drop across emitter resistance $I_E R_E$ will also be increased because,

$$V_{BE} = V_B - V_E$$
$$V_{BE} = V_B - I_E R_E \qquad (6.32)$$

Due to the increment of V_E, the V_{BE} will reduce from Eqn. (6.32). So, I_B will also reduce due to the Eqn. (6.22). Therefore, the increment of I_C will be component by decreasing the value of I_B.

If the selection of R_1 and R_2 in potential divider bias circuit in fine, then the operating point of transistor can be independent of value of β. Hence, we can see that the emitter resistance provides negative feedback which prevents the increment in the collector current.

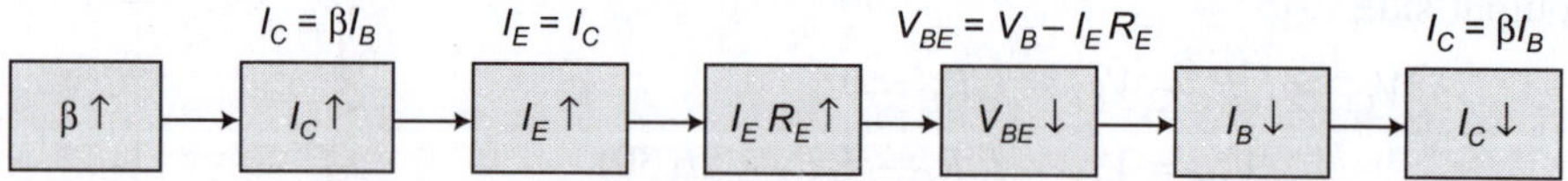

6.3.4.2 Merits and Demerits of Potential Divider Bias Circuit

Merits

   (i) The operating point or *Q*-point is stable.

  (ii) If there is any change in temperature or value of β, then there is no shifting in operating point i.e., independent of change in temperature or value of β.

 (iii) The number of combination of R_1 and R_2 can be used.

 (iv) Only single DC supply is required.

Demerits

In potential divider bias circuit, the AC signal is also given feedback by the emitter resistor R_E. Due to this, the voltage gain of amplifier is reduced. We can overcome this problem by using an electrolytic capacitor in parallel with emitter resistor R_E. This capacitor is called as emitter resistor R_E. This capacitor is also called as emitter bypass capacitor.

Stability Factor of Voltage Divider Bias

To obtain the stability factor, firstly we will find Thevenin's equivalent voltage and resistance for voltage divider bias circuit. The Thevenin's equivalent circuit is shown in Fig. 6.8 (d).

The Thevenin voltage,

$$V_{Th} = \frac{R_2}{R_1 + R_2} \cdot V_{CC} \qquad \text{from Eqn. (6.27)}$$

and Thevenin equivalent resistance,

$$R_{Th} = \frac{R_1 \, R_2}{R_1 + R_2} \qquad \text{from Eqn. (6.28)}$$

Now, applying KVL to the emitter base loop of the Thevenin equivalent circuit of Fig. 6.8 (d).

$$V_{Th} - I_B \, R_{Th} - V_{BE} - I_E \, R_E = 0 \qquad \text{from Eqn. (6.28)}$$

$$I_B \, R_{Th} + (I_C + I_B) \, R_E = V_{Th} - V_{BE} \qquad [\because I_E = I_C + I_B]$$

$$I_B \, (R_{Th} + R_E) + I_C \, R_E = V_{Th} - V_{BE}$$

$$I_B \, (R_{Th} + R_E) = V_{Th} - V_{BE} - I_C \, R_E$$

$$I_B = \frac{V_{Th} - V_{BE} - I_C \, R_E}{R_{Th} + R_E} \qquad (6.33)$$

Differentiating the above Eqn. (6.33) with respect to I_C

$$\frac{dI_B}{dI_C} = \frac{0 - 0 - 1 \cdot R_E}{R_{Th} + R_E} = -\frac{R_E}{R_{Th} + R_E}$$

Since, the stability factor is given by the expression,

$$S = \frac{1 + \beta}{1 - \beta \left(\dfrac{dI_B}{dI_C} \right)} \qquad (6.34)$$

Substituting the value of $\dfrac{dI_B}{dI_C}$ in Eqn. (6.34), we obtain,

$$S = \frac{1 + \beta}{1 - \beta \left(-\dfrac{R_E}{R_{Th} + R_E} \right)} = \frac{1 + \beta}{1 + \beta \left(\dfrac{R_E}{R_{Th} + R_E} \right)}$$

$$S = \frac{(1 + \beta) \, (R_{Th} + R_E)}{(R_{Th} + R_E) + \beta \, R_E} = \frac{(1 + \beta) \left[1 + \dfrac{R_{Th}}{R_E} \right]}{\left[1 + \dfrac{R_{Th}}{R_E} + \beta \right]}$$

$$S = \frac{(1 + \beta)\left(1 + \dfrac{R_{Th}}{R_E}\right)}{1 + \beta + \dfrac{R_{Th}}{R_E}} . \tag{6.35}$$

Example 6.3.4.1   Calculate the values of V_{CE} and I_C in Fig. Ex. 6.3.4.1. Given that $V_{BE} = 0.7$ volt.

Solution   Given that $R_1 = 40$ kΩ, $R_2 = 4.0$ kΩ $R_E = 1.5$ kΩ, $R_C = 15$ kΩ, $\beta = 130$, $V_{CC} = 20$ V, $V_{BE} = 0.7$ volt.

From Eqn. (6.26), we know that,

$$R_{Th} = \frac{R_1 \, R_2}{R_1 + R_2}$$

$$= \frac{40 \text{ k}\Omega \times 4.0 \text{ k}\Omega}{40 \text{ k}\Omega + 4.0 \text{ k}\Omega}$$

$$R_{Th} = \frac{160 \text{ k}\Omega}{44 \text{ k}\Omega}$$

$$R_{Th} = 3.63 \text{ k}\Omega$$

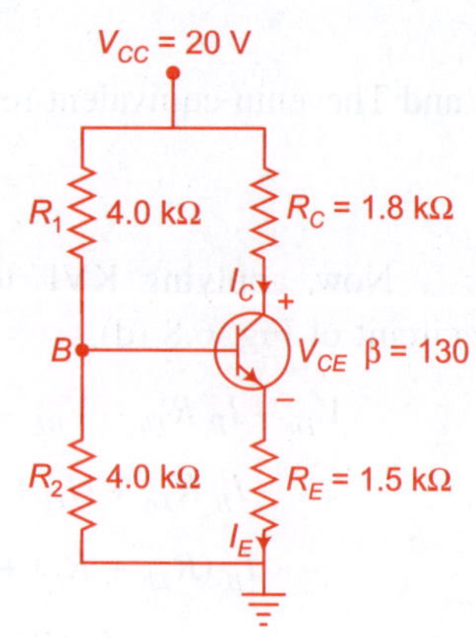

Fig. Ex. 6.3.4.1

From Eqn. (6.27), we know that,

$$V_{Th} = \frac{R_2}{R_1 + R_2} \, V_{CC} = \frac{4}{40 + 4} \times 20$$

$$V_{Th} = \frac{4 \text{ k}\Omega \times 20 \text{V}}{44 \text{ k}\Omega} = 1.8 \text{ V}$$

$$V_{Th} = 1.8 \text{ volt.}$$

From Eqn. (6.29), we know that,

$$I_B = \frac{V_{Th} - V_{BE}}{R_{Th} + (\beta + 1) \, R_E}$$

$$= \frac{1.8 \text{ V} - 0.7 \text{ V}}{3.63 \text{ k}\Omega + (131) \, 1.5 \text{ k}\Omega}$$

$$= \frac{1.1 \text{ V}}{3.63 \text{ k}\Omega + 196.5 \text{ k}\Omega}$$

$$= \frac{1.1 \text{ V}}{200.13 \text{ k}\Omega}$$

$$I_B = 5.5 \text{ }\mu\text{A.}$$

We know that, $I_C = \beta I_B$

$$I_C = 130 \times 5.5 \text{ }\mu\text{A}$$

$$I_C = 0.72 \text{ mA.}$$

From Eqn. (6.32), we know that,

$$V_{CE} = V_{CC} - I_C (R_C + R_E)$$

$$V_{CE} = 20 - 0.72 \text{ mA} (15 \text{ k}\Omega + 1.5 \text{ k}\Omega)$$

$$V_{CE} = 20 - 0.72 \text{ mA} \times 16.5 \text{ k}\Omega$$

$$V_{CE} = 20 \text{ V} - 11.88 \text{ V}$$

$$V_{CE} = 8.12 \text{ volt.}$$

Hence, the location of operating point is,

$$\boxed{I_C = 0.72 \text{ mA}, \; V_{CE} = 8.12 \text{ volt.}}$$

6.3.4.3   *Design of Potential Divider Bias Circuit*

Till now you have done the analysis of different circuits. In these circuits, you have calculated different currents and voltage.

In this section, you will study how to design a circuit. Design of circuit means we know the different values of currents and voltages in the circuit wherein you have to calculate the circuit components. For designing a circuit, it is necessary to know the characteristics of the circuit, and circuit analysis related law like Ohm's law, Kirchhoff's law and so on. Also we need to take some assumptions. So, we can say that the circuit design is more complex than the circuit analysis.

The design problem related to different potential divider circuit is given in next example.

Example 6.3.4.2   In Fig. Ex. 6.3.4.2, if $I_C = 2.5$ mA, $V_{CE} = 1.2$ V, then calculate R_1 and R_C.

Solution   Given that,

$$I_C = 2.5 \text{ mA}, \; V_{CC} = 1.6 \text{ V}, \; R_E = 1.2 \text{ k}\Omega,$$
$$R_L = 1.8 \text{ k}\Omega.$$

We know that,

$$V_E = I_E R_E$$

$$\because \quad I_C = I_E$$

$$V_E = I_C R_E$$

$$= 2.5 \text{ mA} \times 1.2 \text{ k}\Omega$$

$$V_E = 3.0 \text{ V.}$$

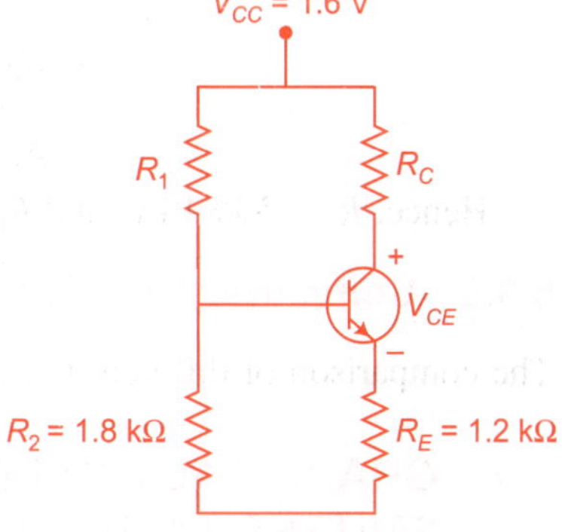

Fig. Ex. 6.3.4.2

Now,

$$V_{BE} = V_B - V_E$$
$$V_B = V_{BE} + V_E$$
$$= 0.7 \text{ V} + 3.0 \text{ V}$$
$$V_B = 3.7 \text{ volt.}$$

Now,

$$V_B = V_{Th} = \frac{R_2}{R_1 + R_2} V_{CC} \qquad \text{from Eqn. (6.27)}$$

$$V_{Th} = \frac{1.8 \text{ k}\Omega}{R_1 + 1.8 \text{ k}\Omega} \times 1.6 \text{ V}$$

$$3.7 \text{ V} = \frac{1.8 \times 1.6}{R_1 + 1.8}$$

$$\Rightarrow \qquad 3.7 R_1 + 66.6 = 288$$
$$3.7 R_1 = 288 - 66.6$$
$$3.7 R_1 = 221.4$$
$$R_1 = \frac{221.4}{3.7} = 59.84 \text{ k}\Omega$$
$$R_1 = 59.84 \text{ k}\Omega.$$

From Eqn. (6.32)

$$V_{CE} = V_{CC} - I_C (R_C + R_E)$$
$$1.2 \text{ V} = 1.6 - 2.5 \text{ mA} (R_C + 1.2 \text{ k}\Omega)$$
$$1.2 \text{ V} = 1.6 - 2.5 R_C - 3 \text{ V}$$
$$2.5 R_C = 1.6 - 3 \text{V} - 1.2 \text{ V}$$
$$2.5 R_C = 1 \text{ V}$$
$$R_C = \frac{1 \text{ V}}{2.5} = 0.4 \text{ k}\Omega$$
$$R_C = 0.4 \text{ k}\Omega$$

Hence, $R_1 = 59.84 \text{ k}\Omega$ and $R_C = 0.4 \text{ k}\Omega$.

6.3.5 Comparison of Biasing Circuit

The comparison of different biasing circuits is shown in Table 6.1.

6.4 GRAPHICAL ANALYSIS OF COMMON EMITTER AMPLIFIER

The common emitter (CE) transistor amplifier is shown in Fig. 6.9 (a). The output characteristics of transistor is shown in Fig. 6.9 (b). The characteristics of transistor have already been mentioned in the previous chapter.

Table 6.1

S.No.	Type of biasing circuit	Complexity	Stability	I_C V_{CE} and I_B	Merits	Demerits
1.	Fixed bias circuit	Simple	Unstable with change in temperature and value of β	$I_C = \beta I_B + I_{CEO}$ $\dot{V}_{CE} = V_{CC} - I_C R_C$ $I_B = V_{CC} - \dfrac{V_{BE}}{R_B}$	Calculations are easy	Q-point shift with change in β and temperature
2.	Collector to base bias	Simple	More stable than fixed bias ckt	$I_C = \beta I_B + I_{CEO}$ $\dot{V}_{CE} = V_{CC} - (I_C + I_B)\, R_C$ $I_B = V_{CE} - \dfrac{V_{BE}}{R_B} + (\beta + 1)\, R_C$	Easy calculation and provides stability	It reduces the voltage gain of amplification
3.	Emitter bias circuit	Simple	More stable than fixed bias	$I_C = \beta I_B + I_{CEO}$ $\dot{V}_{CE} = V_{CC} - I_C\, (R_C + R_E)$ $I_E = \dfrac{V_{CC} - V_{BE}}{R_B + (\beta + 1)\, R_E}$	Easy calculation and provides stability	It requires high voltage battery
4.	Potential divider bias circuit	Complex	More stable than previous ones	$I_C = \beta I_B + I_{CEO}$ $V_{CE} = V_{CC} - I_C\, (R_C + R_B)$ $I_E = \dfrac{V_{Th} - V_{BE}}{R_{Th} + (\beta + 1)\, R_E}$	More stable	Calculations are more complex than previous ones

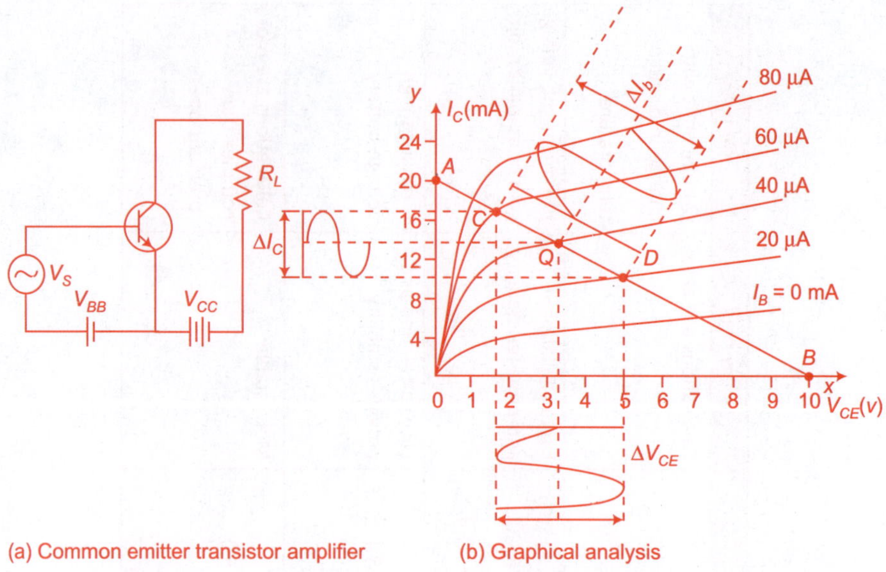

(a) Common emitter transistor amplifier

(b) Graphical analysis

Fig. 6.9 *(a) Common emitter transistor amplifier; (b) Graphical analysis*

Let us assume that $V_{CC} = 10$ V, $R_L = 500$ Ω, $I_B = 40$ μA

For DC load line,

Coordinates on x-axis are B $(V_{CC}, 0)$

 i.e. $(10, 0)$

Coordinates on y-axis are $A\left(0, \dfrac{V_{CC}}{R_L}\right)$

 i.e. $\left(0, \dfrac{10}{500 \ \Omega}\right) \Rightarrow (0, 20 \text{ mA})$.

Hence, load line will pass from point A $(10, 0)$ and B $(0, 20$ mA$)$ as shown in Fig. 6.9 (b). DC load line intersects the output characteristics at point Q. At operating point $V_{CE} = 3.25$ V, $I_C = 13.34$ mA and $I_B = 40$ μA.

AC input signal is applied between base and emitter. Let us assume that the peak value of AC signal is $V_m = 20$ mV. This signal is shown in Fig. 6.10 and denoted by V_S. Due to change in V_S, V_{BE} will also change.

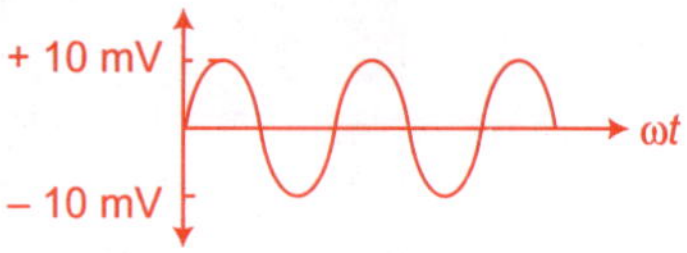

Fig. 6.10 *Input signal V_S*

$$\Delta V_{BE} = 20 \text{ mV} - (-20 \text{ mV})$$

$$\Delta V_{BE} = 40 \text{ mV}.$$

Let us consider that the dynamic input impedance of transistor in $V_i = 1$ kΩ.

Change in I_B due to change in V_{BE},

$$\Delta I_B = \frac{\Delta V_{BE}}{V_i} = \frac{40 \text{ mV}}{1 \text{ k}\Omega} = 40 \text{ }\mu\text{A}$$

$$\Delta I_B = 40 \text{ }\mu\text{A}.$$

$\because$ Change in V_{BE} is ± 20 mV or 40 mV. Hence, change in I_B is ± 20 μA from both side of operating point i.e., the value of I_B will change its own value 40 μA at Q-point to ± 20 μA. So, we can say that it will change between point C ($I_B = 60$ μA) and point D ($I_B = 20$ μA). By drawing perpendicular line at DC load line, it shows the change in base current as shown in Fig. 6.9 (b).

When the value of I_B will change between point A and B, then the value of I_C will also be changed. As we have seen in Fig. 6.9 (b). The value of I_C at point C is 16.5 mA and at point D is 10 mA. Hence, the change in I_C is,

$$\Delta I_C = 16.5 \text{ mA} - 10 \text{ mA} = 6.5 \text{ mA}.$$

Similarly, the value of V_{CE} will also change. The value of V_{CE} at point C is 1.8 V and at point D is 5.0 V. Hence, the change in V_{CE} is,

$$\Delta V_{CE} = 5.0 \text{ V} - 1.8 \text{ V} = 3.2 \text{ V}$$

$$\Delta V_{CE} = 3.2 \text{ V}.$$

With the help of above calculated value we can calculate voltage gain, current gain and power gain of the CE transistor.

(i) Voltage gain,

$$A_V = \frac{\text{Change in } V_{CE}}{\text{Change in } V_{BE}} = \frac{\Delta V_{CE}}{\Delta V_{BE}}$$

$$A_V = \frac{3.2 \text{ V}}{40 \text{ mV}} = 80$$

Hence, $\quad A_V = 80.$

(ii) Current gain,

$$A_i = \frac{\text{Change in } I_C}{\text{Change in } I_B} = \frac{\Delta I_C}{\Delta I_B}$$

$$A_i = \frac{6.5 \text{ mA}}{40 \text{ }\mu\text{A}} = 162.5$$

Hence, $\quad A_i = 162.5.$

(iii) Power gain, $\quad A_p = $ Voltage gain $(A_V) \times$ Current gain (A_i)

$$A_p = 80 \times 162.5$$

$$A_p = 13000.$$

Note The current gain, power gain and voltage gain will be covered in the next section.

6.5 CONCEPT OF VOLTAGE GAIN, CURRENT GAIN AND POWER GAIN

The graphical analysis of common emitter transistor have been discussed earlier. When we see Fig. 6.9 (b), we can state that the voltage gain current gain for common emitter transistor is as below.

Voltage Gain

It is the ratio of the change in output voltage V_{CE} with respect to change in input voltage V_{BE} and is denoted by A_V.

$$A_V = \frac{\Delta V_{CE}}{\Delta V_{BE}}$$

$$= \frac{\text{Change in output voltage}}{\text{Change in input voltage}}. \qquad (6.36)$$

Current Gain

It is the ratio of the change in output current I_C with respect to change in input current I_B and is denoted by A_i.

$$A_i = \frac{\Delta I_C}{\Delta I_B} = \frac{\text{Change in output current}}{\text{Change in input current}}. \qquad (6.37)$$

Power Gain

It is the multiplication of current gain A_i and voltage gain A_V. It is denoted by A_p.

$$A_p = A_i \times A_V. \qquad (6.38)$$

6.6 *h*-PARAMETER MODEL (LOW FREQUENCY)

The hybrid parameters are mostly used in electronic circuit. These are the components of a small signal equivalent model. This hybrid model with all its parameters was the selected model for the educational and industrial communities. The specification sheets provide the hybrid parameters and the hybrid model continues to receive a good measure of attention, especially in constructing models for transistor. The parameter relating the four variables (input voltage V_1, input current I_1, output voltage V_2 and output current I_2) are called *h*-parameters. It is derived from the word "hybrid". This word is chosen because the mixture of variables (V and I) in each equation results in a "hybrid" set of units of measurement for the *h*-parameters.

In this case, voltage of the input port V_i and the current of the output port I_o are expressed in terms of the current of the input port I_i and the voltage of the output port V_o. Due to this reason, these parameters are called as "hybrid" parameters. For the determination of the h-parameters, we consider the transistor as a two port network as shown in Fig. 6.11.

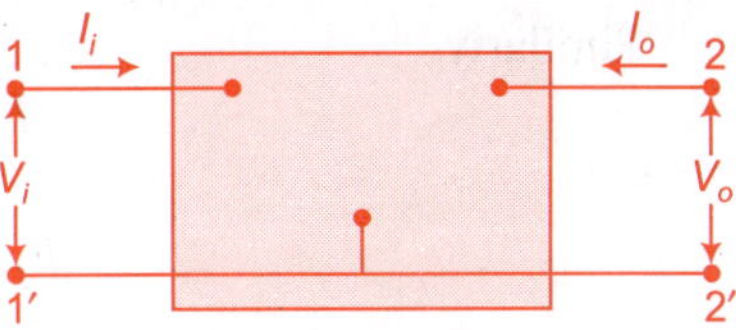

Fig. 6.11   *Two port network*

The equation is,

$$(V_i, I_o) = f(I_i, V_o)$$

$$\begin{bmatrix} V_i \\ I_o \end{bmatrix} = [h] \begin{bmatrix} I_i \\ V_o \end{bmatrix} \tag{6.39}$$

$$\begin{bmatrix} V_i \\ I_o \end{bmatrix} = \underbrace{\begin{bmatrix} h_{11} & h_{12} \\ h_{21} & h_{22} \end{bmatrix}}_{\text{hybrid matrix}} \begin{bmatrix} I_i \\ V_o \end{bmatrix} \tag{6.40}$$

The matrix Eqn. (6.40) defines h-parameters, and h-parameters matrix is known as hybrid matrix.

$$V_i = h_{11} I_i + h_{12} V_o \tag{6.41}$$

$$V_i = h_{21} I_i + h_{22} V_o. \tag{6.42}$$

Determination of h-Parameters

For determination of h-parameters, we consider two cases as follows:

Case I

When output port of Fig. 6.11 is short circuited as shown in Fig. 6.12 (a), then $V_o = 0$.

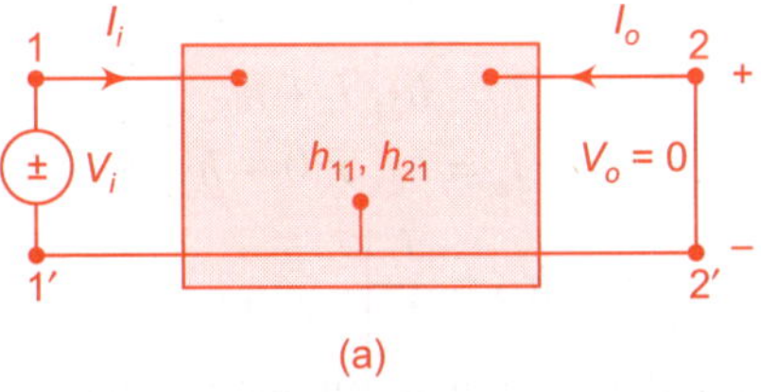

Fig. 6.12   *(a) Two port network when output is short circuited*

Putting the value of V_o in Eqns. (6.41) and (6.42) we obtain,

$$V_i = h_{11} I_i + h_{12} \times 0$$

$$h_{11} = \left. \frac{V_i}{I_i} \right|_{V_o = 0} \tag{6.43}$$

Parameter h_{11} is called forward current gain with the output port short circuited and it is also called as short circuit input-impedance parameter.

Similarly,

$$I_o = h_{21}\,I_i - h_{22} \times 0$$

$$h_{21} = \left.\frac{I_o}{I_i}\right|_{V_o = 0} \tag{6.44}$$

Paramater h_{21} is called as forward current gain with the output port short circuited and it is also called as short circuit forward current gain.

Case II

When input port of Fig. 6.11 is open circuited as shown in Fig. 6.12 (b), then $I_i = 0$

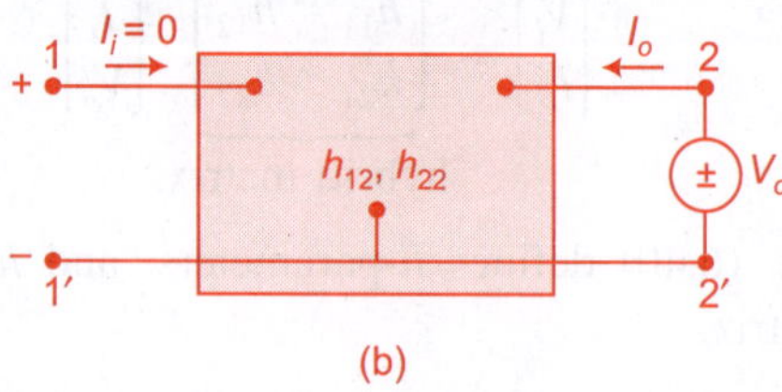

Fig. 6.12   *(b) Two port network when input is open circuited*

Putting the value of I_i in Eqns. (6.41) and (6.42) we obtain,

$$V_i = h_{11}\,I_i + h_{12}\,V_o$$
$$V_i = h_{11} \times 0 + h_{12}\,V_o$$

$$h_{12} = \left.\frac{V_i}{V_o}\right|_{I_i} = 0. \tag{6.45}$$

Parameter h_{12} is called as reverse voltage gain with the input port open circuited and it is also called as open circuit reverse voltage gain parameter.

Similarly,

$$I_o = h_{21}\,I_i + h_{22}\,V_o$$
$$I_o = h_{21} \times 0 + h_{22}\,V_o$$

$$h_{22} = \left.\frac{I_o}{V_o}\right|_{I_i}. \tag{6.46}$$

Parameter h_{22} is called as output admittance with the input port open-circuited and it is also called as open-circuit output admitted parameters.

Since each term of Eqn. (6.41) has the unit volt, let us apply Kirchhoff's voltage law to find a circuit that fits the equation. Performing this operation will result in the circuit of Fig. 6.13. Since the parameter h_{11} has the unit Ohm, it is represented by a resistor in Fig. 6.13. The quantity h_{12} is

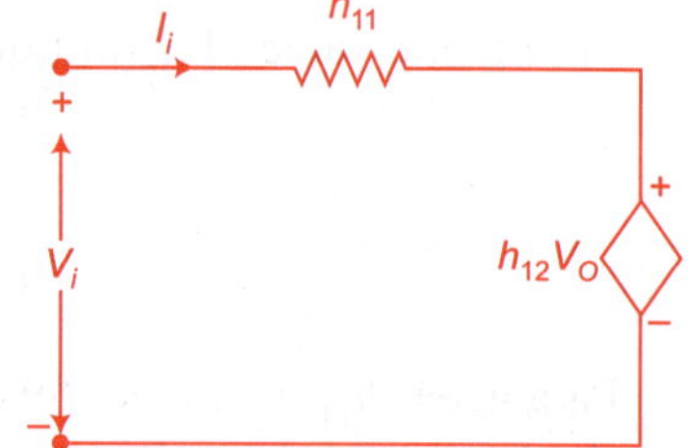

Fig. 6.13   *Hybrid input equtant circuit*

dimensionless and therefore simply appears as a multiplying factor of the "feedback" term in the input circuit.

Since each term of Eqn. (6.42) has the units of current, let us now apply Kirchhoff's current law to obtain the circuit that fits the equation. Performing this operation will result in the circuit of Fig. 6.14. Since the parameter h_{22} has the unit of admittance, for which the transistor model is conductance. It is represented by the resistor symbol. Notice that the resistance in Ohms of this resistor is equal to the reciprocal of conductance ($1/h_{22}$). The quantity h_{21} is dimensionless and therefore simply appears as a multiplying factor of the "feedback" term in the input circuit.

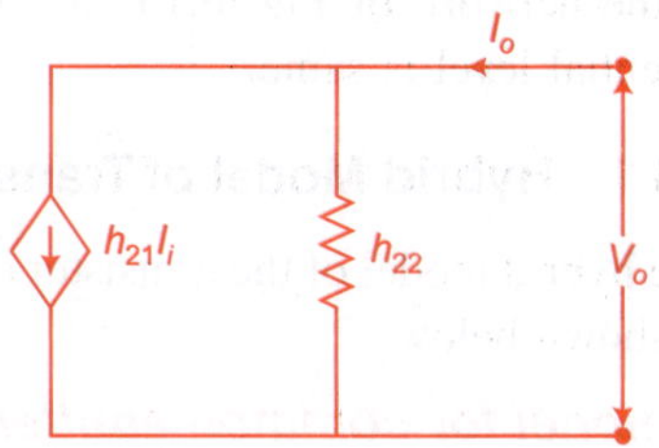

Fig. 6.14 *Hybrid output equivalent model*

The complete AC equivalent circuit for the basic three terminal linear devices are obtained from Figs. 6.13 and 6.14 and shown in Fig. 6.15 with a new set of subscripts for the *h*-parameters.

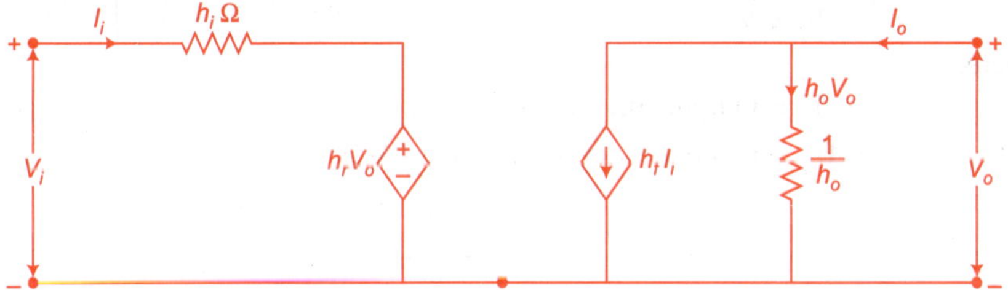

Fig. 6.15 *Hybrid equivalent model*

The notation of Fig. 6.15 is of more practical nature since it relates the *h*-parameters to the resulting ratio obtained in the previous few paragraphs.

The Eqns. (6.41) and (6.42) will be as below,

$$V_i = h_i\, I_i + h_r\, V_o \tag{6.47}$$
$$I_o = h_f\, I_i + h_o\, V_o \tag{6.48}$$

where,

$h_{11} = h_i$ (input resistance)

$h_{12} = h_r$ (reverse transfer voltage ratio)

$h_{21} = h_f$ (forward transfer current ratio)

$h_{22} = h_o$ (output conductance).

The circuit of Fig. 6.15 is applicable to any linear three terminal electronic device or system with no internal independent sources. Therefore, for the transistor, even though it has three basic configuration, they are all three terminal configurations, so that the resulting equivalent circuit will have the same format as shown in Fig. 6.15. In each case, the bottom of the input and output section of the network of Fig. 6.15 can be connected as shown in Fig. 6.16 since the potential level is same.

6.6.1 Hybrid Model of Transistor

The hybrid model of the transistor can be made for all three types of configuration as shown below.

h-model for common emitter configuration

The common emitter configuration is shown in Fig. 6.16 (a) in which,

$$I_i = I_b$$

where RMS value of input current = I_i

$$I_b = \text{value of base current}$$

$$V_i = V_{be},$$

where,

$$V_i = \text{Input AC voltage}$$

$$V_{be} = \text{Base-emitter voltage (AC)}$$

$$V_o = V_{ce}$$

where,

$$V_o = \text{Output AC voltage}$$

and,

$$V_{ce} = \text{Collector emitter voltage (AC).}$$

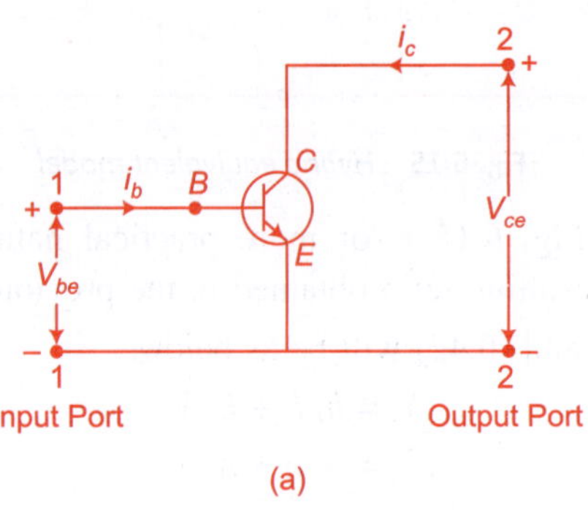

Fig. 6.16 *(a) CE configuration*

Similarly,

$$I_o = I_C$$

where,

$$I_o = \text{Output current (AC)}$$

$$I_s = \text{Collector current (AC).}$$

Note All the values of current and voltages are considered as RMS value.

In common emitter configuration, the emitter terminal is common. So, in *h*-parameter the subscripts '*e*' is added in Eqns. (6.47) and (6.48) and putting

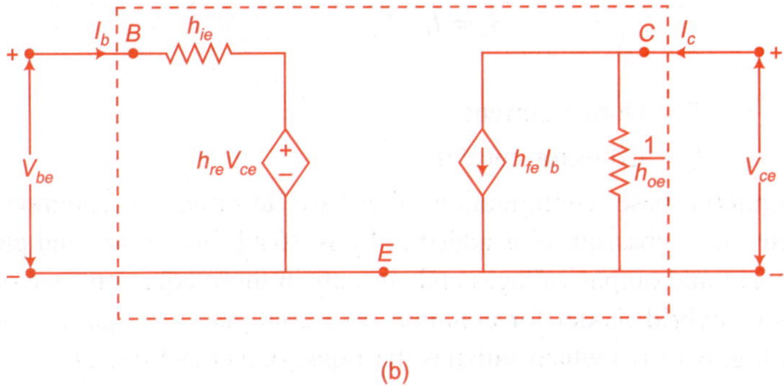

(b)

Fig. 6.16 *(b) Hybrid model of common-emitter transistor*

the value of input and output voltages and currents. We obtain the equations of hybrid model for common emitter configuration transistor. Model is shown in Fig. 6.16 (b) which is satisfies the Eqns. (6.49) and (6.50).

$$V_{be} = h_{ie} I_b + h_{re} V_{ce} \tag{6.49}$$

$$I_c = h_{fe} I_b + h_{oe} V_{ce}. \tag{6.50}$$

h-model for common base configuration

The common base configuration is shown in Fig. 6.17 (a) in which,

$$I_i = I_e$$

where,

$$I_i = \text{Input current}$$

$$I_e = \text{Emitter current}$$

and, $\quad V_i = V_{eb}$

where, $\quad V_i = \text{Input voltage}$

$$V_{eb} = \text{Emitter-base-voltage.}$$

Now, $\quad V_o = V_{cb}$

where,

$$V_o = \text{Output voltage}$$

and, $\quad V_{cb} = \text{Collector base voltage}$

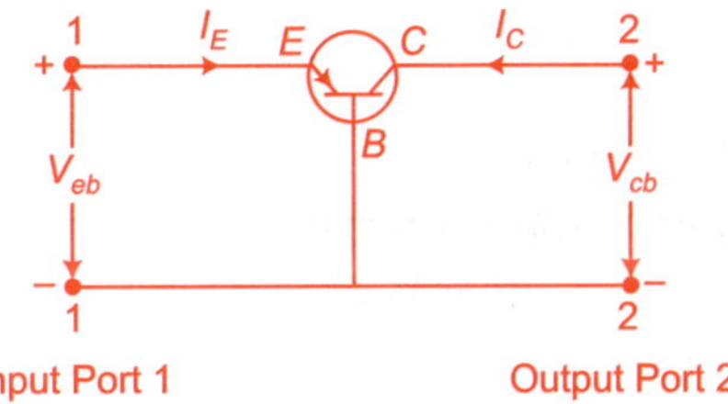

Fig. 6.17 *(a) CB configuration*

$$I_o = I_C$$

where,

I_o = Output current

and, I_C = Collector current

In common base configuration, the base terminal is common. So, in h-parameter, the subscript 'b' is added in Eqns. (6.47) and (6.48) and putting the value of input and output voltages and currents in those equations, we obtain the equations of hybrid model for common base configuration transistor model as shown in Fig. 6.17 (b) which satisfies the Eqns. (6.51) and (6.52).

$$V_{eb} = h_{ib} I_e + h_{rb} V_{cb} \tag{6.51}$$

$$I_C = h_{fb} I_e + h_{ob} V_{cb} \tag{6.52}$$

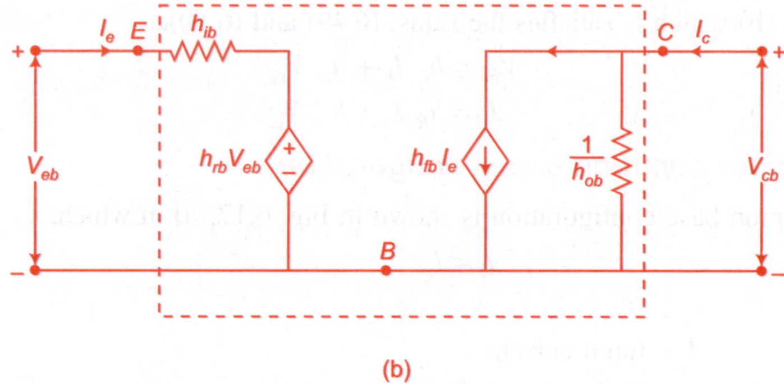

(b)

Fig. 6.17   *(b) Hybrid model of common base transistor*

h-model for common collector configuration

The common collector configuration is shown in Fig. 6.18 (a) in which,

$$I_i = I_b$$

where,

I_i = Input current

I_b = Base current

and, $V_i = V_{bc}$

where,

V_i = Input voltage

V_{bc} = Base collector voltage.

Now, $V_o = V_{eC}$

where,

V_o = Output voltage

V_{ec} = Emitter collector voltage

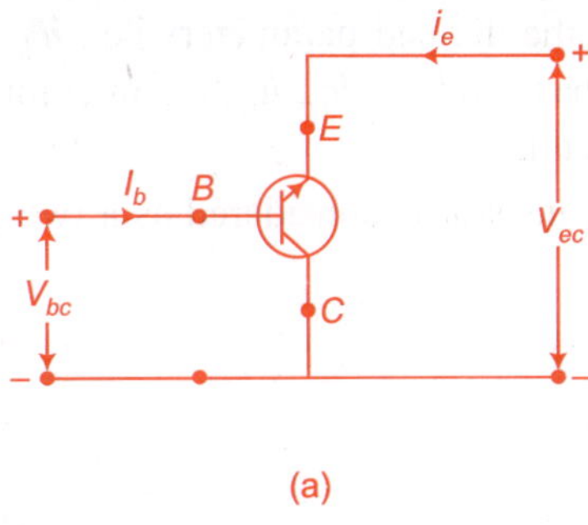

(a)

Fig. 6.18   *(a) Common collector configuration*

and, $i_o = i_e$

where,

i_o = Output current

and, i_e = Emitter current.

In common collector configuration, the collector terminal is common, so in *h*-parameter the subscript '*c*' is added in Eqns. (6.47) and (6.48) and putting the value of input and output voltages and currents in those equations, we obtain the equation of hybrid model for common collector transistor. The hybrid model is shown in Fig. 6.18 (b) which satisfies the Eqns. (6.53) and (6.54).

$$V_{bc} = h_{ic}\, I_b + h_{rc}\, V_{ec} \qquad (6.53)$$
$$I_e = h_{fc}\, I_b + h_{oc}\, V_{ec} \qquad (6.54)$$

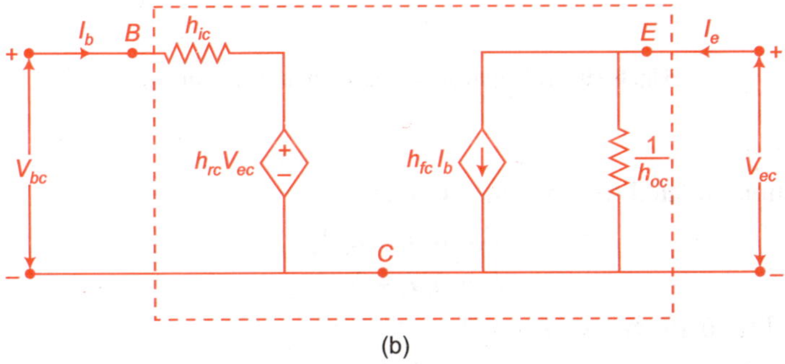

(b)

Fig. 6.18   *(b) Hybrid model of common collector transistor*

6.6.2   Calculation of Transistor Parameters using Hybrid Model

The parameters of transistor like current gain, input impedance, voltage gain, output impedance and power gain can be calculated with the help of hybrid model. As we saw in earlier section that the shape of hybrid equivalent model of transistor for common base, common emitter and common collector are same. The equations we get in this section can be used in all the three configurations.

Only we have to change the defined parameters i.e., h_{fb}, h_{ib}, h_{rb}, h_{ob} in place of h_f, h_i, h_r, h_o in common base and h_{fe}, h_{ie}, h_{re}, h_{oe} in common emitter and h_{fc}, h_{ic}, h_{rc}, h_{oc} in common collector.

In Fig. 6.19 (a) the transistor is considered as a two port network.

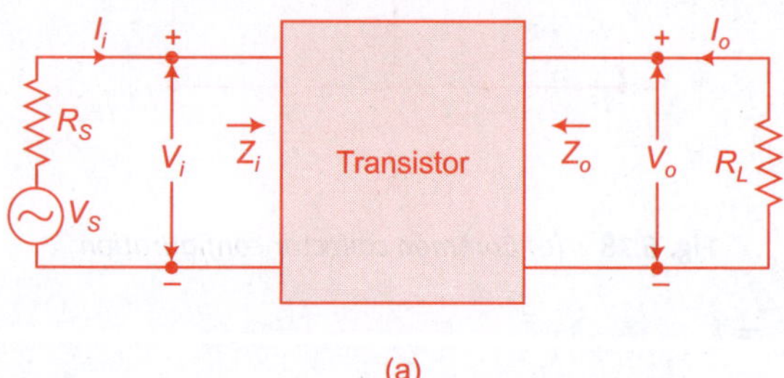

(a)

Fig. 6.19   *(a) Transistor as a two port network*

The equivalent hybrid model for the transistor is shown in Fig. 6.19 (b).

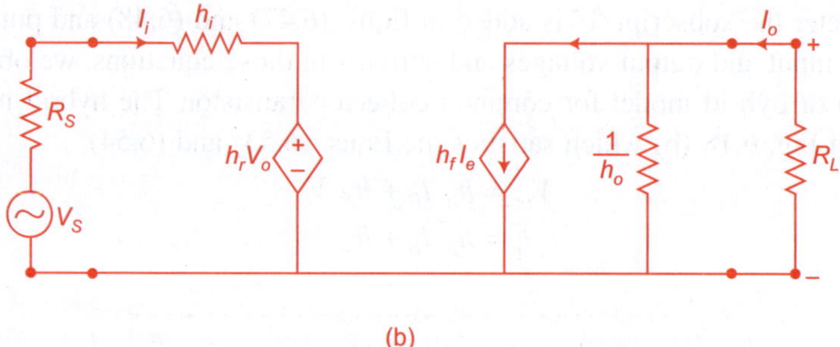

(b)

Fig. 6.19   *(b) Equivalent hybrid model of transistor*

Current Gain

The equations related to h-parameters are:

$$V_i = h_i I_i + h_r V_o \tag{6.55}$$

$$I_o = h_f I_i + h_o V_o \tag{6.56}$$

From Fig. 6.19 (b), we know that,

$$V_o = - I_o R_L \tag{6.57}$$

Putting the value of V_o in Eqn. (6.56), we obtain,

$$I_o = h_f I_i + h_o (- I_o R_L)$$

$$I_o (1 + h_o R_L) = h_f I_i$$

$$\frac{I_o}{I_i} = \frac{h_f}{1 + h_o R_L}$$

i.e.

$$A_i = \frac{I_o}{I_i} = \frac{h_f}{1 + h_o\,R_L}$$

$$A_i = \frac{h_f}{1 + h_o\,R_L} \tag{6.58}$$

If h_o is negligible then,

$$A_i = h_f.$$

Voltage Gain

We know that,

$$V_i = h_i\,I_i + h_r\,V_o$$

From Eqn. (6.58)

$$\frac{I_o}{I_i} = \frac{h_f}{1 + h_o\,R_L}$$

$$h_f\,I_i = I_o\,(1 + h_o\,R_L)$$

$$I_i = \frac{I_o\,(1 + h_o\,R_L)}{h_f} \tag{6.59}$$

Now, from Eqn. (6.57), we know that,

$$V_o = -\,I_o\,R_L$$

$$I_o = -\frac{V_o}{R_L} \tag{6.60}$$

Putting the value of I_o from Eqn. (6.60) into Eqn. (6.59), we obtain,

$$I_i = -\frac{V_o\,(1 + h_o\,R_L)}{R_L \qquad h_f}. \tag{6.61}$$

Now, putting the value of I_i from Eqn. (6.61), to Eqn. (6.55), we obtain,

$$V_i = -\frac{h_i\,(1 + h_0\,R_L)\,V_o}{h_f\,R_L} + h_r\,V_o$$

$$V_i = \left[-\frac{h_i\,(1 + h_o\,R_L)}{h_f\,R_L} + h_r\right]V_o$$

$$V_i = \left[-\frac{(h_i + h_0\,h_i\,R_L) + h_f\,h_r\,R_L}{h_f\,R_L}\right]V_o$$

$$\frac{V_o}{V_i} = \frac{-\,h_f\,R_L}{h_i + (h_i\,h_o - h_f\,h_r)\,R_L}$$

$$A_V = \frac{V_o}{V_i} = \frac{-\,h_f\,R_L}{h_i + (h_i\,h_o - h_f\,h_r)\,R_L}$$

$$A_V = \frac{-h_f R_L}{h_i + (h_i h_o - h_f h_r) R_L} \tag{6.62}$$

if $h_i \gg (h_i h_o - h_f h_r) R_L$ then,

$$A_V = -\frac{h_f R_L}{h_i}.$$

Input Impedance

Putting the value of V_o from Eqn. (6.57) in Eqn. (6.55), we obtain,

$$V_i = h_i I_i + h_r (- I_o R_L). \tag{6.63}$$

We know that,

$$A_i = \frac{I_o}{I_i}$$

Hence,
$$I_o = A_i I_i. \tag{6.64}$$

Putting the value of I_o in Eqn. (6.63), we obtain,

$$V_i = h_i I_i - h_r R_L A_i I_i$$
$$V_i = (h_i - h_r R_L A_i) I_i$$

$$\frac{V_i}{I_i} = h_i - h_r R_L A_i$$

$$Z_i = h_i - h_r R_L A_i$$

$$\because \qquad A_i = \frac{h_f}{1 + h_o R_L}$$

Hence,
$$Z_i = h_i - h_r R_L \frac{h_f}{1 + h_o R_L} \tag{6.65}$$

If the second term in Eqn. (6.65) is negligible then,
$$Z_i = h_i.$$

Power Gain

Power gain is represented by A_p and it is multiplication of voltage gain and current gain.

$$A_p = \text{Voltage gain} \times \text{Current gain}$$

$$A_p = \frac{-h_f R_L}{h_i + (h_i h_o - h_f h_r) R_L} \times \frac{h_f}{1 + h_o R_L}$$

$$A_p = \frac{-h^2_f R_L}{[h_i + (h_i h_o - h_f h_r)] R_L (1 + h_o R_L)}. \tag{6.66}$$

Output Impedance

The output impedance of amplifier is defined as the ratio of output voltage and output current when input source is short circuited.

In Fig. 6.19 (b) the input source V_S is short circuited ($V_S = 0$), then we obtain the Fig. 6.20.

Applying Kirchhoff's voltage law (KVL) at input side,

$$- I_i R_S - h_i I_i - h_f V_o = 0$$

$$- I_i (R_S + h_i) = h_f V_o$$

$$I_i = - \frac{h_r V_o}{R_S + h_i} \tag{6.67}$$

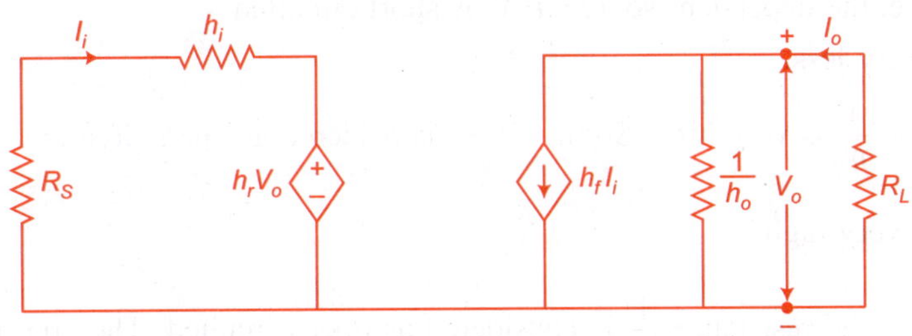

Fig. 6.20   *h-model when input source is short circuited*

Putting the value of I_i in Eqn. (6.56), we obtain,

$$I_o = h_f I_i + h_o V_o$$

$$I_o = h_f \left(- \frac{h_r V_o}{R_S + h_i} \right) + h_o V_o$$

$$I_o = - \frac{h_f h_r V_o}{R_S + h_i} + h_o V_o$$

$$I_o = \left[- \frac{h_f h_r}{R_S + h_i} + h_o \right] V_o$$

$$\frac{V_o}{I_o} = \frac{1}{h_o - \left(\dfrac{h_f h_r}{R_S + h_i} \right)}$$

$$Z_o = \frac{1}{h_o - \left(\dfrac{h_f h_r}{R_S + h_i} \right)} \tag{6.68}$$

if $\dfrac{h_f h_r}{R_S + h_i} \lll h_o$, then,

$$Z_o = \frac{1}{h_o}$$

6.6.3 Approximate Hybrid Model

Generally, the value of h_r and h_o is very less in common base and common emitter configuration. Due to the less value of h_r and h_o, the effect of these h-parameter in calculation of Z_i, Z_o, A_V, A_i is not too much.

$\because\ h_r$ is very less.

Hence, $h_r\,V_o$ is also very less. So, we consider it equal to zero.

$$h_r\,V_o = 0$$

Hence, the dependent source $h_r\,V_o$ is short circuited.

$\because\ h_o$ is very less.

Hence, $\dfrac{1}{h_o}$ is very high. So, it will be considered as open circuited.

$\dfrac{1}{h_o}$ is very high.

Hence, the resistance $\dfrac{1}{h_o}$ is considered as open circuited. The approximated model is shown in Fig. 6.21 after above approximation.

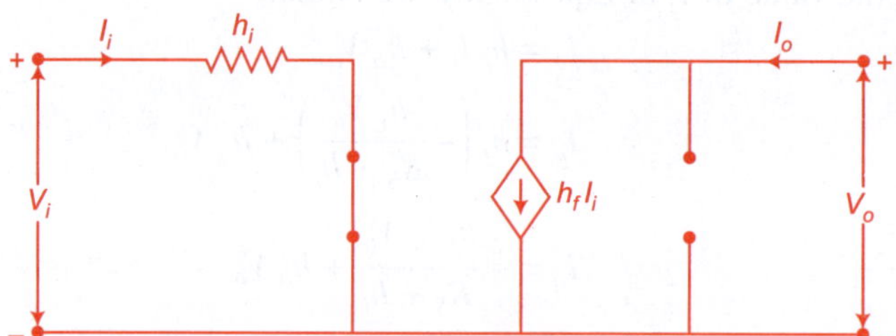

Fig. 6.21 *Approximate hybrid model*

The typical value of h-parameter for common emitter configuration approximate hybrid model is as follows:

$$h_{ie} = 1\ \text{k}\Omega \qquad\qquad h_{re} = 2.5 \times 10^{-4}$$

$$h_{fe} = 50 \qquad\qquad h_{oe} = 25\ \mu\text{S} \quad \text{or} \quad \frac{1}{h_{oe}} = 40\ \text{k}\Omega.$$

Analysis of Common Emitter Amplifier by Approximate Hybrid Model

The approximate model is shown in Fig. 6.22. From Fig. 6.21

Current Gain

$$I_o = h_f\,I_i$$
$$\frac{I_o}{I_i} = h_f$$

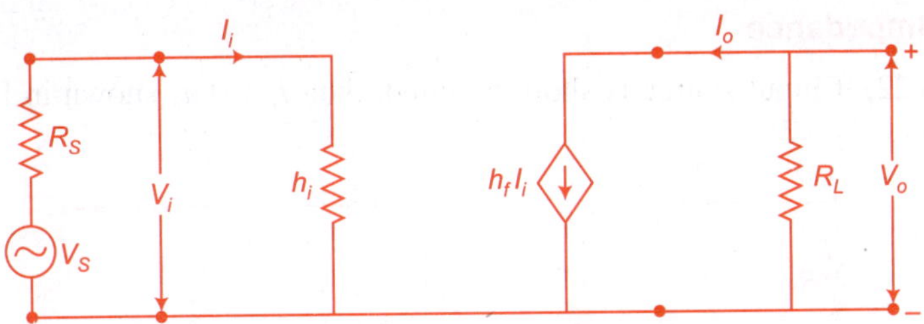

Fig. 6.22 *Equivalent approximate hybrid model*

Hence, $\qquad A_i = h_f.$ $\hfill (6.69)$

Input Impedance

From Fig. 6.22, we know that,

$$V_i = I_i\, h_i$$

$$\frac{V_i}{I_i} = h_i$$

Hence, $\qquad Z_i = h_i$ $\hfill (6.70)$

Voltage Gain

From Fig. 6.22, we know that,

$$V_o = -\, I_o\, R_L \qquad (6.71)$$

$\therefore \qquad I_o = h_f\, I_i$

and, $\qquad I_i = \dfrac{V_i}{h_i}$

Hence, $\qquad I_o = \dfrac{h_f\, V_i}{h_i}.$

Putting the value of I_o in Eqn. (6.71), we obtain,

$$V_o = -\, \frac{h_f\, V_i}{h_i}\, R_L$$

$$\frac{V_o}{V_i} = -\, \frac{h_f\, V_i}{h_i} \Rightarrow A_V = -\, \frac{h_f\, R_L}{h_i} \qquad (6.72)$$

Power Gain

$$A_p = \text{Voltage gain} \times \text{Current gain}$$

$$A_p = -\, \frac{h_f\, R_L}{h_i} \times h_f$$

$$A_p = -\, \frac{h^2_f\, R_L}{h_i}. \qquad (6.73)$$

Output Impedance

In Fig. 6.22, if input source is short circuited, then $I_i = 0$ as shown in Fig. 6.23.

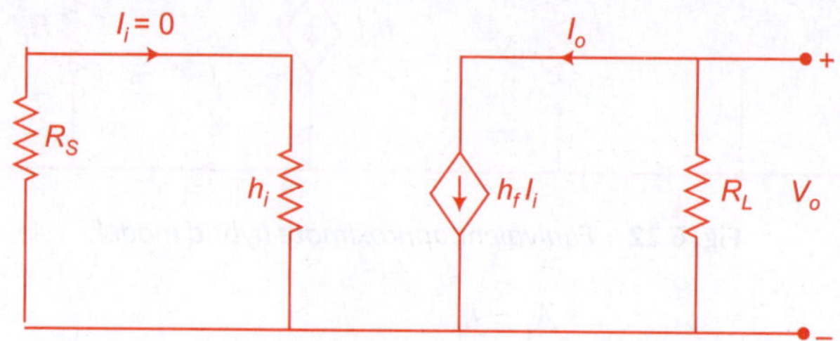

Fig. 6.23 *Circuit for calculation of output impedance*

Hence,

$$h_f I_i = 0$$
$$I_o = 0$$
$$Z_o = \frac{V_o}{I_o}$$
$$Z_o = \infty \tag{6.74}$$
$$Z_o = \frac{1}{h_o - \left(\dfrac{h_f h_r}{R_S + h_i} \right)}$$

if $h_o \gg \dfrac{h_f h_r}{R_S + h_i}$

then,

$$Z_o = \frac{1}{h_o}. \tag{6.75}$$

6.6.4 Determination of *h*-parameters from Transistor Characteristics

h-parameters for common emitter mode can be represented by the following formula:

$$h_{ie} = \frac{\Delta V_i}{\Delta I_i} = \left. \frac{\Delta V_{be}}{\Delta I_o} \right|_{V_{CE} = \text{constant}} \tag{6.76}$$

where,

    h_{ie} = Input impedance in Ohms for common emitter configuration

    Δ = Small change in voltage and current

$$h_{re} = \frac{\Delta V_i}{\Delta V_o} = \left. \frac{\Delta V_{be}}{\Delta V_{ce}} \right|_{I_B = \text{constant}} \tag{6.77}$$

where,

h_{re} = Reverse transfer voltage ratio

$$h_{fe} = \frac{\Delta I_o}{\Delta I_i} = \frac{\Delta I_C}{\Delta I_b}\bigg|_{V_{CE} = \text{constant}} \tag{6.78}$$

where,

h_{fe} = Forward transfer current ratio

and,

$$h_{oe} = \frac{\Delta I_o}{\Delta V_o} = \frac{\Delta I_C}{\Delta V_{ce}}\bigg|_{I_B = \text{constant}} \quad \text{mho} \tag{6.79}$$

where,

h_{oe} = Output conductance in mho.

(i) Calculation of h_{fe}

h_{fe} is the most important parameter out of four parameters. For calculation of h-parameters, we find the location of Q-point. According to Eqn. (6.78), V_{CE} must be constant for calculation of h_{fe}. The change in base and collector current can be read out at vertical line which passes through Q-point because at this line, V_{CE} is constant as shown in Fig. 6.24. h_{fe} can be calculated after division of ΔI_b in ΔI_C.

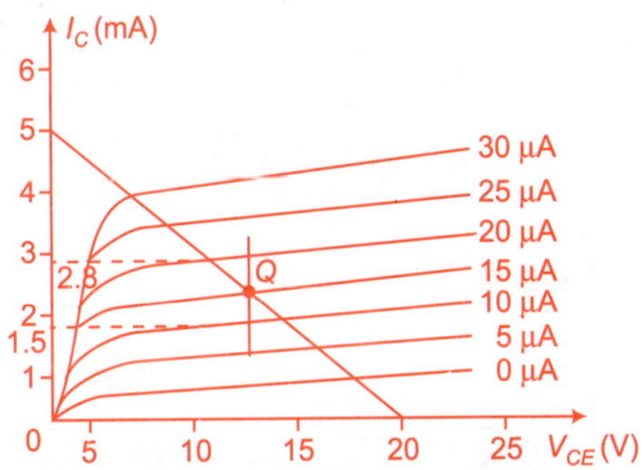

Fig. 6.24 *Calculation of h_{fe} (from output characteristics)*

In Fig. 6.24, the value of I_b changes between 10 μA to 20 μA at V_{CE} = 10 V.

The value of I_C is 1.5 mA and 2.8 mA.

Hence,

$$\Delta I_C = (2.8 - 1.5) \text{ mA}$$

$$\Delta I_C = 1.3 \text{ mA}$$

$$\Delta I_b = (20 \text{ μA} - 10 \text{ μA}) = 10 \text{ μA}.$$

Hence,

$$h_{fe} = \left.\frac{\Delta I_C}{\Delta I_b}\right|_{V_{CE}=\text{constant}} = \frac{1.3 \times 10^{-3}}{10 \times 10^{-6}}$$

$$\boxed{h_{fe} = 130.}$$

(ii) Calculation of h_{oe}

According to Eqn. (6.79), I_B must be constant for calculation of h_{oe}. The change in collector current ΔI_C with respect to change in collector emitter voltage ΔV_{ce} can be read out at horizontal line which passes through Q-point, because at this line, I_B is constant as shown in Fig. 6.25. h_{oe} can be calculated after taking the ratio of ΔI_C and ΔV_{Ce}.

$$I_B = 15 \ \mu A$$
$$\Delta I_C = (2.2 - 2.1) \ mA$$
$$\Delta I_C = 0.1 \ mA$$
$$\Delta V_{CE} = (12 - 8) \ V$$
$$\Delta V_{CE} = 4 \ V.$$

Hence,

$$h_{oe} = \left.\frac{\Delta I_C}{\Delta V_{Ce}}\right|_{I_B=\text{constant}}$$

$$h_{oe} = \frac{0.1}{4}$$

$$\boxed{h_{oe} = 0.025 \ mho.}$$

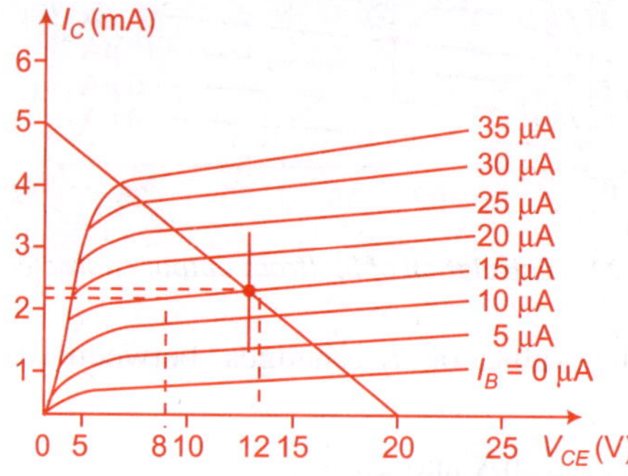

Fig. 6.25   *Calculation of h_{oe} (from output characteristics)*

This parameter is calculated from input characteristics of transistor. According to Eqn. (6.76), h_{ie} can be calculated by the values of ΔV_{be} and ΔI_b at constant value of V_{CE} as shown in Fig. 6.26.

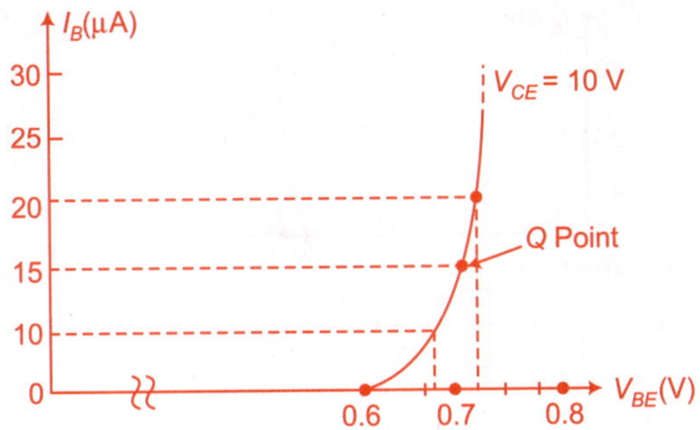

Fig. 6.26 *Calculation of h_{ie} (from input characteristics)*

From Fig. 6.26 at $V_{CE} = 10$ V

$$\Delta V_{be} = (0.72 - 0.68) \text{ V}$$

$$\Delta V_{be} = 0.04 \text{ V}$$

$$\Delta I_b = (20 - 10) \text{ }\mu\text{A}$$

$$\Delta I_b = 10 \text{ }\mu\text{A}.$$

Hence,

$$h_{ic} = \left.\frac{\Delta V_{be}}{\Delta I_b}\right|_{V_{CE}} = 10 \text{ V}$$

$$h_{ic} = \frac{0.04 \text{ V}}{10 \text{ }\mu\text{A}} = 4 \text{ k}\Omega$$

$$\boxed{h_{ic} = 4 \text{ k}\Omega.}$$

(iv) Calculation of h_{re}

This parameter also can be calculated from input characteristics of transistor. If horizontal line is drawn at input characteristics, then I_B is constant at this line as shown in Fig. 6.27.

$$I_B = 15 \text{ }\mu\text{A}$$

$$\Delta V_{be} = (0.72 - 0.68) \text{ V}$$

$$\Delta V_{be} = 0.04 \text{ V}$$

and,

$$\Delta V_{ce} = (20 - 0) \text{ V} = 20 \text{ V}.$$

Hence,

$$h_{re} = \left.\frac{\Delta V_{be}}{\Delta V_{ce}}\right|_{I_B} = \text{constant}$$

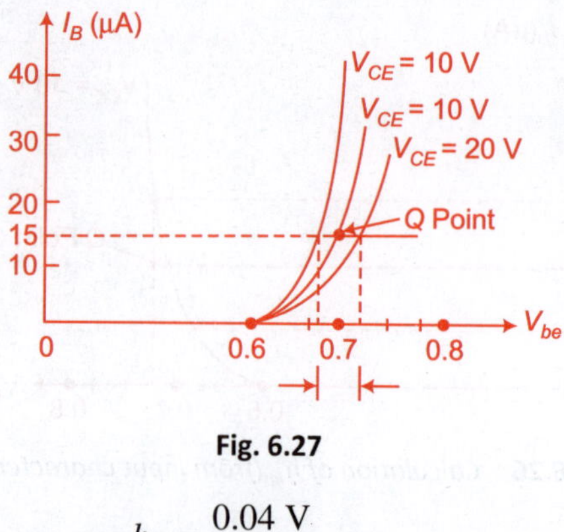

Fig. 6.27

$$h_{re} = \frac{0.04 \text{ V}}{20 \text{ V}}$$

$$\boxed{h_{re} = 2 \times 10^{-3}.}$$

Example 6.6.1   Given that $V_{be} = 0.02$ V, $\Delta I_b = 10$ µA at $V_{CE} = 10$ V, $h_{fe} = 130$, $h_{oe} = 20$ µS and $\Delta V_{be} = 0.04$ and $\Delta V_{ce} = 15$ V at $I_B = 15$ µA. Determine the common emitter hybrid equivalent circuit.

Solution   Given that $\Delta V_{be} = 0.02$ V and $\Delta I_b = 10$ µA

$$h_{ie} = \frac{\Delta V_{be}}{\Delta I_b}\bigg|_{V_{CE} = \text{constant}} = \frac{0.02 \text{ V}}{10 \text{ µA}}$$

$$= 2 \text{ k}\Omega$$

and,

$$h_{re} = \frac{\Delta V_{be}}{\Delta V_{ce}}\bigg|_{I_B = \text{constant}} = \frac{0.04 \text{ V}}{15 \text{ V}}$$

$$h_{re} = 2.6 \times 10^{-3}.$$

The h-model is shown in Fig. Ex. 6.6.1

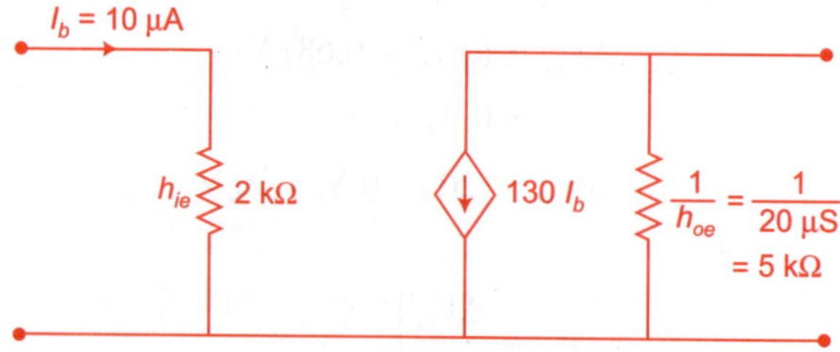

Fig. Ex. 6.6.1   *Common emitter hybrid equivalent model*

Examples

Example 6.1 Calculate the maximum and minimum collector current in the circuit of Fig. Ex. 6.1.

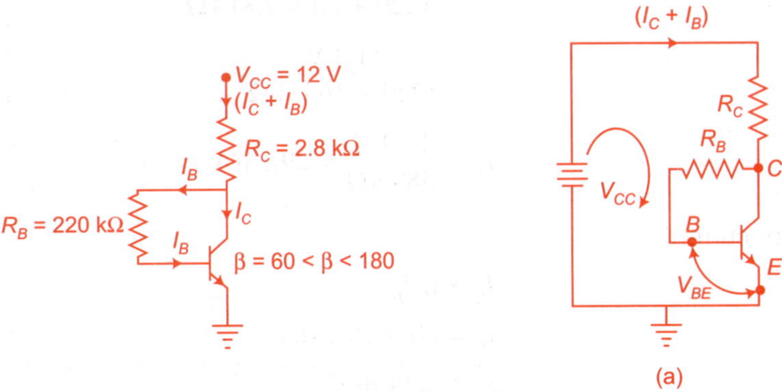

Fig. Ex. 6.1 *Applying KVL input site*

Solution Given that,

$$R_B = 220 \text{ k}\Omega, \quad R_C = 2.8 \text{ k}\Omega$$

$$V_{CC} = 12 \text{ V}$$

Range of β = 60 < β < 180.

Applying KVL at the input side

$$V_{CC} - (I_C + I_B) R_C - I_B R_B - V_{BE} = 0$$

$$V_{CC} = (I_C + I_B) R_C + I_B R_B + V_{BE}$$

$$V_{CC} = I_C R_C + I_B R_C + I_B R_B + V_{BE}$$

$$V_{CC} = I_C R_C + (R_C + R_B) I_B + V_{BE} \quad\quad (6.80)$$

$$\because \quad I_C = \beta I_B.$$

Putting the value of I_C in Eqn. (6.80), we obtain,

$$V_{CC} = \beta I_B R_C + (R_C + R_B) I_B + V_{BE}$$

$$V_{CC} = \beta R_C I_B + R_C I_B + R_B I_B + V_{BE}$$

$$V_{CC} = (\beta + 1) R_C I_B + R_B I_B + V_{BE}$$

$$V_{CC} = [(\beta + 1) R_C + R_B] I_B + V_{BE}$$

$$I_B [(\beta + 1) R_C + R_B] = V_{CC} - V_{BE}$$

$$I_B = \frac{V_{CC} - V_{BE}}{R_B + (\beta + 1) R_C}$$

$\because$ β is much greater than 1 so β + 1 ≃ β.

Hence,
$$I_B = \frac{V_{CC} - V_{BE}}{R_B + \beta R_C}$$

when $\beta = 60$

$$I_B = \frac{(12\ \text{V} - 0.7\ \text{V})}{(220 + 60 \times 2.8)\ \text{k}\Omega}$$

$$= \frac{11.3\ \text{V}}{(220 + 168)\ \text{k}\Omega}$$

$$I_B = \frac{11.3\ \text{V}}{388\ \text{k}\Omega} = 29.1\ \mu\text{A}.$$

Collector current

$$I_C = \beta\, I_B$$
$$I_C = 60 \times 29.1\ \mu\text{A}$$
$$\boxed{I_C = 1.74\ \text{mA}}$$

(**Ans.**)

$$\boxed{I_{C\ \min} = 1.74\ \text{mA}.}$$

(**Ans.**)

when $\beta = 180$

$$I_B = \frac{(12 - 0.7)\ \text{V}}{(220 + 180 \times 2.8)\ \text{k}\Omega}$$

$$= \frac{11.3\ \text{V}}{(220 + 504)\ \text{k}\Omega}$$

$$I_B = \frac{11.3\ \text{V}}{724\ \text{k}\Omega} = 15.6\ \mu\text{A}$$

$$I_B = 15.6\ \mu\text{A}.$$

Collector current

$$I_C = \beta\, I_B$$
$$I_C = 180 \times 15.6\ \mu\text{A}$$
$$I_C = 2.81\ \text{mA}$$
$$\boxed{I_{C\ \max} = 2.81\ \text{mA}.}$$

(**Ans.**)

Example 6.2   Determine I_B, I_C, V_{CE}, V_E, V_B and V_o for the emitter bias network.

Solution   Given that,

$$V_{CC} = 15\ \text{V}, \qquad\qquad R_C = 2.2\ \text{k}\Omega$$
$$R_B = 420\ \text{k}\Omega, \qquad\qquad R_E = 1\ \text{k}\Omega$$
$$\beta = 49$$
$$C_C = 10\ \mu\text{F} \quad \text{and} \quad C_E = 40\ \mu\text{F}$$

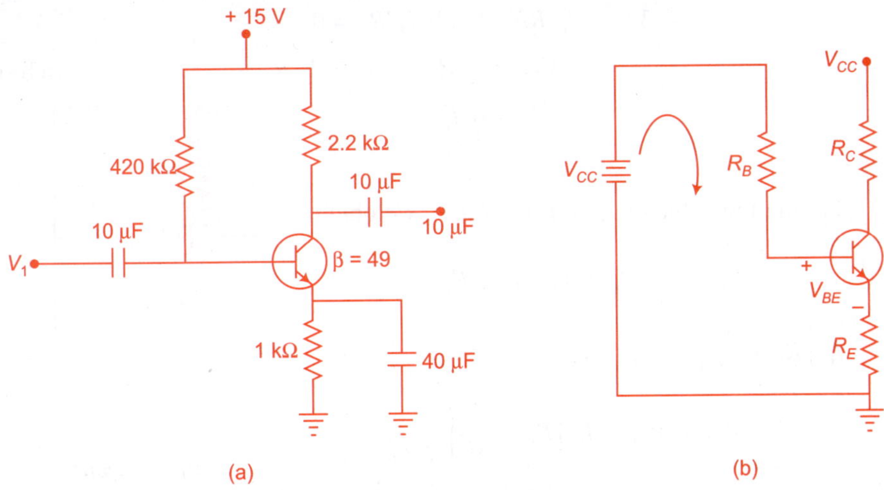

Fig. Ex. 6.2

For DC analysis, the AC input is removed and the circuit will be shown in Fig. Ex. 6.2 (b).

Applying KVL at input side, we obtain,

$$V_{CC} - I_B R_B - V_{BE} - I_E R_E = 0$$

$$V_{CC} = I_B R_B + V_{BE} + I_E R_E \qquad (6.81)$$

$$\therefore \quad I_E = (\beta + 1) I_B.$$

Putting the value of I_E in Eqn. (6.81), we obtain,

$$V_{CC} = I_B R_B + V_{BE} + (\beta + 1) I_B R_E$$

$$I_B [R_B + (\beta + 1) R_E] = V_{CC} - V_{BE}$$

$$I_B = \frac{V_{CC} - V_{BE}}{R_B + (\beta + 1) R_E} \qquad (6.82)$$

Putting the given values in Eqn. (6.82), we obtain,

$$I_B = \frac{(15 - 0.7)\ \text{V}}{420\ \text{k}\Omega + (49 + 1) \times 1\ \text{k}\Omega}$$

$$I_B = \frac{14.3\ \text{V}}{(420 + 50)\ \text{k}\Omega} = \frac{14.3\ \text{V}}{470\ \text{k}\Omega}$$

$$I_B = 0.030\ \text{mA}$$

$$I_B = 30\ \mu\text{A}$$

$$I_C = \beta I_B = 49 \times 30\ \mu\text{A}$$

$$I_C = 1.47\ \text{mA}.$$

Applying KVL at output side in Fig. Ex. 6.2 (c)

$$V_{CC} - I_C R_C - V_{CE} - I_E R_E = 0$$

$$V_{CC} = I_C R_C + V_{CE} + I_E R_E \qquad (6.83)$$

$\because$
$$I_C = \alpha \, I_E$$

$$I_C = \frac{I_C}{\alpha}$$

Putting the value of I_E in Eqn. (6.83), we obtain,

$$V_{CC} = I_C R_C + V_{CE} + \frac{I_C}{\alpha} R_E$$

$$I_C \left[R_C + \frac{R_E}{\alpha} \right] = V_{CC} - V_{CE}$$

$$V_{CE} = V_{CC} - I_C \left[R_C + \frac{R_E}{\alpha} \right]$$

$$V_{CE} = 15 - 1.47 \left[2.2 \text{ k}\Omega + \frac{1 \text{ k}\Omega \, 50}{49} \times 1 \right]$$

$\therefore$
$$\frac{1}{\alpha} = \frac{\beta + 1}{\beta}$$

$$V_{CE} = (15 - 3.38) \text{ V}$$

$$V_{CE} = 11.62 \text{ volt.}$$

We know that,

$$V_E = I_E R_E = 1 \, \frac{I_C R_E}{\alpha}$$

$$V_E = \frac{(\beta + 1)}{\beta} \times I_C R_E$$

$$= \frac{50}{49} \times 1.47 \times 1 \text{ k}\Omega$$

$$V_E = 1.5 \text{ volt}$$

$$V_C = V_{CC} - I_C R_C$$

$$V_C = (15 \text{ V} - 1.47 \text{ mA} \times 2.2 \text{ k}\Omega)$$

$$V_C = (15 \text{ V} - 3.23 \text{ V})$$

$$V_C = 11.77$$

Now,

$$V_E = V_C - V_{CE}$$

$$V_E = (11.77 - 11.62) \text{ V} = 0.15 \text{ V}$$

$$V_E = 0.15 \text{ volt}$$

2.2 kΩ

V_{CC}

V_{CE}

1.8 kΩ

(b)

Fig. Ex. 6.2 (c)

and,

$$V_{BE} = V_B - V_E$$
$$V_B = V_{BE} + V_E$$
$$= 0.7 + 0.15 \text{ V}$$
$$\boxed{V_B = 0.22 \text{ volt.}}$$

(Ans.)

Example 6.3 A common emitter amplifier has the following h-parameters $h_{ie} = 1.2 \text{ k}\Omega$, $h_{re} = 0.30 \times 10^{-3}$, $h_{fe} = 45$, $h_{oe} = 26 \ \mu$ mho, if the load resistance is 1 kΩ and source resistance R_S is also 1 kΩ. Find (i) voltage gain (ii) current gain.

Solution   Given that,

$$h_{ie} = 1.2 \text{ k}\Omega, \qquad h_{re} = 0.30 \times 10^{-3}$$
$$h_{fe} = 45, \qquad h_{oe} = 26 \ \mu \text{ mho},$$
$$R_C = 1 \text{ k}\Omega, \qquad R_S = 1 \text{ k}\Omega.$$

From Eqn. (6.58), we know that,

Current gain
$$A_i = \frac{h_{fe}}{1 + h_{oe} R_L}$$

$$A_i = \frac{45}{1 + 26 \ \mu \text{ mho} \times 1 \text{ k}\Omega}$$

$$= \frac{45}{1 + 26 \times 10^{-3}}$$

$$A_i = \frac{45}{1.026}$$

$$\boxed{A_i = 43.86.}$$

(Ans.)

From Eqn. (6.62), we know that,

Voltage gain $A_V = \dfrac{- h_{fe} R_L}{h_{ie} + (h_{ie} h_{oe} - h_{fe} h_{re}) R_L}$

$$A_V = - \frac{45 \times 1 \text{ k}\Omega}{1.2 \text{ k}\Omega + [(1.2 \text{ k}\Omega \times 26 \ \mu \text{ mho}) - (45 \times 0.3 \times 10^{-3})] \, 1 \text{ k}\Omega}$$

$$A_V = - \frac{45 \text{ k}\Omega}{1.2 \text{ k}\Omega + [(31.2 \times 10^{-3}) - (13.5 \times 10^{-3})] \, 1 \text{ k}\Omega}$$

$$A_V = - \frac{45 \text{ k}\Omega}{1.2 \text{ k}\Omega + [31.2 - 13.5) \times 10^{-3} \times 1 \text{ k}\Omega]}$$

$$A_V = - \frac{45 \text{ k}\Omega}{1.2 \text{ k}\Omega + 17.7 \ \Omega}$$

$$= -\frac{45 \text{ k}\Omega}{1.217 \text{ k}\Omega}$$

$$\boxed{A_V = -36.97.}$$ **(Ans.)**

Example 6.4 A common emitter amplifier uses a voltage source internal resistance $R_S = 750 \ \Omega$ and the load resistance is $R_L = 1 \text{ k}\Omega$. The *h*-parameter are $h_{ie} = 1 \text{ k}\Omega$, $h_{re} = 2 \times 10^{-3}$, $h_{fe} = 50$ and $h_{oe} = 25 \ \mu$ mho. Calculate the current gain A_i, input impedance Z_i and voltage gain A_V.

Solution Given that,

$$R_S = 750 \ \Omega, \qquad R_L = 1 \text{ k}\Omega,$$
$$h_{ie} = 1 \text{ k}\Omega \qquad h_{re} = 2 \times 10^{-4},$$
$$h_{fe} = 50, \qquad h_{oe} = 25 \ \mu \text{ mho}$$

From Eqn. (6.58), we know that,

Current gain
$$A_i = \frac{h_{fe}}{1 + h_{oe} \, R_L}$$

$$= \frac{50}{1 + 25 \ \mu \text{ mho} \times 1 \text{ k}\Omega}$$

$$A_i = \frac{50}{1 + 25 \times 10^{-3}} = \frac{50}{1.025}$$

$$\boxed{A_i = 48.78.}$$ **(Ans.)**

From Eqn. (6.65), we know that,

$$Z_i = h_{ie} - h_{re} \, R_L \, \frac{h_{fe}}{1 + h_{oe} \, R_L}$$

$$\because \qquad \frac{h_{fe}}{1 + h_{oe} \, R_L} = A_i$$

from Eqn. (6.79)

Now,

$$Z_i = h_{ie} - h_{re} \, R_L \, A_i$$

$$Z_i = 1 \text{ k}\Omega - 2 \times 10^{-4} \times 1 \text{ k}\Omega \times 48.78$$

$$Z_i = 1 \text{ k}\Omega - 9.756 \ \Omega$$

$$\boxed{Z_i = 990.24 \ \Omega.}$$ **(Ans.)**

From Eqn. (6.62), we know that,

Voltage gain $A_V = -\dfrac{h_{fe}\,R_L}{h_{ie} + (h_{ie}\,h_{oe} - h_{fe}\,h_{re})\,R_L}$

$$A_V = -\frac{50 \times 1\ \text{k}\Omega}{1\ \text{k}\Omega + [(1\ \text{k}\Omega \times 25\ \mu\ \text{mho}) - (50 \times 2 \times 10^{-3})]\,1\ \text{k}\Omega}$$

$$A_V = -\frac{50\ \text{k}\Omega}{1\text{k}\Omega + [(25 \times 10^{-3}) - (100 \times 10^{-4})]1\text{k}\Omega}$$

$$A_V = -\frac{50\ \text{k}\Omega}{1\ \text{k}\Omega + [(25 - 10) \times 10^{-3} \times 1\ \text{k}\Omega]}$$

$$A_V = -\frac{50\ \text{k}\Omega}{1\ \text{k}\Omega + 15\ \Omega}$$

$$= -\frac{50\ \text{k}\Omega}{1.015\ \text{k}\Omega}$$

$$\boxed{A_V = -\,49.26.}$$
(Ans.)

Example 6.5 The h-parameter for a common emitter configuration are $h_{ie} = 2500\ \Omega$, $h_{fe} = 120$, $h_{re} = 0.02 \times 10^{-2}$ and $h_{oe} = 5\ \mu$ mho. Find h-parameter for common collector configuration.

Solution   We know that,

$$h_{ic} = h_{ie}$$

$$\boxed{h_{ic} = 2.5\ \text{k}\Omega} \qquad\qquad (\textbf{Ans.})$$

$$\boxed{h_{fc} = -\,(1 + h_{fe})} \qquad\qquad (\textbf{Ans.})$$

$$= -\,(1 + 120)$$

$$h_{fc} = -\,121$$

$$h_{rc} = 1 - h_{re}$$

$$h_{rc} = 1 - 0.02 \times 10^{-2}$$

$$\simeq 1$$

$\Rightarrow$
$$\boxed{h_{rc} \simeq 1.}$$

Now,

$$h_{oc} = h_{oe} = 5\ \mu\ \text{mho}$$

$$\boxed{h_{oc} = 5\ \mu\ \text{mho.}} \qquad\qquad (\textbf{Ans.})$$

Example 6.6 h-parameter of common emitter configuration are as follows:

$$h_{ie} = 1.5\ \text{k}\Omega, \qquad\qquad h_{re} = 2 \times 10^{-4}$$
$$h_{fe} = 100, \qquad\qquad h_{oe} = 30\ \mu\text{S}\ (\text{S} = \text{mho})$$
$$R_L = 4.7\ \text{k}\Omega, \qquad\qquad R_S = 1\ \text{k}\Omega$$

Calculate the following:

(a) Current gain

(b) Voltage gain

    (c) Input impedance

    (d) Output impedance

With the help of h-parameter model and approximate h-parameter model.

Solution I-By full hybrid parameter model

(a) From Eqn. (6.58), we know that,

$$A_i = \frac{h_{fe}}{1 + h_{oe} R_L}$$

$$= \frac{100}{1 + (30 \times 10^{-6} \times 4.7 \times 10^3)}$$

$$A_i = \frac{100}{1 + 0.141}$$

$$= \frac{100}{1.141}$$

$$\boxed{A_i = 87.64.}$$ **(Ans.)**

(b) From Eqn. (6.62), we know that,

Voltage gain $A_V = -\dfrac{h_{fe} R_L}{h_{ie} + (h_{ie} h_{oe} - h_{fe} h_{re}) R_L}$

$$A_V = -\frac{100 \times 4.7 \text{ k}\Omega}{1.5 \text{ k}\Omega + [(1.5 \text{ k}\Omega \times 30 \text{ μS}) - (100)(2 \times 10^{-4})] 4.7 \text{ k}\Omega}$$

$$A_V = -\frac{470 \text{ k}\Omega}{1.5 \text{ k}\Omega + [(0.045) - (0.020)] 4.7 \text{ k}\Omega}$$

$$A_V = -\frac{470 \text{ k}\Omega}{1.5 \text{ k}\Omega + [0.025 \times 4.7 \text{ k}\Omega]}$$

$$A_V = \frac{470 \text{ k}\Omega}{1.5 \text{ k}\Omega + 0.12 \text{ k}\Omega} = 470 \text{ k}\Omega$$

$$\boxed{A_V = -290.12.}$$

(c) From Eqn. (6.65), we know that,

Input impedance $Z_i = h_{ie} - \dfrac{h_{fe} h_{re} R_L}{1 + h_{oe} R_L}$

$$Z_i = h_{ie} - h_{re} R_L A_i$$

$\because$ $A_i = \dfrac{h_{fe}}{1 + h_{oe} R_L}$

$$Z_i = 1.5 \text{ k}\Omega - (2 \times 10^{-4}) \times 4.7 \text{ k}\Omega \times 87.64$$

$$Z_i = 1.5 \text{ k}\Omega - 0.08 \text{ k}\Omega$$

$$\boxed{Z_i = 1.42 \text{ k}\Omega.}$$ **(Ans.)**

(d) From Eqn. (6.68), we know that,

$$Z_o = \cfrac{1}{h_{oe} - \left[\cfrac{h_{fe}\,h_{re}}{R_S + h_{ie}}\right]}$$

$$Z_o = \cfrac{1}{30 \times 10^{-6} \text{ mho} - \left[\cfrac{100 \times 2 \times 10^{-4}}{1 \text{ k}\Omega + 1.5 \text{ k}\Omega}\right]}$$

$$Z_o = \cfrac{1}{30 \ \mu S - \cfrac{200 \times 10^{-4}}{2.5 \text{ k}\Omega}}$$

$$Z_o = \cfrac{1}{30 \ \mu S - 80 \times 10^{-7} \text{ S}}$$

$$Z_o = \cfrac{1}{30 \ \mu S - 8\mu S} = \cfrac{1}{22 \ \mu S}$$

$$\boxed{Z_o = 45.45 \text{ k}\Omega.}$$ **(Ans.)**

II By Approximate Hybrid Model

(a) From Eqn. (6.69), we know that,

$$A_i = h_{fe} = 100$$

$$\boxed{A_i = 100.}$$ **(Ans.)**

(b) From Eqn. (6.72), we know that,

$$A_V = -\frac{h_{fe}\,R_L}{h_{ie}} = -\frac{100 \times 4.7 \text{ k}\Omega}{1.5 \text{ k}\Omega}$$

$$A_V = -\frac{470 \text{ k}\Omega}{1.5 \text{ k}\Omega}$$

$$\boxed{A_V = 313.33.}$$ **(Ans.)**

(c) From Eqn. (6.70), we know that,

$$Z_i = h_{ie} = 1.5 \text{ k}\Omega$$

$$\boxed{Z_i = 1.5 \text{ k}\Omega.}$$ **(Ans.)**

(d) From Eqn. (6.74), we know that,

$$Z_o = \frac{1}{h_{oe}} = \frac{1}{30 \ \mu S}$$

$$\boxed{Z_o = 33.33 \text{ k}\Omega.}$$ **(Ans.)**

Example 6.7 Calculate I_B, I_C, V_E and V_{CE} for the following circuit, if $\beta = 130$. **(UPTU 2006-2007)**

Solution Given that,

$$R_C = 9.1 \text{ k}\Omega, \qquad R_E = 7.5 \text{ k}\Omega$$
$$R_1 = 510 \text{ k}\Omega, \qquad R_2 = 510 \text{ k}\Omega$$
$$V_{CC} = 18 \text{ V}, \qquad \beta = 130$$

From Eqn. (6.26), we know that the Thevenin resistance,

$$R_{Th} = \frac{R_1 R_2}{R_1 + R_2} = \frac{510 \times 510}{510 + 510} \text{ k}\Omega$$

$$R_{Th} = \frac{260100}{1020} = 255 \text{ k}\Omega$$

$$R_{Th} = 255 \text{ k}\Omega.$$

From Eqn. (6.27), we know that the Thevenin voltage,

$$V_{Th} = \frac{R_2}{R_1 + R_2} \, V_{CC} = \frac{510}{510 + 510} \times 18$$

$$V_{Th} = \frac{9180}{1020} = 9 \text{ V}$$

$$V_{Th} = 9 \text{ volt}.$$

Now, from Eqn. (6.29), we know that,

$$I_B = \frac{V_{Th} - V_{BE}}{R_{Th} + (\beta + 1) \, R_E}$$

Let $(V_{BE} = 0.7 \text{ V})$

$$I_B = \frac{9 \text{ V} - 0.7 \text{ V}}{255 \text{ k}\Omega + (130 + 1) \times 7.5 \text{ k}\Omega)}$$

$$= \frac{8.3 \text{ V}}{255 \text{ k}\Omega + 982.5 \text{ k}\Omega}$$

$$I_B = \frac{8.3 \text{ V}}{1237.5 \text{ k}\Omega} = 6.7 \times 10^{-6} \text{ A}$$

$$I_B = 6.7 \text{ } \mu\text{A}.$$

We know that,

$$I_C = \beta \, I_B$$
$$I_C = 130 \times 6.7 \text{ } \mu\text{A} = 0.87 \text{ mA}$$
$$I_C = 0.87 \text{ mA}.$$

From Fig. 6.7 (a), we know that,

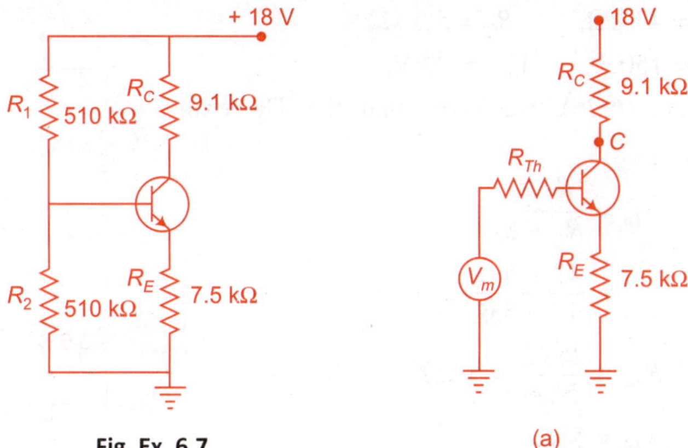

Fig. Ex. 6.7

(a)

Fig. Ex. 6.7 (a)

$$V_C = V_{CC} - I_C R_C$$
$$V_C = 18 - (0.87 \text{ mA} \times 9.1 \text{ k}\Omega)$$
$$V_C = 7.92 \text{ volt.}$$

Now, from Fig. Ex. 6.7(a), the emitter voltage V_E is,

$$V_E = I_E R_E$$

$\because$

$$I_E = (\beta + 1) I_B$$
$$V_E = (\beta + 1) I_B R_E$$
$$V_E = (130 + 1) \times 6.7 \text{ μA} \times 7.5 \text{ k}\Omega$$
$$V_E = 6.58 \text{ V.}$$

Now, we know that,

$$V_{CE} = V_C - V_E$$
$$V_{CE} = 7.92 \text{ V} - 6.58 \text{ V}$$
$$V_{CE} = 1.34 \text{ V}$$

or,

From Eqn. (6.32), we know that,

$$V_{CE} = V_{CC} - I_C (R_C + R_E)$$
$$V_{CE} = 18 - 0.87 \text{ mA} (9.1 + 7.5)\text{k}\Omega$$
$$V_{CE} = (18 \text{ V} - 14.44 \text{ V})$$
$$\boxed{V_{CE} = 3.56 \text{ volt.}}$$

(Ans.)

Example 6.8 For a voltage divider biasing circuit in Fig. Ex. 6.8. Find I_C, V_{CE}, I_B, V_E and V_B.

(UPTU 2007-08)

Solution Given that,

$$R_C = 2.2 \text{ k}\Omega, \qquad R_E = 1 \text{ k}\Omega$$

$$R_1 = 47 \text{ k}\Omega, \qquad R_2 = 5.6 \text{ k}\Omega$$
$$\beta = 150, \qquad V_{CC} = 20 \text{ V}.$$

From Eqn. (6.26), we know that the Thevenin resistance,

$$R_{Th} = \frac{R_1 R_2}{R_1 + R_2}$$

$$= \frac{47 \times 5.6}{47 + 5.6} \text{ k}\Omega$$

$$R_{Th} = \frac{263.2}{52.6} = 5 \text{ k}\Omega$$

$$R_{Th} = 5 \text{ k}\Omega.$$

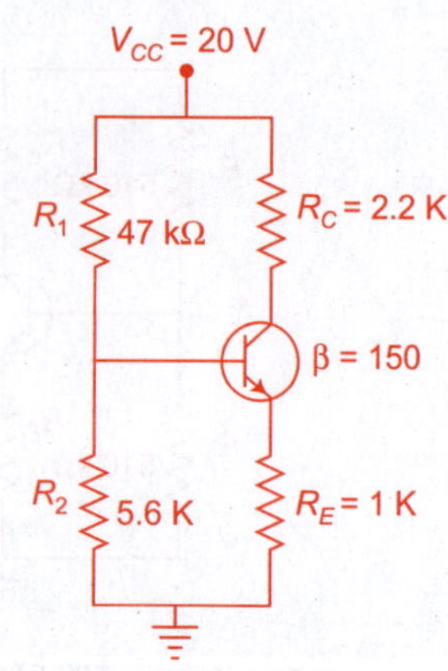

Fig. Ex. 6.8

From Eqn. (6.27), we know that the Thevenin voltage is,

$$V_{Th} = \frac{R_2}{R_1 + R_2} = \frac{5.6 \text{ k}\Omega}{47 \text{ k}\Omega + 5.6 \text{ k}\Omega} \times 20 \text{ V}$$

$$V_{Th} = \frac{112}{52.6} \text{ V} = 2.13 \text{ V}$$

$$V_{Th} = 2.13 \text{ volt}.$$

Now, from Eqn. (6.29), we know that,

$$I_B = \frac{V_{Th} - V_{BE}}{R_{Th} + (\beta + 1) R_E}$$

$$= \frac{(2.13 - 0.7) \text{ V}}{5 \text{ k}\Omega + (150 + 1) \, 1 \text{ k}\Omega}$$

$$I_B = \frac{1.43 \text{ V}}{5 \text{ k}\Omega + 151 \text{ k}\Omega}$$

$$= \frac{1.43}{156 \text{ k}\Omega} = 9.2 \text{ μA}$$

$$I_B = 9.2 \text{ μA}.$$

We know that,

$$I_C = \beta \, I_B = 150 \times 9.2 \text{ μA}$$

$$I_C = 1.38 \text{ mA}.$$

From Eqn. (6.32), we know that,

$$V_{CE} = V_{CC} - I_C \, (R_C + R_E)$$
$$V_{CE} = 20 - 1.39 \text{ mA} \, (2.2 \text{ k}\Omega + 1 \text{ k}\Omega)$$

$$V_{CE} = 20 - (1.39 \text{ mA} \times 3.2 \text{ k}\Omega)$$
$$V_{CE} = (20 - 4.45) \text{ V}$$
$$V_{CE} = 15.6 \text{ V}$$
$$\therefore \quad V_B = V_{Th} = 2.13 \text{ V.}$$

Now,

$$V_{BE} = V_B - V_E$$
$$V_E = V_B - V_{BE}$$
$$V_E = (2.13 - 0.7) \text{ V}$$
$$\boxed{V_E = 1.43 \text{ volt.}} \hspace{3cm} \textbf{(Ans.)}$$

Example 6.9 For a common emitter amplifier circuit with h-parameters $h_{ie} = 2 \text{ k}\Omega$, $h_{re} = 6 \times 10^{-4}$, $h_{fe} = 50$, $h_{oe} = 25 \text{ }\mu \text{ AV}$ and $R_L = 4 \text{ k}\Omega$ and $R_S = 10 \text{ k}\Omega$. Compute A_V, A_i, R_i, R_o. **(UPTU 2008-09)**

Solution Given that,

$$h_{ie} = 2 \text{ k}\Omega, \hspace{2cm} h_{re} = 6 \times 10^{-4}$$
$$h_{fe} = 50, \hspace{2cm} h_{oe} = 25 \text{ }\mu \text{ AV}$$
$$R_L = 4 \text{ k}\Omega, \hspace{2cm} R_S = 10 \text{ k}\Omega.$$

From Eqn. (6.79), we know that the,

Current gain
$$A_i = \frac{h_{fe}}{1 + h_{oe} R_L}$$

$$A_i = \frac{50}{1 + (25 \times 10^{-6} \text{ } \mho \times 4 \text{ k}\Omega)}$$

$$= \frac{50}{1 + 100 \times 10^{-3}}$$

$$A_i = \frac{50}{1.1} = 45.45$$

$$A_i = 45.45$$

From Eqn. (6.65), we know that,

$$R_i = h_{ie} - h_{re} R_L \frac{h_{fe}}{1 + h_{oe} R_L}$$

$$R_i = h_{ie} - h_{re} R_L A_i$$

$$R_i = 2 \text{ k}\Omega - [6 \times 10^{-4} \times 4 \text{ k}\Omega \times 45.45]$$

$$R_i = [2 \text{ k}\Omega - 0.1090.8 \text{ k}\Omega]$$

$$R_i = 1.9 \text{ k}\Omega.$$

From Eqn. (6.62), we know that,

Voltage gain $\quad A_V = -\dfrac{h_{fe}\, R_L}{h_{ie} + (h_{ie}\, h_{oe} - h_{fe}\, h_{re})\, R_L}$

$$A_V = -\frac{50 \times 4\ \text{k}\Omega}{2\ \text{k}\Omega + [(25\ \mu\, A_V \times 2\ \text{k}\Omega) - (50 \times 6 \times 10^{-4})] \times 4\ \text{k}\Omega}$$

$$A_V = -\frac{200\ \text{k}\Omega}{2\ \text{k}\Omega + [50 \times 10^{-3} - 300 \times 10^{-4}] \times 4\ \text{k}\Omega}$$

$$A_V = -\frac{200\ \text{k}\Omega}{2\ \text{k}\Omega + [0.050 - 0.03] \times 4\ \text{k}\Omega}$$

$$A_V = -\frac{-\,200\ \text{k}\Omega}{2\ \text{k}\Omega + 0.08\ \text{k}\Omega} = -\frac{200\ \text{k}\Omega}{2.08\ \text{k}\Omega}$$

$$\boxed{A_V = -\,96.15.} \qquad\qquad \textbf{(Ans.)}$$

Now, from Eqn. (6.68), we know that,

$$R_o = \frac{h_{fe}\, h_{ie}}{R_S + h_{ie}}$$

$$= \frac{1}{25 \times 10^{-6}\ \text{℧} - \left[\dfrac{50 \times 6 \times 10^{-4}}{10\ \text{k}\Omega + 2\ \text{k}\Omega}\right]}$$

$$= \frac{1}{(25 \times 10^{-6}\ \text{℧} - 2.5 \times 10^{-6}\ \text{℧})}$$

$$= \frac{1}{22.5 \times 10^{-6}\ \text{℧}}$$

$$R_0 = 44.4\ \text{k}\Omega.$$

Example 6.10 The potential divider circuit shown in Fig. Ex. 6.10 has the following values: $I_E = 2.1$ mA, $I_B = 50$ μA, $V_{BE} = 0.2$ V, $R_E = 1$ kΩ, $R_2 = 15$ kΩ and $V_{CC} = 12$ V. Find the value of R_1.

Solution Given that,

$$R_2 = 15\ \text{k}\Omega, \qquad R_E = 1\ \text{k}\Omega$$
$$V_{CC} = 12\ \text{V}, \qquad V_{BE} = 0.2\ \text{V}$$
$$I_B = 50\ \mu\text{A}, \qquad I_E = 2.1\ \text{mA}$$

Voltage across R_2,

$$V_2 = V_{BE} + I_E\, R_E$$
$$V_2 = 0.2 + 2\ \text{mA} \times 1\ \text{k}\Omega$$
$$V_2 = 0.2 + 2 = 2.2\ \text{V}$$
$$V_2 = 2.2\ \text{V}$$

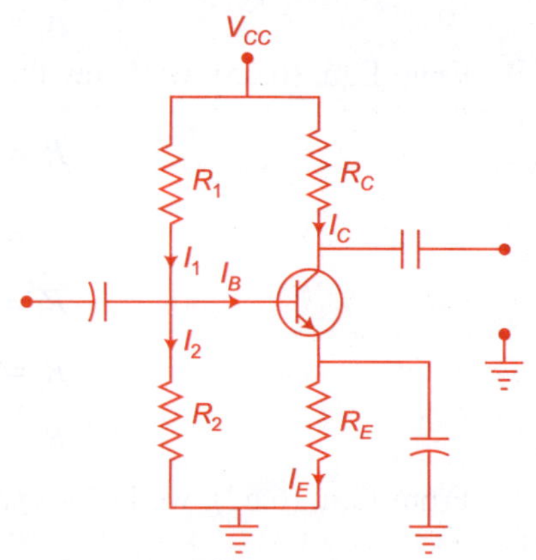

Fig. Ex. 6.10

Current through
$$R_2 = \frac{V_2}{R_2}$$

$$I_2 = \frac{2.2 \text{ V}}{15 \text{ k}\Omega} = 0.15 \text{ mA}$$

$$I_2 = 0.15 \text{ mA}.$$

Current through
$$R_1 \quad I_1 = I_2 + I_B$$

$$= 0.15 \text{ mA} + 0.05 \text{ mA}.$$

$$I_1 = 0.20 \text{ mA}.$$

Voltage across
$$R_1 \quad V_1 = V_{CC} - V_2$$

$$V_1 = (12 - 2.2)\text{V} = 9.8 \text{ V}$$

$$V_1 = 9.8 \text{ V}$$

Now,

$$R_1 = \frac{V_1}{I_1} = \frac{9.8 \text{ V}}{0.20 \text{ mA}}$$

$$\boxed{R_1 = 49 \text{ k}\Omega.} \hspace{3cm} \textbf{(Ans.)}$$

Example 6.11 In the circuit shown in Fig. Ex. 6.11, the Q-point is chosen such that $I_C = 2.1$ mA, $V_{CE} = 3$ V. If $R_C = 2.5$ kΩ, $V_{CC} = 10$ V and $\beta = 49$, determine the values of R_1, R_2 and R_E. Take $V_{BE} = 0.3$ V and $I_2 = 10\,I_B$.

Solution Given that,

$$R_C = 2.5 \text{ k}\Omega, \hspace{2cm} V_{CC} = 10 \text{ V}$$
$$I_C = 2.1 \text{ mA}, \hspace{2cm} V_{CE} = 3 \text{ V}$$
$$\beta = 49, \hspace{2.5cm} V_{BE} = 0.3 \text{ V}$$
$$I_1 = 10\,I_B$$

As $I_B \lll I_1$.

So, we can assume with reasonable accuracy that I_1 flowing through R_1 also flows through R_2.

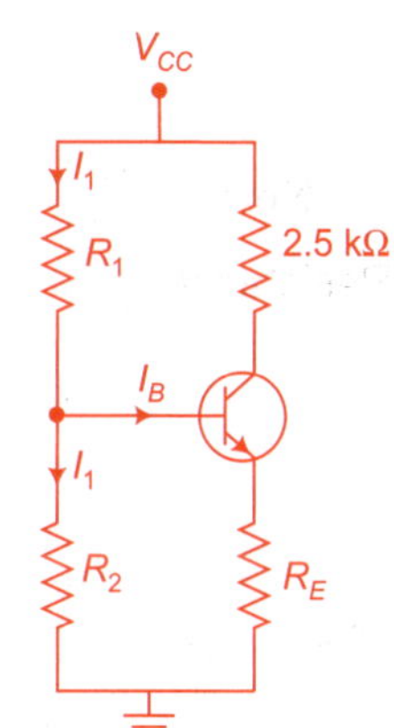

Fig. Ex. 6.11

Base current,
$$I_B = \frac{I_C}{\beta} = \frac{2.1 \text{ mA}}{49}$$

$$I_B = 0.043 \text{ mA}.$$

Current through R_1 and R_2 is,

$$I_1 = 10\,I_B = 10 \times 0.043 \text{ mA}$$
$$I_1 = 0.43 \text{ mA}.$$

Now,
$$I_1 = \frac{V_{CC}}{R_1 + R_2}$$

$$R_1 + R_2 = \frac{V_{CC}}{I_1}$$

$$R_1 + R_2 = \frac{10 \text{ V}}{0.43 \text{ mA}} = 23.26 \text{ k}\Omega$$

$$R_1 + R_2 = 23.26 \text{ k}\Omega.$$

Applying Kirchhoff's voltage law to the collector side of the circuit, we get,

$$V_{CC} - I_C R_C - V_{CE} - I_E R_E = 0$$
$$V_{CC} = I_C R_C + V_{CE} + I_E R_E$$

$\because$
$$I_E = I_C$$
$$10 = (2.1 \text{ mA} \times 2.5 \text{ k}\Omega) + 3 \text{ V} + (2.1 \text{ mA} \times R_E)$$
$$10 = 5.25 \text{ V} + 3 \text{ V} + (2.1 \text{ mA } R_E)$$
$$2.1 \text{ mA } R_E = (10 - 3.25) \text{ V}$$
$$2.1 \text{ mA } R_E = 1.75 \text{ V}$$
$$R_E = 1.75 \text{ V}/2.1 \text{ mA} = 0.833 \text{ k}\Omega$$
$$R_E = 833 \ \Omega.$$

Voltage across R_2

$$V_2 = V_{BE} + V_E$$
$$= V_{BE} + I_E R_E$$
$$= V_{BE} + I_C R_E$$
$$V_2 = 0.3 + (2.1 \text{ mA} \times 0.83 \text{ k}\Omega)$$
$$V_2 = (0.3 + 1.74) \text{ V}$$

Now,
$$V_2 = 2.04 \text{ volt}$$

Resistance

$$R_2 = \frac{V_2}{I_1} = \frac{2.04 \text{ V}}{0.43 \text{ mA}}$$

$$R_2 = 4.75 \text{ k}\Omega$$

and,
$$R_1 = 23.26 \text{ k}\Omega - 4.75 \text{ k}\Omega$$

$$\boxed{R_1 = 18.51 \text{ k}\Omega.}$$ (Ans.)

Example 6.12 A *n-p-n* transistor circuit of Fig. Ex. 6.12 has $\alpha = 0.99$ and $V_{BE} = 0.3$ V if $V_{CC} = 15$ V. Calculate R_1 and R_C to place Q-point at $I_C = 2$ mA, $V_{CE} = 6$ V.

Solution Given that,

$$I_C = 2 \text{ mA}, \qquad V_{CE} = 6 \text{ V},$$
$$V_{CC} = 15 \text{ V} \qquad \alpha = 0.99,$$
$$V_{BE} = 0.3 \text{ V} \qquad R_2 = 20 \text{ k}\Omega,$$
$$R_E = 2 \text{ k}\Omega \qquad R_1 = ?,$$
$$R_C = ?$$

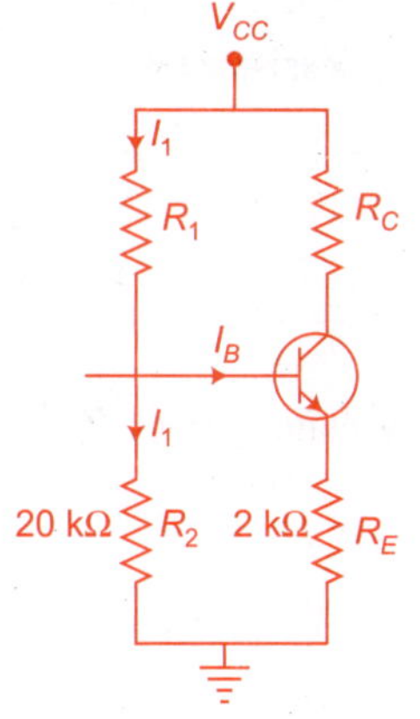

Fig. Ex. 6.12

We know that,

$$\beta = \frac{\alpha}{1 - \alpha} = \frac{.99}{1 - 0.99}$$

$$\beta = \frac{0.99}{0.01} = 99$$

$$\beta = 99.$$

Base Current

$$I_B = \frac{I_C}{\beta} = \frac{2 \text{ mA}}{99} = 0.020 \text{ mA}$$

$$I_B = 20 \text{ μA}.$$

Voltage across R_2

$$V_2 = V_{BE} + V_E$$
$$= V_{BE} + I_E R_E$$
$$= V_{BE} + I_C R_E$$

$\because$

$$I_E = I_C$$
$$V_2 = 0.3 \text{ V} + (2 \text{ mA} \times 2 \text{ k}\Omega)$$
$$V_2 = 0.3 \text{ V} + 4 \text{ V} = 4.3 \text{ V}$$
$$V_2 = 4.3 \text{ V}$$

$\therefore$ Voltage across,

$$R_1 = V_{CC} - V_2 = V_{R_1}$$
$$V_{R_1} = (15 - 4.3) \text{ V}$$
$$V_{R_1} = 10.7 \text{ V}.$$

$\therefore$ Current through R_1 and R_2 is,

$$I_1 = \frac{V_2}{R_2} = \frac{4.3 \text{ V}}{20 \text{ k}\Omega}$$

$$I_1 = 0.215 \text{ mA}$$

$\therefore$ Resistance $\qquad R_1 = $ Voltage across $\dfrac{R_1}{I_1} = \dfrac{V_{R_1}}{I_1}$

$$R_1 = \frac{10.7 \text{ V}}{0.215 \text{ mA}} = 49.76 \text{ k}\Omega$$

$$\boxed{R_1 = 49.76 \text{ k}\Omega.}$$
(Ans.)

Voltage across R_C

$$V_{R_C} = V_{CC} - V_{CE} - V_E$$
$$= V_{CC} - V_{CE} - I_E R_E$$
$$V_{R_C} = 15 - 6 - (2 \text{ mA} \times 2 \text{ k}\Omega)$$
$$V_{R_C} = (15 - 10) \text{ V}$$
$$V_{R_C} = 5 \text{ V}$$

$\therefore$ Collector resistance, $\qquad R_C = \dfrac{V_{R_C}}{I_c}$

$$R_C = \frac{5 \text{ V}}{2 \text{ mA}} = 2.5 \text{ k}\Omega$$

$$\boxed{R_C = 2.5 \text{ k}\Omega.}$$
(Ans.)

Example 6.13 Calculate the exact value of emitter current in the circuit shown in Fig. Ex. 6.13. Assume that the transistor made by Si and $\beta = 99$.

Solution Given that,

$$R_C = 1 \text{ k}\Omega, \qquad R_E = 3 \text{ k}\Omega$$
$$R_1 = 20 \text{ k}\Omega, \qquad R_2 = 15 \text{ k}\Omega$$

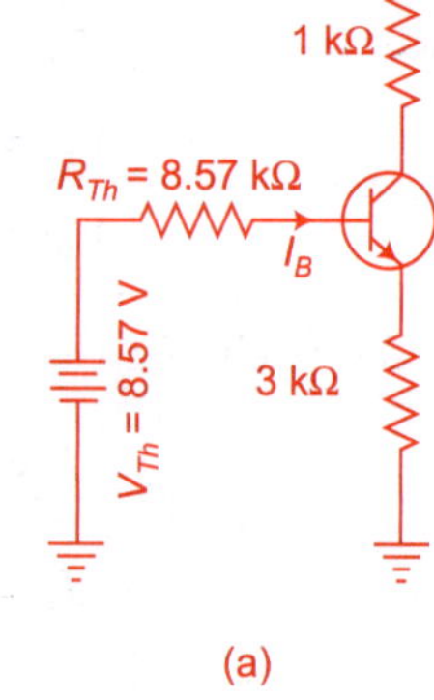

Fig. Ex. 6.13

$$V_{CC} = 20 \text{ V}, \qquad \beta = 99$$

From Eqn. (6.26), we know that the Thevenin resistance,

$$R_{Th} = \frac{R_1 R_2}{R_1 + R_2}$$

$$= \frac{20 \text{ k}\Omega \times 15 \text{ k}\Omega}{(20 + 15) \text{ k}\Omega}$$

$$R_{Th} = \frac{300 \text{ k}^2\Omega^2}{35 \text{ k}\Omega} = \frac{300}{35} \text{ k}\Omega$$

$$R_{Th} = 8.57 \text{ k}\Omega.$$

Now, from Eqn. (6.27), we know that,

$$V_{Th} = \frac{R_2}{R_1 + R_2} V_{CC}$$

$$= \frac{15 \text{ k}\Omega}{20 \text{ k}\Omega + 15 \text{ k}\Omega} \times 20 \text{ V}$$

$$V_{Th} = \frac{300 \text{ kV}\Omega}{35 \text{ k}\Omega}$$

$$\boxed{V_{Th} = 8.57 \text{ volt.}}$$

Now, we will replace the bias portion of Fig. Ex. 6.13 by its Thevenin equivalent of Fig. Ex. 6.13 (a).

Applying Kirchhoff's voltage law to the base emitter loop,

$$V_{Th} - I_B R_{Th} - V_{BE} - I_E R_E = 0$$

$$\because \qquad I_E = I_C$$

Hence,

$$I_B = \frac{I_E}{\beta}$$

$$V_{Th} = \frac{I_E}{\beta} R_{Th} + V_{BE} + I_E R_E$$

$$I_E \left(\frac{R_{TH}}{\beta} + R_E \right) = V_{Th} - V_{BE}$$

$$I_E = \frac{V_{Th} - V_{BE}}{\dfrac{R_{Th}}{\beta} + R_E}$$

$$= \frac{8.57 \text{ V} - 0.7 \text{ V}}{\dfrac{8.57 \text{ k}\Omega}{99} + 3 \text{ k}\Omega}$$

$$I_E = \frac{7.87 \text{ V}}{0.087 \text{ k}\Omega + 3 \text{ k}\Omega}$$

$$= \frac{7.87 \text{ V}}{3.087 \text{ k}\Omega}$$

$$I_E = 2.55 \text{ mA}.$$

Example 6.14 Find the value of I_B for the circuit shown in Fig. Ex. 6.14. Given that β has a range of 100 to 400, when $I_C = 10.10$ mA.

Solution Given that,

$$I_C = 10.10 \text{ mA}$$

Range of $\beta = 100$ to 400.

Voltage across R_L is,

$$V_2 = \frac{V_{CC}}{R_1 + R_2} R_2$$

$$= \frac{10}{1.5 + 0.65 \times 0.65} \times 0.65$$

$$V_2 = \frac{6.5}{2.15} = 3.02$$

$$V_2 = 3.02 \text{ V}.$$

$\therefore$ Emitter current,

$$I_E = \frac{V_2 - V_{BE}}{R_E}$$

$$= \frac{(3.02 - 0.7) \text{ V}}{0.23 \text{ k}\Omega}$$

$$I_E = \frac{2.32}{0.23} = 10.10 \text{ mA}$$

$$I_E = 10.10 \text{ mA}.$$

$\therefore$ Collector current,

$$I_C \simeq I_E$$

$$I_C = 10.10 \text{ mA}.$$

It is given that β has a range of 105 to 410, when Q-point value of I_C is 10.10 mA

$$\beta_{av} = \sqrt{\beta_{min} \times \beta_{max}} = \sqrt{100 \times 400}$$

$$\beta_{av} = 200.$$

Base current,

$$I_E = \frac{I_E}{\beta_{av} + 1} = \frac{10.10 \text{ mA}}{200 + 1}$$

Fig. Ex. 6.14

$$I_B = \frac{10.10}{201} = 50.2 \ \mu A$$

$$\boxed{I_B = 50.2 \ \mu A.}$$

Example 6.15 Fig. Ex. 6.15 shows the circuit of fixed bias using Si transistor. Determine the following:

  (i) base current I_B

 (ii) collector current I_C

(iii) V_{CE}

(iv) V_C and V_B

 (v) V_{CB}

Solution We know that,

(i)
$$I_B = \frac{V_{CC} - V_{BE}}{R_B} \ \text{from Eqn. (6.5)}$$

$$I_B = \frac{18 \ V - 0.7 \ V}{150 \ k\Omega}$$

$$= \frac{17.3 \ V}{150 \ k\Omega} = 0.115 \ mA$$

$$I_B = 115 \ \mu A.$$

(ii) Collector current,

$$I_C = \beta \ I_B$$
$$I_C = 99 \times 115 \ \mu A$$
$$I_C = 11.39 \ mA.$$

(iii) From Eqn. (6.7), we know that,

$$V_{CE} = V_{CC} - I_C \ R_C$$
$$V_{CE} = 18 \ V - 11.39 \ mA \times 1 \ k\Omega$$
$$V_{CE} = 18 \ V - 11.39 \ V$$

$$\boxed{V_{CE} = 6.61 \ \text{volt.}} \hspace{2cm} \textbf{(Ans.)}$$

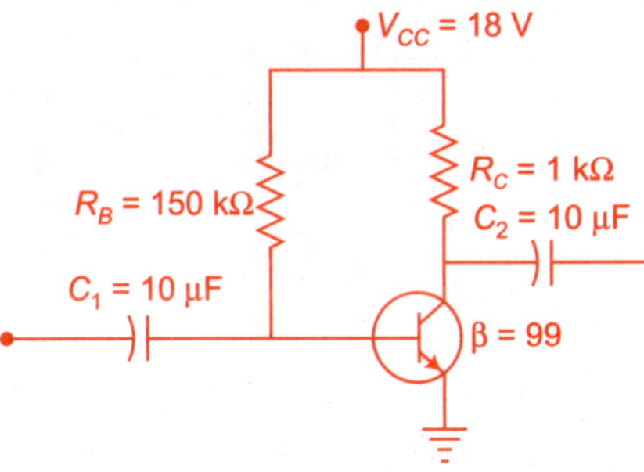

Fig. Ex. 6.15

(iv)
$$V_{CE} = V_C = 6.61 \text{ V}$$
$$\Rightarrow \boxed{V_C = 6.61 \text{ volt.}} \qquad \textbf{(Ans.)}$$
$$\propto V_B = V_{BE} = 0.7 \text{ V}$$
$$\Rightarrow \boxed{V_B = 0.7 \text{ volt.}} \qquad \textbf{(Ans.)}$$

(v)
$$V_{CB} = V_C - V_B$$
$$V_{CB} = (6.61 - 0.7) \text{ V}$$
$$\boxed{V_{CB} = 5.91 \text{ volt.}} \qquad \textbf{(Ans.)}$$

Example 6.16 Figure Ex. 6.16 shows the circuits of fixed bias using a Germanium transistor with $\beta = 90$. Determine the following:

(i) Operating point   (ii) Stability factor

Solution Given that,

$$V_{CC} = 12 \text{ V}$$
$$R_B = 950 \text{ k}\Omega$$
$$R_C = 5 \text{ k}\Omega, \quad \beta = 90$$

Since the transistor is Germanium, therefore,

$$V_{BE} = 0.3 \text{ volt.}$$

The circuit Fig. Ex. 6.16 (a) can be redrawn in Fig. Ex. 6.16 (b).

(i) Applying KVL for the input side, we have,

$$V_{CC} - I_B R_B - V_{BE} = 0$$
$$V_{CC} = I_B R_B + V_{BE}$$
$$I_B R_B = V_{CC} - V_{BE}$$
$$I_B = \frac{V_{CC} - V_{BE}}{R_B}$$
$$= \frac{12 - 0.3}{950 \text{ k}\Omega}$$
$$\boxed{I_B = \frac{11.7 \text{ V}}{950 \text{ k}\Omega} = 12.3 \text{ }\mu A.} \qquad \textbf{(Ans.)}$$

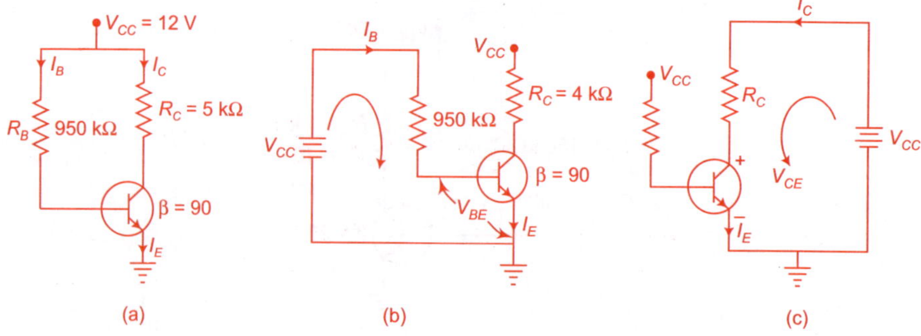

Fig. Ex. 6.16

$$\boxed{I_B = 12.3 \ \mu A.}$$ (**Ans.**)

We know that the collector current is given by,

$$I_C = \beta \, I_B = 90 \times 12.3 \ \mu A$$

$$\boxed{I_C = 1.10 \ mA.}$$ (**Ans.**)

For output loop the circuit of Fig. Ex. 6.16 (a) is redrawn in Fig. 6.16 (c).
Now, applying KVL at the output side, we have,

$$V_{CC} - I_C \, R_C - V_{CE} = 0$$

$$V_{CE} = V_{CC} - I_C \, R_C$$

$$V_{CE} = V_{CC} - I_C \, R_C$$

$$V_{CE} = 12 - 1.10 \ mA \times 5 \times 10^3 \ \Omega$$

$$V_{CE} = 12 \ V - 5.5 \ V$$

$$V_{CE} = 6.5 \ volt.$$

Therefore, the operating $Q = (V_{CE}, I_C) \Rightarrow (6.5 \ V, 1.1 \ mA)$.

(ii) For fixed bias circuit, stability factor is given by,

$$S = \beta + 1$$

$$S = 90 + 1 = 91$$

$$\boxed{S = 91.}$$ (**Ans.**)

Example 6.17 A transistor with $\beta = 100$ is used in common emitter configuration.
The collector circuit resistance is $R_C = 10 \ k\Omega$ and $V_{CC} = 20 \ V$. Assuming $V_{BE} = 0$,
find the value of collector to base resistance, such that Quiescent collector emitter
voltage is 4 V. Also determine the stability factor in this case.

(UPTU Tutorial Question Bank)

Solution From Fig. Ex. 6.17, we observe that,

$$I'_C = I_C + I_B$$

$$\because \qquad I_C \gg I_B$$

Hence, $\qquad I'_C = I_C$ (A)

From Fig. Ex. 6.17, we obtain,

$$V_{CC} = I'_C \, R_C + V_{CE}.$$

Putting the value of I'_C from Eqn. (A), we obtain,

$$V_{CC} = I_C \, R_C + V_{CE}$$

$$20 = I_C \times 1 \ k\Omega + 4$$

$$16 = I_C \times 10^3$$

$$\boxed{I_C = 16 \ mA.}$$ (**Ans.**)

$$\because \qquad I_C = \beta \, I_B$$

$$\Rightarrow \qquad I_B = \frac{I_C}{\beta}$$

Fig. Ex. 6.17

$$I_B = \frac{16 \text{ mA}}{100} = 160 \text{ μA}$$

$$\boxed{I_B = 160 \text{ μA.}} \hspace{3cm} \textbf{(Ans.)}$$

From Fig. 6.17 we observe that,

$$V_{CE} = I_B R_B + V_{BE}$$
$$V_{CE} = I_B R_B + 0$$

$\therefore$
$$V_{BE} = 0$$

$$R_B = \frac{V_{CB}}{I_B}$$

$$R_B = \frac{4}{160 \times 10^{-6}}$$

$$= \frac{4 \times 10^5}{16}$$

$$= 25 \times 10^3 \ \Omega$$

$$\boxed{R_B = 25 \text{ k}\Omega.} \hspace{3cm} \textbf{(Ans.)}$$

The stability factor is given by,

$$S = \frac{\beta + 1}{1 + \dfrac{\beta R_C}{R_B + R_C}}$$

$$= \frac{100 + 1}{1 + \dfrac{100 \times 1 \text{ k}\Omega}{25 + 1}}$$

$$S = \frac{101}{1 + \dfrac{100}{26 \text{ k}\Omega}}$$

$$= \frac{101}{1 + 3.84}$$

$$S = \frac{101}{4.84} = 20.86$$

$$\boxed{S = 20.86.} \hspace{3cm} \textbf{(Ans.)}$$

Example 6.18 In a common emitter circuit, a *n-p-n* transistor having a value of $\beta = 49$ is used with $V_{CC} = 12$ V and $R_C = 2.5$ kΩ. If a 1.5 kΩ resistor is connected between collector and base and $V_{BE} = 0$, determine,

(UPTU Tutorial Question Bank)

(a) Location of Q-point
(b) Stability factor

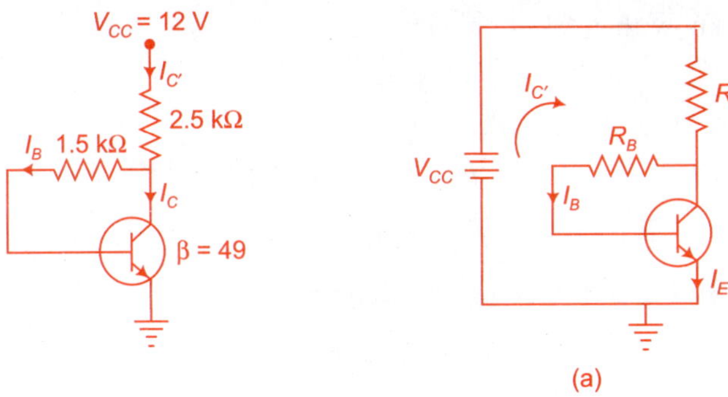

Fig. Ex. 6.18

Solution   From Fig. Ex. 6.18, we observe that,

$$I'_C = I_C + I_B$$

∵

$$I_C \gg I_B$$

∴

$$I'_C \simeq I_C.$$

The circuit of Fig. Ex. 6.18 can be redrawn as shown in Fig. Ex. 6.18 (a).
Now, applying KVL at input side,

$$V_{CC} - I'_C R_C - I_B R_B - V_{BE} = 0$$

$$V_{CC} = I'_C R_C + I_B R_B + V_{BE} = 0$$

∵

$$I'_C = I_C$$

∴

$$V_{CC} = I_C R_C + I_B R_B + V_{BE}$$

∵

$$I_C = \beta I_B \propto V_{BE} = 0$$

∴

$$V_{CC} = I_C R_C + \frac{I_C}{\beta} R_B + 0$$

$$V_{CC} = I_C R_C + \frac{I_C}{\beta} R_B$$

$$V_{CC} = I_C \left[R_C + \frac{R_B}{\beta} \right]$$

$$I_C = \frac{V_{CC}}{\left[R_C + \dfrac{R_B}{\beta} \right]} = \frac{12 \text{ V}}{2.5 \text{ k}\Omega + \dfrac{250 \text{ k}\Omega}{49}}$$

$$I_C = \frac{12 \text{ V}}{2.5 \text{ k}\Omega + 5.1 \text{ k}\Omega} = \frac{12 \text{ V}}{7.60 \text{ k}\Omega}$$

$$I_C = 1.58 \text{ mA.} \qquad\qquad \textbf{(Ans.)}$$

Now, we know that,

$$I_C = \beta \, I_B$$

$$I_B = \frac{I_C}{\beta} = \frac{1.58}{49} = 0.032 \text{ mA}$$

$$\boxed{I_B = 32 \ \mu\text{A.}}$$ **(Ans.)**

Again, from Fig. Ex. 6.18 (a), we can write,

$$V_{CE} = I_B \, R_B + V_{BE}$$

$$\because \qquad V_{BE} = 0$$

$$V_{CE} = I_B \, R_B$$

$$= 32 \times 10^{-6} \times 0.25 \times 10^6$$

$$\boxed{V_{CE} = 8 \text{ volt.}}$$ **(Ans.)**

The location of Q-point $= (V_{CE}, I_C)$

$$= (8 \text{ V}, 1.58 \text{ mA})$$

(ii) The stability factor,

$$S = \frac{\beta + 1}{\dfrac{\beta \ R_C}{R_B + R_C} + 1}$$

$$= \frac{49 + 1}{\dfrac{49 \times 2.5 \text{ k}\Omega}{250 \text{ k}\Omega + 2.5 \text{ k}\Omega} + 1}$$

$$S = \frac{50}{\dfrac{122 \text{ k}\Omega}{252.5 \text{ k}\Omega} + 1} = \frac{50}{1.48}$$

$$\boxed{S = 33.78.}$$ **(Ans.)**

Example 6.19 Determine the value of collector to emitter bias voltage for the given circuit in Fig. Ex. 6.19 if $\beta = 90$ and $I_C = 8$ mA.

Solution Given that,

$$\beta = 90, \qquad\qquad I_C = 8 \text{ mA}$$

$$V_{CC} = 18 \text{ V}, \qquad\qquad R_C = 2.5 \text{ k}\Omega$$

$$R_B = 100 \text{ k}\Omega$$

We know that,

$$I_C = \beta \, I_B$$

$$I_B = \frac{I_C}{\beta} = \frac{8 \text{ mA}}{90} = 88.9 \ \mu\text{A}$$

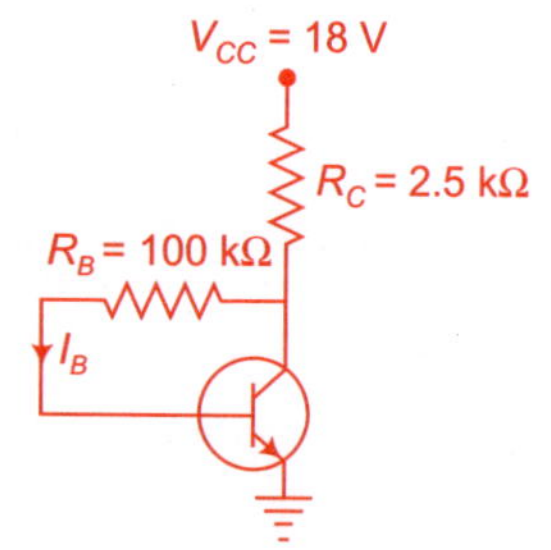

Fig. Ex. 6.19

$$I_B = 88.9 \ \mu A$$

$\because \qquad V_{CE} = I_B \, R_B + V_{BE}.$

Let us consider that transistor is made of *Si*

i.e. $\qquad V_{BE} = 0.7 \text{ V}$

Hence,

$$V_{CE} = 88.9 \ \mu A \times 100 \ k\Omega + 0.7 \text{ V}$$

$$V_{CE} = 8.89 \text{ V} + 0.7 \text{ V}$$

$$\boxed{V_{CE} = 9.59 \text{ volt.}} \qquad \textbf{(Ans.)}$$

Example 6.20 In a *n-p-n* transistor amplifier stage, a 9 V battery supply is to be used with collector to base biasing. Determine the value of R_B and R_C if, I_{CBO} is negligible, $\beta = 100$ and if the Q-point is specified by $I_C = 0.2$ A and $V_{CE} = 5$ V. **(UPTU Tutorial Question Bank)**

Solution   Data given that,

$$I_{CBO} \simeq 0$$

$$I_C = 0.2 \text{ A}$$

and, $\qquad V_{CE} = 5 \text{ V}.$

Applying KVL at output side we obtain,

$$V_{CC} - I'_C \, R_C - V_{CE} = 0$$

$$V_{CC} = I'_C \, R_C + V_{CE}$$

$\because \qquad I'_C = I_C + I_B$

from Fig. Ex. 6.20,

Now,

$$V_{CC} = (I_C + I_B) \, R_C + V_{CE}$$

But, $\qquad I_e = \ >> I_B$

$\therefore \qquad V_{CC} = I_C \, R_C + V_{CE}$

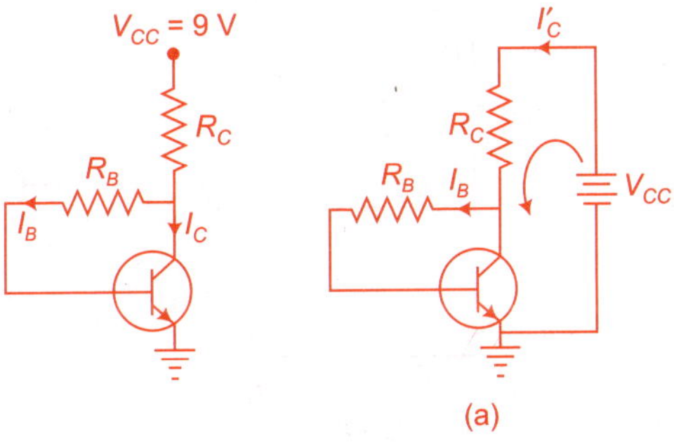

Fig. Ex. 6.20

$$I_C R_C = V_{CC} - V_{CE}$$

$$I_C R_C = (9 - 5)\ \text{V} = 4\ \text{V}$$

$$R_C = \frac{4\ \text{V}}{I_C} = \frac{4\ \text{V}}{0.2\ \text{A}}$$

$$\boxed{R_C = 20\ \Omega.}$$ **(Ans.)**

We know that,

$$I_C = \beta\, I_B$$

$$I_B = \frac{I_C}{\beta} = \frac{0.2}{100} = 2\ \text{mA}$$

From Fig. Ex. 6.20 (a)

$$V_{CE} = I_B R_B + V_{BE}$$

Assume,

$$V_{BE} = 0$$

$$V_{CE} = I_B R_B$$

$$R_B = \frac{V_{CE}}{I_B} = \frac{5\ \text{V}}{2\ \text{mA}}$$

$$\boxed{R_B = 2.5\ \text{k}\Omega.}$$ **(Ans.)**

Example 6.21   Find I_C and V_{CE} for the following circuit in Fig. Ex. 6.21. What will happen to V_{CE}, if β increases due to temperature?   **(UPTU II Sem. 2001)**

Solution   Data given that,

$$R_B = 510\ \text{k}\Omega, \qquad R_C = 2.2\ \text{k}\Omega$$

$$V_{CC} = 18\ \text{V}, \qquad \beta = 90$$

$$R_E = 1.8\ \text{k}\Omega.$$

Applying KVC at input side in Fig. Ex. 6.21(a).

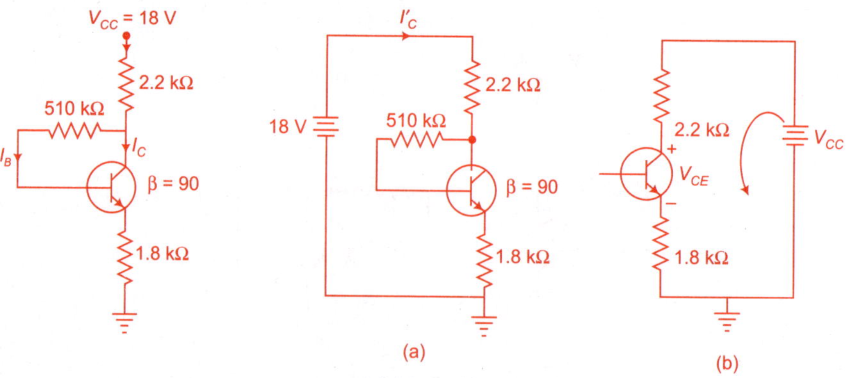

Fig. Ex. 6.21

$$V_{CC} - I'_C R_C - I_B R_B - V_{BE} - I_E R_E = 0$$

$$V_{CC} = I'_C R_C + I_B R_B + V_{BE} + I_E R_E$$

$$18 = 2.2 \times 10^3 \, I'_C + 510 \times 10^3 \, I_B + 0 + 1.8 \times 10^3 \, I_E$$

$$\frac{18}{10^3} = 2.2 \, I_C + 510 \, I_B + 1.8 \, I_E$$

$$\because \quad I_E \simeq I'_C$$

$$\frac{18}{1000} = 2.2 \, I'_C + 1.8 \, I_C + 510 \, I_B$$

$$\frac{18}{1000} = 4 \, I'_C + 510 \, I_B$$

$$\because \quad I'_C = I_C + I_B$$

$$0.018 = 4 \, (I_C + I_B) + 510 \, I_B \tag{A}$$

But,
$$\beta = \frac{I_C}{I_B} = 90$$

$$I_C = 90 \, I_B \tag{B}$$

Putting the value of I_C in Eqn. (A), we obtain,

$$0.018 = 4 \, (90 \, I_B + I_B) + 510 \, I_B$$

$$= 364 \, I_B + 510 \, I_B = 874 \, I_B$$

$$I_B = \frac{0.018}{874} = 20.59 \; \mu A$$

$$\boxed{I_B = 20.59 \; \mu A.} \tag{Ans.}$$

From Eqn. (B), we obtain,

$$I_C = 90 \times 20.59 \; \mu A$$

$$\boxed{I_C = 1.85 \; mA.} \tag{Ans.}$$

Again applying KVL at output side in Fig. Ex. 6.21(b), we obtain,

$$V_{CC} - I'_C R_C - V_{CE} - I_E R_E = 0$$

$$V_{CE} = V_{CC} - I'_C R_C - I_E R_E$$

$$V_{CE} = 18 - 2.2 \times 10^3 \, I'_C - 1.8 \times 10^3 \, I_E$$

$$\because \quad I'_C \simeq I_E$$

$$\therefore \quad V_{CE} = 18 - 2.2 \times 10^3 \, I_E - 1.8 \times 10^3 \, I_E$$

$$V_{CE} = 18 - 4 \times 10^3 \, I_E$$

$$\because \quad I_E = I_C + I_B$$

$$\therefore \quad V_{CE} = 18 - 4 \times 10^3 \, (I_C + I_B)$$

$$V_{CE} = 18 - 4 \times 10^3 \, (1.85 \times 10^{-3} + 20.59 \times 10^{-6})$$

$$V_{CE} = 18 - 7.48$$

$$\boxed{V_{CE} = 10.52 \text{ volt.}} \qquad \textbf{(Ans.)}$$

If β increases due to temperature, then V_{CE} will decrease.

Example 6.22 A transistor with $\beta = 50$ is used with collector to base resistor (R_B) biasing, with Q value of 8 V for V_{CE}, if $V_{CC} = 20$ V, $R_C = 15$ kΩ and $R_E = 250$ Ω. Find the value of (i) R (ii) S.   **(UPTU Tutorial Question Bank)**

Solution The circuit of Fig. Ex. 6.22 can be redrawn and is shown in Fig. Ex. 6.22 (a).

Applying KVL at output side as shown in Fig. Ex. 6.22 (a).

$$V_{CC} - (I_C + I_B) \, R_C - V_{CE} - I_E \, R_E = 0$$

$$V_{CC} = (I_C + I_B) \, R_C + I_E \, R_E + V_{CE}$$

$\because \qquad I_E = I_C + I_B$

$\therefore \qquad V_{CC} = (I_C + I_B) \, R_C + (I_C + I_B) \, R_E + V_{CE}$

$$V_{CC} = (I_C + I_B) \, (R_C + R_E) + V_{CE}$$

$\because \qquad I_C = \beta \, I_B$

$\therefore \qquad V_{CC} = (\beta I_B + I_B) \, [R_C + R_E] + V_{CE}$

$$V_{CC} = (\beta + 1) \, I_B \, (R_C + R_E) + V_{CE}$$

$$I_B = \frac{V_{CC} - V_{CE}}{(\beta + 1) \, (R_C + R_E)} \qquad\qquad \text{(A)}$$

Substituting the given value in Eqn. (A), we obtain,

$$I_B = \frac{(20 - 8) \text{ V}}{(50 + 1) \, (15 \text{ k}\Omega + 250 \ \Omega)} = \frac{12 \text{ V}}{51 \, (15 \text{ k}\Omega + 0.350 \text{ k}\Omega)}$$

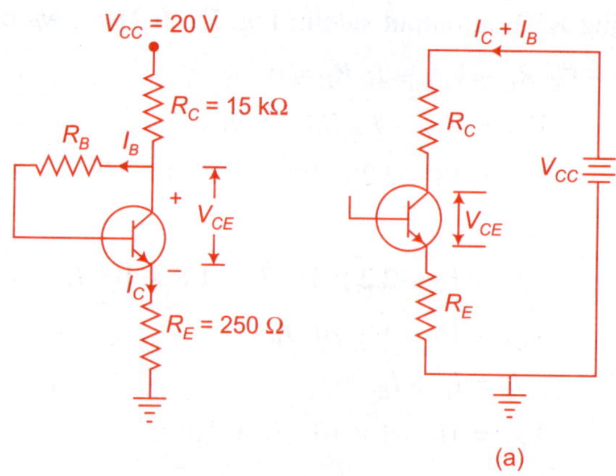

Fig. Ex. 6.22

$$I_B = \frac{12 \text{ V}}{777.75 \text{ k}\Omega} = 0.015 \text{ mA}$$

$$I_B = 15 \text{ }\mu\text{A}.$$

From circuit Fig. Ex. 6.22, we can write,

$$V_{CE} = V_{BE} + I_B R_B$$

$$I_B R_B = V_{CE} - V_{BE}$$

$$R_B = \frac{V_{CE} - V_{BE}}{I_B} = \frac{(8 - 0.7) \text{ V}}{15 \text{ }\mu\text{A}}$$

$$R_B = \frac{7.3 \text{ V}}{15 \text{ }\mu\text{A}} = 486.7 \text{ k}\Omega$$

$$\boxed{R_B = 486 \text{ k}\Omega.}$$ **(Ans.)**

(ii) Stability factor

$$S = \frac{\beta + 1}{1 + \dfrac{\beta R_E}{R_E + R_B}} = \frac{50 + 1}{1 + \dfrac{50 \times 250 \text{ }\Omega}{250 \text{ }\Omega + 486 \text{ k}\Omega}}$$

$$S = \frac{51}{1 + \dfrac{12.5 \text{ k}\Omega}{486.25 \text{ k}\Omega}} = \frac{51}{1 + 0.026}$$

$$S = \frac{51}{1.026} = 49.7$$

$$\boxed{S = 49.7.}$$ **(Ans.)**

Example 6.23 In a CE Germanium transistor amplifier, self bias used various parameters are $V_{CC} = 18$ V, $R_C = 2$ kΩ, $R_E = 1.5$ kΩ, $R_1 = 55$ kΩ, $R_2 = 20$ kΩ and $\alpha = 0.98$. Determine the following:

(i) Q-point

(ii) The stability factor.

Solution Given that,

$$\alpha = 0.98$$

so

$$\beta = \frac{\alpha}{1 - \alpha} = \frac{0.98}{1 - 0.98}$$

$$\beta = \frac{0.98}{0.02} = 49.$$

Since given transistor is Ge so,

$$V_{BE} = 0.3 \text{ V}.$$

By voltage dividing rule, the voltage across R_2 is,

$$V_B = \frac{R_2}{R_1 + R_2} \times V_{CC}$$

$$= \frac{20 \text{ k}\Omega}{(20 + 55) \text{ k}\Omega} \times 18 \text{ V}$$

$$V_B = 0.266 \times 18 \text{ V} = 4.8 \text{ V}$$

$$V_B = 4.8 \text{ V}.$$

From Fig. Ex. 6.23, the emitter current is given by,

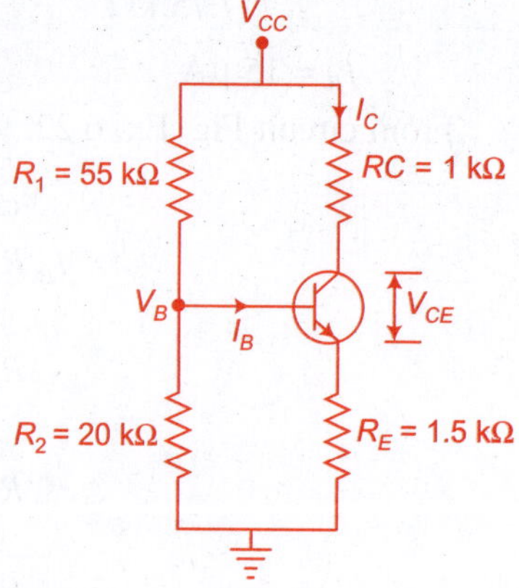

Fig. Ex. 6.23

$$I_E = \frac{V_B - V_{BE}}{R_E} = \frac{(4.8 - 0.3) \text{ V}}{1.5 \text{ k}\Omega}$$

$$I_E = \frac{4.5 \text{ V}}{15 \text{ k}\Omega} = 0.3 \text{ mA}$$

$\because \qquad I_E = I_C$

$\therefore \qquad I_C = 0.3 \text{ mA}.$

From Eqn. (6.32) we know that,

$$V_{CE} = V_{CC} - I_C (R_C + R_E)$$

$$V_{CE} = 18 \text{ V} - 0.3 \text{ mA} (2 \text{ k}\Omega + 1.5 \text{ k}\Omega)$$

$$V_{CE} = 18 \text{ V} - 0.3 \text{ mA} \times 3.5 \text{ k}\Omega$$

$$V_{CE} = 18 \text{ V} - 1.05 \text{ V}$$

$$V_{CE} = 16.95 \text{ volt.}$$

Therefore, the operating point,

$$Q = (V_{CE}, I_C) = (16.95 \text{ V}, 0.3 \text{ mA}).$$

(ii) The stability factor,

$$S = \frac{1 + \beta \left(1 + \dfrac{R_{Th}}{R_E}\right)}{1 + \beta + \dfrac{R_{Th}}{R_E}} \qquad \text{(A)}$$

$\because \qquad$
$$R_{Th} = \frac{R_1 R_2}{R_1 + R_2}$$

$$= \frac{55 \text{ k}\Omega \times 20 \text{ k}\Omega}{55 \text{ k}\Omega + 20 \text{ k}\Omega} = \frac{1100 \text{ k}\Omega}{75}$$

$$R_{Th} = 14.67 \text{ k}\Omega.$$

Substituting the value of R_{Th} in Eqn. (A)

$$S = \frac{1 + 49\left(1 + \dfrac{14.67\ k\Omega}{1.5\ k\Omega}\right)}{1 + 49 + \dfrac{14.67\ k\Omega}{1.5\ k\Omega}}$$

$$S = \frac{1 + 49\,(1 + 9.78)}{1 + 49 + 9.78}$$

$$= \frac{1 + 528.11}{59.78}$$

$$S = \frac{529.11}{59.78} = 8.85$$

$$\boxed{S = 8.85.}$$
(Ans.)

Example 6.24   In a single stage *CE* amplifier $V_{CC} = 20$ V, $\beta = 45$, $R_E = 250\ \Omega$, $R_1 = 55\ k\Omega$ and $R_2 = 35\ k\Omega$. Determine the DC voltage across R_E.

Solution   Data given,

$$V_{CC} = 20\ V, \qquad \beta = 45, \qquad R_E = 250\ \Omega$$
$$R_1 = 55\ k\Omega, \qquad R_2 = 35\ k\Omega.$$

By voltage dividing rule, the voltage across R_2 is given,

$$V_B = \frac{R_2}{R_1 + R_2}\ V_{CC}$$

$$V_B = \frac{35\ k\Omega}{(35 + 55)\ k\Omega} \times 20\ V$$

$$V_B = \frac{35\ k\Omega}{90\ k\Omega} \times 20\ V$$

$$\boxed{V_B = 7.37\ \text{volt.}}$$
(Ans.)

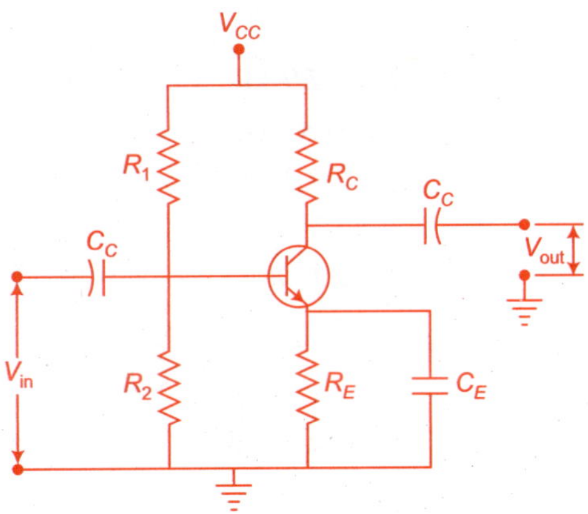

Fig. Ex. 6.24

The voltage across emitter resistor is given by,

$$V_E = V_B - V_{BE}$$

Let, $\qquad\qquad\qquad V_{BE} = 0.7 \text{ V}$

Hence, $\qquad\qquad V_E = 7.37 \text{ V} - 0.7 \text{ V}$

$$\boxed{V_E = 6.67 \text{ volt.}} \qquad\qquad\qquad \textbf{(Ans.)}$$

Example 6.25 A *Si* transistor with $V_{BE} = 0.75$ V, $h_{fe} = 100$, $V_{CE_{sat}} = 0.3$ V, is used in the circuit as shown in Fig. Ex. 6.25. Find the minimum value of R_C for which the transistor reaches in saturation.

Solution From Fig. Ex. 6.25, the base current is given by,

$$I_B = \frac{V_{BB}}{R_B} = \frac{5 \text{ V}}{150 \text{ k}\Omega}$$

$$I_B = 33.3 \text{ }\mu\text{A} \qquad\qquad \textbf{(Ans.)}$$

Collector current,

$$I_C = \beta I_B = 100 \times 33.3$$

$$I_C = 3.33 \text{ mA}$$

We know that,

$$V_{CE} = V_{CC} - I_C R_C$$

$$0.2 \text{ V} = 12 \text{ V} - 3.33 \text{ mA } R_C$$

$$3.33 \text{ mA} \times R_C = (12 - 0.2) \text{ V}$$

$$R_C = \frac{(12 - 0.2) \text{ V}}{3.33 \text{ mA}} = \frac{11.8 \text{ V}}{3.33 \text{ mA}}$$

$$\boxed{R_{C \text{ (min)}} = 3.54 \text{ k}\Omega.} \qquad\qquad\qquad \textbf{(Ans.)}$$

Fig. Ex. 6.25

Example 6.26 Find the value of β, V_{CC} and R_B of the following circuit of Fig. Ex. 6.26. $\qquad\qquad$ **(UPTU II Sem. 2000-01)**

Solution Data given that,

$$R_C = 2.7 \text{ k}\Omega, V_E = 2.1 \text{ V}$$

$$I_B = 20 \text{ }\mu\text{A}, R_E = 0.68 \text{ V}$$

$$V_{CE} = 7.3 \text{ V}.$$

We know that,

$$V_E = I_E R_E$$

$$I_E = \frac{V_E}{R_E} = \frac{2.1 \text{ V}}{0.68 \text{ k}\Omega}$$

$$\boxed{I_E = 3.08 \text{ mA.}} \qquad\qquad\qquad \textbf{(Ans.)}$$

But we know that,

$$I_C \simeq I_E$$

$$\therefore \qquad\qquad I_C = 3.08 \text{ mA.}$$

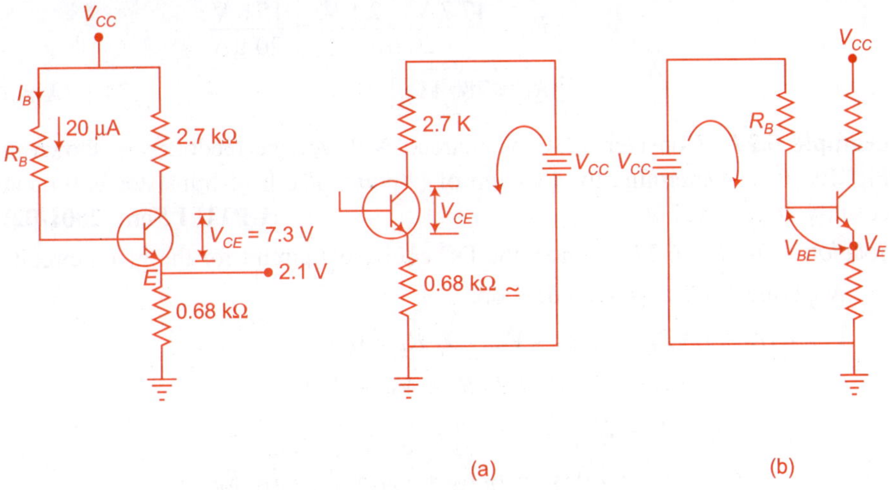

Fig. Ex. 6.26

Also we know that $I_C = \beta\, I_B$

$$\beta = \frac{I_C}{I_B} = \frac{3.08 \text{ mA}}{20 \text{ µA}}$$

$$\boxed{\beta = 154.}$$ (Ans.)

Applying KVL at the output side loop, we obtain,

$$V_{CC} - I_C\, R_C - V_{CE} - I_E\, R_E = 0$$

$$V_{CC} = V_{CE} + I_C\, R_C + I_E\, R_E$$

$$V_{CC} = V_{CE} + I_C\, (R_C + R_E)$$

$$\therefore \quad I_C \simeq I_E$$

$$V_{CC} = 7.3 \text{ V} + 3.08 \text{ mA} \, (2.7 + 0.68) \text{ k}\Omega$$

$$V_{CC} = 7.3 \text{ V} + 3.08 \text{ mA} \, (3.38) \text{ k}\Omega$$

$$V_{CC} = 7.3 \text{ V} + 10.41 \text{ V}$$

$$\boxed{V_{CC} = 17.7 \text{ volt.}}$$ (Ans.)

Again applying KVL at input side loop,

$$V_{CC} - I_B\, R_B - V_{BE} - V_E = 0$$

$$V_{CC} = I_B\, R_B + V_{BE} + V_E$$

$$I_B\, R_B = V_{CC} - V_E - V_{BE}$$

$$R_B = V_{CC} - \frac{V_E}{I_B}$$

neglecting $V_{BE} = 0$

$$R_B = \frac{17.7\ V - 2.1\ V}{20\ \mu A} = \frac{15.6\ V}{20\ \mu A}$$

$$\boxed{R_B = 780\ k\Omega.}$$ **(Ans.)**

Example 6.27 Consider a DC bias circuit with voltage feedback as shown in Fig. Ex. 6.27. Determine the location of Q-point. The β of transistor is 90 and cut-in-voltage is 0.7 V. **(UPTU I Sem. 2001-02)**

Solution Fig. Ex. 6.27 (a) show the DC equivalent circuit for the given circuit.

Applying KVL at input side loop,

$$V_{CC} - (I_C + I_B)\,R_C - I_B\,R_B - V_{BE} - I_E\,R_E = 0$$

$$V_{CC} = (I_C + I_B)\,R_C + I_B\,R_B + V_{BE} + I_E\,R_E$$

$$\because \quad I_E = I_C + I_B$$

$$\therefore \quad V_{CC} = (I_C + I_B)\,R_C + I_B\,R_B + V_{BE} + (I_C + I_B)\,R_E$$

$$V_{CC} = (I_C + I_B)\,(R_C + R_E) + I_B\,R_B + V_{BE}$$

$$\because \quad I_B = \frac{I_C}{\beta}.$$

We have,

$$V_{CC} = \left[I_C + \frac{I_C}{\beta} \right] (R_C + R_E) + \frac{I_C}{\beta} . R_B + V_{BE}$$

$$V_{CC} = I_C \left[\left(1 + \frac{1}{\beta} \right) (R_C + R_E) + \frac{R_B}{\beta} \right] + V_{BE}$$

Hence,

$$V_{CC} = I_C \left[\left(1 + \frac{1}{\beta} \right)(R_C + R_E) + \frac{R_B}{\beta} \right] + V_{BE}$$

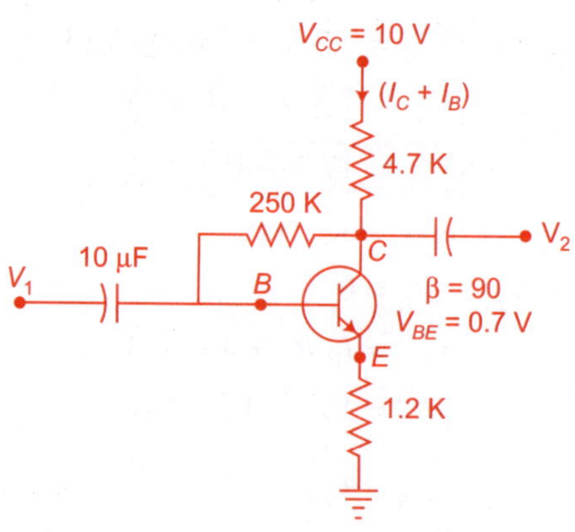

Fig. Ex. 6.27

$$I_C = \frac{(V_{CC} - V_{BE})}{\left(1 + \dfrac{1}{\beta}\right)(R_C + R_E) + \dfrac{R_B}{\beta}} = \frac{\beta(V_{CC} - V_{BE})}{(\beta + 1)(R_C + R_E) + R_B}$$

$$I_C = \frac{90\,(10 - 0.7)\ \text{V}}{(1 + 90)\,(4.7\ \text{k}\Omega + 1.2\ \text{k}\Omega) + 0.250\ \text{k}\Omega}$$

$$I_C = \frac{837\ \text{V}}{91\,(5.9\ \text{k}\Omega) + 0.250\ \text{k}\Omega} = \frac{837\ \text{V}}{536.9\ \text{k}\Omega + 0.250\ \text{k}\Omega}$$

$$I_C = \frac{837\ \text{V}}{537.15\ \text{k}\Omega} = 1.5\ \text{mA}$$

$$\boxed{I_C = 1.5\ \text{mA.}}$$
   (Ans.)

Fig. Ex. 6.27 (a) and (b)

Further, applying KVL to output side, we get,

$$V_{CC} - I_C R_C - V_{CE} - I_E R_E = 0$$

$$V_{CC} = I_C R_C + V_{CE} + I_E R_E$$

$$V_{CE} = V_{CC} - I_C R_C - I_E R_E$$

$$\because \quad I_E \simeq I_C$$

$$\therefore \quad V_{CE} = V_{CC} - I_C R_C - I_C R_E$$

$$V_{CE} = V_{CC} - I_C (R_C + R_E)$$

$$V_{CE} = 10\ \text{V} - 1.5\ \text{mA}\,(4.7 + 1.2)\ \text{k}\Omega$$

$$V_{CE} = 10\ \text{V} - 1.5\ \text{mA} \times 5.9\ \text{k}\Omega$$

$$V_{CE} = 10\ \text{V} - 8.85\ \text{V}$$

$$\boxed{V_{CE} = 1.15\ \text{volt.}}$$
   (Ans.)

Example 6.28 Determine V_C and V_B for the circuit shown in Fig. Ex. 6.28. Given that $\beta = 50$ and $V_{BE} = 0.7$ V.

Solution From Fig. Ex. 6.28,

Applying KVL at input side, we get,

$$V_{EE} + I_B R_B - V_{BE} = 0$$

$$V_{EE} - V_{BE} = - I_B R_B$$

$$I_B = - \frac{V_{EE} - V_{BE}}{R_B} = \frac{(10 - 0.7)\ \text{V}}{90\ \text{k}\Omega}$$

$$I_B = - \frac{9.3\ \text{V}}{90\ \text{k}\Omega} = - 103\ \mu\text{A}$$

$$I_B = - 103\ \mu\text{A}.$$

The collector current is given by,

$$I_C = \beta I_B$$

$$I_C = 90 \times - 103 \times 10^{-6} = - 9.29\ \text{mA}$$

$$\boxed{I_C = - 9.29\ \text{mA.}} \qquad \textbf{(Ans.)}$$

$$\therefore \quad V_C = I_C R_C = (- 9.29 \times 10^{-3}) \times 1 \times 10^3$$

$$\boxed{V_C = - 9.29\ \text{volt.}} \qquad \textbf{(Ans.)}$$

Also,

$$V_B = - I_B R_B = - 103 \times 10^{-6} \times 90 \times 10^3$$

$$\boxed{V_B = 9.2\ \text{volt.}} \qquad \textbf{(Ans.)}$$

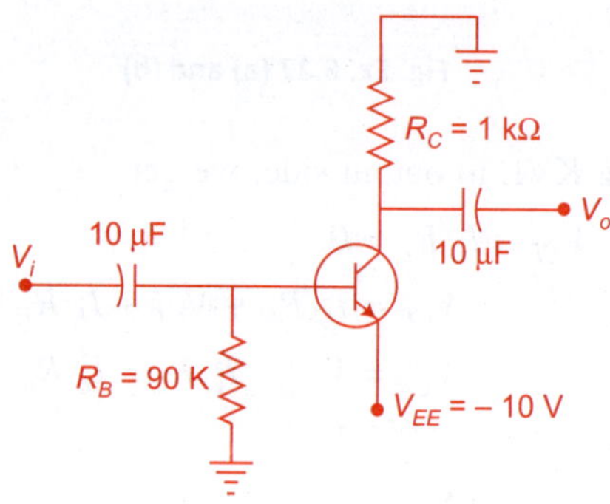

Fig. Ex. 6.28

Example 6.29 For the emitter bias circuit shown in Fig. Ex. 6.29, determine I_B, I_C, V_{CE}, V_E, V_B and V_C. **(UPTU 2002-03)**

Solution The circuit of Fig. Ex. 6.29 is redrawn in Fig. Ex. 6.29 (a). It is obtained by making the capacitors open circuited for DC analysis.

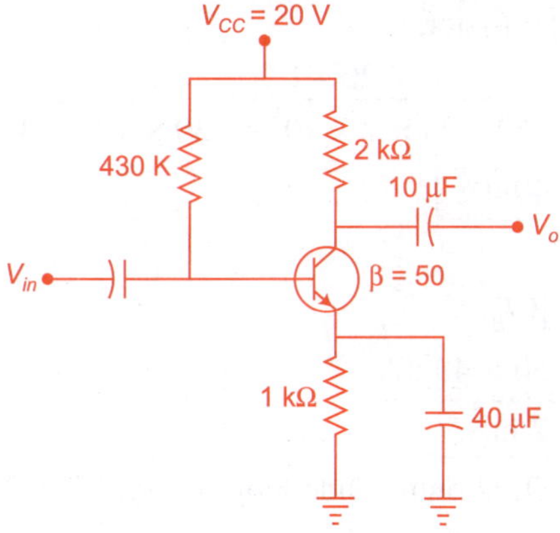

Fig. Ex. 6.29

Applying KVL at input side loop, we get,

$$V_{CC} - I_B R_B - V_{BE} - I_E R_E = 0 \qquad \text{(A)}$$

$$\therefore \qquad I_E = I_C + I_B$$

and,

$$I_C = \beta \, I_B.$$

Now, substituting the above value in Eqn. (A), we obtain,

$$V_{CC} = I_B R_B + V_{BE} + (I_C + I_B) R_E$$

$$V_{CC} = I_B R_B + V_{BE} + (\beta \, I_B + I_B) R_E$$

$$V_{CC} = I_B ((\beta + 1) R_E + R_B] + V_{BE}$$

$$I_B ((\beta + 1) R_E + R_B] = V_{CC} - V_{BE}$$

$$I_B = \frac{V_{CC} - V_{BE}}{(\beta + 1) R_E + R_B}$$

Fig. Ex. 6.29 (a) and (b)

$\because V_{BE} = 0.7$ V for *Si* transistor,

$$I_B = \frac{20 - 0.7}{(50 + 1) \times 1 \times 10^3 + 430 \times 10^3} = \frac{19.3 \text{ V}}{4.81 \times 10^3 \; \Omega}$$

$$\boxed{I_B = 40 \; \mu A.}$$

We know that,

$$I_C = \beta \, I_B$$
$$I_C = 50 \times 40 \; \mu A$$
$$\boxed{I_C = 2 \text{ mA.}}$$

Now, applying KVL at output side loop of Fig. Ex. 6.29 (b), we get,

$$V_{CC} - I_C R_C - V_{CE} - I_E R_E = 0$$
$$V_{CC} = I_C R_C + V_{CE} + I_E R_E$$
$$\because \qquad I_E = I_C + I_B$$
$$V_{CE} = V_{CC} - I_C R_C - (I_C + I_B) R_E$$
$$V_{CE} = 20 \text{ V} - 2 \times 10^{-3} \times 2 \times 10^3 - (2 \times 10^{-3} + 40 \times 10^{-6}) \times 1 \times 10^3$$
$$V_{CE} = 20 \text{ V} - 4 \text{ V} - 2.04 \text{ V}$$
$$V_{CE} = 20 \text{ V} - 6.04 \text{ V.}$$
$$\boxed{V_{CE} = 13.96 \text{ volt.}} \qquad \text{(Ans.)}$$

$$\because \qquad V_E = I_E R_E = (I_C + I_B) R_E$$
$$V_E = (2 \times 10^{-3} + 40 \times 10^{-6}) \times 1 \times 10^3 \text{ V}$$
$$\boxed{V_E = 2.04 \text{ volt.}} \qquad \text{(Ans.)}$$

Again,

$$V_B = V_{CC} - I_B R_B$$
$$V_B = 20 - 40 \times 10^{-6} \times 430 \times 10^3$$
$$V_B = (20 - 17.2) \text{ V}$$
$$\boxed{V_B = 2.8 \text{ volt.}}$$

$$\therefore \qquad V_C = V_{CE} + V_E$$
$$V_C = 13.96 \text{ V} + 2.04 \text{ V}$$
$$\boxed{V_C = 16 \text{ volt.}}$$

Example 6.30 In the circuit shown in Fig. Ex. 6.30, $h_{fe} = 100$, $V_{BE} = 0.8$ mt $V_{CE} = 0.2$ V. Determine whether or not the silicon transistor is in saturation and find I_B and I_C.

(UPTU 2002-03)

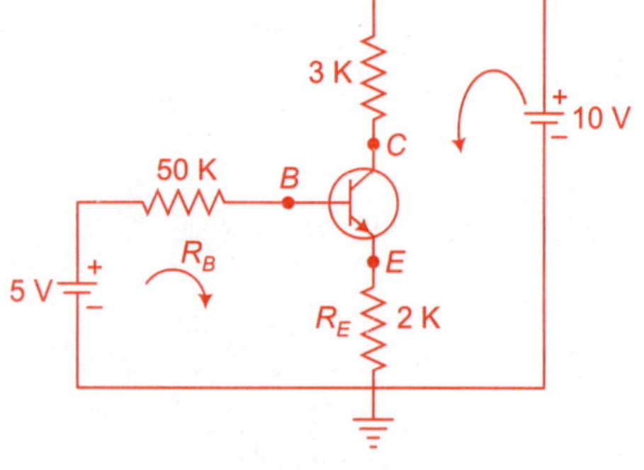

Fig. Ex. 6.30

Solution   As $V_{CE} = 0.2$ V for the S_i transistor which is in saturation state.

Applying KVL to input side, we have,

$$V_{BB} - I_B R_B - V_{BE} - I_E R_E = 0$$

$$V_{BB} = I_B R_B + V_{BE} + I_E R_E$$

$$V_{BB} = I_B R_B + V_{BE} + (I_C + I_B) R_E$$

$\because \qquad I_E = I_C + I_B$

$$5 = 50 \times 10^3 I_B + 0.8 + (\beta I_B + I_B) R_E$$

$\because \qquad I_C = \beta I_B = h_{fe} I_B$

$$I_C = 100 I_B$$

$$5 = 50 \times 10^3 I_B + 0.8 + I_B (100 + 1) \times 2 \times 10^3$$

$$5 = I_B [50 \times 10^3 + 202 \times 10^3] + 0.8$$

$$I_B = \frac{(5 - 0.8) \text{ V}}{252 \times 10^3} = \frac{4.2 \text{ V}}{252 \times 10^3 \text{ } \Omega}$$

$$\boxed{I_B = 16.67 \text{ } \mu A.}$$ **(Ans.)**

Now,

$$I_C = \beta I_B = 100 I_B$$

$$I_C = 100 \times 16.67 \text{ } \mu A$$

$$\boxed{I_C = 1.67 \text{ mA.}}$$ **(Ans.)**

Now, applying KVL to output side, we obtain,

$$V_{CC} - I_C R_C - V_{CE} - I_E R_E = 0$$

$$V_{CC} = I_C R_C + V_{CE} + I_E R_E$$

$$V_{CE} = V_{CC} - I_C R_C - (I_C + I_B) R_E$$

$\because \qquad I_E = I_C + I_B$

$$V_{CE} = 10 - 1.67 \times 10^{-3} \times 3 \times 10^3$$

$$- (1.67 \times 10^{-3} + 16.67 \times 10^{-6}) \times 2 \times 10^3$$

$$V_{CE} = 10 - 5.01 - 3.374$$

$$V_{CE} = 10 - 8.38$$

$$\boxed{V_{CE} = 1.62 \text{ volt.}}$$ **(Ans.)**

Thus, in the given circuit, we have,

$$\boxed{V_{CE} = 1.62 > 0.2 \text{ volt.}}$$ **(Ans.)**

Therefore, the transistor is not working in saturation region, rather it is in active region.

Example 6.31 A silicon transistor with $(V_{BE})_{\text{sat}} = 0.8$ V

$\beta = h_{fe} = 100$, $V_{CE(\text{sat})} = 0.2$ V is used in the circuit shown in Fig. Ex. 6.31. Find the minimum value of R_C for which the transistor remains in saturation.

(UPTU 2002-03)

Solution Applying KVL to input side, we have,

$$V_{BB} - I_B R_B - V_{BE} = 0$$

$$V_{BB} = (I_B)_{\text{sat}} R_B + (V_{BE})_{\text{sat}}$$

$$5 = (I_B)_{\text{sat}} \times 200 \times 10^3 + 0.8$$

$$(I_B)_{\text{sat}} = \frac{5 - 0.8}{200 \times 10^3}$$

$$(I_B)_{\text{sat}} = \frac{4.2}{200 \times 10^3} = 21 \times 10^{-6} \text{ amp}$$

$$(I_B)_{\text{sat}} = 21 \ \mu A$$

Now, we have

$$(I_C)_{\text{sat}} = \beta(I_B)_{\text{sat}} = 100 \times 21 \times 10^{-6} \text{ amp}$$

$$(I_C)_{\text{sat}} = 2.1 \text{ mA.}$$

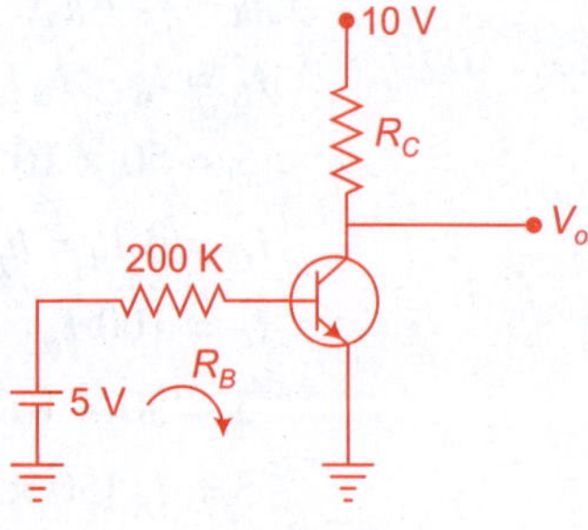

Fig. Ex. 6.31

Again using KVL to output side loop, we have,

$$V_{CC} - I_C R_C - V_{CE} = 0$$

$$I_C R_C = V_{CC} - V_{CE}$$

$$R_C = \frac{V_{CC} - V_{CE}}{I_C} = \frac{10 - 0.2}{2.1 \times 10^{-3}}$$

$$R_C = \frac{9.8}{2.1 \times 10^{-3}} = 4.67 \text{ k}\Omega$$

$$\boxed{R_C = 4.67 \text{ k}\Omega.}$$

(Ans.)

Example 6.32 Find V_{CE} and I_E in Fig. Ex. 6.32.

Solution Given that,

$$R_C = 3.5 \text{ k}\Omega, \ R_E = 1 \text{ k}\Omega$$

$$R_1 = 15 \text{ k}\Omega, \ R_2 = 2.5 \text{ k}\Omega$$

$$V_{CC} = 12 \text{ V}, \ \beta = 150.$$

From Eqn. (6.26), we know that the Thevenin resistance,

$$R_{Th} = \frac{R_1 R_2}{R_1 + R_2} = \frac{15 \text{ k}\Omega \times 2.5 \text{ k}\Omega}{(15 + 2.5) \text{ k}\Omega}$$

$$R_{Th} = \frac{37.5}{17.5} \text{ k}\Omega = 2.14 \text{ k}\Omega$$

$$\boxed{R_{Th} = 2.14 \text{ k}\Omega.}$$

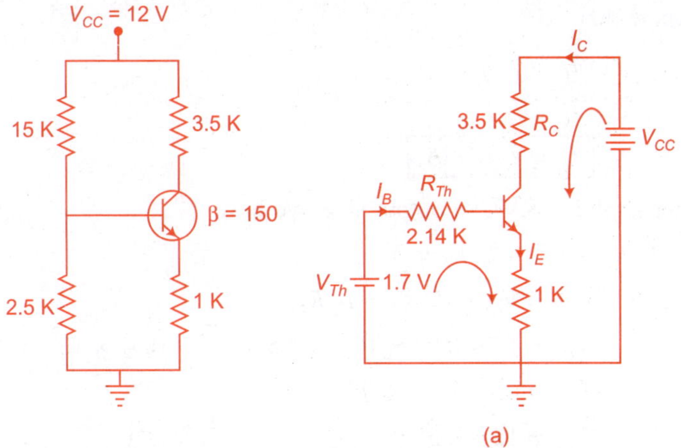

Fig. Ex. 6.32

Now, from Eqn. (6.27), we know that the Thevenin equivalent voltage is,

$$V_{Th} = \frac{R_2}{R_1 + R_2} \times V_{CC} = \frac{2.5 \text{ k}\Omega}{(15 + 2.5) \text{ k}\Omega} \times 12 \text{ V}$$

$$V_{Th} = \frac{2.5 \text{ k}\Omega}{17.5 \text{ k}\Omega} = 12 \text{ V}$$

$$\boxed{V_{Th} = 1.71 \text{ volt.}}$$

Now, the Thevenin equivalent circuit of Fig. Ex. 6.32 is shown in Fig. Ex. 6.32 (a).

Applying KVL to input side of Fig. Ex. 6.32 (a), we get,

$$V_{Th} - I_B R_{Th} - V_{BE} - I_E R_E = 0$$

$\because$

$$I_E = (\beta + 1) I_B$$

$$V_{Th} = \frac{I_E}{(\beta + 1)} R_{Th} + V_{BE} + I_E R_E$$

$$I_E \left[\frac{R_{Th}}{\beta + 1} + R_E \right] = V_{Th} - V_{BE}$$

$$I_E = \frac{V_{Th} - V_{BE}}{\dfrac{R_{Th}}{\beta + 1} + R_E} = \frac{1.7 - 0.7}{\dfrac{2.14 \text{ k}\Omega}{101} + 1 \text{ k}\Omega}$$

$$I_E = \frac{1}{21.2 + 1000} = \frac{1}{1021.2}$$

$$\boxed{I_E = 0.97 \text{ mA.}} \qquad \qquad \textbf{(Ans.)}$$

We know that,

$$I_C = \beta \, I_B$$

$$I_C \simeq I_E$$

$$\boxed{I_C = 0.97 \text{ mA.}}$$

Further applying KVL to output side, we have,

$$V_{CC} - I_C \, R_C - V_{CE} - I_E \, R_E = 0$$

$$V_{CE} = V_{CC} - I_C \, R_C - I_E \, R_E$$

$$V_{CE} = 12 - 0.97 \times 10^{-3} \times 3.5 \times 10^3 - 0.97 \times 10^{-3} \times 1 \times 10^3$$

$$V_{CE} = 12 - 3.4 - 0.97$$

$$V_{CE} = 12 - 4.37$$

$$\boxed{V_{CE} = 7.63 \text{ volt.}} \hspace{3cm} \textbf{(Ans.)}$$

Example 6.33 Determine V_C and I_B for the following circuit shown in Fig. Ex. 6.33. **(UPTU Special Exam, 2001)**

Solution From Eqn. (6.26), the Thevenin resistance is,

$$R_{Th} = \frac{R_1 \, R_2}{R_1 + R_2} = \frac{82 \, k \times 16 \, k}{(82 + 16) \, k}$$

$$R_{Th} = \frac{1312 \text{ k}\Omega}{98} = 13.39 \text{ k}$$

$$\boxed{R_{Th} = 13.39 \text{ k.}}$$

Now, from Eqn. (6.27) the Thevenin voltage is,

$$V_{Th} = \frac{R_2}{R_1 + R_2} \, V_{CC} = \frac{16 \, k}{16 + 82 \, k} \times 22 \text{ V}$$

$$V_{Th} = 0.163 \, k \times 22 \text{ V}$$

$$\boxed{V_{Th} = 3.59 \text{ volt.}}$$

Thevenin's equivalent circuit has been drawn in Fig. Ex. 6.33(a).

Applying KVL to input side loop of Fig. Ex. 6.33 (a), we have,

$$V_{Th} - I_B \, R_{Th} - V_{BE} - I_E \, R_E = 0$$

$$\because \hspace{2cm} I_E = (\beta + 1) \, I_B$$

$$V_{Th} = (\beta + 1) \, I_B \, R_E + I_B \, R_{Th} + V_{BE}$$

$$V_{Th} = [(\beta + 1) \, R_E + R_{Th}] \, I_B + V_{BE}$$

$$I_B = \frac{V_{Th} - V_{BE}}{(\beta + 1) \, R_E + R_{Th}} = \frac{3.59 - 0.7}{(200 + 1) \times 750 + 13.39 \times 10^3}$$

$$I_B = \frac{2.89 \text{ V}}{150 \text{ k}\Omega + 13.39 \text{ k}\Omega} = \frac{2.89 \text{ V}}{163.39 \text{ k}\Omega}$$

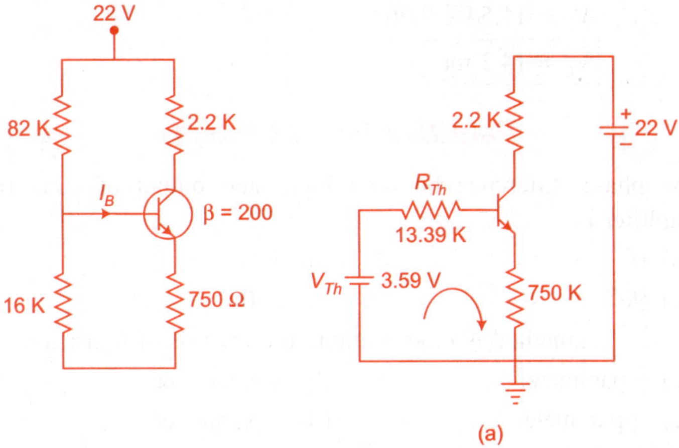

Fig. Ex. 6.33

$$\boxed{I_B = 17.69 \ \mu A.}$$ (**Ans.**)

We know that,

$$I_C = \beta \ I_B$$

$$I_C = 200 \times 17.69 \ \mu A$$

$$\boxed{I_C = 3.54 \ mA.}$$ (**Ans.**)

Now, applying KVL to output side loop, we have,

$$V_{CC} - I_C \ R_C - V_{CE} - I_E \ R_E = 0$$

$$\because \qquad I_E = I_C + I_B$$

$$V_{CE} = V_{CC} - I_C \ R_C - (I_C + I_B) \ R_E$$

$$V_{CE} = 22 - 3.54 \times 10^{-3} \times 2.2 \times 10^3 - (3.54 \times 10^{-3} + 17.69$$
$$\times 10^{-6}) \times 750$$

$$V_{CE} = 22 - 7.79 - 2.67$$

$$V_{CE} = 22 - 10.46 = 11.54$$

$$\boxed{V_{CE} = 11.54 \ volt.}$$

We know that,

$$V_{CE} = V_C - V_E$$

$$V_C = V_{CE} + V_E$$ (A)

$$V_C = V_{CE} + I_E \ R_E$$

$$\because \qquad I_E \simeq I_C$$

$$V_C = V_{CE} + I_C \ R_E$$

$$V_C = 11.54 + 3.54 \times 10^{-3} \times 750$$

$$V_C = 11.54 + 2.66$$

$$\boxed{V_C = 14.2 \text{ mt.}}$$

(Ans.)

Objective Type Questions

1. The phase difference between input and output of common emitter amplifier is ....
   - (a) $0°$
   - (b) $180°$
   - (c) $90°$
   - (d) $360°$

2. ....method is more suitable for analysis of transistor.
   - (a) r-parameter
   - (b) z-parameter
   - (c) h-parameter
   - (d) y-parameter

3. Transistor biasing represents .... conditions.
   - (a) AC
   - (b) DC
   - (c) Both AC and DC
   - (d) None of these

4. If biasing is not done in an amplifier circuit, it results in ....
   - (a) Unfaithful amplification
   - (b) Decrease in I_B
   - (c) Excessive collector bias
   - (d) None of these

5. DC and AC load lines are ....
   - (a) Parallel
   - (b) Perpendicular
   - (c) Passes through Q-point
   - (d) None of these

6. What are the units of h_{oe} and h_{ie} respectively?
   - (a) Ω, Ω
   - (b) $\mho, \Omega$
   - (c) $\Omega, \mho$
   - (d) $\mho, \mho$

7. Operating point represents ....
   - (a) Value of I_C and V_{CE} when signal is applied
   - (b) The magnitude of signal
   - (c) Zero signal values of I_C and V_{CE}
   - (d) None of the above

8. Transistor biasing is generally provided by a ....
   - (a) Bias battery
   - (b) Biasing circuit
   - (c) Diode
   - (d) None of these

9. Which parameter is approximately equal to β?
   - (a) h_{ie}
   - (b) h_{fe}
   - (c) h_{oe}
   - (d) h_{re}

10. The value of h_{fe} may be
    - (a) Approximatey 100
    - (b) 1
    - (c) 0.99
    - (d) -1

11. The circuit that provides the best stabilization of operating point is,
    - (a) Base resistor bias
    - (b) Collector to base bias circuit
    - (c) Potential divider bias
    - (d) None of these

12. An ideal value of stability factor is
    - (a) 100
    - (b) 1
    - (c) 200
    - (d) more than 200

13. The coupling capacitor used in amplifier works as
    - (a) Prevent AC
    - (b) Prevent DC
    - (c) Prevents both AC and DC
    - (d) Used for transistor biasing

14. The disadvantage of base resistor method of transistor biasing is that it ....
    - (a) is complicated
    - (b) Provides high stability
    - (c) is sensitive to change in β
    - (d) None of these

15. The biasing circuit has a stability factor of 50. If due to temperature change, I_{CBO} changes by 1 μA, then I_C will change by
    - (a) 50 μA
    - (b) 100 μA
    - (c) 20 μA
    - (d) 25 μA

16. An amplifier circuit whose voltage gain is 100. If the output voltage is 2 V, what will be the input voltage?
    - (a) 200 V
    - (b) 20 mV
    - (c) 50 V
    - (d) 2 mV

17. Which biasing circuit has maximum stability factor?
    - (a) Fixed bias
    - (b) Collector to base bias
    - (c) Potential divider bias
    - (d) Self biased

18. Think about following h-parameter (a) h_{fe} (b) h_{oe} (c) h_{ie} (d) (h_{re}. Out of these, which one has no unit?
    - (a) a and d
    - (b) only c
    - (c) a and b
    - (d) a, b and c

19. The operating point is also called the ....
    - (a) Cut-off point
    - (b) Quiescent point
    - (c) Saturation point
    - (d) None of these

20. The operating point .... on the *AC* load line.
    - (a) Does not lie
    - (b) Also lie
    - (c) May or may not lie
    - (d) Data insufficient

21. The base resistor method is generally used in ....
    - (a) Switching circuit
    - (b) Amplifier circuits
    - (c) Rectifier circuit
    - (d) None of these

22. In a base resistor method, if the value of β changes by 45, then collector current I_C will change by a factor of ....
    - (a) 25
    - (b) 90
    - (c) 45
    - (d) 180

23. The stability factor of a collector feedback bias circuit is .... that of base resistor bias.
    - (a) The same as
    - (b) More than
    - (c) Less than
    - (d) None of these

24. The value of stability factor for a base resistor,
    - (a) $R_B (\beta + 1)$
    - (b) $(\beta + 1) R_E$
    - (c) $1 - \beta$
    - (d) $\beta + 1$

25. In practical biasing circuit, the value of R_E is about ....
    - (a) 10 kΩ
    - (b) 1 mΩ
    - (c) 100 kΩ
    - (d) 800 Ω

26. For good stabilization in voltage divider bias, the current I_1 flowing through R_1 and R_2 should be equal to or greater than ....
    - (a) $3 I_B$
    - (b) $10 I_B$
    - (c) $2 I_B$
    - (d) $4 I_B$

27. The disadvantage of voltage divider bias is that it has ....
    - (a) High stability factor
    - (b) Low base current
    - (c) Many resistor
    - (d) None of these

28. In voltage divider bias, Q-point is 3 V, 2 mA. If $V_{CC} = 9$ V, $R_C = 2.2$ kΩ, what is the value of R_E?
    - (a) 2000 Ω
    - (b) 1400 Ω
    - (c) 800 Ω
    - (d) 1600 Ω

29. In voltage divider bias, $V_{CC} = 25$ V, $R_1 = 10$ kΩ, $R_2 = 2.2$ kΩ, $R_C = 3.6$ kΩ and $R_E = 1$ kΩ. What is the emitter voltage?
    - (a) 3.8 V
    - (b) 5.5 V
    - (c) 6.5 V
    - (d) 4.2 V

30. In the above question, what is the collector voltage?
    - (a) 12.3 V
    - (b) 8.7 V
    - (c) 11.2 V
    - (d) 15.2 V

ANSWERS

1. (b)	2. (c)	3. (b)	4. (a)	5. (c)	6. (b)	7. (c)
8. (b)	9. (b)	10. (a)	11. (c)	12. (b)	13. (b)	14. (c)
15. (a)	16. (b)	17. (a)	18. (a)	19. (a)	20. (b)	21. (a)
22. (c)	23. (c)	24. (d)	25. (d)	26. (b)	27. (c)	28. (c)
29. (a)	30. (a)					

Exercise

6.1. What do you understand by transistor biasing? What is its need?

6.2. What are the essentials of biasing circuit?

6.3. Write short notes on the following,

   (a) Stabilization of Q-point   (b) Operating point

6.4. Derive the voltage gain, current gain expressions for a potential divider biased common emitter using h-parameter?

6.5. For the following circuit, calculate Z_i, A_i, A_V and Z_0.

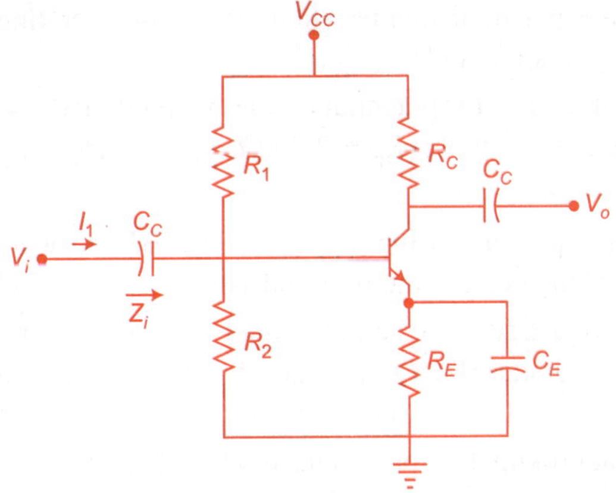

Fig. E. 6.5

6.6. How operating point is selected from amplification in *CE* mode using graphical method Explain?

6.7. A silicon transistor has $I_{CBO} = 0.02$ μA at 27°C. The leakage current doubles for every 6°C rise in temperature. Calculate the base current at 57°C, when the emitter current is 1 mA. Given that $\alpha = 0.99$.

$$\textbf{(Ans.}\ I_B = 9.4\ \mu A\textbf{)}$$

6.8. Draw a fixed bias circuit and obtain the value of collector current in the circuit.

6.9. Determine whether or not the circuit shown in Fig. E. 6.9 is mid-point biased.

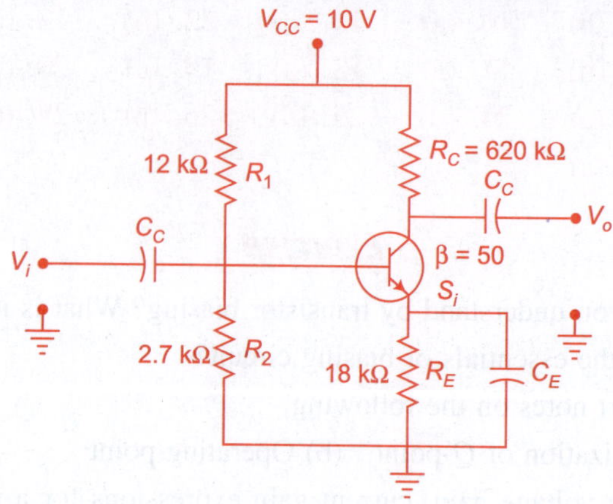

Fig. E. 6.9

(**Ans.** $I_C = I_E = 6.33$ mA, $V_{CE} = 4.94$ V) circuit is mid-point biased.

6.10. Describe the potential divider stabilization of operating point which is achieved by this method?

6.11. Find the value of I_C for potential divider method, if $V_{CC} = 9$ V, $R_E = 1$ kΩ, $R_1 = 39$ kΩ, $R_2 = 10$ kΩ, $R_C = 2.7$ kΩ, $V_{BE} = 0.15$ V and $\beta = 90$.

(**Ans.** 1.5 mA)

6.12. How many types of biasing are done on a BJT to work properly as an amplifier? Which one is the best and why? **(UPTU 2006-07)**

6.13. Explain why a transistor action cannot be achieved by connecting two back-to-back diodes? In a transistor why emitter region is heavily doped, base width is small and collector area is large? **(UPTU 2006-07)**

6.14. Using the approximate h-parameter model, obtain the expression for a *CE* circuit for (a) A_i (b) R_i (c) A_V (d) R_0. **(UPTU 2007-08)**

6.15. Sketch a voltage divider bias circuit using a *n-p-n* transistor. Show all voltage polarities and current directions. Explain the operation of the circuit and write the approximate equations for V_B, I_E, I_C and V_{CE}.

(UPTU 2008-09)

7.1 INTRODUCTION

We have already discussed in chapter 5 about bipolar junction transistor. Basically, the bipolar transistors are current amplifying or current regulating device, that controls the amount of current flowing through them in proportion to the amount of biasing current applied to the base current. So, we can say that BJT is a current controlled device, whereas in field effect transistor, the flow of current through the device is controlled by an applied electric field across the same conduction region. Since the current is carried by majority carriers, only the Field Effect Transistor (FET) is said to be a unipolar device. Whereas in bipolar junction transistor, the current is carried by majority and minority carriers, so it is called as bipolar device.

Field effect transistor is a three terminal device which uses the same applications as bipolar junction transistor. The basic diagram of BJT and FET are shown in Figs. 7.1(a) and (b) respectively.

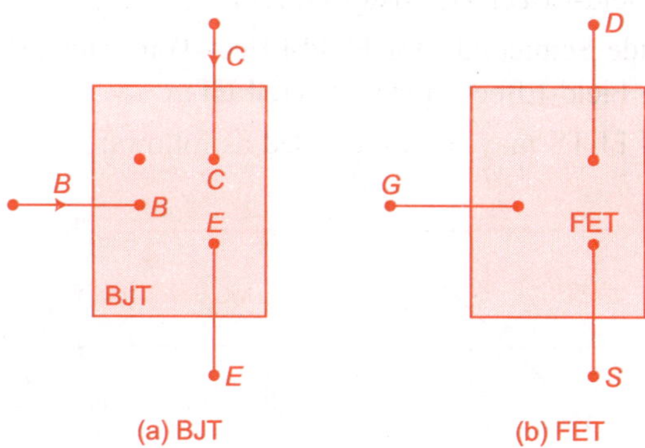

Fig. 7.1

As we can see in Fig. 7.1 that BJP and FET both have three terminal devices named as base, emitter and collector which is named as gate, drain and source in filed effect transistor. Although the bipolar junction transistor and field effect transistor has many differences but they have many similarities also.

Differences in BJT and FET

1. BJT is a current controlled device i.e., the about characteristics of BJT is controlled by the base current (input current) whereas the FET is a voltage controlled device. This means that the output characteristics of FET are controlled by applied input voltage (Gate voltage).

2. FET is a unipolar device in which the current flows through the majority carriers only i.e., one type of charge carriers, whereas in BJT the current flows through it by the majority and minority carriers so it is called bipolar device.

3. The input impedance of FET's are high. It varies from 1M Ω to many hundred MΩ. So, we can say that the input impedance of FET (10^8 to 10^{12})Ω is very high as compared to input impedance of BJT (10^2 to 10^6) Ω.

4. The BJT is more sensitive for applied input voltage as compared to FET. Hence, equal change in applied voltage of BJT and FET, the output current of BJT changes more as compared to the output current of FET. Due to this, the AC voltage gain of BJT is large as compared to FET.

5. The temperature stability of FET is greater than that of BJT. Due to the size of FET, it is small as compared to BJT.

6. FET is a type of semiconductor device in which the width of conducting channel can be controlled by the electric field.

In this chapter, there are two main types of field effect transistor as follows:

(i) Junction Field-Effect Transistor (JFET)

(ii) Metal Oxide Semiconductor Field-Effect Transistor (MOSFET) or Insulated Gate Field-Effect Transistor (IGFET).

Both types of FETS may be sub-divided as follows:

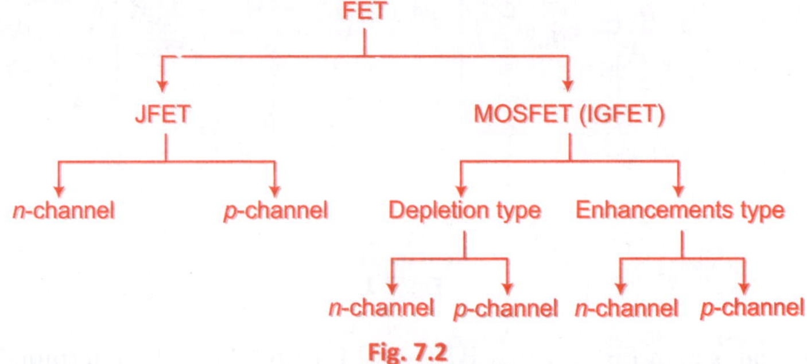

Fig. 7.2

7.2 JUNCTION FIELD EFFECT TRANSISTOR (JFET)

As we can see in Fig. 7.2, the junction field effect transistor may be divided into two parts, one is *n*-channel and second one *p*-channel. The construction of both types of JFET are same and the only difference is that the channel in *n*-channel JFET is made with *n*-type and the channel in *p*-channel JFET is made with *p*-type material.

7.2.1 Construction of *n*-channel JFET

n-channel JFET is shown in Fig. 7.3(a). In this figure, we see that one bar or stick is used of *n*-type material with two *p*-types heavily doped region diffused on opposite side of its middle part i.e., Ohmic contacts are made on both sides of semiconductor bar. The *p*-type region from two *pn*-junctions. The space between the junctions is known as a channel and it is made with *n*-type of semiconductor material. So, it is called as *n*-channel and the JFET which consist *n*-channel is called *n*-channel JFET. Both *p*-type regions are connected internally and a single wire is taken out. This terminal is called as gate (G).

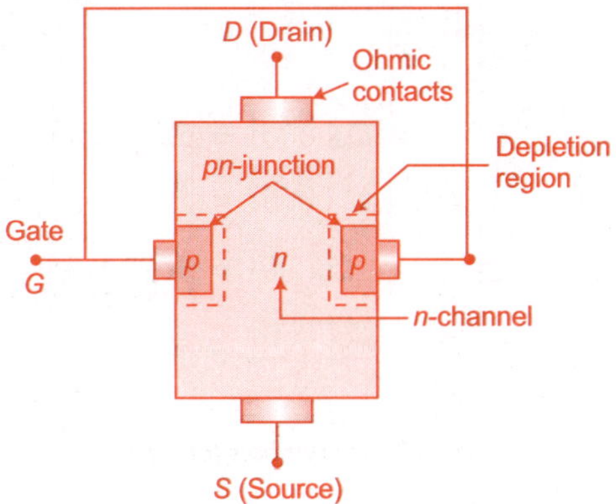

Fig. 7.3   *(a) n-channel JFET*

The Ohmic contacts are made to both ends of the *n*-types semiconductor and are taken out in the form of two terminals known as Source (S) and Drain (D).

   (i) **Source (S)**   In the *n*-channel JFET, the terminal is connected to the negative terminal of the battery. Electrons are majority carriers in *n*-type bar and enter the bar through this terminal.

   (ii) **Drain (D)**   In the *n*-channel JFET, the terminal is connected to the positive terminal of the battery. The majority carriers leave the bar through this terminal.

Gate (g)

On both sides of *n*-type silicon bar, layers of heavily doped *p*-types silicon are formed by diffusion or by other methods by which *pn*-junctions are formed. These layers are joined together and they are called the gate G. Thus, two *pn*-junctions are formed between the *p*-type gate and the *n*-type bar. Since the gate regions are heavily doped, the depletion regions at the junctions are entirely confined to the *n*-type silicon bar.

Channel

The region of the *n*-type bar between the depletion region is called the Channel. Majority carriers move from the source to drain through channel, when a potential difference is applied between the source and the drain.

Symbol for JFET

The symbols for JFETS are shown in Fig. 7.3 (b) symbol for *n*-channel JFET is shown Fig. 7.3 (b) (i) in which we can see that, the arrow points towards the vertical line i.e., channel. The vertical line represents the *n*-channel. On the other hand, the symbol for *n*-channel JFET is shown in Fig. 7.3 (b) (ii) in which, the arrow points away from the vertical line. In this case, the vertical line represents the *p*-channel.

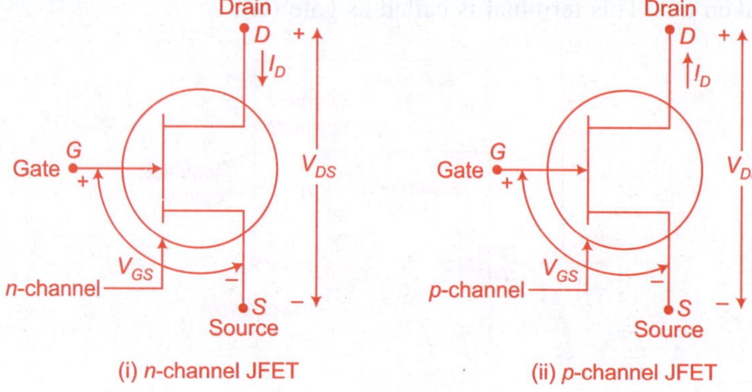

(i) *n*-channel JFET   (ii) *p*-channel JFET

Fig. 7.3   *(b) symbols for JFET*

7.2.2   Formation of Depletion Region in JFET

The *n*-channel JFET is shown in Fig. 7.4 (a) in which *p*-types and *n*-channel makes a *p-n* junction. In the absence of any applied potential, the JFET has two *p-n* junction i.e., under no-bias conditions.

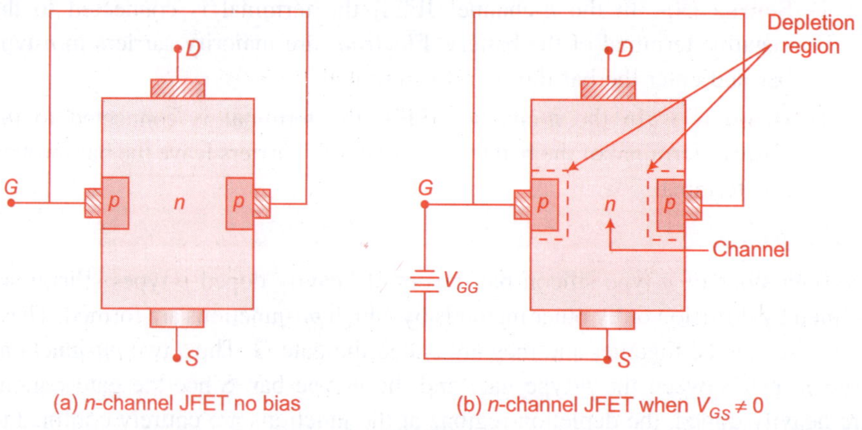

(a) *n*-channel JFET no bias   (b) *n*-channel JFET when $V_{GS} \neq 0$

Fig. 7.4

In n-channel JFET, the p-n junction always keeps reverse biased as shown in Fig. 7.4 (b) in which we can see that the positive terminal of battery V_{GG} is connected to source (negative) and the negative terminal of the battery is connected to the gate (positive).

As a matter of fact, the gate source junction of JFET must never be allowed to become forward biased. The gate material is not constructed to handle any significant amount of current. If the junction is allowed to become forward bias, current will start to flow in the gate circuit. Under most circumstances, this current will destroy the JFET.

The general relationship which can be applied to the DC analysis of all FET amplifier are:

$$I_G \cong 0 \text{ i.e., when gate-source junction is reverse biased i.e., } V_{GS} = 0$$

and, $\qquad I_D = I_S.$

Whenever the p-n junction is reversed bias i.e., $V_{GS} < 0$ (negative), the electrons and holes diffuses across the junction and after diffusion the electrons leave the positive ions on the n-side i.e., towards channel and the holes leave the negative ions on the p-side i.e., towards the gate. The region containing these imobiles ions, is called Depletion region. In other words, we can say that the region near the junction, which contains uncompensated or uncovered acceptor and donor ions is called depletion region. The thickness of depletion region is dependent upon the applied reverse voltage. If applied reverse voltage is increased, the thickness of depletion region will also increase. If both the p-side and n-side of the junction are equally doped, the depletion region will increase equally in both the regions.

Width of depletion region

The width of depletion region is depenedent upon the applied reverse voltage between gate terminal and source and depends on the doping levels of p-side and n-side material. If p-side and n-side of the junction are equally doped, the depletion region will extend equally in both the regions. If one side of the junction is heavily doped as compared to other side, the depletion region extends more into the region of lower doping. In n-channel JFET, the p-region of an n-channel JFET is heavily doped as compared to the n-channel, the depletion region extends less into the p-region and deeper into the n-channel as shown in Fig. 7.4 (b).

When there is no voltage applied between the drain and source, the width of depletion region is symmetrical around the junction and the conductivity of depletion region is zero because there is no mobile charge carriers in this region, due to the absence of voltage between drain and source. So, the effective width of the *n*-channel is reduced. When the reverse voltage is increased, then the effective width of *n*-channel further reduces.

Shape of channel

The *n*-channel JFET is shown in Fig. 7.5, in which the gate terminal is open and the reverse bias across the gate source junction of JFET can also be achieved simply by applying a voltage across the drain and source terminal as shown in Fig. 7.5

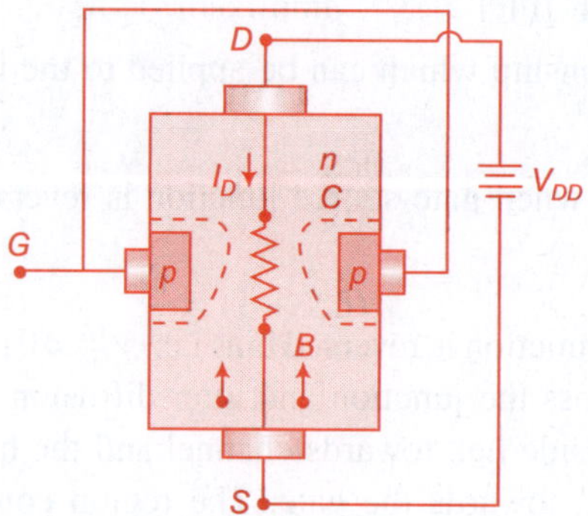

Fig. 7.5 *Effect of V_{DS} on channel shape*

In Fig. 7.5, we can see that the drain (D) is connected to the positive terminal of the DC supply (V_{DD}) and source (*s*) is connected to the negative terminal. In the presence of positive voltage supply (V_{DD}), the electrons flow from sources to drain through *n*-channel and constitute a current known as drain current (I_D) flow from drain to source through the device. The drain current causes a voltage drop across the resistance which has the effect of reverse biasing the gate to source junction. It creates the depletion region within the channel as shown in Fig. 7.5. But the depletion region is not symmetrical around the gate to source junction. The width of depletion region is high at point C, where as it reduces when moves towards point B. It is due to the fact that voltage drop across channel resistance at point C is greater than the voltage drop at point B. Thus, the reverse bias voltage is higher near the drain end of the channel as compared to the source end.

7.2.3 Working and Characteristics of *n*-channel JFET

In Fig. 7.6 (a) working of JFET is shown as follows:

(i) If $V_{DS} = 0$ and $V_{GS} < 0$.

When $V_{DS} = 0$ and $V_{GS} < 0$, in this case both *p-n* junction of *n*-channel JFET are reverse biased and due to this, the width of depletion region will increase as shown in Fig. 7.6 (a). As we have discussed in earlier section, when V_{GS} increases more towards the negative value, then the reverse bias at junctions increases and the width of depletion region increases continuously with increasing the reverse bias. At particular negative value of V_{GS}, the both layers of depletion region come in contact with each other. In this position, the *n*-channel is cut-off at point *p* as shown in Fig. 7.6 (b) the value of V_{GS} at which the channel is cut-off is called cut-off voltage or pinch-off voltage V_p.

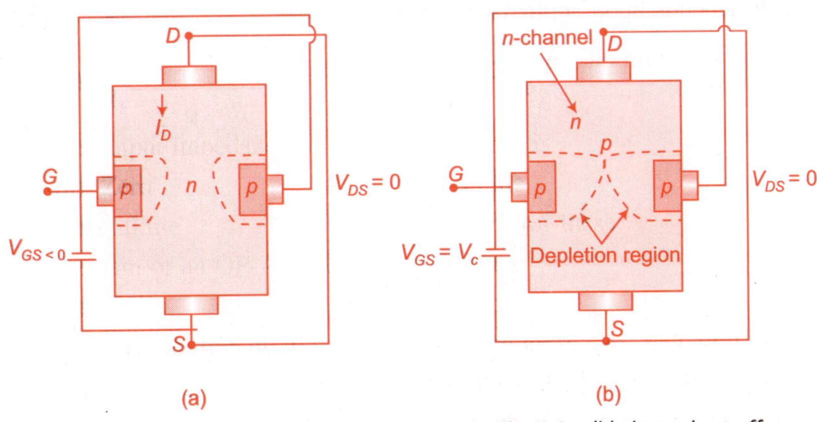

(a)   (b)

Fig. 7.6   *(a) n-channel JFET*   **Fig. 7.6**   *(b) channel cut-off*

(ii) If $V_{GS} = 0$ and V_{DS} has some positive value

In Fig. 7.6 (c), V_{DS} is applied with some positive voltage $V_{DS} > 0$ i.e., the positive terminal of the battery V_{DD} is connected to the drain terminal and the negative terminal of the battery is connected to the source terminal. The gate and source terminals are short circuited $V_{GS} = 0$.

Since voltage is applied between drain source, therefore the electrons move from sources to drain as depicted in Fig. 7.6 (c). The direction of conventional drain current will flow in the opposite direction of electrons, hence, from drain to source. The drain current is represented by I_D.

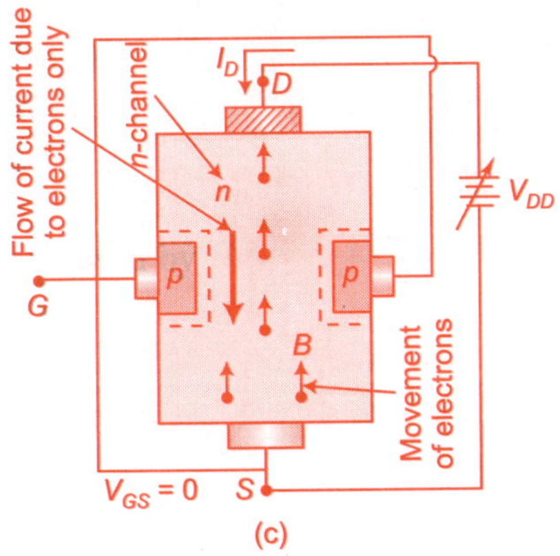

(c)

Fig. 7.6   *(c) n-channel JFET when $V_{DS} > 0$*

The value of drain current depends upon the following factors:

(a) It depends upon the number of majority carriers (number of electrons in the channel) i.e., on the resistivity 'ρ' of the channel

(b) It depends upon the channel

(c) It depends upon cross-sectional area 'A' of the channel

(d) It depends upon applied voltage V_{DS}.

Hence, the drain current,

$$I_D = \frac{V_{DS}}{R} \qquad \qquad \because \quad R = \rho \frac{l}{A}$$

$$I_D = \frac{V_{DS}}{\rho l/A} = \frac{V_{DS} \, A}{\rho l}$$

$$\boxed{I_D = V_{DS} \frac{A}{\rho l}.} \qquad\qquad (7.1)$$

From Eqn. (7.1), we can say that the value of drain current depends upon the resistance of the channel and applied voltage V_{DS}.

(iii) If $V_{GS} = 0$ and $V_{DS} > 0$, Drain characteristics

This condition is already explained in the previous point. We have seen that in this condition that electrons move from source to drain and drain current will flow from drain to source. Due to the flow of drain current, the width of channel will differ at different places of channel. Hence, the shape differs at different places of channel. Hence, the shape of channel is wedged.

Now, if we increase V_{DS}, then the drain current I_D will also increase. The graph between V_{DS} and drain current I_D is shown in Fig. 7.7.

In the Fig. 7.7, we can see that at the low values of V_{DS}, the characteristics are shown as state line (*OA* line). The *OA* region in Fig. 7.7 represents the Ohmic region i.e., the value of channel resistance at low value of V_{DS} is approximately constant.

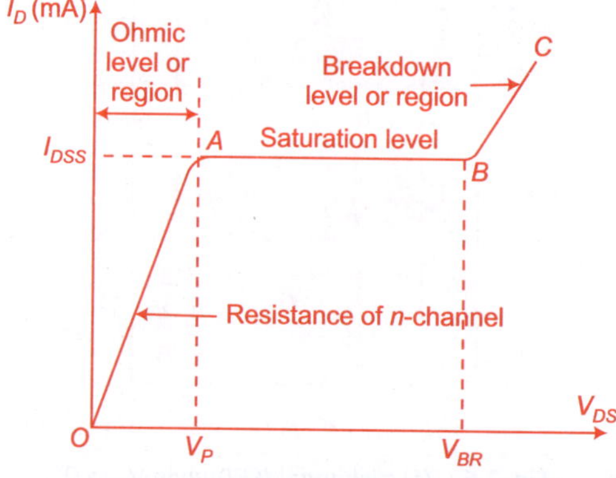

Fig. 7.7 *Drain characteristics with $V_{GS} = 0$*

When V_{DS} is further increased, the reverse bias is also increased and due to this the width of depletion region will increase. Accordingly, increment in width of depletion region, the channel becomes narrow and area of the channel will reduce. Therefore, the resistance of channel will increase.

If the value of V_{DS} is increased from pinch-off voltage V_p, then the raise in drain current will stop. Hence, current is constant due to the constant of depletion region at pinch-off voltage V_p.

When we see in Fig. 7.6 (b), at pinch-off voltage channel is narrow and depletion region comes in contact but it seems that at pinch-off voltage the drain current is zero, due to the blockage of channel. If drain current becomes zero, then the voltage drop at every point of channel is also zero. The reverse bias is same at complete channel i.e., the width of channel at every point will be the same.

Hence, we can state that by increasing the V_{DS} above V_p, the drain current I_D is not zero. The drain current is constant at its maximum value. The maximum value of drain current is I_{DSS}.

Where, I_{DSS} = Maximum drain to source current when gate and source terminals are short circuited.

In Fig. 7.7, region AB represents saturation region. Hence, I_{DSS} flows in n-channel JFET when $V_{Gs} = 0$ and $V_{DS} > V_p$.

If $V_{DS} >>> V_p$, then the covalent bond will break in depletion region between gate and drain of JFET. Due to the breaking of covalent bond avalanche multiplication will occur. Therefore, the breakdown will occur and drain current I_D suddenly increases. This effect is due to the avalanche multiplication of electrons caused by breaking of covalent bond. The BC region in Fig. 7.7 represents the breakdown occur which is called as Breakdown Voltage and represented by V_{BR}.

(iv) If $V_{GS} < 0$ and $V_{DS} > 0$ i.e., V_{DS} increased from zero

As we have studied earlier that in BJT, the base current is the controlling current, at different value of I_B, The characteristics is drawn between V_{CE} and I_C. Similarly, in JFET, V_{GS} is a controlling voltage, Now, we draw characteristics between V_{DS} and I_D at different values of V_{GS}. In drain characteristics of JFET, the V_{GS} is increased towards more negative values and the values of I_D are measured at difference value of V_{DS}. The graph is drawn between V_{DS} and I_D at different values of V_{GS} as shown in Fig. 7.8. When more negative value of V_{GS} is applied between gate and source, the width of depletion region will increase. Due to this, less current will flow at the same value of V_{DS}, when V_{GS} is zero.

It is clear from the Fig. 7.8 that the saturation value of drain current is reduced with increasing negative value of V_{GS} at low values of V_{DS}. If $V_{GS} = -V_p$, then the

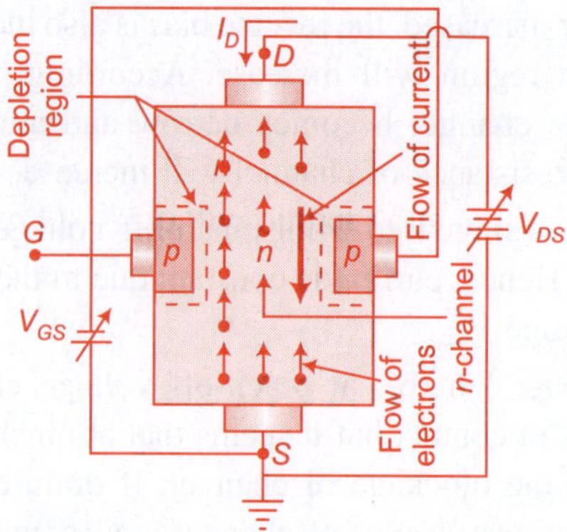

Fig. 7.8   *(a) n-channel JFET when $V_{GS} < 0$ and V_{DS} increased from zero*

saturation current is zero even though we increase V_{DS}. In this condition, JFET will be in 'off' state.

The *n*-channel JFET with its drain characteristic in shown in Fig. 7.8 in which we can see that the region from the left of V_p is known as Ohmic region and the region to the right side of V_p is shown as saturation region.

From the characteristics, it is clear that the drain current I_D is controlled by the electric field that extends into the channel due to the reverse biased voltage applied to the gate. Hence, this device has been given the name "Field Effect Transistor".

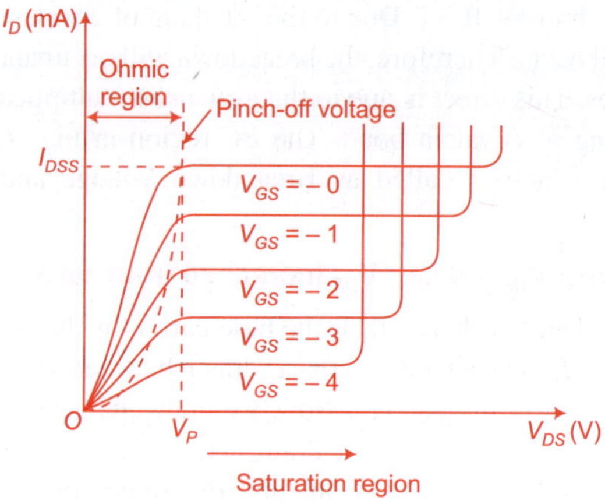

Fig. 7.8   *(b) n-channel JFET drain characteristics*

7.2.4   Transfer Characteristics of *n*-channel JFET

Transfer characteristic is shown in Fig. 7.9 for a *n*-channel JFET, it gives us the relationship between drain current (I_D) and gate to source voltage (V_{GS}) for a constant value of drain to source voltage (V_{DS}). It also known as

Transconductance Curve. The circuit arrangement for transfer characteristics is shown in Fig. 7.10.

In circuit, firstly we set the drain to source voltage at some suitable value and then we increase the gate to source voltage in negative direction in small steps.

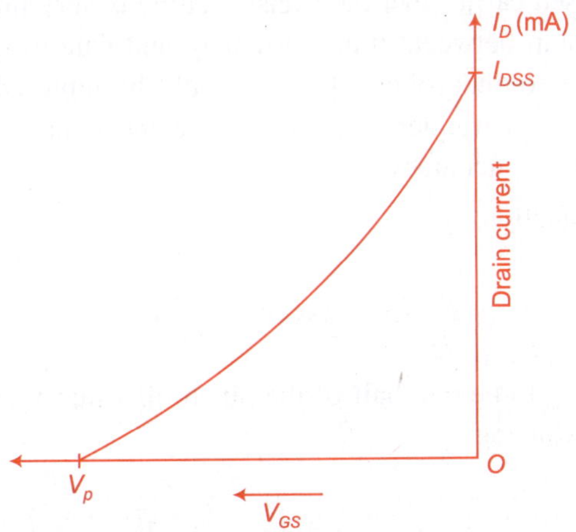

Fig. 7.9   *Transfer characteristic for n-channel JFET*

We note the value of drain current at each step. Now, we plot a graph with gate to source voltage (V_{GS}) and drain current (I_D) as shown in Fig. 7.9.

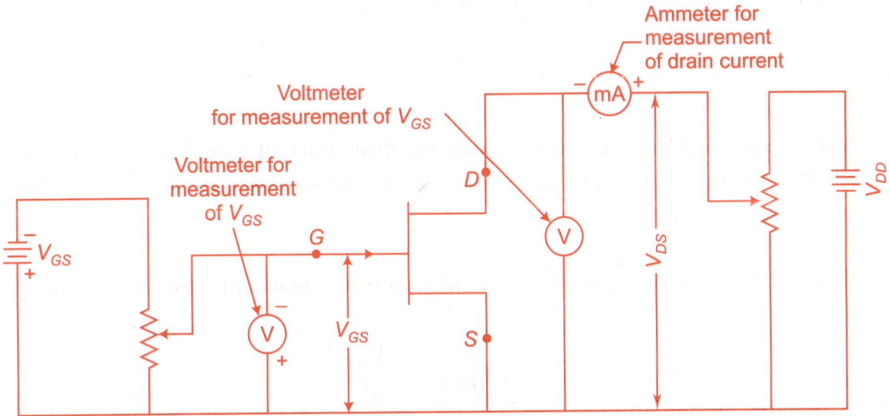

Fig. 7.10   *The circuit arrangement for n-channel JFET characteristics*

In JFET, the relation between I_D and V_{GS} can be expressed by the Shockley equation as shown below:

$$I_D = I_{DSS}\left(1 - \frac{V_{GS}}{V_p}\right)^2. \tag{7.2}$$

From Eqn. (7.2), it is clear that the relationship between I_D and V_{GS} is non-linear because equation has squared term. It is clear from transfer characteristic that

(i) if $V_{GS} = 0$ then, value of $I_D = I_{DSS}$ (maximum)

(ii) If $V_{GS} = -V_p$, then drain current $I_D = 0$.

7.2.5   Method for Drawing Transfer Curve

As we have discussed earlier that the transfer curve is very important because it shows the relationship between drain current I_D and gate to source voltage V_{GS} at constant drain to source voltage V_{DS}. It would be quite advantageous if the transfer curve is plotted frequently in most efficient manner while maintaining an acceptable degree of accuracy.

By Shockley equation:

$$I_D = I_{DSS}\left(1 - \frac{V_{GS}}{V_P}\right)^2 \tag{7.2a}$$

If we specify V_{GS} to be one-half of the pinch-off value V_p, the resulting level of I_D will be the following:

$$I_D = I_{DSS}\left(1 - \frac{V_{p/2}}{V_p}\right)^2 = I_{DSS}\left(1 - \frac{1}{2}\right)^2$$

$$I_D = I_{DSS}(0.5)^2$$

$$I_D = 0.25\, I_{DSS}$$

i.e.
$$I_D = \left.\frac{I_{DSS}}{4}\right|_{V_{GS} = V_p/2} \tag{7.2b}$$

From Eqn. (7.2(b)), we can state that the drain current I_D will always be one-fourth of the saturation level I_{DSS} as long as the gate to source voltage is one-half of the pinch-off value.

Now, if we assume $I_D = \dfrac{I_{DSS}}{2}$ and substitute in equation below, we obtain,

$$I_D = I_{DSS}\left(1 - \frac{V_{GS}}{V_P}\right)^2$$

$$\left(1 - \frac{V_{GS}}{V_P}\right)^2 = \frac{I_D}{I_{DSS}}$$

$$\left(1 - \frac{V_{GS}}{V_P}\right) = \sqrt{\frac{I_D}{I_{DSS}}}$$

$$\frac{V_{GS}}{V_P} = 1 - \sqrt{\frac{I_D}{I_{DSS}}}$$

$$V_{GS} = V_P \left[1 - \sqrt{\frac{I_D}{I_{DSS}}} \right] \tag{7.2c}$$

Now, putting the value of $I_D = \dfrac{I_{DSS}}{2}$ in Eqn. (7.2(c)), we obtain,

$$V_{GS} = V_P \left[1 - \sqrt{\frac{\frac{I_{DSS}}{2}}{I_{DSS}}} \right]$$

$$= V_P[1 - \sqrt{0.5}]$$

$$V_{GS} = 0.293\ V_P$$

Hence, $\qquad V_{GS} \cong 0.3\ V_P\Big|_{I_D = \frac{I_{DSS}}{2}}$ $\qquad\qquad$ (7.2d)

Some other point can be determined, but the transfer curve can be drawn to a satisfactory level of accuracy by using these four points as shown in Table 7.1

Table 7.1

Points	V_{GS}	I_D
A	0	I_{DSS}
D	0.3 V_p	$\dfrac{I_{DSS}}{2}$
C	0.5 V_p	$\dfrac{I_{DSS}}{4}$
B	V_p	0 mA

7.2.5.1 Construction of p-channel JFET

p-channel JFET is similar in construction to the *n*-channel JFET. The only difference is that *p*-channel JFET has the channel made with *p*-type material and gate is made with *n*-type material as shown in Fig. 7.11. The direction of drain current in *p*-channel JFET is in opposite direction of drain current in *n*-channel JFET i.e., from source

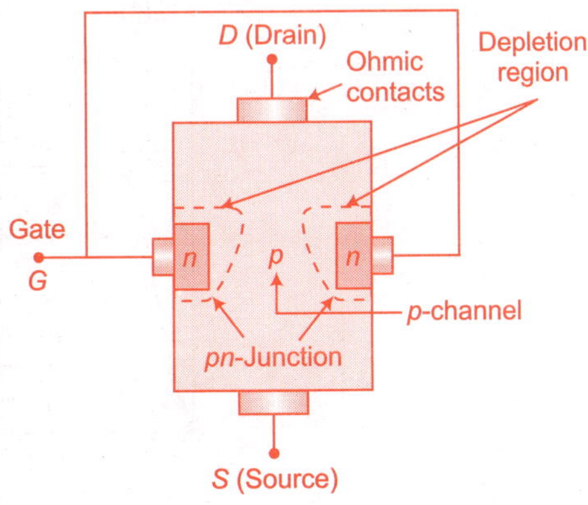

Fig. 7.11 *p-channel JFET*

to drain because current flows due to the holes. Similarly, the polarities of V_{GS} and V_{DS} are opposite of n-channel JFET.

7.2.5.2 Working and characteristics of p-channel JFET

Working of p-channel JFET is similar to the working of n-channel JFET, excluding some differences such as, we need positive gate voltage for keeping p-channel narrow whereas in n-channel JFET, we use negative gate voltage. The gate is made with n-types in p-channel JFET and it is reverse biased with positive voltage.

Figure 7.12 shows a p-channel JFET. Here, we take V_{DS} is negative that means the voltage of source is greater than the voltage of drain. The positive terminal of battery V_{DD} is connected to the source and negative terminal of battery is connected to the drain.

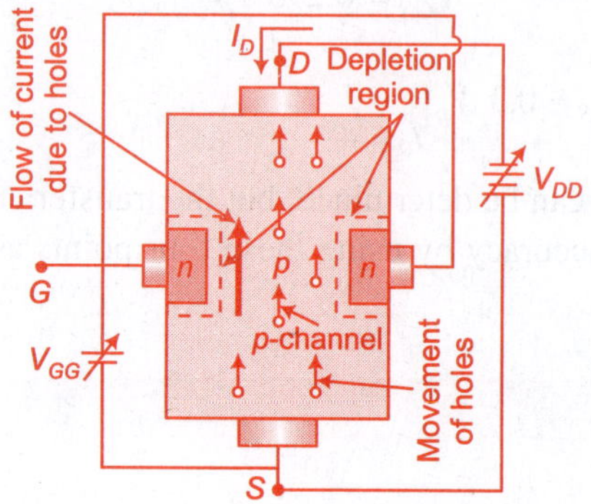

Fig. 7.12 *p-channel JFET, when $V_{GS} > 0$ and V_{DS} decreased from zero*

The drain characteristics of p-channel JFET is shown in Fig. 7.13. In drain characteristics of p-channel JFET, V_{GS} is increased towards more positive value and the values of V_{DS} and I_D are measured at different values of V_{GS} and graph is drawn

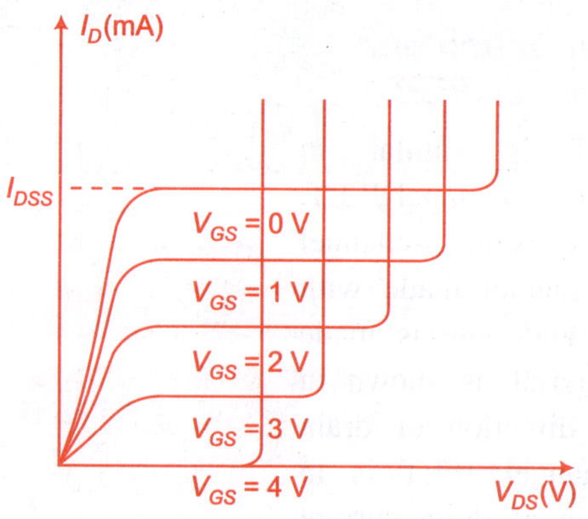

Fig. 7.13 *Drain characteristics of p-channel JFET*

between V_{DS} and I_D at different values of V_{GS}. When more positive value of V_{GS} is applied between gate and source, the width of the depletion region will increase. Due to this, less current will flow at the same value of V_{DS} when V_{GS} is 0.

7.3 PARAMETERS OF JFET

In JFET, the value of drain current I_D is dependent upon the value of drain to source voltage V_{DS} and gate to source voltage V_{GS}. We can find the relation between any two quantities out of three quantities (I_D, V_{DS}, V_{GS}) when the third one is fixed. This relation can be obtained by following three parameters.

(i) Mutual Conductance or Transconductance

It is the ratio of the a small change in the drain current I_D to the small change in the gate to source voltage V_{GS} at constant drain to source voltage V_{DS}.

It has a unit of conduction i.e., mA/volt or Amp/volt (mho). In other words, it is called slope of the transfer characteristics. It is denoted by g_m, and is defined by

$$g_m = \left. \frac{\Delta I_D}{\Delta V_{GS}} \right|_{V_{DS}\,=\,\text{constant}} \text{(mho)}. \tag{7.3}$$

It is the ratio of a small change in the drain to source voltage V_{DS} to small change in the drain current I_D at a constant gate to source voltage V_{GS}. It has unit of resistance i.e., volt/Amp (ohm). It is denoted by r_d, and is defined by

$$r_d = \left. \frac{\Delta V_{DS}}{\Delta I_D} \right|_{V_{GS}\,=\,\text{constant}} \text{(ohm)}. \tag{7.4}$$

(iii) Amplification factor μ

Amplification factor is the ratio of small change in the drain to source voltage V_{DS} to the small change in the gate to source voltage V_{GS} at a constant drain current. It has no unit and is denoted by μ.

$$\mu = -\left(\frac{\Delta V_{DS}}{\Delta V_{GS}} \right)_{I_D\,=\,\text{constant}} \tag{7.5}$$

The negative sign shows that when gate to source voltage V_{GS} is increased, drain to source voltage V_{DS} must be decreased for I_D to remain constant.

(iv) Relation between three parameters of JFET

The multiplication of Transconductance g_m and drain resistance r_d will be equal to the amplification factor μ

i.e. $$\mu = g_m \times r_d. \tag{7.6}$$

7.3.1 Procedure to Obtain a IDQ and VGSQ with the Help of Transfer Curve and Bias Line Curve

1. First obtain the gate to source voltage V_{GS}.

   i.e.
   $$V_{GS} = V_G - V_S$$
   $$V_{GS} = V_G - I_D R_S \quad \text{(A)} \qquad (\because V_S = I_D R_S)$$

2. Now Arbitrary choose the value of drain current and obtain the value of V_{GS} at particular drain current I_D.

3. Now draw a bias curve by using point of drain current and gate to source voltage obtained in previous point.

4. After drawing the bias line curve, calculate the points for transfer curve. For drawing the transfer curve, we need at least four points for satisfactory curve. These points are named as A, B, C and D.

   Point A: When $I_D = I_{DSS}$ mA and $V_{GS} = 0V$

   Point B: When $I_D = 0$mA and $V_{GS} = V_P$

   Point C: When $I_D = \dfrac{I_{DSS}}{4}$ and $V_{GS} = \dfrac{V_P}{2}$

   Point D: When $I_D = \dfrac{I_{DSS}}{2}$ and $V_{GS} = 0.3 \ V_P$.

5. After determining point A, B, C and D of transfer curve, locate these points on V_{GS} and I_D axis. You will find the transfer curve.

6. Now superimpose the bias line curve, a transfer curve and the intersection point of these two curves which is known as Q-point or Operating point.

7. The coordinate of Q-point is I_{DQ} and V_{GSQ}.

8. Further calculate the required parameter of FET with the help of V_{GSQ} and I_{DQ}.

7.4 SPECIFICATION OF JFET

The specifications of JFET is given in its specification sheet. The important specifications for the JFET are as follows:

(i) Maximum rating

    (a) Maximum drain to source voltage V_{DS}

    (b) Maximum gate to source voltage V_{GS}

    (c) Maximum drain current I_D

    (d) Maximum power dissipation P_D

    (e) Temperature range of junction T_j

(ii) Identification of terminals

Terminals of JFET can be identified according to Fig. 7.14 and is also given a specification sheet.

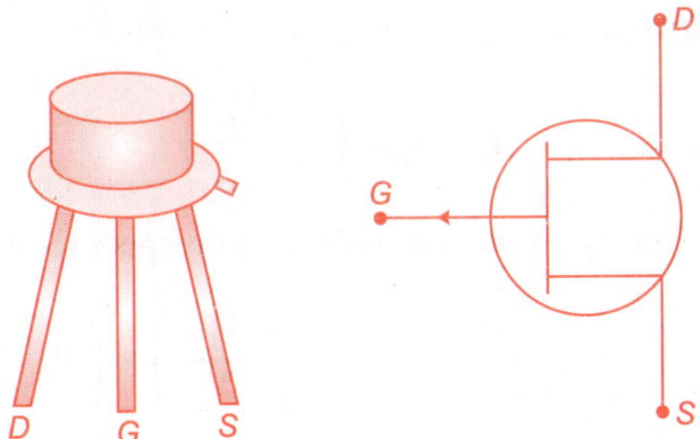

Fig. 7.14   *2N2 844 p-channel JFET*

7.5 COMPARISON BETWEEN FET AND BJT

The comparison between field effect transistor and bipolar junction transistor is shown in Table 7.2.

Table 7.2

S.No	Field effect Transistor (FET)	Bipolar Junction Transistor BJT
1.	FET is a unipolar device i.e., the current flow in the device is only due to the majority carriers. (electrons in *n*-channel and hole in *p*-channel).	It is a bipolar device i.e., the current flow in device is due to the electrons and holes i.e., due to majority and minority carriers.
2.	FET is a voltage controlled device, i.e., gate voltage controls the amount of drain current.	It is a current controlled device i.e., the base current controls the amount of collector current.
3.	FET has a negative temperature coefficient at high current level, i.e., the drain current decreases as the temperature increases. It prevents the FET from thermal breakdown.	It has a positive temperature coefficient at high current level i.e., the collector current increases with the increase in temperature. It leads BJT to thermal breakdown.
4.	JFET has high input resistance in the order of megaohms.	Its resistance is very low and is the order of few kΩ.
5.	JFET has high switching speed and cut-off frequencies because it does not suffer from minority carrier storage effect.	It has low switching speed and cut-off frequencies due to the effect of minority carrier storage.
6.	JFET is easily fabricated in the form of *IC* and occupies a less space on *IC*	It is comparatively difficult to fabricate it as an integrated circuit (*IC*) and occupies more space on *IC* than that of FET.
7.	JFET is less noisy than a BJT or vacuum. Thus tubes is more suitable as an input amplifier for low level signals.	It is comparatively more noisy.

7.6   MATHEMATICAL EXPRESSION FOR TRANSCONDUCTANCE

As we have discussed earlier, the drain current is expressed by the relation given in Eqn. (7.2) as follows:

$$I_D = I_{DSS} \left(1 - \frac{V_{GS}}{V_p} \right)^2 \tag{7.7}$$

Differentiating Eqn. (7.7) on both sides with respect to V_{GS}, we have,

$$\frac{dI_D}{dV_{GS}} = 2I_{DSS} \left(1 - \frac{V_{GS}}{V_p} \right) \times \left(0 - \frac{1}{V_p} \right)$$

$$\frac{dI_D}{dV_{GS}} = - \frac{2I_{DSS}}{V_p} \left(1 - \frac{V_{GS}}{V_p} \right) \tag{7.8}$$

From Eqn. (7.3), we know that,

$$g_m = \frac{\Delta I_D}{\Delta V_{GS}} \tag{7.9}$$

Substituting the Eqn. (7.9) in Eqn. (7.8), we obtain,

$$g_m = - \frac{2I_{DSS}}{V_P} \left(1 - \frac{V_{GS}}{V_p} \right) \tag{7.10}$$

Now, substituting $V_{GS} = 0$ in above Eqn. we get,

$$g_m = - \frac{2I_{DSS}}{V_p} \bigg|_{V_{GS} = 0} = g_{m0}$$

The value of g_m at $V_{GS} = 0$ is the maximum and equal to $- \dfrac{2I_{DSS}}{V_p}$. Thus, the value of g_m is designated as g_{mo}. The Eqn. (7.8) can now be re-written as,

$$g_m = g_{m0} \left(1 - \frac{V_{GS}}{V_p} \right). \tag{7.11}$$

Example 7.6.1   The following information is given on the data sheet for n-channel JFET

$$I_{DSS} = 15 \text{ mA } V_p = -6 \text{ volt and } g_{ms} = 4000 \text{ } \mu s.$$

Find the values of the drain current and transconduction at $V_{GS} = -3$ volts.

Solution   Given that,

$$I_{DSS} = 15 \text{ mA}, V_p = -6V, g_{mo} = 4000 \text{ } \mu s$$

and, $V_{GS} = -3$ volt.

Now, from Eqn. (7.2), we know that,

$$I_D = I_{DSS}\left(1 - \frac{V_{GS}}{V_p}\right)^2$$

$$I_D = 15\text{mA} \times \left(1 - \frac{-2}{-2}\right)^2$$

$$I_D = 15\text{mA} \times (0.5)^2 = 3.75\ \text{mA}$$

$$\boxed{I_D = 3.75\ \text{mA.}} \hspace{4cm} \textbf{(Ans.)}$$

Now, from Eqn. (7.11), we know that,

$$g_m = g_{mo}\left(1 - \frac{V_{GS}}{V_p}\right) = 4000\ \mu\text{s}\left(1 - \frac{-3}{-6}\right)$$

$$g_m = 4000\ \mu\text{s} \times 0.5$$

$$\boxed{g_m = 2000\ \mu\text{s.}} \hspace{4cm} \textbf{(Ans.)}$$

Example 7.6.2   As n-channel JFET has $I_{DSS} = 6\text{mA}$ and $V_p = -3\text{V}$. Find the minimum value of drain to source voltage V_{DS} for pinch-off region and the drain current I_D for $V_{GS} = -2\text{V}$ in the pinch-off region.

Solution   The minimum value of V_{DS} for pinch-off to occur for $V_{GS} = -2\text{V}$ is,

$$V_{DS}(\text{min}) = V_{GS} - V_p$$

$$= -2 - (-3) = 1\text{V}$$

$$I_D = I_{DSS}\left[1 - \frac{V_{GS}}{V_p}\right]^2$$

$$I_D = 6\ \text{mA} \times \left(1 - \frac{-2\text{v}}{-3\text{v}}\right)^2 = 6\ \text{mA} \times (0.33)^2$$

$$\boxed{I_D = 0.653\ \text{mA.}} \hspace{4cm} \textbf{(Ans.)}$$

7.7 SOME IMPORTANT APPLICATIONS OF FET

As we have discussed in Table 7.2 the comparison between FET and BJT. We found that FET has high input impedance, low noise level and low output impedance. Therefore, the FET is superior to BJT. Some important applications of FET are explained below.

7.7.1 JFET Amplifier

The circuit for JFET amplifier is shown in Fig. 7.15 and has following components:

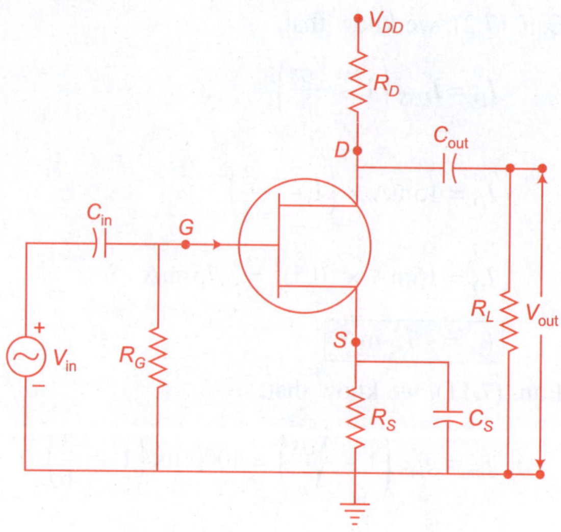

Fig. 7.15 *n-channel JFET Amplifier*

(i) V_{DD} is a drain supply voltage.

(ii) R_G is a gate resistance. It is used for proper biasing of gate.

(iii) C_{in} and C_{out} are coupling capacitors which are used to block DC to input and output side respectively.

(iv) R_S and R_D are source and drain resistances.

(v) C_s is a bypass capacitor which is used to bypass the AC to the ground.

(vi) V_{in} and V_{out} are input AC voltage and amplified output voltage respectively.

(vii) R_L is a load resistance across which amplified output is obtained.

Working of JFET amplifier

DC biasing is used in the circuit with battery V_{DD}. Due to biasing, drain current I_D flows in the circuit. The drain current flow in the source resistance R_S. Due to the drain current, the voltage drop $I_D R_S$ occurs across source resistance R_S. This voltage drop produces reverse voltage V_{GS} between gate and sources because $V_G = 0$ and $V_S = I_D R_S$.

Hence,
$$V_{GS} = V_G - V_S$$

$$V_{GS} = 0 - I_D R_S = - I_D R_S$$

$$V_{GS} = - I_D R_S. \tag{7.12}$$

In Eqn. (7.12), the negative sign indicates that the gate is reverse biased. So, we can apply suitable reverse voltage to the gate by changing V_{DD} and R_S.

When we apply AC voltage at the gate, then this AC voltage is superimposed on the DC voltage (reverse bias). Due to this, the value of V_{GS} changes with the change in input AC signal. Due to the change in V_{GS}, the drain current will change. The varying drain current produces an amplified AC voltage across the load resistance R_L and thus the circuit acts as an amplifier.

7.7.2 As a Buffer Amplifier

A buffer amplifier is shown in Fig. 7.16. Basically, the buffer amplifier is used to isolate the preceding stage from the following stage. Since, FET has a high input impedance (in the order of MΩ) and low output impedance (in the order of few hundered Ω), it acts as an excellent buffer amplifier. The high input impedance means high load on the preceding stage. The low output impedance means that the buffer can drive heavy loads.

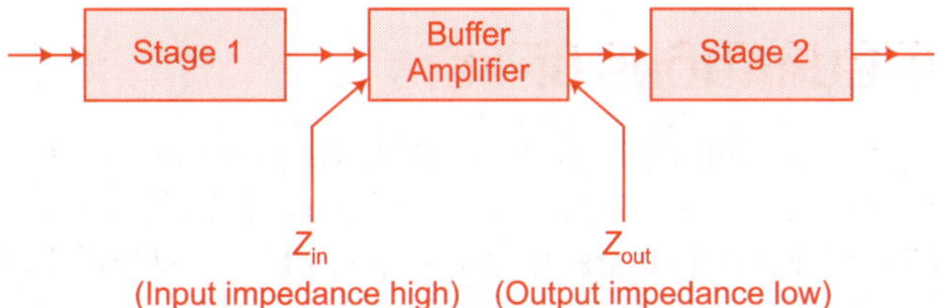

Fig. 7.16 *The buffer amplifier stage*

7.7.3 As Phase Shift Oscillator

Generally all the oscillators work with FETS. But it is valuable especially in phase shift oscillator to minimize loading due to the high input impedance. The circuit diagram of a phase shift oscillator using *n*-channel JFET is shown in Fig. 7.17.

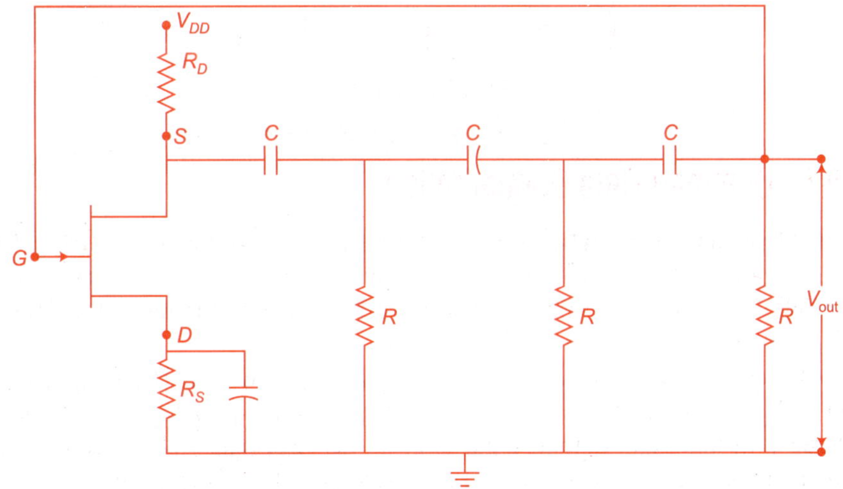

Fig. 7.17 *Phase shift oscillator using n-channel JFET*

7.7.4 Low Noise Amplifier

As we have discussed earlier that JFET is a low noise device. Therefore, it is preferred to use it near the front end of the receivers and other electronic equipments. As we know that noise is an unwanted signal which is superimposed upon the wanted signal or information signal. Basically the wanted signal contains the desired information. As the noise level increase in the wanted signal, the information becomes inferior.

For instance, the noise in radio receiver, produces hissing; sever noise may completely block the voice or music. Similarly, the noise in television receivers produces white spots on the picture. A heavy noise may wipe-off the picture. Noise is independent of the signal strength because it exists even when the signal is off.

7.8 CONFIGURATIONS OF FET

FET is also used as an amplifier similar to BJT. We give input to the amplifier and it gives amplified output, so we need two terminals for input and two terminals for output. Total four terminal are needed in amplifier as shown in Fig. 7.18. but a transistor has three terminals is common in input and output side. So, we get four terminals effectively. So it is necessary to take one terminal as a common terminal, it has three configuration.

   (i) Common Gate configuration (CG configuration)

  (ii) Common Drain (CD) configuration

 (iii) Common Source (CS) configuration

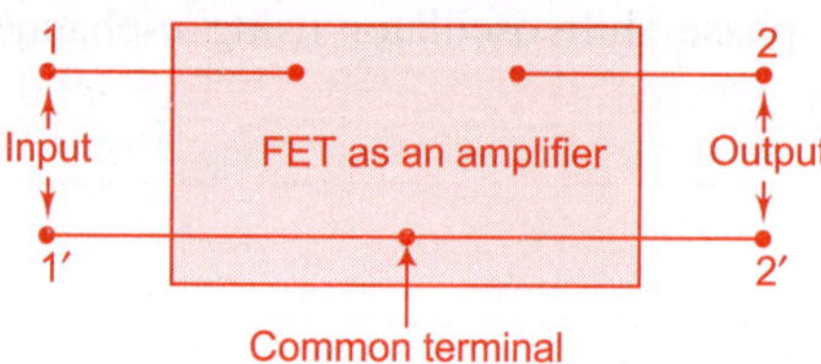

Fig. 7.18 *FET as an amplifier*

7.8.1 Common Gate Configuration

A common gate configuration is one of three basic configurations, of field effect transistor to topologies, typically used as a current buffer or voltage amplifier. In this circuit, the source terminal of the transistor serves as the input, the drain is the output and the gate is common to both.

The common gate configuration is shown in Figs. 7.19 (a) and (b) for *n*-channel and *p*-channel JFET respectively. This type of configuration gate is common or reference to input and output. Input is given between gate and source terminal and output is taken between gate and drain terminal. So, it is called common gate configuration.

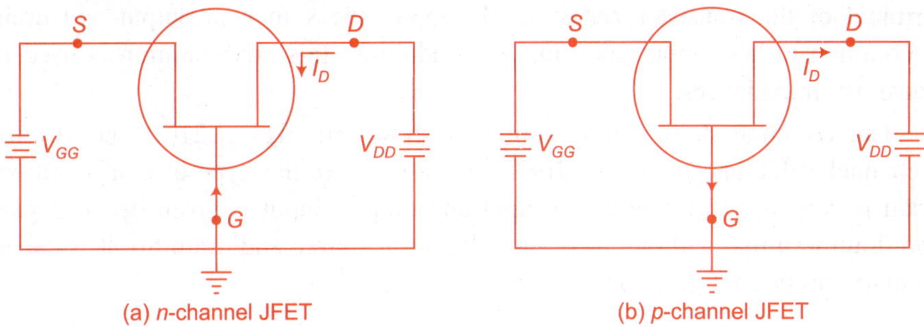

(a) n-channel JFET   (b) p-channel JFET

Fig. 7.19   Common gate configuration

7.8.2 Common Source (CS) Configuration

A common source configuration is one of three basic configurations of field effect transistor topologies, typically used as a voltage or transconductance amplifier.

The common source configuration is shown in Figs. 7.20 (a) and (b) for *n*-channel and *p*-channel JFET respectively.

This type of configuration is widely used. In this type of configuration, source is common in reference to input and output. Input is given between gate and source terminal and output is taken between source and drain. So, it is called common source configuration.

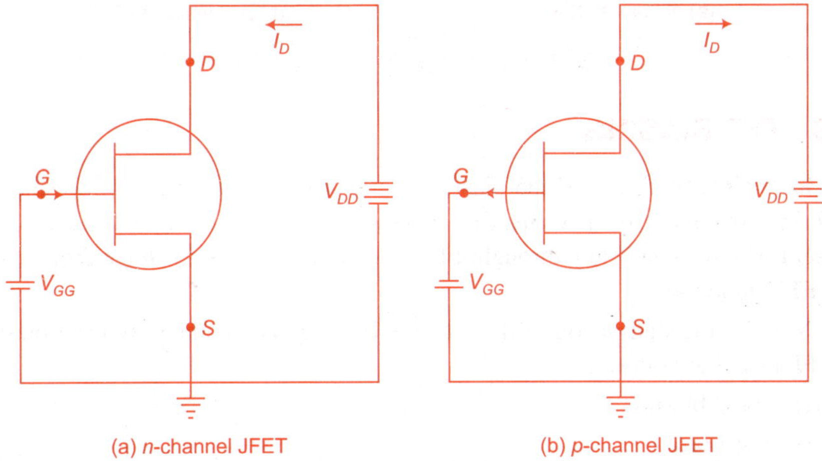

(a) n-channel JFET   (b) p-channel JFET

Fig. 7.20   Common source configuration

7.8.3 Common Drain (CD) Configuration

A Common Drain configuration is also used as an amplifier, also known as a Source Follower. It is one of three basic configurations of field effect transistor topologies, typically used as a voltage buffer. In this configuration, the gate terminal of the transistor serves as the input, the source is output and drain is common to both input and output. In addition, this configuration is used to transform impedances.

The common drain configuration is shown in Figs. 7.21(a) and (b) for *n*-channel JFET and *p*-channel JFET respectively. In this type of configuration, drain is common in reference to input and output. Input is given between gate and drain terminal and output is taken between source and drain So, it is called common drain configuration.

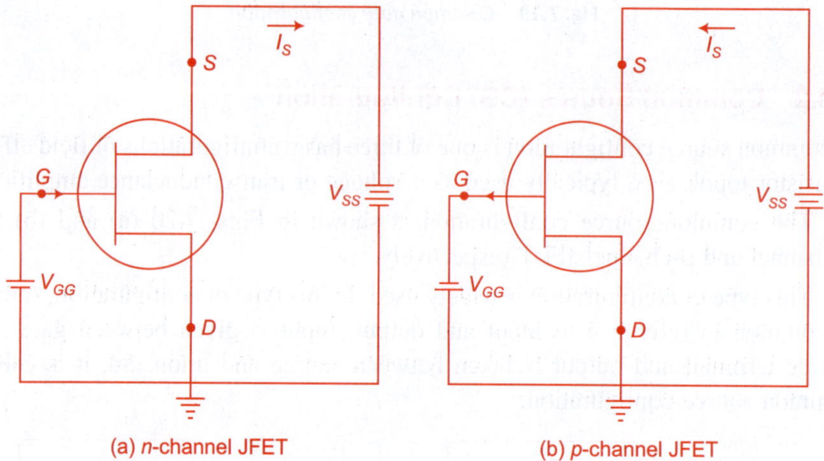

Fig. 7.21 *Common drain configuration*

7.9 FET BIASING

We have already discussed the biasing of transistor and the need of biasing for BJT in previous chapter. Generally, biasing deals with setting a fixed level of current which should flow through. FET with a desired fixed voltage drop across the FET Junction.

Now let us, study about different FET biasing. The commonly used biasing for FETs are as follows:

  (i) Fixed biasing

 (ii) Self biasing

(iii) Potential divider bias or voltage bias

As we have studied earlier section, the gate source junction of a JFET will always be reverse biased i.e., in all methods of biasing, we keep gate source junction in reverse bias. We have seen earlier that when the gate and source terminals are short i.e., $V_{GS} = 0$, the drain current (I_D) will cause the gate source junction to be reverse biased.

7.9.1 Fixed Bias Circuit

The fixed bias circuit for JFET is shown in Fig. 7.22. It is also known that as gate bias circuit for *n*-channel JFET. The battery of V_{GS} is used to ensure the gate source junction is reverse biased i.e., the positive terminal of battery is connected to the source (negative) and the negative terminal of battery is connected to the gate (positive). Since there is no gate current, hence there is no voltage drop across gate resistor R_G.

Analysis of the Circuit

The DC analysis of the circuit is divided into two parts one is input side and another one is output side as shown in Figs. 7.22 (a) and (b).

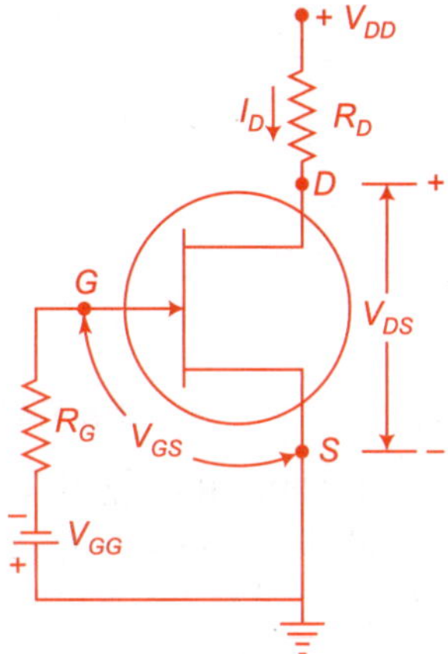

Fig. 7.22 *Fixed bias circuit for n-channel JFET*

(i) Analysis of input side or gate source side

The circuit for analysis of input side is shown in Fig. 7.22 (a)

We know that for DC analysis,

$$I_G = 0 \tag{7.13}$$

Now, applying KVL to input side or gate circuit side loop,

$$V_{GG} + V_{GS} = 0$$

$$V_{GS} = -V_{GG} \tag{7.14}$$

The expression for the drain current I_D is as under:

$$I_D = I_{DSS}\left[1 - \frac{V_{GS}}{V_p}\right]^2 \tag{7.15}$$

The Eqn. (7.15) is nothing but Shockley's equation as discussed in previous section.

(ii) Analysis of output side loop

The circuit for analysis of output side is shown in Fig. 7.22 (b)

Applying KVL at output side loop we obtain,

$$V_{DD} - I_D R_D - V_{DS} = 0$$

$$V_{DS} = V_{DD} - I_D R_D \qquad\qquad (7.16)$$

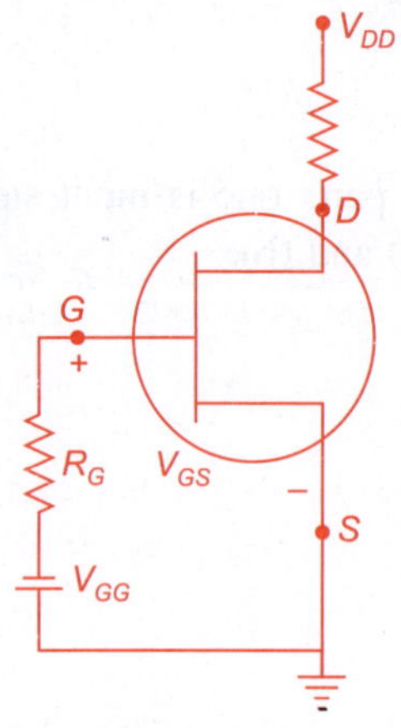

Fig. 7.22   (a) Input side loop

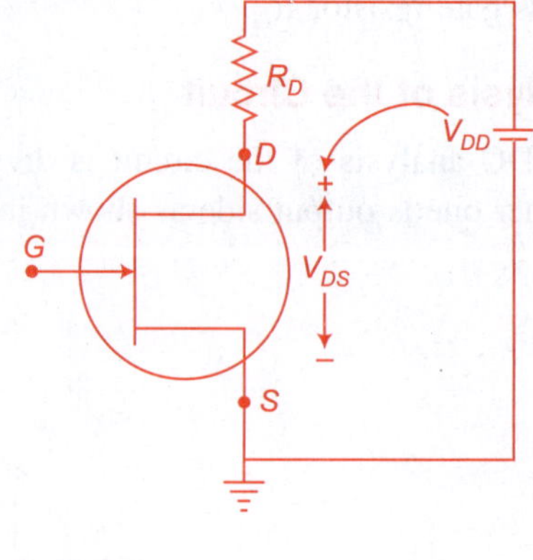

Fig. 7.22   (b) Output side loop

Example 7.9.1   Fig. Ex. 7.9.1 show the fixed bias circuit for *n*-channel JFET. Find the values of I_D and V_{DS}. Assume $I_{DSS} = 6$mA and $V_p = -5$V.

Solution   Data given that,

$$I_{DSS} = 6\text{mA}; \qquad V_p = -5\text{V}$$

$$V_{GG} = +3\text{V}, \qquad V_{DD} = 15\text{V}$$

$$R_D = 3\text{ k}\Omega, \qquad R_G = 1\text{M}\Omega$$

Appling KVL to input loop, we obtain,

$$V_{GG} + I_G R_G + V_{GS} = 0$$

$$3 + (1 \times 10^6) I_G + V_{GS} = 0. \qquad (A)$$

But for DC analysis we know that $I_G = 0$.

Substituting the value of I_G in equation *A*, we obtain,

$$3 + (0 \times 1 \times 10^6) + V_{GS} = 0$$

$$V_{GS} = -3\text{V}.$$

Now, using Shockley equation, we get,

$$I_D = I_{DSS}\left[1 - \frac{V_{GS}}{V_P}\right]^2$$

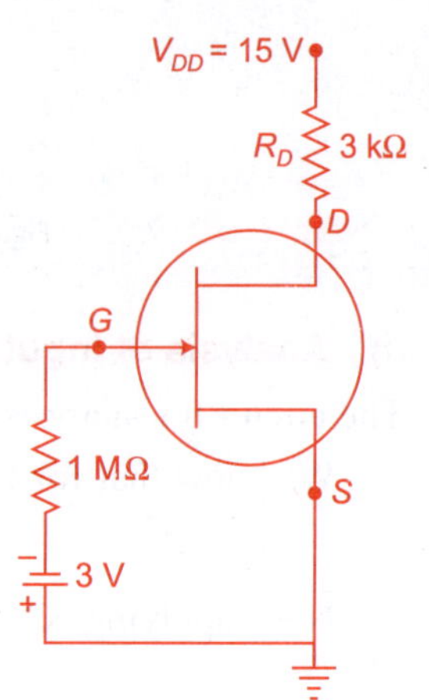

Fig. Ex. 7.9.1

Substituting the values in above equation,

$$I_D = 6 \times 10^{-3} \left[1 - \left(\frac{-3}{-5}\right)\right]^2$$

$$I_D = 6 \times 10^{-3} \left[1 - 0.6\right]^2 = 6 \times 10^{-3} \left[0.4\right]^2$$

$$I_D = 6 \times 0.16 \times 10^{-3} \, A$$

$$\boxed{I_D = 0.96 \text{ mA.}} \hspace{3cm} \textbf{(Ans.)}$$

Now, applying KVL to output side, we obtain from Eqn. (7.16)

$$V_{DD} - I_D R_D - V_{DS} = 0$$

$$V_{DS} = V_{DD} - I_D R_D$$

$$\boxed{V_{DS} = 12.12 \text{ volt.}} \hspace{3cm} \textbf{(Ans.)}$$

The fixed bias circuit cannot provide a stable Q-point from one JFET to another. The reason for this is for a given JFET, there is a range of values for $V_{GS \text{ (off)}}$ and I_{DSS} which can lead to a major problem with fixed bias circuits. As we know that there is no gate current but still the circuit contains R_G. R_G is used for AC operation purposes. Thus, R_G has little to do with DC operation of the amplifier, but it is a vital part of the AC operation of the circuit.

7.9.2 Self-Bias Circuit

Self-bias circuit is shown for in Fig. 7.23. In this Fig., we can see that only drain supply is used and there is no need for the gate supply. The gate terminal is connected to ground through gate resistance R_G. Similarly, the source resistance is connected to the ground through source resistance R_S. When the drain voltage is applied by V_{DD}, a drain current flows even if there is no gate current. The voltage drop V_S produced across sources resistance R_S is due to the drain current and is given by,

$$V_S = I_D R_S \hspace{2cm} (7.17)$$

The voltage drop V_S produces the gate to sources reverse voltage required for a FET operation. The resistor R_G is known as feedback resistor and is used to prevent any variation in the FET drain current I_D which can be understood from the following discussion:

Firstly, let us consider an increase in the drain current, which will increase the voltage drop V_S across the source resistor R_S. The increased voltage drop increases the reverse gate to source

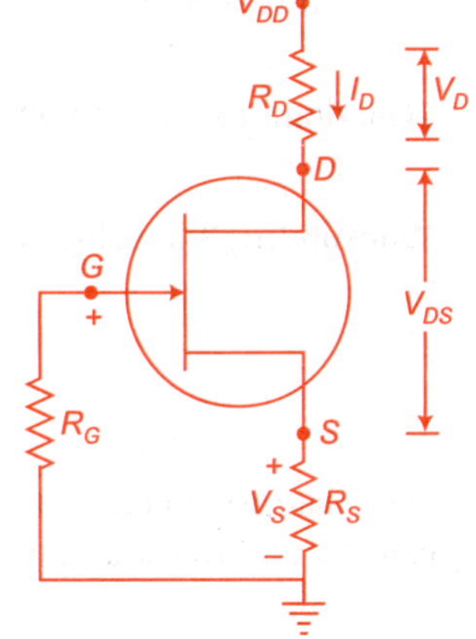

Fig. 7.23 *Self-bias circuit for n-channel JFET*

voltage, which decreases the effective width of the channel W_{effc}. It reduces the value of drain current.

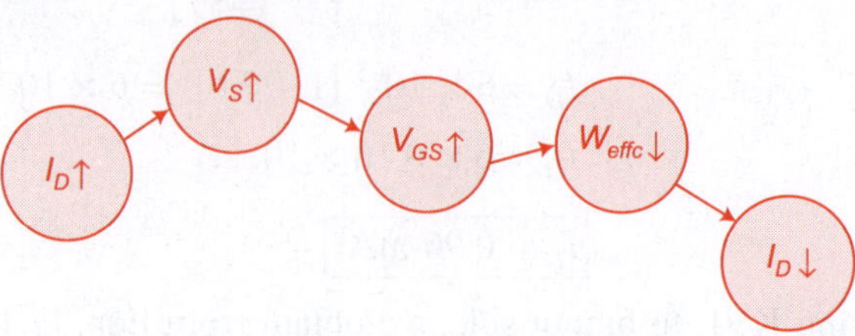

Now, if drain current decreases, then reverse action takes place i.e., it reduces gate to source voltage, which in turn increases the effective width of the channel by increasing the value of drain current.

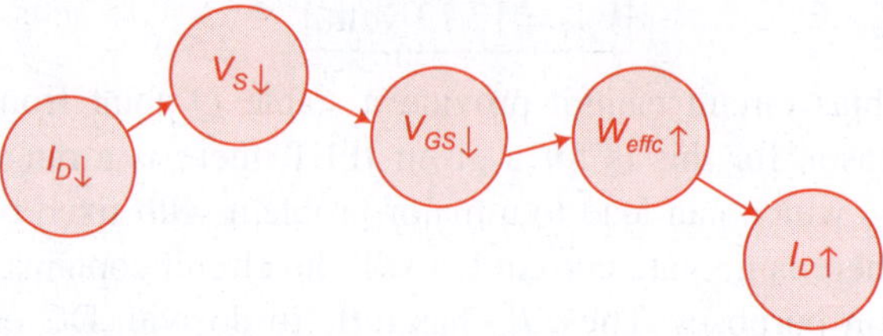

Where, W_{effc} = effective width of channel.

As we saw earlier that there is no current flow through the gate terminal to the ground because the gate to source junction is reverse biased. But practically a small current flows through gate to ground and is known as Reverse leakage current I_{GSS}. But its value is very small and can be neglected. Therefore, we shall assume that the gate terminal is at zero voltage i.e., $V_G = 0$.

We know that,

$V_S = I_D R_S$ from Eqn. (7.17), the drain voltage is given by,

$$V_D = V_{DD} - I_D R_D \qquad (7.18)$$

The drain to source voltage V_{DS} is given by,

$$V_{DS} = V_D - V_S \qquad (7.19)$$

Substituting the values of V_D and V_S in Eqn. (7.19) we obtain,

$$V_{DS} = V_{DD} - I_D R_D - I_D R_S$$
$$V_{DS} = V_{DD} - I_D (R_D + R_S)$$
$$V_{DS} = V_{DD} - I_D (R_D + R_S). \qquad (7.20)$$

The gate to source voltage V_{GS} is equal to the difference between the gate voltage V_G and the source voltage V_S.

$$V_{GS} = V_G - V_S = 0 - I_D R_S$$
$$V_{GS} = - I_D R_S \qquad (7.21)$$

From Eqn. (7.21), it is clear that the gate to source voltage is equal to the negative of the voltage across the source resistor. Hence, it is obvious that greater the value of I_D, more negative will be the gate to source voltage.

From Eqn. (7.21), we can say,

$$I_D = -\frac{V_{GS}}{R_S}.$$

(7.22)

Now, if we plot a graph using the value of drain current obtained from Eqn. (7.22) for a constant value resistor (R_S) against V_{GS}, we get a straight line which is known as is self-bias line.

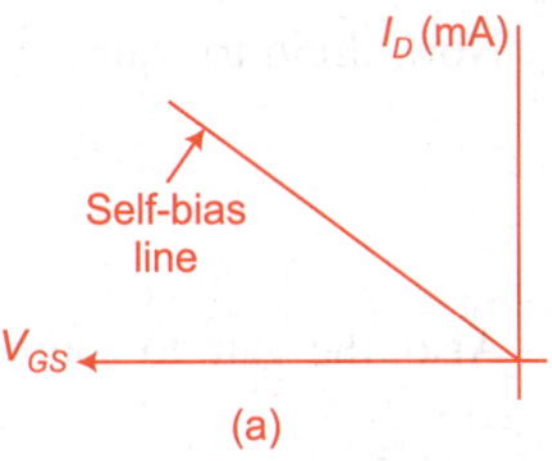

Fig. 7.23 (a)

Example 7.9.2 Determine the values of drain to source voltage V_{DS} and the gate to source voltage V_{GS} for the circuit shown in Fig. Ex. 7.9.2. Give that $I_D = 6$ mA.

Solution Given that,

$$I_D = 6 \text{ mA}, \quad R_D = 1 \text{ k}\Omega$$
$$R_S = 550 \text{ }\Omega, \quad V_{DD} = 15 \text{ V}.$$

We know that,

$$V_S = I_D R_S$$
$$V_S = 6 \times 10^{-3} \times 550$$
$$V_S = 3300 \times 10^{-3} \text{ V}$$
$$V_S = 3.3 \text{ volt}.$$

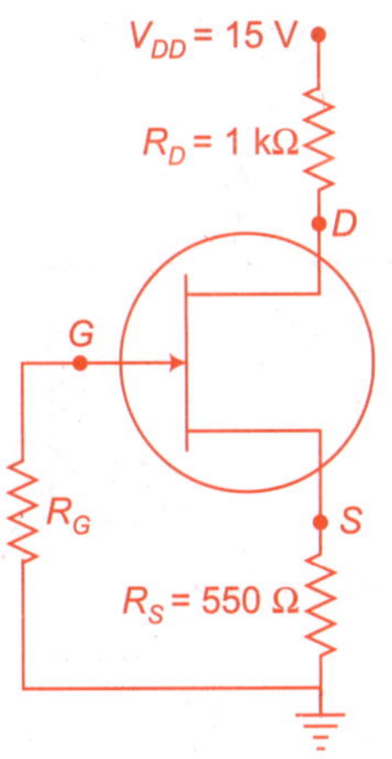

Fig. Ex. 7.9.2

Drain voltage is given by,

$$V_{DD} = V_D + I_D R_D \Rightarrow V_D = V_{DD} - I_D R_D$$

$$V_D = 15 - 6 \times 10^{-3} \times 1 \times 10^3$$

$$V_D = 15 - 6$$

$$V_D = 9 \text{ volt.}$$

Now, drain to source voltage,

$$V_{DS} = V_D - V_S = 9 \text{ V} - 3.3 \text{ V}$$

$$\boxed{V_{DS} = 5.7 \text{ volts.}}$$ (Ans.)

Also, the gate to source voltage is given by,

$$V_{GS} = V_G - V_S = 0 - V_S$$

$$V_{GS} = -V_S$$

$$\boxed{V_{GS} = -3.3 \text{ mt.}}$$ (Ans.)

7.9.3 Voltage Divider Bias

The potential divider bias circuit is shown in Fig. 7.24. In this type of circuit, three resistance R_1, R_2 and R_S are used. R_D is a drain resistance. The resistor R_1 and R_2 are connected on the gate side between V_{DD} supply and ground. Assuming the gate current to be zero i.e., $I_G = 0$, the voltage V_G is given by the voltage dividing rule,

$$V_G = \left(\frac{R_2}{R_1 + R_2} \right) V_{DD} \tag{7.23}$$

where, V_G is the voltage across R_2 and is calculated by the voltage dividing rule.

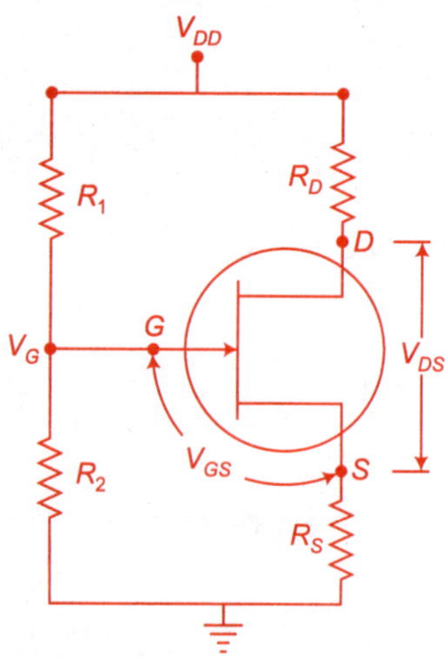

Fig. 7.24 *Voltage divider bias circuit*

The Thevenin resistance is given by the Eqn. (6.26)

$$R_{Th} = R_1 \parallel R_2 = \frac{R_1 R_2}{R_1 + R_2} \tag{7.24}$$

Now, the Thevenin equivalent circuit of Fig. 7.23 is shown in Fig. 7.23 (a). Now, applying KVL at input side in Fig. 7.23 (a), we obtain,

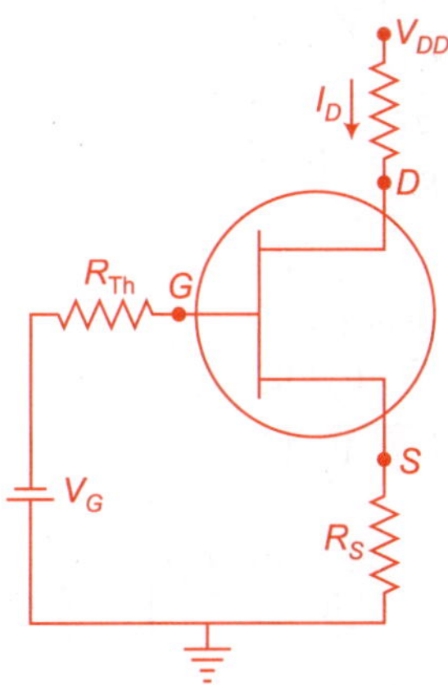

Fig. 7.24 *(a) simplified circuit*

$$V_G - I_G R_{Th} - V_{GS} - I_D R_S = 0$$

$$\therefore \quad I_G = 0$$

$$V_G - (0 \times R_{Th}) - V_{GS} - I_D R_S = 0$$

$$I_D R_S = V_G - V_{GS}$$

$$I_D = \frac{V_G - V_{GS}}{R_S}. \tag{7.25}$$

DC voltage from drain to ground is given as:

$$V_D = V_{DD} - I_D R_D \tag{7.26}$$

Now, if $V_G >>> V_{GS}$, then the drain current is approximately constant for any JFET. But it is found that the potential divider bias is less effective with JFET than BJT. It is because of the fact that in BJT the base to emitter voltage V_{BE} is approximately 0.7 V with only minor variation from one transistor to another. However, in a JFET, the gate to source voltage (V_{GS}) can vary several volts from one JFET to another. Due to this, it is difficult to make gate voltage (V_G) larger than the gate to source voltage (V_{GS}).

Example 7.9.3 A JFET amplifier with stabilized biasing circuit shown in Fig. 7.9.3, has following parameters:

$$V_p = -2V, \quad I_{DSS} = 5mA, \quad R_L = 910\ \Omega, \quad R_S = 2.29\ k\Omega$$

$$R_1 = 12\ M\Omega \quad R_2 = 8.57\ M\Omega \quad \text{and} \quad V_{DD} = 24V.$$

Determine the value of drain current I_D at the operating point also, verify that FET will operate in pinch-off region.

Solution Now, we know that gate voltage is given by,

$$V_G = \left(\frac{R_2}{R_1 + R_2}\right) V_{DD}$$

$$V_G = \left(\frac{8.57}{12 + 8.57}\right) \times 24$$

$$V_G = 10 \text{ volt.}$$

The drain current I_D, is given by,

$$I_D = I_{DSS} \left[1 - \frac{V_{GS}}{V_p} \right]^2$$

$$I_D = 5 \times 10^{-3} \left[1 - \frac{(10 - 2.29 \times 10^3\ I_D)}{-2} \right]^2 \qquad \because V_{GS} = V_G - I_D R_S$$

$$I_D = 5 \times 10^{-3} \left[1 + 5 - 1.145\ I_D \right]^2$$

$$I_D = 0.005 \left[6 - 1.145\ I_D \right]^2$$

By solving above equations, we get,

$$\boxed{I_D = 4.45 \text{ mA.}}$$

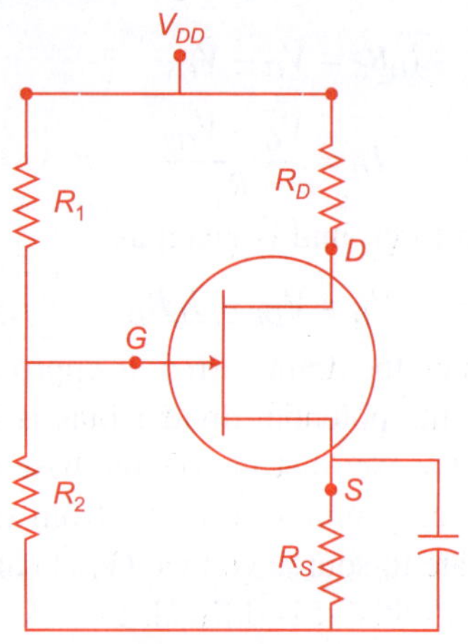

Fig. Ex. 7.9.3

Gate source voltage V_{GS} is given by,

$$V_{GS} = V_G - V_S = V_G - I_D R_S \qquad \because \quad V_S = I_D R_S$$

$$V_{GS} = 10 - (2.29 \times 10^3 \times 4.45 \times 10^{-3})$$

$$\boxed{V_{GS} = -0.213.}$$

∴

Current at operation point is given by Eqn. (7.25), we obtain,

$$I_D = \frac{V_G - V_{GS}}{R_S} = \frac{10 - 0.213}{2.29 \times 10^3}$$

$$\boxed{I_D = 4.26 \text{ mA.}}$$ (**Ans.**)

It may be noted that the value of I_D at operating point is almost equal to previously computed value of I_D. So, we can say that the JFET is operational in pinch-off region.

7.10 SETTING OF A *Q*-POINT

As we have discussed in the previous chapter that the operating point is defined, when zero signal is applied at transistor input i.e., there is no input given and only DC biasing is used. The value of drain current I_D for a desired value of source voltage V_{GS} or vice versa, gives one point on the load line. This point is known as Operating point or *Q*-point. However, if JFET includes transfer characteristics curve, then the procedure for determining the *Q*-point is as follows:

(i) Firstly, select a convenient value of drain current I_D. Generally, its value is taken half of the I_{DSS}

i.e. $$I_D = \frac{I_{DSS}}{2} \qquad (7.27)$$

where, I_{DSS} = Maximum possible value of drain current.

(ii) Find the voltage drop across source resistor R_S

$$V_S = I_D R_S \qquad (7.28)$$

and calculate gate-to-source voltage from the expression,

$$V_{GS} = -V_S. \qquad (7.29)$$

(iii) Now, plot the drain current I_D and the corresponding gate-to-source voltage V_{GS} on the transfer characteristic curve as shown in Fig. 7.25 (a)

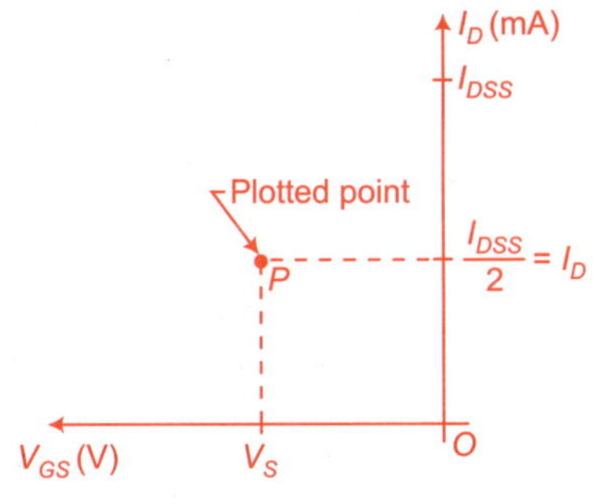

Fig. 7.25 (a)

(iv) Draw a line through the plotted point 'P' and the origin 'O' as shown in Fig. 7.25(b).

(v) Now, the point of intersection of the self bias line and the curve is given at the desired Q-point Read the coordinate of Q-point as shown in Fig. 7.25(c).

It is required to set Q-point near the mid-point of Q-point near the mid-point of the transfer characteristic curve of a JFET. The mid-point bias allows a maximum amount of drain current swing between the values of I_{DSS} and O.

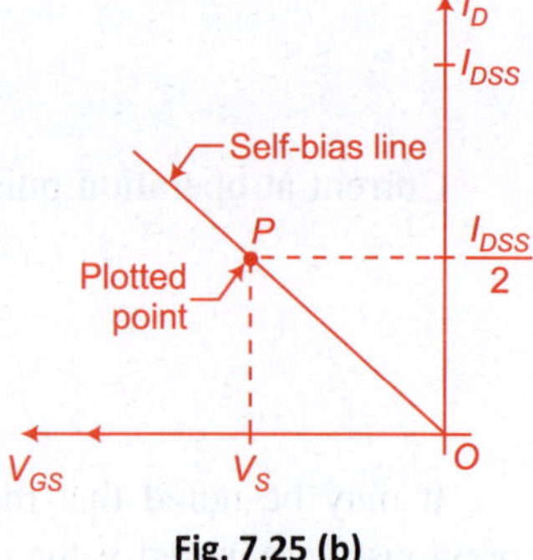

Fig. 7.25 (b)

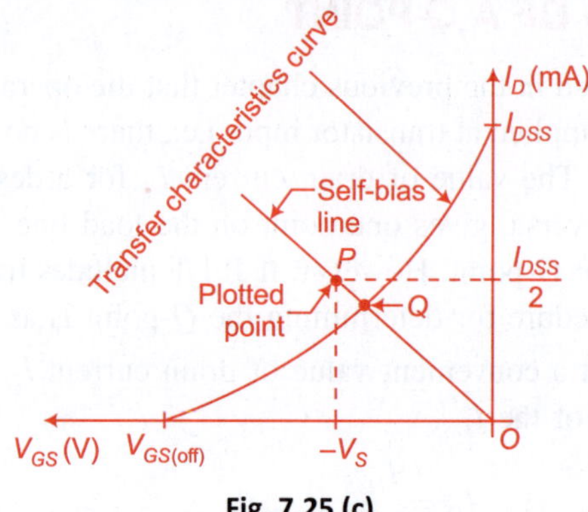

Fig. 7.25 (c)

Example 7.10.1   Figure Ex. 7.10.1 (a) shows that circuit of a self biased JFET amplifier and Fig. Ex. 7.10.1 (b) shows the transfer characteristics curve of the JFET.

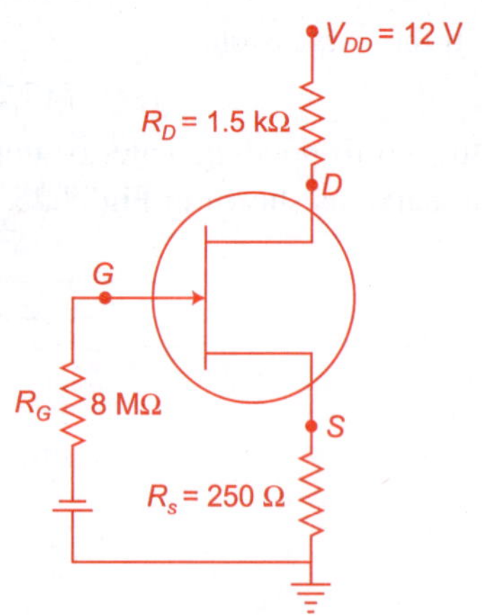

Fig. Ex. 7.10.1 (a)

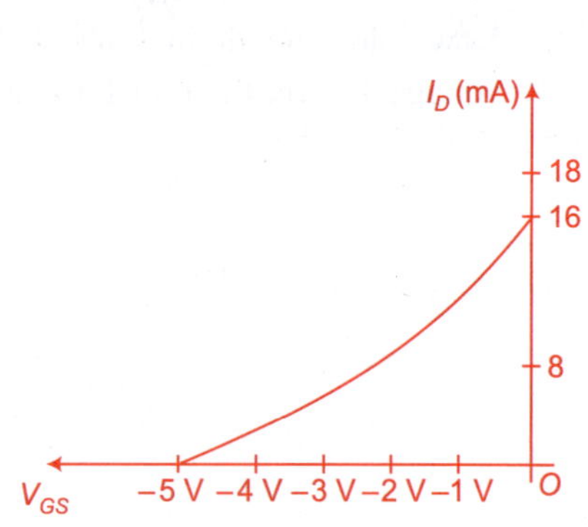

Fig. Ex. 7.10.1 (b)

Determine the Q-point values of I_D and V_{GS}. Compute the values of DC voltage between drain and the ground V_D.

Solution   Given that,

$$R_D = 1.5 \text{ k}\Omega, \quad R_S = 250\Omega$$

$$R_G = 8 \text{ M}\Omega, \quad I_{DSS} = 16 \text{ mA.}$$

(i) Convenient value of drain current I_D

$$I_D = \frac{I_{DSS}}{2} = \frac{16\text{mA}}{2} = 8 \text{ mA}$$

$$\boxed{I_D = 8 \text{ mA.}}$$

(ii) Voltage drop across source resistor R_S

$$V_S = I_D R_S = 8 \times 10^{-3} \times 250 \ \Omega$$

$$\boxed{V_S = 2 \text{ volt.}}$$

And gate to source voltage V_{GS}

$$V_{GS} = -V_S$$

$$\boxed{V_{GS} = -2.}$$

(iii) Now, Plot the values of drain current I_D and the corresponding value of voltage V_{GS} on the graph of transfer characteristics curve as shown in Fig. Ex. 7.10.1(c).

(iv) Now, Draw a line through the plotted point P and the origin O as shown in Fig. Ex. 7.10.1(c).

(v) Now, the point of intersection of the self-bias line and the curve gives at the desired Q-point and the coordinate of Q-point is as follows:

and $\boxed{\begin{array}{c} I_D = 6.8 \text{ mA} \\ V_{GS} = -1.75 \text{ volt.} \end{array}}$ **(Ans.)**

Fig. Ex. 7.10.1 (c)

Now, we know that the drain voltage is given by,

$$V_D = V_{DD} - I_D R_D$$

$$V_D = 12 \text{ V} - (6.8 \times 10^{-3}) \times (1.5 \times 10^{+3})$$

$$V_D = 12\text{V} - 10.2$$

$$\boxed{V_D = 1.8 \text{ volt.}}$$
(Ans.)

7.11 DC LOAD LINE OF JFET

As we have discussed in previous chapter about DC load line and its importance. The DC load line can be drawn easily on the drain characteristics as shown in Fig. 7.26 (a). The values of drain current (I_D) and drain to source voltage (V_{DS}) at the upper and lower ends of the DC load can be obtained by using the expression,

$$V_{DS} = V_{DD} - I_D (R_D + R_S) \tag{7.30}$$

As we can see in Fig. 7.26 (a) the value of drain to source voltage at point A is zero. So, substituting $V_{GS} = 0$ in Eqn. (7.30), we obtain,

$$0 = V_{DD} - I_D (R_D + R_S)$$

$$I_D (R_D + R_S) = V_{DD}$$

$$I_D = \frac{V_{DD}}{R_D + R_S} \tag{7.31}$$

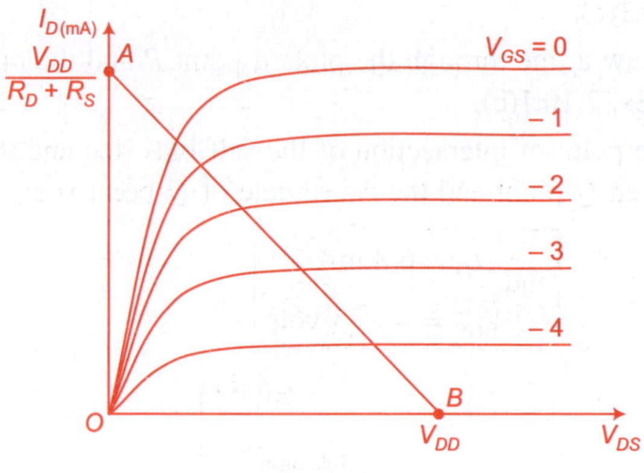

Fig. 7.26 (a)

Similarly, at point B, the value of drain current is zero, so substituting $I_D = 0$ in Eqn. (7.30), we obtain,

$$V_{DS} = V_{DD} - 0 \times (R_S + R_D)$$

$$V_{DS} = V_{DD} \tag{7.32}$$

Hence, point A and B of load lines are given by,

$$I_D = \frac{V_{DD}}{R_D + R_S} \text{ and } V_{DS} = 0 \text{ (point } A)$$

$$V_{DS} = V_{DD} \text{ and } I_D = 0 \text{ (point } B).$$

The Q-point is set at mid-point of DC load line as shown in Fig. 7.26 (b). In this case, the value of drain current at Q-point will be given by,

$$I_{DQ} = \frac{V_{DD}}{2(R_D + R_S)} \tag{7.33}$$

the drain to source voltage at Q-point is given by,

$$V_{DSQ} = \frac{V_{DD}}{2}. \tag{7.34}$$

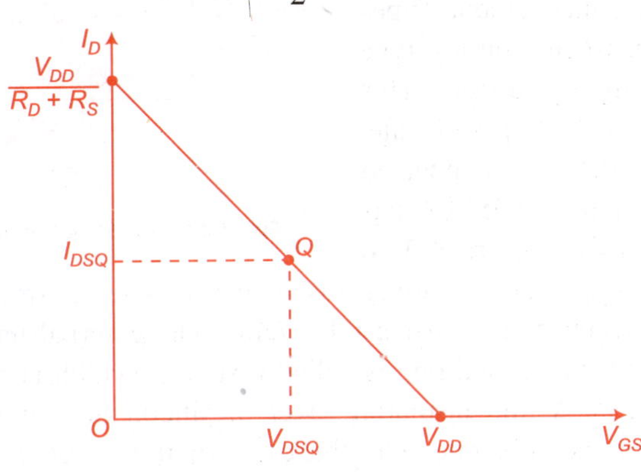

Fig. 7.26 (b)

7.12 METALOXIDE SEMICONDUCTOR FIELD EFFECT TRANSISTOR (MOSFET)

MOSFET is a three terminal device like FET and BJT. Current conduction in such device is only due to one carrier like FET and unlike BJT. Thus, it is a unipolar device. The type of carrier involved in current conduction may be electron or hole depending upon the type of device. Current conduction in MOSFET depends on the control of majority carriers, which are available in a channel, by an applied electric field. Thus, the device behaves as a voltage controlled current source.

The MOSFET is most widely used device in the fabrication of large scale integrated circuits.

Basically, MOSFETs are of two types like a JFET.

(i) Enhancement type MOSFET (E-MOSFET)

(ii) Depletion type MOSFET (D-MOSFET)

Depending on the types of majority carrier present in the channel, a MOSFET is further identified as a *n*-channel or *p*-channel device. We consider the construction and the working of each types in the following sections.

7.12.1   *n*-channel Enhancement Types MOSFET (E-NMOSFET)

The enhancement types MOSFET is widely used device in the fabrication of Large Scale Integration (LSI) circuits.

The enhancement types MOSFET is shown in Fig. 7.27 (a) which is *n*-channel MOSFET, abbreviation E–NMOSFET.

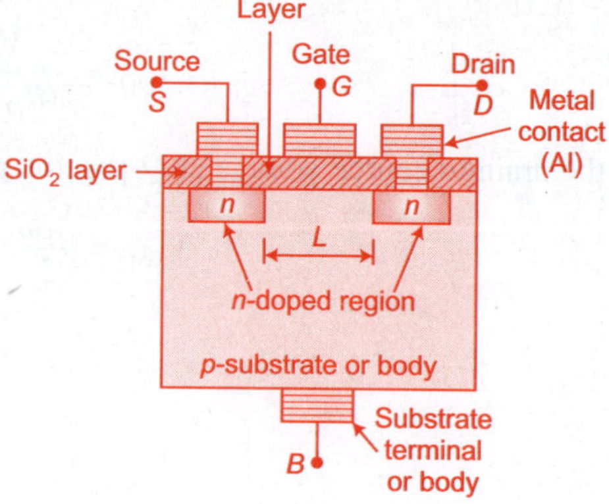

Fig. 7.27   *(a) Struction of E-NMOSFET*

The *n*-channel enhancement types MOSFET is constructed on a *p*-type slab which is lightly doped. This *p*-type slab is called *p*-substrate. Generally, substrate is connected to a source but some MOSFET has separate terminal for substrate. Two heavily doped *n*-type regions are diffused in a *p*-type substrate or body. The doping level concentration is at least 10^{18}/cm³. The external terminal of this is called a drain and the second one is called a source, which is brought out with metal (*Al*) contacts. A thin insulating layer of silicon dioxide (SiO₂) is grown on the surface of the substrate. The thickness of this layer is about 40 nm to 100 nm. Two windows are open on *n*-region for getting metal contact of source and drain. Now, a layer of metal (*Al*) is deposited on the oxide insulator and the third terminal between drain and source is called the gate (*G*). It is brought out formation of the transistor with the metal (*Al*) gate, the insulating oxide layer (SiO₂) and the semiconductor below the oxide layer gives rise to metal oxide semiconductor field effect transistor. Two important points are derived from following construction.

(i)   There is no direct electrical connection between the gate terminal and substrate because there is a SiO₂ layer between them.

(ii)   The input impedance of MOSFET is high due to SiO₂ layer.

7.12.1.2   *Working of Enhancement Type NMOSFET*

When the gate terminal is open i.e., zero bias condition, the body, source and drain forms two *p-n* junction connected back to back. The device is normally off, with negligible current, the gate is normally off. The gate is insulated from source and drain by oxide layer. The resistance between the gate and drain and gate and source is of order 10^{10} Ω to 10^{15} Ω. Due to this reason, the MOSFET is sometimes called as the Insulated Gate Field Effect Transistor (IGFET).

(i) If $V_{GS} = 0$ and the voltage between drain and source is positive i.e., $V_{DS} > 0$, the drain current in NMOSFET is zero due to the absence of conducting channel (*n*-channel). Whereas the current flows in JFET due to the presence of channel.

i.e. $V_{GS} = 0$ and $V_{DS} > 0$

then $I_D = 0.$

(ii) Connecting body and source together and applying a positive gate voltage establishes an electric field between the gate and body i.e., $V_{GS} > 0$ (positive). This field repels the majority holes towards the *p*-substrate under the gate region and causes a depletion region. There are lack of holes near SiO_2 layer. The minority carriers (electrons) of *p*-substrate will attract towards the gate and accumulate beneath SiO_2 layer as the electrons cannot cross the layer. As the gate voltage V_{GS} *is* increased, the increasing field attracts electrons from two n^+ region of the source and drain into the region directly below the oxide layer, i.e., the concentration of electron will increase with increasing the gate voltage, for sufficient number of electrons to accumulate under the gate terminal to form a conducting channel between the source and the drain.

The minimum positive gate voltage required for the formation of conducting path in the NMOSFET is called the Threshold voltage V_T. The induced channel has a length L (below 1μm to 6 μm) as shown in Fig. 7.27 (a). Width of the channel is up to 500 μm.

(iii) If $V_{DS} > 0$ i.e., then drain is kept at slightly positive voltage relative to the source, electron from the source end reaches the drain causing positive current flow from drain to source.

If gate voltage is increased above the threshold voltage V_T, the number of induced electrons in the channel increases and the channel becomes deeper as shown in Fig. 7.27 (b).

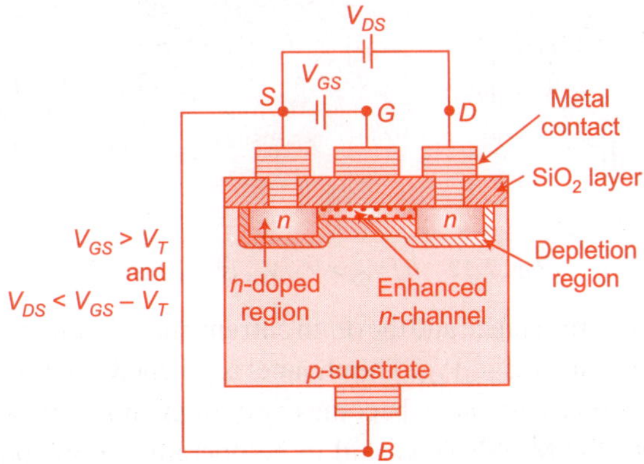

Fig. 7.27   *(b) n-channel formation*

i.e. $$V_{GS} > V_T$$

and, $$V_{DS} < V_{GS} - V_T$$

where, V_T = Threshold voltage.

The channel and drain current, are thus enhanced for positive gate voltage. The device, is therefore called an Enhancement-Type MOSFET, where an increase in the gate voltage cause an innercase is the drain current for small drain voltage, is called Ohmic region, the voltage controlled resistance region or linear region.

(iv) If drain voltage increases while the gate voltage is kept constant at or above threshold value V_T, then the voltage between gate and body near the drain decreases

i.e. $$V_{GD} = V_{GS} - V_{DS} \tag{7.35}$$

where, V_{GD} = Voltage between gate and drain

V_{GS} = Voltage between gate and source

V_{DS} = Voltage between drain and source.

From Eqn. (7.35), we can say that when V_{DS} is increased at constant V_{GS}, then the voltage between gate and drain will reduce. Due to this, the conducting channel becomes shallow near the drain-channel junction as shown in Fig. 7.27 (c). If there is further increase in the drain voltage V_{DS}, it causes a shrink in the channel near the drain. Since the gate to source voltage V_{GS} is kept constant, the channel depth does not change near the source end.

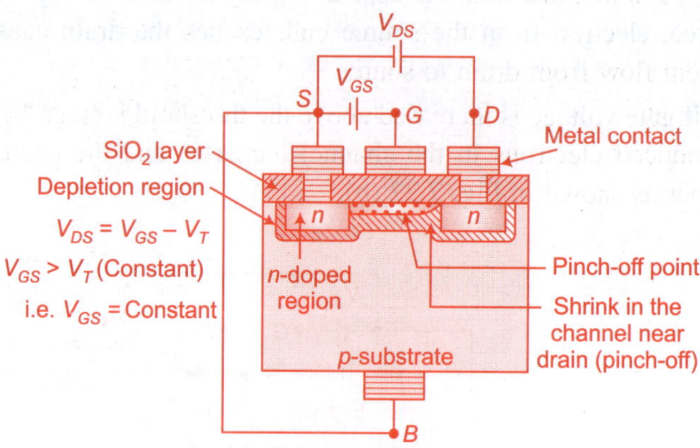

Fig. 7.27 *(c) $V_{GS} > V_T$, V_{DS} is at pinch-off*

The channel resistance and the drain current, therefore increases non-linearly with the drain voltage V_{DS}. The channel is pinched-off near the drain and the drain to source resistance becomes very high about 300 kΩ to 500 kΩ. At this point, the MOSFET is said to be operating in the pinch-off mode.

Channel pinch-off occurs when,

gate to drain voltage, $V_{GD} = V_{GS} - V_{DS}$

$$\because \quad V_{GS} - V_{DS} < V_T$$

$$\therefore \quad V_{GD} < V_T.$$

The voltage V_{GD} at the drain end of the channel is below the threshold voltage V_T which is needed to form the channel.

(v) When the drain voltage increases above pinch-off

i.e. $\quad V_{DS} > V_{GS} - V_T.$

The depletion region near the drain end extends towards the sources due to which the channel length (L) is reduced as shown in Fig. 7.27 (d). But the number of electrons arriving from the source and the drain current, remains the same. Electrons leaving the source are accelerated towards the drain while their number remains the same. The drain current saturates at the values reached, at the pinch-off.

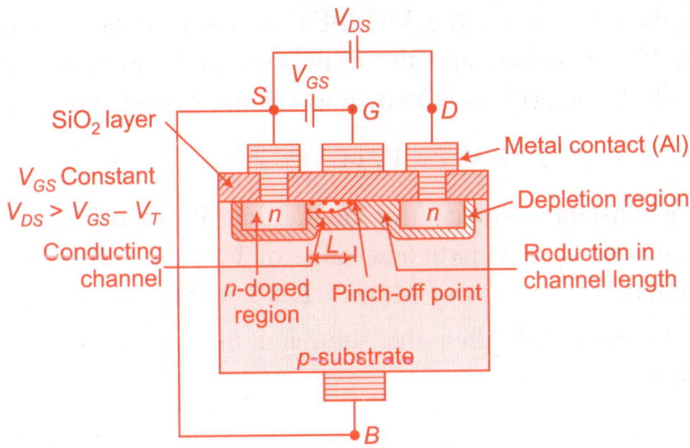

Fig. 7.27   *(d) $V_{GS} > V_T$, V_{DS} is above pinch-off*

In this region, the drain current is constant for a given drain voltage, so this region is called the saturation region or pinch-off region.

(vi) Drain voltage further increases about 20 V. The depletion region near the drain extends completely to the source. Now, the channel becomes narrow and gate looses its control on the drain current. So, the drain current increases rapidly. This breakdown is called punch-through of MOSFET. When drain voltage increases about 50 V to 100 V, the large reverse bias causes avalanche breakdown of the *pn*-junction. Due to the breakdown, holes from the substrate form a drift component of current, leading to a rapid increase in the drain current. It is severe when the gate voltage exceeds the insulating oxide breakdown voltage wherein MOSFET will be damaged.

7.12.1.3 *Channel length modulation*

The drain current increases with increase in drain voltage V_{DS} after pinch-off. When V_{DS} increases above $V_{GS} - V_T$, the depletion region extends near the drain and pinch-off point extends towards the source. Due to this, the effective channel length will reduce. The reduction in channel length is known as Channel Length Modulation as shown in Fig. 7.27 (d).

7.12.1.4 *Characteristics of NMOSFET*

As we saw in the earlier section that the induced channel in E-NMOSFET is not formed until the gate to source voltage V_{GS} reaches the threshold voltage V_T.

The characteristics of drain current I_D and drain to source voltage V_{DS} is shown in Fig. 7.28. From the above discussion, we can say that *n*-channel MOSFET characteristic can be mentioned as following:

(i)   If gate to source voltage V_{GS} is less than the threshold voltage V_T, then the drain current I_D does not flow in MOSFET.

(ii) For a given $V_{GS} > V_T$, the MOSFET begins to conduct when the drain voltage V_{DS} increases and the drain current I_D increases linearly. This region of characteristics is known as Ohmic region and is defined by:

$$V_{DS} \leq V_{GS} - V_T \text{ or } V_{GD} \geq V_T.$$

(iii) If V_{GS} is constant at value V_T or greater than V_T, then I_{DS} is increased with increasing the V_{DS}. A particular value of V_{DS}, the values of drain current is saturated. This region of MOSFET is known as saturated region.

(iv) If V_{GS} is increased, then the saturated magnitude of drain current is increased.

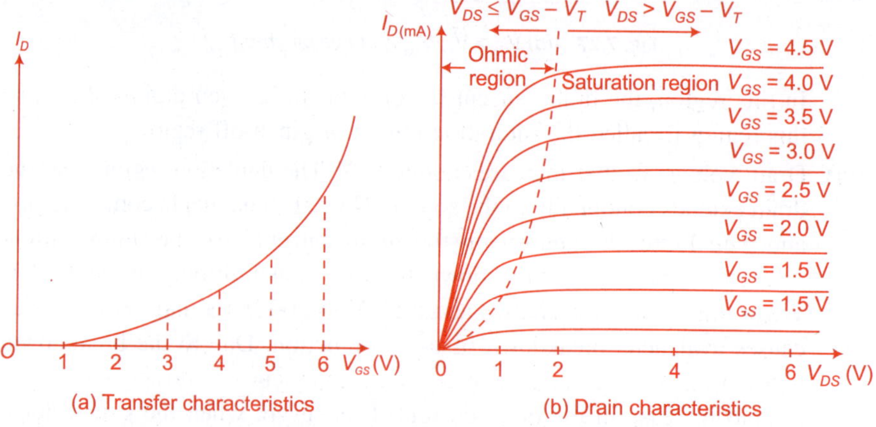

Fig. 7.28 *V_I characteristics of E-NMOSFET*

7.12.1.5   *Mathematical expression for drain current*

The drain current in Ohmic region is given by,

$$I_D = \frac{\text{Average induced charge}}{\text{Average transit time}} = \frac{Q_c}{\tau}. \qquad (7.36)$$

Since I_D is the drift current due to the field from source to drain,

$$\text{Transit time } \tau = \frac{\text{Channel length}}{\text{Drift velocity}} = \frac{L}{v_d} \qquad (7.37)$$

where the drift velocity is given by,

$$V_d = \mu E.$$

For *n*-channel,

$$V_d = \mu_n E = \mu_n \frac{V_{DS}}{L} \qquad (7.38)$$

$$\therefore \quad E = \frac{dV}{dx}$$

where, μ_n = Mobility of electrons

E = Uniform electric field along the length of the channel length L.

Now, substituting the value of V_d from Eqn. (7.38) in Eqn. (7.37), we obtain,

$$\tau = \frac{L}{\dfrac{\mu_n V_{DS}}{L}} = \frac{L^2}{\mu_n V_{DS}} \qquad (7.39)$$

Now, substituting the value of τ in Eqn. (7.36), we obtain,

$$I_D = \frac{Q_c}{L^2/\mu_n V_{DS}} = \frac{Q_c \mu_n V_{DS}}{L^2}$$

$$I_D = \frac{Q_C \mu_n V_{DS}}{L^2}. \qquad (7.40)$$

The gate and body of MOSFET acts as parallel plates of capacitor and the size acts as dielectric. Hence, the change Q_C stored in the channel is given by,

$$Q_c = V_G C_g = V_G \left(\frac{A\epsilon}{t_{ox}}\right) = V_G WL \left(\frac{\epsilon}{t_{ox}}\right)$$

$$Q_c = V_G WL \left(\frac{\epsilon}{t_{ox}}\right) \qquad (7.41)$$

$$\therefore \quad A = WL$$

where, C_g = Gate to channel capacitance

A = Area of the plate

t_{ox} = Thickness of oxide

$\in$ = Permitivity of gate oxide

V_G = Gate to channel voltage or plate voltage.

We know that channel is formed when,

$$V_{GS} > V_T.$$

Hence, plate voltage $V_G = V_{GS} - V_T$ for $V_{DS} = 0$.

The source to channel voltage increases along the length from 0, at the source to V_{DS} at the drain. As a result, V_G varies from $V_{GS} - V_T$ to $V_{GS} - V_T - V_{DS}$ along the length of the plate.

We assume that the channel voltage increases linearly, the average, or effective, plate voltage becomes,

$$V_G = V_{GS} - V_T - \frac{V_{DS}}{2}.$$

Putting the value of V_G in Eqn. (7.41), we obtain,

$$Q_C = \left(V_{GS} - V_T - \frac{V_{DS}}{2}\right) WL \left(\frac{\in}{t_{ox}}\right) \tag{7.42}$$

Now, putting the value Q_C in Eqn. (7.40), we obtain,

$$I_D = \left(V_{GS} - V_T - \frac{V_{DS}}{2}\right) W_L \left(\frac{\in}{t_{ox}}\right) \frac{\mu_n V_{DS}}{L^2}$$

$$I_D = \left(\frac{1}{2}\right) \left(\frac{W}{L}\right) \left(\frac{\mu_n \in}{t_{ox}}\right) [2(V_{GS} - V_T) V_{DS} - V_{DS}^2]$$

$$I_D = \left(\frac{1}{2}\right) \left(\frac{W}{L}\right) \mu_n C_{ox} [2 (V_{GS} - V_T) V_{DS} - V_{DS}^2]$$

where, $C_{ox} = \dfrac{\in}{t_{ox}}$ is gate capacitance per unit area

$$\boxed{I_D = k \, [2(V_{GS} - V_T) V_{DS} - V_{DS}^2]} \tag{7.43}$$

$$V_{DS} \leq V_{GS} - V_T \text{ and } V_{GS} > V_T$$

where, $k = \dfrac{1}{2} \mu_n c_{ox} \left(\dfrac{W}{L}\right)$ Amp/Volt^2

= Conduction parameter.

In the above equation for a given fabrication process, the oxide thickness is constant. Hence, conduction parameter k depends on width or to length or aspect ratio $\dfrac{W}{L}$.

Hence,
$$k = \left(\frac{1}{2}\right) k' \left(\frac{W}{L}\right)$$

where,
$$k' = \mu_n c_{ox}$$

For saturation region or pinch-off region,

$$V_{DS} \geq V_{GS} - V_T \quad \text{and} \quad V_{GS} > V_T.$$

Now, substituting the value V_{DS} in Eqn. (7.43) we obtain

$$I_D = k \left[2(V_G - V_T)(V_{GS} - V_T) - (V_{GS} - V_T)^2\right]$$

$$I_D = k \left[2(V_{GS} - V_T)^2 - (V_{GS} - V_T)^2\right]$$

$$\boxed{I_D = k (V_{GS} - V_T)^2.} \tag{7.44}$$

The Eqn. (7.44) is for saturation drain current in the pinch-off region.

The effect of channel length modulation, i.e., is the reduced effective channel length in pinch-off region.

The drain current is given by,

$$I_D = k (V_{GS} - V_T)^2 (1 + \lambda V_{DS}) \tag{7.45}$$

where, λ = Channel length modulation factor.

7.12.1.6   *Circuit symbol for enhancement types NMOSFET*

Figure 7.29(a) shows the circuit symbol for *n*-channel enhancement types MOSFET when body terminal is given separately and Fig. 7.29 (b) shows the circuit symbol for *n*-channel enhancement types MOSFET for which body is connected to the source or most negative potential terminal. In this figure, the broken line indicates that there is no conducting channel between drain and source when $V_{GS} = 0$. Due to this, the device is also known as "normally off MOSFET". The arrow points are the direction of channel.

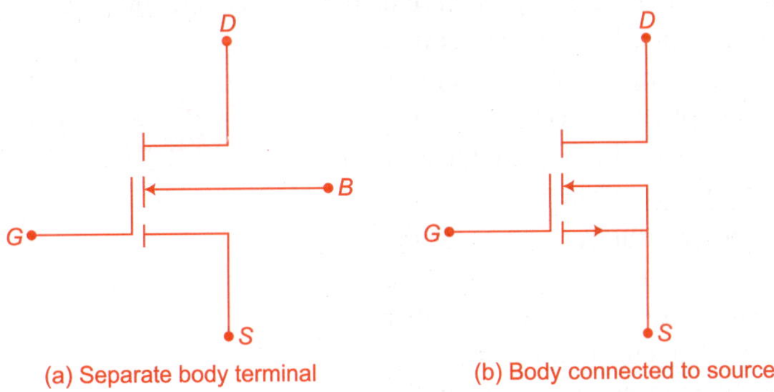

(a) Separate body terminal   (b) Body connected to source

Fig. 7.29   *Circuit symbols for n-channel enhancement type MOSFET*

7.12.2 *n*-channel Depletion Type MOSFET (D-N MOSFET)

The depletion types MOSFET is shown in Fig. 7.30, which is a *n*-channel MOSFET abbreviated D-NMOSFET.

7.12.2.1 Construction

The construction of depletion types MOSFET is shown in Fig. 7.30. The structure of depletion type *n*`-channel MOSFT is similar to that of an enhancement type *N*-MOSFET. The only difference is that it has a lightly doped *n*-channel implanted between the source and the drain at the time of fabrication.

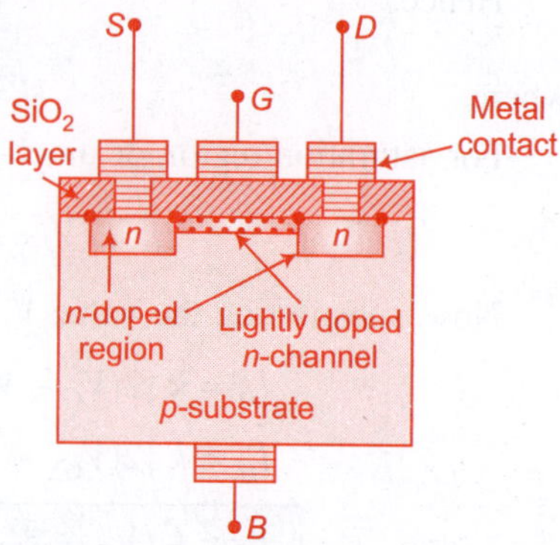

Fig. 7.30   *n-channel depletion type MOSFET (D-NMOSFET)*

7.12.2.2 Working of depletion type NMOSFET

It is clear from the structure of depletion types NMOSFET that *n*-types channel is already formed between source and drain.

(i) If gate and source terminals are short circuited ($V_{GS} = 0$) and positive voltage is applied between drain source, then the electrons attract towards the drain. Similarly, the JFET i.e., the electron are flown from source to drain, hence drain current flows from drain to source. The maximum value of drain current I_D at $V_{GS} = 0$ and is called saturated drain current I_{DSS}. The current conduction in depletion type MOSFET is shown in Fig. 7.31.

(ii) If $V_{GS} < 0$ i.e., gate voltage is negative, then this negative will induce positive charge in channel. In other words, we can say that this negative voltage will repel the electrons of channel towards the *p*-substrate while attracting the holes from *p*-substrate towards the channel. Due to this, there is a recombination of electrons which are moving towards the substrate and holes which comes from substrate. Hence, there is a lack of free electrons in the channel and the conductivity of *n*-channel is reduced.

When we further increase the negative gate voltage, then the conductivity of channel is further reduced and the drain current will also be reduced. At one point, the drain current will be zero.

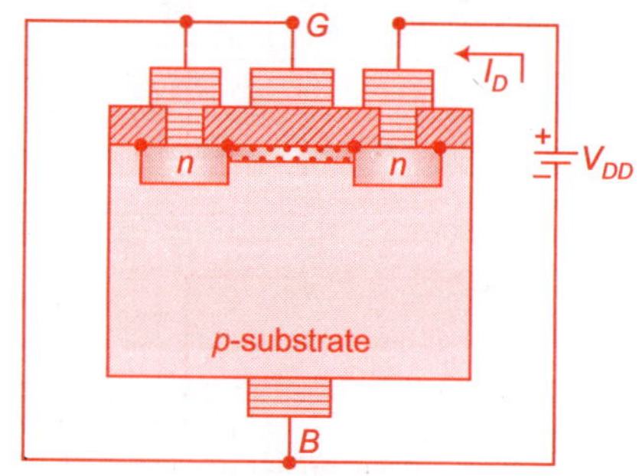

Fig. 7.31   *Current condition in n-channel D-MOSFET*

(iii) If positive voltage is applied on gate i.e., $V_{GS} > 0$. Due to this, the number of free electrons in the channel will increase. Therefore, the conductivity of the channel will increase and hence the drain current increases rapidly.

7.12.2.3 *Characteristics of depletion type NMOSFET*

The drain and transfer characteristics of depletion type NMOSFET shown are Figs. 7.32 (a) and (b) respectively. From Fig. 7.32, it is clear that,

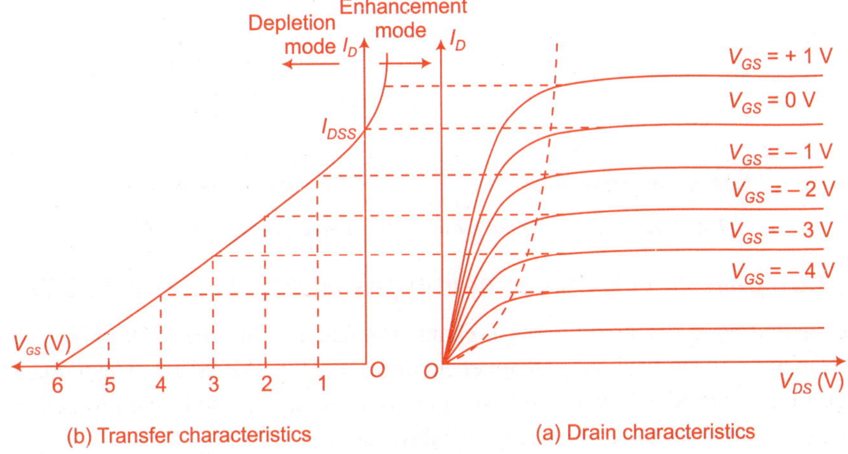

Fig. 7.32 *n-channel depletion type MOSFET characteristics*

(i) If $V_{GS} = 0$ and V_{DS} increased from zero, then the drain current increases initially and at pinch-off voltage V_p, the drain current will be at its maximum value I_{DSS}. After pinch-off if V_{DS} is further increased, the drain current will be constant.

(ii) If $V_{GS} < 0$ i.e., negative, then the conductance of the channel will reduce and due to this the channel will also decrease. The channel will be pinch-off at low value and the pinch-off value at $V_{GS} = 0$. If the gate voltage further increases in negative direction, then at one value of V_{GS} the drain current will be zero.

(iii) If $V_{GS} > 0$ i.e., positive and V_{DS} is increased from zero, then the drain current increases rapidly.

The Shockley equation which has been used for JFET can also be used for depletion types MOSFET.

7.12.2.4 *Circuit symbol for depletion type NMOSFET*

Figure 7.33 (a) shows the circuit symbol for *n*-channel depletion type MOSFET when body terminal is given separately and Fig. 7.33 (b) shows the circuit symbol for *n*-channel depletion type MOSFET for which body connected to the source is most negative potential terminal. In the figure the continuous line of channel indicates that availability of conduction channel between drain and source when $V_{GS} = 0$. Due to this, the device is also known as "normally on MOSFET".

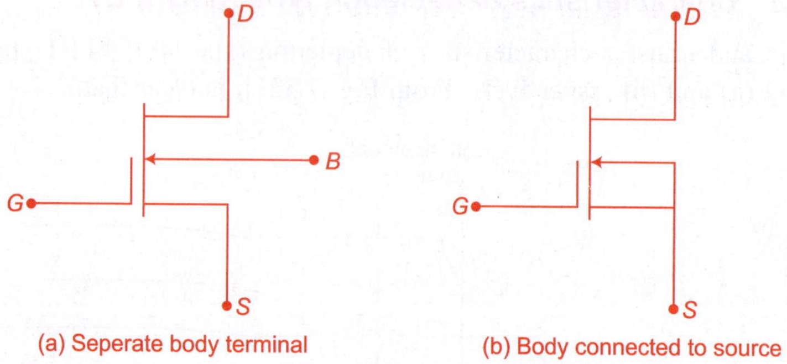

(a) Seperate body terminal   (b) Body connected to source

Fig. 7.33   *Circuit symbol for n-channel depletion type MOSFET*

7.12.3   *p*-channel Enhancement-types MOSFET (E-PMOSFET)

We have already discussed about *n*-channel enhancement type MOSFET. In this section, we will study about *p*-channel enhancement type MOSFET. The *p*-channel MOSFET (PMOSFET) was the first to be used in large scale integration circuits because of their relative simplicity in fabrication.

7.12.3.1   *Construction*

The *p*-channel enhancement MOSFET is similar in construction as *n*-channel enhancement MOSFET. The difference is that, the polarity of voltage is opposite as that of *n*-channel enhancement MOSFET. The *p*-channel enhancement types MOSFET is shown in Fig. 7.33. The *p*-channel enhancement types MOSFET is constructed on a *n*-type slab. This *n*-type slab is called *n*-substrate. Two heavily doped *p*-type regions are diffused in a *n*-type substrate, or body. One external terminal of this is called as drain and second one is called as a source which was brought out with metal (*Al*) contacts.

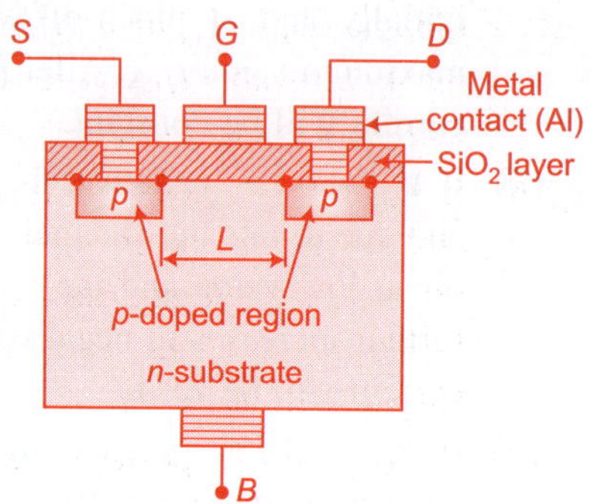

Fig. 7.33   *Structure of E-PMOSFET*

7.12.3.2   *Working of Enhancement Type PMOSFET*

To form an induced channel of holes in the region between the drain and the source, the gate must be at lower potential than the source. Hence, the threshold voltage V_T is negative for *p*-channel MOSFET. The drain must be at a lower voltage than the source so that it can draw holes from the channel to conduct drain current. The channel is pinched-off at the drain end.

$$\text{when,} \quad V_{GD} \geq V_T$$

$$\text{or,} \quad V_{SD} \geq V_{SG} + V_T \tag{7.46}$$

Then the channel is pinched-off and the drain current is in saturation.

Condition for pinch-off is,

$$V_{SD} \geq V_{SG} - |V_T|$$

$$\because \qquad V_T \text{ is negative i.e., } V_T < 0$$

$$\therefore \qquad V_{SD} \geq V_{SG} + V_T. \tag{7.47}$$

The current in the Ohmic region is given by,

$$I_D = k\ [2(V_{SG} + V_T)\ V_{SD} - V_{SD}^2)\quad \text{from Eqn. (7.43)}$$

i.e.

$$I_D = k\ [2(V_{SG} + V_T)V_{SD} - V_{SD}^2] \tag{7.48}$$

for,

$$V_{SD} \leq V_{SG} + V_T \text{ and } V_{SG} > |V_T|.$$

The current in the pinch-off region is given by,

$$I_D = k\ (V_{SG} + V_T)^2 \tag{7.49}$$

for,

$$V_{SD} \geq V_{SG} + V_T \text{ and } V_{SG} > V_T.$$

These current equations are similar to that of NMOSFET with the reversal of terminal where only polarity will differ. Since the mobility of holes is only about half as much as that of electrons, the conduction parameter of a NMOSFET is approximately half of NMOSFET for the same aspect ratio (W/L).

7.12.3.3 Characteristics of PMOSFET

The V-I characteristics of PMOSFET shown in Fig. 7.34 (b) and transfer characteristics in Fig. 7.34 (a) are similar to those of NMOSFET.

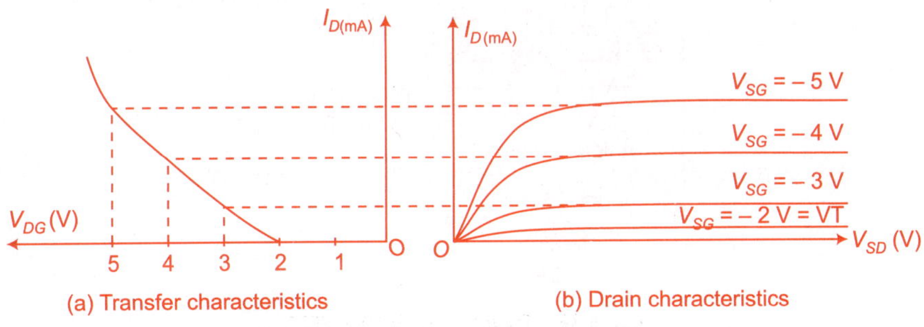

Fig. 7.34 *VI Characteristics of E-PMOSFET*

7.12.3.4 *Circuit symbol for enchancement type PMOSFET*

Figure 7.35 (a) shows the circuit symbol for *p*-channel enhancement type MOSFET when body terminal is separate and Fig. 7.35 (b) shows the circuit symbol for *p*-channel enhancement type MOSFET for which body is connected to the source.

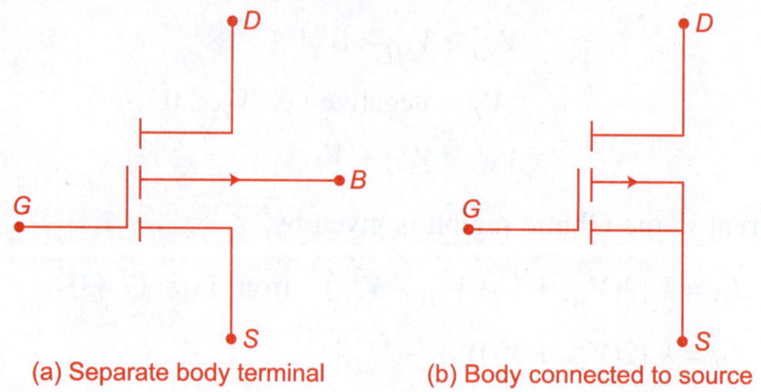

Fig. 7.35 *Circuit Symbols for p-channel enhancement type MOSFET*

7.12.4 *p*-Channel Depletion Types MOSFT (D-PMOSFET)

We have already discussed about *n*-channel depletion types MOSFET. In this section, we will study about *p*-channel depletion type MOSFET.

7.12.4.1 *Construction*

The *p*-channel depletion type MOSFET is similar in construction as *n*-channel depletion type MOSFET. Only the polarity of voltage differs. The *p*-channel depletion type MOSFET is shown in Fig. 7.36. *P*-channel depletion type MOSFET is mostly similar with *p*-channel enhancement type MOSFET. The only difference is that the conducting channel is implanted at the time of fabrication in depletion type MOSFET.

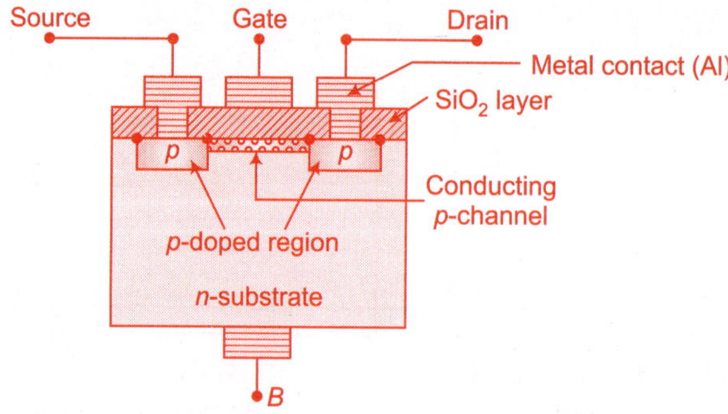

Fig. 7.36 *Structure of D-PMOSFET*

7.12.4.2 *Working of p-channel depletion type MOSFET*

The working of *p*-channel depletion type MOSFET is similar to *n*-channel depletion MOSFET. The only difference is that the channel in *p*-type and the polarities of voltages are opposite to that of *n*-channel MOSFET. Hence, drain is kept at negative and gate is kept at positive. To control the flow of current into the channel, the positive voltage is given to the gate.

7.12.4.3 *Characteristics of D-PMOSFET*

The *VI* Characteristics of D-PMOSFET is shown in Fig. 7.37 (b) and transfer characteristics is shown in Fig. 7.37(a) which are similar to those of NMOSFET.

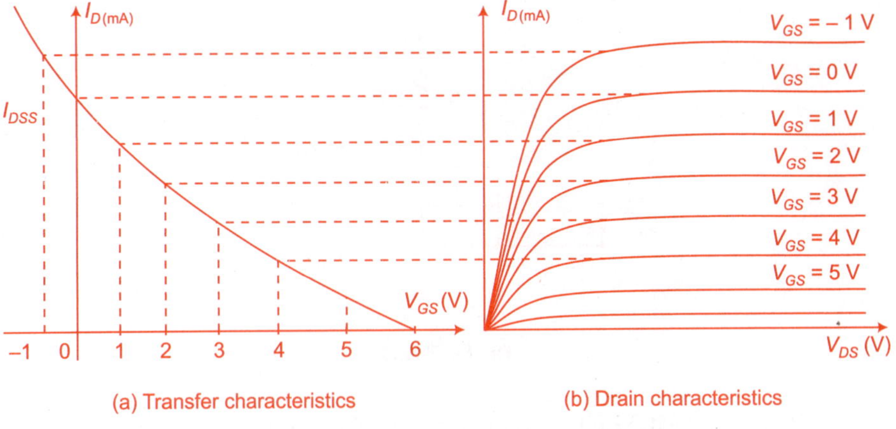

(a) Transfer characteristics   (b) Drain characteristics

Fig. 7.37   *VI characteristics of D-PMOSFET*

7.12.4.4 *Circuit symbols for depletion types MOSFET*

Figure 7.38 (a) Shows the circuit symbol for *p*-channel depletion types MOSFET when body terminal is separate and Fig. 7.38 (b) shows the circuit symbol for *p*-channel depletion types MOSFET.

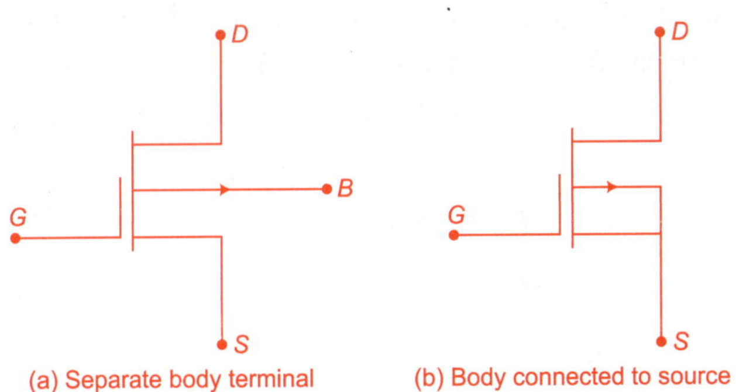

(a) Separate body terminal   (b) Body connected to source

Fig. 7.38   *Circuit symbols for p-channel depletion type MOSFET*

Example 7.12.1 Determine the drain current I_D and drain voltage V_D for the circuit given in Fig. Ex. 7.12.1 for an input voltage V_{GS} of,
(a) 0.6V (b) 2.5 V (c) 5V. What is the voltage V_{GS} at which $V_{DS} = V_i$?

Solution Given that,

$$R_D = 5 \text{ k}\Omega, \quad V_D = 10 \text{ V}$$
$$V_T = 1\text{V}, \qquad k = 100 \text{ }\mu A/V^2$$

(a) Given that,

$$V_G = 0.5\text{V}$$
$$\therefore \qquad V_{GS} < V_T.$$

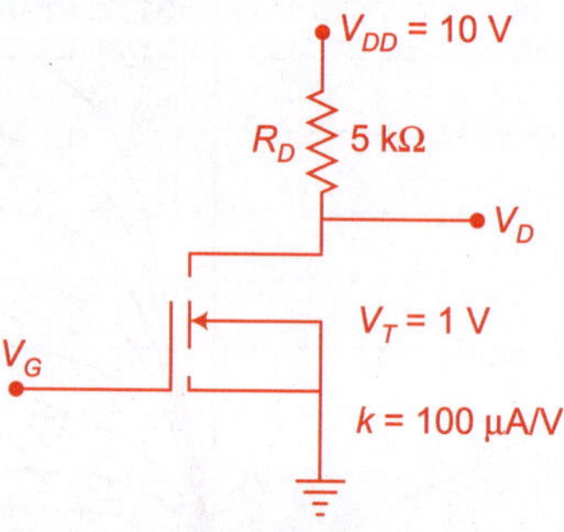

Fig. Ex. 7.12.1

Hence, the MOSFET is in off mode so the drain current $I_D = 0$

$$\therefore \qquad V_D = V_{DS} = V_{DD} = 10\text{V}$$

$$\boxed{V_D = 10 \text{ volt.}} \qquad \qquad \text{(Ans.)}$$

(b) Given that $V_{GS} = 2.5$ V

$\because$ $V_{GS} > V_T$, so the MOSFET will conduct or be in 'ON' state and the gate voltage will be greater than V_T. So, we assume that MOSFET is operating in saturation. The drain current is given by,

$$I_D = k \, (V_{GS} - V_T)^2 = 100 \text{ }\mu A/V^2 \, (2.5 - 1)$$
$$I_D = 100 \text{ }\mu A/V^2 \times 2.25 \text{ V}^2$$
$$I_D = 225 \text{ }\mu A$$

$$\boxed{I_D = 0.2 \text{ mA.}} \qquad \qquad \text{(Ans.)}$$

The drain voltage is,

$$V_{DS} = V_{DD} - I_D R_D \quad \text{from Fig. Ex. 7.12.1}$$
$$V_{DS} = 10 \text{ V} - (0.2 \times 10^{-3})\text{A} \times (4) \text{ k}\Omega$$

$$V_{DS} = 10 \text{ V} - 0.8\text{V}$$

$$\boxed{V_{DS} = 9.2 \text{ volt.}} \qquad \textbf{(Ans.)}$$

(c) Given that $V_{GS} = 50$ V

At this large value of $V_{GS} = V_{DD} =$ maximum possible V_{DS}, the device must be in the Ohmic mode of operation. The drain current is given by

$$I_D = k\,[2(V_{GS} - V_T)\,V_{DS} - V_{DS}^2]$$
$$I_D = 100 \text{ }\mu A/V^2 \,[2(5-1)V_{DS} - V_{DS}^2]$$
$$I_D = (800\,V_{DS} - 100\,V_{DS}^2) \times 10^{-6}$$

Hence,

$$V_{DS} = V_{DD} - I_D R_D$$
$$V_{DS} = 5 - 5 \times 10^3\,(800\,V_{DS} - 100\,V_{DS}^2) \times 10^{-6}$$
$$V_{DS} = 5 - 5\,(0.8\,V_{DS} - 0.1\,V_{DS}^2)$$
$$V_{DS}^2 - 10\,V_{DS} + 10 = 0.$$

After solving above equation, we obtain,

$$V_{DS} = 1.13 \text{ V and } 8.87 \text{ V}$$

But, $\qquad V_{DS}$ cannot be 8.87 V

Hence, $\qquad \boxed{V_{DS} = 1.13\text{V}}$

$\because \qquad V_{DS} < V_{GS} - V_T = 4 \text{ V.}$

The drain current I_D is,

$$I_D = \frac{V_{DD} - V_{DS}}{R_D} = \frac{5 - 1.13}{5 \times 10^3}$$

$$\boxed{I_D = 0.77 \text{ mA.}}$$

(D) For $V_{GS} = V_{DS}$, we have $V_{DS} > V_{GS} - V_T$

Hence, the MOSFET is operating in the saturation region.

The drain current I_D is given by,

$$I_D = k\,(V_{GS} - V_T)^2 = k\,(V_{DS} - V_T)^2.$$

The drain voltage is,

$$V_{DS} = V_{DD} - I_D R_D$$
$$V_{DS} = 5 - 5\,(V_{DS} - V_T)^2 \times 100 \text{ }\mu A/V^2$$
$$V_{DS} = 5 - 0.5\,(V_{DS} - 1)^2$$

Solving above equation we obtain,

$$V_{DS} = V_{GS} = 3V.$$

Then, $\boxed{I_D = 0.4 \text{ mA.}}$ **(Ans.)**

Example 7.12.2 For the circuit shown in Fig. Ex. 7.12.2 determine (a) the drain voltage V_{DS} for $V_{GS} = 0$, (b) the input voltage V_G, when the drain is at 1V and the MOSFET parameters are $k = 0.4 \text{ mA/V}^2$ and $V_T = -2V$.

Solution (a) For $V_G = 0$, let us assume saturation mode of operation for the MOSFET.

The drain current is,

$$I_D = k\,(V_{GS} - V_T)^2$$
$$I_D = 0.4\,(0 - (2))^2 \times 10^{-3} \text{ A}$$

$$\boxed{I_D = 1.6 \text{ mA.}}$$

$$I_D = 1.6 \text{ mA}$$

Now, the drain voltage V_{DS} is given by,

$$V_{DS} = V_{DD} - I_D R_D$$
$$V_{DS} = 8 - 1.6 \times 10^{-3} \times 3k$$
$$V_{DS} = (8 - 4.8) \text{ V}$$

$$\boxed{V_{DS} = 3.2 \text{ volt.}}$$ **(Ans.)**

Fig. Ex. 7.12.2

Since $V_{DS} > V_{GS} - V_T = 2V$, the MOSFET is operating in the saturation mode as assumed earlier.

(b) At the small voltage of $V_{DS} = 1V$, we may assume operation of the device in the Ohmic region then,

$$I_D = k[2(V_{GS} - V_T)\,V_{DS} - V_{DS}^2]$$
$$= 0.4 \times 10^{-3}\,[\,2(V_{GS} - (-2)) \times 1 - 1^2\,]$$
$$I_D = [0.8\,V_{GS} + 1.6 - 0.8] \times 10^{-3} \text{ A}$$
$$I_D = (0.8\,V_{GS} + 0.8) \text{ mA} \qquad\qquad \text{(A)}$$

From the circuit,

$$I_D = \frac{V_{DD} - V_{DS}}{R_D} = \frac{8 - 1}{3 \text{ k}\Omega}$$

$$\boxed{I_D = 2.3 \text{ mA.}}$$

Now, putting the value of I_D in Eqn. (A), we obtain,

$$2.3 = 0.8\, V_{GS} + 0.8$$

$$0.8\, VGS = 2.3 - 0.8$$

$$V_{GS} = \frac{1.5}{0.8} = 1.9 \text{ V}$$

$$\boxed{V_{GS} = 1.9 \text{ volt.}}$$ (Ans.)

Example 7.12.3 Analyze the PMOSFET circuit shown in Fig. Ex. 7.12.3 and determine (a) V_{DS} and drain current I_D for $V_G = 0$, (b) V_{DS} and I_D for $V_G = 4V$, and (c) V_G for $V_{DS} = 1V$. The transistor has $V_T = -1V$ and $k = 50\mu A/V^2$.

Solution Data given that,

$$k = 50 \ \mu/V^2, \quad R_D = 5k\Omega$$

(a) For $V_{GS} = 0$, $V_{SG} = V_S - V_G$

$$V_{SG} = 5V - 0 = 5 \text{ volt.}$$

At this high voltage (much above the threshold voltage) the device conducts heavily and $V_{SD} = V_S - V_D$, V_{SD} is likely to be small. Hence, assuming Ohmic mode of operation, we have,

$$I_D = \frac{V_D}{R_D} = \frac{V_{SS} - V_{SD}}{R_D} = 0.05 \times 10^{-3}\,[2(5-1)V_{SD} - V_{SD}^2]$$

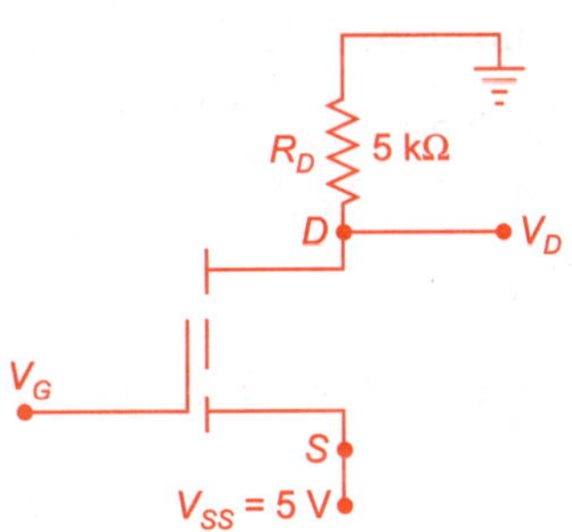

Fig. Ex. 7.12.3

After solving the above equation, we get,

$$V_{SD} = 2 \text{ V and } 10 \text{ V}$$

i.e. $$V_{SD} = 2 \text{ V is possible}$$

$$I_D = \frac{V_{SS} - V_{SD}}{R_D} = \frac{5 - 2}{5} \text{ mA}$$

$$\boxed{I_D = 0.6 \text{ mA.}}$$

The drain voltage,

$$V_{SD} = I_D R_D = 0.6 \times 5 = 3 \text{ V}$$

$$\boxed{V_{SD} = 3 \text{ volt.}}$$

Hence, the calculated voltage will satisfy the Ohmic mode.

(b) At

$$V_G = 4V,$$

$$V_{SG} = V_S - V_G = 5 - 4 = 1V.$$

Since the channel is not formed, hence drain current $I_D = 0$

$$V_{SD} \cong 5V \text{ and } V_D = 0$$

(c) At,

$$V_D = 1V$$

$$V_{SD} = V_S - V_D = 5 - 1 = 4V$$

and,

$$I_D = \frac{V_D}{R_D} = \frac{1V}{5k\Omega}$$

$$\boxed{I_D = 0.2 \text{ mA.}} \hspace{2cm} \textbf{(Ans.)}$$

Assuming pinch-off mode of operation, we have,

$$I_D = 0.05 \times 10^{-3} (5 - V_G - 1)^2$$

$$\boxed{I_D = 0.2 \text{ mA.}}$$

Hence $V_G = 2V$, for situation

$$V_{SD} = 4 > V_{SG} - |V_T|$$

$$4 > 5 - 2 - 1$$

$$V_{SD} = 4 > 3.$$

7.13 COMPARISON OF JFET AND MOSFET

S.No	JFET	MOSFET
1.	In JFET, there is no insulation between gate and channel.	There is no direct electrical connection between gate and channel *i.e.;* there is insulation between gate and channel.
2.	It is a two type one is *n*-channel and second one is *p*-channel.	MOSFET is of two types one is enhancement and second one is depletion types and both can be *n*-channel or *p*-channel.
3.	Its conduction in depletion mode only.	Depletion MOSFET can conduct both in enhancement and depletion mode but enhancement mode can conduct in enhancement mode only.
4.	The input impedance of JFET is high (about > 1MΩ).	The input is very high as compared to JFET (approx > 10,000 MR).
5.	The resistance is high.	Drain resistance is low.
6.	Generally, used in amplifiers.	These are used in digital circuits.

7.14 ADVANTAGES OF MOSFET OVER BJT

MOSFETS have following advantage over BJTS.

   (i) MOSFETS have one third size of BJTs i.e., MOSFET has small size.

 (ii) The fabrication of MOSFET circuits requires fewer and less complex steps than the fabrication of BJT circuits.

(iii) Circuit using MOSFET are less expensive to manufacture and has higher density per chip area.

(iv) MOSFET consumes less power as compared to BJTs.

7.14.1 Limitation of MOSFET

  (i) A major limitation of MOSFET logic device is their relatively low speed of operation, due to large lumped capacitance at the output node.

 (ii) The second limitation of MOSFET circuits arises from their low current-driving capability.

7.14.2 Disadvantages of *p*-channel MOSFET Over *n*-channel MOSFET

The switching speed of *p*-channel MOSFET is slow due to the lower mobility of the majority hole which is a disadvantage as compared to NMOSFET.

7.15 MOSFET AS A RESISTOR

The important application of MOSFET is that it can be used as a resistor, capacitor, amplifier and as a switch. If we use MOSFET as a resistor, it makes the design of electronic circuits very simple because complete circuit consist only MOSFETS and no other component. Generally, the MOSFETS are also used as a resistor in memory circuits and microprocessor.

The use of MOSFET as a resistor is shown in Fig 7.38 in which Fig. 7.38 (a) shows the enhancement type MOSFET as a resistor in which the gate terminal is connected to the drain. In Fig. 7.38 (b) the depletion type MOSFET is used as a resistor and its gate terminal is connected to the source i.e.; $V_{GS} = 0$.

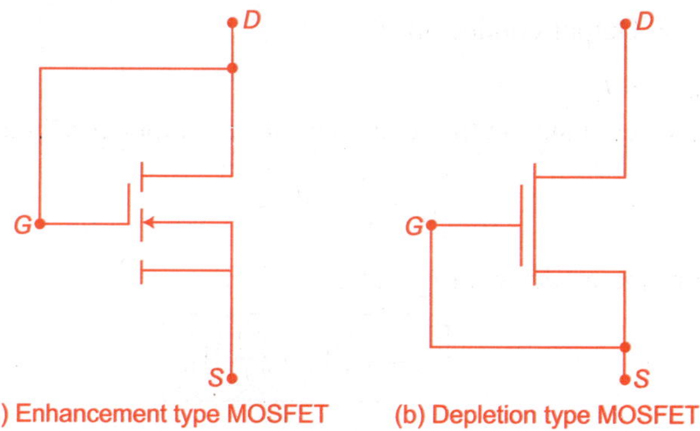

(a) Enhancement type MOSFET   (b) Depletion type MOSFET

Fig. 7.38

7.16 SMALL SIGNAL FET MODEL

The small signal FET model in the AC domain can be constructed. It is used to relate small changes in FET current and voltage about the Q-point. This model is valid both for JFET and MOSFET. The model is different at low and high frequencies. In this section, we will study both the models and FET shall be considered in common source configuration.

7.16.1 Small-signal Low-frequency FET Model

In this model, the control of i_d by V_{gs} is included as a current source $g_m V_{gs}$ connected from drain to source as shown in Fig. 7.39. In Fig. 7.39, we see that the gate to source junction V_{gs} is represented by an open circuit and no current is drawn by the input terminals of the FET, because the input impedance of FET is very high.

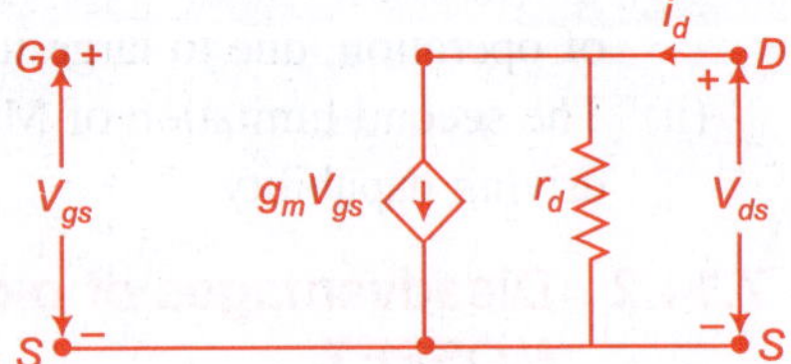

Fig. 7.39   *Low frequency FET model*

Although the gate source junction appears as an open circuit, yet the gate to source voltage affects the value of drain current. The source used in FET model is a dependent source. Basically it is a voltage controlled current source ($g_m v_{gs}$), The typical values of g_m are from 0.5 mA/V to 10 mA/V for JFET and 0.5 mA/V to 20 mA/V for MOSFET.

The FET output resistance is represented by r_d and connected in parallel with dependent current source. The typical values of drain resistance r_d are from 100 kΩ to 1 MΩ for JFETS and 1 kΩ to 50 kΩ for MOSFET.

We can also find the change in drain current due to an increment in the drain source voltage,

$$g_o = \left.\frac{\Delta i_D}{\Delta V_{DS}}\right|_Q = \frac{1}{2}\,\mu_n\,C_{ox}\left(\frac{W}{L}\right)(V_{GS} - V_T)^2\,\lambda_n \tag{7.50}$$

where, λ_n = Channel length modulation factor for n-channel MOSFET.

g_o = Output conductance

$$g_o \equiv \lambda_n I_D. \tag{7.51}$$

The output resistance is the reciprocal of the output conductance.

Hence, $$r_d = \frac{1}{g_o} = \frac{1}{\lambda_n I_D} \tag{7.52}$$

Now, from the model with r_d added,

$$i_d = g_m V_{gs} + \left(\frac{1}{r_d}\right)V_{ds}. \tag{7.53}$$

It is possible to convert voltage controlled current source in to its equivalents voltage controlled voltage source as shown in Fig. 7.40 (a). The equivalent circuit consists of a voltage controlled voltage source in series with drain resistance (output resistance). The Fig. 7.40 (b) shows another form of the low frequency model of FET.

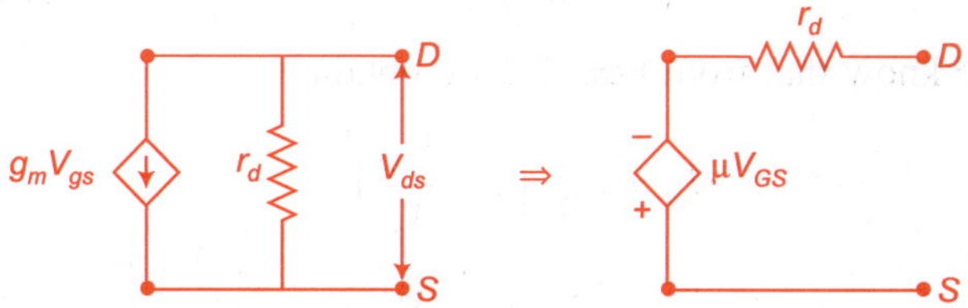

Fig. 7.40   *(a) Conversion of voltage controlled current source into voltage controlled voltage source*

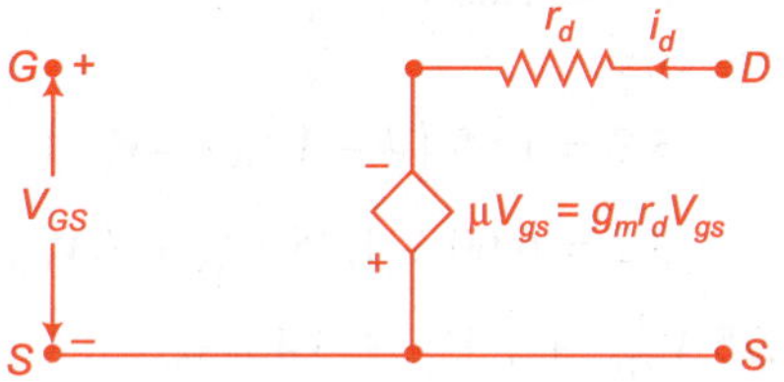

Fig. 7.40   *(b) Low frequency FET model with voltage controlled voltage source*

where, $\mu = g_m \cdot r_d$

= Amplification factor of the FET.

The small signal high frequency FET model is shown in Fig. 7.41 which is similar to the low frequency model except the addition of capacitances which consider between each pair of terminals. Basically, when we are working with high frequency signal, then there is an effect of capacitances between each pair of terminals. The capacitor C_{gs} represents the barrier capacitance between the gate and source. Its typical values are from 1 pF to 10 pF for FETS. The capacitor c_{gd} represents the barriers capacitance between the gate drain. Its typical value is also 1 pF to 10 pF for both JFET and MOSFET. Similarly, the capacitor c_{ds} represents the barring capacitance between the drain to source. Its trypical value is 0.1 pF to 1 pF.

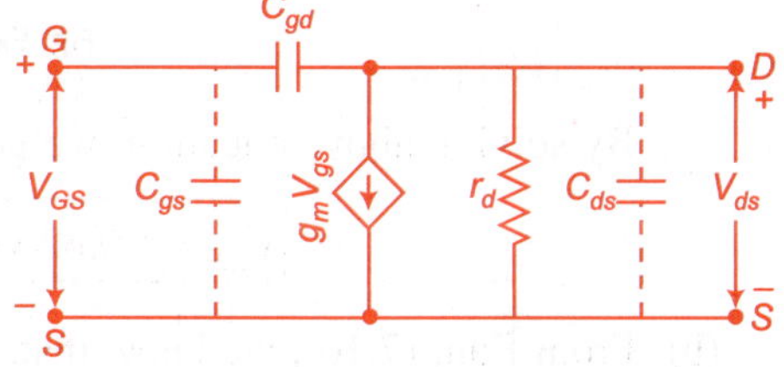

Fig. 7.41   *High frequency FET model*

Examples

Example 7.1   The amplifier of Fig. Ex. 7.1 utilizes a *n*-channel FET for which $V_p = -2$ V and $I_{DSS} = 1.65$ mA. It is desired to bias the circuit at $I_D = 0.8$ mA using $V_{DD} = 20$ V find,

   (a) V_{GS} (b) g_m (c) R_S

Solution   We know that from Eqn. 7.2 we obtain,

$$I_D = I_{DSS} \left(1 - \frac{V_{GS}}{V_p} \right)^2$$

$$0.8 = 1.65 \left(1 - \frac{V_{GS}}{-2} \right)^2$$

$$0.8 = 1.65 \left(\frac{2 + V_{GS}}{2} \right)^2$$

$$3.2 = 1.65 \left(4 + V_{GS}^2 + 4V_{GS} \right)$$

$$3.2 = 6.60 + 1.75\, V_{GS}^2 + 7\, V_{GS}$$

$$1.75\, V_{GS}^2 + 7\, V_{GS} + 3.4 = 0$$

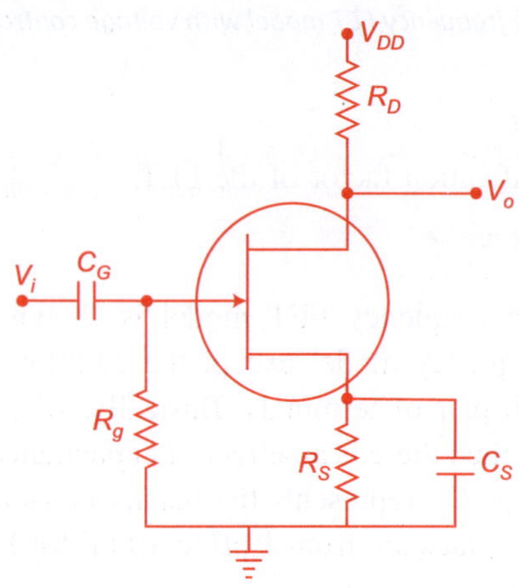

Fig. Ex. 7.1

By solving above equation, we get

$$\boxed{V_{GS} = -0.62 \text{ volt.}}$$ **(Ans.)**

(b)   From Eqn. (7.10), we know that,

$$g_m = -\frac{2\, I_{DSS}}{V_p} \left(1 - \frac{V_{GS}}{V_p} \right)$$

$$g_m = -\frac{2\,(1.65)}{-2}\left[1 - \frac{0.62}{-2}\right]$$

$$g_m = 1.65\,[1 + .31] = 1.65\,(1.31)$$

$$\boxed{g_m = 2.16 \text{ mA/V.}}$$ (Ans.)

(c) We know that,

$$R_S = -\frac{V_{GS}}{I_D} = \frac{0.62\text{V}}{0.8 \text{ mA}}$$

$$R_S = 0.77 \text{ k}\Omega$$

$$\boxed{R_S = 770 \ \Omega.}$$ (Ans.)

Example 7.2   For *n*-channel JFET for which $V_P = -5\text{V}$ and $I_{DSS} = 12$ mA. Find the drain current for $V_{GS} = 0\text{V}, -1\text{V}$ and -4 V.

Solution   Given that,

$$I_{DSS} = 12 \text{ mA}, \quad V_p = -5 \text{ V}$$

we know that for drain current,

$$I_D = I_{DSS}\left[1 - \frac{V_{GS}}{V_P}\right]^2$$

(i) If $V_{GS} = 0\text{V}$, then,

$$I_D = 12 \times 10^{-3}\left[1 - \frac{0}{-5}\right] = 12 \times 10^{-3} \times 1$$

$$\boxed{I_D = 12 \text{ mA.}}$$ (Ans.)

(ii) If $V_{GS} = -1\text{V}$, then,

$$I_D = 12 \times 10^{-3}\left[1 - \frac{1\text{V}}{5\text{V}}\right]^2 = 12 \times 10^{-3}\,(0.8)^2$$

$$I_D = 7.68 \times 10^{-3}\text{A}$$

$$\boxed{I_D = 7.68 \text{ mA.}}$$ (Ans.)

(iii) If $V_{GS} = -4\text{V}$, then,

$$I_D = 12 \times 10^{-3}\left[1 - \frac{4}{5}\right]^2 = 12 \times 10^{-3}(0.2)^2$$

$$\boxed{I_D = 4.8 \text{ mA.}}$$ (Ans.)

Example 7.3   A JFET has parameters of $V_P = 18$ V and I_{DSS} is 15 mA. Plot the transconductance curve for the device using V_{GS} values of 0 V, -3 V -6 V, -9 V, -12 V, -15 V, -18 V.

Solution   Given that,

$$I_{DSS} = 15 \text{ mA} \quad \text{and} \quad V_p = -20 \text{ V}$$

We know that for drain current,

$$I_D = I_{DSS} \left[1 - \frac{V_{GS}}{V_p} \right]^2$$

(i) if $V_{GS} = 0V$, then,

$$I_D = 15 \left[1 - \frac{0}{-20} \right] = 15 \text{ mA}.$$

(ii) If $V_{GS} = -3V$, then,

$$I_D = 15 \left[1 - \frac{3}{18} \right]^2 = 15 \, [0.83]^2$$

$$= 15 \times 0.628 = 10.33 \text{ mA}.$$

(iii) If $V_{GS} = -6$ V, then,

$$I_D = 15 \left[1 - \frac{6}{18} \right]^2 = 15 \, (0.67)^2$$

$$= 15 \times 0.45 = 6.73 \text{ mA}.$$

(iv) If $V_{GS} = -9$ V, then,

$$I_D = 15 \left[1 - \frac{9}{18} \right] = 15(0.5)^2$$

$$I_D = 15 \times 0.25 = 3.75 \text{ mA}.$$

(v) If $V_{GS} = -12V$, then,

$$I_D = 15 \left[1 - \frac{12}{18} \right] = 15 \, (0.33)^2$$

$$I_D = 15 \times 0.11 = 1.65 \text{ mA}.$$

(vi) If $V_{GS} = -15$ V, then,

$$I_D = 15 \left[1 - \frac{15}{18} \right]^2 = 15 \, (0.17)^2$$

$$I_D = 15 \times 0.029$$

$$\boxed{I_D = 0.43 \text{ mA.}}$$

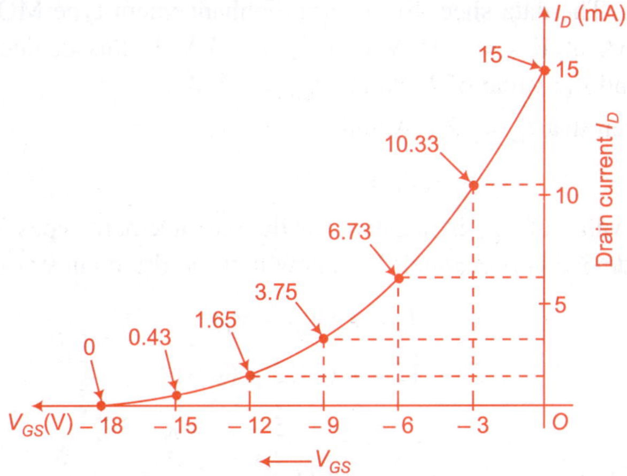

Fig. Ex. 7.3 *Transconductance curve*

(vii) If $V_{GS} = -18$ V, then

$$I_D = 15\left[1 - \frac{18}{18}\right]^2$$

$$I_D = 0 \text{ mA}.$$

Now, plotting the values of V_{GS} and I_D we obtain a curve known as transconductance curve as shown in Fig. Ex. 7.3.

Example 7.4 Determine the value of transconductance of a FET, when the drain current charge from 1mA to 1.8mA with a change in gate to source voltage from –3.5V to –3.0 V.

Solution Given that,

Change in drain current

$$\Delta I_D = I_{D_2} - I_{D_1}$$
$$= 1.8 \text{ mA} - 1.0 \text{mA} = 0.8 \text{ mA}.$$

Change in gate to source voltage,

$$\Delta V_{GS} = V_{GS2} - V_{GS1}$$
$$= +3.5 - 3.0 \text{ volt}$$
$$= 0.5 \text{ volt}.$$

Now, transconductance given by Eqn. (7.9), we obtain

$$g_m = \frac{\Delta I_D}{\Delta V_{GS}} = \frac{0.8\text{mA}}{0.5\text{V}}$$

$$\boxed{g_m = 1.6 \text{ mA/V.}}$$ **(Ans.)**

Example 7.5 The data sheet for a certain enhancement-type MOSFET reveals that $I_D = 12$mA at $V_{GS} = -15$ V and $V_T = -3$ V. Is this device *n*-channel or *p*-channel? Find the value of I_D when $V_{GS} = -7$ V.

Solution Given that $I_D = 12$ mA and $V_{GS} = -15$ V

$$V_T = -3 \text{ V.}$$

Since the value of V_{GS} is negative for the enhancement types MOSFET, this indicates the device is *p*-channel. We know that the drain current is given by,

$$I_D = k\,(V_{GS} - V_T)^2$$

$$12 = k\,(-15 - (-3))^2$$

$$= k\,(-15 + 3) = k\,(-12)^2$$

$$= k\,144$$

$$k = \frac{12}{144} = 0.083 \text{ mA/V.}$$

Now, the drain current for $V_{GS} = -7$ V

$$I_D = k\,(V_{GS} - V_T)^2$$

$$I_D = 0.083\,(-7 + 3)^2 = 0.083\,(-4)^2$$

$$\boxed{I_D = 1.33 \text{ mA.}} \hspace{4cm} \textbf{(Ans.)}$$

Example 7.6 For the given measurement $V_S = 1.7$ V for the network as shown in Fig. Ex. 7.6. Determine,

(i) I_{DQ} (ii) V_{GSQ} (iii) I_{DSS} (iv) V_D (v) V_{DS} **(UPTU 2006-07)**

Solution Given that,

$$V_S = 1.7 \text{ V}, \; V_{DD} = 18 \text{ V}$$

$$R_S = 510 \; \Omega, \; R_D = 2 \text{ k}\Omega$$

$$R_G = 1 \text{ M}\Omega, \; V_P = -4 \text{ V}$$

(i) We know that,

$$V_S = I_{DRS}$$

$$I_D = \frac{V_S}{R_S} = \frac{1.7 \text{ V}}{510 \; \Omega}$$

$$\boxed{I_{DQ} = 3.33 \text{ mA.}} \hspace{4cm} \textbf{(Ans.)}$$

(ii) $V_{GSQ} = V_G - V_S$

$$V_{GSQ} = 0 - 1.7 = -1.7 \text{ V}$$

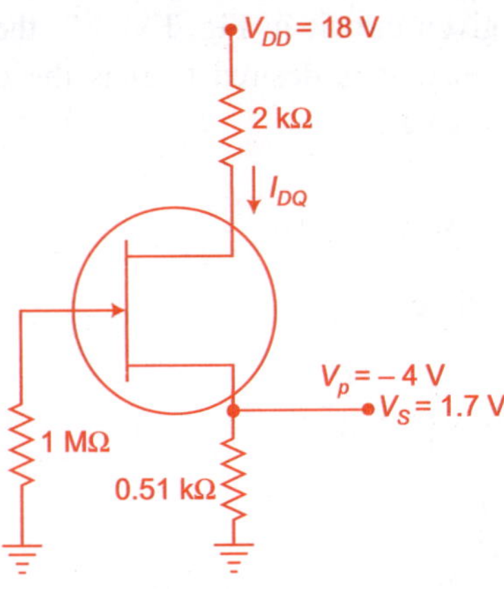

Fig. Ex. 7.6

$$\boxed{V_{GSQ} = -\,1.7 \text{ volt.}}$$ (Ans.)

(iii) We know that,

$$I_D = I_{DSS}\left[1 - \frac{V_{GS}}{V_p}\right]^2$$

$$I_{DSS} = \frac{I_D}{\left[1 - \dfrac{V_{GS}}{V_p}\right]^2} = \frac{3.33}{\left[1 - \dfrac{1.7}{4}\right]^2}$$

$$I_{DSS} = \frac{3.33}{(1 - 0.425)^2} = \frac{3.33}{0.33}$$

$$\boxed{I_{DSS} = 10.09 \text{ mA.}}$$ (Ans.)

(iv) $$V_{DS} = V_D - V_S$$

$$V_D = V_{DS} - V_S$$

(v) We know that,

$$V_{DS} = V_{DD} = I_D\,(R_D + R_S)$$

$$V_{DS} = 18 \text{ V} - 3.33\,(2 + 0.51)\ \text{k}\Omega$$

$$V_{DS} = 9.64 \text{ V.}$$

Hence,

$$V_D = V_{DS} - V_S = 9.64 - (-\,1.7)$$

$$V_D = 9.64 + 1.7$$

$$\boxed{V_D = 11.34 \text{ volt.}}$$ (Ans.)

Example 7.7 For the given circuit in Fig. Ex. 7.7, the data sheet is given as $V_p = -2$V, $I_{DSS} = 1.65$ mA. It is desired to bias the circuit at $I_D = 0.8$ mA, $V_{DD} = 24$ V. Find (i) V_{GS} (ii) g_m.

Solution   Given that,

$$V_{DD} = 24 \text{ V}, \quad V_p = -2\text{V}$$

$$I_{DSS} = 1.65 \text{ mA}$$

(i) We know that,

$$I_D = I_{DSS}\left[1 - \frac{V_{GS}}{V_p}\right]^2$$

$$\left[1 - \frac{V_{GS}}{V_p}\right]^2 = \frac{I_D}{I_{DSS}}$$

$$1 - \frac{V_{GS}}{V_p} = \sqrt{\frac{I_D}{I_{DSS}}}$$

$$\frac{V_{GS}}{V_p} = 1 - \sqrt{\frac{I_D}{I_{DSS}}}$$

$$V_{GS} = V_p\left[1 - \sqrt{\frac{I_D}{I_{DSS}}}\right]$$

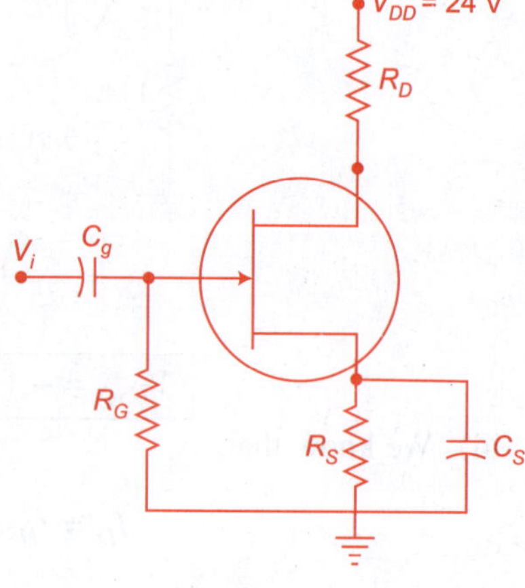

Fig. Ex. 7.7

Now substitute the value in the above equation,

$$V_{GS} = -2\left[1 - \sqrt{\frac{0.8 \times 10^{-3}}{1.65 \times 10^{-3}}}\right]$$

$$V_{GS} = -0.61 \text{ V}.$$

(ii) We know that,

$$g_m = -\frac{2 I_{DSS}}{V_p}\left[1 - \frac{V_{GS}}{V_p}\right] = \frac{-2 \times 1.65 \times 10^{-3}}{-2}\left(1 - \frac{0.61}{2}\right)$$

$$= 1.65 \times 10^{-3} \times 0.696$$

$$\boxed{g_m = 1.15 \text{ mA/V.}} \tag{Ans.}$$

Example 7.8 An n-channel JFET having $V_p = -4$V and $I_{DSS} = 10$ mA, is used in the circuit of Fig. Ex. 7.8(a). The parameter values are $V_{DD} = 18$ V, $R_S = 2$ kΩ, $R_D = 2$ kΩ $R_1 = 450$ kΩ and $R_2 = 90$ kΩ. Determine I_D and V_{DS}.

(UPTU 2007-2008)

Solution Given that,

$$V_p = -4 \text{ V}, \quad I_{DSS} = 10\text{mA}$$

$$V_{DD} = 18 \text{ V}, \quad R_S = 2 \text{ k}\Omega$$

$$R_D = 2 \text{ k}\Omega, \quad R_1 = 450 \text{ k}\Omega$$

$$R_2 = 90 \text{ k}\Omega, \quad I_D = ?$$

$$V_{DS} = ?$$

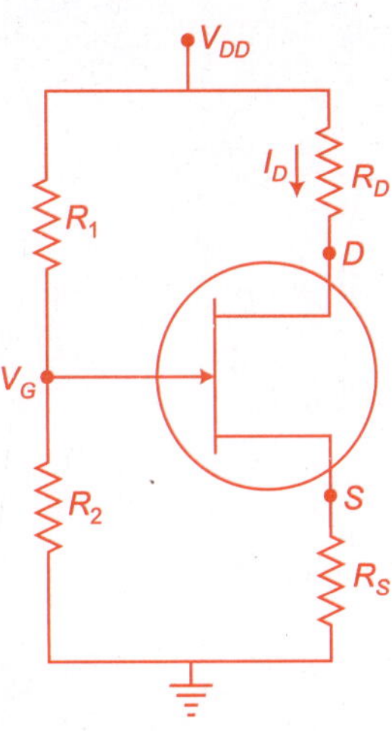

Fig. Ex. 7.8(a)

By Voltage dividing rule,

$$V_G = \frac{R_2}{R_1 + R_2} \times V_{DD}$$

$$V_G = \frac{90 \text{ k}\Omega}{(90 + 450) \text{ k}\Omega} \times 18 \text{ V}$$

$$V_G = \frac{1620}{540} \text{ V} = 3\text{V}.$$

The Thevenin resistance is,

$$R_{Th} = \frac{R_1 R_2}{R_1 + R_2} = \frac{90 \times 450}{90 + 450} \text{ k}\Omega$$

$$R_{Th} = \frac{40500}{540} = 75 \text{ k}\Omega.$$

Thevenin equivalent circuit of Fig. 7.8 (a) is shown in Fig. Ex. 7.8 (b).

Now, applying KVL to input loop,

$$V_G - I_G R_{Th} - V_{GS} - I_D R_S = 0$$

$$\because \qquad I_G = 0$$

$$\therefore \qquad V_G - V_{GS} - I_D R_S = 0$$

$$I_D = \frac{V_G - V_{GS}}{R_S}$$

If, $\qquad V_{GS} = 0$

$$I_D = \frac{V_G}{R_S} = \frac{3\text{V}}{2 \text{ k}\Omega} = 1.5 \text{ mA}$$

$$\boxed{I_D = 1.5 \text{ mA.}} \qquad\qquad \textbf{(Ans.)}$$

Now, applying KVL to output side,

$$V_{DD} - I_D R_D - V_{DS} - I_D R_S = 0$$

$$V_{DS} = V_{DD} - I_D (R_D + R_S)$$

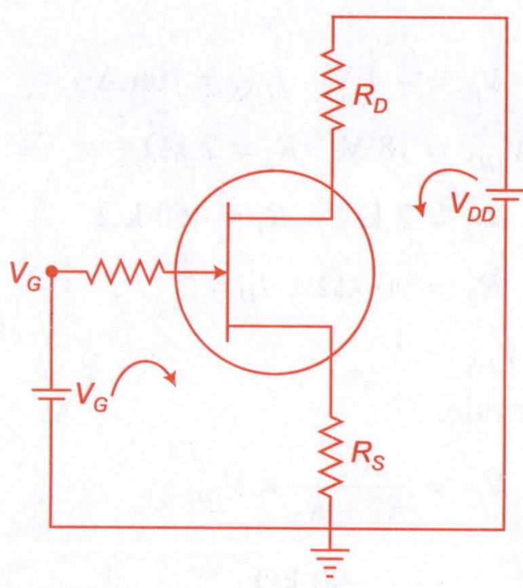

Fig. Ex. 7.8(b)

$$V_{DS} = 18\text{V} - 1.5 \text{ mA} \,(2 + 2) \text{ k}\Omega$$

$$V_{DS} = 18 \text{ V} - 6\text{V}$$

$$\boxed{V_{DS} = 12 \text{ volt.}}$$

(Ans.)

Example 7.9 The pinch-off voltage of a *p*-channel junction FET is $V_p = 5$V and the drain to source saturation current $I_{DSS} = -40$ mA. The value of drain-source voltage V_{DS} is such that the transistor operates in the saturation region. The drain current is given as $I_D = -15$mA. Determine the gate source voltage V_{GS}.

Solution Given that,

(Gate 2001)

$$V_p = 5\text{V}, \ I_{DSS} = -40 \text{ mA}$$

$$I_D = -15 \text{ mA}.$$

We know that the drain current in saturation,

$$I_D = I_{DSS}\left(1 + \frac{V_{GS}}{V_P}\right)^2$$

$$-15 \text{ mA} = -40 \text{ mA}\left(1 + \frac{V_{GS}}{5}\right)^2$$

$$\left(1 + \frac{V_{GS}}{5}\right)^2 = \frac{-15}{-40}$$

$$\left(1 + \frac{V_{GS}}{5}\right)^2 = 0.375$$

$$1 + \frac{V_{GS}}{5} = \sqrt{0.375} = 0.612$$

$$\frac{V_{GS}}{5} = 0.612 - 1 = -0.3876$$

$$V_{GS} = -3876 \times 5$$

$$\boxed{V_{GS} = -1.938.}$$ (Ans.)

Example 7.10 Determine the value of transconductance of a FET, when the drain current changes from 1mA to 2.8 mA with a change in gate-source voltage from –3.8 V to 2.9 V.

Solution Data given that,

Change in drain current $\Delta I_D = I_{D_2} - I_{D_1}$

$$\Delta I_D = (2.8 - 1.0)\text{mA}$$

$$\Delta I_D = 1.8 \text{ mA.}$$

Change in gate to source voltage $\Delta V_{GS} = V_{GS2} - V_{GS1}$

$$\Delta V_{GS} = -2.9 - (-3.8)$$

$$\Delta V_{GS} = -2.9 + 3.8 = 0.9 \text{ V.}$$

Now, the transconductance is given by,

$$g_m = \frac{\Delta I_D}{\Delta V_{GS}} = \frac{1.8 \text{ mA}}{0.9\text{V}}$$

$$g_m = 2\text{mA/V}$$

$$\boxed{g_m = 2000 \text{ µS.}}$$ (Ans.)

Example 7.11 Consider the *n*-channel enhancement MOSFET as shown in Fig. Ex. 7.11. If $V_T = 4$ V and $I_D = 1.28$ mA at $V_{DS} = 12$ V and $I_D = 1.28$ mA at $V_{DS} = 12$ V. Determine the value of R_D.

Solution Given that,

$$V_T = 4\text{V}, \quad I_D = 1.28 \text{ mA}$$

$$V_{DS} = 12\text{V}, \quad R_D = ?, \quad V_{DD} = 24 \text{ V.}$$

We know that from Fig. Ex. 7.11,

$$V_{DS} = V_{DD} - I_D R_D$$

$$12 = 24 - (1.28 \text{ mA}) R_D$$

$$1.28 \text{ mA } R_D = 12$$

$$R_D = \frac{12 \text{ V}}{1.28 \times 10^{-3}\text{A}} = 9.38 \text{ k}\Omega$$

$$\boxed{R_D = 9.38 \text{ k}\Omega.}$$ (Ans.)

At pinch-off,
$$V_{GS} = -V_p$$
$$I_D = 0.$$

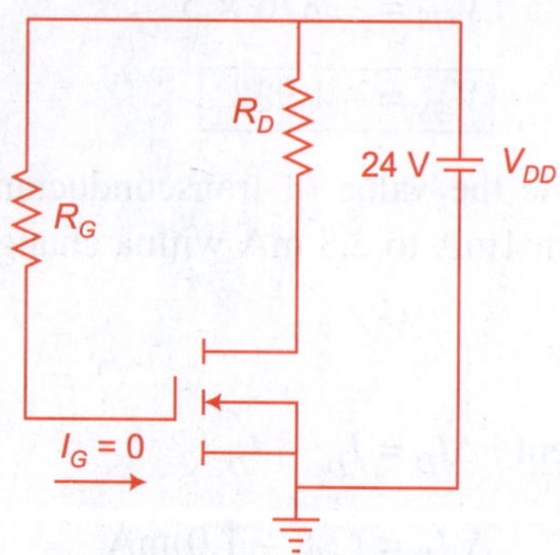

Fig. Ex. 7.11

Example 7.12 Calculate the self-bias operating point for the FET circuit shown in Fig. Ex. 7.12. Also find the values of resistance R_D and R_S to obtain this bias condition, given the maximum values of drain current as 15 mA and $V_{GS} = -2.5$ V at $I_D = 6$ mA.

Solution Given that,

$$I_{DSS} = 15 \text{ mA}, \quad V_{GS} = -2.5 \text{ V}$$

$$V_{DD} = 24 \text{ V}$$

We know that the value of drain current at Q-point is given by,

$$I_{DQ} = \frac{I_{DSS}}{2} = \frac{15 \text{ mA}}{2} = 7.5 \text{ mA and the drain to source voltage is given as:}$$

$$V_{DSQ} = \frac{V_{DD}}{2} = \frac{24}{2} = 12 \text{ V.}$$

Therefore, the operating point will be given as,

$$I_D = 7.5 \text{ mA} \quad \text{and} \quad V_{DS} = 12 \text{ V}$$

Now, $$V_{GS} = -2.5 \text{ V}$$

and, $$I_D = 6 \text{ mA.}$$

The drain to source voltage VDS is given,

$$V_{DS} = V_{DD} - I_D R_D$$

$$12 = 24 - (6 \times 10^{-3}) R_D$$

$$(6 \times 10^{-3}) R_D = 12$$

$$R_D = \frac{12}{6 \times 10^{-3}} = 2\ k\Omega$$

$$\boxed{R_D = 2\ k\Omega.}$$ (**Ans.**)

The source voltage is given by,

$$V_{GS} = V_G - V_S$$

$\therefore$
$$V_G = 0$$

$$V_S = -V_{GS} = -(-2.5\ V)$$

$$V_S = 2.5\ V.$$

The voltage drop across the source resistor,

$$V_S = I_D R_S$$

$$R_S = \frac{V_s}{I_D} = \frac{2.5\ V}{6 \times 10^{-3}}\ A$$

$$R_S = 0.416\ k\Omega$$

$$\boxed{R_S = 416\ \Omega.}$$ (**Ans.**)

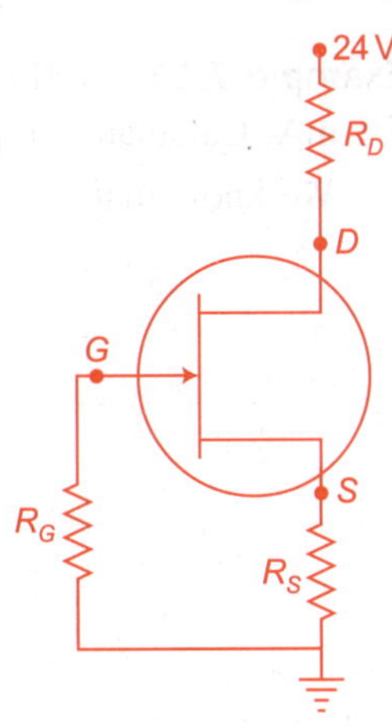

Fig. Ex. 7.12

Example 7.13 For a FET, following parameter is given $I_{DSS} = 9.2$ mA, $V_p = -2.5$ V. What is the value of I_D for $V_{GS} = -1.5$ V? Find g_m at this point.

Solution Data given that,

$$I_{DSS} = 9.2\ mA, \quad V_{GS} = -1.5\ V$$

$$V_p = -2.5\ V.$$

We know that the value of drain current in saturation region is,

$$I_D = I_{DSS}\left[1 - \frac{V_{GS}}{V_p}\right]^2$$

$$I_D = 9.2\ mA\left[1 - \left(\frac{-1.5}{-2.5}\right)\right]^2$$

$$I_D = 9.2\ mA\ [1 - 0.6]^2$$

$$I_D = 9.2 \times 10^{-3} \times 0.4$$

$$I_D = 3.68 \times 10^{-3}\ A = 3.7\ mA$$

$$\boxed{I_D = 3.7\ mA.}$$ (**Ans.**)

Transconductance for $V_{GS} = 0$ is given by,

$$g_m = -\frac{2\ I_{DSS}}{V_p}\left(1 - \frac{V_{GS}}{V_p}\right)$$

$$= \frac{-2 \times 9.2 \text{ mA}}{-2.5} \left(1 - \frac{1.5}{2.5}\right) = 7.36 \text{ mA}(1 - 0.6)$$

$$\boxed{g_m = 2.9 \text{ mA/volt.}} \qquad \text{(Ans.)}$$

Example 7.14 A JFET has $g_m = 15$ mA/V, the maximum situation current is 7.5 mA. Calculate the pinch-off voltage V_p.

We know that,

$$g_m = g_{m_o} = -2 \frac{I_{DSS}}{V_p}$$

$$15 \times 10^{-3} = \frac{-2 \times 7.5 \times 10^{-3}}{V_p}$$

$$V_p = \frac{-2 \times 7.5 \times 10^{-3}}{15 \times 10^{-3}}$$

$$V_p = -1\,\text{V}$$

$$\boxed{V_p = -1 \text{ volt.}} \qquad \text{(Ans.)}$$

Example 7.15 A n-channel JFET has a pinch-off voltage of -3.5 V and maximum saturation current is 8 mA. At what value of V_{GS} will I_D be equal to 2 mA? what is its g_m at this I_D? **(A.M.I.E. Examination)**

Solution Given that,

$$V_p = -3.5, \ I_{DSS} = 8 \text{ mA}, \ I_D = 2 \text{ mA}$$

From, Shockley's equation, we know that,

$$I_D = I_{DSS} \left[1 - \frac{V_{GS}}{V_p}\right]^2$$

$$\left[1 - \frac{V_{GS}}{V_p}\right]^2 = \frac{I_D}{I_{DSS}}$$

$$1 - \frac{V_{GS}}{V_p} = \sqrt{\frac{I_D}{I_{DSS}}}$$

$$\frac{V_{GS}}{V_p} = 1 - \sqrt{\frac{I_D}{I_{DSS}}}$$

$$V_{GS} = V_p \left[1 - \sqrt{\frac{I_D}{I_{DSS}}}\right]$$

$$V_{GS} = -3.5 \times \left[1 - \sqrt{\frac{2 \times 10^{-3}}{8 \times 10^{-3}}}\right]$$

$$V_{GS} = -3.5 \left[1 - \sqrt{\frac{1}{4}} \right] = -3.5 \times [1 - 0.5]$$

$$\boxed{V_{GS} = -1.75 \text{ volt.}}$$ **(Ans.)**

Transconductance gm for $I_D = 2$ mA for which $V_{GS} = -1.75$ V

$$g_m = \frac{-2 \, I_{DSS}}{V_p} \left[1 - \frac{V_{GS}}{V_p} \right] = \frac{-2 \times 8 \times 10^{-3}}{-3.5} \left(1 - \frac{1.75}{3.5} \right)$$

$$g_m = +4.57(1 - 0.5) \times 10^{-3}$$

$$g_m = +2.29 \text{ mA/V}$$

$$\boxed{g_m = 2.29 \text{ ms.}}$$ **(Ans.)**

Example 7.16   The JFET used in Fig. Ex. 7.16 has $V_p = -2.5$V, $I_{DSS} = 1.75$ mA. It is desired to bias the circuit at $I_D = 0.9$mA. If $V_{DD} = 20$ V and $r_d \gg R_D$. Find

    (a) V_{GS} (b) g_m (c) R_S

Solution   Given that,

$$V_p = -2.5 \text{ V}, I_{DSS} = 1.75 \text{ mA}$$

$$I_D = 0.9 \text{ mA}, V_{DD} = 20 \text{ V}.$$

We know that,

$$I_D = I_{DSS} \left[1 - \frac{V_{GS}}{V_p} \right]^2$$

or,

$$V_{GS} = V_p \left[1 - \sqrt{\frac{I_D}{I_{DSS}}} \right]$$

$$V_{GS} = -2.5 \left[1 - \sqrt{\frac{0.9\text{mA}}{1.75\text{mA}}} \right]$$

$$V_{GS} = -2.5 \left[1 - \sqrt{\frac{1}{2}} \right]$$

$$V_{GS} = -2.5 \, [1 - 0.71]$$

$$V_{GS} = -2.5 \times 0.29$$

$$\boxed{V_{GS} = -0.73 \text{ volt.}}$$ **(Ans.)**

    (b)   We know that,

$$g_m = \frac{-2 \, I_{DSS}}{V_p} \left[1 - \frac{V_{GS}}{V_p} \right]$$

$$g_m = \frac{-2 \times 1.75 \times 10^{-3}}{-2.5}\left[1 - \frac{0.73V}{2.5}\right]$$

$$g_m = \frac{3.5 \times 10^{-3}}{2.5}[1 - 0.29] = 1.4 \times 10^{-3} \times 0.71$$

$$\boxed{g_m = 0.99 \text{ mA/V.}} \qquad \text{(Ans.)}$$

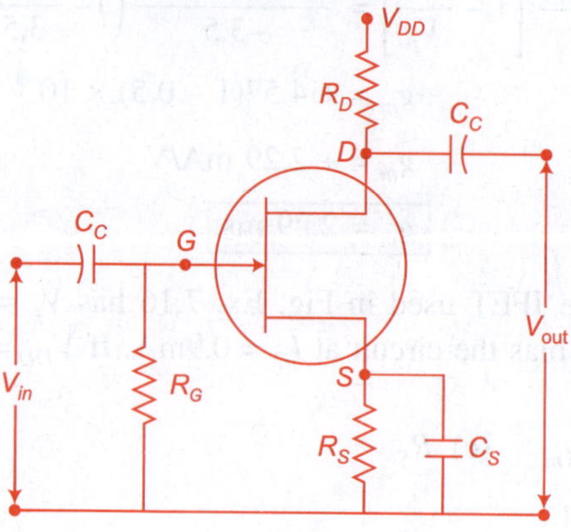

Fig. Ex. 7.16

(c) Appling the KVL to input side,

$$V_{GS} = V_G - V_S$$

$\therefore$

$$V_G = o$$

$$V_{GS} = -V_S = -I_D R_S$$

$$R_S = -\frac{V_{GS}}{I_D} = \frac{-(-0.73V)}{0.9 \text{ mA}}$$

$$R_S = 0.811 \text{ k}\Omega$$

$$\boxed{R_S = 811 \ \Omega.} \qquad \text{(Ans.)}$$

Example 7.17   Sketch the transfer curve defined by $I_{DSS} = 12$ mA and $V_P = -6V$.

Solution   Two point of transfer curve are defined by,

Point A: $I_{DSS} = 12$ mA and $V_{GS} = 0V$

Point B: $I_D = 0mA$ and $V_{GS} = V_p$

Point C: At $V_{GS} = \dfrac{V_p}{2} = \dfrac{-6}{2} = -3V$

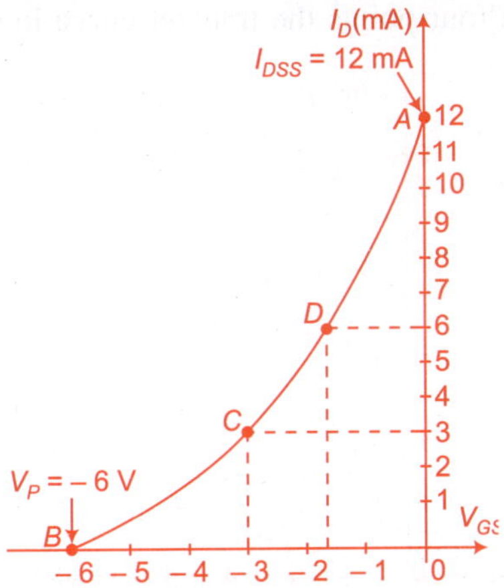

Fig. Ex. 7.17   *Tranfer curve*

The drain current,

$$I_D = \frac{I_{DSS}}{4} = \frac{12 \text{ mA}}{4} = 3 \text{ mA.}$$

Point D: At $I_D = \dfrac{I_{DSS}}{2} = \dfrac{12 \text{ mA}}{2} = 6 \text{ mA.}$

The gate to source voltage,

$$V_{GS} \equiv 0.3 V_P = 0.3 \,(-6)$$

$$= -1.8 \text{ V.}$$

Now, we plot these four points the characteristics is shown in Fig. Ex. 7.17.

Example 7.18   Sketch the transfer curve for a *p*-channel device with I_{DSS} = 6 mA and V_p = 2V.

Solution   Two points of transfer curve is defined by

Point A: I_{DSS} = 6 mA and V_{GS} = 0V

Point B: $I_D = 0$ and $V_{GS} = +\, V_p$

Now,

Point C: At $V_{GS} = \dfrac{V_p}{2} = \dfrac{2}{2} = 1$V and the drain current $I_D = \dfrac{I_{DSS}}{4} = \dfrac{6 \text{ mA}}{4} = $

1.5 mA.

Point D: At $I_D = \dfrac{I_{DSS}}{2} = \dfrac{6 \text{ mA}}{2} = 6 \text{ mA}$ and $= V_{GS} = 0.3\, V_p = 0.3\,(2)$

$$V_{GS} = 0.6 \text{ V.}$$

Now, we plot these four points the transfer curve in shown in Fig. Ex. 7.18.

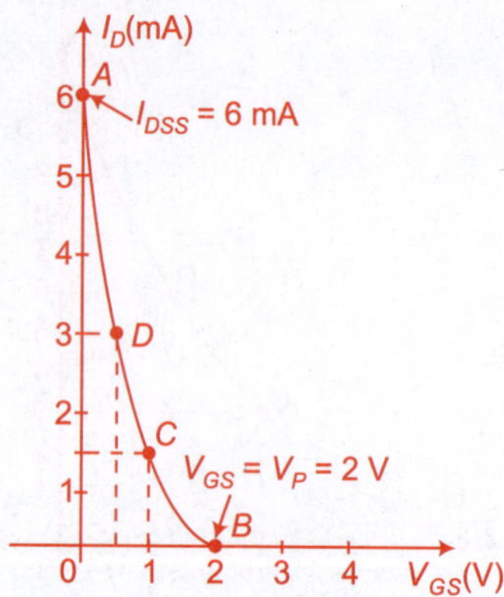

Fig. Ex. 7.18 *Transfer curve*

Example 7.19 The data given on specification sheet is as follows

$$I_D = 5\text{mA},\ V_{GS} = 12\text{V}$$

Determine $\qquad V_T = 3\text{V}.$

(a) The resulting value of k for the MOSFET

(b) The transfer characteristics at $V_{GS} = 5\text{V},\ 8\text{V},\ 10,\ 12\text{V},\ 14\text{V}.$

Solution (a) From Eqn. (7.49), we know that,

$$I_D = k\,(V_{GS} - V_T)^2$$

$$k = \frac{I_D}{(V_{GS} - V_T)^2} = \frac{5\text{mA}}{(12-3)^2} = \frac{5\text{mA}}{(9V)^2}$$

$$k = \frac{5\text{mA}}{81\ V^2} = 0.062\ \text{mA/V}^2$$

$$\boxed{k = 62\ \mu\text{A/V}^2.}\qquad\qquad\text{(Ans.)}$$

Now, drain current equation becomes,

$$I_D = 62 \times 10^{-6}\,(V_{GS} - 3\text{V})^2$$

for, $\qquad\qquad V_{GS} = 3\text{V}$

$$I_D = 0.$$

for, $\qquad\qquad V_{GS} = 5\text{V}$

$$I_D = 62 \times 10^{-6}\,(5-3)^2 = 62 \times 10^{-6} \times (4)\ \text{A}$$

$$I_D = 0.25\ \text{mA}.$$

for,
$$V_{GS} = 8V$$
$$I_D = 62 \times 10^{-6} (5 - 8)^2 = 62 \times 10^{-6} \times (9)\ A$$
$$I_D = 0.56\ mA.$$

for,
$$V_{GS} = 10\ V$$
$$I_D = 62 \times 10^{-6} (5 - 10)^2 = 62 \times 10^{-6} (25)A$$
$$= 1.55\ mA.$$

for,
$$V_{GS} = 12V$$
$$I_D = 62 \times 10^{-6} (5 - 12)^2 = 62 \times 10^{-6} (49)\ A$$
$$I_D = 3.04\ mA.$$

for,
$$V_{GS} = 14V$$
$$I_D = 62 \times 10^{-6} (5 - 14)^2 = 62 \times 10^{-6} (81)\ A$$
$$I_D = 5.02\ mA.$$

Now, the transfer curve is sketched by above calculated point and shown in Fig. Ex. 7.19

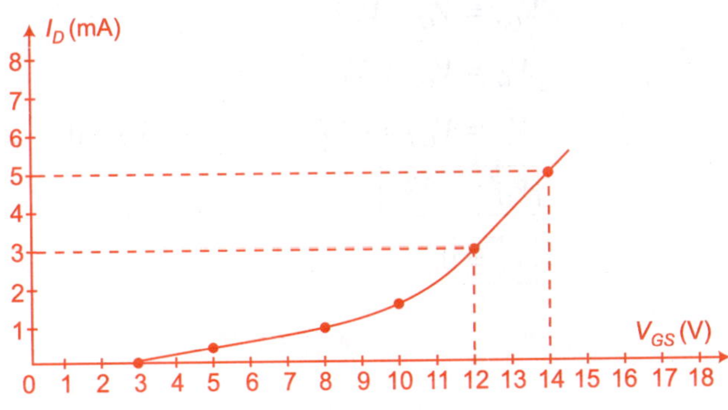

Fig. Ex. 7.19

Example 7.20 Determine the following for the Fig. Ex. 7.20.

(a) V_{GSQ} (b) I_{DQ}

(c) V_{DS} (d) V_D

(e) V_G (f) V_S

Solution   Data given as,

$$R_D = 2.5\ k\Omega, \quad R_G = 1\ M\Omega$$
$$V_{GG} = 3V, \quad V_p = -6V$$
$$I_{DSS} = 12\ mA, \quad V_{DD} = 18\ V.$$

(a)
$$\boxed{V_{GSQ} = -V_{GG} = -3\ volt.}$$

(b)
$$I_{DQ} = I_{DSS}\left(1 - \frac{V_{GS}}{V_p}\right)^2 = 12\text{ mA}\left(1 - \frac{-3}{-6}\right)^2$$

$$= 12\text{ mA }(1 - 0.5)^2 = 12\text{ mA }(0.5)^2$$

$$= 12\text{ mA }(0.25)$$

$$\boxed{I_{DQ} = 3\text{ mA.}}$$ **(Ans.)**

(c) From Fig. Ex. 7.20, we obtain,

$$V_{DS} = V_{DD} - I_D R_D = 18\text{ V} - (3\text{mA} \times 2.5\text{ k}\Omega)$$

$$V_{DS} = 18\text{V} - 7.5\text{V}$$

$$V_{DS} = 10.5\text{ volt.}$$

(d) $\qquad V_{DS} = V_D - V_S \qquad\qquad \because\ \ V_S = 0$

$\therefore \qquad\qquad V_{DS} = V_D = 10.5\text{ V}$

$$\boxed{V_{DS} = 10.5\text{ volt.}}$$ **(Ans.)**

(e) $\qquad\qquad V_{GS} = V_G - V_S$

$$V_G = V_{GS} + V_S$$

$$V_G = V_{GS} = -3\text{V} \qquad\qquad \therefore\ \ V_S = 0$$

$$\boxed{V_G = -3\text{V}}$$ **(Ans.)**

(f) $\qquad\qquad\quad \boxed{V_S = 0.}$ **(Ans.)**

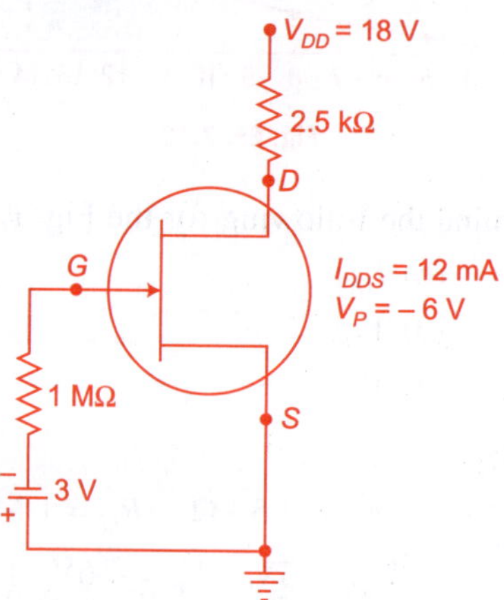

Fig. Ex. 7.20

Example 7.21 Determine the following for given network of Fig. Ex. 7.21

  (a) V_{GSQ} (b) I_{DQ}

  (c) V_{DS} (d) V_S

  (e) V_G (f) V_D

Solution (a) The gate to source voltage is determined by,

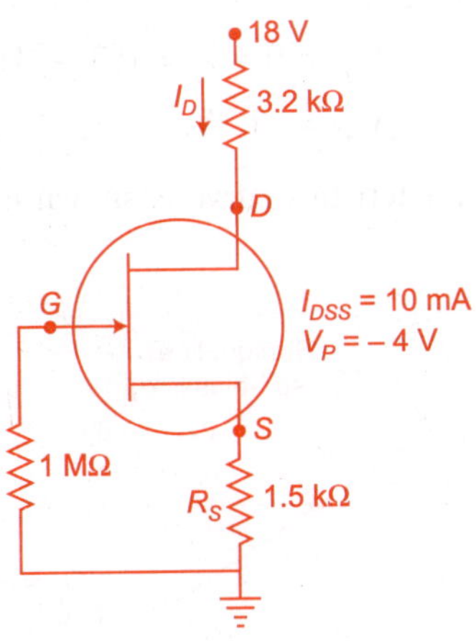

Fig. Ex. 7.21

$$V_{GS} = V_G - V_S$$

Arbitratry choosing $I_D = 3 \text{ mA}$

$$V_{GS} = V_G - I_D R_S$$

$\because$ $V_G = 0$

$\therefore$ $V_{GS} = - I_D R_S = - (3\text{mA}) \times 1.5 \text{ k}\Omega$

$$\boxed{V_{GS} = - 4.5 \text{ volt.}}$$ **(Ans.)**

Let, $I_D = 6 \text{ mA}$ then the resulting value of V_{GS}

$$V_{GS} = - (6 \text{ mA}) \times 1.5 \text{ k}\Omega$$

$$V_{GS} = - 9 \text{ volt.}$$

The values V_{GS} for appropriate values of drain current will provide straight line as shown in Fig. 7.21(a).

Now, from the points for transfer curve, we can determine,

Point A: $I_{DSS} = 10 \text{ mA}$ and $V_{GS} = 0 \text{ V}$

Point B: $I_D = 0$ and $V_{GS} = V_p = - 4 \text{ V}$

  Now,

Point C: $V_{GS} = \dfrac{V_p}{2} = \dfrac{-4}{2} \text{ V} = -2 \text{ V}$

and the drain current,

$$I_D = \frac{I_{DSS}}{4} = \frac{10 \text{mA}}{4}$$

$$I_D = 2.5 \text{ mA}.$$

Point D: At $I_D = \dfrac{I_{DSS}}{2} = \dfrac{10}{2} = 5\text{mA}$

and,

$$V_{GS} = 0.3 \, V_p = 0.3 \, (-4)$$

$$V_{GS} = -1.2\text{V}.$$

Now, the transfer characteristics curve is shown in Fig. Ex. 7.21 (b).

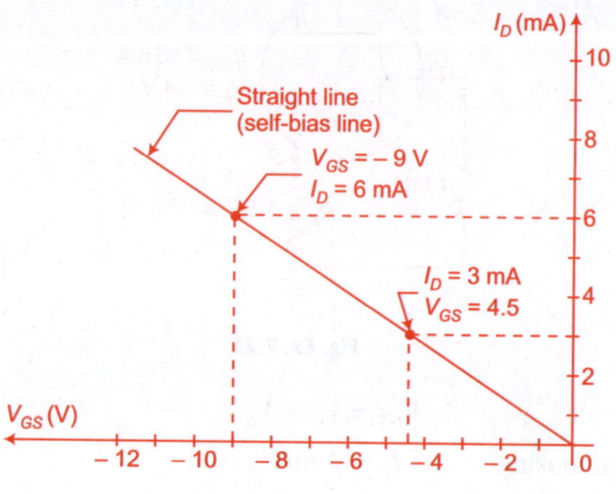

Fig. Ex. 7.21(a)

Now, superimposing the self-bias line curve as shown in Fig. Ex. 7.21(a) on a transfer curve as shown in Fig. 7.21(b). The as resultant curve is shown in Fig. Ex. 7.21(c). In Fig. Ex. 7.21(c) we see that the intersection of transfer curve and self bias line is operating point. From Fig. Ex. 7.21(c).

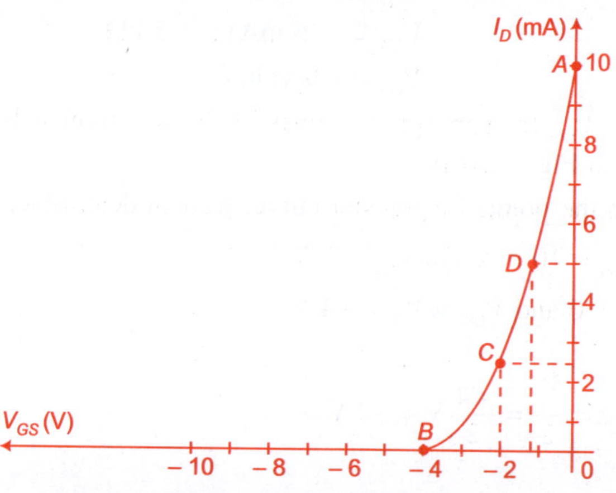

Fig. Ex. 7.21(b)   Transfer curve of Ex. 7.21

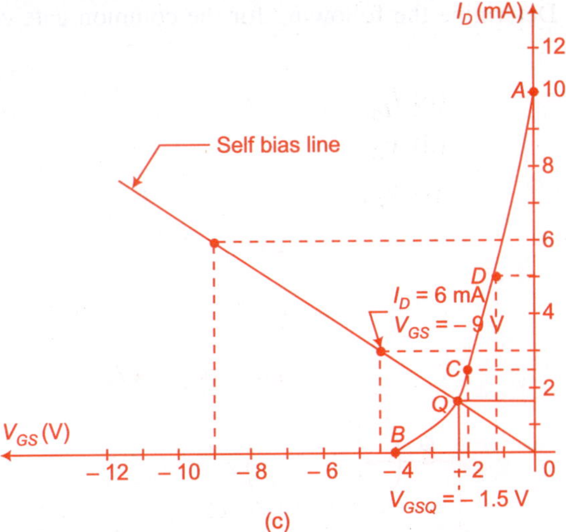

Fig. Ex. 7.21(c)

(a) V_{GSQ} = 2.6 mA

(b) I_{DQ} = 1.6 mA

(c) From Fig. Ex. 7.21

$$V_{DS} = V_{DD} - I_D (R_D + R_S)$$

$$V_{DS} = 18V - (1.6 \text{ mA}) (3.2 \text{ k}\Omega + 1.5 \text{ k}\Omega)$$

$$V_{DS} = 18 \text{ V} - (1.6 \text{ mA}) (4.7 \text{ k}\Omega)$$

$$V_{DS} = 18 \text{ V} - 7.52 \text{ V}$$

$$\boxed{V_{DS} = 10.48 \text{ volt.}}$$ (**Ans.**)

(d) We know that,

$$V_S = I_D R_S = 1.6 \text{ mA} \times 1.5 \text{ k}\Omega$$

$$\boxed{V_S = 2.4 \text{ volt.}}$$ (**Ans.**)

(e) From Fig. Ex. 7.21

$$\boxed{V_G = 0}$$ (**Ans.**)

We know that,

$$V_{DS} = V_D - V_S$$

$$V_D = V_{DS} + V_S$$

$$V_D = (10.48 + 2.4)\text{V}$$

$$\boxed{V_D = 12.52 \text{ volt.}}$$ (**Ans.**)

Example 7.22 Determine the following for the common gate configuration of Fig. Ex. 7.22

    (a) V_{GSQ} (b) I_{DQ}

    (c) V_D (d) V_G

    (e) V_S (f) V_{DS}

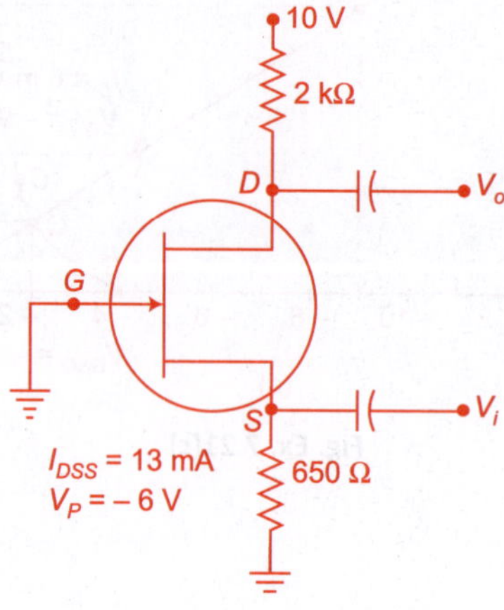

Fig. Ex. 7.22

Solution (a) The gate to source voltage is determined by,

$$V_{GS} = V_G - V_S$$

arbitrary choosing, $I_D = 5\text{mA}$

Now, $V_{GS} = 0 - I_D R_S$ ($\because$ $V_G = 0$)

$$V_{GS} = - (5 \text{ mA} \times 650 \text{ } \Omega)$$

$$V_{GS} = - 3.25\text{V}$$

Similarly choosing, $I_D = 10 \text{ mA}$

$$V_{GS} = - (10 \text{ mA} \times 650 \text{ } \Omega) = - 6.5\text{V}.$$

The self bias line is shown in Fig. Ex. 7.22(a)

Now, the points from transfer curve, we can determine,

Point A: $I_{DSS} = 13 \text{ mA}$ and $V_{GS} = 0\text{V}$

Point B: $V_{GS} = V_p = - 6\text{V}$ and $I_D = 0$

Point C: When $V_{GS} = \dfrac{V_p}{2} = - \dfrac{6V}{2} = - 3V$

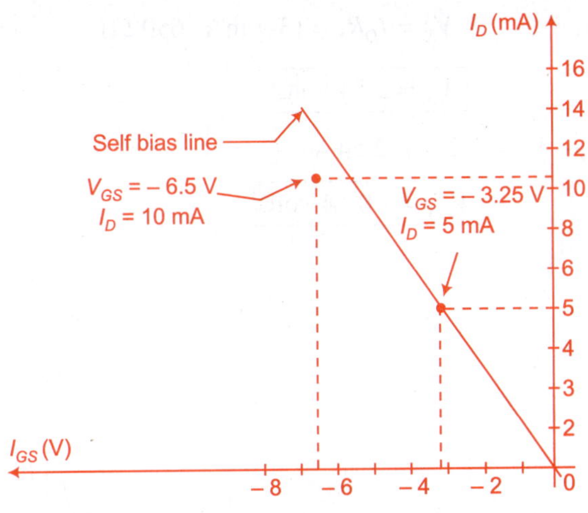

Fig. Ex. 7.22(a)

and the drain current,

$$I_D = \frac{I_{DSS}}{4} = \frac{13 \text{ mA}}{4} = 3.25 \text{ mA}.$$

Point D: At $I_D = \dfrac{I_{DSS}}{2} = \dfrac{13 \text{ mA}}{2} = 6.5$ mA

and,

$$V_{GS} = 0.3 \, V_P = 0.3 \, (-6 \text{ V})$$

$$V_{GS} = -1.8\text{V}$$

Now, the transfer curve is shown in Fig. Ex. 7.22(b).

Now, superimposing the self bias line curve as shown in Fig. Ex. 7.22(a) on a transfer curve which is shown in Fig. Ex. 7.22(b), the resultant curve is shown in Fig. Ex. 7.22(c). In Fig. 7.22 (c) we see that the intersection of transfer curve and self bias line is called as operating point. From Fig. Ex. 7.22(c)

(a) $\boxed{V_{GSQ} = -2.6 \text{ volt.}}$ (Ans.)

(b) $\boxed{I_{DQ} = +3.9 \text{ mA.}}$ (Ans.)

(c) $V_D = V_{DD} - I_D R_D$

$$V_D = 10 - (3.9 \text{ mA}) (2 \text{ k}\Omega)$$

$$V_D = (10 - 7.8) \text{ V} = 2.2\text{V}$$

$$\boxed{V_D = 2.2 \text{ volt.}}$$ (Ans.)

(d) $\boxed{V_G = 0.}$ (Ans.)

(e) We know that,

$$V_S = I_D R_S = (3.9 \text{ mA } (650 \text{ }\Omega)$$

$$\boxed{V_S = 2.54 \text{ volt.}}$$
(Ans.)

(f) $V_{DS} = V_D - V_S = 2.2 \text{ V} - 2.54 \text{ V}$

$$\boxed{V_{DS} = -0.34 \text{ volt.}}$$
(Ans.)

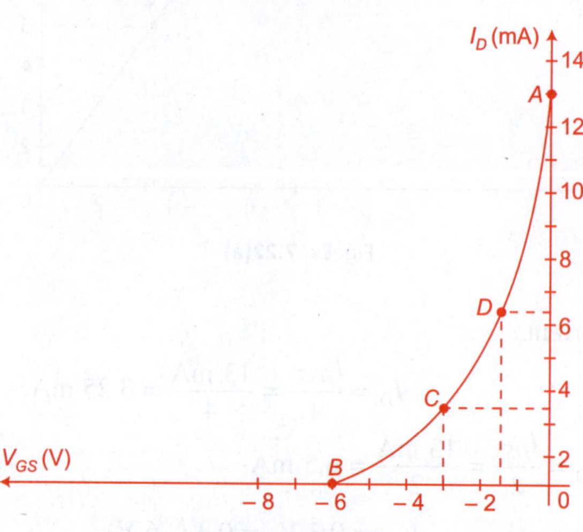

Fig. Ex. 7.22(b)

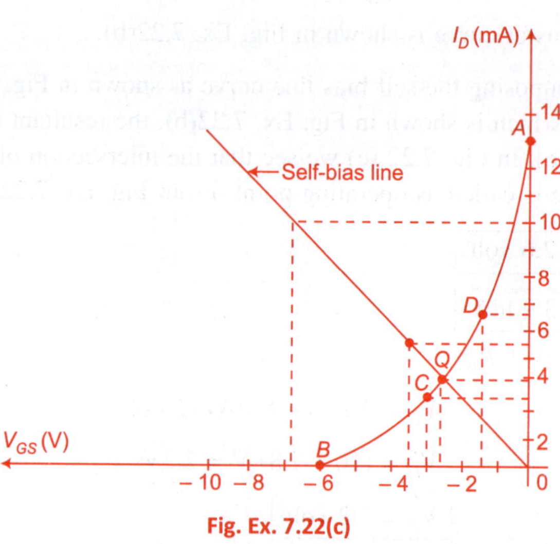

Fig. Ex. 7.22(c)

Example 7.23 The FET shown in Fig. Ex. 7.23 has the following parameters $I_{DSS} = 6$ mA, $V_p = 3$V if V_{in} and V_{out} are constant, then determine,

(a) V_{out} when $V_{in} = 0$

(b) V_{out} when $V_{in} = 10V$

(c) V_{in} when $V_{out} = 0V$

Solution Applying KVL to input side loop,

$$10 - I_D R_S + V_{GS} = 0$$

$$V_{GS} = I_D R_S - 10$$

$$V_{GS} = I_D R_S - 10 \qquad\qquad (A)$$

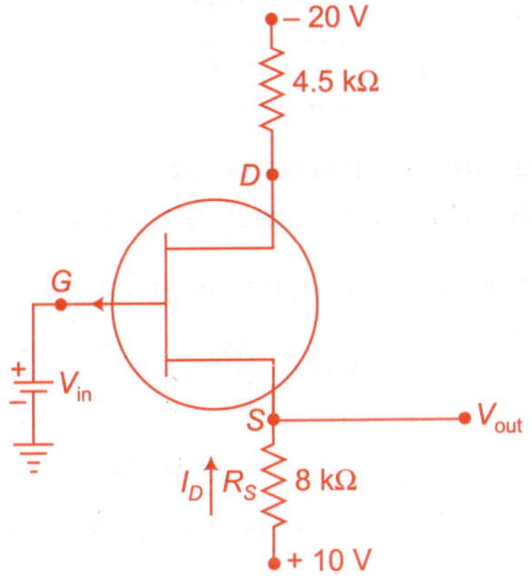

Fig. Ex. 7.23

we know that,

$$I_D = I_{DSS}\left[1 - \frac{V_{GS}}{V_p}\right]^2 = 6\ \text{mA}\left[1 - \frac{V_{GS}}{3V}\right]^2$$

Now, substituting the value of V_{GS} from Eqn. (A)

$$I_D = 6 \times 10^{-3}\left[1 - \frac{8\,I_D + 10}{3}\right]^2$$

$$= 6 \times 10^{-3}\left[\frac{3 - 8_D + 10}{3}\right]^2 = 6 \times 10^{-3}\left[\frac{7 - 8I_D}{3}\right]^2$$

$$= 6 \times 10^{-3}\left[\frac{49 + 64\,I_D^2 - 112\,I_D}{9}\right]$$

$$9\,I_D = 0.294 + 384 \times 10^3\,I_D^2 - 0.672\,I_D$$

$$384 \times 10^3\,I_D^2 - 9.672\,I_D + 0.294 = 0.$$

By solving the above equation,

$$I_D \simeq 1.5 \text{ mA}$$

$$V_{GS} = I_D R_S - 10\text{V}$$

$$= 1.5 \text{ mA} \times 8 \text{ k}\Omega - 10 \text{ V}$$

$$= 12 \text{ V} - 10\text{V}$$

$$\boxed{V_{GS} = 2 \text{ volt.}} \qquad \text{(Ans.)}$$

The output voltage will be,

$$V_{\text{out}} = 10 - I_D R_S$$

$$\boxed{V_{\text{out}} = -2 \text{ volt.}} \qquad \text{(Ans.)}$$

Example 7.24 The maximum transconductance of a certain *n*-channel JFET is 3.5 mS. If I_{DSS} is 25 mA, what is the pinch-off voltage?

Solution The maximum transconductance is given by,

$$g_{mo} = -2 \frac{I_{DSS}}{V_p} \Rightarrow V_p = \frac{-2 \, I_{DSS}}{g_{mo}}$$

$$V_p = \frac{-2 \times 25 \times 10^{-3}}{3.5 \times 10^{-3} \text{ A/V}} = \frac{-50}{3.5} \text{ V}$$

$$\boxed{V_p = -14.29 \text{ volt.}} \qquad \text{(Ans.)}$$

Example 7.25 For *n*-channel JFET the maximum saturation drain current is 8.2 mA, pinch-off voltage is -4V and gate to source voltage is -1V. Find the values of (i) I_D (ii) g_{mo} (iii) g_m.

Solution Data given that,

$$I_{DSS} = 8.2 \text{ mA}, \quad V_p = -4 \text{ V}$$

$$V_{GS} = -1\text{V.}$$

We know that the drain current,

(i)
$$I_D = I_{DSS}\left(1 - \frac{V_{GS}}{V_p}\right)^2$$

$$I_D = 8.2 \times 10^{-3}\left[1 - \left(\frac{-1}{-4}\right)^2\right] = 8.2 \times 10^{-3} \, (1 - 0.25)^2$$

$$I_D = 8.2 \times 10^{-3} \, (0.75)^2$$

$$I_D = 8.2 \times 0.562 \times 10^{-3} \text{ A}$$

$$I_D = 4.61 \text{ mA.} \qquad \textbf{(Ans.)}$$

(ii) Transconductance for $V_{GS} = 0$ i.e.

$$g_{mo} = \frac{-2\,I_{DSS}}{V_p} = \frac{-2 \times 8.2\text{mA}}{-4}$$

$$\boxed{g_{mo} = 4.1 \text{ mA/V.}} \qquad \textbf{(Ans.)}$$

(iii) Transconductance, g_m

$$g_m = g_{mo}\left(1 - \frac{V_{GS}}{V_p}\right)$$

$$g_m = 4.1 \times 10^{-3}\left(1 - \left(\frac{-1}{-4}\right)\right)^2$$

$$g_m = 4.1 \times 0.562 \times 10^{-3}$$

$$\boxed{g_m = 2.30 \text{ mA/V.}} \qquad \textbf{(Ans.)}$$

Example 7.26 Determine the following for the given circuit in Fig. Ex. 7.26.

(a) I_{DQ} and V_{GSQ}

(b) V_D (c) V_S

(d) V_{DS} (e) V_{DD}

Solution The gate to source voltage is determined,

$$V_{GS} = V_G - V_S \qquad (A)$$

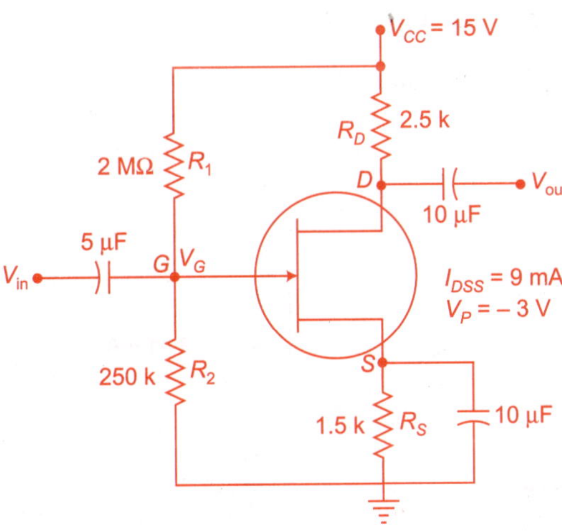

Fig. Ex. 7.26

From Fig. Ex. 7.26 the gate voltage is given by Thevenin equation

$$V_G = \frac{R_2}{R_1 + R_2} \times V_{DD} = \frac{(250\ \text{k}\Omega)\,(15\text{V})}{(250\ \text{k} + 2\text{M}\Omega)}$$

$$V_G = \frac{3750\ \text{k}\Omega\ \text{V}}{2.25\ \text{M}\Omega} = 1.67\text{V}$$

$$V_G = 1.67\ \text{V}.$$

Now, putting the value of V_G in Eqn. (A), we obtain,

$$V_{GS} = 1.67\ \text{V} - V_S$$

$\because$
$$V_S = I_D R_S$$

$\therefore$
$$V_{GS} = 1.67\ \text{V} - I_D R_S$$

$$V_{GS} = 1.67\ \text{V} - \text{I}\,(1.5\ \text{k}\Omega)$$

arbitrary choosing drain current,

where, $\quad I_D = 0$

$$V_{GS} = 1.67\ \text{V}$$

when $V_{GS} = 0$, then drain current is given by,

$$I_D = \frac{1.67\ \text{V}}{1.5\ \text{k}\Omega} = 1.11\ \text{mA}$$

$$I_D = 1.11\ \text{mA}$$

The bias line is shown in Fig. Ex. 7.26 (a).

Now, from the point for transfer characteristics curve we can determine,

Point A: $\quad I_{DSS} = 9$ mA and $V_{GS} = 0V$

Point B: $\quad V_{GS} = V_p = -3V \quad$ and $\quad I_D = 0$

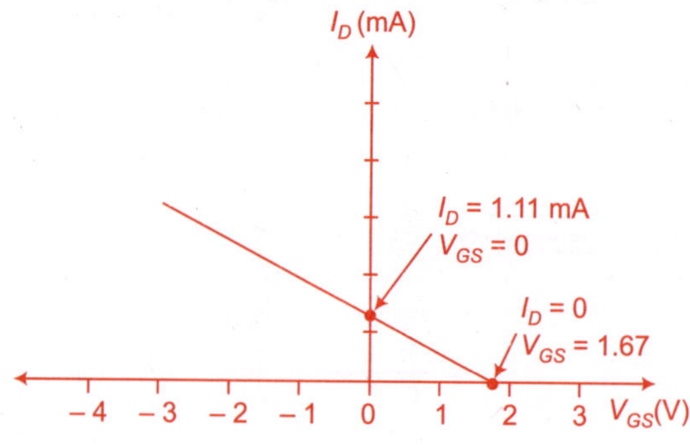

Fig. Ex. 7.26 *(a) bias line*

Point C: When $V_{GS} = \dfrac{V_P}{2} = \dfrac{-3}{2} = -1.5\text{V}$ and the drain current,

$$I_D = \frac{I_{DSS}}{4} = \frac{9\text{ mA}}{4} = 2.25\text{ mA}.$$

Point D: At $I_D = \dfrac{I_{DSS}}{2} = \dfrac{9\text{ mA}}{2} = 4.5\text{ mA}$

and,

$$V_{GS} = 0.3V_p = 0.3\,(-3\text{V})$$

$$V_{GS} = -0.9\text{ V.}$$

Now, the transfer curve is drawn and shown in Fig. Ex. 7.26 (b).

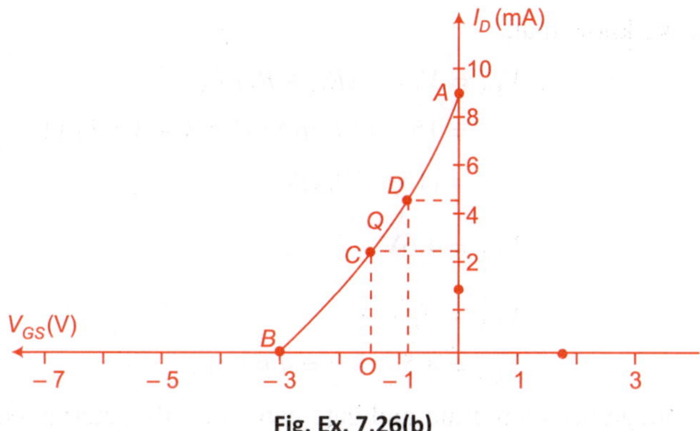

Fig. Ex. 7.26(b)

Now, superimposing the bias line curve on a transfer curve the resultant is shown in Fig. Ex. 7.26(c). In Fig. 7.26 (c), we can see that the intersection of transfer curve and bias line is called as Q-point from Fig. Ex. 7.26(c)

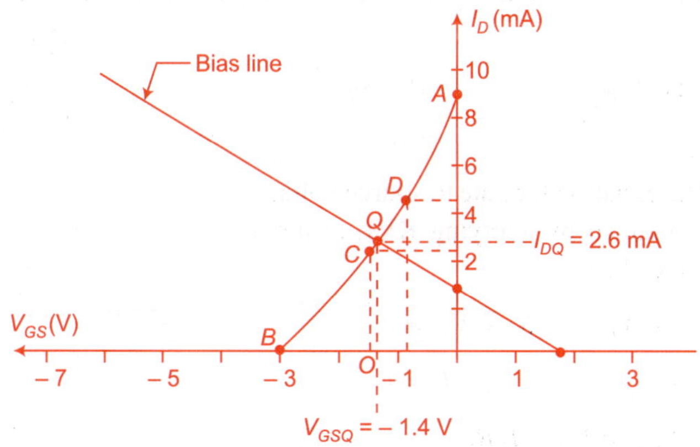

Fig. Ex. 7.26(c)

(a) $V_{GSQ} = -1.4\text{V}$

and, $\qquad I_{DQ} = 2.6$ mA.

(b) Drain voltage is given by,

$$V_D = V_{DD} - I_D R_D = 15 - (2.6 \text{ mA})(2.5 \text{ k}\Omega)$$

$$V_D = (15 - 6.5)\text{V}$$

$$\boxed{V_D = 8.5 \text{ volt.}} \qquad \textbf{(Ans.)}$$

(c) We know that,

$$V_S = I_D R_D = 2.6 \text{ mA} \times 1.5 \text{ k}\Omega$$

$$\boxed{V_S = 3.9 \text{ volt.}} \qquad \textbf{(Ans.)}$$

(d) Now, we know that,

$$V_{DS} = V_{DD} - (R_D + R_S) I_D$$

$$= 15 - (2.6 \text{ mA})(2.5 \text{ k} + 1.5 \text{ k})\,\Omega$$

$$= (15 - 10.4)\text{V}$$

$$V_{DS} = 4.6\text{V}$$

or, $\qquad V_{DS} = V_D - V_S$

$$V_{DS} = 8.5 - 3.9 = 4.6 \text{ volt.}$$

(e) The voltage between drain and gate can be easily determined by,

$$V_{DG} = V_D - V_G = (8.5 - 1.67) \text{ V}$$

$$\boxed{V_{DG} = 6.83 \text{ volt.}} \qquad \textbf{(Ans.)}$$

Example 7.27 Determine the following for the given circuit of Fig. Ex. 7.27.

(UPTU 2006-07)

(a) I_{DQ} and V_{GSQ} (b) V_{DS}

(c) V_D (d) V_S

Solution The equation for gate to source voltage V_{GS} can be obtained by applying KVL at input side of Fig. Ex. 7.27.

$$-V_{GS} - I_S R_S + V_{SS} = 0$$

$$\because \qquad I_S = I_D$$

$$\therefore \qquad V_{GS} = V_{SS} - I_D R_S$$

$$V_{GS} = 10 \text{ V} - I_D (1.5 \text{ k}\Omega) \quad \text{(A)}$$

Arbitrary choosing drain current,

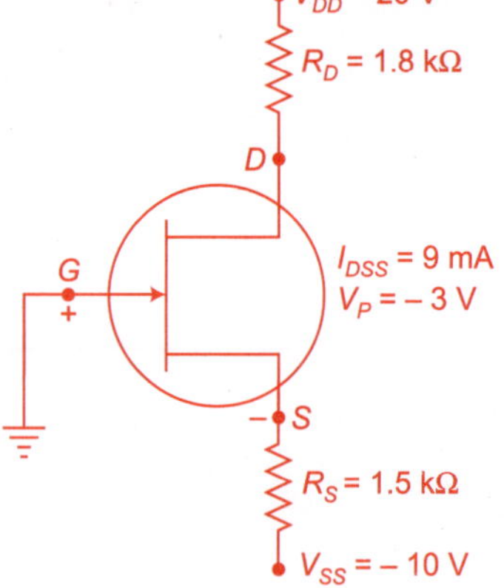

Fig. Ex. 7.27

when, $\qquad I_D = 0\text{mA}$

$\qquad\qquad V_{GS} = 10\text{V}.$

For, $\qquad V_{GS} = 0\text{V},$

$$0 = 10\text{V} - I_D\,(1.5\ \text{k}\Omega)$$

$$I_D = \frac{10\text{V}}{1.5\ \text{k}\Omega} = 6.67\ \text{mA}$$

$$I_D = 6.67\text{mA}.$$

The bias line plot is shown in Fig. Ex. 7.27(a)

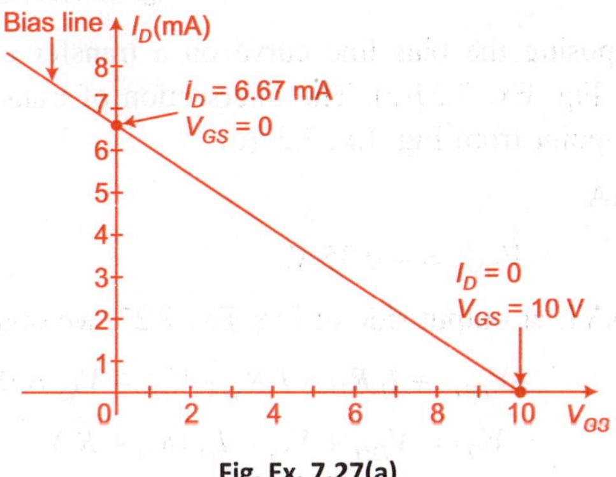

Fig. Ex. 7.27(a)

Now, the point for transfer curve, Fig. Ex. 7.27(a) we can determine,

Point A: $\quad I_{DSS} = g\ \text{mA}$ and,

$$V_{GS} = 0\text{V}$$

Point B: $\quad V_{GS} = V_p = -3\text{V}\quad$ and $\quad I_D = 0\ \text{mA}$

Point C: When $V_{GS} = \dfrac{V_p}{2} = \dfrac{-3}{2} = -1.5\text{V}$

and the drain current,

$$I_D = \frac{I_{DSS}}{4} = \frac{g\ \text{mA}}{4} = 2.25\ \text{mA}.$$

Point D: At $I_D = \dfrac{I_{DSS}}{2} = \dfrac{g\ \text{mA}}{2} = 4.5\ \text{mA}$

$$V_{GS} = 0.3\ V_P$$

$$V_{GS} = 0.3\ (-3) = -9\text{V}$$

$$V_{GS} = -0.9\text{V}.$$

The transfer curve is shown in Fig. 7.27 (b).

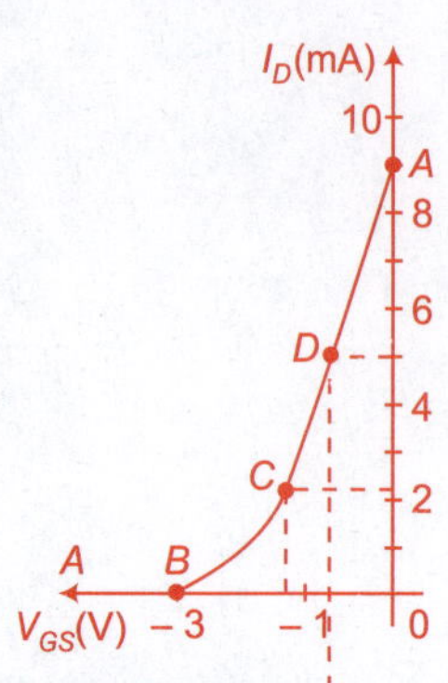

Fig. Ex. 7.27(b)

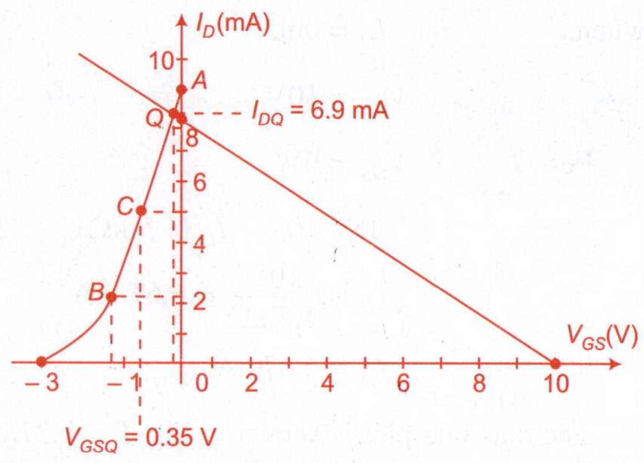

Fig. Ex. 7.27(c)

Now, superimposing the bias line curve on a transfer curve, the resultant curve is shown in Fig. Ex. 7.27(c). The intersection of transfer curve and bias line is called as Q-point from Fig. Ex. 7.27(c).

(a) $I_{DQ} = 6.9 \text{mA}$

$$V_{GSQ} = -0.35 \text{ V.}$$

(b) Applying KVL at output side of Fig. Ex. 7.27, we obtain,

$$-V_{DD} + I_D R_D + I_S R_S + V_{DS} - V_{SS} = 0$$

$$V_{DS} = V_{DD} + V_{SS} - I_D (R_D + R_S)$$

$$V_{DS} = 20\text{V} + 10\text{V} - (6.9 \text{ mA}) (1.8 \text{ k}\Omega + 1.5 \text{ k}\Omega)$$

$$V_{DS} = 30\text{V} - 22.77\text{V}$$

$$\boxed{V_{DS} = 7.23 \text{ volt.}} \qquad \textbf{(Ans.)}$$

(c) $V_D = V_{DD} - I_D R_D$

$$= 20 - (6.9 \text{ mA}) (1.8 \text{ k}\Omega)$$

$$= 20 - 12.42\text{V}$$

$$\boxed{V_D = 7.58 \text{ volt.}} \qquad \textbf{(Ans.)}$$

(d) We know that,

$$V_{DS} = V_D - V_S$$

$$V_S = -V_{DS} + V_D \Rightarrow V_D - V_{DS}$$

$$V_S = 7.58\text{V} - 7.23\text{V}$$

$$\boxed{V_S = 0.35 \text{ volt.}} \qquad \textbf{(Ans.)}$$

Example 7.28 For n-channel depletion-type MOSFET of circuit in Fig. Ex. 7.28. Determine,

(a) I_{DQ} and V_{GSQ}

(b) V_{DS}

Solution The equation for gate to source voltage V_{GS} can be determined by,

$$V_{GS} = V_G - V_S$$

$\therefore$

$$V_G = \frac{R_2}{R_1 + R_2} \times V_{DD}$$

$$= \frac{10 \text{ m}\Omega \times 15 \text{ V}}{(10 + 100) \text{ M}\Omega}$$

$$V_G = \frac{10 \text{ M}\Omega \times 15\text{V}}{110\text{M}\Omega} = 1.36 \text{ V}$$

$$\boxed{V_G = 1.4 \text{ volt.}}$$

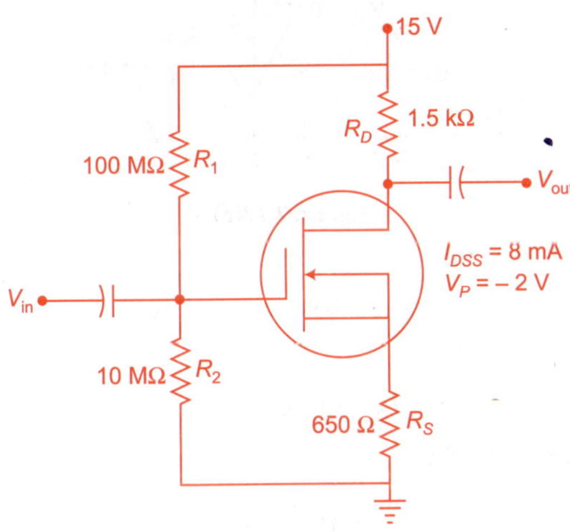

Fig. Ex. 7.28

Now,

$$V_{GS} = 1.4 \text{ V} - I_D R_S$$
$$= 1.4 \text{ V} - I_D (650 \text{ }\Omega). \qquad \text{(A)}$$

Arbitrary choosing drain current,

when,

$$I_D = 0 \text{ mA results in}$$

$$\boxed{V_{GS} = 1.4\text{V.}}$$

Now,

$$V_{GS} = 0$$

$$0 = 1.4\text{V} - I_D (750 \text{ }\Omega)$$

$$I_D = \frac{1.4\text{V}}{750} = 1.8 \text{ mA}$$

$$\boxed{I_D = 1.8 \text{ mA.}}$$

Plotting the points and bias line as shown in Fig. Ex. 7.28(a). Now, the points for transfer curve, we can determine,

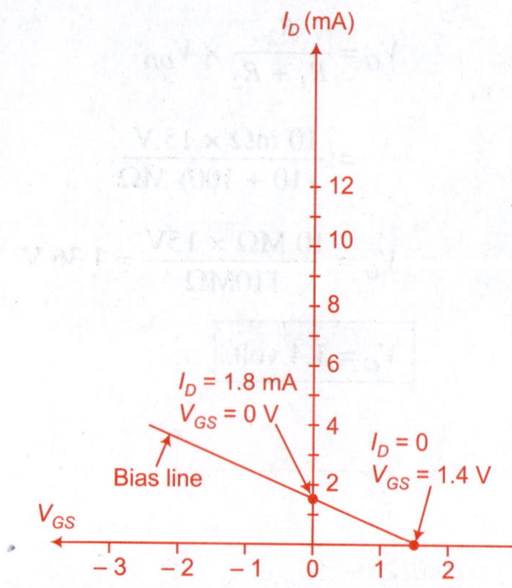

Fig. Ex. 7.28(a)

Point A: $I_{DSS} = 8$ mA and

$$V_{GS} = 0V$$

Point B: $V_{GS} = V_P = -2V$

and,

$$I_D = 0 \text{mA}$$

Point C: $I_D = \dfrac{I_{DSS}}{4} = \dfrac{8 \text{ mA}}{4} = 2 \text{ mA}$

and,

$$V_{GS} = \frac{V_P}{2} = \frac{-2}{2} = -1V$$

Point D: Shockley equation yields,

$$I_D = I_{DSS}\left(1 - \frac{V_{GS}}{V_P}\right)^2$$

$$= (8 \text{ mA})\left(1 - \frac{1V}{-2V}\right)^2 = 8 \text{ mA } (1.5)^2$$

$$= 18 \text{ mA}$$

i.e. The curve rises more rapidly as V_{GS} becomes more positive.

The transfer curve is shown in Fig. Ex. 7.28(b).

Now, superimposing the bias line curve on a transfer curve, the resultant curve is show in Fig. Ex. 7.28(c). The intersection point of two curves is called as Q-point. From Fig. Ex. 7.28(c)

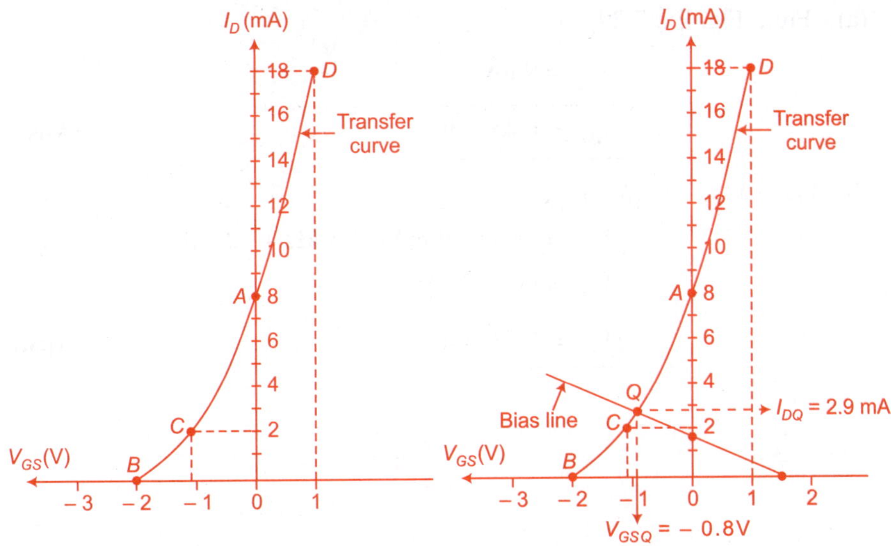

(a) $I_{DQ} = 2.9$ mA

$$V_{GSQ} = -0.8 \text{ mA.}$$

(b) $V_{DS} = V_{DD} - I_D (R_D + R_S)$

$$= 15\text{V} - (2.9 \text{ mA}) (1.5 \text{ k}\Omega + 0.65 \text{ k}\Omega)$$

$$= 15\text{V} - 6.24 \text{ V}$$

$$\boxed{V_{DS} \cong 8.76 \text{ volt.}} \hspace{3cm} \textbf{(Ans.)}$$

Example 7.29   Repeat the Example 7.28 for $R_S = 120 \ \Omega$.

Solution   The equation for gate to source voltage is similar to Example 7.28

$$V_{GS} = 1.4 - I_D R_S$$

$\because$
$$R_S = 120 \ \Omega$$
$$V_{GS} = 1.4 - I_D (120 \ \Omega)$$

where, $I_D = 0$ mA

$$V_{GS} = 1.4\text{V}$$

when, $V_{GS} = 0$V yields

$$0 = 1.4 - I_D (120 \ \Omega)$$

$$I_D = \frac{1.4 \text{ V}}{120 \ \Omega} = 11.7 \text{ mA}$$

The transfer curve is similar to Example 7.28. The plot in Fig. 7.29 shows the resultant plot.

(a) From Fig. Ex. 7.29

$$I_{DQ} = 9\text{mA}$$

$$\boxed{V_{GSQ} = 0.45 \text{ volt.}}$$ **(Ans.)**

(b) $V_{DS} = V_{DD} - I_D (R_D + R_S)$

$$V_{DS} = 15 - (9.0 \text{ mA}) (1.5 \text{ k}\Omega + 120 \text{ }\Omega)$$

$$V_{DS} = (15 - 14.58)\text{V}$$

$$\boxed{V_{DS} = 0.42 \text{ volt.}}$$ **(Ans.)**

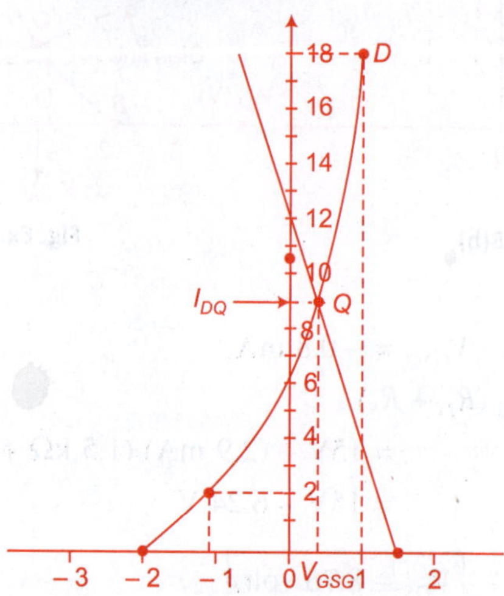

Fig. Ex. 7.29

Example 7.30 Determine V_{DS} for circuit in Fig. Ex. 7.30

Solution From Fig. Ex. 7.30, given,

$$V_{GS} = 0\text{V}$$

Since V_{GS} is found at 0V, the drain current must be I_{DSS}.

i.e.

$$V_{GSQ} = 0\text{V}$$

$$I_{DQ} = I_{DSS} = 8 \text{ mA}$$

Therefore, there is no need to draw the transfer curve and,

$$V_D = V_{DD} - I_D R_D = 0$$

$$V_D = 18 - (8\text{mA}) (1.6 \text{ k}\Omega)$$

$$V_D = (18 - 12.8)\text{V}$$

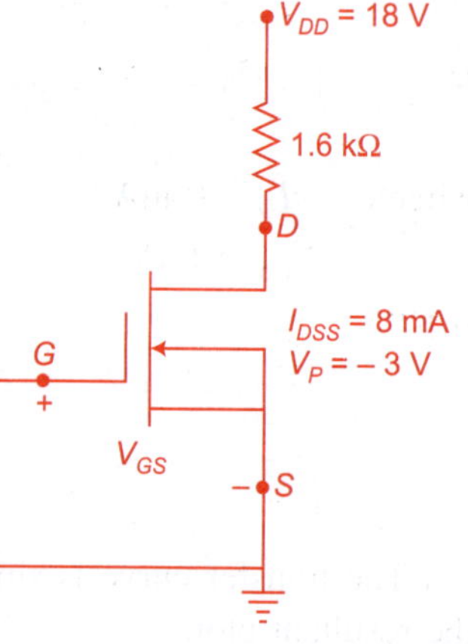

Fig. Ex. 7.30

$$V_D = 5.2\text{V}$$

Now, $V_{DS} = V_D - V_S \quad \because \quad V_S = 0$

$\therefore$ $V_{DS} = V_D = 5.2 \text{ V}$

$$\boxed{V_{DS} = 5.2 \text{ volt.}}$$ **(Ans.)**

Example 7.31   Determine the following for given circuit in Fig. Ex. 7.31.

(a) I_{DQ} and V_{GSQ}

(b) V_D

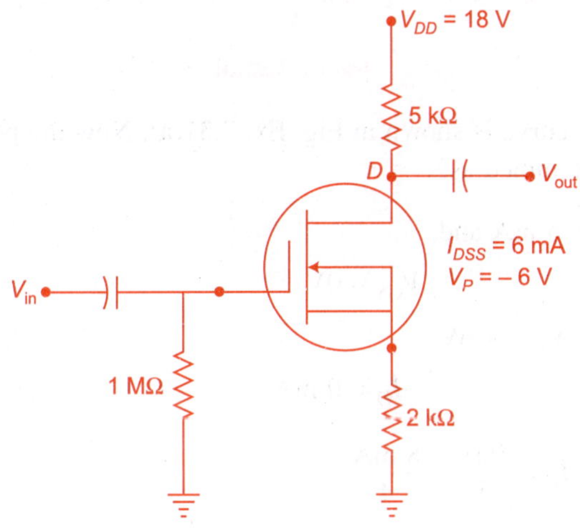

Fig. Ex. 7.31

Solution   The equation for gate to source voltage V_{GS} can be determined,

$$V_{GS} = V_G - V_S$$
$$V_{GS} = V_G - I_D R_S$$

$\because$ $V_G = 0$

$\therefore$ $V_{GS} = - I_D R_S.$

Arbitrary choosing drain current,

when, $I_D = 0\text{mA}$ results is

$$V_{GS} = 0\text{V}$$

when, $V_{GS} = - 4\text{V}$

then, $I_D = - \dfrac{- 4\text{V}}{2.0 \text{ k}\Omega} = 2 \text{ mA}$

$$I_D = 2 \text{ mA}.$$

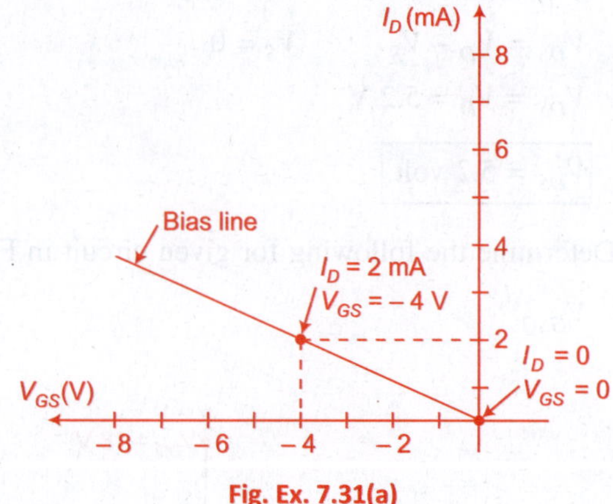

Fig. Ex. 7.31(a)

The bias line curve is shown in Fig. Ex. 7.31(a). Now the points for transfer curve can be determined as:

Point A: $I_{DSS} = 6$ mA and,

$$V_{GS} = 0\text{V}$$

Point B: $V_{GS} = V_P = -6\text{V}$

$$I_D = 0 \text{ mA}$$

Point C: when $I_D = \dfrac{I_{DSS}}{4} = \dfrac{8 \text{ mA}}{4}$

$$I_D = 2 \text{ mA}$$

and,

$$V_{GS} = \dfrac{V_p}{2} = \dfrac{-6}{2} = -3\text{volt.}$$

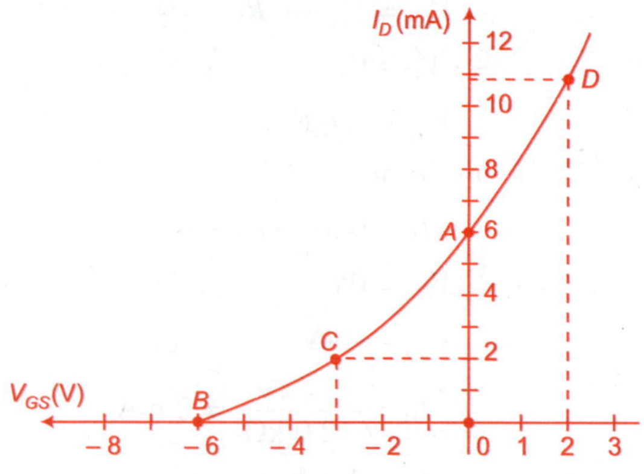

Fig. Ex. 7.31(b)

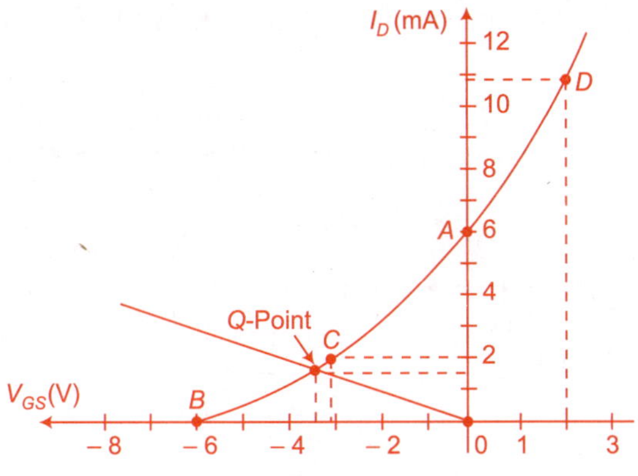

Fig. Ex. 7.31(c)

Point D: $V_{GS} > 0$ i.e. $V_{GS} = 2\text{V}$

$$I_D = I_{DSS}\left(\frac{1 - V_{GS}}{V_P}\right)^2$$

$$I_D = 6\ \text{mA}\left(1 - \frac{2}{-6}\right)^2$$

$$I_D = 10.61\ \text{mA}.$$

Now, superimposing the bias line curve on a transfer line curve which is shown in Fig. 7.31(c). It will give the Q-point.

From Fig. Ex. 7.31(c),

(a) $I_{DQ} = 1.55\ \text{mA}$

$$V_{GSQ} = -3.4\text{V}$$

(b) $V_D = V_{DD} - I_D R_D$

$$V_D = 18 - (1.55\ \text{mA})(5\ \text{k}\Omega)$$

$$V_D = (18 - 7.75)\text{V}$$

$$\boxed{V_D = 10.25\ \text{volt.}}$$
(Ans.)

Example 7.32 Determine I_{DQ} and V_{GSQ} for the enhancement-types MOSFET of Fig. Ex. 7.32.

Solution   We know the equation,

$$I_D = k(V_{GS} - V_T)^2$$

$$k = \frac{I_D}{(V_{GS} - V_T)^2}$$

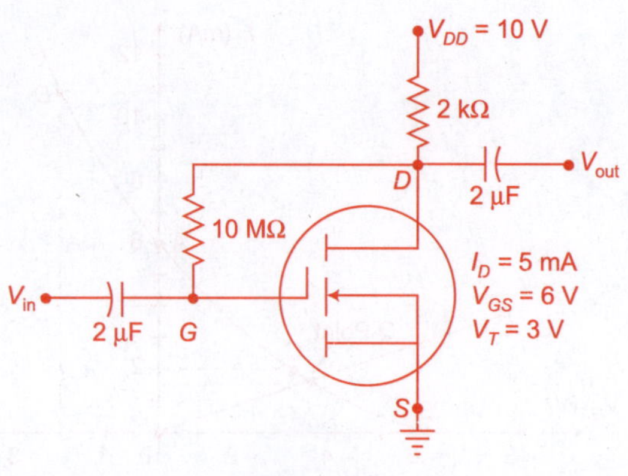

Fig. Ex. 7.32

$$k = \frac{5 \text{ mA}}{(6-3)^2} = \frac{5 \text{ mA}}{9}$$

$$k = 0.56 \text{ mA/V}$$

For, $\qquad V_{GS} = 5\text{V}$ (select between $3 \times 6\text{V}$)

$$I_D = 0.56 \times 10^{-3} (5-3)^2 = 2.22 \text{ mA}$$

$$\boxed{I_D = 2.2 \text{ mA.}}$$

For, $\qquad V_{GS} = 8\text{V}$ (slightly greater than V_T)

$$I_D = 0.56 \times 10^{-3} (8-3)^2 = 0.56 \times 10^{-3} (25)$$

$$I_D = 14 \text{ mA.}$$

The transfer curve is shown in Fig. Ex. 7.32(a)

If, $\qquad\qquad V_{GS} = V_T = 3\text{V}$

$$I_D = 0\text{mA}$$

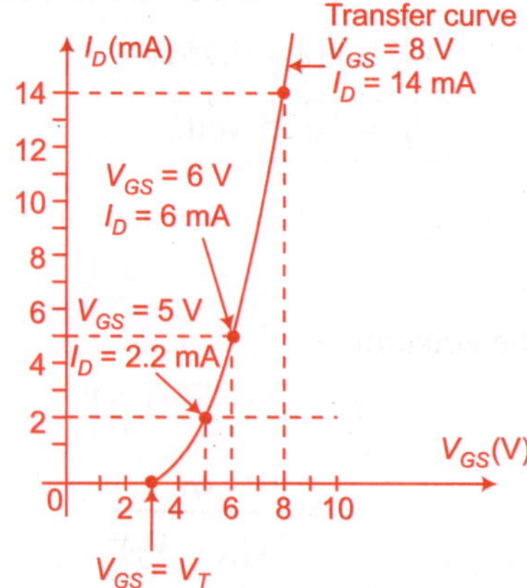

Fig. Ex. 7.32(a)

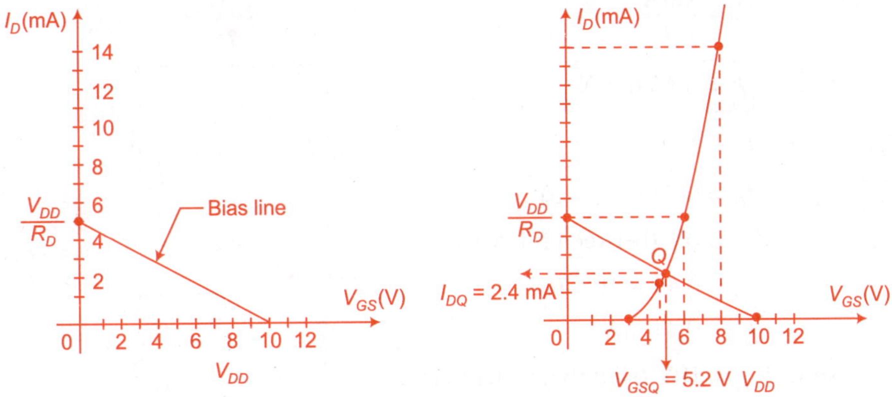

Fig. Ex. 7.32(b)   **Fig. Ex. 7.32(c)**

For bias line,

$$V_{GS} = V_{DD} - I_D R_D$$

$$V_{GS} = 10V - I_D\,(2\text{ k}\Omega) \tag{A}$$

when,

$$I_D = 0$$

$$V_{GS} = 10V$$

when,

$$V_{GS} = 0$$

$$I_D = \frac{10V}{2\text{ k}\Omega} \text{ from Eqn. (A)}$$

$$I_D = 5\text{mA}.$$

The resulting bias line is shown in Fig. 7.32 (b) and the resultant is shown in Fig. Ex. 7.32(c). In which the Transfer curve is superimposed on a bias line and the intersection point of both is called as Q-point.

$$V_{GSQ} = 5.2V$$

$$I_{DQ} = 2.4\text{ mA}$$

and,

$$\boxed{V_{DSQ} = V_{GSQ} = 5.2 \text{ volt.}} \qquad \textbf{(Ans.)}$$

Example 7.33 Determine I_{DQ}, V_{GSQ}, and V_{DS} for given circuit of Fig. Ex. 7.33.

Solution   We know the Eqn.:

$$I_D = k\,(V_{GS} - V_T)^2$$

$$k = \frac{I_D}{(V_{GS} - V_T)^2} = \frac{4\text{mA}}{(8 - 5)^2}$$

$$k = \frac{4\text{mA}}{9}$$

$$\boxed{k = 0.44 \text{ mA/V.}}$$

i.e.

$$I_D = k\,(V_{GS} - V_T)^2$$

$$I_D = 0.44\,(V_{GS} - 5)^2$$

For, $V_{GS} = 6\text{V}$ (between 5 k·8 V)

$$I_D = 0.44 \times 10^{-3}\,(6 - 5)^2$$

$$I_D = 0.44 \text{ mA.}$$

Now, $V_{GS} = 10\text{V}$ (slightly greater than V_T)

$$I_D = 0.44 \times 10^{-3}\,(10 - 5)^2$$

$$I_D = 11 \cdot 11$$

mA

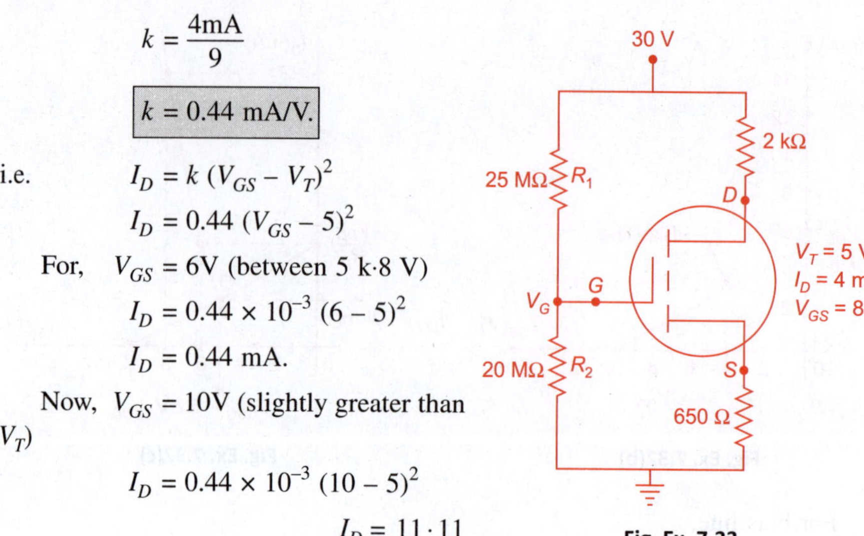

Fig. Ex. 7.33

If, $V_{GS} = V_T$, then $I_D = 0$.

Now, determining the point for plotting transfer curve.

We know that,

$$V_G = \frac{R_2}{R_1 + R_2}\,V_{DD} = \frac{20 \text{ M}\Omega \times (30 \text{ V})}{(20 + 25) \text{ M}\Omega}$$

$$\boxed{V_G = 13.33\text{V.}}$$

Now,

$$V_{GS} = V_G - V_S = V_G - I_D R_S$$

$$V_{GS} = 13.33 - I_D\,(650 \ \Omega) \tag{A}$$

when, $I_D \doteq 0 \text{ mA}$

$$V_{GS} = 13.33 \text{ V}$$

when, $V_{GS} = 0\text{V}$

$$0 = 13.33 - I_D\,(650 \ \Omega)$$

$$I_D = \frac{13.33}{650\Omega} = 20.5 \text{ mA.}$$

Now, plotting the transfer curve and bias line on the same scale as shown in Fig. Ex. 7.33(a).

From Fig. 7.33 (a),

$$I_{DQ} = 6.5 \text{ mA}$$

$$V_{GSQ} = 8.5 \text{ V} \qquad\qquad \textbf{(Ans.)}$$

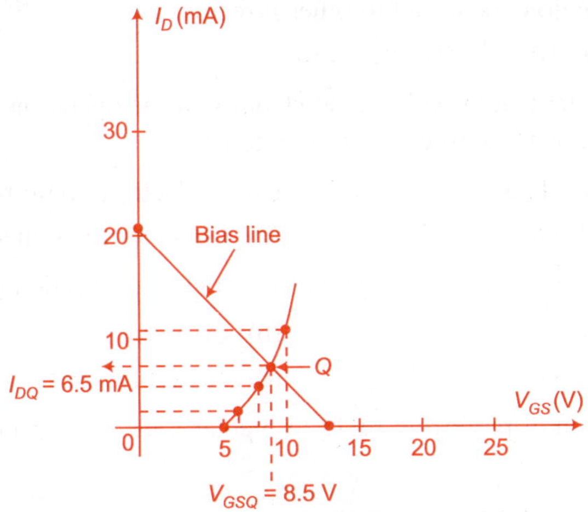

Fig. Ex. 7.33(a)

Now,
$$V_{DS} = V_{DD} - I_D (R_S + R_D)$$

$$V_{DS} = 30 - (6.5 \text{ mA}) (2 \text{ k}\Omega + 0.65 \text{ k}\Omega)$$

$$V_{DS} = 30 - (6.5 \text{ mA}) (2.65 \text{ k}\Omega)$$

$$V_{DS} = (30 - 17.23) \text{ V}$$

$$\boxed{V_{DS} = 12.77 \text{ volt.}}$$ **(Ans.)**

Objective Type Question

1. The metal from which the electrons are emitted is called

   (a) An emitter
   (b) Base
   (c) Collector
   (d) None of these

2. An emitter is generally called

   (a) Anode
   (b) Cathode
   (c) Drain
   (d) Grid

3. The work function of metal is generally expressed in

   (a) Volts
   (b) Joules
   (c) Electron volts
   (d) Electrons

4. A transistor is a combination of two *pn*-junction with their

   (a) *p*-region connected together
   (b) *n*-region connected together

(c) n-region connected to other p-region

(d) Both (A) and (B) are correct

5. The biasing that provides good stability of the operation point but fails to provide good amplification is said to be

    (a) Fixed bias

    (b) Collector to base bias

    (c) Self bias

    (d) Collector to emitter bias

6. The arrow in a transistor symbol indicates the direction of current is

    (a) Base

    (b) Collector

    (c) Emitter

    (d) None of these

7. The reason of preferring FETs to bipolar transistors in integrated circuits is that

    (a) It has become a tradition

    (b) They occupy very less space

    (c) Fabrication of circuit is easy

    (d) It is cheaper

8. FET is a device which has

    (a) High input impedance and is voltage controlled

    (b) High Z_i and is current controlled

    (c) Low Z_i and is voltage controlled

    (d) Low Z_i and is current controlled

9. FET is a .... controlled device.

    (a) Current

    (b) Impedance

    (c) Voltage

    (d) None of these

10. The terminals in FET are named as

    (a) Emitter, base, collector

    (b) Drain, source, gate

    (c) Anode, cathode, gate

    (d) None of these

11. FET is a .... device.

    (a) Bipolar

    (b) Unipolar

    (c) Both (a) & (b)

    (d) None of these

12. Current conduct in FET is due to

    (a) Majority carriers only

    (b) Minority carries only

    (c) Majority and minority carriers both

    (d) None of these

13. The input impedance of FET is an order of
    - (a) kΩ
    - (b) MΩ
    - (c) Many hundred MΩ
    - (d) None of these

14. FET is .... sensitive for applied input voltage
    - (a) More
    - (b) Less
    - (c) Not
    - (d) None of these

15. The temperatue stabiltity of .... is greater
    - (a) Diode
    - (b) BJT
    - (c) FET
    - (d) None of these

16. MOSFET is also called as...........
    - (a) BJT
    - (b) FET
    - (c) IGFT
    - (d) None of these

17. In FETs electrons emits from .... terminal
    - (a) Drain
    - (b) Source
    - (c) Gate
    - (d) None of these

18. The region of the NMOSFET between the depletion region is called
    - (a) Substrate
    - (b) Channel
    - (c) Bar
    - (d) None of these

19. The width of depletion region is dependent upon the ....
    - (a) Drain to source voltage
    - (b) Gate voltage
    - (c) Reverse voltage between gate terminal & source
    - (d) None of these

20. The value of V_{GS} at which the channel is cut-off is called
    - (a) Threshold
    - (b) Pinch-off
    - (c) Threshold or pinch-off
    - (d) None of these

21. The output characteristics of JFET is drawn between
    - (a) V_{GS} and I_D
    - (b) V_{DS} and I_D
    - (c) V_{GS} and V_{DS}
    - (d) None of these

22. In JFET, the region is called as .... when drain to source voltage is greater than pinch-off voltage.
    - (a) Ohmic region
    - (b) Linear region
    - (c) Saturation region
    - (d) None of these

23. JFET is future classified as

    (a) *p*-channel, *n*-channel

    (b) Enhancement type, depletion type

    (c) Both (a) & (b)

    (d) None of these

24. Depletion MOSFET can conduct in .... mode.

    (a) Enhancement

    (b) Depletion

    (c) Both (a) & (b)

    (d) None of these

25. Enhancement MOSFET can conduct in .... mode.

    (a) Enhancement

    (b) Depletion

    (c) Both (a) & (b)

    (d) None of these

26. In current equation of MOSFET and NMOSFET. The .... differs.

    (a) Voltages

    (b) Current

    (c) Polarity of voltage

    (d) None of these

27. The drain resistance is given by

    (a) $\dfrac{\Delta V_{GS}}{\Delta I_D}$

    (b) $\dfrac{\Delta I_D}{\Delta V_{DS}}$

    (c) $\dfrac{\Delta V_{DS}}{\Delta I_D}$

    (d) None of these

28. Amplification factor of FET is given by

    (a) $\left(\dfrac{\Delta V_{GS}}{\Delta V_{DS}}\right)_{I_G\,=\,constant}$

    (b) $\left(\dfrac{\Delta V_{DS}}{\Delta V_{GS}}\right)_{I_D\,=\,constant}$

    (c) $\left(\dfrac{\Delta V_{DS}}{\Delta V_{GS}}\right)_{I_G\,=\,constant}$

    (d) None of these

29. A .... configuration is also known as source follower:

    (a) Common drain

    (b) Common gate

    (c) Common source

    (d) None of these

30. The point on a load that gives drain current I_D for desired values of V_{GS} or vice versa is called as

    (a) Operating point

    (b) *Q*-point

    (c) Both (a) & (b)

    (d) None of these

31. The Q-point is set at .... point of the DC load line
    (a) Upper end   (b) Mid
    (c) Lower end   (d) None of these

32. In PMOSFET, the substrate is made with ....
    (a) p-type   (b) n-type
    (c) Both (a) & (b)   (d) None of these

33. In NMOSFET, the substance is made with ....
    (a) p-type   (b) n-type
    (c) Both (a) & (b)   (d) None of these

34. In NMOSFET, the two regions of n-type are,
    (a) Heavily doped   (b) Lightly doped
    (c) No doping   (d) None of these

35. When V_{DS} increases after $V_{GS} - V_T$, then the effective channel length,
    (a) Increases   (b) Decreases
    (c) No effect   (d) None of these

36. The drain current in saturation region is given by
    (a) $I_D = k \left[2(V_{GS} - V_T)V_{DS} - V_{DS}^2 \right]$
    (b) $I_D = k (V_{GS} - V_T)^2$
    (c) $I_D = k (2V_{GS} - V_T)^2$
    (d) None of these

37. In NMOSFET, the conduction parameter k is given by,
    (a) $\dfrac{1}{2} \mu_n c_{ox} \left(\dfrac{W}{L} \right) \text{A/V}^2$ (b) $\dfrac{1}{2} \mu_n co_x^2 \left(\dfrac{W}{L} \right)$

    (c) $\dfrac{1}{2} \mu_n c_{ox} \left(\dfrac{L}{W} \right) \text{A/V}^2$ (d) None of these

38 The circuit symbol for E-N MOSFET is,

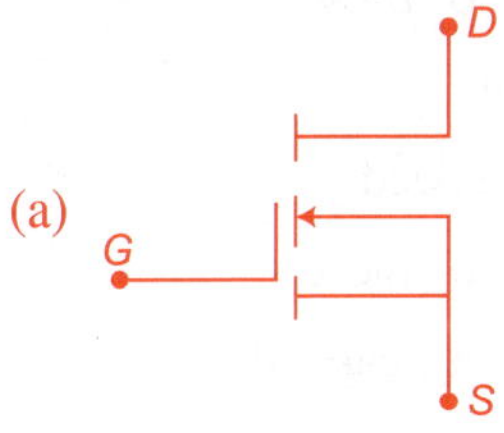

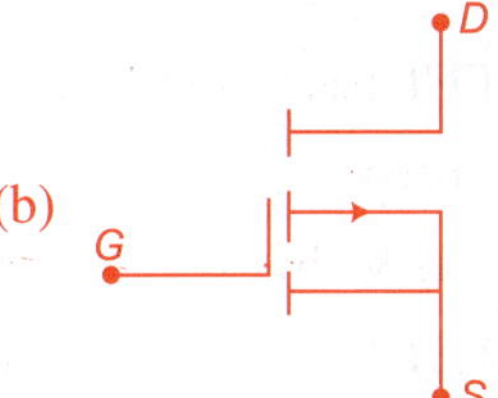

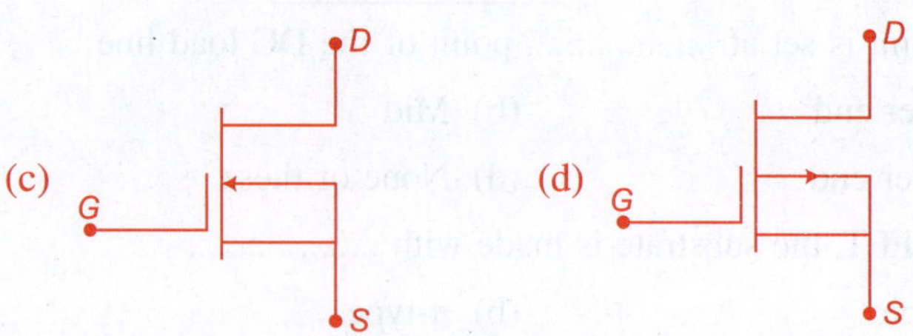

39. The circuit symbol for D-N MOSFET is copy above diagram

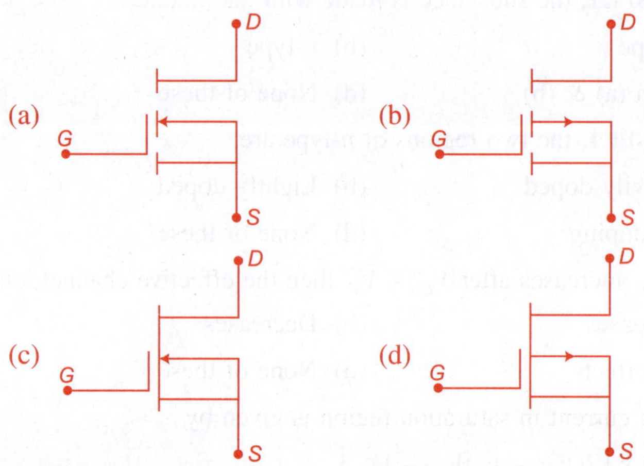

40. MOSFET requires .... size of BJT.

   (a) Twice
   (b) Thrice
   (c) One third
   (d) Two third

41. Circuit using MOSFET have .... density per chip area.

   (a) Lower
   (b) Higher
   (c) Equal
   (d) None of these

42. MOSFET consumes .... power, as of BJT.

   (a) Less
   (b) More
   (c) Equal
   (d) None of these

43. The MOSFET has .... of operation.

   (a) Low speed
   (b) High speed
   (c) Both (a) & (b)
   (d) None of these

44. The MOSFET has .... current driving capacity.

   (a) Low
   (b) High
   (c) No
   (d) None of these

45. The .... of PMOSFET is low as compared to NMOSFET.
    - (a) Switching speed
    - (b) Current driving capacity
    - (c) Both (a) & (b)
    - (d) None of these

46. Input Impedance of MOSFET is
    - (a) Less than that of FET but more than BJT
    - (b) More than that of FET and BJT
    - (c) More than that of FET but less than BJT
    - (d) Less than that of FET and BJT

47. The pinch-off voltage for JFET is 4V, when $V_{GS} = 1$, the pinch-off occurs for V_{DS} is equal to
    - (a) 5V
    - (b) 3V
    - (c) 2V
    - (d) 1 V

48. V_p is the pinch-off voltage for $V_{GS} = 0$ is a FET. When the gate is reverse biased by V_{GS} the pinch-off voltage
    - (a) Will be less than V_p
    - (b) Will be more than V_p
    - (c) Same as V_p
    - (d) Does not depend on V_{GS}

49. Which of the following is expected to have the higher input impedance?
    - (a) MOSFET
    - (b) CE transistor
    - (c) CC transistor
    - (d) JFET

50. The pinch-off voltage of a JFET is 4V. Its cut-off voltage is
    - (a) $(4)\frac{1}{2}$V
    - (b) 3V
    - (c) 8V
    - (d) 2V

51. In a JFET, V_{GS} off equals 4V and $I_{DSS} = 8$ mA. If V_{GS} is 2V, the drain current will be
    - (a) 1 mA
    - (b) 2 mA
    - (c) 4 mA
    - (d) 6 mA

52. In a JFET, drain current is maximum, when V_{GS} is
    - (a) Zero
    - (b) Negative
    - (c) Positive
    - (d) Equal to V_p

53. After V_{DS} reaches pinch-off values V_P in a JFET, drain current I_D becomes
    - (a) Zero
    - (b) Low
    - (c) Saturated
    - (d) Reversed

54. In common source configuration the input resistance of JFET will be approximately
    - (a) $10^{15}\,\Omega$
    - (b) $500\,\Omega$
    - (c) $100\,\Omega$
    - (d) $3\,M\Omega$

55. Self bias cannot be used in
    - (a) BJT circuit
    - (b) JFET
    - (c) EMOSFET
    - (d) Depletion mode

56. Pinch-off voltage in a FET is
    - (a) The drain voltage that gives zero drain current
    - (b) The gate to source voltage that gives unity I_D
    - (c) The V_{GS} that gives zero drain current
    - (d) The drain voltage that gives infinite drain current.

ANSWERS

1 (a)	2 (b)	3 (c)	4 (d)	5 (b)	6 (c)	7 (a)
8 (a)	9 (c)	10 (b)	11 (b)	12 (a)	13 (c)	14 (b)
15 (c)	16 (c)	17 (b)	18 (b)	19 (c)	20 (b)	21 (b)
22 (c)	23 (a)	24 (c)	25 (a)	26 (c)	27 (c)	28 (c)
29 (a)	30 (c)	31 (b)	32 (b)	33 (a)	34 (a)	35 (b)
36 (b)	37 (a)	38 (a)	39 (c)	40 (c)	41 (b)	42 (a)
43 (a)	44 (a)	45 (a)	46 (b)	47 (b)	48 (d)	49 (b)
50 (d)	51 (b)	52 (d)	53 (c)	54 (d)	55 (a)	56 (c)

Exercise

7.1. Why there is no gate current in the operation of a MOSFET? Explain.

7.2. Sketch the structure of a *n*-channel JFET and show its circuit symbol.

7.3. Draw the circuit of fixed biased FET and explain it.   **(UPTU 2002-03)**

7.4. What is the channel length modulation in a MOSFET?

7.5. Derive the expression for A_V, R_i and R_o of a JFET source follower.

7.6. Define the aspect ratio of a MOSFET. What is its significance?

7.7. For the circuit shown in Fig. E 7.1, determine,
    - (a) V_G and V_D at which the MOSFET is at the boundary between the pinch-off and Ohmic mode,
    - (b) V_D for $V_G = 5V$ assume $V_T = 1V$ and $k = 100\ \mu/V^2$

7.8. Write short notes on MOSFET.

7.9. What is the significant difference between the construction of an enhancement type and depletion types MOSFET.

7.10. What do you understand by pinch-off voltage and cut-off voltage as applied to FET?

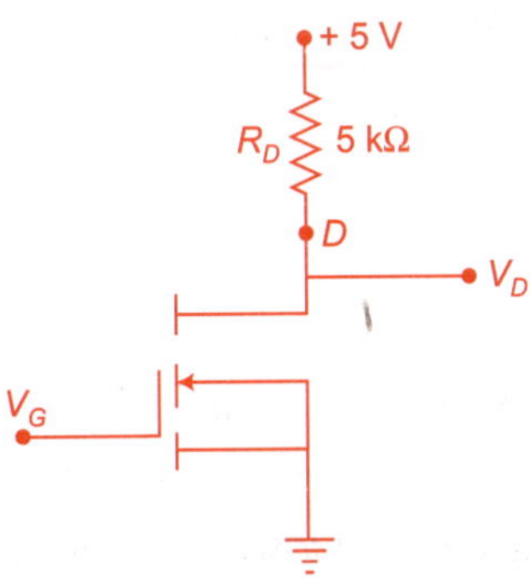

Fig. E 7.1

7.11. Sketch the small signal high frequency circuit of a CS amplifier. Also derive the expression for voltage gain.   **[UPTU 2003-04, 05]**

7.12. Obtain the approximate equivalent model for JFET as an amplifer.

7.13. How a FET can be used as a voltage controlled resistor? Explain.

7.14. Write short notes on the following:

   (a) Transfer characteristics of JFET

   (b) Pinch-off voltage   **(UPTU 2005-06)**

7.16. A FET has a drive current of mA. IF I_{DSS} = 8 mA and V_{GS} (off) = – 6V, find the values of V_{GS} and V_p.   **[Ans.** V_{GS} = – 1.76 V, V_p = 6V]

7.17. Determine the value of transconductance of a FET, when the drain current changes from 1mA to 1.5 mA with a change in gate voltage from –2.125 V to – 2V.   **[Ans.** 4m℧]

7.18. Explain the construction and operation of enhancement MOSFET with characteristics.

7.19. Explain the construction and operation of depletion MOSFET with characteristics.

7.20. Write short notes on depletion type MOSFET.   **(UPTU 2004-05)**

7.21. Explain JFET as an amplifier.   **(UPTU 2004-05)**

7.22. Compare BJT and FET.

7.23. Compare JFET and MOSFET.

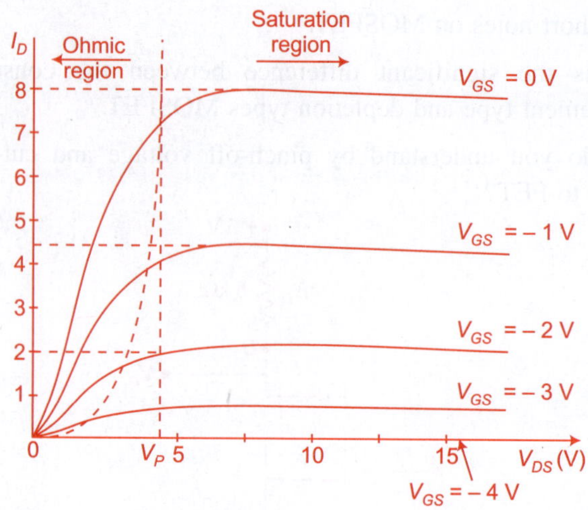

Fig. E 7.2 *NMOSFET characteristics*

7.24. Derive the expression for input and output impedances of a single stage FET amplifier. **(UPTU 2004-05)**

7.25. What is the difference in physical structure between E-NMOSFET and D-NMOSFET? How is the channel in E-MOSFET formed?

7.26. What are the regions of operation of a MOSFET? How are they defined?

7.27. Why are NMOSFETS preferred over PMOSFET?

7.28. How does saturation of drain current occur in a NMOSFET?

7.29. Sketch the transfer and drain characteristics of a D-NMOSFET with $I_{DSS} = 12$ mA and $V_p = -8$V for a range of $V_{GS} = -V_p$ to $V_{GS} = 1$V.

7.30. Given $I_D = 14$ mA and $V_{GS} = 1$V, determine V_p if $I_{DSS} = 9.5$ mA for a D-MOSFET.

7.31. In what ways the construction of a depletion type MOSFET is similar to that of a JFET? In what ways does it differ?

7.32. (a) Determine V_{DS} for $V_{GS} = 0$ and $I_D = 6$ mA using the characteristic of Fig. E 7.2.

(b) Using the result of part (a), Calculate the resistance of JFET for the region $I_D = 0$ to 6 mA for $V_{GS} = 0$ V. **[Ans. $r_d = 233.33$ Ω]**

7.33. Given $I_{DSS} = 9$ mA and $V_p = -3.5$ V determine I_D, where (a) $V_{GS} = 0$ V (b) $V_{GS} = -2$V (c) $V_{GS} = -3.5$V (d) $V_{GS} = -5$V. **[Ans.** (a) 9 mA (b) 1.65 mA (c) 0 mA (d) 0 mA]

7.34. Given a Q-point of $I_{DQ} = 3$ mA and $V_{GS} = -3$V, determine I_{DSS} if $V_p = -6$ V. **[Ans. $I_{DSS} = 12$ mA]**

7.35. Design a self bias network using a JFET transistor with $I_{DSS} = 8$ mA and $V_p = -6$V to have a Q-point at $I_{DQ} = 4$ mA using a supply of 14V. Assume that $R_D = 3\ R_S$ and use standard value. $\qquad$ [$R_S = 0.43$ kΩ, $R_D = 1.3$ kΩ]

7.36. How is FET used as a voltage variable resistance? Define (a) Transconductance gm (b) Drain resistance r_d (c) Amplification factor of FETμ

(UPTU 2007-08)

Operational Amplifier

8

8.1 INTRODUCTION

As in the previous chapters we have studied about the application of the transistor as an amplifier, which are used to amplify input signal i.e., when the weak signal is applied at the input of the amplifier, it will produce strong signal or amplified output signal. In this chapter, we will study about operational amplifier. The operational amplifier is made with one or more differential amplifier (will be discussed later). It is a direct-coupled high-gain amplifier. The operational amplifier can amplify signals having frequency ranging from 0 Hz to 1 MHz. It can amplify AC as well as DC signals. This amplifier was designed for mathematical operation like addition, substraction, multiplication and integration so it is named as operational amplifier (OP-amp.). Now-a-days operational amplifier is available as a single Integrated Circuit (IC) package. An IC operational Amplifier is made with large number of transistors, resistors and one or two capacitors.

8.2 BLOCK DIAGRAM REPRESENTATION OF AN OP-AMP

As we have discussed earlier that the OP-Amp. is a multistage amplifier because it is made with large number of transistors. So, it can be represented by a block diagram as shown in Fig. 8.1. As we see in Fig 8.1, it has 4 stages named as input stage, intermediate stage, level shifting stage and output stage.

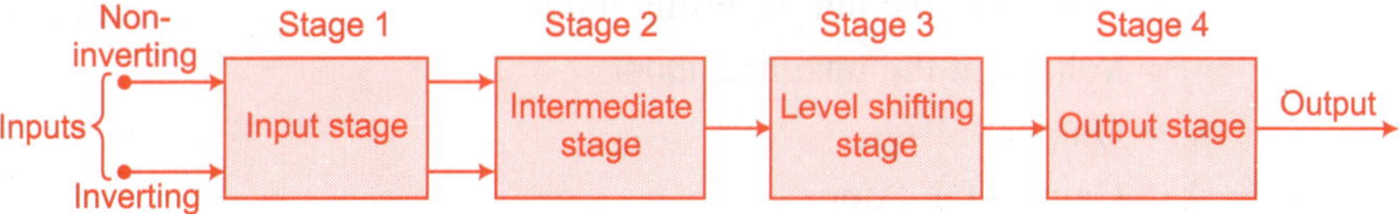

Fig. 8.1 *Block diagram of an OP-Amp*

Stage 1 Input stage is a dual input balanced output differential amplifier because two inputs are given to this stage. (one is non-inverting and second one is inverting). This stage provides high voltage gain of the amplifier and also provides the input resistance of the OP-Amp. The outputs of this stage are balanced.

Stage 2 This stage is also a dual input but unbalanced output differential amplifier. This stage is driven by the output of first stage i.e., stage 1. and is named as intermediate stage i.e., the inputs of this stage are two but the output is single ended. The DC voltage of stage 2 is above the ground potential because direct coupling is used. Intermediate stage increases the overall gain of the OP-Amp.

Stage 3 This stage is a level shifting (translator) amplifier. This stage is driven by the output of intermediate stage i.e., stage 2. This stage is used to shift the DC level at the output of intermediate stage downwards to 0 volts with respect to ground. Therefore, it is named as level shifting stage. Usually this stage is an emitter follower using constant current source.

Stage 4 Output stage is a push-pull complementary amplifier. This stage increases the output voltage swing and raises the current supply capability of the OP-Amp. This stage also provides low output resistance.

8.3 SCHEMATIC SYMBOL OF OPERATIONAL AMPLIFIER

The schematic symbol of OP-Amp. is shown in Fig. 8.2. In Fig. 8.2, we can see that there are of two inputs named as inverting input labled as '−' terminal and second one is non-inverting input labled as '+' terminal. In Fig. 8.2 for simplicity, power supply and other pin connections are not displayed.

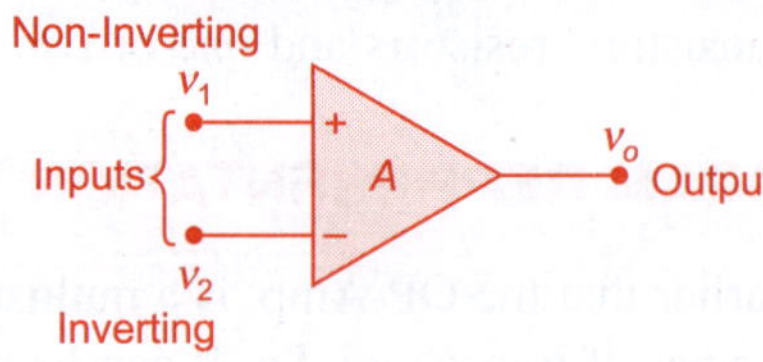

Fig. 8.2 *Schematic symbol for the OP-Amp*

where,

v_1 = Voltage at the non-inverting input

v_2 = Voltage at the inverting input

v_0 = Output voltage

A = Gain of OP-Amp.

When the DC or AC signal applied on the non-inverting input i.e., on '+' input, then the OP-Amp produces output which is in the same phase or polarity with the input. On the other hand, when the AC or DC input applied at the inverting input i.e., on '−' input, then the OP-Amp produces the output which is 180° out of phase or opposite polarity with the input.

8.4 INTEGRATED CIRCUIT PACKAGE OF OP-AMP

As we have told earlier that the OP-Amp is also available in IC package. The Fig. 8.3(a) shows an OP-Amp in IC packages. Basically, the OP-Amp is available in two packages, one is Dual-In-Line (DIP) package and second one is metal can package.

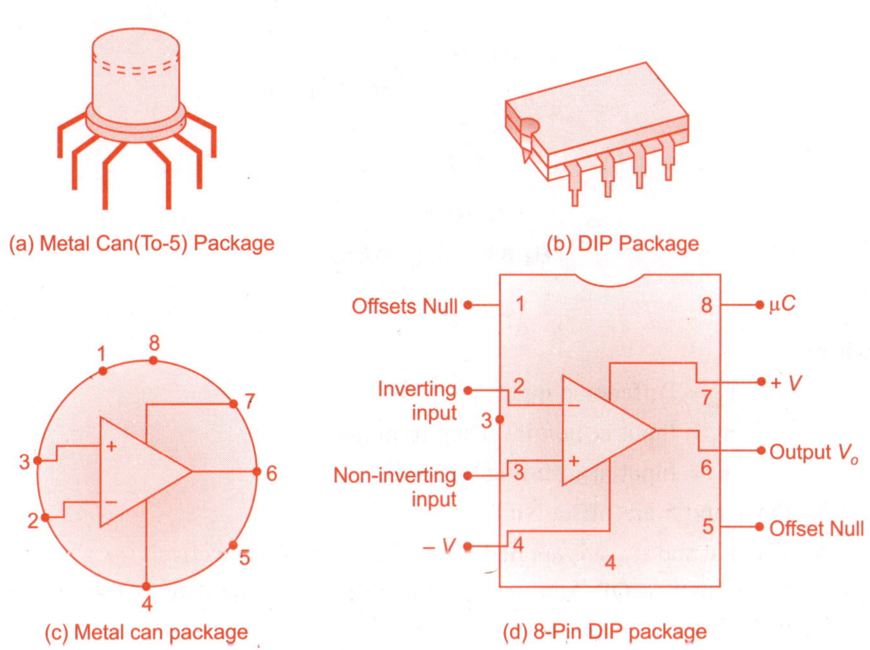

Fig. 8.3 *Packaging and pin outs of OP-Amp*

Pin 1: Off set null

Pin 2: Inverting input

Pin 3: Non-inverting input

Pin 4: $-v$ (volts)

Pin 5: Offset null

Pin 6: Output

Pin 7: $+v$ (volts)

Pin 8: NC.

8.4.1 Circuit Symbol of 741C IC of OP-Amp

The IC 741C OP-Amp is widely used, so the circuit symbol of 741C is shown in Fig. 8.4. This is a 8-pin IC. Its operating range is 0°C to 70°C.

1. As we can see in Fig. 8.4, it provides two inputs at pin no. 2 and 3 named and inverting input and non-inverting input respectively.

2. Basically, the OP-Amp amplifiers the difference between the voltage applied is at two inputs. In Fig. 8.4, we see that V_1 is the input applied at the non-inverting terminal and V_2 is the input applied at the inverting terminal. Then, the difference of two inputs act as an input to the OP-Amp.

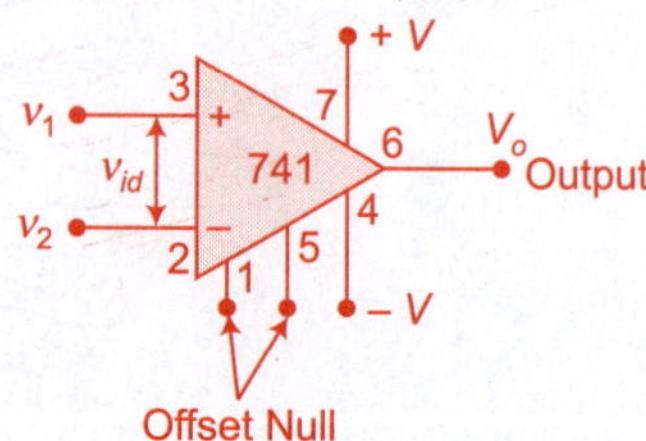

Fig. 8.4   *741C OP-Amp IC*

$$v_1 - v_2 = v_{id} \tag{8.1}$$

where,

v_{id} = Difference input

v_1 = Input at non-inverting terminal

v_2 = Input at inverting terminal.

3. Pin 1 and 5 are offset Null.

4. The $+ v$ and $- v$ are applied at pin 7 and pin 4 respectively.

5. The output of OP-Amp is available on pin 6. If the gain of the OP-Amp is A, then,

$$A = \frac{v_o}{v_{id}}$$

i.e.

$$v_o = A \, v_{id}. \tag{8.2}$$

From Eqn. (8.2) we can state that the output of an OP-Amp is directly proportional to the algebraic difference between two input voltages i.e., we can say that a the OP-Amp amplifies the difference is between two input voltages but does not amplify the input voltage themselves.

8.4.2   Electrical Parameters of 741C OP-Amp

The electrical specifications of 741C OP-Amp are as follows:

(i)   *Input offset voltage*

When the output of an OP-Amp is zero or null, then the input applied to the OP-Amp at this output, is known as input offset voltage and is denoted by v_{io}, as shown in Fig. 8.5

$$v_{io} = v_1 - v_2 = \text{input offset voltage} \tag{8.3}$$
$$R_s = \text{source resistance} \leq 10 \text{ k}\Omega$$
$$v_{io} = 150 \ \mu v \ (\text{maximum})$$

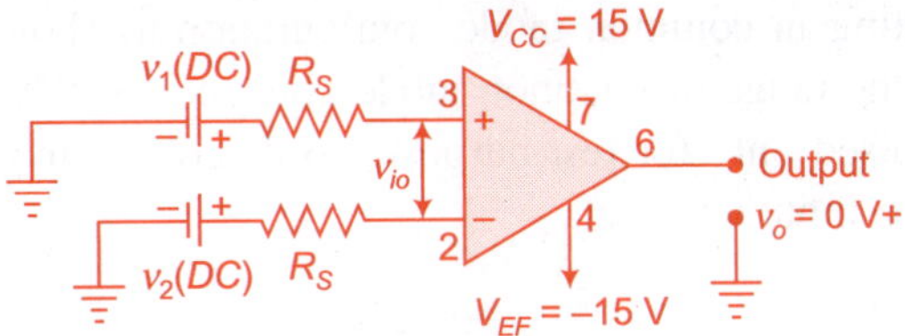

Fig. 8.5   *Circuit for input offset voltage*

(ii)   Input offset current

The input offset current is defined as the algebraic sum of the input current at non-inverting and inverting terminals

$$I_{io} = |I_1 - I_2| \qquad (8.4)$$

where,

I_{io} = Input offset current

I_1 = Current at non-inverting terminal

I_2 = Current at inverting terminal.

The input offset current I_{io} for 741C is 200 µA (maximum).

(iii)   Input resistance

Input resistance is the equivalent resistance that can be measured at either the non-inverting or inverting input terminal with the other terminal connected to ground. The input resistance for 741C is relatively high by 2 MΩ approximately.

(iv)   Offset voltage adjustment range

As we can see in Fig. 8.3 (d) that 741C IC has two pins 1 and 5 for offset null voltage. Basically, the 741 family of IC has an offset voltage null capability. In Fig. 8.6 we see that the potentiometer of 10 kΩ is connected between pin 1 and 5 and the wiper of potentiometer is connected to the negative supply $-V_{EE}$ terminal. By varying the potentiometer the output voltage can be reduced to zero volts without any input applied. So, we can say that the offset voltage adjustment range is the range through which the input offset voltage can be adjusted by varying the Potentiometer.

(v)   Common mode voltage

When the same voltage is applied to both input terminals of OP-Amp, then the voltage is called Common Mode Voltage and is denoted by V_{cm} and the OP-Amp is

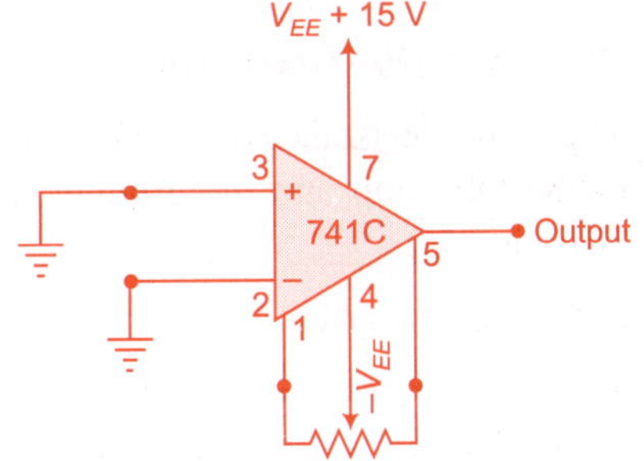

Fig. 8.6   *Offset voltage adjustment range*

said to be operating in common mode configuration as shown in Fig. 8.7. For 741C OP-Amp, the range of common mode voltage is ± 13V (maximum). This configuration is used only for test purposes to determine the degree of making bottom of the two input terminals.

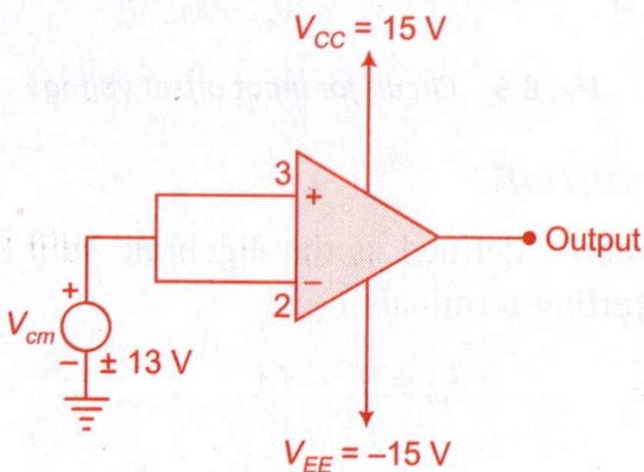

Fig. 8.7 *Common mode configuration*

(vi) *Common mode rejection ratio*

It is defined as the ratio of differential voltage gain A_d to the common mode voltage gain A_{cm} that is,

$$CMRR = \frac{A_d}{A_{cm}} \tag{8.5}$$

where,

A_d = Differential voltage gain is same as large signal voltage gain A

A_{cm} = Common mode voltage gain

$$A_{cm} = \frac{V_{ocm}}{V_{cm}} \tag{8.6}$$

where,

V_{ocm} = Common mode output voltage

V_{cm} = Input common mode voltage

A_{cm} = Common mode voltage gain.

(vii) *Output resistance*

The output resistance of an OP-Amp can be measured between the output terminal of the OP-Amp and the ground. It is denoted by R_o. The output resistance for 741C is 75Ω.

(viii) *Slew rate*

The slew rate (SR) of an OP-Amp is defined as the maximum rate of change of output voltage v_0 per unit of time.

i.e.

$$SR = \left.\frac{dv_0}{dt}\right|_{maximum} v/\mu s. \tag{8.7}$$

8.5 THE IDEAL OPERATIONAL AMPLIFIER

The ideal OP-Amp should meet the following electrical properties:

1. The input resistance R_i of an OP-Amp must be infinite. Due to the infinite input resistance, any signal source can drive it and there is no loading effect of the preceding stage.

   i.e. $\qquad R_i = \infty.$ $\hfill$ (8.8)

2. The output resistance R_o of an OP-Amp must be zero. Due to zero output resistance, the output of an OP-Amp can derive an infinite number of other devices

   i.e. $\qquad R_o = 0.$ $\hfill$ (8.9)

3. The gain of an OP-Amp is infinite

   i.e. $\qquad A = \infty$

   Because, $\qquad A = \dfrac{v_o}{v_i} = \dfrac{I_o R_o}{I_i R_i}$

   $\because \qquad R_i = \infty \quad$ and $\quad R_o = 0$

   $\because \qquad A = \dfrac{o}{\infty} = \infty$

   i.e. $\qquad A = \infty.$ $\hfill$ (8.10)

4. When input voltage of an OP-Amp is zero, then the output voltage is zero.

   i.e. $\qquad v_0 = o|_{v_i = o}.$ $\hfill$ (8.11)

5. The bandwidth of an OP-Amp must be infinite so that any frequency signal from 0 to ∞Hz can be amplified without attenuation.

6. The Common Mode Rejection Ratio (*CMRR*) must be infinite so that the output common mode noise voltage is zero.

7. An ideal OP-Amp should have infinite slew rate so that output voltage changes occur simultaneously with input voltage changes.

8.5.1 Equivalent-Circuit of an Ideal OP-Amp

For understanding the equivalent circuit of an ideal OP-Amp, first we should know few points about an ideal OP-Amp as follows:

1. The ideal OP-Amp does not draw input current because the input resistance R_i is infinite

   i.e. $\qquad v = IR$

   $\qquad\qquad I = \dfrac{v}{R}$

$$\because \qquad R_i = \infty \text{ for ideal OP-Amp}$$

$$\therefore \qquad I_i = \frac{V_i}{R_i} = \frac{v_i}{\infty} = 0.$$

Hence, there is zero input current which means there is an open circuit between two terminals.

2. For an ideal OP-Amp, the output resistance R_o is zero i.e., output voltage v_o does not depend on the load resistance connected across the output. Since the output resistance is zero, therefore it is replaced by a short circuit in the output side of the equivalent circuit.

Hence, the equivalent circuit of an ideal OP-Amp is drawn with the help of above explained points and is shown in Fig. 8.8.

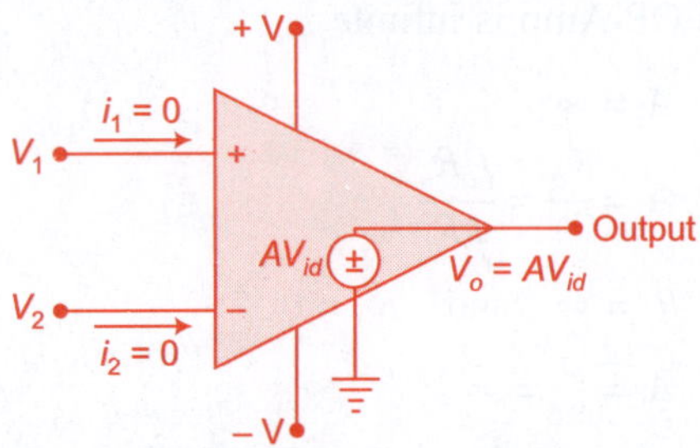

Fig 8.8 *Equivalent circuit of an ideal OP-Amp*

8.5.2 Voltage Transfer Curve of Ideal OP-Amp

The Eqn. (8.2) is the basic OP-Amp equation in which the output offset voltage is assumed to be zero. This equation of OP-Amp is useful in studying the OP-Amp's characteristics. It is used to analyze the circuit configurations that employ feedback.

$$v_0 = Av_{id}.$$

The graphical representation of the above equation is shown in Fig. 8.9. In the Fig., we can see that the output v_o is plotted against input difference voltage v_{id} when gain A is constant.

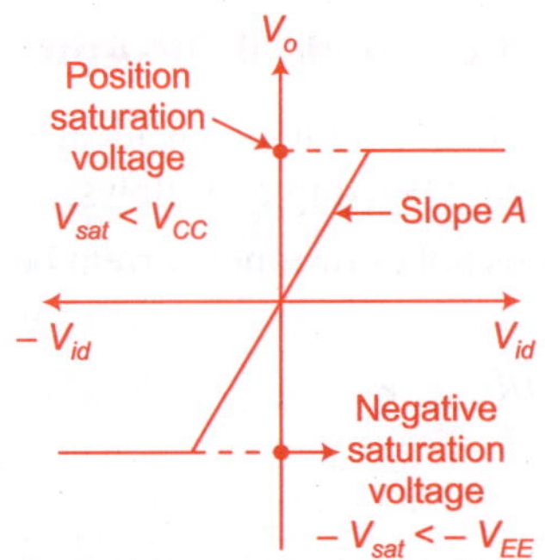

Fig. 8.9 *Ideal voltage transfer curve*

From the curve, it is clear that the output voltage v_o cannot exceed the positive and negative saturation voltages (+ v_{sat} and – v_{sat}). These positive and negative saturation voltages are specified by an output voltage swing rating of the OP-Amp for given values of supply voltages. i.e., output voltage is directly proportional to the input difference voltage v_{id} only, until it reaches the saturation voltages. After that the output voltage remain constant.

8.6 EQUIVALENT CIRCUIT OF A PRACTICAL OP-AMP

The Equivalent circuit of a practical OP-Amp is shown in Fig. 8.10. This equivalent circuit is useful in analyzing the basic operating principles of OP-Amp and observes the effect of feedback arrangement.

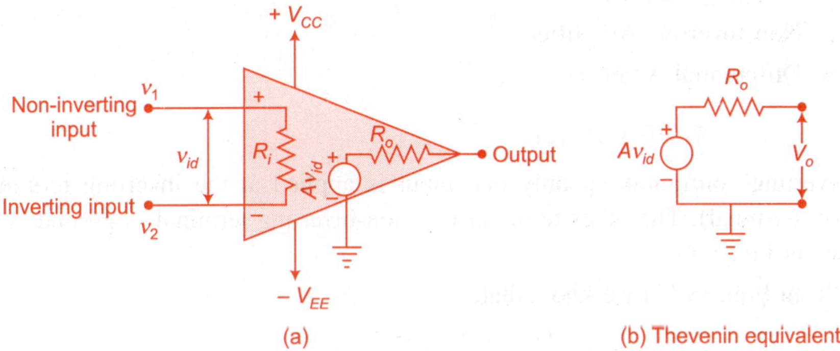

Fig. 8.10 *Equivalent circuiting an OP-Amp*

When we see in Fig. 8.10, the resistance R_i is the input resistance that appears between two terminals of OP-Amp. The resistance R_s is the output resistance and Av_{id} is the equivalent voltage source that appears at the output terminal of an OP-Amp. The output resistance R_o and voltage source Av_{id} makes a Thevenin equivalent circuit. So, the resistance R_o is called as Thevenin resistance that is looking back into the output terminal of an OP-Amp and voltage source. Av_{id} is called as Thevenin voltage source and both are connected in series, so the network is called as Thevenin network as shown in Fig. 8.10 (b) which looks back into the output terminal.

For the circuit shown in Fig. 8.10, the output voltage is,

$$v_0 = Av_{id} = A\,(v_1 - v_2) \tag{8.12}$$

where,

A = Large signal voltage gain

v_{id} = Difference input voltage

v_1 = Voltage at non-inverting terminal w.r.t. ground

v_2 = Voltage at inverting terminal w.r.t. ground.

The Eqn. (8.12) shows that the output voltage v_o is directly proportional to the algebraic difference between two input voltages. In other words, we can say that the OP-Amp amplifies the difference between two input and not the input voltages themselves.

8.7 OPEN-LOOP OP-AMP CONFIGURATIONS

Open-loop OP-Amp means there is no connection between output and input of OP-Amp directly or indirectly. That is the output signal is not feedback in any form as a part of input signal and the loop formed by feedback is open.

When OP-Amp is connected in open-loop, then it works as a high gain amplifier. There are three types of open-loop configuration

1. Inverting Amplifier
2. Non-Inverting Amplifier
3. Differential Amplifier.

8.7.1 Inverting Amplifier

In inverting configuration, only one input is applied at the inverting terminal (– 've' terminal). The other terminal i.e., non-inverting terminal is grounded as shown in Fig. 8.11

From Eqn. (8.12) we know that,

$$v_0 = A\, v_{id} = A\,(v_1 - v_2)$$

$$\because \qquad v_1 = 0 \text{ Volt} \quad \text{and} \quad v_2 = v_{in}$$

If source resistance R_{in} is neglected,

$$\therefore \qquad v_0 = A(0 - v_{in})$$

$$v_0 = -A\, v_{in} \tag{8.13}$$

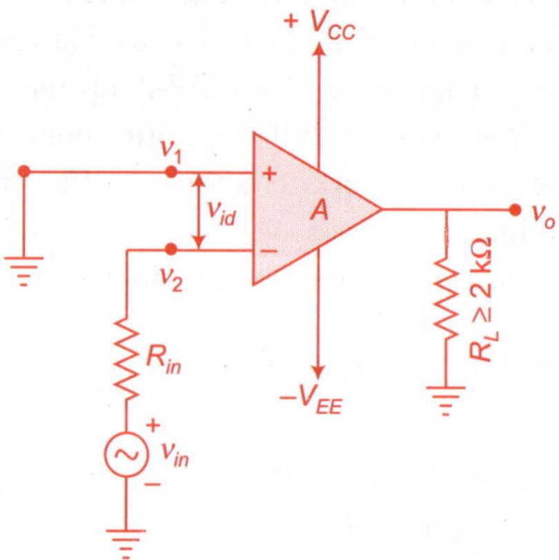

Fig. 8.11 *Inverting amplifier*

When we see the Eqn. (8.13), we observe that the input signal v_{in} is amplified and the output signal is larger. A times of input signal. The negative sign indicates that the output is opposite in phase with input i.e., 180° out of phase.

8.7.2 Non-inverting Amplifier

In non-inverting configuration, only one input is applied at the non-inverting terminal ('+ *ve*' terminal). The other terminal i.e., inverting terminal is grounded as shown in Fig. 8.12.

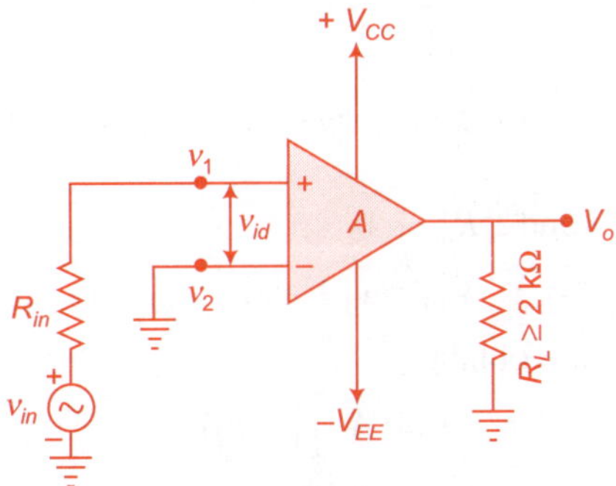

Fig. 8.12 *Non-inverting amplifier*

From Eqn. (8.12), we know that,

$$v_0 = A\, v_{id} = A\, (v_1 - v_2)$$

$\because$
$$v_2 = 0 \quad \text{and} \quad v_1 = v_{in}$$

if source resistance R_{in} is neglected,

$\therefore$
$$v_0 = A\, (v_{in} - 0)$$

$$v_0 = A\, v_{in}. \tag{8.14}$$

From Eqn. (8.14), we observe that the output voltage is large than the input voltage by gain A and is in phase with the input signal.

8.7.3 Differential Amplifier

In differential configuration, the inputs are applied at both inverting and non-inverting terminal. As we have discussed earlier that when the input is applied at both terminals of OP-Amp, then it amplifies the difference of two inputs and not themselves. That is why it is called as Differential Amplifier.

The source voltage and resistance applied at non-inverting terminal are v_{in1} and R_{in1} respectively. Similarly, the source voltage and resistance applied at inverting terminal are v_{in2} and R_{in2} respectively as shown in Fig. 8.13.

The source resistance R_{in1} and R_{in2} is very small as compared to input resistance R_i so we can neglect them. Therefore, the voltage drop across the resistor can be assumed to be zero.

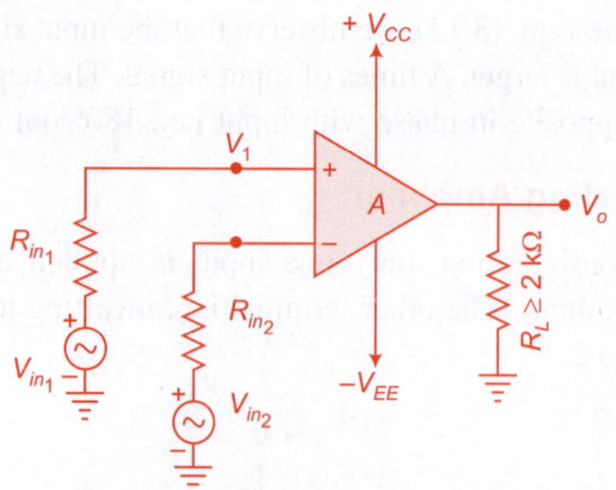

Fig. 8.13 *Differential amplifier*

$$\because \qquad R_{in1} \quad \text{and} \quad R_{in2} \simeq 0$$

$$\therefore \qquad v_{in1} = v_1 \quad \text{and} \quad v_{in2} = v_2.$$

From Eqn. (8.12), are obtain,

$$v_0 = A\,(v_1 - v_2)$$

$$v_0 = A\,(v_{in1} - v_{in2}). \tag{8.15}$$

From Eqn. (8.15), we observe that the output voltage is A times larger than the difference between two inputs. The polarity of the output is dependent upon the polarity of the difference voltage $(v_{in1} - v_{in2})$.

8.7.4 Limitations of Open-Loop OP-Amp Configuration

Open-loop configuration of OP-Amp may not be used in linear application i.e., there is no linear relationship between input and output because the gain of open-loop OP-Amp is very large. So, when the input signal is slightly increased above zero, then the output v_o reaches up to saturation level as observed in ideal characteristics of an OP-Amp. Hence, we can say that when OP-Amp is operated in open-loop configuration, the output is either negative saturation or positive saturation or switcher between both saturation levels. Basically, it is used in non-linear applications like stable multivibrators, square wave generator etc.

Example 8.7.1   Determine the output voltage in each of the following cases for the open-loop differential amplifier of Fig. Ex. 8.7.1.

(a) $\qquad\qquad v_{in1} = 6\ \mu V\ DC, \qquad\qquad\qquad v_{in2} = -\,8\ \mu V\ DC$

(b) $\qquad\qquad v_{in2} = 9mV\ rms, \qquad\qquad\qquad v_{in2} = 18\ mV\ rms$

The OP-Amp is 741C with the following specification:

$$A = 1.5 \times 10^5, \qquad\qquad R_i = 2\ M\Omega$$

$$R_o = 70\Omega, \quad V_{cc} = 18\,\text{Volt}, \qquad v_{EE} = -\,18\,\text{Volt}$$

and output voltage swing $= \pm\,15v,$

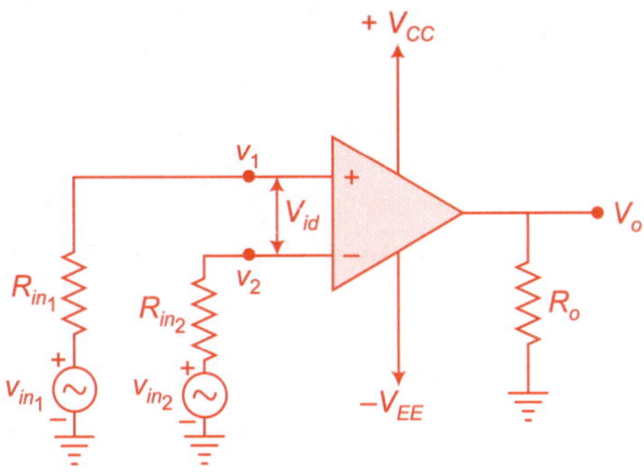

Fig. Ex. 8.7.1

Solution   Given that,

(a)
$$v_{in1} = 6\ \mu\ \text{Volt DC and } v_{in2} = -8\ \mu\ \text{Volt DC}$$

From Eqn. (8.12), we know that,

$$v_0 = A\ (v_{in1} - v_{in2}) \quad \because v_1 = v_{in1} \text{ and } v_2 = v_{in2}$$
$$v_0 = 1.5 \times 10^5\ [6 \times 10^{-6} - (-8 \times 10^{-6})]$$
$$v_0 = 1.5 \times 10^5 \times 10^{-6}\ [6 + 8]$$
$$v_0 = (1.5 \times 1.4)\ \text{Volt}$$

$$\boxed{v_0 = 2.1 \text{ volt DC.}} \tag{Ans.}$$

(b)
$$v_{in1} = 9\ \text{mV } rms \quad \text{and} \quad v_{in2} = 18\ \text{mV } rms$$
$$v_o = A\ (v_1 - v_2) = 1.5 \times 10^5\ [(9 \times 10^{-3}) - (18 \times 10^{-3})]$$
$$= 1.5 \times 10^{-3} \times 10^5\ [9 - 18]$$
$$= -13.5 \times 10^2\ v\ rms$$

$$\boxed{v_0 = 1350 \text{ volt } rms.} \tag{Ans.}$$

The wave shape for input voltage is shown in Fig. Ex. 8.7.2 (a) and (b).

$$v_{in1} = 10\ \text{mV } rms$$
$$v_{in1}(\text{max}) = \pm 10 \times \sqrt{2}\ \text{mV}$$
$$= \pm 14.1\ \text{mV}$$

and,
$$v_{in2} = 20\ \text{mV } rms$$
$$v_{in2}(\text{max}) = \pm 20 \times \sqrt{2}\ \text{mV}$$
$$= \pm 28.2\ \text{mV}.$$

The waveforms are shown in Fig. Ex. 8.7.2(a) and (b) and the output waveform is shown in Fig. Ex. 8.7.2(c), in which the output waveform swings between ± 15V as given in examples. However, the OP-Amp saturates at ± 15V. Therefore, the actual output waveform will be clipped as shown in Fig. Ex. 8.7.2(c).

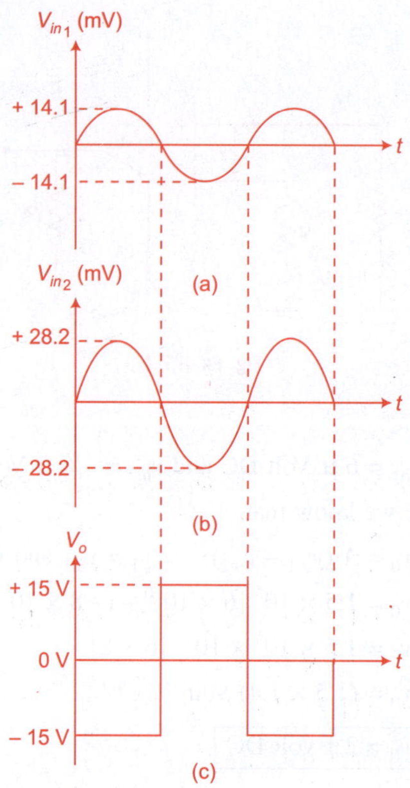

Fig. Ex. 8.7.2   *Input and output waveform*

8.8   CLOSED-LOOP OP-AMP CONFIGURATIONS

In earlier section, we studied about open-loop OP-Amp but the limitations of open-loop OP-Amp is that it cannot be used and for linear applications. And we have seen that the output is clipped in open-loop configuration when output exceeds the saturation levels of the OP-Amp because the gain of open-loop OP-Amp is very high. Therefore, only smaller signals (order of μ volts or less) having very low frequency (less than 5 Hz) may be amplified with acceptable accuracy without distortion. Due to these drawbacks of open loop OP-Amp, the feedback is introduced in open-loop OP-Amp for effective use of OP-Amp in linear applications. So, the OP-Amp with feedback is called as closed-loop OP-Amp. Feedback means that a fraction of the output is applied back to the input. If the signal feedback is out of phase by 180° with respect to the input, then the feedback is called negative feedback. On the other hand, if the signal feedback is in phase with respect to the input, then the feedback is called positive feedback.

Most widely used closed-loop configuration are:

(i) Inverting Amplifier

(ii) Non-Inverting Amplifier.

8.8.1 Inverting Amplifier

As we have discussed earlier that the input is applied at the inverting terminal of an OP-Amp. The output of inverting amplifier is out of phase by 180° with respect to input. The fraction of inverted output is feedback through feedback resistor R_F. Figure 8.14 shows the circuit diagram of closed-loop inverting Amplifier.

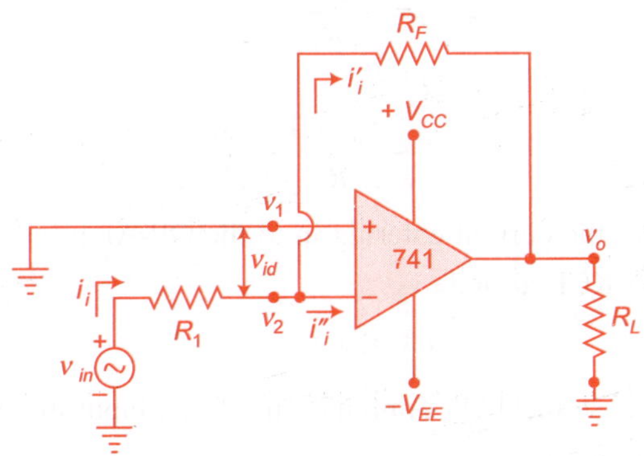

Fig. 8.14 *Closed-loop inverting Amplifier*

Here, we assume that the OP-Amp is ideal and it produces finite output. Therefore, the open-loop gain will be neatly infinite. For an OP-Amp, we know that,

Gain, $$A = \frac{v_0}{v_{id}}$$

∵ $$v_{id} = v_1 - v_2$$

Now, $$A = \frac{v_0}{v_1 - v_2}$$

i.e. $$v_1 - v_2 = \frac{v_0}{A} \qquad \because \ A = \text{Gain of an open-loop}$$

$$v_1 - v_2 = \frac{v_0}{\infty} \qquad\qquad\qquad \text{OP-Amp}$$

$$= \infty$$

$$v_1 - v_2 = 0$$

$$v_1 = v_2. \tag{8.16}$$

From Eqn. (8.16), it is clear that the potential difference the between two terminals of an OP-Amp is zero i.e., both terminals are at the same potential. So, we can say that virtual ground exists between two terminals. This is called as virtual ground, since we are not shorting the two input terminals. In other words, we can say that whatever is the voltage at non-inverting terminal, it will automatically appear at the inverting terminal due to the infinite voltage gain A.

In Fig. 8.14, $v_1 = 0$ (Because at ground)

So from Eqn. (8.16), we have,

$$v_2 = 0. \tag{8.17}$$

From Eqn. (8.17), we can say that the potential at inverting terminal is zero and it is called virtual ground. Virtual ground means that the terminal i is not connected to the ground physically. The current i_i through resistor R_1, we have,

$$i_i = \frac{v_{in} - v_2}{R_1}$$

$\because$ $\qquad\qquad v_2 = 0$

$\therefore$ $$i_i = \frac{v_{in} - 0}{R_1} = \frac{v_{in}}{R_1}. \tag{8.18}$$

Let us consider the current entering in to the OP-Amp is i_i''

Now, applying KCL at node v_2

$$i_i = i_i' + i_i''. \tag{8.19}$$

But $i_i'' \simeq 0$ because ideal OP-Amp has an infinite input impedance and hence draws zero current.

Now, $\qquad\qquad i_i = i_i' + 0$

i.e. $\qquad\qquad i_i = i_i'$

But, $\qquad\qquad i_i = \dfrac{v_{in}}{R_1}$ $\qquad\qquad$ from Eqn. (8.18)

therefore, $\qquad\qquad i_i = i_i' = \dfrac{v_{in}}{R_1}.$ $\qquad\qquad$ (8.19)

From Fig. 8.14, we know that,

$$v_2 = i_i' R_F + v_0$$

$$v_0 = v_2 - i_i' R_F$$

$\because$ $\qquad\qquad v_2 = 0$

$\therefore$ $\qquad\qquad v_0 = - i_i' R_F$

$\because$ $\qquad\qquad i_i' = \dfrac{v_{in}}{R_1}$ $\qquad\qquad$ from Eqn. (8.19)

Now, $\qquad\qquad v_0 = - \dfrac{v_{in} R_F}{R_1}$ $\qquad\qquad$ (8.20)

$$\frac{v_0}{v_{in}} = - \frac{R_F}{R_1}$$

$$\boxed{A_F = - \frac{R_F}{R_1}.} \tag{8.20a}$$

Where, A_F = closed loop gain

$$= \frac{v_0}{v_{in}}.$$

From Eqn. (8.20), it is clear that the gain of closed-loop OP-Amp is the ratio of the two resistance R_1 and R_F. The negative sign indicates that closed-loop OP-Amp provides signal inversion.

8.8.2   Non-inverting Amplifier

As we have discussed earlier, that in non-inverting amplifier only one input is applied at the non-inverting terminal (+ ve terminal) and the inverting terminal is grounded. The output of non-inverting amplifier is in phase with input. The fraction of output is feedback through feedback resistor R_F. Figure 8.15 shows the circuit diagram of closed-loop non-inverting amplifier.

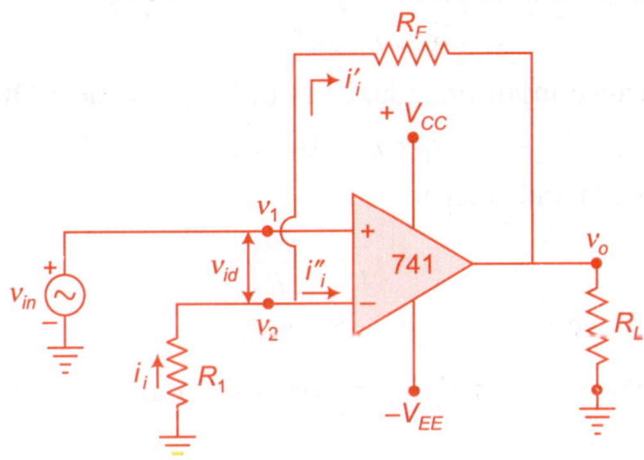

Fig. 8.15   *Closed-loop non-inverting amplifier*

The closed-loop gain,

$$A_F = \frac{v_0}{v_{in}}. \tag{8.21}$$

Here, we assume that the OP-Amp is ideal and it produces finite output. Therefore, the open-loop gain will be nearly infinite i.e., open-loop gain A is,

$$A = \frac{v_0}{v_{id}}.$$

$$= \frac{v_0}{v_1 - v_2}$$

$$\Rightarrow \qquad v_1 - v_2 = \frac{v_0}{A}$$

$$v_1 - v_2 = \frac{v_0}{\infty} = 0$$

$$v_1 = v_2. \tag{8.22}$$

But in Fig. 8.15, $\qquad v_1 = v_{in}$

so, $\qquad\qquad v_2 = v_1 = v_{in}$ $\qquad$ from Eqn. (8.22) $\qquad$ (8.23)

The current i_i through resistor R_1, we have,

$$v_2 + i_i R_1 = 0$$

i.e. $$v_2 = -i_i R_1$$

$$i_i = -\frac{v_2}{R_1}$$

$\because$ $\qquad\qquad v_2 = v_{in}$ $\qquad$ from Eqn. (8.23)

i.e. $$i_i = -\frac{v_{in}}{R_1}.$$ $\qquad$ (8.24)

Let us consider the current entering into the OP-Amp is i''_i

Now, applying KCL at node v_2

$$i_i = i'_i + i''_i.$$ $\qquad$ (8.25)

But $i''_i = 0$ because input impedance is infinite for ideal OP-Amp

$$i_i = i'_i + 0 = i_i.$$ $\qquad$ (8.26)

From Eqns. (8.24) and (8.26)

$$i_i = i'_i = -\frac{v_{in}}{R_1}.$$ $\qquad$ (8.27)

From Fig. 8.15, we have, $v_2 = i'_i R_F + v_0$

Now, putting the value of i'_i from Eqn. (8.27) and v_2 from Eqn. (8.14), we get,

$$v_{in} = -\frac{v_{in}}{R_1} R_F + v_0$$

$$v_0 = v_{in} + \frac{v_{in}}{R_1} R_F$$

$$v_0 = v_{in}\left(1 + \frac{R_F}{R_1}\right)$$

$$\frac{v_0}{v_{in}} = 1 + \frac{R_F}{R_1}$$

i.e. $$\boxed{A_F = 1 + \frac{R_F}{R_1}.}$$ $\qquad$ (8.28)

Where, $\qquad\qquad A_F$ = closed-loop gain

$$= \frac{v_0}{v_{in}}.$$

8.8.3 Unity Gain Configuration

In the previous section, we have studied about non-inverting configuration. If in a closed-loop non-inverting amplifier, feedback resistor R_F is shorted i.e., $R_F = 0$ and resistor R_1 is open circuited i.e., $R_1 = \infty$, then the configuration is called as because the output will be equal to the input (for all value of input). So, when these two conditions are applied at Fig. 8.15, then a circuit which is formed is called as as shown in Fig. 8.16.

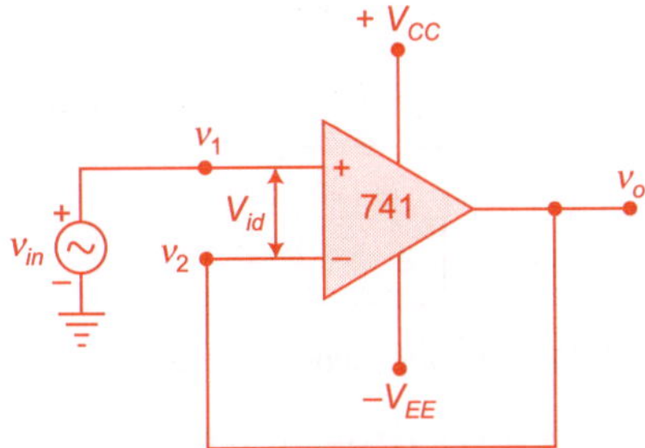

Fig. 8.16   *Unity gain amplifier*

From Eqn. (8.28), we know that,

$$A_F = 1 + \frac{R_F}{R_1}$$

For Fig. 8.16, we have,

$$R_F = 0 \text{ and } R_1 = \infty$$

putting the values in above equation, we obtain,

$$A_F = 1 + \frac{0}{\infty} = 1 + 0$$

$$A_F = 1$$

i.e.
$$\frac{v_0}{v_{in}} = 1 \Rightarrow v_0 = v_{in}.$$

8.9   APPLICATIONS OF OP-AMPS

As discussed earlier, open-loop OP-Amp cannot be used for linear applications and the closed-loop OP-Amps is widely used in linear applications such as Buffer, voltage follower, Adder, Difference Amplifier, Integrator, Differentiator, Convertors and Phase Shifter etc. The important applications of closed-loop OP-Amp are discussed below.

8.9.1   Buffer Amplifier or Voltage Follower

Buffer amplifier is used for isolating the input from the load i.e., buffer amplifier can be used as an isolating unit. Due to the high input resistance, unity gain and low output resistance, the unity gain amplifier is used as a buffer amplifier. The circuit of buffer amplifier is shown in Fig. 8.17.

The unity gain amplifier is also used as a voltage follower because in unity gain amplifier, the output is equal and in same phase with the input. In other words, we can say that the output voltage of voltage follower follows the input. So, the circuit for buffer amplifier and voltage follower is shown in Fig. 8.17.

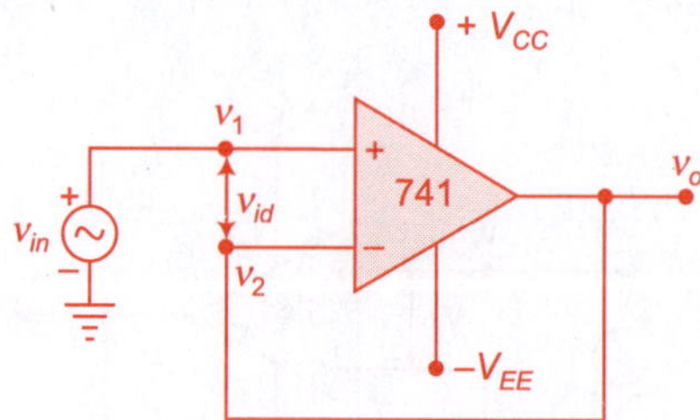

Fig. 8.17 *Voltage follower or buffer amplifier*

8.9.2 OP-Amp as an Adder (Summer)

OP-Amp can be used as an adder. Basically, an adder is a circuit which is used to add or sum the number of inputs and gives an output as a sum of inputs. In other words, we can say that this circuit adds electronically two or more inputs. The circuit for adder is shown in Fig. 8.18.

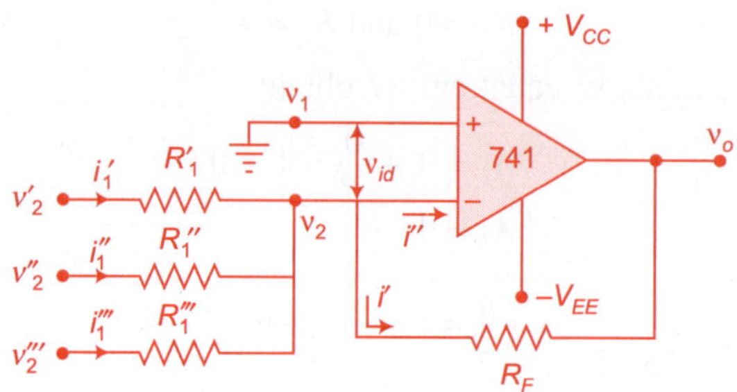

Fig. 8.18 *A summer*

Let us consider that three voltages v_2', v_2'' and v_2'' are applied to the inverting terminal of the OP-Amp.

Assuming that OP-Amp is ideal, we have,

$$A \cong \infty$$

we know that,

$$A = \frac{v_0}{v_{id}}$$

$$\therefore \qquad v_{id} = \frac{v_0}{A} = \frac{v_0}{\infty} = 0$$

$$v_{id} = 0$$

i.e. $\qquad v_1 - v_2 = 0 \Rightarrow v_1 = v_2$

Now, from Fig. 8.18, $\qquad v_1 = 0$

$\therefore \qquad v_2 = 0$

Now, current i_1' is given as,

$$i_1' = \frac{v_2' - v_2}{R_1'}$$

$\because \qquad v_2 = 0$

$$\therefore \qquad i_1' = \frac{v_2'}{R_1'}$$

Similarly, the current i_1'' is,

$$i_1'' = \frac{v_2''}{R_1''}$$

and, $\qquad i_1''' = \dfrac{v_2'''}{R_1'''}$

Now, applying KCL at node v_2, we obtain,

$$i_1' + i_1'' + i_1''' = i_i' + i_i'' \tag{8.29}$$

But $i_i'' = 0$ due to infinite input impedance for ideal OP-Amp.

From Eqn. (8.29), we obtain,

$$i_1' + i_1'' + i_1''' = i_i' \tag{8.30}$$

Now, putting the values of i_1', i_1'', i_1''' and i_i' in above equation, we obtain,

$$\frac{v_2'}{R_1'} + \frac{v_2''}{R_1''} + \frac{v_2'''}{R_1'''} = \frac{v_2 - v_0}{R_F}$$

$\because \qquad v_2 = 0$

$$\therefore \qquad \frac{v_2'}{R_1'} + \frac{v_2''}{R_1''} + \frac{v_2'''}{R_1'''} = \frac{0 - v_0}{R_F} = -\frac{v_0}{R_F}$$

$$v_0 = -\left(\frac{R_F}{R_1'} v_2' + \frac{R_F}{R_1''} v_2'' + \frac{R_F}{R_1'''} v_2''' \right)$$

Let us assume, $\qquad R_1' = R_1'' = R_1''' = R_F \tag{8.31}$

Then, $\qquad v_0 = -(v_2' + v_2'' + v_2''') \tag{8.32}$

Therefore from Eqn. (8.32), we can say that the output is a sum of three inputs.

8.9.3 OP-Amp as an Integrator

OP-Amp can also be used as an integrator i.e., an integrator is a circuit for which the output voltage is integral of the input voltage. The circuit for integrator is shown in Fig. 8.19 in which the fraction of output is feedback to input through feedback capacitor C_F.

Assuming that OP-Amp is ideal, we have,

$$A \cong \infty$$

we know that,
$$A = \frac{v_0}{v_{id}}$$

$\therefore$
$$v_{id} = \frac{v_0}{A} = \frac{v_0}{\infty} = 0$$

i.e.
$$v_1 - v_2 = 0$$
$$v_1 = v_2$$

$\because$
$$v_1 = 0 \qquad \text{from Fig. 8.19}$$

$\therefore$
$$v_2 = 0.$$

From Fig. 8.19, we can determine the current i_i
$$i_i = \frac{v_{in} - v_2}{R_1}.$$

$\because$ OP-Amp is ideal, so it will draw zero current

$\therefore$
$$i_i'' = 0$$

i.e.
$$i_i = i_i'$$

$\therefore$
$$i_i = i_i = \frac{v_{in} - o}{R_1}$$

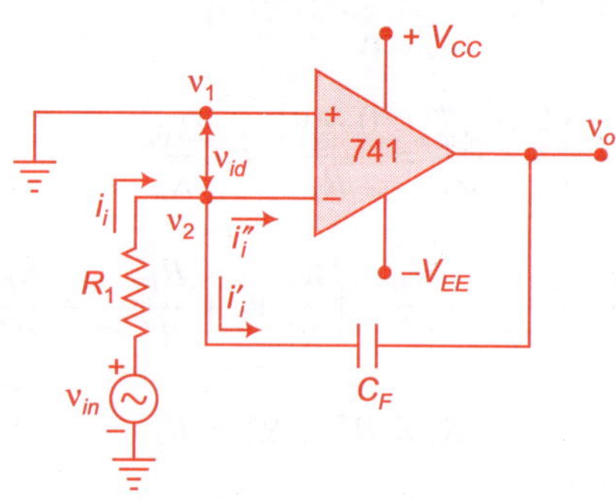

Fig. 8.19 *Integrator*

$$i'_i = \frac{v_{in}}{R_1} \tag{8.33}$$

where i'_i = current through capacitor, we know that the current through a capacitor may also be expressed as,

$$i'_i = C_F \frac{dv_c}{dt} \tag{8.34}$$

equating Eqns. (8.33) and (8.34), we obtain,

$$\frac{C_F dv_c}{dt} = \frac{v_{in}}{R_1} \tag{8.35}$$

in Eqn. (8.35), v_c is expressed as voltage across capacitor,

$$v_c = v_2 - v_0$$

and substituting the value of v_c in Eqn. (8.35), we obtain,

$$C_F \frac{d\,[v_2 - v_0]}{dt} = \frac{v_{in}}{R_1}$$

$$\because \qquad v_2 = 0$$

$$\therefore \qquad C_F \frac{d\,[-v_0]}{dt} = \frac{v_{in}}{R_1}\,. \tag{8.36}$$

Integrating on both sides of Eqn. (8.36), with respect to it, we obtain,

$$\int_0^t \frac{v_{in}}{R_1}\, dt = C_F\,[-v_0] + A$$

$$\boxed{v_0 = -\frac{1}{R_1 C_F} \int_0^t v_{in}\, dt + A.} \tag{8.37}$$

Where,

$$A \;=\; \text{Integration constant.}$$

From Eqn. (8.37), it is clear that the output voltage v_0 is equal to the integration of input voltage v_{in}.

8.9.4 OP-Amp as a Differentiator

OP-Amp can be used as a differentiator i.e., the differentiator is a circuit for which the output voltage is the differential of the input voltage. The circuit for differentiator is shown in Fig. 8.20 in which the fraction of output is feedback through feedback resistor R_F.

Assuming that OP-Amp is ideal, we have open-loop gain $A \cong \infty$ we know that,

$$A = \frac{v_0}{v_{id}}$$

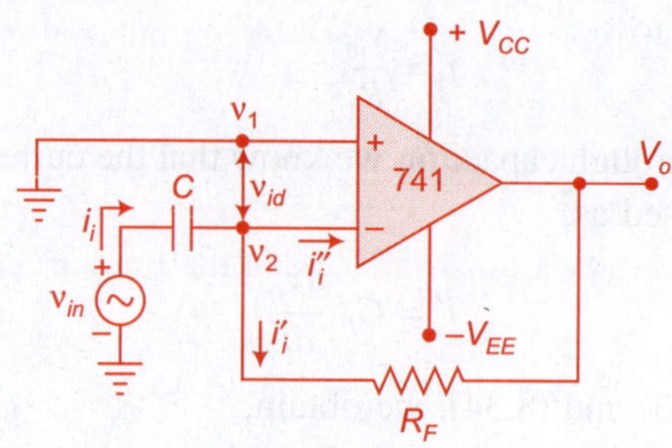

Fig. 8.20 *Differentiator*

$$\therefore \quad v_{id} = \frac{v_0}{A} = \frac{v_0}{\infty}$$

$$v_{id} = v_1 - v_2 = 0$$

$$v_1 = v_2$$

$$\because \quad v_1 = 0 \qquad \text{from Fig. 8.20}$$

$$\therefore \quad v_2 = 0$$

$\because$ OP-Amp is ideal, so it will draw zero current,

$$\therefore \quad i_i'' = 0$$

i.e. $\qquad i_i' = i_i''$

From Fig. 8.20, we can determine the current,

$$i_i = C \frac{dv_c}{dt} \tag{8.38}$$

and, $\qquad\qquad i_i' = \frac{v_2 - v_0}{R_F}. \tag{8.39}$

Equating Eqns. (8.38) and (8.39), we have,

$$C \frac{dv_c}{dt} = \frac{v_2 - v_0}{R_F}$$

$$\because \quad v_2 = 0$$

$$\therefore \quad C \frac{dv_c}{dt} = 0 \; \frac{-v_0}{R_F}$$

$$C \frac{d}{dt}(v_{in} - v_2) = -\frac{v_0}{R_F}$$

$$C \frac{d}{dt}(v_{in}) = -\frac{v_0}{R_F}$$

$$\boxed{v_0 = -R_F \, C \, \frac{d}{dt}(v_{in}).} \tag{8.40}$$

From Eqn. (8.40), it is clear that the output voltage v_0 is equal to the differential of input voltage v_{in}.

Example 8.9.1   As an example for an integrator, if input is a DC voltage, then the output will be a linearly increasing ramp waveform, and similarly, for a differentiator, if input is a ramp signal, then the output is a constant DC voltage as shown in Fig. 8.21.

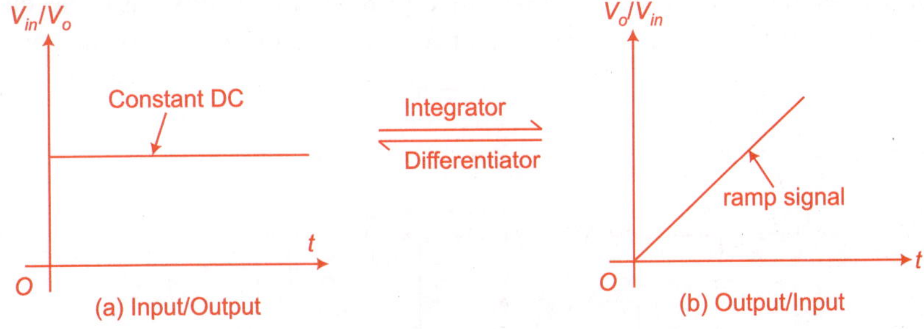

Fig. 8.21   *Input and output waveform for Integrator and differentiator*

8.9.5   OP-Amp as a Differential Amplifier

Another application of OP-Amp is that, it can be used as a differential amplifier or difference amplifier i.e., the differential amplifier is a circuit which amplifies the difference between two input voltages. In such type of amplifier, voltages are applied at both OP-Amp input terminals i.e., inverting and non-inverting terminals. The circuit for difference amplifier is shown in Fig. 8.22 in which the fraction of output is feedback to the input through feedback resistor R_F.

The circuit for difference amplifier is a linear circuit. So, it can be analyzed with the help of superposition principal.

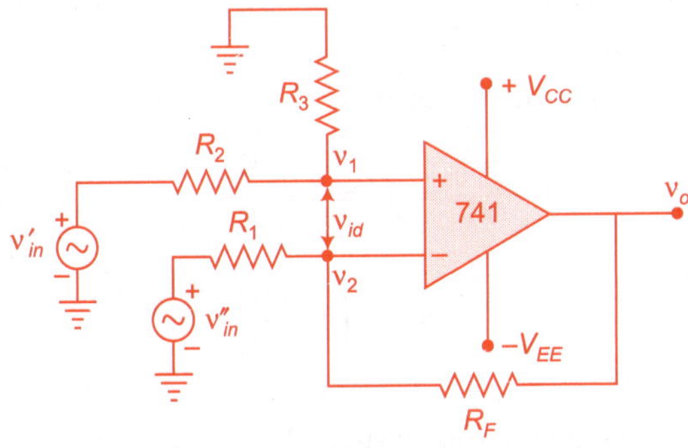

Fig. 8.22   *A difference amplifier*

Super position principal

"In an active linear network containing two or more sources (Including dependent sources), the overall response in any branch in the network equals the algebraic sum of the response of each individual source considered separately with all other sources, replaced by their internal resistances or impedances"

Case 1   Firstly, considering only one source v''_{in} is active and v'_{in} reduced to zero. The equivalent circuit is shown in Fig. 8.22 (a). In this case, the circuit works as inverting amplifier. The resistor R_2 and R_3 will draw zero current.

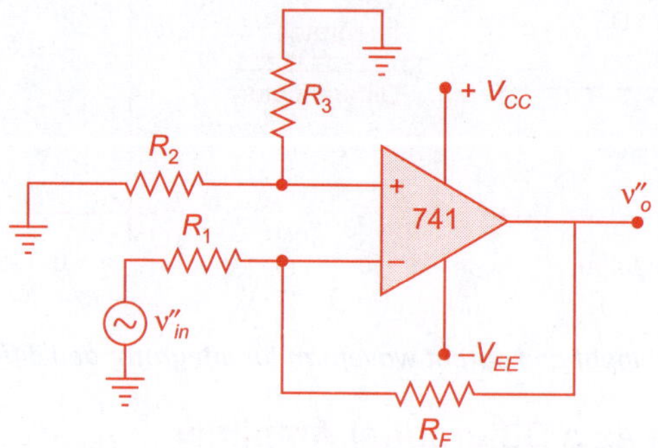

Fig. 8.22   *(a) Equivalent circuit when $v'_n = 0$*

Let the output voltage in this case be v''_0,

$$A_F = \frac{v''_0}{v''_{in}} = -\frac{R_F}{R_1} \qquad \text{from Eqn. (8.20)}$$

$$v''_0 = -\frac{R_F}{R_1} v''_{in}. \qquad (8.41)$$

Case 2   Secondly, considering the source v'_{in} is active and v''_{in} reduced to zero. The equivalent circuit is shown in Fig. 8.22 (b). In this case, the circuit works as non-inverting amplifier.

Let the output voltage in this case be v'_0,

$$\frac{v'_0}{v_{R_3}} = A_F = 1 + \frac{R_F}{R_1} \qquad \text{from Eqn. (8.28)}$$

$$v'_0 = v_{R_3}\left(1 + \frac{R_F}{R_1}\right) \qquad (8.42)$$

where,

$$v_{R_3} = \text{Voltage across resistor } R_3.$$

Hence in Eqn. (8.38) v_{R_3} is voltage across resistor R_3, for this case ($v''_{in} = 0$) voltage across R_4

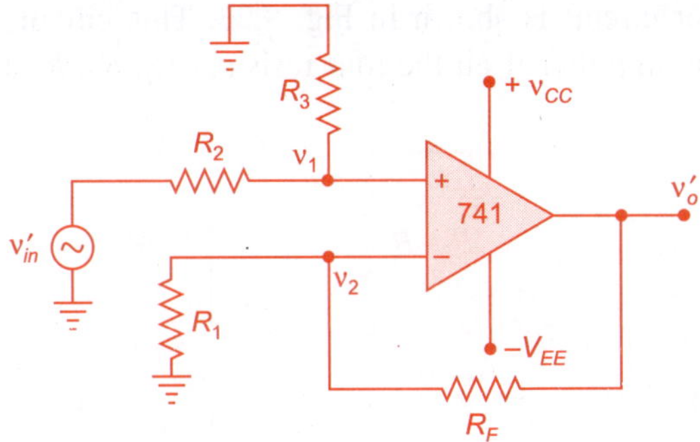

Fig. 8.22 *(b) Equivalent circuit when $v''_{in} = 0$*

$$v_1 = v_{R_3} = \frac{R_3}{R_2 + R_3}\, v'_{in}$$

using Eqn. (8.38), we get,

$$v'_0 = \frac{R_3}{R_2 + R_3} \cdot v'_{in} \cdot \left(1 + \frac{R_F}{R_1}\right). \tag{8.43}$$

According to superposition principal, we have output voltage,

$$v_0 = v'_0 + v''_0. \tag{8.43a}$$

Putting the values of v'_0 and v''_0 from Eqns. (8.43) and (8.35) respectively.

$$v_0 = \frac{R_3}{R_2 + R_3}\, v'_{in} \left(1 + \frac{R_F}{R_1}\right) + \left(-\frac{R_F}{R_1}\, v''_{in}\right)$$

$$= \frac{R_3/R_2}{1 + R_3/R_2}\, v'_{in} \left(1 + \frac{R_F}{R_1}\right) + \left(-\frac{R_F}{R_1}\, v''_{in}\right).$$

Now, if,

$$\frac{R_F}{R_1} = \frac{R_3}{R_2}$$

then,

$$v_0 = \frac{R_F/R_1}{1 + R_F/R_1}\, v'_{in} \left(1 + \frac{R_F}{R_1}\right) - \frac{R_F}{R_1}\, v''_{in}$$

$$= \frac{R_F}{R_1}\, v'_{in} - \frac{R_F}{R_1}\, v''_{in}$$

$$v_0 = \frac{R_F}{R_1}\, (v'_{in} - v''_{in}). \tag{8.44}$$

From Eqn. (8.40), it is clear that the output of difference amplifier is the amplified output of the difference input of amplifier.

8.9.6 OP-Amp as a Subtractor

The OP-Amp also can be used as a subtractor i.e., the subtractor is a circuit which subtract one voltage from another voltage.

The subtractor circuit is shown in Fig. 8.23. This circuit is obtained from differential circuit amplifier if all the four resistors R_1, R_2, R_3 and R_F are equal.

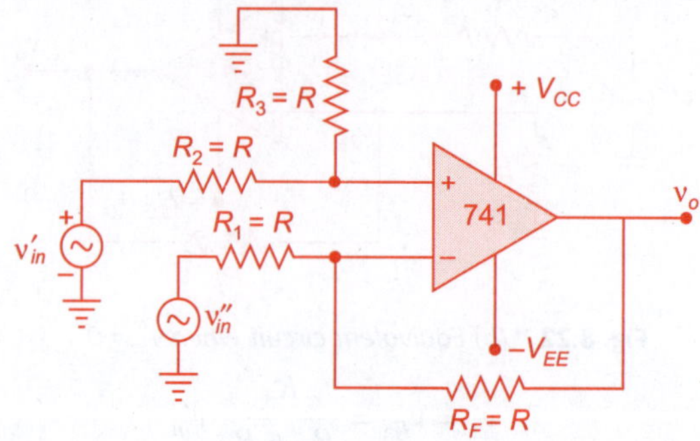

Fig. 8.23 *A substractor*

i.e. Let $$R_1 = R_2 = R_3 = R_F = R.$$

We know that the output of a differential amplifier is expressed as,

$$v_0 = R_F/R_1 \ (v'_{in} - v''_{in})$$

$$\because \qquad R_F = R_1 = R$$

$$\therefore \qquad v_0 = (v'_{in} - v''_{in}). \tag{8.45}$$

8.9.7 OP-Amp as a Comparator

The OP-Amp can be used as a comparator i.e., the comparator is a circuit which compares two voltages and gives an output that indicates the relationship between these two voltages.

The circuit for comparator is shown in Fig. 8.24 in which we observe that there is no feedback used in the circuit. The input voltage is applied to the non-

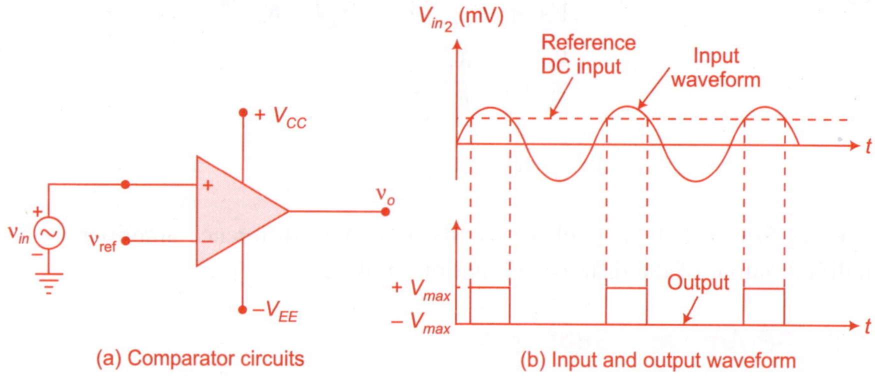

(a) Comparator circuits

(b) Input and output waveform

Fig. 8.24 *(a) Comparator circuit (b) Input and output waveform*

inverting terminal and a set DC reference voltage v_{ref} is applied to the inverting terminal of the OP-Amp.

The waveform of input and output is shown in Fig. 8.24 (b).

Examples

Example 8.1   Calculate the output voltage of an OP-Amp summing amplifier for the following sets of voltages and resistors. And R_F is 1 mΩ **(UPTU 2006-07)**

(a) $\qquad\qquad v_2' = 1v, \qquad\qquad v_2'' = 2v, \qquad\qquad v_2''' = 3v$

$\qquad\qquad R_1' = 400\ k\Omega, \qquad R_1'' = 1\ M\Omega, \qquad R_1''' = 1\ M\Omega$

(b) $\qquad\qquad v_2' = -1v, \qquad\qquad v_2'' = 2v, \qquad\qquad v_2''' = -3v$

$\qquad\qquad R_1' = 250\ k\Omega, \qquad R_1'' = 500\ k\Omega, \qquad R_3''' = 1\ M\Omega$

Solution   (a) From Eqn. (8.24), we know that,

$$v_0 = -R_F\left[\frac{v_2'}{R_1'} + \frac{v_2''}{R_1''} + \frac{v_2'''}{R_1'''}\right]$$

$$v_0 = -1\times10^6\left[\frac{1v}{400\times10^3} + \frac{2v}{1\times10^6} + \frac{3v}{1\times10^6}\right]$$

$$v_0 = -[2.5 + 2 + 3]$$

$$\boxed{v_0 = -7.5\ \text{volt.}}\qquad\qquad\text{(Ans.)}$$

(b)
$$v_0 = -1\times10^6\left[\frac{-1v}{250\times10^3} + \frac{2v}{500\times10^3} + \frac{-3v}{1\times10^6}\right]$$

$$v_0 = -[-4 + 4 - 3]$$

$$\boxed{v_0 = +3\ \text{volt.}}\qquad\qquad\text{(Ans.)}$$

Example 8.2   What is the range of the voltage gain adjustment in the following circuit Fig. Ex. 8.2?   **(UPTU 2006-07)**

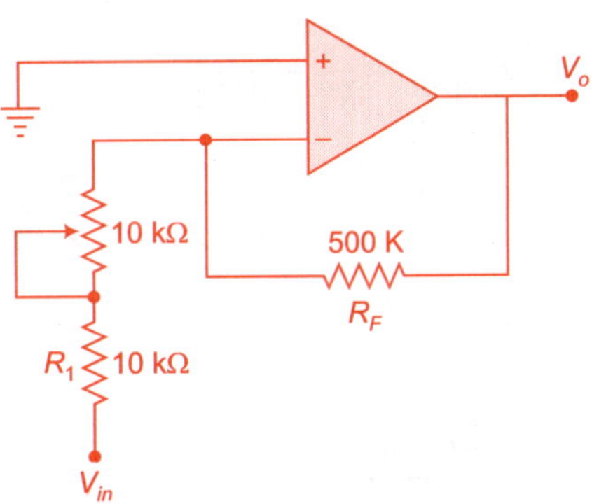

Fig. Ex 8.2

Solution Given that,

$$R_1 = 10 \text{ k}\Omega, \quad R_F = 500 \text{ k}\Omega$$

The gain of inverting amplifier is given by,

$$A = -\left(\frac{R_F}{R_1 + R_p}\right)$$

where, R_p = resistance of potentiometer.

Case (i) when 10 kΩ pot is at,

or,

then,

$$A = -\frac{R_F}{R_1 + 0} = -\frac{R_F}{R_1}$$

$$A = -\frac{500 \text{ k}\Omega}{10 \text{ k}\Omega} = -50$$

$$\boxed{A = -50.}$$

Case (ii) when 10 k is at 10 kΩ

then,

$$A = -\left(\frac{R_F}{R_1 + 10 \text{ k}}\right) = -\left(\frac{500 \text{ k}\Omega}{10 \text{ k}\Omega + 10 \text{ k}\Omega}\right)$$

$$\boxed{A = -25.} \quad \textbf{(Ans.)}$$

Hence, the range of voltage gain is -25 to -50.

Example 8.3 Find the output voltage of the following circuit of Fig. Ex. 8.3.

(UPTU 2006-07, 2008-09)

Solution Given that,

$$v_2' = 0.2 \, v \qquad v_2'' = -0.5 \, v \qquad v_2''' = 0.8 \, v$$
$$R_1' = 33 \text{ k}\Omega \qquad R_1'' = 22 \text{ k}\Omega \qquad R_1''' = 12 \text{ k}\Omega$$
$$R_F = 68 \text{ k}\Omega.$$

From Eqn. (8.20), we have,

$$v_0 = -R_F\left[\frac{v_2'}{R_1'} + \frac{v_2''}{R_1''} + \frac{v_2'''}{R_1'''}\right]$$

$$v_0 = -68 \times 10^3\left[-\frac{0.2}{33 \times 10^3} + \frac{-0.5}{22 \times 10^3} + \frac{0.8}{12 \times 10^3}\right]$$

$$v_0 = \times 0.41 + 1.55 - 4.53$$

$$v_0 = -4.94 + 1.55$$

$$\boxed{v_0 = -3.39 \text{ volt.}} \quad \textbf{(Ans.)}$$

Example 8.4   For a given OP-Amp, $CMRR = 10^4$ and differential gain $A_d = 10^5$. Determine the common mode gain A_{cm}, of the OP-Amp.   **(UPTU 2004-05)**

Solution   Given that,

$$A_d = 10^4, \quad CMRR = 10^4$$

we know that,

$$CMRR = \frac{A_d}{A_{cm}} \qquad \text{from Eqn. (8.5)}$$

$$A_{cm} = \frac{A_d}{CMRR} = \frac{10^5}{10^4}$$

$$\boxed{A_{cm} = 10.}$$ **(Ans.)**

Example 8.5   In an OP-Amp used as differentiator $R_F = 1$ MΩ and $C = 2$ μF. If a signal $v_{in} = 5 \sin 500 \, \pi t \, (m_v)$ is used as input voltage, calculate the output voltage.

Solution   Given that,

$$R_F = 1 \text{ M}\Omega, \quad C = 2 \, \mu F$$

$$v_{in} = 5 \sin 500 \, \pi t \, (m_v)$$

We know that the output voltage for differentiator is given,

$$v_0 = - R_F \, C \, \frac{d}{dt} \, (v_{in}) \qquad \text{from Eqn. (8.40)}$$

$$v_0 = - 1 \times 10^6 \times 2 \times 10^{-6} \, \frac{d}{dt} \, (5 \sin 500 \, \pi t) \, m_v$$

$$v_0 = - 2 \times 5 \, (500 \, \pi) \cos 500 \, \pi t] \times 10^{-3} \text{ volt}$$

$$\boxed{v_0 = - 15.7 \cos (500 \, \pi t) \text{ volt.}}$$ **(Ans.)**

Example 8.6   If the circuit of Fig. Ex. 8.6 has $R_1 = 80$ kΩ and $R_F = 400$ kΩ, what output voltage results for an input of $v_{in} = 5 \, v$?

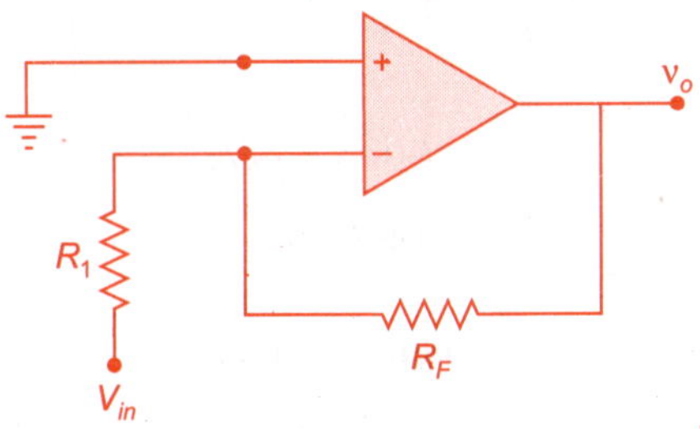

Fig. Ex. 8.6

Solution   Given that,

$$R_1 = 80 \text{ k}\Omega, \quad R_F = 400 \text{ k}\Omega$$

$$v_{in} = 5 \, v.$$

The output voltage for inverting amplifier is given as,

$$v_0 = -\frac{R_F}{R_1} v_{in} \qquad \text{from Eqn. (8.20)}$$

$$v_0 = -\frac{400}{80} \times 5 \ v$$

$$\boxed{v_0 = -25 \text{ volt.}} \qquad \textbf{(Ans.)}$$

Example 8.7 The output voltage of a certain OP-Amp circuit changes by 24 V in 3 μ sec. What is the slew rate?

Solution Given that,

$$\Delta v_0 = 24 \text{ Volt}$$

$$\Delta t = 3 \ \mu \text{ sec.}$$

From Eqn. (8.7), we know that,

$$SR = \frac{\Delta v_0}{\Delta t} = \frac{24 \text{ Volt}}{3 \ \mu s}$$

$$\boxed{SR = 8 \text{ V/μ sec.}} \qquad \textbf{(Ans.)}$$

Example 8.8 For the OP-Amp circuit shown in Fig. Ex. 8.8. Find the values of R_1 and R_2 for the output to be $v_0 = -5 \ v_a + 3 \ v_b$. **(UPTU 2007-08)**

Solution From Fig. Ex. 8.8

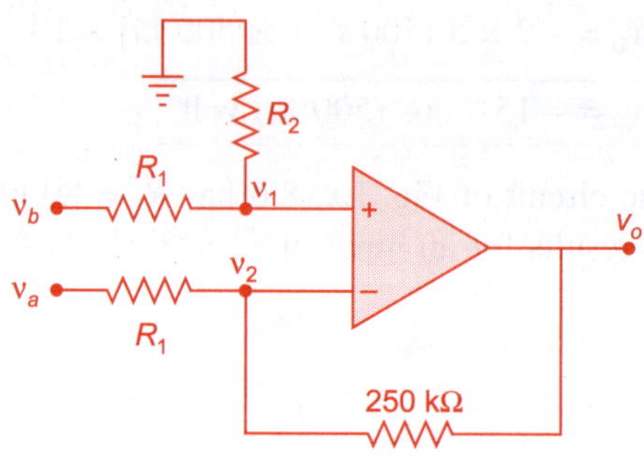

Fig. Ex. 8.8

Case (i) Let $v_b = 0$ and $v_a \neq 0$ then the output,

$$v_{0_1} = -\frac{R_F}{R_1} v_a \qquad \text{from Eqn. (8.37)}$$

i.e.

$$v_{0_1} = -\frac{250 \ k}{R_1} v_a.$$

Case (ii) Let $v_a = 0$ and $v_b \neq 0$ then the output,

$$v_{0_2} = v_1 \left(1 + \frac{R_F}{R_1}\right) \qquad \text{from Eqn. (8.42)}$$

$$v_{0_2} = \frac{R_2}{R_1 + R_2}\, v_b \left(1 + \frac{250\,k}{R_1}\right)$$

The output of Fig. Ex. 8.8 is,

$$v_0 = v_{0_1} + v_{0_2}$$

$$v_0 = -\frac{250\,k}{R_1}\, v_a + \frac{R_2}{R_1 + R_2}\, v_b \left(1 + \frac{250\,k}{R_1}\right)$$

$$v_0 = -\frac{250\, v_a}{R_1} + \frac{R_2(R_1 + 250)\, v_b}{R_1\,(R_1 + R_2)} \qquad (A)$$

Now, comparing Eqn. (A) with $v_0 = -5\, v_a + 3\, v_b$, we obtain,

$$-\frac{250}{R_1} = -5$$

$$\boxed{R_1 = 50\ \text{k}\Omega.} \qquad \textbf{(Ans.)}$$

and,

$$\frac{R_2\,(R_1 + 250)}{R_1\,(R_1 + R_2)} = 3$$

$$\frac{R_2\,(50 + 250)}{50\,(50 + R_2)} = 3$$

$$\frac{300\, R_2}{50\,(50 + R_2)} = 3$$

$$100\, R_2 = 50\,(50 + R_2)$$

$$2\, R_2 = 50 + R_2$$

$$\boxed{R_2 = 50\ \text{k}\Omega.} \qquad \textbf{(Ans.)}$$

Example 8.9 A 5 mV, 1 kHz sinusoidal voltage is applied at the input of an OP-Amp integrator for which $R_1 = 100\ \text{k}\Omega$ and $C_F = 1\ \mu F$. Calculate the output voltage. **(Agra University 1999)**

Solution Given that,

$$v_{in} = 5 \sin \omega t \ \text{mV}$$

$$= 5 \sin 2\pi f t \ \text{mV}$$

$$\because \qquad \omega = 2\pi f$$

$$= 5 \sin 2\pi \, (1000)t \text{ mV}$$

$$\therefore \qquad f = 1 \text{ kHz}$$

$$v_{in} = 5 \sin 2000 \, \pi t \text{ mV}$$

and,
$$R_1 = 100 \text{ k}\Omega \quad \text{and} \quad C_F = 1 \, \mu F.$$

From Eqn. (8.37), we know that,

$$v_0 = -\frac{1}{R_1 C_F} \int_0^t v_{in} \, dt$$

$$= \frac{1}{100 \times 10^3 \times 1 \times 10^{-6}} \int_0^t 5 \sin 2000 \, \pi f \, dt$$

$$v_0 = -\frac{1}{0.1} \int_0^t 5 \sin 2000 \, \pi t \, dt$$

$$v_0 = -10 \left[\frac{-5 \cos 2000 \, \pi t}{2000 \, \pi} \right]_o^t$$

$$v_0 = -\frac{1}{40 \, \pi} \left[\cos 2000 \, \pi t \right]_o^t$$

$$\boxed{v_0 = -\frac{1}{4 \, \pi} \left[\cos 2000 \, \pi + 1 \right] \text{ mV.}} \qquad \textbf{(Ans.)}$$

Example 8.10 Design a non-inverting amplifier circuit, which is capable of providing a voltage gain of 15. Assume an ideal OP-Amp.

Solution The closed-loop gain in given as:

$$A_F = 1 + \frac{R_F}{R_1}$$

Given that,
$$A_F = 15$$

$$\therefore \qquad 15 = 1 + \frac{R_F}{R_1}$$

$$14 = \frac{R_F}{R_1}$$

$$R_F = 14 \, R_1$$

Let,
$$R_1 = 2 \text{ k}\Omega$$

then,
$$R_F = 14 \times 2 = 28 \text{ k}\Omega$$

i.e.
$$\boxed{\begin{aligned} R_1 &= 2 \text{ k}\Omega \\ R_F &= 28 \text{ k}\Omega. \end{aligned}} \qquad \textbf{(Ans.)}$$

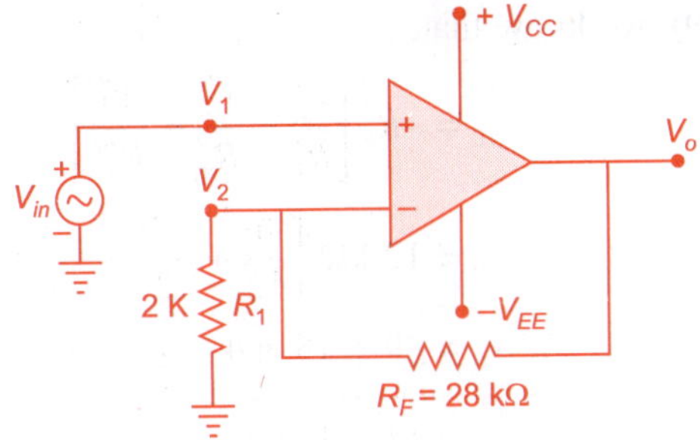

Fig. Ex. 8.10 *Shows the required non-inverting amplifier*

Example 8.11 For an OP-Amp having slew rate of $SR = 2.5$ V/µs, what is the maximum closed-loop voltage gain A_F that can be used when input signal varies by 0.4 V in 20 µs?

Solution Given that,

$$\Delta v_{in} = 0.4 \text{ V} \quad \text{and} \quad \Delta t = 20 \text{ µs}$$

$$SR = 2.5 \text{ v/µ s}.$$

We know that. the closed-loop gain,

$$A_F = \frac{v_0}{v_{in}}$$

i.e.

$$v_0 = A_F \, v_{in}$$

$$\frac{\Delta v_0}{\Delta t} = A_F \frac{\Delta v_{in}}{\Delta t}$$

$$A_F = \frac{\Delta v_0/\Delta t}{\Delta v_{in}/\Delta t} = \frac{SR}{\Delta v_{in}/\Delta t}$$

$$A_F = \frac{2.5 \text{ V/µs}}{0.4 \text{ V/20 µs}} = \frac{2.5 \text{ V/µs}}{0.02 \text{ V/µs}}$$

$$\boxed{A_F = 125.} \qquad\qquad\qquad \textbf{(Ans.)}$$

Example 8.12 An OP-Amp has feedback resistor $R_F = 12$ kΩ and resistance in the input sides are $R_1' = 12$ kΩ, $R_2'' = 2$ kΩ, $R_1''' = 3$ kΩ. The corresponding input voltage $v_{in}' = 9$ V, $v_{in}'' = -3$ V and $V_{in}''' = -1$V. Non-inverting terminal is grounded. Calculate the output voltage. **(UPTU Question Back)**

Solution Given that,

$$R_1' = 12 \text{ kΩ}, \qquad R_1'' = 2 \text{ kΩ}, \qquad R_1''' = 3 \text{ kΩ}$$
$$v_{in}' = 9 \text{ V}, \qquad v_{in}'' = -3 \text{ V}, \qquad v_{in}''' = -1 \text{ V}$$
$$R_F = 12 \text{ kΩ}$$

From Eqn. (8.20), we know that,

$$v_0 = -R_F \left[\frac{v_2'}{R_1'} + \frac{v_2''}{R_1''} + \frac{v_2'''}{R_1'''} \right]$$

$$v_0 = 12 \text{ k}\Omega \left[\frac{9}{12} + \frac{-2}{2} + \frac{-1}{3} \right]$$

$$v_0 = -9 + 18 + 4$$

$$\boxed{v_0 = 13 \text{ volt.}}$$

(Ans.)

Example 8.13   For the circuit shown in Fig. Ex. 8.13. Find out voltage v_0.

(UPTU 2007-08)

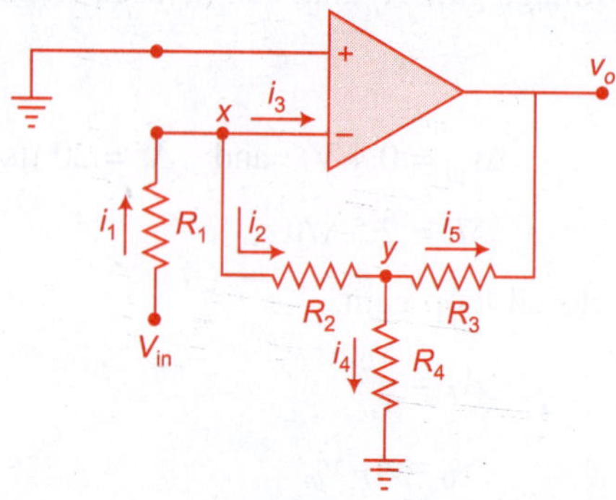

Fig. Ex 8.13

Solution   Let us consider the voltage at node x is v_a and at node y is v_b and the current flowing through R_1 is i_1, through R_2 is i_2, through R_3 and R_4 is i_5 respectively.

Applying *KCL* at node x, we obtain,

$$i_1 = i_2 + i_3$$

$$\because \qquad i_3 = 0 \quad \text{For ideal OP-Amp}$$

$$\therefore \qquad i_1 = i_2$$

$$\frac{v_{in} - v_a}{R_1} = \frac{v_a - v_b}{R_2}$$

using virtual ground concept. We know that $v_a = 0$.

Hence,

$$\frac{v_{in} - 0}{R_1} = 0 - \frac{v_b}{R_2}$$

$$\frac{v_{in}}{R_1} = -\frac{v_b}{R_2}$$

(A)

Now, applying *KCL* at node *y*, are obtain,

$$i_2 = i_4 + i_5$$

$$\frac{v_a - v_b}{R_2} = \frac{v_b}{R_4} + \frac{v_b - v_0}{R_3}$$

Since, $$v_a = 0$$

$$\therefore \quad -\frac{v_b}{R_2} = \frac{v_b}{R_4} + \frac{v_b}{R_3} - \frac{v_0}{R_3}$$

$$\frac{v_0}{R_3} = v_b \left[\frac{1}{R_2} + \frac{1}{R_3} + \frac{1}{R_4} \right]$$

$$v_0 = v_b R_3 \left[\frac{R_3 R_4 + R_2 R_4 + R_2 R_3}{R_2 R_3 R_4} \right]$$

$$v_0 = v_b \left[\frac{R_3 R_4 + R_2 R_4 + R_2 R_3}{R_2 R_4} \right]$$

Now, putting the value of v_b, we obtain,

$$v_0 = -\frac{R_2}{R_1} v_{in} \left[\frac{R_3 R_4 + R_2 R_4 + R_2 R_3}{R_2 R_4} \right]$$

$$\boxed{v_0 = -v_{in} \left[\frac{R_3 R_4 + R_2 R_4 + R_2 R_3}{R_1 R_4} \right].}$$ **(Ans.)**

Example 8.14   In the given Fig. Ex. 8.14 the variable resistance varies from zero to 100 kΩ. Find out the maximum and minimum closed-loop voltage gain.

(UPTU 2000-01 even seen)

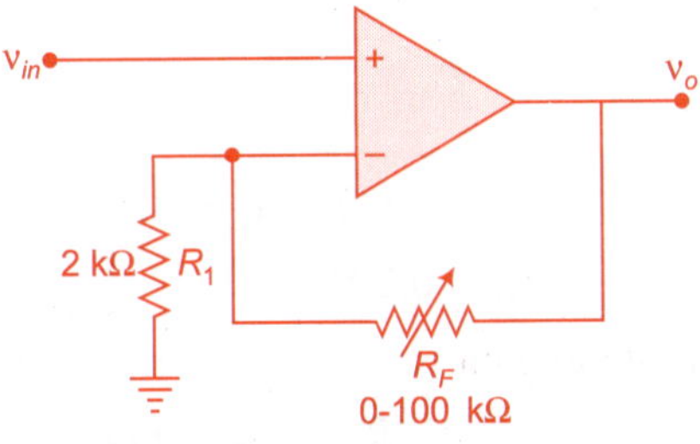

Fig. Ex 8.14

Solution   Given that,

$$R_1 = 2 \text{ k}\Omega \quad \text{and} \quad R_F = 0 - 100 \text{ k}\Omega$$

The given circuit is a non-inverting amplifier, therefore,

$$A_F = 1 + \frac{R_F}{R_1}$$

$$\because \quad A_F = \frac{v_0}{v_{in}}$$

$$\therefore \quad \frac{v_0}{v_{in}} = 1 + \frac{R_F}{R_1}$$

Hence,

$$v_0 = v_{in}\left[1 + \frac{R_F}{R_1}\right]$$

Case (i) When forward resistance is maximum i.e., $R_F = 0\Omega$

then,

$$v_0 = v_{in}\left[1 + \frac{0}{2 \text{ k}\Omega}\right]$$

$$v_0 = v_{in}$$

$$\frac{v_0}{v_{in}} = 1$$

i.e.

$$\boxed{A_F|_{min} = 1.}$$

(**Ans.**)

Case (ii) When forward resistance is maximum i.e., $R_F = 100 \text{ k}\Omega$

then,

$$v_0 = v_{in}\left[1 + \frac{100 \text{ k}\Omega}{2 \text{ k}\Omega}\right]$$

$$v_0 = v_{in}\,[1 + 50]$$

$$\frac{v_0}{v_{in}} = 51$$

Hence,

$$\boxed{A_F|_{max} = 51.}$$

(**Ans.**)

Example 8.15 An inverting amplifier has feedback resistor $R_F = 450 \text{ k}\Omega$ and $R_1 = 5 \text{ k}\Omega$. Find the amplifier circuit voltage gain, input resistance. Also find the output voltage and input current, if the input voltage is given as 0.2 V. Assume the OP-Amp to be ideal.

Solution Given that,

$$R_1 = 5 \text{ k}\Omega, \quad R_F = 450 \text{ k}\Omega$$

$$v_{in} = 0.2 \text{ V}.$$

We know that the closed-loop gain is,

$$A_F = -\frac{R_F}{R_1} = -\frac{450 \text{ k}\Omega}{5 \text{ k}\Omega}$$

$$\boxed{A_F = -90.}$$

(**Ans.**)

The input resistance of an amplifier,

$$R_1 = R_{in} = 5 \text{ k}\Omega$$

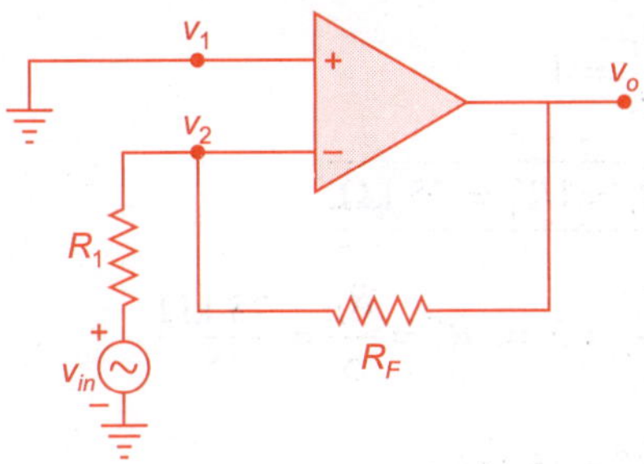

Fig. Ex. 8.15 *Inverting Amplifier*

The output resistance of an amplifier is,

$$R_0 = 0 \ \Omega.$$

The output voltage of an amplifier is given as,

$$A_F = \frac{v_0}{v_{in}}$$

$$v_0 = A_F \, v_{in} = -90 \times 0.2$$

$$\boxed{v_0 = -18 \text{ volt.}}$$ **(Ans.)**

The input current is given as,

$$i_{in} = \frac{v_{in}}{R_1} = \frac{0.2 \text{ V}}{5 \times 10^3 \ \Omega}$$

$$\boxed{i_{in} = 0.04 \text{ mA.}}$$ **(Ans.)**

Example 8.16 Design an adder circuit using an OP-Amp to get the output expression as given below,

$$v_0 = -\left(v_2' + 5v_2'' + 80\,v_2'''\right)$$

where v_2', v_2'' and v_2''' are the inputs. Feedback resistance R_F is given as 75 kΩ.

Solutions Given that,

$$R_F = 75 \text{ k}\Omega$$

From Eqn. (8.31), we know that,

$$v_0 = -\left(\frac{R_F}{R_1'} v_2' + \frac{R_F}{R_1''} v_2'' + \frac{R_F}{R_1'''} v_2'''\right). \qquad \text{(A)}$$

The given expression is,

$$v_0 = -\left(v_2' + 8v_2'' + 150\,v_2''\right). \qquad \text{(B)}$$

Now, comparing expression (A) and (B) for the output, we obtain,

$$\frac{R_F}{R_1'} = 1$$

i.e.
$$\boxed{R_F = 1\,R_1' = 75 \text{ k}\Omega.}$$
(**Ans.**)

Similarly,
$$\frac{R_F}{R_1''} = 5 \Rightarrow R_1'' = \frac{R_F}{5} = \frac{75 \text{ k}\Omega}{5}$$

$$\boxed{R_1'' = 15 \text{ k}\Omega.}$$
(**Ans.**)

Similarly,
$$\frac{R_F}{R_1'''} = 150$$

$$R_1''' = \frac{R_F}{150} = \frac{75}{150}$$

$$\boxed{R_1''' = 0.5 \text{ k}\Omega.}$$
(**Ans.**)

Example 8.17 An OP-Amp has feedback resistor R_F is 10 kΩ and the resistances in the input sides are $R_1' = 8$ kΩ, $R_1'' = 6$ kΩ, $R_1''' = 4$ kΩ. The corresponding inputs are $v_2' = 6$ V, $v_2'' = -4$ V and $v_2''' = 2$ V. Non-inverting terminal is grounded. Calculate the output voltage and tell about the action of OP-Amp.

Solution From Eqn. (8.31), we know that,

$$v_0 = -\left(\frac{R_F}{R_1'}\, v_2' + \frac{R_F}{R_1''}\, v_2'' + \frac{R_F}{R_1'''}\, v_2'''\right)$$
(A)

Given that,

$$R_F = 10 \text{ k}\Omega \quad \text{and} \quad R_1' = 8 \text{ k}\Omega, \quad R_1'' = 6 \text{ k}\Omega, \quad R_1''' = 4 \text{ k}\Omega$$

$$v_2' = 6 \text{ V}, \quad v_2'' = -4 \text{ V}, \quad v_2''' = 2 \text{ V}.$$

Now, substituting the above values in Eqn. (A) we obtain

$$v_0 = -\left[\frac{10}{8} \times 6 + \frac{10}{6} \times (-4) + \frac{10}{2}\,(2)\right]\text{V}$$

$$v_0 = -(7.5 \text{ V} - 6.67 \text{ V} + 10 \text{ V})$$

$$v_0 = -17.5 \text{ V} + 6.67 \text{ V}$$

$$\boxed{v_0 = -10.8 \text{ volt.}}$$
(**Ans.**)

Example 8.18 Fig. Ex. 8.18, shows a non-inverting OP-Amp summer with $v_1' = 5$ V and $v_1'' = -2$ V. Calculate the output voltage V_0.

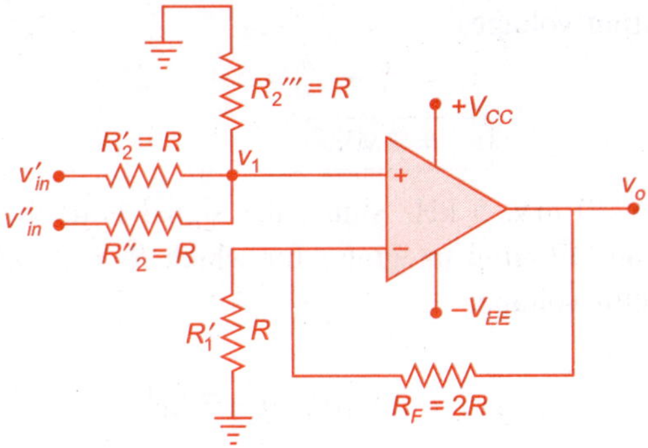

Fig. Ex. 8.18

Solution   Given that,

$$v'_{in} = 5 \text{ V}, \quad v''_{in} = -2 \text{ V}$$

$$R_F = R_1 = R'_2 = R''_2 = R'''_2 = R.$$

According to principal of superposition, we have $v_0 = v'_0 + v''_0$.

From Eqn. (8.43(a)).

where v'_0 = output produced by OP-Amp due to v'_{in} and v''_{in} is reduced to zero.

Case (i)   when $v'_{in} = 5$ V and v''_{in} is reduced to zero, then the input at non-inverting terminal is,

$$v'_1 = \frac{R''_2 \| R'_2}{R''_2 + R'_2 \| R'''_2} \, v'_{in} = \frac{R \| R}{R + R \| R} \, v'_{in}$$

$$v'_1 = \frac{\dfrac{R \times R}{R + R}}{R + \dfrac{R \times R}{R + R}} = \frac{\dfrac{R}{2}}{R + \dfrac{R}{2}} \, v'_{in}$$

$$v'_1 = \frac{5}{3} \text{ volt.}$$

i.e.
$$v'_0 = v'_1 \left(1 + \frac{R_F}{R_1}\right) = \frac{5}{3}\left(1 + \frac{2R}{R}\right) = 5 \text{ volt.}$$

Case (ii)   when $v''_{in} = -2$ V and v'_{in} is reduced to zero, then the input at non-inverting terminal is,

$$v'_1 = \frac{R \| R}{R + R \| R} \, v''_{in} = -\frac{2}{3} \text{ volt}$$

i.e.
$$v''_0 = v''_1 \left(1 + \frac{R_F}{R_1}\right) = -\frac{2}{3}\left(1 + \frac{2R}{R}\right)$$

$$= -2 \text{ volt.}$$

Hence, the output voltage,

$$v_0 = v'_0 + v''_0 = 5 + (-2)$$

$$\boxed{v_0 = 3 \text{ volt.}} \qquad \textbf{(Ans.)}$$

Example 8.19 A 20 mV, 5 kHz sinusoidal signal is applied to the inverting input terminal of an OP-Amp integrator for which $R_1 = 20$ kΩ and $C_F = 3$ μF. Determine the output voltage.

Solution Given that,

$$v_{rms} = 20 \text{ mV}, \quad f = 5 \text{ kHz}$$

i.e. the input signal is,

$$v_{in}(t) = v_{max} \sin \omega t$$

$$v_{in}(t) = \sqrt{2}\, v_{rms} \sin 2\pi f t \qquad \because \quad \frac{v_{max}}{\sqrt{2}} = v_{rms}$$

$$v_{in}(t) = \sqrt{2} \times 20 \sin 2 \times 5 \times 10^3\, \pi t \text{ mV}$$

$$= 20 \times 1.41 \sin 10{,}000\, \pi t \text{ mV}$$

$$= 28.2 \sin 10{,}000\, \pi t \text{ mV}$$

and, $\qquad R_1 = 20$ kΩ and $C_F = 3$ μF.

The output voltage for integrator, we have,

$$v_0 = -\frac{1}{R_1 C_F} \int_0^t v_{in}(t)\, dt$$

$$= -\frac{1}{20 \times 10^3 \times 3 \times 10^{-6}} \int_0^t [28.2 \sin 10{,}000\, \pi t]\, dt$$

$$= -16.67 \times 28.2 \left[\frac{\cos 10{,}000\, \pi t}{10{,}000\, \pi}\right]_0^t$$

$$\boxed{V_0 = -469.9 \,(\cos 10{,}000\, \pi t - 1) \text{ mV.}} \qquad \textbf{(Ans.)}$$

Example 8.20 The input to an OP-Amp differentiator circuit is a sinusoidal voltage of peak value 15 μV and frequency 3 kHz. Determine the output voltage, if $R_F = 30$ kΩ and $C = 1$ μF. Determine the output voltage.

Solution Given that,

$$V_{max} = 15 \text{ }\mu\text{V}, f = 3 \text{ kHz}$$

$$R_F = 30 \text{ k}\Omega \quad \text{and} \quad C = 1 \text{ }\mu\text{F}.$$

The input signal is,

$$V_{in} = V_{max} \sin \omega t$$

$$V_{in} = V_{max} \sin 2\pi f t$$

$$V_{in} = 15 \sin 2 \times 3000\, \pi t$$

$$V_{in} = 15 \sin 6000\, \pi t.$$

The output voltage for differentiator, we have,

$$V_0 = - R_F\, C\, \frac{d}{dt}\,(V_{in})$$

$$= - 30 \times 10^3 \times 1 \times 10^{-6}\, \frac{d}{dt}\,(15 \sin 6000\,\pi t)$$

$$= - 0.03 \times 15\, \frac{d}{dt}\,(\sin 6000\,\pi t)$$

$$= - 0.45\,[\cos 6000\,\pi t] \times 6000\,\pi$$

$$= - 2700\,\pi \cos 6000\,\pi t\ \mu V$$

$$\boxed{V_0 = - 2.7\,\pi \cos 6000\,\pi t\ mV.}$$
   (Ans.)

Example 8.21 Consider an ideal OP-Amp circuit as shown in Fig. Ex. 8.21. Given that $R_1 = 20$ kΩ, $R_F = 50$ kΩ $v_2' = 5$ V, Find V_0.

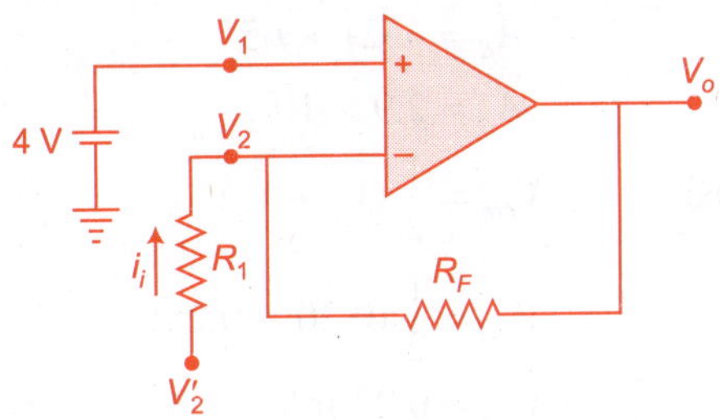

Fig. Ex. 8.21

Solution From Fig. Ex. 8.21,

$$V_1 = V_2 = 4\ V$$

The current i_i is,

$$i_i = \frac{V_2' - V_2}{R_1}$$

$$= \frac{5\ V - 4\ V}{20\ k\Omega}$$

$$= \frac{1}{20 \times 10^3}\ Amp$$

$$= 0.05\ mA.$$

The output V_0 is,

$$V_0 = - i_i\, R_F + V_2$$

$$V_0 = (- 0.05 \times 50 \times 10^3 \times 10^{-3}) + (4\ V)$$

$$V_0 = - 2.5 + 4\ V$$

$$\boxed{V_0 = 1.5\ volt.}$$
   (Ans.)

Example 8.22   A differential amplifier has a typical common mode gain of 65 *CMMR* of 4521. Find the output voltage (V_0), when input voltages are 0.20 mV and 0.25 mV.

Solution   Given that,

Common mode gain, $A_{cm} = 65$

$$CMRR = 4521$$

input voltages are,

$$V_1 = 0.20 \text{ mV} \quad \text{and} \quad V_2 = 0.25 \text{ mV}$$

we know that,

$$CMRR = \frac{A_d}{A_C}$$

differential gain, $A_d = CMRR \times A_C$

$$A_d = 4521 \times 65$$

$$A_d = 2.9 \times 10^5.$$

Common mode signal, $V_{cm} = \dfrac{1}{2}(V_1 + V_2)$

$$V_{cm} = \frac{1}{2}(0.20 + 0.25)$$

$$V_{cm} = 0.23 \text{ mV}$$

differential mode signal, $V_d = V_2 - V_1$

$$= (0.25 - 0.20) \text{ mV}$$

$$= 0.05 \text{ mV}.$$

Hence output voltage,

$$V_0 = A_{cm} V_{cm} + A_d V_d$$

$$= 65 \times 0.23 \text{ mV} + 2.9 \times 10^5 \times 0.05 \times 1 \text{ mV}$$

$$= 14.95 \text{ mV} + 14.5 \text{ mV}$$

$$\boxed{V_0 = 29.45 \text{ mV.}}$$

(Ans.)

Example 8.23   The summing amplifier as shown in Fig. Ex 8.23 has $R_F = 3$ kΩ, $R'_1 = 8$ kΩ, $R''_1 = 3.5$ kΩ and $R'''_1 = 3.2$ kΩ if $V'_2 = 5$ V, $V''_2 = -4$ V and $V'''_2 = -0.1$ V find V_0.

Solution   We know that from Eqn. (8.31) is,

$$V_0 = -\left[\frac{R_F}{R'_1} V'_2 + \frac{R_F}{R''_1} V''_2 + \frac{R_F}{R'''_1} V'''_2\right]$$

$$V_0 = -\left[\frac{3}{8} \times (5) + \frac{3}{35} \times (-4) + \frac{3}{3.2}(-0.1)\right]$$

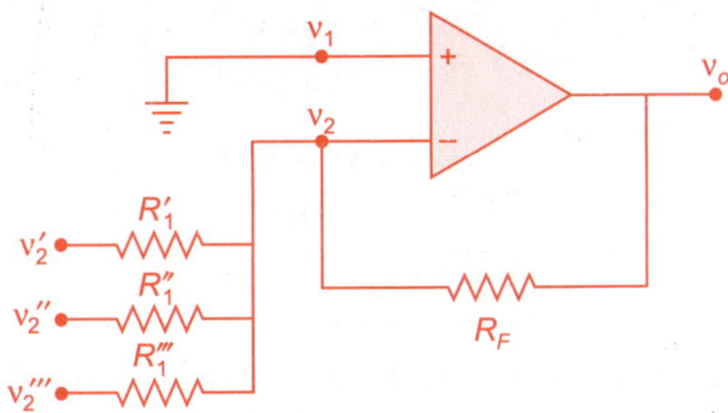

Fig. Ex. 8.23

$$= -[1.88 - 3.43 - 0.09]$$

$$= -[1.88 - 3.52]$$

$$\boxed{V_0 = 1.64.}$$ **(Ans.)**

Example 8.24 In the circuit of Fig. Ex 8.23, $V_2' = 2$ V, $V_2'' = 3$ V, $V_2''' = 5$ V, $R_1' = R_1'' = R_1''' = 2$ kΩ and $R_F = 1$ kΩ the supply voltage is ± 18 V. Assuming that the OP-Amp is initially nulled. Determine the output voltage.

Solution From Eqn. (8.32), we know that,

$$v_0 = -\frac{R_F}{R_1}\,(v_2' + v_2'' + v_2''')$$

when,

$$R_1' = R_1'' = R_1''' = R_1$$

Now,

$$v_0 = -\frac{1\ \text{k}\Omega}{2\ \text{k}\Omega}\,(2\ \text{V} + 3\ \text{V} + 5\ \text{V})$$

$$= -\frac{11\ \text{V}}{2}$$

$$\boxed{v_0 = -5.5\ \text{volt.}}$$ **(Ans.)**

Example 8.25 In the circuit of Fig. Ex. 8.23, $V_2' = 5$ V $V_2'' = -3$ V and $V_2''' = 2$ V. The resistances are $R_1' = 1$ kΩ, $R_1'' = 2$ kΩ, $R_1''' = 4$ kΩ. The output voltage is -8 V. Determine the feedback resistances R_F.

Solution From Eqn. (8.31), we know that,

$$v_0 = -R_F\left(\frac{1}{R_1'}\,v_2' + \frac{1}{R_1''}\,v_2'' + \frac{1}{R_2'''}\,v_2'''\right)$$

Now, substituting the values,

$$-8\text{V} = -R_F\left(\frac{5\ \text{V}}{1\ \text{k}\Omega} + \frac{-3\ \text{V}}{2\ \text{k}\Omega} + \frac{2\ \text{V}}{4\ \text{k}\Omega}\right)$$

$$-8\text{V} = -R_F\left(\frac{20\ \text{V} - 6\ \text{V} + 2\ \text{V}}{4\ \text{k}\Omega}\right)$$

$$-8\text{ V} = -R_F\left(\frac{16\text{ V}}{4\text{ k}\Omega}\right)$$

$$R_F = +\frac{8\text{ V} \times 4\text{ k}\Omega}{16\text{ V}}$$

$$\boxed{R_F = 2\text{ k}\Omega.} \tag{Ans.}$$

Example 8.26 For the inverting amplifier as shown in Fig. Ex. 8.26. Determine the output offset voltage due to (a) Input offset voltage and (b) due to input bias current. The input offset voltage $V_{io} = 6$ mV DC and input bias current I_B is 500 nA × DC.

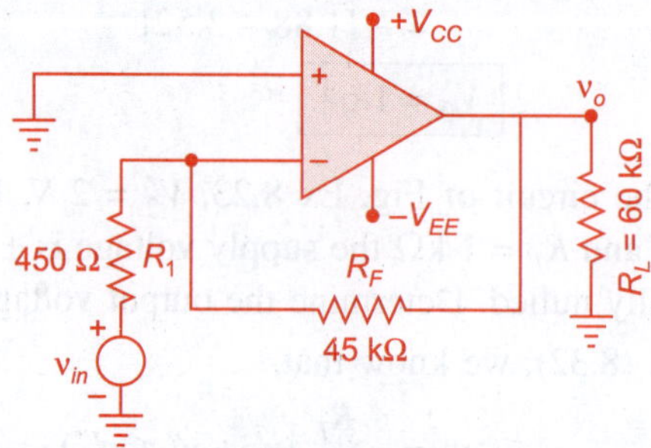

Fig. Ex. 8.26

Solution The output offset voltage due to input offset voltage is given as,

$$V_{00} = \left(1 + \frac{R_F}{R_1}\right) V_{io}$$

$$V_{00} = \left(1 + \frac{45\text{ k}\Omega}{450\text{ }\Omega}\right)(6\text{ mV})$$

$$\boxed{V_{00} = 606\text{ mV DC.}}$$

where, $V_{00} =$ output offset voltage.

The output offset voltage due to bias current is,

$$V_{00} = R_F I_B$$

$$V_{00} = (45\text{ k}\Omega)(500\text{ nA})$$

$$\boxed{V_{00} = 22.5\text{ mV DC.}} \tag{Ans.}$$

Example 8.27 Consider an ideal OP-Amp circuit as shown in Fig. Ex 8.27. Given that $R_1 = 8$ kΩ, $R_F = 20$ kΩ and $V_{in} = 15$ V. Find V_0.

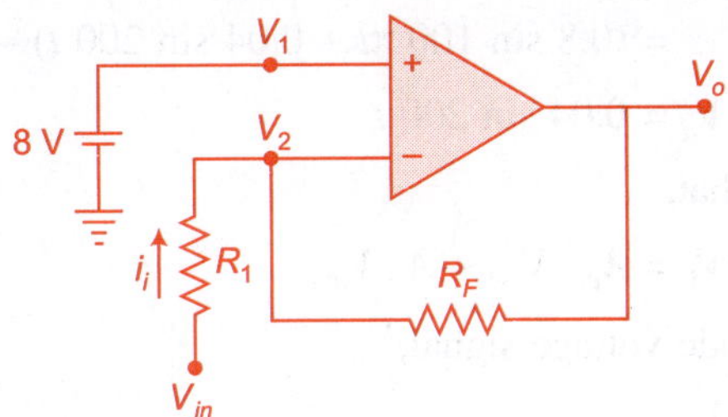

Fig. Ex. 8.27

Solution   From Fig. Ex. 8.27,

$$V_1 = V_2 = 8 \text{ V}$$

The current i_i is,

$$i_i = \frac{V_{in} - V_2}{R_1}$$

$$= \frac{15 \text{ V} - 8 \text{ V}}{3 \text{ k}\Omega} = \frac{7 \text{ V}}{8 \text{ k}\Omega}$$

$$i_i = 0.88 \text{ mA}.$$

The output voltage V_0 is,

$$V_0 = - i_i R_F + v_2$$

$$V_0 = (- 0.88 \times 10^{-3} \times 20 \times 10^3 + 8 \text{ V})$$

$$V_0 = - 17.5 + 8 \text{ V}$$

$$\boxed{V_0 = - 9.5 \text{ volt.}} \qquad \text{(Ans.)}$$

Example 8.28   The inputs of an amplifier are $V_2 = 0.8 \sin 100\,\pi t + 0.04 \sin 200\,t$ and $V_1 = 0.8 \sin (100\,\pi t)$. The differential gain of the amplifier is 250 and the *CMRR* is 1200. What will be the output signal?

Solution   Given that,

$$A_d = 250, \quad CMRR = 1200$$

$$V_1 = 0.8 \sin 100\,\pi t$$

$$V_2 = 0.8 \sin 100\,\pi t + 0.04 \sin 200\,t.$$

We know that,

common mode gain, $\qquad A_C = \dfrac{\text{differential gain } A_d}{CMRR}$

$$A_{cm} = \frac{250}{1200} = 0.21.$$

The differential voltage V_d,

$$V_d = V_2 - V_1$$

$$= (0.8 \sin 100 \, \pi t + 0.04 \sin 200 \, t) - (0.8 \sin 100 \, \pi t)$$

$$V_d = 0.04 \sin 200 \, t.$$

Now, we know that,

$$V_0 = A_{cm} \, V_{cm} + A_d \, V_d. \tag{A}$$

The common mode voltage signal,

$$V_{cm} = \frac{1}{2} \, (V_1 + V_2)$$

$$= \frac{1}{2} \, [(0.8 \sin 100 \, \pi t + 0.04 \sin 200 \, t) + (0.8 \sin 100 \, \pi t)]$$

$$= \frac{1}{2} \, [1.6 \sin 100 \, \pi t + 0.04 \sin 200 \, t]$$

$$V_{cm} = 0.8 \sin 100 \, \pi t + 0.02 \sin 200 \, t.$$

Now, substituting the calculated values in Eqn. (A), we obtain,

$$V_0 = 0.21 \, (0.8 \sin 100 \, \pi t + 0.02 \sin 200 \, t) + 250 \, (0.04 \sin 200 \, t)$$

$$= 0.17 \sin 100 \, \pi + 0.42 \sin 200 \, t + 10 \sin 200 \, t$$

$$\boxed{V_0 = 0.17 \sin 100 \, \pi + 10.42 \sin 200 \, t.} \qquad \textbf{(Ans.)}$$

Example 8.29   What will be the range of potentiometer for the range of voltage gain adjustment $- 10$ to $- 60$ for the given circuit of Fig. Ex. 8.29? Given that $R_1 = 15 \text{ k}\Omega$ and $R_F = 1000 \text{ k}\Omega$.

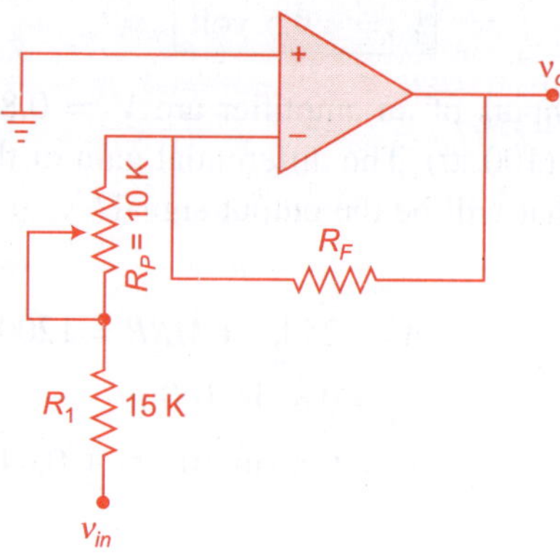

Fig. Ex. 8.29

Solution   The gain of inverting amplifier is given by,

$$A = - \left(\frac{R_F}{R_1 + R_p} \right)$$

where, R_p = Resistance of potentiometer.

When gain is –10, then pot is at,

$$A = -\left(\frac{R_F}{R_1 + R_p}\right) \Rightarrow -10 = -\left(\frac{600 \text{ k}\Omega}{15 \text{ k}\Omega + R_p}\right)$$

$$150 \text{ k}\Omega + 10\, R_p = 600 \text{ k}\Omega$$

$$10\, R_p = (1000 - 150)$$

$$10\, R_p = 850 \text{ k}\Omega$$

$$R_p = \frac{850 \text{ k}\Omega}{10} = 85 \text{ k}\Omega$$

$$R_p = 45 \text{ k}\Omega.$$

Case (ii)   When gain is – 60, then pot is at

$$60 = \left(\frac{1000 \text{ k}\Omega}{15 \text{ k}\Omega + R_p}\right)$$

$$900 \text{ k}\Omega + 60\, R_p = 1000 \text{ k}\Omega$$

$$60\, R_p = 1000 \text{ k}\Omega - 900 \text{ k}\Omega$$

$$60\, R_p = 100 \text{ k}\Omega$$

$$R_p = \frac{1000 \text{ k}\Omega}{60} = 1.6 \text{ k}\Omega$$

$$\boxed{R_p = 1.6 \text{ k}\Omega.}$$
  (Ans.)

Example 8.30   For a given OP-Amp circuit the differential gain is 12×10^5 and the common mode gain is 100. Determine the common mode rejection ratio.

Solution   We know that,

$$CMRR = \frac{A_d}{A_{cm}}$$

Given that,

$$A_d = 12 \times 10^5$$

$$A_{cm} = 100$$

$\therefore$

$$CMRR = \frac{12 \times 10^5}{100}$$

$$\boxed{CMRR = 12 \times 10^3.}$$
  (Ans.)

Example 8.31   Determine the output voltage for the open-loop differential amplifier for given data as $V_{in_1} = 5 \text{ }\mu A$, $V_{in_2} = - 6 \text{ }\mu A$ The OP-Amp is 741 with the following specification.

$$A = 2 \times 10^6 \quad R_i = 2 \text{ M}\Omega, \quad R_0 = 50 \ \Omega$$

$$V_{cc} = 15 \text{ V}, \quad V_{EE} = -15 \text{ V}$$

we know that the open loop gain is,

$$A = \frac{v_0}{v_{in}}$$

$$v_0 = A \ v_{in} = A(v_{in_1} - v_{in_2})$$

$$v_0 = 2 \times 10^6 \ [5 \times 10^{-6} - (-6 \times 10^{-6})]$$

$$v_0 = 2 \times 10^6 \times 10^{-6} \ [5 + 6]$$

$$\boxed{V_0 = 22 \text{ volt.}}$$

(**Ans.**)

Example 8.32 If the circuit of Fig. Ex 8.32, has $R_1 = 50$ kΩ and output voltage is -80 V for the input voltage of 8 V. Determine the required feedback resistance.

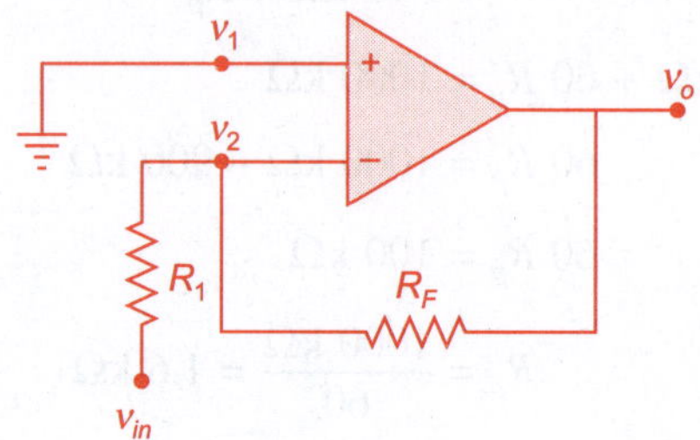

Fig. Ex. 8.32

Solution   Given that,

$$v_0 = -80 \text{ volt}$$

$$R_1 = 50 \text{ k}\Omega$$

$$R_F = ?, \quad v_{in} = 8 \text{ V}$$

we know that,

$$v_0 = -\frac{R_F}{R_1} v_{in}$$

$$R_F = -\frac{v_0 R_1}{V_{in}} = -\left(\frac{-80 \times 50 \text{ k}\Omega}{8}\right)$$

$$\boxed{R_F = 500 \text{ k}\Omega.}$$

(**Ans.**)

Example 8.33 The output voltage of certain OP-Amp at 2 µs is 15 V and 6 µs is 60 V. What is the slew rate?

Solution   Given that,

$$v_{01} = 15 \text{ V} \quad \text{at} \quad t_1 = 2 \text{ µs}$$

and, $v_{02} = 60$ V   at $t_2 = 6$ μs

The change in output voltage is,

$$\Delta v_0 = v_{02} - v_{01} = 60 \text{ V} - 15 \text{ V}$$

$$\Delta v_0 = 45 \text{ V}.$$

The change in time is,

$$\Delta t = t_2 - t_1 = 6 \text{ μs} - 2 \text{ μs}$$

$$\Delta t = 4 \text{ μs}.$$

Now, from Eqn. (8.7), we know that,

$$SR = \frac{\Delta v_0}{\Delta t} = \frac{45 \text{ V}}{4 \text{ μs}}$$

$$\boxed{SR = 11.25 \text{ V/μs}.}$$ **(Ans.)**

Example 8.34   In the given Fig. Ex. 8.34, the variable resistance varies from 1 kΩ to 10 kΩ. Find the maximum and the minimum closed-loop voltage gain.

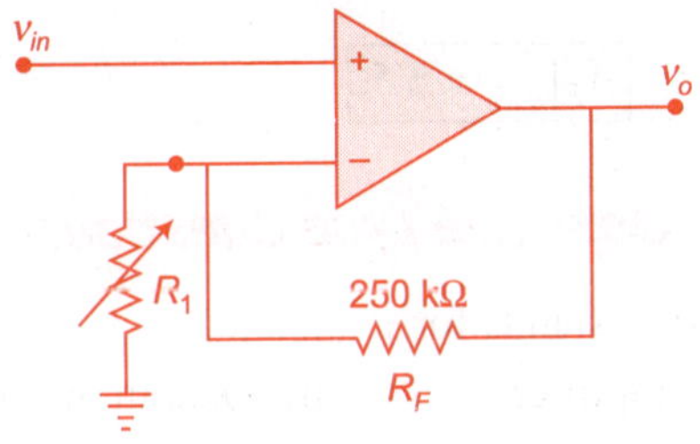

Fig. Ex. 8.34

Solution   Given that,

$$R_F = 250 \text{ kΩ}$$

$$R_1 = 1 \text{ kΩ to } 100 \text{ kΩ}.$$

The given circuit is a non-inverting amplifier

∴ $A_F = 1 + \dfrac{R_F}{R_1}$

∴ $A_F = \dfrac{v_0}{v_{in}}$

Hence, $v_0 = \left(1 + \dfrac{R_F}{R_1}\right) v_{in}$

Case (i)   when resistance R_1 is minimum.

i.e. $R_1 = 1 \text{ kΩ}$

then,
$$v_0 = \left(1 + \frac{250 \text{ k}\Omega}{2 \text{ k}\Omega}\right)$$

$$v_0 = 251 \, v_{in}$$

$$A_F = \frac{v_0}{v_{in}} = 251$$

$$\boxed{A_F|_{max} = 251.}$$

(Ans.)

Case (ii)   when resistance R_1 is maximum

i.e.
$$R_1 = 100 \text{ k}\Omega$$

then,
$$v_0 = \left(1 + \frac{250 \text{ k}\Omega}{100 \text{ k}\Omega}\right) v_{in}$$

$$\frac{v_0}{v_{in}} = (1 + 2.5)$$

$$A_F|_{min} = 3.5$$

$$\boxed{A_F|_{min} = 3.5.}$$

(Ans.)

Objective Type Questions

1. The word OP-Amp stands for ....
   - (a) Optimum Amplifier
   - (b) Operational Ampere
   - (c) Operational Amplifier
   - (d) None of these

2. For the circuit shown in Fig. OT 8, the output voltage V_0 in given by:

   (a) $v_0 = - RC \int_0^t v_{in}(t) \, dt$

   (b) $v_0 = - \dfrac{1}{RC} \dfrac{d}{dt} v_{in}(t)$

   (c) $v_0 = - RC \dfrac{d}{dt} v_{in}(t)$

   (d) $v_0 = - \dfrac{1}{RC} \int_0^t v_{in}(t) \, dt$

   Fig. OT 8.1

3. The input signal of an amplifier is given as $\cos (314 \, t)$, the output signal for a gain of 200 will be
   - (a) $200 \cos (314 \, t + 90°)$
   - (b) $200 \cos (314 \, t + 180°)$
   - (c) $200 \cos (314 \, t - 90°)$
   - (d) $100 \cos (314 \, t + 180°)$

4. The input impedance of an ideal OP-Amp is
   - (a) Finite
   - (b) Zero
   - (c) Infinite
   - (d) Unity

5. The gain of an inverting amplifier is given as
   - (a) $\dfrac{R_F}{R_1}$
   - (b) $-\dfrac{R_F}{R_1}$
   - (c) $\dfrac{R_1}{R_F}$
   - (d) $\dfrac{R_1}{R_1 + R_F}$

6. The output impedance of an ideal OP-Amp is
   - (a) Zero
   - (b) Finite
   - (c) Infinite
   - (d) None of these

7. The gain of an OP-Amp voltage follower is
   - (a) Zero
   - (b) Unity
   - (c) Infinite
   - (d) Very high

8. In the circuit, V_0 is given by
   - (a) $2V_s$
   - (b) $3V_s$
   - (c) $4V_s$
   - (d) V_s

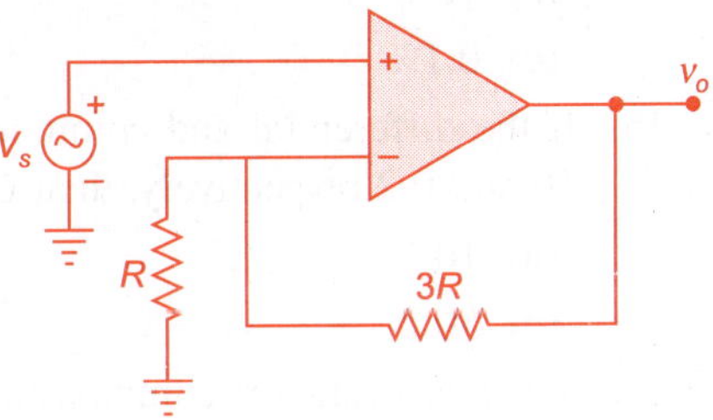

9. The gain of an ideal OP-Amp is
   - (a) Zero
   - (b) Infinite
   - (c) Finite
   - (d) None of these

10. In the OP-Amp circuit shown in Fig. OT 8.2, V_0 is given by
    - (a) $3V_{s1} - 6V_{s2}$
    - (b) $2V_{s1} - 3V_{s2}$
    - (c) $2V_{s1} - 2V_{s2}$
    - (d) $3V_{s1} - 2V_{s2}$

11. The value of V_0 is given by
    - (a) $-3V_1 + V_2/2$
    - (b) $3V_2/2 - 9V_1/4$
    - (c) $-3V_2$
    - (d) $2V_2 - 3V_1$

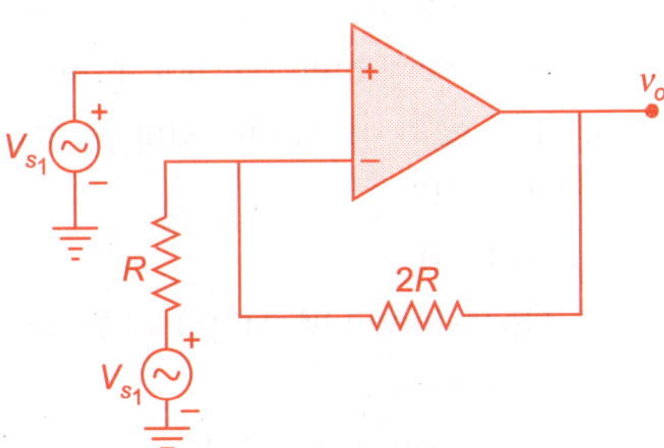

Fig. OT 8.2

12. For an ideal deference amplifier, the *CMRR* should be

    (a) As high as possible

    (b) As low as possible

    (c) Constant

    (d) None of these

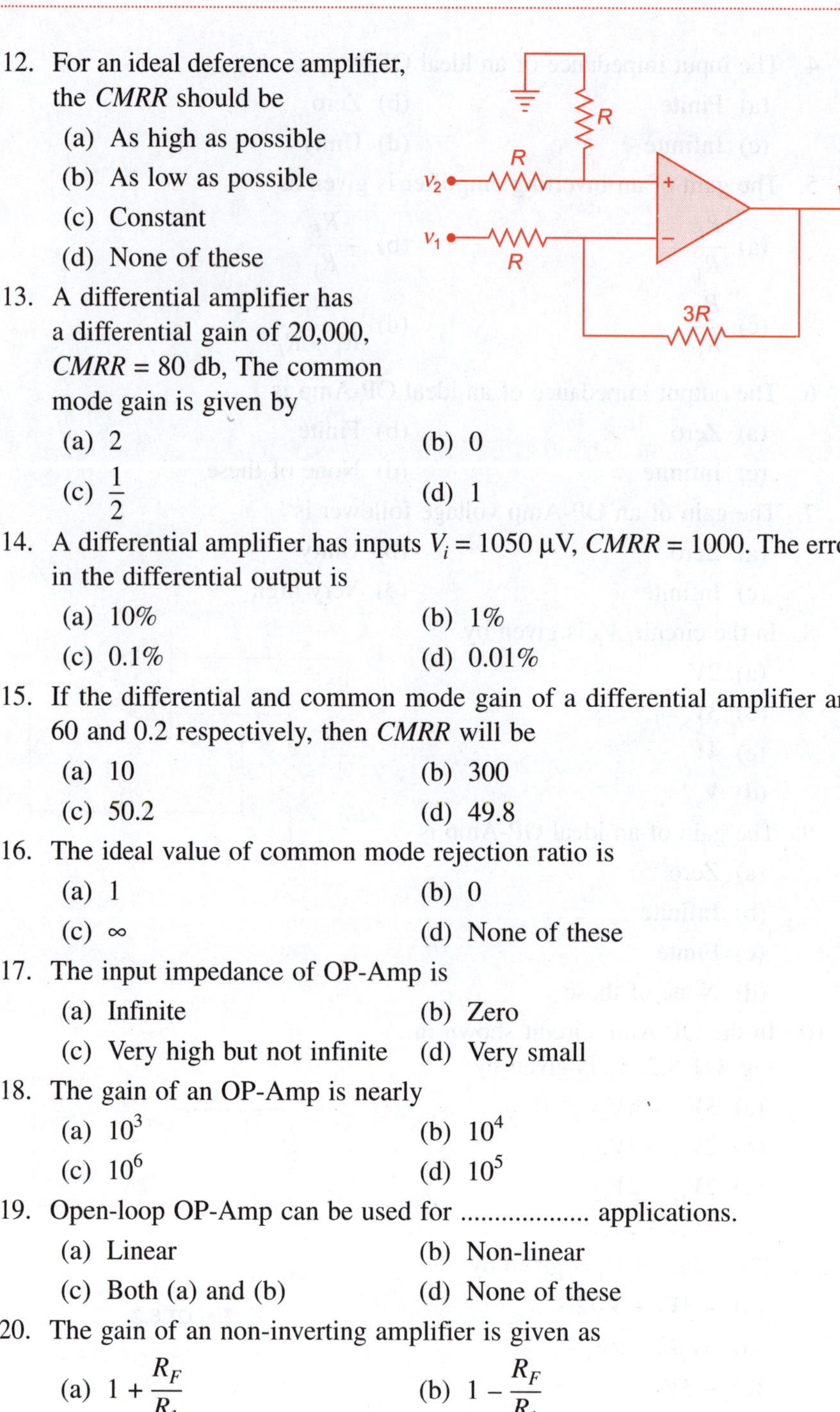

13. A differential amplifier has a differential gain of 20,000, *CMRR* = 80 db, The common mode gain is given by

    (a) 2   (b) 0

    (c) $\dfrac{1}{2}$ (d) 1

14. A differential amplifier has inputs $V_i = 1050\ \mu V$, *CMRR* = 1000. The error in the differential output is

    (a) 10%   (b) 1%

    (c) 0.1%   (d) 0.01%

15. If the differential and common mode gain of a differential amplifier are 60 and 0.2 respectively, then *CMRR* will be

    (a) 10   (b) 300

    (c) 50.2   (d) 49.8

16. The ideal value of common mode rejection ratio is

    (a) 1   (b) 0

    (c) ∞ (d) None of these

17. The input impedance of OP-Amp is

    (a) Infinite   (b) Zero

    (c) Very high but not infinite   (d) Very small

18. The gain of an OP-Amp is nearly

    (a) 10^3 (b) 10^4

    (c) 10^6 (d) 10^5

19. Open-loop OP-Amp can be used for .... applications.

    (a) Linear   (b) Non-linear

    (c) Both (a) and (b)   (d) None of these

20. The gain of an non-inverting amplifier is given as

    (a) $1 + \dfrac{R_F}{R_1}$ (b) $1 - \dfrac{R_F}{R_1}$

    (c) $1 + \dfrac{R_1}{R_F}$ (d) $1 - \dfrac{R_1}{R_F}$

21. The OP-Amp amplifies the .... of two input voltage applied at input terminal.
    - (a) Addition
    - (b) Difference
    - (c) Multiplication
    - (d) None of these

22. When the output of OP-Amp is zero at applied input voltage, then the input voltage is named as
    - (a) Offset voltage
    - (b) Input voltage
    - (c) Input offset voltage
    - (d) None of these

23. *CMMR* is defined as the ratio of differential voltage gain to the ....
    - (a) Common mode gain
    - (b) Differential gain
    - (c) Common mode output voltage
    - (d) None of these

24. When the input voltage of an ideal OP-Amp is zero, then the output voltage is ....
    - (a) Finite
    - (b) Infinite
    - (c) One
    - (d) Zero

25. The bandwidth of an ideal OP-Amp must be ....
    - (a) Zero
    - (b) Finite
    - (c) Infinite
    - (d) None of these

26. The *CMRR* of an ideal OP-Amp must be
    - (a) Zero
    - (b) Infinite
    - (c) Finite
    - (d) None of these

27. The slew rate of rate of an OP-Amp is defined as
    - (a) $\left.\dfrac{dv_o}{dv_i}\right|_{max}$
    - (b) $\left.\dfrac{dv_i}{dv_o}\right|_{max}$
    - (c) $\left.\dfrac{dv_o}{dt}\right|_{max}$
    - (d) $\left.\dfrac{dv_i}{dt}\right|_{max}$

28. The output voltage for an open-loop inverting amplifier is ....
    - (a) $-\dfrac{A}{v_{in}}$
    - (b) $-Av_{in}$
    - (c) $\dfrac{v_{in}}{A}$
    - (d) $-Av_{in}$

29. The output voltage for an open-loop non-inverting amplifier is ....

(a) $-\dfrac{A}{V_{in}}$ (b) $-AV_{in}$

(c) $\dfrac{V_{in}}{A}$ (d) $-AV_{in}$

30. For a unity gain Op-Amp, the output voltage is ....
    (a) Equal to input   (b) One
    (c) Infinite   (d) Finite

31. For the circuit shown in Fig. OT 8.3, the output voltage V_0 is given by

(a) $-\left(\dfrac{R_F}{R'_1}\,v'_2 + \dfrac{R_F}{R''_1}\,v''_2\right)$

(b) $-\dfrac{R_F}{R'_1}\,(v'_2 + v''_2)$

(c) $-\left(\dfrac{R_F}{R'_1} + \dfrac{R_F}{R'_2}\right)(v'_2 + R''_1)$

(d) None of these

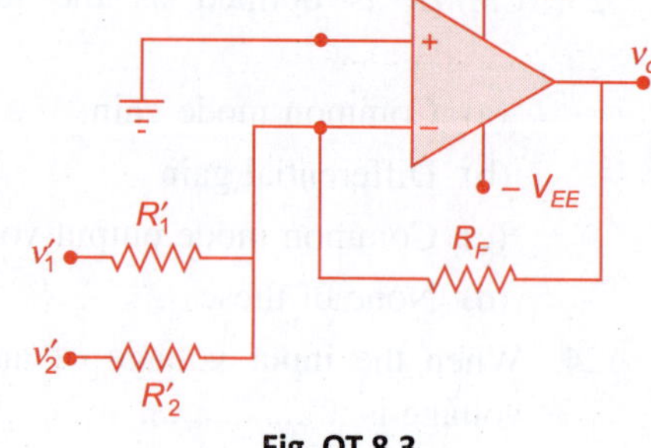
Fig. OT 8.3

32. For the circuit shown in Fig. OT 8.3, is named as
    (a) Integrator   (b) Summer
    (c) Differentiator   (d) None of these

33. For the circuit shown in Fig. OT 8.4, the output voltage V_0 is given as

(a) $-\dfrac{R_F}{C}\dfrac{d}{dt}\,(v_{in})$

(b) $-R_F\,C\displaystyle\int_0^t (v_{in})\,dt$

(c) $-R_F\,C\dfrac{d}{dt}\,(v_{in})$

(d) None of these

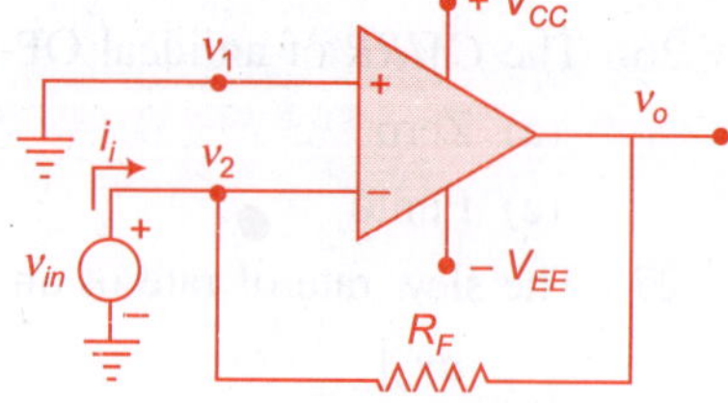
Fig. OT 8.4

34. For the circuit shown in Fig. OT 8.4 is named as ....
    (a) Integrator
    (b) Differentiator
    (c) Summer
    (d) None of these

35. For the circuit shown in Fig. OT 8.5, the output voltage V_0 is given as

(a) $\dfrac{R_F}{R_1}\,(v'_{in} - v''_{in})$

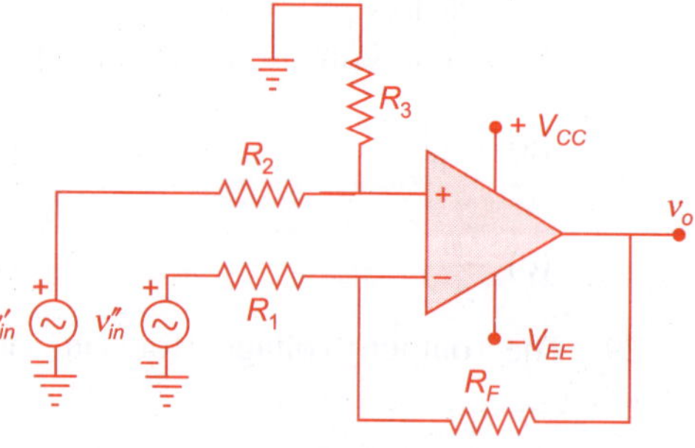
Fig. OT 8. 5

(b) $\dfrac{R_1}{R_F}\,(v'_{in} - v''_{in})$

(c) $\dfrac{R_F}{R_1}\,(v''_{in} - v''_{in})$ (d) $R_1\,R_F\,(v'_{in} - v''_{in})$

36. For the circuit shown in Fig. OT 8.5 is named as
    (a) Differential Amplifier (b) Integrator
    (c) Differentiator (d) None of these

37. For the circuit shown in Fig. OT 8.6, the output voltage V_0 is given as
    (a) $R\,(v'_{in} - v''_{in})$
    (b) $\dfrac{1}{R}\,(v'_{in} - v''_{in})$
    (c) $(v'_{in} - v''_{in})$
    (d) None of these

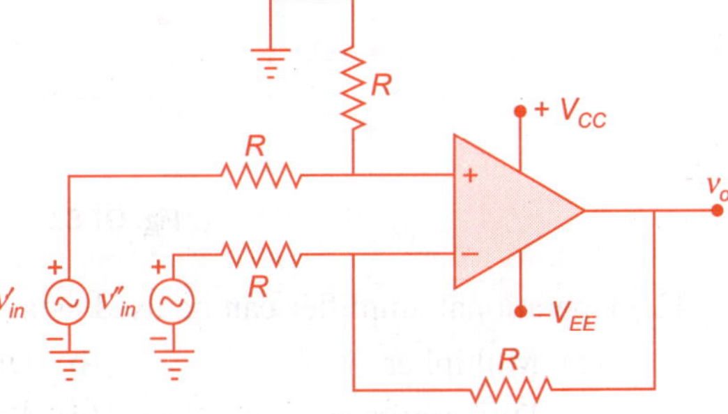

Fig. OT 8.6

38. For the circuit shown in Fig. OT 8.6 is named as
    (a) Differential Amplifier (b) Substractor
    (c) Summer (d) Integrator

39. For the circuit shown in Fig. OT 8.7, the circuit is named as

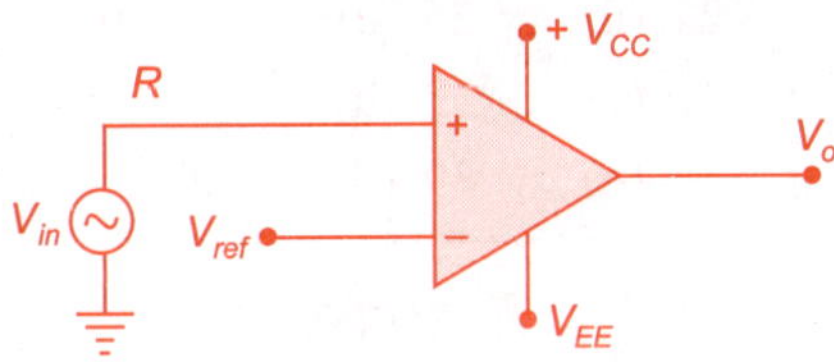

Fig. OT 8.7

    (a) Differentiator (b) Integrator
    (c) Substractor (d) Comparator

40. From the circuit given in Fig. OT 8.8 the value of output voltage V_0 is
    ....
    (a) V_{in}
    (b) $0.5\,V_{in}$
    (c) $2V_{in}$
    (d) $\dfrac{V_{in}}{2}$

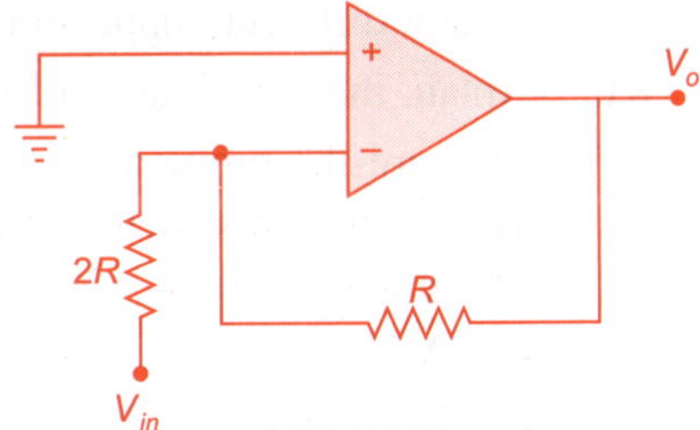

Fig. OT 8.8

41. The circuit shown in Fig. OT 8.9 acts *S* as a .... For the given inputs, its output voltage is .... *V*.

    (a) Integrator, – 1 V (b) Integrator, – 2 V

    (c) Summer, – 1 V (d) Summer, – 2 V

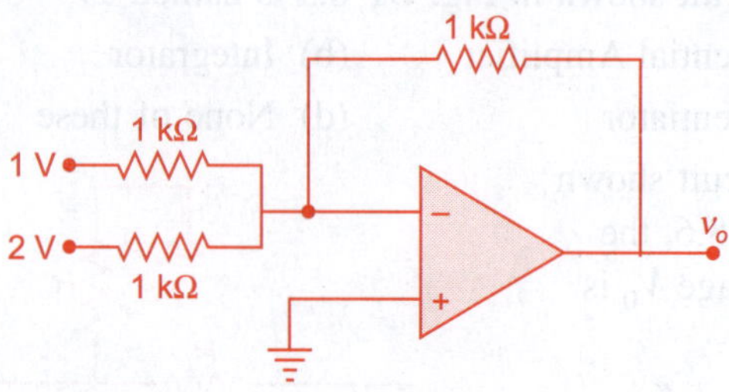

Fig. OT 8.9

42. Operational amplifier can be used as a

    (a) Multiplier (b) Divider

    (c) Differentiator (d) None of these.

ANSWERS

1 (c)	2 (d)	3 (b)	4 (c)	5 (b)	6 (a)	7 (b)
8 (c)	9 (b)	10 (a)	11 (c)	12 (d)	13 (b)	14 (c)
15 (b)	16 (c)	17 (c)	18 (d)	19 (b)	20 (a)	21 (b)
22 (c)	23 (a)	24 (d)	25 (c)	26 (b)	27 (c)	28 (d)
29 (b)	30 (a)	31 (a)	32 (b)	33 (c)	34 (b)	35 (a)
36 (a)	37 (c)	38 (b)	39 (d)	40 (b)	41 (c)	42 (c)

Exercise

8.1. What is an Operational Amplifier? Draw the block diagram of Op-Amp and explain it.

8.2. Draw and explain the ideal Op-Amp.

8.3. Explain open-loop Op-Amp configuration.

8.4. What are the limitations of open-loop Op-Amp configurations?

8.5. Explain different applications of Op-Amp.

8.6. Explain the significance of virtual round in a basic inverting amplifier.

8.7. What are the properties of an ideal Op-Amp used in measurement and instrumentation system? Explain with the help of circuit diagram, how it is used as? **(UPTU 2006-07)**

    (a) Adder (b) Substractor

    (c) Integrator (d) Differentiator

8.8. Write short notes on:

    (a) Inverting and non-inverting amplifier

    (b) *CMMR* **(UPTU 2007-08)**

8.9. Draw the circuits of inverting amplifier, non-inverting amplifier and difference amplifier using Op-Amp. Derive the expression for output voltage. **(UPTU 2007-08)**

8.10. In an Op-Amp used as an integrator $R_1 = 10^6$ Ω and $C = 0.5$ mF. If the input voltage is 0.4 mt, find the value of output voltage.

(Ans. $0.4\ t^2$ mt)

8.11. List the ideal characteristics of an Op-Amp. Why OP-Amp is called Operational amplifier? Find the voltage output of the following circuit of Fig. E 8.11 **(UPTU 2006-07)**

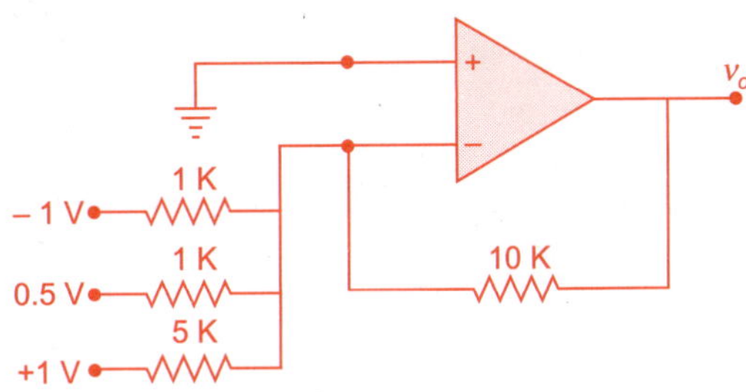

Fig. E 8.11

(Ans. $V_0 = 12$ V)

8.12. Discuss an integrator and show that an integrator is a circuit in which the output voltage waveform in the integral of the input voltage waveform.

8.13. Define *CMMR* of a differential amplifier.

8.14. Draw the circuit of integrator and differentiator using Op-Amp. Derive the expression for output. **(UPTU 2007-08)**

8.15. Define and explain the term *CMRR* and virtual ground in an Op-Amp. Why integrators are preferred over differentiator?

(UPTU 2008-09)

8.16. What is Op-Amp? How it is used as an integrator and summer?

(UPTU 2008-09)

8.17. Calculate the output voltage of a non-inverting amplifier for which the values are $V_1 = 2$ V, $R_F = 500$ kΩ and $R_1 = 100$ kΩ.

(Ans. $V_0 = 12$ V)

8.18. Explain closed-loop Op-Amp configurations.

8.19. Draw the circuit of a substractor using Op-Amp and explain its working.

8.20. Draw the circuit diagram of unity gain amplifier. Where is it and why?

Switching Theory and Logic Design

9

9.1 INTRODUCTION

In previous chapters we have studied about analog electronics i.e., which deals with analog input and produces analog output. In this chapter, we will study about digital logic circuit and design, but before knowing about digital logic circuit, we should know about digital signal because the digital logic circuits are the circuits which deals with digital input and produces digital output. We should also understand the basic difference between analog and digital signal:

"A signal which is defined at every instant of time (continuous range of time) or a signal that has infinite number of levels over a defined interval of time is called analog signal i.e., continuous signal as shown in Fig. 9.1(a)".

"The signal which is defined at particular instant of time is called digital signal or non-continuous signal as shown in Fig. 9.1(b)".

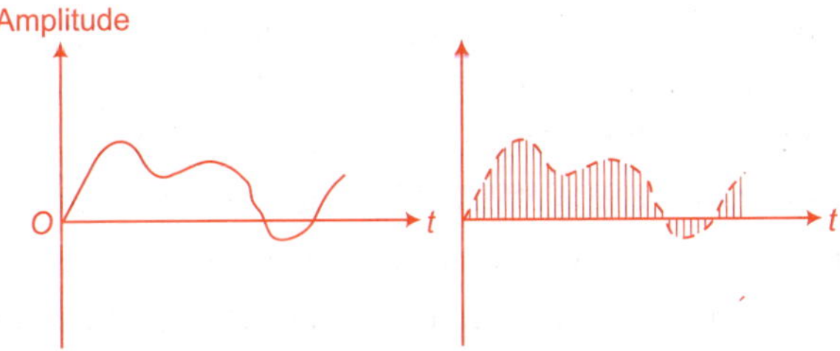

Fig. 9.1   *(a) Continuous signal (b) Non continuous signal*

Widely used digital signal has two levels: level 1 or level 0 which is also called as level high and level low as shown in Fig. 9.1(c).

There is large development in the field of information technology in the recent years. This development is possible due to digital communication system for transmission and reception of the signal.

In other words, we can define analog and digital system as follows,

"An analog system contains devices that process (manipulates) physical quantities which are represented in analog form. In this system, the quantities can vary over a continuous range of values".

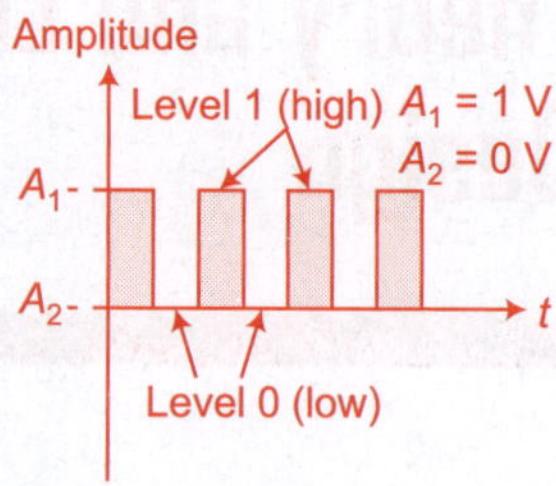

Fig. 9.1 (c) *Two level digital signal*

"A digital system contains devices which are designed to process (or manipulate) physical quantities or information that are represented in digital form".

9.2 ADVANTAGES AND APPLICATIONS OF DIGITAL LOGIC DESIGN

Basically most of the units of the world are analog but most of these units are converted in digital form before processing. The process is done by the digital circuits due to the advantages following advantages:

(i) Digital systems are easier to design because switching circuits are used in these systems. In switching circuits, the actual value of the signal is not important but the level of the signal (high or low) is important.

(ii) Digital information can easily be stored in digital systems.

(iii) The accuracy and precision of digital systems are greater and can also be improved by increasing switching circuits.

(iv) The operation of digital circuits can be programmed.

(v) Digital systems are less effected by noise and distortion.

(vi) Digital circuitry can be fabricated on integrated circuit chip.

(vii) Error detection and correction is possible in digital circuits.

(viii) Secure communication is possible due to decryption/encryption and wide spectrum provides anti-jamming facility.

(ix) Data compression is possible so it can save the memory.

(x) Due to the VLSI technology, the size and weight of transceivers is small, handy and mobile.

(xi) Digital circuits are more reliable.

9.3 NUMBER SYSTEMS USED IN DIGITAL ELECTRONICS

As you are aware of decimal number system because this number system is used in our daily life. Some other systems are also available which, we will study in this section.

One important term which is used in number system is base or radix. The base or radix means the number of symbols or digits which are used in particular number system. The number systems are as follows:

1. Decimal number system

   Base or radix: 10

   Symbols or digits: 0, 1, 2, 3, 4, 5, 6, 7, 8, 9

2. Binary number system

   Base or radix: 2

   Symbols or digits: 0, 1

3. Octal number system

   Base or radix: 8

   Symbols or digits: 0, 1, 2, 3, 4, 5, 6, 7

4. Hexadecimal number system

   Base or radix: 16

   Symbols or digits: 0, 1, 2, 3, 4, 5, 6, 7, 8, 9, A, B, C, D, E, F

9.3.1 Decimal Number System

Decimal number system is made with ten different symbols or digits i.e., 0, 1, 2, 3, 4, 5, 6, 7, 8, 9. It is also called as base 10 system. The decimal number has evolved naturally because there are ten fingers in human hands and he can count with the help of fingers. Basically, the digit words come from latin word which is used for finger. The process of writing the decimal numbers after digit 9 consists of writing the second digit 1 first and then following it up with other digits from 0 to 9 one by one to get the next ten numbers from 10 to 19. The next ten numbers are obtained by writing the third digit 2 first and then following it up with other digits from 0 to 9 one by one to get next ten numbers from 20 to 29. This process continues until we have used all two digit combinations and we have reached up to '99'. After that we begin with three digits combinations and start from 100 and reach up to 999. This process goes on endlessly as shown in Fig. 9.2.

0	10	20	90	100
1	11	21	91	101
2	12	22	92	–
3	13	23	–	–
4	14	24	–	–
5	15	25	–	–
6	16	26	–	–
7	17	27	–	–
8	18	28	–	–
9	19	29	99	110

Fig. 9.2 *Decimal counting*

In decimal number, every digit has its positional value. This positional value of digit is called as weight of the digit. The value or magnitude of a given decimal number can be expressed as a sum of various digits multiplied by their place values or weights.

The weight can be expressed as follows:

For Integer part,

$$\text{weight} = (\text{base})^n$$

where,   base = 10 for decimal

$$n = \text{position of digit from 0 to } n$$

i.e. $\text{weight} = 10^n$

For fractional part,

$$\text{weight} = (\text{base})^{-n}$$

where,

$$\text{base} = 10 \text{ for decimal number}$$

$$n = \text{position of digit after decimal point from } -1 \text{ to } -n$$

i.e. $\text{weight} = 10^{-n}$.

Considering the example of decimal number 26329.247. The decimal number and the weights of different digits may be tabulated as in Fig. 9.3

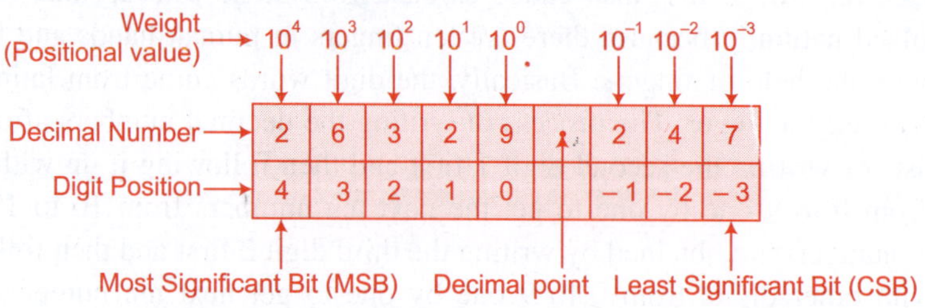

Fig. 9.3 *Representation of positional value in 10 to the power*

$$(2 \times 10^4) + (6 \times 10^3) + (3 \times 10^2) + (2 \times 10^1) + (9 \times 10^0)$$
$$+ (2 \times 10^{-1}) + (4 \times 10^{-2}) + (7 \times 10^{-3})$$

$$20000 + 6000 + 300 + 20 + 9 + 0.2 + 0.04 + 0.007 = (26329.247)_{10}.$$

The bit which has more weight is called as Most Significant Bit (MSB) and the bit which has minimum weight is called as Least Significant Bit (LSB) as shown in Fig. 9.3.

9.3.2 Binary Number System

Unfortunately, the decimal number system cannot be implemented in digital system because for decimal number system, such type of electronic devices are necessary to design which are capable to work on ten different voltage levels. So, such type of systems are difficult to implement because the hardware must have ten discrete steps or states representing each of the digit 0 to 9.

Binary number system is made with two different symbols or digits i.e., 0, 1. It is also called as base 2 system. The binary number system is much simpler, faster, more reliable and less expensive. All binary numbers consist of a string of 0s and 1s. Like 11, 101, 011 etc. Like the decimal number system, binary number system is also positionally weighted. The value or magnitude of a given binary number can be expressed as a sum of various bits multiplied by their place values or weight.

The weight can be expressed as follows:

For Integer part

$$\text{weight} = (\text{base})^n$$

where,   base = 2 for binary

n = position of bit from 0 to n

i.e.   weight = 2^n.

For fractional part

$$\text{weight} = (\text{base})^{-n}$$

where,   base = 2 for binary

n = position of bit after binary point from -1 to $-n$.

i.e.   weight = 2^{-n}.

Considering the example of binary number 1001.110. The binary number and the weight of different bit may be tabulated as in Fig. 9.4.

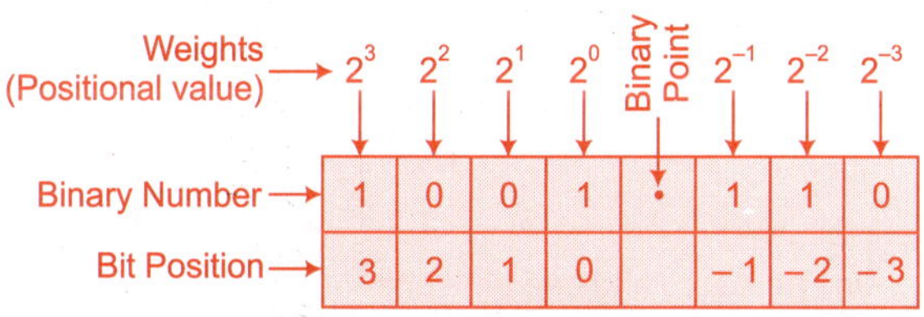

Fig. 9.4   *Here binary digit is abbreviated as bit*

The process of writing binary number after bit 1 consists of writing the second bit 1 first and then following it up with other bit 0 and 1 one by one to get the next two numbers. The binary counting is shown in Fig. 9.5. The binary counting also can be understood with the help of Table 9.1.

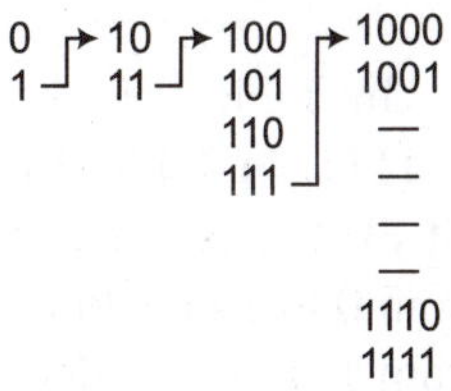

Fig. 9.5   *Binary counting*

Table 9.1

Weight or positional value →	$2^3 = 8$	$2^2 = 4$	$2^1 = 2$	$2^0 = 1$	Decimal Equivalent
	0	0	0	0	0
	0	0	0	1	1
	0	0	1	0	2
	0	0	1	1	3
	0	1	0	0	4
	0	1	0	1	5
	0	1	1	0	6
	0	1	1	1	7
	1	0	0	0	8
	1	0	0	1	9
	1	0	1	0	10
	1	0	1	1	11
	1	1	0	0	12
	1	1	0	1	13
	1	1	1	0	14
	1	1	1	1	15

$T \rightarrow$ Toggle

$2^0 = 1$ i.e. One time zero and one time one

$2^1 = 2$ i.e. Two times zero and two times one

$2^2 = 4$ i.e. 4 times zero and 4 times one

$2^3 = 8$ i.e. 8 times zero and 8 times one

In Table 9.1, 4-bit binary counting is shown. Counting starts from 0000 to 1111. In Table 9.1 we can see that counting starts from 0000. After that, the bit of 2^0 position toggle at every count. Here, toggle means when bit changes from 0 to 1 or 1 to 0. The 2^1 position bit toggle when the 2^0 position bit changes from 1 to 0. i.e., decimal count changes from 1 to 2 and 3 to 4. The 2^2 position bit toggle when 2^1 position bit changes from 1 to 0. i.e., decimal count changes from 3 to 4 and 7 to 8. Similarly, 2^3 position bit toggle when 2^2 position bit changes from 1 to 0 i.e., decimal count changes from 7 to 8. In Table 9.1, the toggle is marked with T and an arrow is indicated when positional bit changes from 1 to 0. In Table 9.1, we can see that the toggle (T) occurs in 2^n positional bit when the arrow ($\leftarrow$) occurs at 2^{n-1} positional bit.

In other words, from Table 9.1, we can say that LSB changes at every count, 2^1 positional bit changes after every two count, 2^2 positional bit changes after every four count and 2^3 positional bit changes after every eight count.

Octal number system is made with eight different symbols or digits i.e., 0, 1, 2, 3, 4, 5, 6, 7. It is also called as base 8 system. All higher digits are expressed as a combination of these or the same pattern as followed in the case of binary and decimal number systems. Like the decimal and binary number system, octal number system is also positionally weighted. The value of magnitude of a given octal number can be expressed as a sum of various digits multiplied by their place values or weight.

The weight can be expressed as follows:

For Integer part,

$$\text{weight} = (\text{base})^n$$
$$= 8^n$$

where,   base = 8 for octal

$$n = \text{position of digit from 0 to } n$$

i.e.   weight $= 8^n$

For fractional part

$$\text{weight} = (\text{base})^{-3}$$

where,   base = 8 for octal

$$n = \text{position of digit after octal point from} - 1 \text{ to} - n$$

i.e.   weight $= 8^{-n}$.

Considering the example of octal number 742.468. The octal number and the weight of different digits may be tabulated as in Fig. 9.6

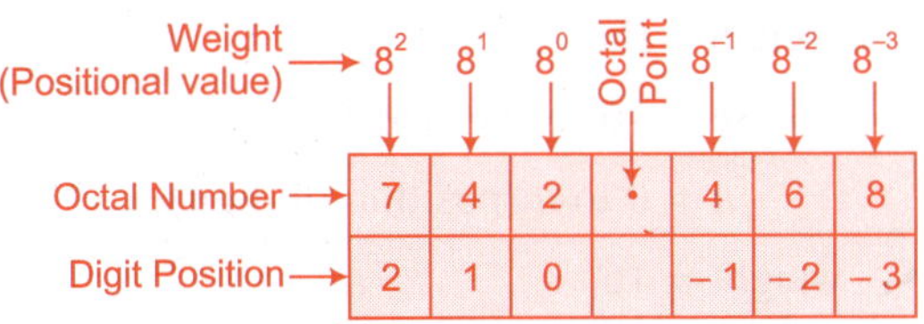

Fig. 9.6 *Representation of positional value in 8 to the power*

The process of writing the octal numbers after digit 7 consist of writing the second digit 1 first and then following it up with other digits from 0 to 7 one by one to get the next 8 numbers from 10 to 17. The next eight numbers are obtained by writing the third digit 2 first and then following it up with other digits from 0 to 7 one by one to get next eight numbers from 20 to 27. This process continues until we have used all two digit combinations and we have reached up to 77. After that, we begin with three digit combinations and start from 100 and reach up to 777. This process goes on endlessly as shown in Fig. 9.7.

```
0    10    20 –    – 90    100
1    11    21 –    – 91    101
2    12    22 –    – 92    –
3    13    23 –    – 93    –
4    14    24 –    – 94    –
5    15    25 –    – 95    –
6    16    26 –    – 96    106
7    17    27 –    – 97    107
```

Fig. 9.7 *Octal counting*

$$(742.468)_8 = (7 \times 8^2) + (4 \times 8^1) + (2 \times 8^0) + (4 \times 8^{-1}) + (6 \times 8^{-2})$$
$$+ (8 \times 8^{-2})$$
$$= (7 \times 64) + (4 \times 8) + (2 \times 1) + \left(4 \times \frac{1}{8}\right) + \left(6 \times \frac{1}{64}\right)$$
$$+ \left(8 \times \frac{1}{64}\right)$$
$$= 448 + 32 + 2 + 0.5 + 0.09 + 0.125$$
$$= (482.715)_{10}.$$

9.3.4 Hexadecimal Number System

Hexadecimal number system is made with sixteen different symbols or digits i.e., 0, 1, 2, 3, 4, 5, 6, 7, 8, 9, A, B, C, D, E, F. It is also called as base 16 system. All higher digits are expressed as a combination of these on the same pattern as followed in the case of octal number systems. Like the octal number system, hexadecimal number system is also positionally weighted. The value of magnitude of a given hexadecimal number can be expressed as a sum of various digits multiplied by their place value or weight.

The weight can be expressed as follows:

For Integer part weight = $(\text{base})^n$

where,   base = 16 for hexadecimal

$n = 0$ to n (position of digit)

i.e.   weight = 16^n

For fractional part

$$\text{weight} = (\text{base})^{-n}$$

where, $\text{base} = 16$ for hexadecimal

n = position of digit after hexadecimal point from -1 to $-n$.

i.e. $\text{weight} = 16^{-n}$.

Considering the example of hexadecimal number $79A2 \cdot 3FB$. The hexadecimal number and the weight of different digits may be tabulated as in Fig. 9.8.

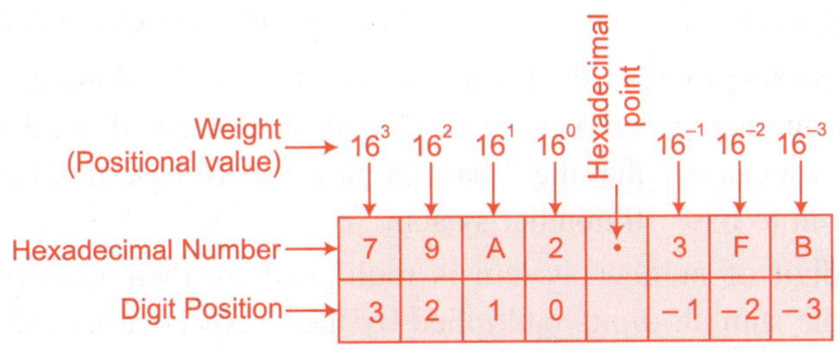

Fig. 9.8 *Representation of positional value in 16 to the power*

The process of writing the hexadecimal number after digit F consists of writing the second digit 1 first and then following it up with other digits from 0 to F one by one to get the next 16 numbers from 10 to $1F$. The next sixteen numbers are obtained by writing the third digit 2 first and then following it up with other digits from 0 to F one by one to get next sixteen numbers from 20 to $2F$. This process continues until we have used all two digit combinations and we have reached up to FF. After that, we begin with three digit combination and start from 100 and reach up to FFF. This process goes on endlessly as shown in Fig. 9.9.

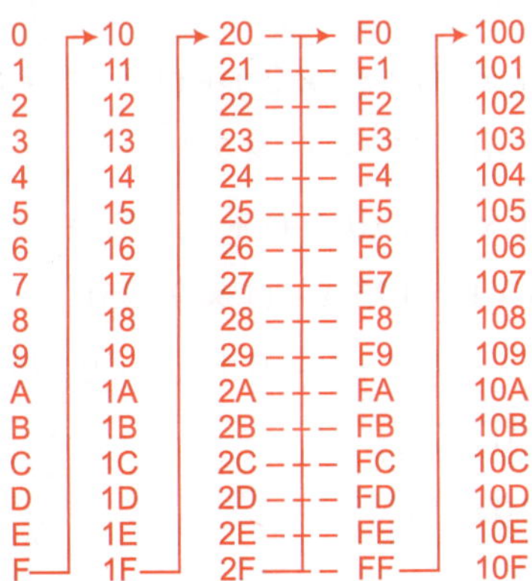

Fig. 9.9 *Hexadecimal counting*

$$(79A2 \cdot 3FB)_{16} = (7 \times 16^3) + (9 \times 16^2) + (A \times 16^1) + (2 \times 16^0)$$

$$+ (3 \times 16^{-1}) + (F \times 16^{-2}) + (B \times 16^{-3})$$

$$= 28672 + 2304 + 160 + 2 + 0.188 + 0.059$$

$$+ 0.003 = (31138.250)_{10}.$$

9.4 CONVERSION OF A NUMBER FROM ONE NUMBER SYSTEM TO ANOTHER

By following some rules, we can convert one number system to another.

Rule 1   According to this rule, binary number system, octal number system and hexadecimal number system can be converted in to decimal number system. In other words, we can say that the 2 base, 8 base and 16 base number system can be converted in to base 10 number system.

"Every digit of number system is multiplied by their positional value or weight and the sum all digits multiplied by their respective weights. In case of mixed numbers, the integer and fractional parts should be handled separately"

(i) Binary to decimal conversion (Rule 1).

Example 9.4.1   Convert $(11011)_2$ and (1010.101) to decimal.

Solution

Weights   Binary digit of number system

$$(11011)_2 = (1 \times 2^4) + (1 \times 2^3) + (0 \times 2^2) + (1 \times 2^1) + (1 \times 2^0)$$

$$= (1 \times 16) + (1 \times 8) + (0 \times 4) + (1 \times 2) + (1 \times 1)$$

$$= 16 + 8 + 0 + 2 + 1$$

$$(11011)_2 = (27)_{10} \hspace{4cm} \textbf{(Ans.)}$$

$$(1010.101)_2 = (1 \times 2^3) + (0 \times 2^2) + (1 \times 2^1) + (0 \times 2^0) + (1 \times 2^{-1})$$

$$+ (0 \times 2^{-2}) + (1 \times 2^{-3})$$

$$= (1 \times 8) + (0 \times 4) + (1 \times 2) + (0 \times 1) + (1 \times 0.5)$$

$$+ (0 \times 0.25^1) + (1 \times 0.125)$$

$$= 8 + 0 + 2 + 0 + 0.5 + 0 + 0.125$$

$$= (10.625)_{10}$$

i.e. $(1010.101)_2 = (10.625)_{10}$ \hspace{4cm} **(Ans.)**

(ii) Octal to decimal conversion (Rule 1).

Example 9.4.2   Convert $(1072)_8$ and $(2105.732)_8$ to decimal.

Solution

$$(1072)_8 = (1 \times 8^3) + (0 \times 8^2) + (7 \times 8^1) + (2 \times 8^0)$$
$$= (1 \times 512) + (0 \times 64) + (7 \times 8) + (2 \times 1)$$
$$= 512 + 0 + 56 + 2$$
$$= (570)_{10} \qquad \textbf{(Ans.)}$$

$$(2105.732)_8 = (2 \times 8^3) + (1 \times 8^2) + (0 \times 8^1) + (5 \times 8^0)$$
$$+ (7 \times 8^{-1}) + (3 \times 8^{-2}) + (2 \times 8^{-3})$$
$$= (2 \times 512) + (1 \times 64) + (0 \times 8) + (5 \times 1) + (7 \times 0.125)$$
$$+ (3 \times 0.016) + (2 \times 0.002)$$
$$= 1024 + 64 + 0 + 5 + 0.875 + 0.048 + 0.004$$
$$= (1093.927)_{10} \qquad \textbf{(Ans.)}$$

(iii) Hexadecimal to decimal conversion (Rule 1).

Example 9.4.3   Convert $(26BC)_{16}$ and $(23A.17)_{16}$ to decimal.

Solution

$$(26BC)_{16} = (2 \times 16^3) + (6 \times 16^2) + (11 \times 16^1) + (12 \times 16^0)$$
$$= (2 \times 4096) + (6 \times 256) + (11 \times 16) + (12 \times 1)$$
$$= 8192 + 1536 + 176 + 12$$
$$= (9916)_{10} \qquad \textbf{(Ans.)}$$

$$(23A.17)_{16} = (2 \times 16^2) + (3 \times 16^1) + (10 \times 16^0) + (1 \times 16^{-1})$$
$$+ (7 \times 16^{-2})$$
$$= (2 \times 256) + (3 \times 16) + (10 \times 1) + (1 \times 0.0625)$$
$$+ (7 \times 0.004)$$
$$= 512 + 48 + 10 + 0.0625 + 0.028$$
$$= (570.\,905)_{10}. \qquad \textbf{(Ans.)}$$

Rule 2   According to this rule, the decimal number system can be converted in to other number system i.e., binary, octal and hexadecimal number system. In other words we can say that the 10 base number system can be converted in to 2 base, 8 base and 16 base number system.

"According to this rule, the given decimal number is divided continuously by base of other number system (i.e., 2, 8, 16) and the remainders must be written in reverse order i.e., first remainder will be LSD and last remainder will be MSD. This process is called Double Dabble".

"If the decimal number also has a fractional part, then the fractional part is multiplied by base continuously and the left part of decimal point of multiple must be noted. (i.e., called carry). The carry digits are written in forward order".

Note If the process is endless, then we use repeated multiplication of fractional part with base. If we cannot get zero, then we can stop multiplication after four or five steps.

(iv) Decimal to binary conversion (Rule 2).

Example 9.4.4 Convert $(82)_{10}$ to binary number.

Solution

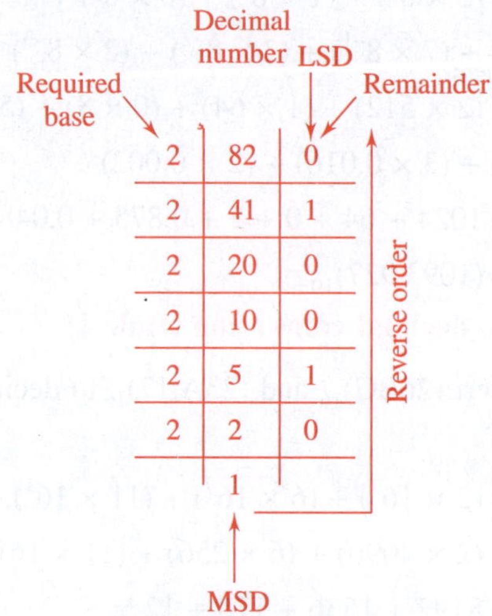

Remainder in reverse order = 1010010

i.e. $(82)_{10} = (1010010)_2$. **(Ans.)**

Example 9.4.5 Convert $(0.625)_{10}$ to binary number.

Solution

Integer is forward order = 10 00

i.e. $(0.625)_{10} = (10\,00)_2$ **(Ans.)**

if the process is endless, then we use repeated multiplication of fractional part with base. If we do not get zero, then we can stop multiplication after four or five steps.

(v) Decimal to octal conversion (Rule 2).

Example 9.4.6   Convert $(672)_{10}$ to octal number.

Solution

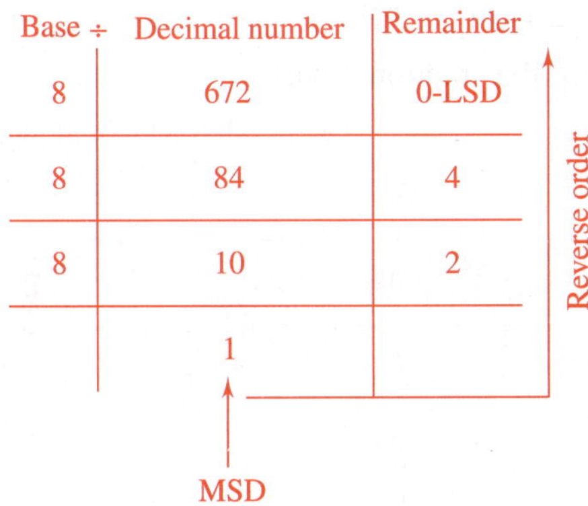

Remainder in reverse order = 1240

$$\text{i.e. } (672)_{10} = (1240)_8.$$ **(Ans.)**

Example 9.4.7   Convert $(0.68)_{10}$ to octal number.

Solution

Integer in forward order = 5341

i.e. $\qquad (0.68)_{10} = (0.5341)_8.$ **(Ans.)**

(vi) Decimal to hexadecimal conversion (Rule 2)

Example 9.4.8   Convert $(250.85)_{10}$ to hexadecimal number.

Solution   Firstly, converting integer part

Base ÷	Decimal number	Remainder
16	250	$10 = A$
16	15	$15 = F$
	0	

i.e. Remainder = $(FA)_{16}$

Now, converting the fractional part

Base ×	Fractional part	Integer of multiple
16	0.85	$13 = D$
16	0.6	9
16	0.6	9
	0.6	

(forward order)

i.e. Integer in forward order = $(D99)_{16}$.

The final result is,

$$(250 \cdot 85)_{10} = (FA \cdot D99)_{16}.$$

(Ans.)

Rule 3 According to this rule, binary number can be converted in to octal or hexadecimal number system. In other words, we can say that the 2 base number system can be converted into 8 base and 16 base number system.

"While converting from binary to octal number, group the bits in group of three and write the correct octal number for each group. Similarly, while converting from binary to hexadecimal number, group the bits in group of four and write the correct hexadecimal number for each group".

Note For completing the group of integer part place zero in the left if necessary and for fractional part place zero in the right of factors part if necessary.

(vii) Binary to octal conversion (Rule 3)

Example 9.4.9 Convert $(1101011)_2$ and $(1110011 \cdot 11001)_2$ to octal number.

Solution Making a group of three from left to right

$$(\underline{001} \ \underline{101} \ \underline{011})_2 = (153)_8$$

(Ans.)

$$(1110011 \cdot 11001)_2 = (\underline{001} \ \underline{110} \ \underline{011} \cdot \underline{110} \ \underline{010})_2$$

$$= (163 \cdot 62)_8.$$

Since octal number has base 8 which is equal to 2^3 i.e., $n = 3$, so we make group of 3-bits.

(viii) Binary to hexadecimal conversion. (Rule 3).

Example 9.4.10   Convert $(1001010 \cdot 01110)_2$ to hexadecimal.

Solution   Making a group of four from left to right

$$(1001010 \cdot 01110)_2 = (\underline{0100}\ \ \underline{1010}\ \cdot\ \underline{0111}\ \ \underline{0000})$$
$$= 4A \cdot 70$$
$$= (4A \cdot 70)_{16}.$$

Since hexadecimal number has base 16 which is equal to 2^4 i.e., $n = 4$, therefore we make group of 4-bits.

Rule 4   According to this rule, octal number and hexadecimal number system can be converted in to binary number system. In other words, we can say that 8 base and 16 base number can be converted in to 2 base number system.

"According to this rule, when we convert octal number system in to binary number, then we convert each digit into its 3-bit binary equivalent. Similarly, when we convert hexadecimal number system into binary number, then we convert each digit into its 4-bit binary equivalent".

(ix) Octal to binary conversion (Rule 4).

Example 9.4.11   Convert $(357)_8$ to binary number and also convert $(37 \cdot 67)_8$ to binary number.

Solution   Octal   3   5   7

   Binary   011   101   111

   i.e. $(357)_8 = (011101111)_2$ **(Ans.)**

   Similarly,

   Octal   3   7   ·   6   7

   Binary   011   111   ·   101   111

   i.e. $(37 \times 67)_8 = (011\ 111 \cdot 101111)_2.$ **(Ans.)**

(x) Hexadecimal to binary conversion (Rule 4).

Example 9.4.12   Convert $(8B5 \cdot C4)_{16}$ to binary.

Solution   Hexadecimal   8   *B*   5   ·   *C*   4

   Binary   1000   1010   0101   ·   1100   0100

   i.e. $(8B5.C4)_{16} = (100010100101 \cdot 11000100)_2.$ **(Ans.)**

Rule 5   According to this rule, octal number can be converted into hexadecimal number system and vice versa. In other words, we can say that base 8 system can be converted into base 16 number system and vise versa.

"According to this rule, when converting from octal to hexadecimal number, (or vice versa) first convert into binary and then convert in the binary into desired number system. i.e., this rule is indirect, before applying this rule first, we must apply rule 4 and then rule 3.

(xi) Octal to hexadecimal conversion (Rule 5).

Example 9.4.13 Convert $(74.5)_8$ to hexadecimal number.

Solution

Applying rule 4

Octal   7   4   .   5

Binary   111   100   .   101

i.e. $(74.5)_8 = 111100 \cdot 101)_2$

Now, applying rule 3

making group of 4-bit,

$$(111100.101)_2 = \underline{0011} \ \underline{1100} \cdot \underline{1010}$$
$$= 3 \ C \cdot A$$
$$= (3C \cdot A)_{16}. \qquad \text{(Ans.)}$$

(xii) Hexadecimal to octal conversion (Rule 5).

Example 9.4.14 Convert $(5C9 \cdot A3)_{16}$ to octal number apply rule 4.

Solution

Hexadecimal   5 C 9   .   A   3

Binary   0101   1100   1001   .   1010   0011

i.e. $(5C9 \times A3)_{16} = (010111001001 \cdot 10100011)_2$

Now, applying rule 3

making group of 3-bits

$$(010111001001 \cdot 10100011)_2 = \underline{010} \ \underline{111} \ \underline{001} \ \underline{001} \cdot \underline{101} \ \underline{000} \ \underline{110}$$
$$\qquad\qquad 2 \quad 7 \quad 1 \quad 1 \ . \ 5 \quad 0 \quad 6$$
$$= 506$$
$$= (2711 \times 506)_8. \qquad \text{(Ans.)}$$

9.5   BINARY OPERATIONS

The arithmetic operations also can be done in binary numbers. There are four binary operations that can be done as given below.

1. Binary addition
2. Binary subtraction
3. Binary multiplication
4. Binary division.

9.5.1 Binary Addition

Binary addition is accomplished in a same manner that in decimal. The following rules must be followed at the time of addition in binary:

Rule 1A: $0 + 0 = 0$

Rule 2A: $0 + 1 = 1$

Rule 3A: $1 + 0 = 1$

Rule 4A: $1 + 1 = 10$ (1 is carry)

From rule 4A, it is clear that the addition of $1 + 1 = 10$ i.e., the addition of binary number 1 with binary number 1 is 10. Because in binary 2 is written as 10. In other words, we can state that in addition of $1 + 1$, sum is 0 and 1 is carry.

We start addition in the same manner as in decimal i.e., we start with LSB and if we obtain, then we carry it in the next higher position. For example, adding binary number 1110 and 1011 as follows:

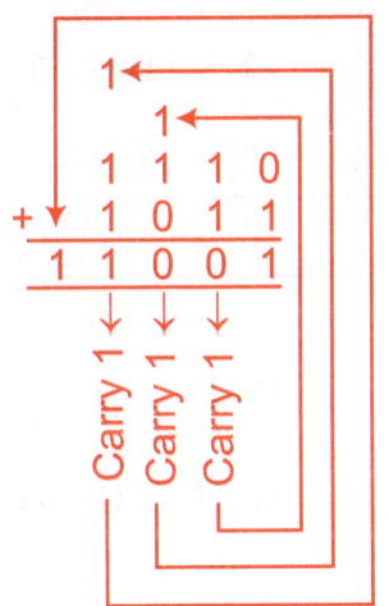

Example 9.5.1   Add the binary number $(10110)_2$ and $(10100)_2$.

Solution

$$\begin{array}{r} 1 \\ 10110 \\ +\ 10100 \\ \hline 101010 \end{array}$$

Hence, $10110 + 10100 = 101010.$ **(Ans.)**

Example 9.5.2   Add the binary number $(1101 \cdot 01)_2$ and $(1011 \cdot 11)_2$.

Solution

$$\begin{array}{r} 1\ 1\ 1\ 1\ \ 1 \\ 1\ 1\ 0\ 1 \cdot 0\ 1 \\ 1\ 0\ 1\ 1 \cdot 1\ 1 \\ \hline 1\ 1\ 0\ 0\ 1 \cdot 0\ 0 \end{array}$$

Hence, $1101 \cdot 01 + 1011 \cdot 11 = 11001 \cdot 00.$ **(Ans.)**

9.5.2 Binary Subtraction

Binary subtraction is accomplished in the same manner which is followed in decimal subtraction. The following rules must be followed at the time of subtraction in binary.

Rule 1S: $0 - 0 = 0$

Rule 2S: $0 - 1 = 1$ (Borrow = 1)

Rule 3S: $1 - 0 = 1$

Rule 4S: $1 - 1 = 0.$

From rule 2S, we can state that if we subtract 1 from 0 i.e., higher bit from lower bit, then we borrow from next higher position and 1 placed before bit 0 i.e., it will be 10 which is equal to 2 in decimal and we subtract 1 from it, then we get result 1. i.e., $10 - 1 = 1$.

For example, subtracting $(1011)_2$ from $(1101)_2$

$$
\begin{array}{cccc}
 & 0 \nearrow 1 & \text{borrow} & \\
1 & 1 & 0 & 1 \\
-1 & 0 & 1 & 1 \\
\hline
0 & 0 & 1 & 0 \\
\hline
\end{array}
$$

Hence, $1101 - 1011 = 0010.$ **(Ans.)**

Example 9.5.3   Subtract $(100.1)_2$ from $(110.01)_2.$

Solution

$$
\begin{array}{cccccc}
 & 0\nearrow 1 \nearrow 1 & & & & \\
1 & 1 & 0 \bullet & 0 & 1 \\
-1 & 0 & 0 \bullet & 1 & 0 \\
\hline
0 & 0 & 1 \bullet & 1 & 1 \\
\hline
\end{array}
$$

Hence, $(110.01)_2 - (100.10)_2 = (001.11)_2.$ **(Ans.)**

Example 9.5.4   Subtract $(10000.01)_2$ from $(00111.01)_2.$

Solution

$$
\begin{array}{ccccccccc}
 & \nearrow 1 \nearrow 1 \nearrow 1 \nearrow 1 & & & & & & & \\
1 & 0 & 0 & 0 & 0 \bullet & 0 & 1 \\
-0 & 0 & 1 & 1 & 1 \bullet & 0 & 1 \\
\hline
0 & 1 & 0 & 0 & 1 \bullet & 0 & 0 \\
\hline
\end{array}
$$

Hence, $(10000.01)_2 - (00111.01)_2 = (01001.00)_2$.   **(Ans.)**

9.5.3 Binary Multiplication

Multiplication in binary is accomplished in the same manner that in decimal. The following rules must be followed at the time of multiplication in binary.

Rule 1M: $0 \times 0 = 0$

Rule 2M: $0 \times 1 = 0$

Rule 3M: $1 \times 0 = 0$

Rule 4M: $1 \times 1 = 1$.

Example 9.5.5 Multiply binary number 1011 and 101.

Solution

$$
\begin{array}{r}
1\ 0\ 1\ 1 \quad \text{multiplicant} \\
\times\ 1\ 0\ 1 \quad \text{multiplier} \\
\hline
1\ 0\ 1\ 1 \\
0\ 0\ 0\ 0\ \times \\
1\ 0\ 1\ 1\ \times \\
\hline
1\ 1\ 0\ 1\ 1\ 1 \quad \text{product}
\end{array}
$$

Hence, $(1011)_2 \times (101)_2 = (110111)_2$.   **(Ans.)**

9.5.4 Binary Division

Binary division is same as division of decimal numbers. Two rules must be followed at the time of division in binary.

Rule 1D $0 \div 1 = 0$ (remainder 1)

Rule 2D $1 \div 0 = 1$ (remainder 0).

Note that the division cannot be done with zero.

Example 9.5.6 Divide binary number 1100111 by 101.

Solution

$$
\begin{array}{r}
1\ 0\ 1\ 0\ 0 \quad \rightarrow \text{quotient} \\
101\ \overline{)\ 1\ 1\ 0\ 0\ 1\ 1\ 1} \\
1\ 0\ 1 \\
\hline
0\ 1\ 0 \\
0\ 0\ 0 \\
\hline
1\ 0\ 1 \\
1\ 0\ 1 \\
\hline
0\ 0\ 1 \\
0\ 0\ 0 \\
\hline
0\ 1\ 1 \quad \rightarrow \text{remainder.}
\end{array}
$$

Hence, $\qquad$ $1100111 \div 101 = 10100$ and remainder is 11.

Example 9.5.7 Divide binary number $(1100.11)_2$ by $(10)_2$.

Solution

$$
\begin{array}{r}
110 \cdot 01 \rightarrow \text{quotient} \\[2pt]
10\,\overline{)\,1\,1\,0\,0 \cdot 0\,0} \\
\underline{1\,0} \\
1\,0 \\
\underline{1\,0} \\
0\,0 \\
\underline{0\,0} \\
1\,1 \\
\underline{1\,0} \\
1 \rightarrow \text{remainder.}
\end{array}
$$

Hence, $\quad (1100.11)_2 \div (10)_2 = (110.01)_2$ and remainder is 1.

9.6 COMPLEMENT OF A NUMBER TO REPRESENT NEGATIVE NUMBERS

Generally we represent a negative sign (–) before the decimal number if the number is negative. But in digital circuits, the negative number is stored in the form of complement. The advantage of this technique is that the operation of subtraction can be performed by the same circuitary which is used for addition. In binary number system, there are two types of complements: 1's complement and 2's complement.

In general, there are two types of complement for any number system.

1. Radix complement (or base complement)

2. (Radix–1) complement or (base-1) complement

For example, in decimal number system there are two types of complement:

(i) Radix complement i.e., 10's complement

(ii) (Radix–1) complement i.e., $10 - 1 = 9$'s complement

In binary number system,

(i) Radix complement i.e., 2's complement

(ii) (Radix–1) complement i.e., $2 - 1 = 1$'s complement etc.

9.6.1 9's Complement (Radix–1, Complement in Decimal Number)

For getting the 9's complement of decimal number, each digit of a decimal number is subtracted from 9.

Example 9.6.1 Find the 9's complement of each of the following decimal numbers.

(a) 15   (b) 32   (c) 215

Solution As we know that to get 9's complement each digit of decimal number is subtracted from 9.

(a) $\quad$ 99

$\quad\quad\underline{-\ 15}\quad\rightarrow$ decimal number whose 9's complement to be find

$\quad\quad\ \ 84\quad\rightarrow$ 9's complement

$\quad\quad$ Hence 9's complement of 15 is 84.   **(Ans.)**

(b) $\quad$ 99

$\quad\quad\underline{-\ 32}\quad\rightarrow$ Decimal number whose 9's complement to be find

$\quad\quad\ \ 67\quad\rightarrow$ 9's complement

$\quad\quad$ Hence 9's complement of 32 is 67.   **(Ans.)**

(c) $\quad$ 999

$\quad\quad\underline{-\ 215}\quad\rightarrow$ Decimal number whose 9's complement to be find

$\quad\quad\ \ 784\quad\rightarrow$ 9's complement

$\quad\quad$ Hence 9's complement of 215 is 784.   **(Ans.)**

9.6.2 9's Complement Subtraction

With the help of 9's complement, we can subtract smaller number from a larger number. By this method, the 9's complement of small number (Subtrahend) is to be added in larger number (minuend) and adding the carry to the result. When we subtract larger number from small number, there is no carry and hence the result is in 9's complement form and negative.

Example 9.6.2 Subtract 5 from 13 using 9's complement. and verify the result by direct subtraction.

Solution The subtrahend is 05 (smaller than minuend).

The 9's complement of subtrahend is

$$\begin{array}{r} 99 \\ \underline{-\ 05} \\ 94 \end{array} \rightarrow \text{9's complement of 05.}$$

The minuend is 13 (larger than subtrahend)

i.e.

$$\begin{array}{r} 13 \\ \underline{+\ 94} \end{array} \rightarrow \text{minuend} \atop \rightarrow \text{9's complement of subtrahend}$$

carry $\longleftarrow$ ①07

$\quad\quad\quad\quad\quad\ \underline{+\ 1}\ \rightarrow$ Adding carry to result

$\quad\quad\quad\quad\quad\quad\ 08$

Direct subtraction

$$
\begin{array}{r}
13 \\
-\ 5 \\
\hline
08 \\
\hline
\end{array}
$$

(Ans.)

i.e., result obtained by 9's complement and direct subtraction is same, so the result is verified.

Example 9.6.3 Subtract 35 from 21 using 9's complement and verify result by direct subtraction.

Solution The subtrahend is 35 (larger than minuend) 9's complement of subtrahend is,

$$
\begin{array}{r}
99 \\
-\ 35 \\
\hline
64 \\
\hline
\end{array}
\quad \rightarrow \text{9's complement of 35.}
$$

The minuend is 21 (smaller than subtrahend)

i.e.

$$
\begin{array}{r}
21 \quad \text{minuend} \\
+\ 64 \quad \rightarrow \text{9's complement of subtrahend} \\
\hline
85 \quad \rightarrow \text{No carry means that answer is negative} \\
\hline
\end{array}
$$
and in complemented form.

Hence,

$$
\begin{array}{r}
99 \\
85 \\
\hline
-\ 14 \\
\hline
\end{array}
$$

(Ans.)

Direct subtraction

$$
\begin{array}{r}
21 \\
-\ 35 \\
\hline
-\ 14 \\
\hline
\end{array}
$$

(Ans.)

Hence, the result obtained by 9's complement and direct subtraction is same so the result is verified.

9.6.3 10's Complement (Radix, Complement in Decimal Number)

For getting the 10's complement of decimal number, first we find 9's complement, then we add 1 in 9's complement.

Example 9.6.4 Find the 10's complement of each of the following decimal number

(a) 14 (b) 39 (c) 129

Solution   (a) First we find 9's complement

$$
\begin{array}{l}
99 \\
-\ 14 \quad \rightarrow \text{Decimal number whose 9's complement to be find} \\
\hline
\ \ 85 \quad \rightarrow \text{9's complement of 14} \\
\hline
\end{array}
$$

Now, adding 1 is 9's complement

$$
\begin{array}{l}
\ \ 85 \quad -\ \text{9's complement} \\
+\ 1 \\
\hline
\ \ 86 \quad -\ \text{10's complement} \\
\hline
\end{array}
$$

Hence, 10's complement of 14 is 86.   **(Ans.)**

(b) First we find 9's complement

$$
\begin{array}{l}
99 \\
-\ 39 \quad \rightarrow \text{Decimal number whose 9's complement to be find} \\
\hline
\ \ 60 \quad \rightarrow \text{9's complement of 39} \\
\hline
\end{array}
$$

Now, adding 1 in 9's complement

$$
\begin{array}{l}
\ \ 60 \quad \rightarrow \text{9's complement} \\
+\ 1 \\
\hline
\ \ 61 \quad \rightarrow \text{10's complement} \\
\hline
\end{array}
$$

Hence, 10's complement of 39 is 61.   **(Ans.)**

(c) First we find 9's complement

$$
\begin{array}{l}
999 \\
-\ 129 \quad \rightarrow \text{Decimal number whose 9's complement to be find} \\
\hline
\ \ 870 \quad \rightarrow \text{9's complement of 129} \\
\hline
\end{array}
$$

Now, adding 1 M 9's complement

$$
\begin{array}{l}
\ \ 870 \quad \rightarrow \text{9's complement} \\
+\ 1 \\
\hline
\ \ 871 \quad \rightarrow \text{10's complement} \\
\hline
\end{array}
$$

Hence, 10's complement of 129 is 871.   **(Ans.)**

9.6.4   10's Complement Subtraction

Subtraction also can be performed with the help of 10's complement. By this method, we find the 10's complement of subtrahend and add it with minuend and neglect the carry.

Example 9.6.5 Subtract 22 from 85 using 10's complement and verify the result by direct subtraction.

Solution The subtrahend is 22

Now, we will find the 10's complement of 22.

$$
\begin{array}{r}
99 \\
-\,22 \\
\hline
77 \\
\hline
\end{array}
\quad\rightarrow \text{subtrahend}
$$

$$77 \rightarrow 9\text{'s complement}$$

Now, adding 1 in 1's complement

$$
\begin{array}{r}
77 \\
+\,1 \\
\hline
78 \\
\hline
\end{array}
$$

$77 \rightarrow 9\text{'s complement}$

$78 \rightarrow 10\text{'s complement}$

The minuend is 85, now adding minuend with 10's complement of subtrahend.

$$
\begin{array}{r}
85 \\
+\,78 \\
\hline
163 \\
\hline
\end{array}
$$

neglecting carry

Hence, the result is 63.

Direct subtraction

$$
\begin{array}{r}
85 \\
-\,22 \\
\hline
63 \\
\hline
\end{array}
$$

Hence, result obtained by 10's complement and direct subtraction is same, so the result is verified.

9.6.5   1's Complement (Radix-1, Complement in Binary Number)

1's complement of binary number is same as 9's complement in decimal number. 1's complement can be achieved, by changing each bit of binary number from 1 to 0 and 0 to 1.

i.e.

$$1 \xrightarrow{\text{changes to}} 0$$

$$0 \xrightarrow{\text{changes to}} 1.$$

Example 9.6.6   Find 1's complement of binary number 1011010.

Solution   Binary number $\rightarrow$ 1011010

1's complement $\rightarrow$ 0100101

i.e., is complement of binary number 1011010 is 0100101.   **(Ans.)**

9.6.6 1's Complement Subtraction

With the help of 1's complement, we can subtract smaller number from a larger number. By this method, the 1's complement of small number (subtrahend) is to be added in larger number (minuend) and adding the carry to the result. When we subtract larger number (subtrahend) from small number (minuend), there is no carry and hence the result is in 1's complement form and negative.

Example 9.6.7 Subtract $(10101)_2$ from $(11001)_2$ using is complement and verify the result by direct subtraction.

Solution   The subtrahend is $(10101)_2$ i.e., smaller than minuend.
The 1's complement of subtrahend is

$$\text{subtrahend} \to 1\ 0\ 1\ 0\ 1$$
$$\text{1's complement} \to 0\ 1\ 0\ 1\ 0$$

The minuend is $(11001)_2$ i.e., larger than subtrahend.

i.e.

$$1\ 1\ 0\ 0\ 1 \to \text{Minuend}$$
$$+\ 0\ 1\ 0\ 1\ 0 \to \text{1's complement of subtrahend}$$

carry $\longleftarrow$ ①$0\ 0\ 0\ 1\ 1$

$+\ 1 \to$ Adding carry to result

$$0\ 0\ 1\ 0\ 0$$

Direct subtraction

$$1\ 1\ 1\ 0\ 0\ 1$$
$$-\ 1\ 0\ 1\ 0\ 1$$
$$\overline{0\ 0\ 1\ 0\ 0}$$

i.e., result obtained by 1's complement and direct subtraction is same, so the result is verified.

Example 9.6.8 Subtract $(10110)_2$ from $(10001)_2$ using 1's complement and verify the result by direct subtraction.

Solution   The subtrahend is $(10110)_2$ i.e., larger than minuend.
The 1's complement of subtrahend is

$$\text{subtrahend} \to 10110$$

i.e.

$$\text{complement} \to 01001$$

The minuend is $(10001)_2$ i.e., smaller than subtrahend.

$$\text{i.e., } 1\ 0\ 0\ 0\ 1 \to \text{minuend}$$
$$+\ 0\ 1\ 0\ 0\ 1 \to \text{1's complement of subtrahend}$$
$$\overline{1\ 1\ 0\ 1\ 0} \to \text{No carry means that answer is negative and}$$
$$\text{in complement form.}$$

Hence, 00101.   **(Ans.)**

Direct subtraction

$$
\begin{array}{r}
1\ 0\ 0\ 0\ 1 \\
-\ 1\ 0\ 1\ 1\ 0 \\
\hline
-\ 0\ 0\ 1\ 0\ 1
\end{array}
$$
(Ans.)

i.e., result obtained by 1's complement and direct subtraction is same, so the result is verified.

9.6.7   2's Complement (Radix, Complement of Binary Number)

The 2's complement in binary number is similar to the 10's complement in decimal number. For getting 2's complement, first we find 1's complement, then we add 1 in 1's complement.

Example 9.6.9   Find the 2's complement of each of the following binary number.

   (a) 10110   (b) 110100

Solution   (a) First we find 1's complement

$$
\begin{array}{l}
\text{Binary number} \rightarrow 1\ 0\ 1\ 1\ 0 \\
\text{1's complement} \rightarrow 0\ 1\ 0\ 0\ 1 \\
\hline
\qquad\qquad\qquad\qquad +\ 1 \quad \text{adding 1} \\
\hline
\qquad\qquad\qquad 0\ 1\ 0\ 1\ 0
\end{array}
$$

       Hence 2's complement of 10110 is 01010.   **(Ans.)**

  (b) First we find 2's complement

$$
\begin{array}{l}
\text{Binary number} \leftarrow \quad 1\ 1\ 0\ 1\ 0\ 0 \\
\text{1's complement} \leftarrow \quad 0\ 0\ 1\ 0\ 1\ 1 \\
\text{Adding} \leftarrow +\ \underline{\qquad\qquad 1} \\
\qquad\qquad\qquad\qquad 0\ 0\ 1\ 1\ 0\ 0
\end{array}
$$

Hence 2's complement of 110100 is 1100.   **(Ans.)**

9.6.8   2's Complement Subtraction

Subtraction also can be performed with the help of 2's complement. By this method, we find the 2's complement of subtrahend, add it with minuend and neglect the carry.

Example 9.6.10   Subtract $(11001)_2$ from $(11101)_2$ using 2's complement and also verify the result with direct subtraction.

Solution   The subtrahend is 11001.

    Now, we will find the 2's complement of 11001.

$$1\;1\;0\;0\;1 \leftarrow \text{subtrahend}$$
$$0\;0\;1\;1\;0 \leftarrow \text{1's complement of subtrahend}$$
$$\underline{+\;1} \leftarrow \text{adding 1}$$
$$\underline{0\;0\;1\;1\;1} \leftarrow \text{2's complement of subtrahend}$$

The minuend is $(11101)_2$, now adding 2's complement of subtrahend with minuend.

$$1\;1\;1\;1$$
$$1\;1\;1\;0\;1 \leftarrow \text{minuend}$$
$$\underline{+\;0\;0\;1\;1\;1} \leftarrow \text{2's complement of subtrahend}$$
$$\text{neglecting carry} \leftarrow \textcircled{1}\;0\;0\;1\;0\;0$$

Hence, the result is 00100. **(Ans.)**

Direct subtraction

$$1\;1\;1\;0\;1$$
$$\underline{-\;1\;1\;0\;0\;1}$$
$$\underline{0\;0\;1\;0\;0}$$

(Ans.)

Hence, result obtained by 2's complement and direct subtraction is same, so the result is verified.

9.7   BCD NUMBERS

In the previous section, you have seen that we can convert decimal number into its equivalent binary. As we know that binary number is a group of 1's and 0's. In other words, we can say that binary number is one type of code to represent a decimal number.

"When letters, numbers or words are represented by a special group of symbols, this process of representation is called encoding and the group of symbols is called a code".

If decimal number represents its equivalent binary, then it is called as straight binary code or simply binary code.

When the decimal number is very large, then the conversion of decimal number into its equivalent binary is long and complicated. So, for the conversion of decimal number into binary, some other codes are used. BCD (Binary Coded Decimal) is one of them.

"When each digit of decimal number is represents its 4-bit binary equivalent number, then this coding is called Binary Coded Decimal (BCD) number". The Table 9.2 represents the decimal number at its equivalent 4-bit binary number. For example, $(748)_{10}$ can be represented in BCD as follows:

$$7 \qquad 4 \qquad 8 \qquad \text{(decimal)}$$
$$0111 \qquad 0100 \qquad 1000 \qquad \text{(BCD)}$$

Hence, $(748)_{10} = 0111\ 0100\ 1000$ (BCD).

Table 9.2

Decimal Number	BCD Code
	8 4 2 1 ← weight
0	0 0 0 0
1	0 0 0 1
2	0 0 1 0
3	0 0 1 1
4	0 1 0 0
5	0 1 0 1
6	0 1 1 0
7	0 1 1 1
8	1 0 0 0
9	1 0 0 1

From Table 9.2, it is clear that BCD code is also called as 8421 code. When we see in Table 9.2, we observe that each bit of BCD code has some weight. The LSB of BCD code is weight 1, next bit has weight 2, next bit from it has weight 4 and MSB of BCD code has weight 8. That's why it is also called as 8421 code. If we find the equivalent decimal number of BCD, then we add the weight of those bits of BCD code whose binary is 1.

For example, $(1001)_2$ can be represented in decimal as follows:

$$1\ 0\ 0\ 1 \quad \leftarrow \text{BCD code}$$
$$8 + 1 \quad \leftarrow \text{weight}$$
$$= 9 \text{ decimal number.}$$

Binary coded decimal is 8421 code, so it is also called as weighted binary code. It is necessary to understand that BCD number is not similar to binary, octal, decimal and hexadecimal number system. It is only a process to code the decimal number system. The process of BCD conversion and binary conversion are different.

Example 9.6.11 Convert decimal number $(8472)_{10}$ into BCD number

Solution

$$8 \qquad 4 \qquad 7 \qquad 2 \qquad \leftarrow \quad \text{decimal number}$$
$$1000 \quad 0100 \quad 0111 \quad 0010 \quad \leftarrow \quad \text{BCD code}$$

Hence, the BCD number for $(8472)_{10}$ is 1000 0100 0111 0010.

Example 9.6.12 Convert $(245)_{10}$ into binary and BCD and compare both the results.

Solution   Binary conversion

2	245	1
2	122	0
2	61	1
2	30	0
2	15	1
2	7	1
2	3	1
	1	

$$(245)_{10} = (11110101)_2$$

(ii) BCD conversion

     2   4   5 decimal number

  0010   0100   0101   BCD code

Hence, $(245)_{10}$ = 0010 0100 0101 (BCD).

From the above conversions we see that, 8-bits are needed for conversion of decimal number 245 in to binary number while 12-bits are needed for conversion of decimal number 245 into BCD number.

The circuitry required for conversion of BCD in to decimal and decimal number to BCD number are very simple.

9.7.1   Limitations of BCD Numbers

As we know from the previous section that BCD code is a weighted code. From Table 9.2, we can state that in BCD code, there are sixteen possible group of 4-bit from 0000 to 1111. Out of these 16 combinations, only 10 groups (0000 to 1001) are used because remaining six groups such as 1010, 1011, 1100, 1101, 1110 and 1111 of BCD code are not valid code, due to unavailability of equivalent decimal digit of these codes. Due to this limitation, there is a problem to find the complement of BCD code.

For example, 1's complement of binary number 0010 (2) is 1101 (13) but this is not valid in BCD.

9.8   LOGIC GATES

Logic gates are based on Boolean algebra. Basically, the Boolean algebra is a mathematical logic system and is different from general algebra and binary number system. The basic Boolean operations are AND, OR and NOT. In this chapter, we will consider the positive logic i.e., 0 is low state and 1 is high state.

9.8.1 AND Gates

The AND gate is shown in Fig. 9.9 (a) and its electrical equivalent is shown in Fig. 9.9 (b). It consists two inputs A and B and Y is output of the gate.

A AND $B = A \times B = Y$

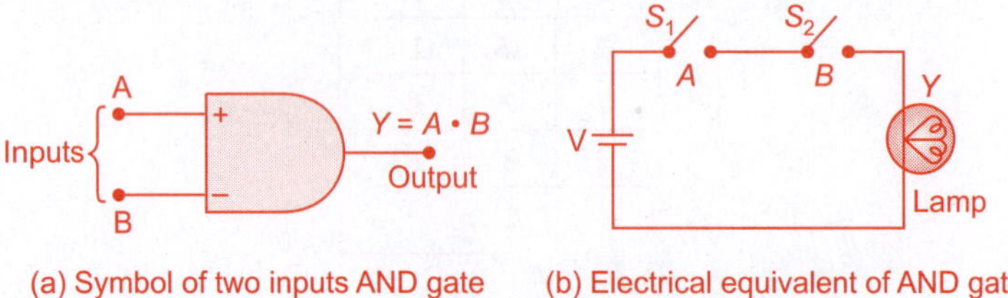

(a) Symbol of two inputs AND gate   (b) Electrical equivalent of AND gate

Fig. 9.9

In Fig. 9.9 (b) the two switches S_1 and S_2 are used as the input A and B and also called series input switch when it is close, then input is 1 and when it is open their input is 0. The Lamp indicates the output Y. The truth table for AND gate is shown in Table 9.3. In Table 9.3, we see that when both the inputs at AND gate are 1, then the output is 1. If any one input or both the inputs have logic 0, then the output is 0.

Table 9.3 Truth Table of AND Gate

Input		Output
A	B	Y
0	0	0
0	1	0
1	0	0
1	1	1

9.8.2 OR Gate

The OR gate is shown in Fig. 9.10 (a) and its electrical equivalent is shown in Fig. 9.10 (b). It also consists of two inputs A and B and Y is the output of the gate.

$$A \text{ OR } B = A + B = Y$$

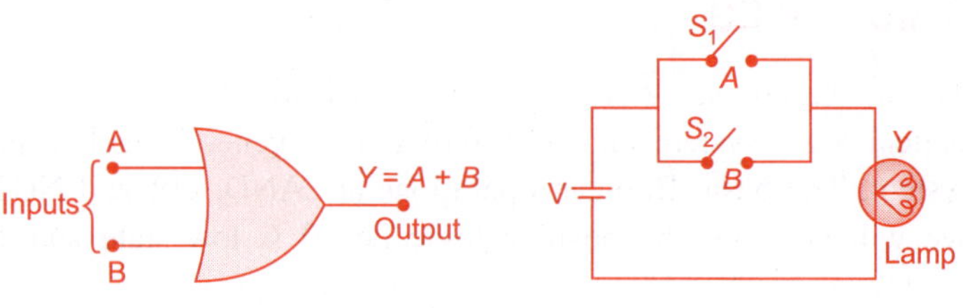

(a) Symbol of two inputs OR gate   (b) Electrical equivalent of OR gate

Fig. 9.10

In Fig. 9.10 (b) the two switches S_1 and S_2 are used as input A and B and also called parallel input. When the switch is close, then the input is equal to 1 and when the switch is open, then the input is equal to 0. Y is the output of the OR gate. The truth table for OR gate is shown in Table 9.4. In this table, we see when both the inputs at OR gate is 0, then the output is 0. If any one input or both the inputs have logic 1, then the output is logic 1.

Table 9.4 Truth Table of OR Gate

Input		Output
A	B	Y
0	0	0
0	1	1
1	0	1
1	1	1

9.8.3 NOT Gate

The NOT gate is shown in Fig. 9.11(a) and its electrical equivalent is shown in Fig. 9.11(b). It has single input and single output. This gate is indicated by the bar of the input. It is also called complementation, negation or inversion. In Fig. 9.11(b). When the switch is open i.e., $A = 0$, then the output is 1 i.e., bulb is on.

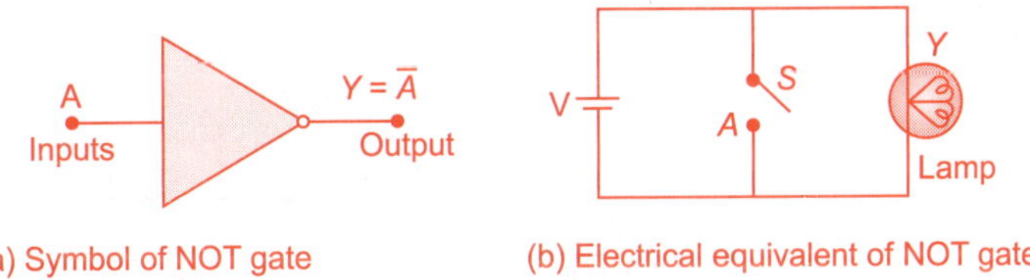

(a) Symbol of NOT gate   (b) Electrical equivalent of NOT gate

Fig. 9.11

If the switch is close i.e., $A = 1$, then the output is 0 i.e., bulb is off. The truth table for NOT gate is shown in Table 9.5. In this table, we see that when input is logic 0, then output is logic high and vice versa.

Table 9.5 Truth Table of NOT Gate

Input	Output
A	Y
0	1
1	0

9.8.4 Exclusive OR Gate (EX-OR Gate)

The Ex-OR gate is shown in Fig. 9.12 (a). It consists two inputs A and B and single output Y. When any one input out of two is 1, then output is 1 otherwise it is 0.

$$A \text{ Ex-OR } B = A \oplus B = Y$$

The truth table for Ex-Or gate is shown in Table 9.6.

Table 9.6 Truth Table of Ex-OR Gate

Input		Output
A	B	Y
0	0	0
0	1	1
1	0	1
1	1	0

The electrical equivalent circuit of Ex-OR gate is shown in Fig. 9.12 (b). The verification of truth table through electrical equivalent circuit when both the switches i.e., S_1 and S_2 are open, then the bulb is off.

i.e.
$$S_1 = 0 \text{ (open)}$$
$$Y = 0 \text{ (OFF)}$$
$$S_2 = 0 \text{ (open)}$$

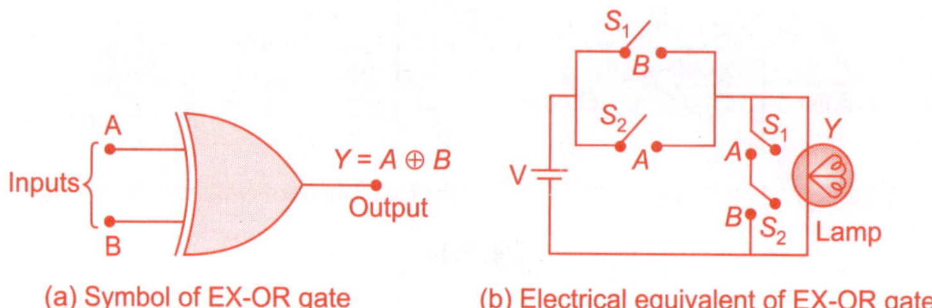

(a) Symbol of EX-OR gate   (b) Electrical equivalent of EX-OR gate

Fig. 9.12

When anyone switch out of two is close, then the bulb is on.

i.e.
$$S_1 = 0 \text{ (open)}$$
$$S_2 = 1 \text{ (close)}$$
$$Y = 0 \text{ (ON)}$$

and,
$$S_1 = 1 \text{ (close)}$$
$$S_2 = 0 \text{ (open)}$$
$$Y = 1 \text{ (ON)}$$

When both the switches are close, then the bulb is OFF.

i.e.
$$S_1 = 1 \text{ (close)}$$
$$S_2 = 1 \text{ (close)}$$
$$Y = 0 \text{ (OFF)}.$$

9.8.5   Universal Gates

In the previous section, we have studied about different logic gates like AND, OR and NOT. Some other gates are available that can perform all logic action. In other words, we can say that by these gates we can take any action like AND, OR and NOT, these gates are called universal gate and named as NAND and NOR gate.

9.8.5.1   NAND gate

The NAND gate is made with the help of AND and NOT gate as shown in Fig. 9.13 (a) and the symbolic representation of NAND gate is shown in Fig. 9.13 (b)

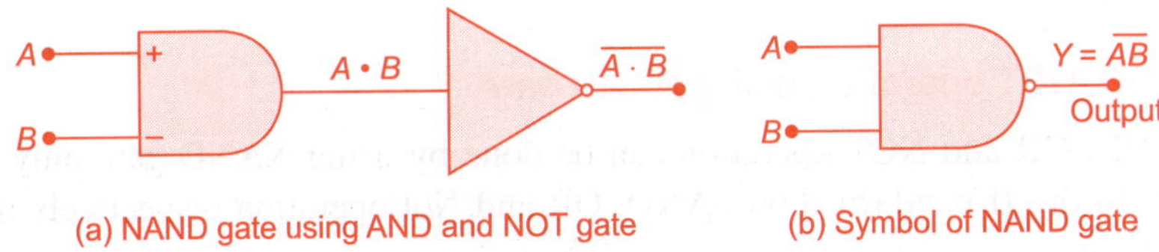

(a) NAND gate using AND and NOT gate   (b) Symbol of NAND gate

Fig. 9.13   *NAND gate*

The truth table for NAND gate is shown in Table 9.7.

Table 9.7   Truth Table of NAND Gate

Input		Output
A	*B*	*Y*
0	0	1
0	1	1
1	0	1
1	1	0

The output of the NAND gate is 0, when both inputs are 1. When both the inputs are 0 and any one input of two is 1, then the output is 1, as we can see in truth table.

The electrical equivalent of NAND gate is shown in Fig. 9.13 (c), when both switches are open i.e.,

$$S_1 = 0 \text{ (open)}$$
$$S_2 = 0 \text{ (open)} \qquad Y = 1 \text{ (ON)}$$

When anyone out of two switches are close i.e.,

i.e.
$$S_1 = 0 \text{ (open)}$$
$$S_2 = 1 \text{ (close)} \qquad Y = 1 \text{ (ON)}$$

and,
$$S_1 = 1 \text{ (close)}$$
$$S_2 = 0 \text{ (open)} \qquad Y = 1 \text{ (ON)}$$

If both the switches are close, then the output is 0

i.e.,
$$S_1 = 1 \text{ (close)}$$
$$S_2 = 1 \text{ (close)}.$$
$$Y = 0 \text{ (OFF)}$$

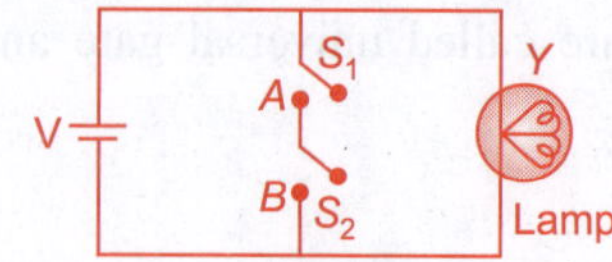

(c) Electrical equivalent of NAND gate

Fig. 9.13(c)

AND, OR, NOT operation using NAND gate

The AND, OR and NOT operation can be done by using NAND gate only. The Figs. 9.14 (a), (b) and (c) shows AND, OR and Not operation respectively using NAND gate.

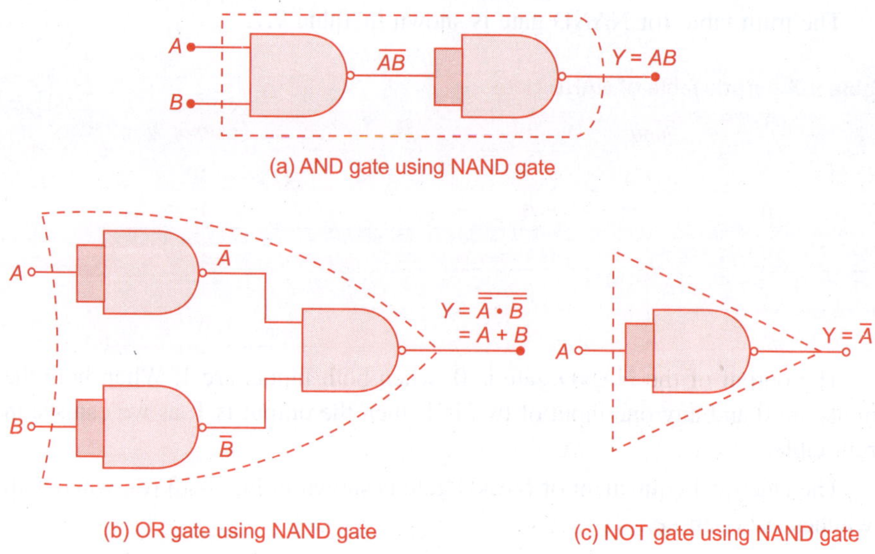

(a) AND gate using NAND gate

(b) OR gate using NAND gate

(c) NOT gate using NAND gate

Fig. 9.14   *AND, OR and NOT operations using NAND gate*

Ex-OR Gate Using NAND Gate

The Ex-OR operation can be done by using NAND gate only. The Fig. 9.14 (d) shows Ex-OR operation using NAND gate.

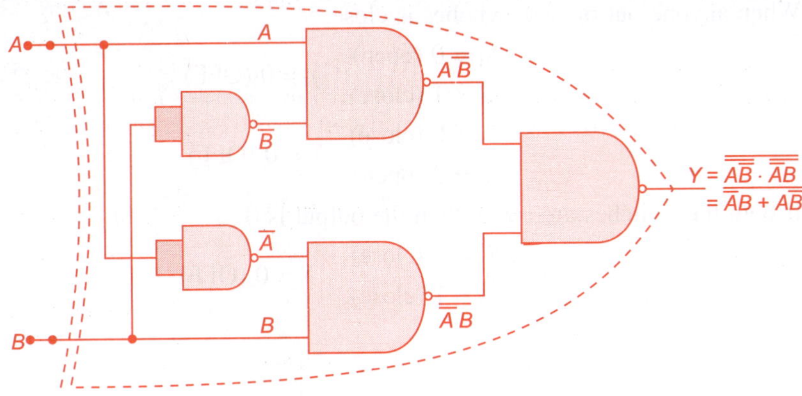

Fig. 9.14(d)   *Ex-OR gate using NAND gate*

9.8.5.2   NOR gate

The NOR gate is made with the help of OR and NOT gate as shown in Fig. 9.15 (a) and the symbolic representation of NOR gate is shown in Fig. 9.15 (b)

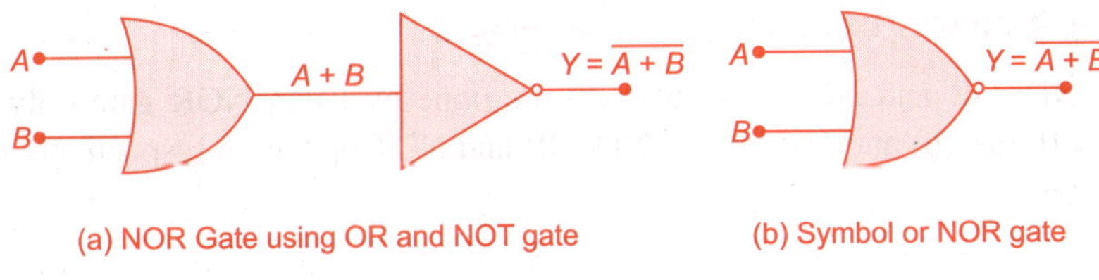

(a) NOR Gate using OR and NOT gate   (b) Symbol or NOR gate

Fig. 9.15   (a) and (b)

The truth table for NOR gate is shown in Table 9.8. From the truth table, we can see that the output of the NOR gate is 1 when both inputs are 0. When any one input or both inputs are 1, then the output is 0. The electrical equivalent of NOR gate is shown in Fig. 9.14 (c). When the both switches are open i.e.,

Table 9.8   Truth Table of NOR Gate

Input		Output
A	B	Y
0	0	1
0	1	0
1	0	0
1	1	0

i.e. $S_1 = 0$ (open) $Y = 1$ (ON)
             $S_2 = 0$ (open)

When anyone out of two switches is close,

i.e.

$$S_1 = 0 \text{ (open)}$$
$$S_2 = 1 \text{ (close)}$$
$$Y = 0 \text{ (OFF)}$$

and,

$$S_1 = 0 \text{ (close)}$$
$$S_2 = 0 \text{ (open)}$$
$$Y = 0 \text{ (OFF)}$$

If both the switches are close, then the output is 0

i.e.

$$S_1 = 1 \text{ (close)}$$
$$S_2 = 1 \text{ (close)}.$$
$$Y = 0 \text{ (OFF)}$$

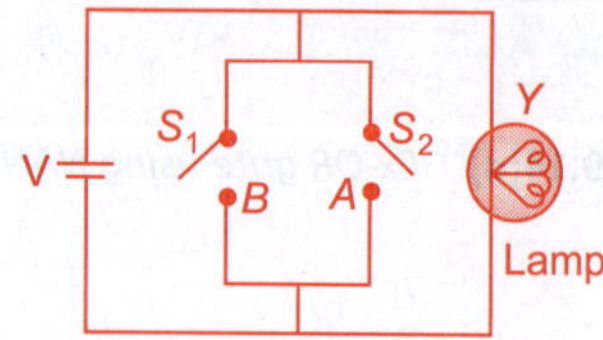

(c) Electrical equivalent of NOR gate

Fig. 9.15(c)

AND, OR, NOT operation using NOR gate

The AND, OR and NOT operation can be done by using NOR gate only. The Figs. 9.16 (a), (b) and (c) shows AND, OR and NOT operation respectively using NOR gate.

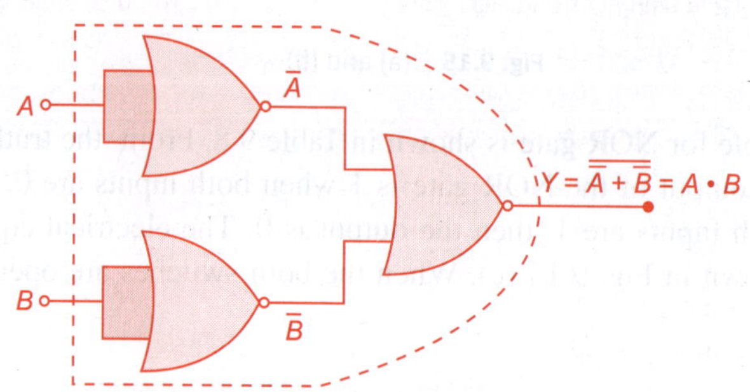

(a) AND gate using NOR gate

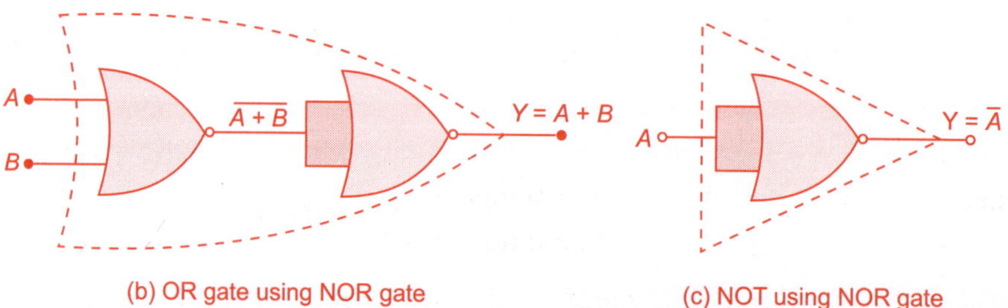

(b) OR gate using NOR gate

(c) NOT using NOR gate

Fig. 9.16   *AND, OR and NOT operation using NOR gate*

Ex-OR Gate Using NOR Gate

The Ex-OR operation can be done by using NOR gate only. The Fig. 9.16 (d) shows Ex-OR operation using NOR gate.

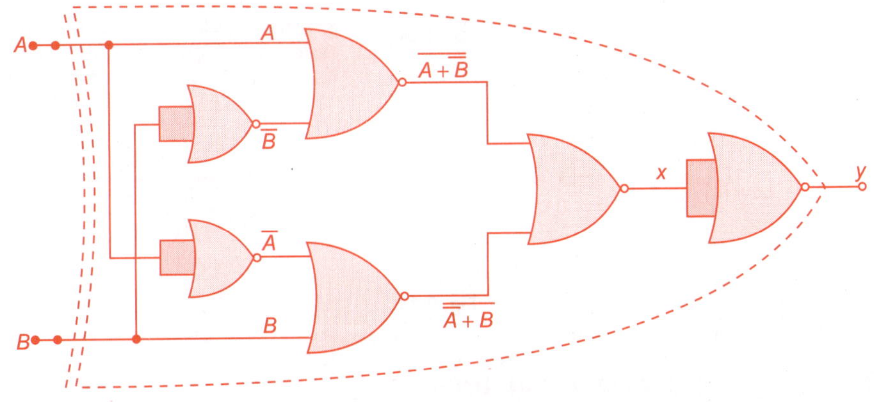

(d) Ex-OR gate wing NOR gate

Fig. 9.16(d) *Ex-OR gate using NOR gate*

where,

$$x = \overline{(A + \overline{B}) + (\overline{A} + B)}$$

$$y = \overline{\overline{(A + \overline{B}) + (\overline{A} + B)}} = (A + \overline{B}) + (\overline{A} + B)$$

$$= \overline{A} \cdot \overline{\overline{B}} + \overline{\overline{A}} \cdot \overline{B}$$

$$= \overline{A}B + A\overline{B}.$$

9.9 BOOLEAN ALGEBRA

In Boolean algebra, the operations are based on logic. Boolean algebra was differs from both ordinary algebra and the binary number system. This algebra was developed by George Boole. This is an algebra of logic and called Boolean algebra by name of George Boole.

When we compare the ordinary algebra, then we can say that in ordinary algebra the variables may have any value such as $4x + 8y$, the variables x and y may have any value. Whereas in Boolean Algebra, the variables used in any expression have a unique value they can have one out of two values i.e., 1 or 0.

In Boolean algebra, two different types of levels occur within digital logic circuit. These levels are indicated by 'High' (1) and 'low' (0). The two levels of Boolean algebra is shown in Figs. 9.17 (a) and (b). In Boolean algebra, the complement of variable (letter) for example, complement of x is represented by $\bar{x}$.

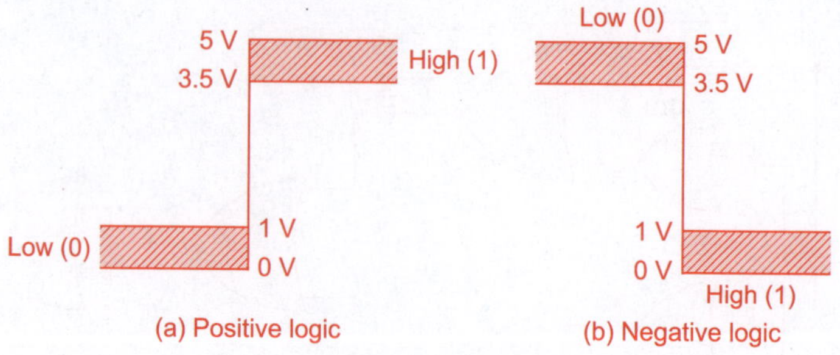

Fig. 9.17

9.9.1 Laws of Boolean Algebra

Basically, the boolean algebra is a mathematical logic system and it is different from general algebra. This algebra consists of some unproved postulates and six theorems. Commonly postulates and theorems are as follows:

Postulate 2: If A is any object then,

   (a) $A + 0 = A$

   (b) $A \cdot 1 = A$

Postulate 3: If A and B are two objects then,

   (a) $A + B = B + A$

   (b) $A \cdot B = B \cdot A$

   These are also called as commutative laws.

Postulate 4: If A, B and C are certain objects then,

   (a) $A \cdot (B + C) = A \cdot B + A \cdot C$

   (b) $A + (B \cdot C) = (A + B) \cdot (A + C)$

   (c) $A + (\bar{A} \cdot B) = A + B$

   These are also called as distributive laws.

Postulate 5: If A is any object and $\bar{A}$ is a inversion of A then,

   (a) $A + \bar{A} = 1$

   (b) $A \cdot \bar{A} = 0$

Theorem 1: If A is any object, then,

   (a) $A + A = A$

   (b) $A \cdot A = A$

Theorem 2:   If A is any object then,

(a) $A + 1 = 1$

(b) $A \cdot 0 = 0$

Theorem 3:   If A is any object and $\overline{A}$ is inversion of A then,

$$\overline{\overline{A}} = A$$

It is also called as Involution Law.

Theorem 4:   If A, B and C are certain objects then,

(a) $A + (B + C) = (A + B) + C$

(b) $A \cdot (B \cdot C) = (A \cdot B) \cdot C$

These are also called as Associative Laws.

Theorem 5:   If A and B are certain objects then,

$$\overline{(A + B)} = \overline{A} \cdot \overline{B} \quad \text{or} \quad \overline{(A \cdot B)} = \overline{A} + \overline{B}$$

It is also called as DeMorgan theorem. According to it, "Break the bar and change the sign i.e., (+) in (·) and (.) in to (+)".

Theorem 6:   If A and B are two certain objects, then,

(a) $A + AB = A$

(b) $A \cdot (A + B) = A$

(c) $A \cdot (\overline{A} + B) = AB$

These are also called as Absorptive Laws.

9.10   KARNAUGH MAP (k-MAP)

The k-map is used to minimize a Boolean function. Basically, the complexity of the digital logic circuit is directly related to the complexity of the Boolean function. So, we can say that, when the Boolean function has more variables, then the complexity of the function will increase and it needs more number of logic gates to implement the function. The k-map is a graphical method by which we can minimise the Boolean function that required minimum number of logic gates. We can also simplify the Boolean function by algebraic means (these methods are out of scope of this book) but this procedure of minimization is awkward. The map method provides a simple straight forward procedure for minimizing Boolean functions. This map method or graphical method was developed by a switch and was modified by Karnaugh. So, it is known as "Veitch diagram" or Karnaugh map.

A.   *Procedure for placing a Boolean function (SOP form) in map or matrix*

1. The map is a diagram made up in the form of matrix. Each element of matrix represents one minterm (m) as shown in Fig. 9.18.

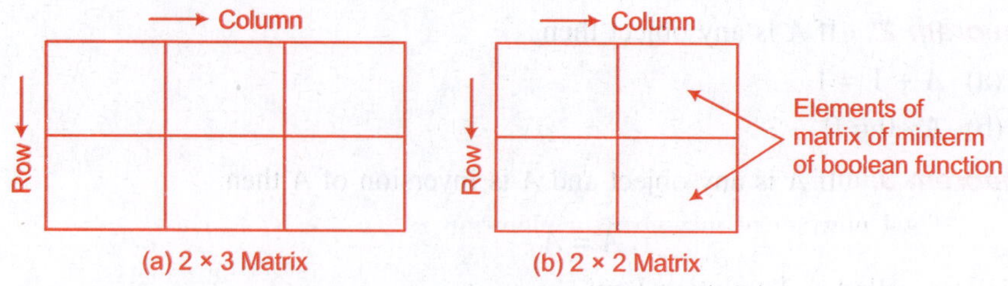

Fig. 9.18

2. The Boolean function may be in Sum of Product form (SOP form) or Product of Sum form (POS form) as shown below,

$$f_1(AB) = \overline{A}B + A\overline{B} + \overline{A}\,\overline{B} \rightarrow \text{Sum of Product form (SOP form)}$$

$$f_2(AB) = (A + B)(\overline{A} + \overline{B})(\overline{A} + B) \rightarrow \text{Product of Sum form (POS form)}$$

3. One line is made at left upper corner of the matrix that will separate the row and column variables as shown in Fig. 9.19

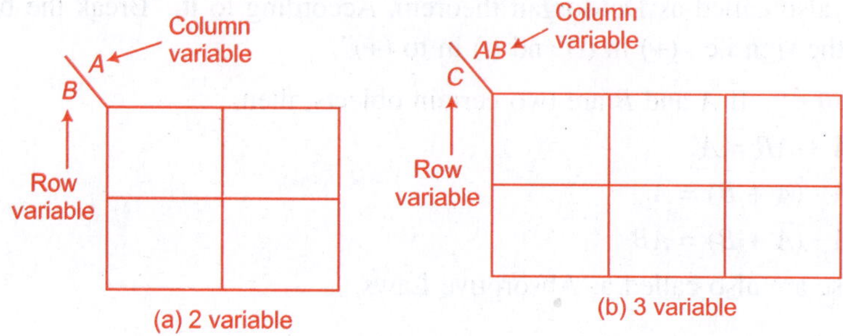

Fig. 9.19

4. The size of matrix is based on the number of variables. The number of row and columns and matrix are defined by the formula given below,

$$\text{Number of rows} = 2^{n_r} \tag{9.1}$$

where n_r = number of row variable

$$\text{Number of columns} = 2^{n_c} \tag{9.2}$$

where n_c = number of column variable.

The total number of minterms element in the matrix.

$$= 2^{n_r} \times 2^{n_c} \tag{9.3}$$

For example of Fig. 9.19

For Fig. 9.19 (a) $n_r = 1$ and $n_c = 1$

The number of rows = $2^1 = 2$

The number of columns = $2^1 = 2$.

The total number of minterms or element of matrix.

$$= 2 \times 2 = 4.$$

For Fig. 9.19 (b)

$$n_r = 1 \quad \text{and} \quad n_c = 2.$$

The number of rows = $2^1 = 2$

The number of columns = $2^2 = 4$.

Total number of minterms or elements = $2 \times 4 = 8$.

5. Now, we will assign some number to the rows and columns. The number of columns are given in the upper side of matrix and the number of rows are on the left side of the matrix and the numbers are in gray code as shown in Figs. 9.20 (a) and (b).

Decimal number	0	1	2	3	4	5	6	7	8
Gray code	0000	0001	0011	0010	0110	0111	0101	0100	1100

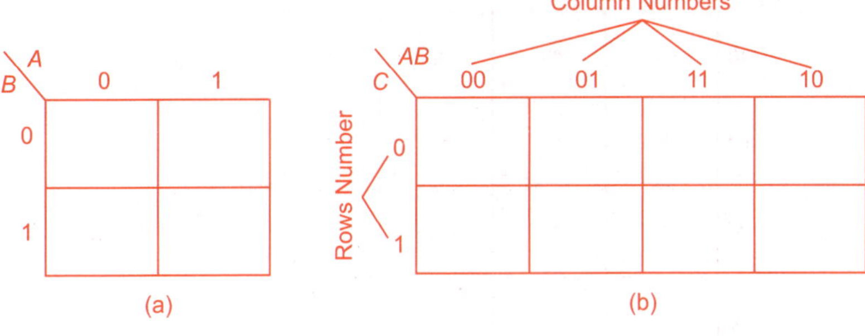

Fig. 9.20

These numbers of rows and columns represent the minterms of Boolean functions.

6. Each Boolean function contain minterms like,

$$f_1(A, B, C) = \overline{A}BC + AB\overline{C} + A\overline{B}\,\overline{C} + ABC$$

$$\quad\;\; m_0 \qquad m_1 \qquad m_2 \qquad m_3$$

$$\underbrace{}_{\text{minterms}}$$

In sum of product form,

$$\overline{A} = 0$$

and, $\qquad A = 1$

i.e., the variable with 'des' or 'bar' represents 0 and the variable without 'bar' or 'des' represents 1.

7. Some Boolean function represents mintern numbers only like,

$$f(A, B, C) = \Sigma\,(3, 4, 6, 7)$$

The matrix has predesided minterms numbers as shown in Figs. 9.21(a) and (b). The minterm that exists in function will place 1 in the matrix as shown in Fig. 9.21(a)

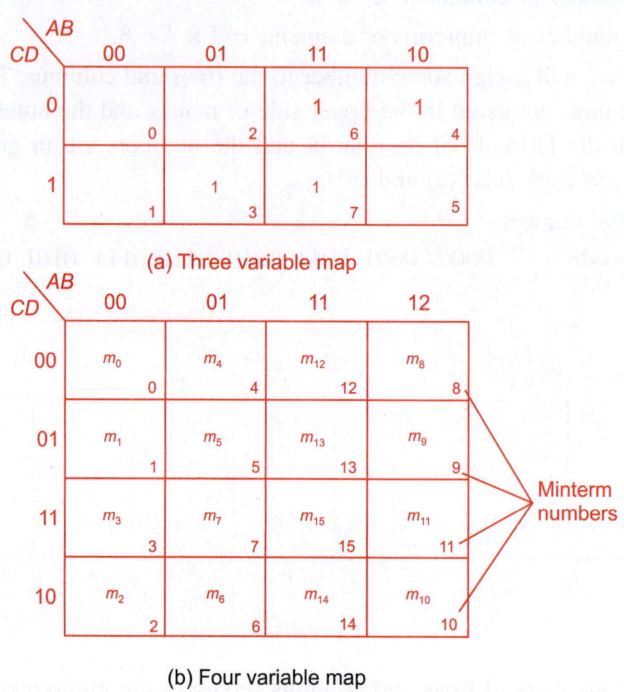

Fig. 9.21

8. The Boolean function which consists minterms place in matrix.

   Assume Boolean function,

$$f(A, B, C) = \overset{\text{minterm}}{\overbrace{\overline{A}BC}} + \overset{\text{minterm}}{\overbrace{ABC}} + \overset{\text{minterm}}{\overbrace{A\overline{B}\,\overline{C}}} + \overset{\text{minterm}}{\overbrace{A\overline{B}C}}$$

$$\overline{A}BC = \overset{C_2}{\overbrace{01}}\ \overset{R_2}{\overbrace{1}} \qquad A\overline{B}\,\overline{C} = \overset{C_4}{\overbrace{10}}\ \overset{R_1}{\overbrace{0}}$$

$$ABC = \overset{C_3}{\overbrace{11}}\ \overset{R_2}{\overbrace{1}} \qquad A\overline{B}C = \overset{C_4}{\overbrace{10}}\ \overset{R_2}{\overbrace{1}}.$$

The minterms that exist in Boolean function will place 1 at the place of minterm in the matrix as shown in Figs. 9.22 (a) and (b)

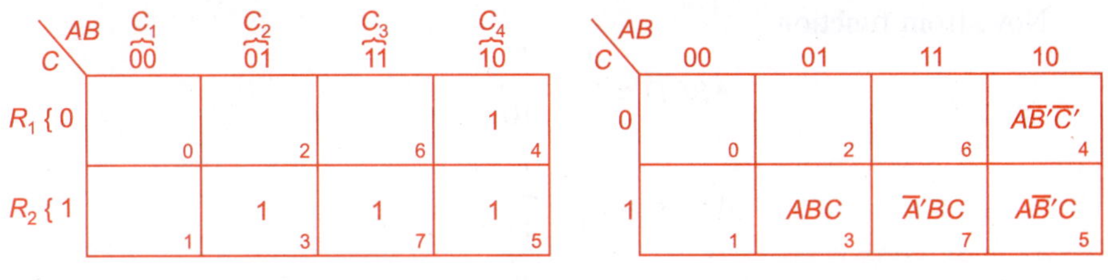

(a) After placing minterms   (b) After placing 1 at the place of minterm

Fig. 9.22

Example 9.10.1   Place the minterms into the matrix for following function,
$$F(A, B, C) = \Sigma\,(1, 4, 5, 6, 7).$$

Solution   Let row variable be C.

Column variables are A and B.

$$\text{The number of rows} = 2^{n_r}$$
$$= 2^1 = 2.$$
$$\text{The number of columns} = 2^{n_c}$$
$$= 2^2 = 4.$$

So the matrix size is 2×4.

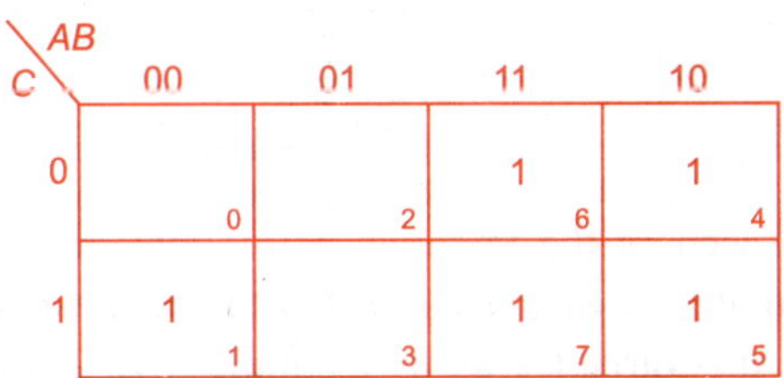

Example 9.10.2   Place the minterms in to the matrix for the following Boolean function,

$$F(A, B, C, D) = ABC\overline{D} + \overline{AB}CD + \overline{A}BCD + AB\overline{CD}$$
$$\downarrow \qquad \downarrow \qquad \downarrow \qquad \downarrow$$
$$m_0 \qquad m_1 \qquad m_2 \qquad m_3$$

Solution   Let column variables $= A, B$

row variable $= C, D$

i.e. $n_r = 2$ and $n_c = 2.$

$$\text{The size of matrix} = 2^{n_r} \times 2^{n_c}$$
$$= 2^2 \times 2^2$$
$$= 4 \times 4.$$

Now, from function,

$$ABC\overline{D} = \overset{C_3}{\overbrace{11}}\ \overset{R_4}{\overbrace{10}}$$

$$\overline{A}BCD = \overset{C_1}{\overbrace{00}}\ \overset{R_3}{\overbrace{11}}$$

$$\overline{A}BCD = \overset{C_2}{\overbrace{01}}\ \overset{R_3}{\overbrace{11}}$$

$$AB\overline{C}\overline{D} = \overset{C_3}{\overbrace{11}}\ \overset{R_1}{\overbrace{00}}$$

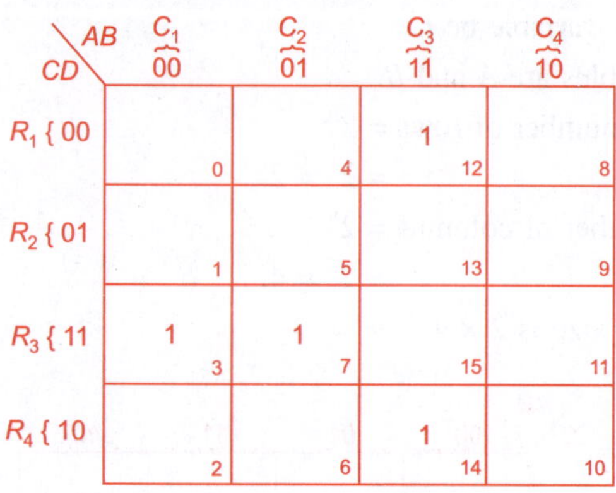

9. As we have seen in the previous points that the Boolean function has number of minterms each 1 place into the matrix is due to one minterm. If it has all the variables as in Example 9.10.2, the function has four minterms m_0, m_1, m_2 and m_3, and all the minterms have all variables. So, the four 1 will be place into the matrix.

10. In some Boolean functions some variables will be missing in some minterm. The number of 1 will be placed into the matrix according to the missing variable and given by the following formula,

The number of 1's which will be placed into the matrix = $2^{n_{v_m}}$ (9.4)

where $n_{v_m} = 0$ (number of missing variable).

Example 9.10.3   Place the minterms into the matrix for the following Boolean function.

Solution
$$F(A, B, C, D) = \underset{\downarrow}{\overline{A}BC\overline{D}} + \underset{\downarrow}{\overline{A}BC} + \underset{\downarrow}{AC} + \underset{\downarrow}{\overline{B}}$$
$$ m_0 \quad\quad m_1 \quad\quad m_2 \quad\quad m_3$$

For minterm m_0 $(\overline{A}BC\overline{D})$

$$n_{v_m} = 0 \text{ (no missing variable).}$$

The number of 1 due to minterm $m_0 = 2^{n_{v_m}} = 2^0 = 1$

For minterm $m_1\ (\overline{A}BC)$

$$n_{v_m} = 1 \text{ (one missing variable, } D)$$

The number of 1 due to minterm $m_1 = 2^{n_{v_m}}$

$$= 2^1 = 2$$

For minterm $m_2(AC)$

$$n_{v_m} = 2 \text{ (Two missing variables; } B, D)$$

The number of 1 due to minterm $m_2 = 2^{n_{v_m}}$

$$= 2^2 = 4.$$

For minterm $m_3(\overline{B})$

$$n_{v_m} = 3 \text{ (Three missing variables; } A, C, D)$$

The number of 1 due to minterms $(m_3) = 2^{n_{v_m}}$

$$= 2^3 = 8.$$

The total number of 1's = Due to m_0 + Due to m_1 + Due to m_2 + Due to m_3

$$= 1 + 2 + 4 + 8$$

$$= 15.$$

More than single 1 may come at a particular place in the matrix but we will place a single 1 at that place.

11. Now, we will find the minterms of missing variables. for Example, 9.10.3 we have,

    The number of 1's due to minterm $m_0 = 1$

    $$ABC\overline{D} = 0110\}\tag{9.5}$$

    The number of 1's due to minterm $m_1 = 2$

    $\because\quad D$ is missing, it may be D or $\overline{D}$

    So, $\qquad\quad \overline{A}BCD = 0111 \atop \overline{A}\,BC\overline{D} = 0110 \Big\}$ $\tag{9.6}$

    The number of 1's due to minterm $m_2 = 4.$

    $\because\quad B$ and D are missing, it may be $B, D, \overline{B}, \overline{D}$

    $$AC$$
    $$\downarrow$$

    $$\begin{aligned} ABCD &= 1\,1\,1\,1 \\ A\overline{B}CD &= 1\,0\,1\,1 \\ ABC\overline{D} &= 1\,1\,1\,0 \\ A\overline{B}C\overline{D} &= 1\,0\,1\,0 \end{aligned}\Bigg\}\tag{9.7}$$

    The number of 1's due to minterm $m_3 = 8$

    $\because\quad A, C$ and D are missing, it may be $A, \overline{A}, C$ and $D, \overline{D}$

$$\begin{matrix}
A\overline{B}CD = 1\ 0\ 1\ 1 \\
\overline{A}BCD = 0\ 0\ 1\ 1 \\
\overline{A}\,\overline{B}CD = 0\ 0\ 0\ 1 \\
\overline{A}\,\overline{B}\,\overline{C}\,\overline{D} = 0\ 0\ 0\ 0 \\
\overline{A}\,\overline{B}C\overline{D} = 0\ 0\ 1\ 0 \\
A\overline{B}\,\overline{C}D = 1\ 0\ 0\ 1 \\
A\overline{B}C\overline{D} = 1\ 0\ 1\ 0 \\
A\overline{B}\,\overline{C}\,\overline{D} = 1\ 0\ 0\ 0
\end{matrix}$$

$$\overset{\overline{B}}{\downarrow}$$

(9.8)

Alternative Method

First make the truth table for the given variable. For Example 9.10.3, there are four variables A, B, C and D. So the truth table is,

Table 9.9

2^3	2^2	2^1	2^0		Minterms
A	B	C	D	Y	
0	0	0	1	1	0
0	0	0	1	1	1
0	0	1	0	1	2
0	0	1	1	1	3
0	1	0	0	0	4
0	1	0	0	0	5
0	1	1	0	1	6
0	1	1	1	1	7
1	0	0	0	1	8
1	0	0	1	1	9
1	0	1	0	1	10
1	0	1	1	1	11
1	1	0	0	0	12
1	1	0	0	0	13
1	1	1	0	1	14
1	1	1	1	1	15

First minterm m_0

$$\overline{A}BC\overline{D} = 0\ 1\ 1\ 0 \text{ minterm 6 in truth table}$$

So, place 1 at minterm 6.

Second minterm m_1

$$\overline{A}BC = 0\ 1\ 1\ X \text{ where } X = \text{don't care it may be '0' or '1'}$$

at minterm 6 and 7 in truth table. So, place 1 at minterm 6 and 7, but at No. 6 already 1 is placed.

Third minterm m_2

$$AC = 1\ X\ 1\ X \text{ This is at minterms 10, 11, 14 and 15 in}$$

truth table, so place 1 at these minterms.

Fourth minterm, m_3

$$\overline{B} = X\ 0\ X\ X.$$

This is at minterms 0, 1, 2, 3, 8, 9, 10 and 11 in truth table, so place 1 at these minterms.

Now, the function will be in the form.

$$F(A, B, C, D) = \Sigma\ (0, 1, 2, 3, 6, 7, 8, 9, 10, 11, 14, 15)$$

This matrix is,

CD \ AB	00	01	11	10
00	1 [0]	[4]	[12]	1 [8]
01	1 [1]	[5]	[13]	1 [9]
11	1 [3]	1 [7]	1 [15]	1 [11]
10	1 [2]	1 [6]	1 [14]	1 [10]

To understand the usefulness of the map for simplifying Boolean functions. Any two adjacent minterms or squares in the matrix or map differ by only one variable, which is primed[0] in one square and unprimed[1] in the other. For example, in Fig. 9.21(a), 3 and 7 lies in two adjacent squares. Variable B is primed in 3 and unprimed in 7, whereas the other two variables are the same in both squares.

In sum of product form (SOP), if the variable is complemented, then the bit is 0 and if uncomplemented, then the bit is 1 and the algebraic expression represents the minterms (m_i) whereas in product of sum (POS) form, if the variables are complemented, then the bit is 1 and if uncomplemented, then the bit is 0. The Boolean function represents the minterms (M_j) where j represents the binary number of the term.

Example 9.10.4 $F(A, B, C) = \overline{A}\,\overline{B}\,\overline{C} + \overline{A}BC + A\overline{B}C + ABC$

Solution $m_j = m_0 + m_2 + m_5 + m_7$

$F(A, B, C) = \Sigma\, m(0, 2, 5, 7)$

and $F(A, B, C) = (A + B + \overline{C}) \cdot (A + \overline{B} + \overline{C}) \times (\overline{A} + B + C) \cdot (\overline{A} + \overline{B} + C)$

$= (M_1 \times M_3 \times M_4 \times M_6)$

$F(A, B, C) = \Pi M(1, 3, 4, 6).$

B. Rules for mapping the square of the matrix

(i) We can use the same 1 more than once in the overlapping or mapping to get the largest group.

(ii) Try to overlap the group the maximum number of 1's if possible because it will eliminate more variables. The number of 1's in a group must be 2^n where $(n = 0, 1, 2, 3, \ldots)$.

As we can see in Fig. 9.23, there are two overlappings one contains two 1's and second one contains eight 1's.

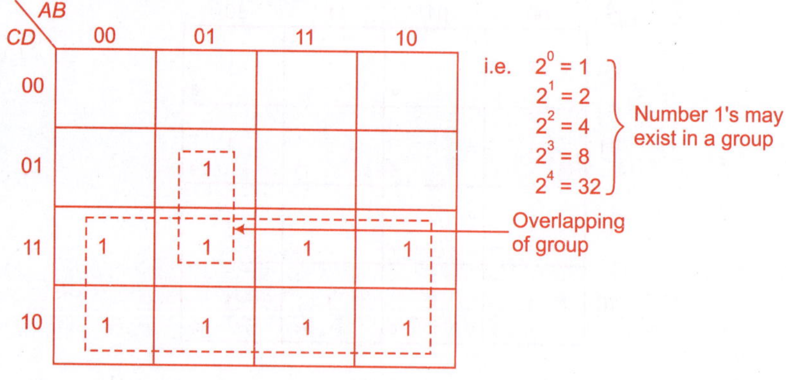

Fig. 9.23

(iii) The minimum number of 1's in a group is one as shown in Fig. 9.24. There are three groups. One contains four 1's second contains single 1 and third contains single 1.

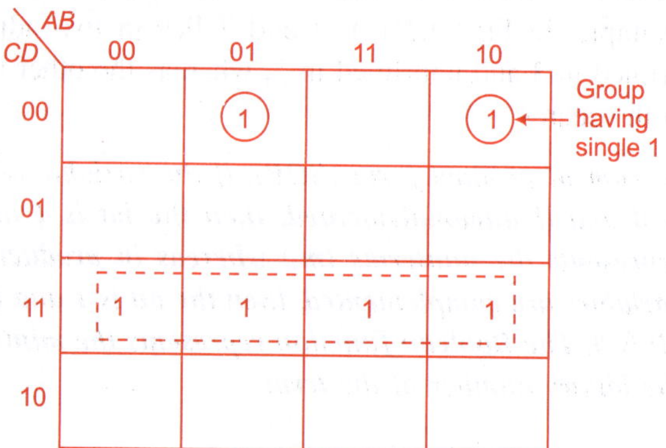

Fig. 9.24

(iv) If we roll the map in such a way that the leftside touches the right side. We see that two pairs actually form a quad. To indicate this, draw half circles around each pair, as shown in Figs. 9.25 (a) and (b).

(v) When we finish encircling groups, then we eliminate redundant group. Basically, the reductant group is a group whose all 1's are already used by other groups, as shown in Fig. 9.26 (a) and (b).

(vi) If some minterms or maxterm have 'don't care' condition, it will be marked as X in a square and may be considered as 0 or 1 according to requirement, as shown in Fig. 9.27

$$F(A, B, C, D) = \Sigma\, m(7) + d(\overline{3}, 10, 11, 12, 13, 15)$$

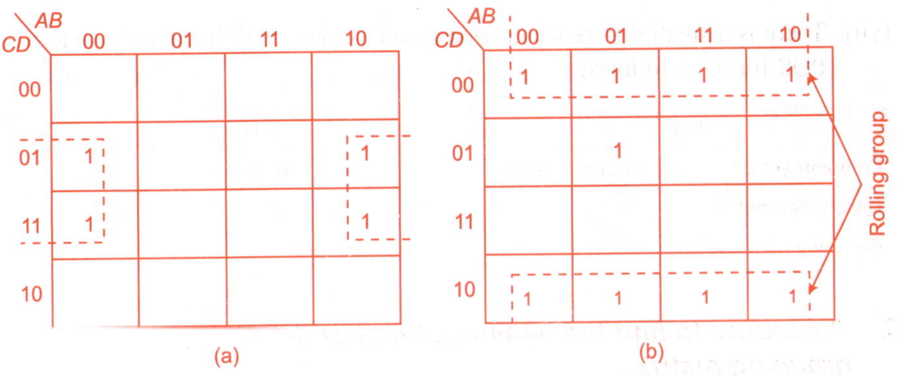

(a)　　　　　(b)

Fig. 9.25

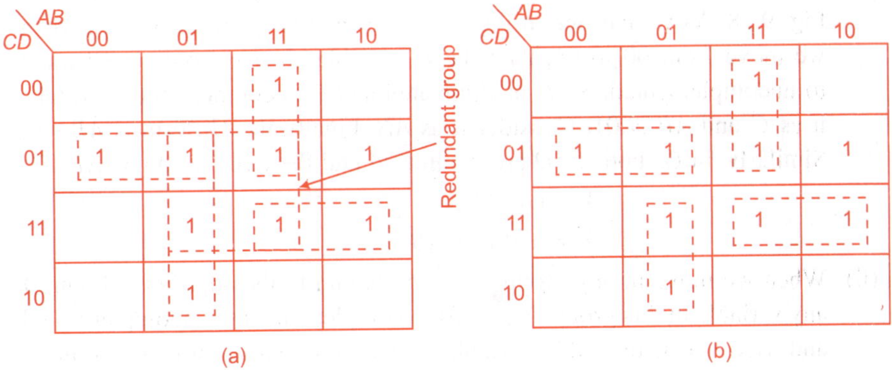

(a)　　　　　(b)

Fig. 9.26

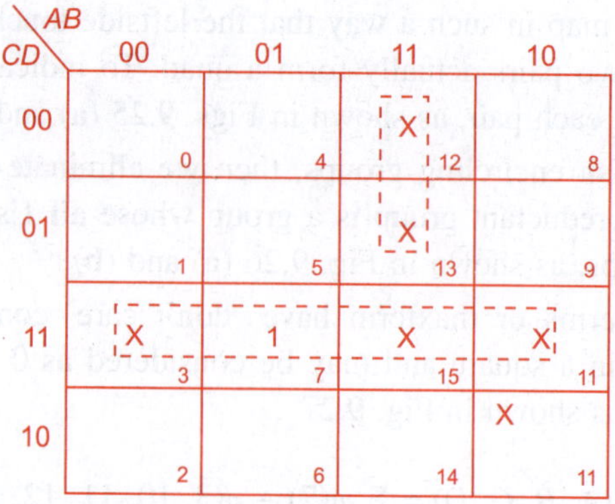

<table>
<tr><td rowspan="2"></td><td colspan="4" align="center">AB</td></tr>
<tr><td align="center">00</td><td align="center">01</td><td align="center">11</td><td align="center">10</td></tr>
<tr><td>CD
00</td><td align="right">0</td><td align="right">4</td><td align="center">X

12</td><td align="right">8</td></tr>
<tr><td>01</td><td align="right">1</td><td align="right">5</td><td align="center">X
13</td><td align="right">9</td></tr>
<tr><td>11</td><td align="center">X
3</td><td align="center">1
7</td><td align="center">X
15</td><td align="center">X
11</td></tr>
<tr><td>10</td><td align="right">2</td><td align="right">6</td><td align="right">14</td><td align="center">X
11</td></tr>
</table>

Fig. 9.27

(vii) There is a small difference between minterm i.e., SOP form and maxterm POS form, as follows:

SOP		POS	
Complement 0	i.e. $A = 1, \overline{A} = 0$	1	i.e. $A = 0, \overline{A} = 1$
Uncomplement 1		0	
Grouping of 1's		Grouping of 0's	

C. Procedure to find the minimize function from grouping matrix

(i) When we move in any group from one row to adjacent row, if any variable in this group change from complemented to uncomplemented and vice versa, then this variable will be eliminated from minterm as shown in Fig. 9.28. As we can see in Fig. 9.28, we see that in group 1 (G_1), when we move from 00 to 01, then the D variable change from complement to uncomplemented. So D is eliminated and C is complemented, so take it as $\overline{C}$ and AB is 01. Consider it as $\overline{A}B$. Finally the G_1 term is $\overline{A}B \cdot \overline{C}$. Similarly, in G_2 both C, D are eliminated and the term is 10 i.e., $A\overline{B}$

$$Y = G_1 + G_2$$
$$Y = \overline{A}\,B\,\overline{C} + A\overline{B}$$

(ii) When we move in any group from on column to the adjacent column, if any variable in this group change from complemented to uncomplemented and vice versa, then this variable will be eliminated from minterm as shown in Fig. 9.29

$$Y = G_1 + G_2$$
$$Y = B\overline{C}D + C\overline{D}$$

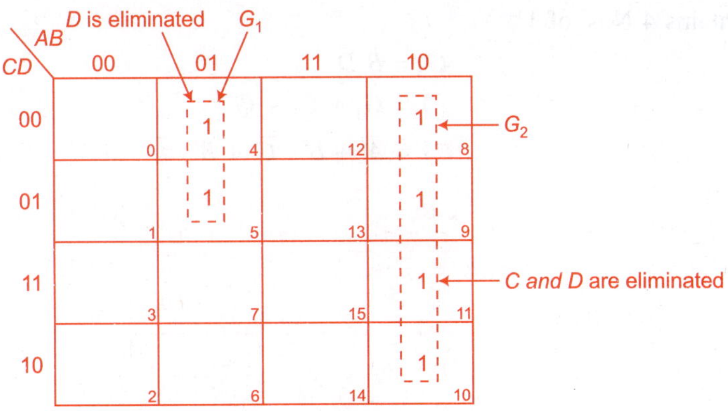

Fig. 9.28

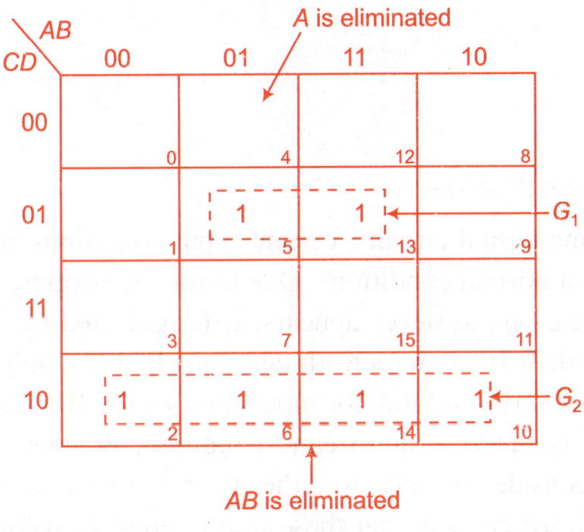

Fig. 9.29

In sum of product form (SOP) A = 1 and $\overline{A}$ = 0 and we will place dot (•) between the variable and plus (+) between the groups whereas in product of sum form (POS), A = 0 and $\overline{A}$ = 1 and we will place plus (+) between the variable and dot (•) between the groups.

Example 9.10.5   Simplify the Boolean function using K-map in SOP form.

$$F(A, B, C, D) = \Sigma\, m(0, 1, 2, 3, 4, 5, 6, 7, 8, 9, 10)$$

Solution   There are four groups G_1, G_2, G_3, G_4

G_1 contains 8 Nos. of 1

$$G_1 = \overline{A}$$

G_2 contains 4 Nos. of 1

$$G_2 = \overline{B}\,\overline{C}$$

G_3 contains 4 Nos. of 1

$$G_3 = \overline{B}\,\overline{D}$$

i.e.
$$Y = G_1 + G_2 + G_3$$

$$Y = \overline{A}' + \overline{B}' \cdot \overline{C}' + \overline{B}' \cdot \overline{D}' \qquad \textbf{(Ans.)}$$

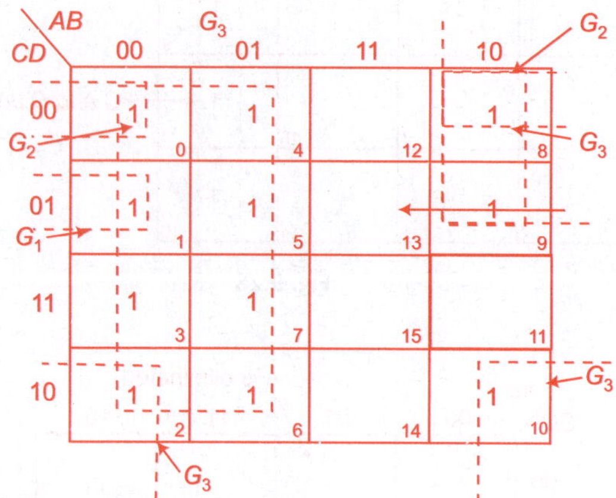

'Don't care' conditions

Generally in some digital circuits, certain input conditions never occur when the circuit operates at normal conditions. Due to this, the corresponding output never appears. Since the output never appears, it is indicated by X in the truth table. For instance, Table 9.10 shows a truth table in which the output is high for inputs entries from 000 to 011 and low for inputs entries for 100 and 101 and X for 110 and 111. The X is called a 'don't care' condition. Whenever you see X in truth table, you can consider it equal to either 0 or 1, whichever produces a simpler logic circuit. So, we can say that these don't cares stand for whatever you like. In k-map, we can consider them as 1's if we get largest group of 1's for SOP form and as 0's, for getting largest group of 0's in POS form. The representation of 'don't care' conditions are as follows for Truth Table 9.10

$$F(A, B, C) = \Sigma\, m(0, 1, 2, 3) + d(6, 7)$$

Table 9.10

A	B	C	Y	Minterms
0	0	0	1	0
0	0	1	1	1
0	1	0	1	2
0	1	1	1	3
1	0	0	0	4
1	0	1	0	5
1	1	0	X ⎫ don't care conditions	6
1	1	1	X ⎭	7

Example 9.10.6   Simplifying the Boolean function,

$$F(A, B, C, D) = \Sigma\, m(0, 1, 2) + d(8, 9, 10, 11)$$

Solution   At the place of 'don't care' matrix.

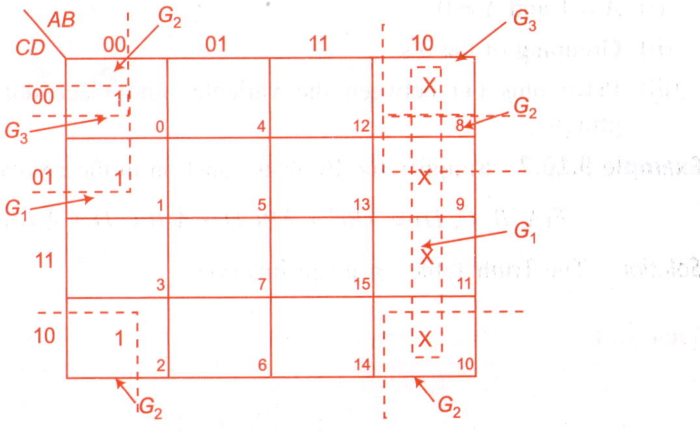

$$G_1 = A\overline{B} \text{ and } G_2 = \overline{B}\overline{D}$$
$$G_3 = \overline{B}\,\overline{C}\,.$$

So the output,

$$Y = G_1 + G_2 + G_3$$
$$Y = A\overline{B} + \overline{D}\overline{B} + \overline{B}\overline{C} \qquad \text{(\textbf{Ans.})}$$

The minimise output is,

$$Y = A\overline{B} + \overline{D}\,\overline{B} + \overline{B}\overline{C}.$$

The logic circuit using gates is shown in Fig. Ex. 9.10.6

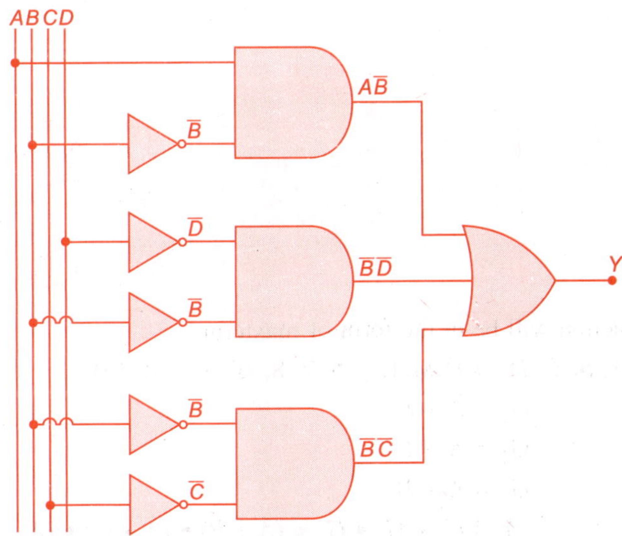

Fig. Ex. 9.10.6

Product of sum form

The procedure for product of sum form is much similar to the sum of product form. Following differences are there:

(i) $\overline{A} = 1$ and $A = 0$

(ii) Grouping of the 0's

(iii) Place plus (+) between the variables and place dot (•) between the groups.

Example 9.10.7 Simplify the Boolean function in the product of sum form

$$F(A, B, C, D) = ABC + \overline{A}BC\overline{D} + \overline{A}\,\overline{B}\,\overline{C}\,\overline{D} + \overline{A}\,\overline{B}\,\overline{C}\,D + A\overline{D}$$

Solution The Truth Table for given function

Table 9.11

Variables					
A	B	C	D	Y	Minterm
0	0	0	0	1	0
0	0	0	1	0	1
0	0	1	0	1	2
0	0	1	1	0	3
0	1	0	0	1	4
0	1	0	1	0	5
0	1	1	0	1	6
0	1	1	1	0	7
1	0	0	0	0	8
1	0	0	1	0	9
1	0	1	0	0	10
1	0	1	1	0	11
1	1	0	0	0	12
1	1	0	1	0	13
1	1	1	0	1	14
1	1	1	1	1	15

The function will be in the form of maxterm.

$$F(A, B, C, D) = \Pi\, M(1, 3, 5, 7, 8, 10, 11, 12, 13)$$
$$G_1 = \overline{A} + B$$
$$G_2 = \overline{A} + C$$
$$G_3 = A + \overline{D}$$
$$Y = G_1 \bullet G_2 \bullet G_3 = (\overline{A} + B) \bullet (\overline{A} + C) \bullet (A + \overline{D}) \quad \textbf{(Ans.)}$$

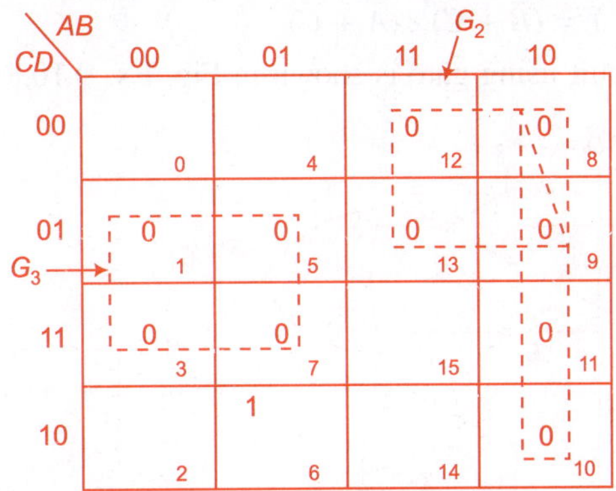

Example 9.10.8 Simplify the Boolean function in product of sum form and implement the logic using gates,

$$F(A, B, C) = (\overline{A} + B + C) \times (A + \overline{B} + \overline{C}) \times (\overline{A} + \overline{C}).$$

Solution The function is given in product of sum form. The truth table for that

Table 9.12

A	B	C	Output Y	Maxterm
0	0	0	1	0
0	0	1	1	1
0	1	0	1	2
0	1	1	0	3
1	0	0	0	4
1	0	1	0	5
1	1	0	1	6
1	1	1	0	7

The function is,

$$F(A, B, C) = \Pi\, M(3, 4, 5, 7)$$
$$G_1 = \overline{A} + B$$
$$G_2 = \overline{B} + \overline{C}$$
$$Y = G_1 \cdot G_2$$
$$Y = (\overline{A} + B) \bullet (\overline{B} + \overline{C}) \qquad \textbf{(Ans.)}$$

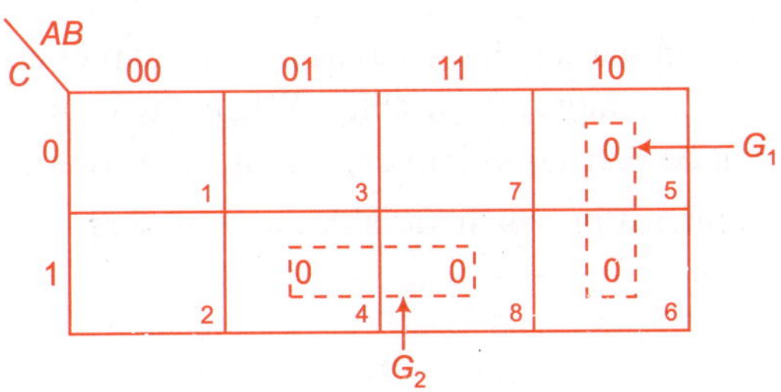

The output is $Y = (\overline{B} + C) \cdot (\overline{A} + \overline{C})$

The logic circuit using gate is shown in Fig. Ex. 9.10.8.

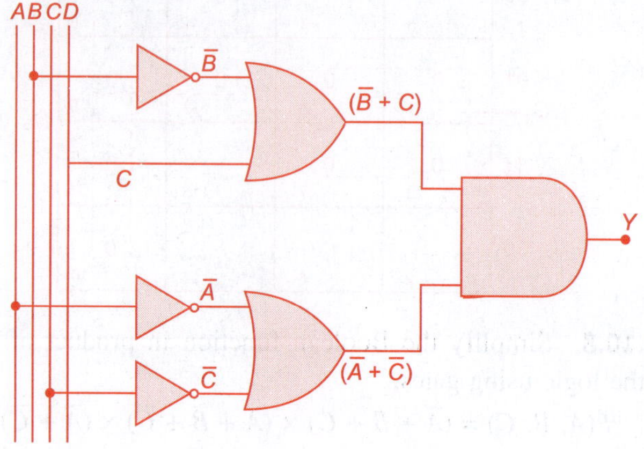

Fig. Ex. 9.10.8

9.11   CANONICAL FORM OF A LOGIC EXPRESSION

Logic expression is called in canonical form when each term of expression contains all variables. When a product of sum (POS) form of logic expression is in canonical form, each term is called maxterm and each maxterm contains all variables. So, it is also called as maxterm canonical form or standard product of sum form. Same way when a sum of product (SOP) form of logic expression is in canonical form, each term is called minterm and each minterm contains all variables. So, it is also called as minterm canonical formor standard sum of product form.

Following are the examples of the canonical form of sum of the products and product of sum canonical forms.

(a) $Y = AB\overline{C} + A\overline{B}C + ABC$

(b) $Y = ABCD + \overline{A}BC\overline{D} + AB\overline{C}\,\overline{D} + \overline{A}\,\overline{B}\,\overline{C}\,\overline{D}$

(c) $Y = (A + B) \cdot (A + \overline{B})$

(d) $Y = (\overline{A} + B + \overline{C})(\overline{A} + \overline{B} + \overline{C})(A + B + C)(A + \overline{B} + C)$

9.11.1   Conversion of SOP Form into Canonical Form

As we have discussed earlier that canonical form expression is such a type in which each minterm contains all variables. When any expression which is not in canonical form can be converted into canonical form by following procedures.

1. Find the minterm in which variables are missing.

2. Multiply the term by the sum of missing variable and complement of missing variable. For example, in any term variable B is missing, then we multiply that particular minterm with $(B + \overline{B}) = 1$.

3. After multiplying the $B + \overline{B}$ in term, we will get the expression in canonical form.

Example 9.11.1 The given expression has three variables Convert this expression into canonical form

$$Y(A, B, C) = AB + B + AB\overline{C}.$$

Solution   In given expression first term has only two variables A and B but C is missing. So first term is multiplied by $(C + \overline{C})$.

The second term has only one variable B i.e., A and C variables are missing. So the second term is multiplied by $(A + \overline{A})$ and $(C + \overline{C})$.

We obtain,

$$Y(A, B, C) = AB \cdot (C + \overline{C}) + B \cdot (A + \overline{A}) \cdot (C + \overline{C}) + AB\overline{C}$$

$$Y(A, B, C) = ABC + AB\overline{C} + B(AC + A\overline{C} + \overline{A}C + \overline{A}\,\overline{C}) + AB\overline{C}$$

$$Y(A, B, C) = ABC + AB\overline{C} + BAC + BA\overline{C} + B\overline{A}C + B\overline{A}\,\overline{C} + AB\overline{C} \quad \textbf{(Ans.)}$$

9.11.2   Conversion of POS Form into Canonical Form

As we know that canonical form expression is such a type in which each maxterm contains all variables. When any expression which is not in canonical form can be converted into canonical form by the following procedure:

1. Find the maxterm in which variables are missing.

2. Add the term by the product of missing variable and complement of missing variable. For example, in any term variable A is missing, then we add $(A\overline{A})$ in particular maxterm.

3. After adding the $A\overline{A} = 0$, in maxterm, we will get the expression in canonical form.

Example 9.11.2   Convert the following expression in canonical form.

Solution $Y(A, B, C) = (A + B) \cdot (B + \overline{C}) \cdot$

$$Y(A, B, C) = (A + B + C\overline{C}) \cdot (A\overline{A} + B + \overline{C})$$

$$Y(A, B, C) = (A + B + C)(A + B + \overline{C})(A + B + \overline{C})(\overline{A} + B + \overline{C}).$$

(Ans.)

Examples

Example 9.1   Convert the following numbers as indicated,   **(UPTU 2005-06)**

(a) $(1000)_8 \quad = (\ )_2$

(b) $(2CCD)_{16} = (\ )_5$

(c) $(0.45)_{10} = (\quad)_8$

(d) $(345)_8 = (\quad)_{10}$

(e) $(7841)_9 = (\quad)_{10}.$

Solution (a) Given number is in octal and required number is in binary. So apply Rule-4 i.e., octal to binary conversion.

Octal,	1	0	0	0
Binary,	001	000	000	000

i.e. $(1000)_8 = (001000000000)_2$

$$= (1000000000)_2. \qquad \textbf{(Ans.)}$$

(b) The given number is in hexadecimal and required number is base 5 number.

First, apply Rule-1 i.e., hexadecimal to decimal conversion.

$$(2CCD)_{16} = 2 \times 16^3 + C \times 16^2 + C \times 16^1 + D \times 16^0$$

$$= 2 \times 16^3 + 12 \times 16^2 + 12 \times 16^1 + 13 \times 16^0$$

$$= (2 \times 4096) + (12 \times 256) + (12 \times 16) + (13 \times 1)$$

$$= 8192 + 3072 + 192 + 13$$

$$(2CCD)_{16} = (11469)_{10}.$$

Now, apply Rule-2 i.e., decimal number to other number system.

Base ÷	Decimal no.	Remainder
5	11469	4
5	2203	3
5	458	3
5	91	1
5	18	3
	3	

i.e. $(11469)_{10} = (331334)_5$

Hence, $(2CCD)_{16} = (11469)_{10} = (331334)_5.$ **(Ans.)**

(c) Given number is in fractional binary and required number is in octal. So apply Rule-2 i.e., from decimal to other number system.

$$(0.45)_{10} = (\quad)_8$$

Base	× Fractional part	Integer of product
8	0.45	3
8	0.60	4
8	0.8	6
8	0.4	3
	0.2	

i.e. $\quad (0.45)_{10} = (0.3463)_8.$ **(Ans.)**

(d) The given number is in octal and required number is decimal, so we apply Rule-1 other number to decimal number system.

$$(345)_8 = (3 \times 8^2) + (4 \times 8^1) + (5 \times 8^0)$$
$$= 192 + 32 + 5$$
$$= (229)_{10}$$

i.e. $\quad (345)_8 = (229)_{10}.$ **(Ans.)**

(c) In this problem, Rule-1 is applicable i.e., from other number system to decimal number system.

$$(7841)_9 = (7 \times 9^3) + (8 \times 9^2) + (4 \times 9^1) + (1 \times 9^0)$$
$$= (7 \times 729) + (8 \times 81) + (4 \times 9) + (1 \times 1)$$
$$= 5103 + 648 + 36 + 1$$
$$= (5788)_{10}$$

i.c. $\quad (7841)_9 = (5788)_{10}.$ **(Ans.)**

Example 9.2 Convert the following in to decimal.

(a) $(10110 \cdot 1101)_2$

(b) $(110101)_2$

(c) $(10111)_2$

Solution Given number are in binary and need to be converted in decimal. So apply Rule-1 i.e., from other to decimal number.

(a) $(10110.1101)_2 = 1 \times 2^4 + 0 \times 2^3 + 1 \times 2^2 + 1 \times 2^1 + 0 \times 2^0$
$$+ 1 \times 2^{-1} + 1 \times 2^{-2} + 0 \times 2^{-3} + 1 \times 2^{-4}$$
$$= (1 \times 16) + (0) + (1 \times 4) + (1 \times 2) + (0 \times 1)$$
$$+ (1 \times 0.5) + (1 \times 0.25) + (0 \times 0.125) + (1 \times 0.0625)$$
$$= 16 + 4 + 2 + 0.5 + 0.25 + 0.625$$
$$= (23.375)_{10}$$

i.e. $\quad (10110.1101)_1 = (23.375)_{10}.$ **(Ans.)**

(b) $(110101)_2 = (1 \times 2^5) + (1 \times 2^4) + (0 \times 2^3) + (1 \times 2^2) + (0 \times 2^1)$
$\qquad\qquad\quad + (1 \times 2^0)$
$\qquad\qquad = (1 \times 32) + (1 \times 16) + (0 \times 8) + (1 \times 4) + (0 \times 2) + (1 \times 1)$
$\qquad\qquad = 32 + 16 + 4 + 1$
$\qquad\qquad = (53)_{10}$

i.e. $(110101)_2 = (53)_{10}.$ **(Ans.)**

(c) $(10111)_2 = (1 \times 2^4) + (0 \times 2^3) + (1 \times 2^2) + (1 \times 2^1) + (1 \times 2^0)$
$\qquad\qquad = (1 \times 16) + (0 \times 8) + (1 \times 4) + (1 \times 2) + (1 \times 1)$
$\qquad\qquad = 16 + 0 + 4 + 2 + 1$
$\qquad\qquad = (23)_{10}$

i.e. $(10111)_2 = (23)_{10}.$ **(Ans.)**

Example 9.3 Convert the following numbers in the desired bases.

(a) $(4521)_{10} = (\ )_2$
(b) $(1201203)_3 = (\ )_{10}$

Solution (a) The given number is in decimal and required number is in binary Now, apply Rule-2 i.e., conversion from decimal to other number system.

Required base ÷	Decimal number	Remainder
2	4521	1
2	2260	0
2	1130	0
2	565	1
2	282	0
2	141	1
2	70	0
2	35	1
2	17	1
2	8	0
2	4	0
2	2	0
	1	

$$= 1000110101001$$

i.e. $\qquad (4521)_{10} = (1000110101001)_2.$ **(Ans.)**

(b) The given number is in base 3 number and required number is decimal. So, apply Rule-1 i.e., other number to decimal number.

$$(1201203)_3 = (1 \times 3^6) + (2 \times 3^5) + (0 \times 3^4) + (1 \times 3^3) + (2 \times 3^2)$$
$$+ (0 \times 3^1) + (3 \times 3^0)$$
$$= (1 \times 729) + (2 \times 243) + (0 \times 81)$$
$$+ (1 \times 27) + (2 \times 9) + (0 \times 3) + (3 \times 1)$$
$$= 729 + 486 + 0 + 27 + 18 + 0 + 3$$
$$= (1263)_{10}$$

i.e. $(1201203)_3 = (1263)_{10}$.   **(Ans.)**

Example 9.4   Add the following binary number

(a) $(101011)_2$ and $(110101)_2$

(b) $(11101)_2$ and $(110)_2$

Solution

(a)

$$
\begin{array}{r}
1\ 1\ 1\ 1\ 1 \leftarrow \text{carry} \quad \because\ 1 + 1 = 0 \text{ with carry } 1\\
1\ 0\ 1\ 0\ 1\ 1\\
+\ 1\ 1\ 0\ 1\ 0\ 1\\
\hline
1\ 1\ 0\ 0\ 0\ 0\ 0
\end{array}
$$

i.e. $(101011)_2 + (110101)_2 = (1100000)_2$.   **(Ans.)**

(b)

$$
\begin{array}{r}
1\ 1 \leftarrow \text{carry}\\
1\ 1\ 1\ 0\ 1\\
+\ 1\ 1\ 0\\
\hline
1\ 0\ 0\ 0\ 1\ 1
\end{array}
$$

i.e. $(11101)_2 + (110)_2 = (100011)_2$.   **(Ans.)**

Example 9.5   Divide the following,

(a) $(110110)_2$ by $(101)_2$

(b) $(11001)_2$ by $(110)_2$

Solution   (a)

$$
\begin{array}{r}
101 \leftarrow \text{Quotient}\\
110\ \overline{)\ 1\ 1\ 0\ 1\ 1\ 0}\\
1\ 0\ 1\\
\hline
1\ 1\ 1\\
1\ 0\ 1\\
\hline
1\ 0\ 0 \quad \text{— Remainder}
\end{array}
$$

Hence, $(110110)_2 \div (101)_2 = 101$ and remainder 100.   **(Ans.)**

(b)

$$110 \overline{\smash{\big)}\, 11001} \; \big(100$$
$$\underline{110}$$
$$01$$

Hence, $(11001)_2 \div (110)_2 = 100$ and remainder is 01.   **(Ans.)**

Example 9.6 Find the 9's complement of each of the following decimal number

(a) 16   (b) 45

(c) 459   (d) 2543

Solution As we know that to get a 9's complement, each digit of decimal number is subtracted from 9.

(a)   99
      $-\ 16$
     ————
      83 $\leftarrow$ 9's complement

(b)   99
      $-\ 45$
     ————
      54 $\leftarrow$ 9's complement

(c)   999
      $-\ 459$
     ————
      540 $\leftarrow$ 9's complement

(d)   9999
      $-\ 2543$
     ————
      7456 $\leftarrow$ 9's complement

Example 9.7 Subtrahend 245 from 158 using 9's complement and verify result

Solution The subtrahend is 245 (larger than minuend) 9's complement of subtrahend is

      999
    $-\ 245$
    ————
      754   $\leftarrow$ 9's complement of 245

The minuend is 158 (smaller than subtrahend)

      158 $\leftarrow$ minuend
    $+\ 754$ $\leftarrow$ 9's complement of subtrahend
    ————
      912 $\rightarrow$ No carry means that answer is negative and is complement form.

Hence,   999
                  $-\ 912$
                  ————
                  $-\ 87$ **(Ans.)**

Direct subtraction

$$
\begin{array}{r}
158 \\
-\ 245 \\
\hline
-\ 87 \\
\hline
\end{array}
$$

Hence, result obtained by 9's complement and direct subtraction is same so the result in verified.

Example 9.8   Convert the following bases:

(a) $(11011.011)_2 = (\ )_{16}$

(b) $(2AC9)_{16} = (\ )_7$

Solution   For conversion of binary to hexadecimal, Rule-3 is applicable.

(a) Making a group of four bits from left to right.

i.e. $(11011.011)_2 = \underline{0001}\ \underline{1011}\ .\ \underline{0110}$

$$= (1\ B \times 6)_{16}. \qquad \textbf{(Ans.)}$$

(b) First we will convert hexadecimal into decimal number by Rule-1

$$
\begin{aligned}
(2AC9)_{16} &= (2 \times 16^3) + (A \times 16^2) + (C \times 16^1) + (9 \times 16^0) \\
&= (2 \times 4096) + (10 \times 256) + (12 \times 16) + (9 \times 1) \\
&= 8192 + 2560 + 192 + 9 \\
&= (10953)_{10}
\end{aligned}
$$

i.e. $(2AC9)_{16} = (10953)_{10}.$

Now we will convert decimal number into base 7 by Rule-2.

Required base $\div$ Decimal number Remainder

7	10953	5 remainder
7	1564	3
7	223	6
7	31	3
	4	

i.e. $(10953)_{10} = (43635)_7$

Hence, $(2AC9)_{16} = (10953)_{10} = (43635)_7.$ \qquad **(Ans.)**

Example 9.9   Convert the following numbers.   **(UPTU 2006-07 (even))**

(a) $(6089.25)_{10} = (\ )_8$

(b) $(A6B.F5)_{16} = (\ )_2$

(c) $(375.37)_8 = (\ )_2$

Solution   (a) Now, to convert decimal to octal number, apply Rule-2.

For integer part of decimal number.

Base ÷ Decimal no. Remainder

8	6089	1
8	761	1
8	95	7
8	11	3
	1	

i.e. $(6089)_{10} = (13711)_8$.

For fractional part of decimal number

Base	Fractional no.	Integer of product
8	0.25	2
	0.00	

i.e. $(0.25)_{10} = (0.2)_8$.

Hence, $(6089.25)_{10} = (13711.2)_8$. **(Ans.)**

(b) For conversion of hexadecimal to binary, apply Rule-4.

$(A6B \cdot F5)_{16} = (\ )_2$

Hexadecimal no. → A 6 B · F 5

Binary No. →   1010   0110   1011   ·   1111   0101

i.e. $(A\,6\,B \times F\,5)_{16} = (1010\,0110\,1011 \cdot 1111\,0101)_2$. **(Ans.)**

(c) For conversion of octal into binary, apply Rule-4.

Octal no. →   3   5   7   ·   3   7

Binary no. →   011   101   111   ·   011   111

i.e. $(357.37)_8 = (011\,101\,111 \cdot 011\,111)_2$. **(Ans.)**

Example 7.10 Simplify the following numbers **(UPTU 2006-07 (even))**

(a) $(A4F \cdot EF)_{16} + (3FD \cdot AB)_{16}$

(b) $(6488 \cdot 43)_9 - (3837 \cdot 78)_9$

(c) $(FA12 \cdot 35)_{16} - (9BCD \cdot EC)_{16}$

Solution

(a) $$(A4F \cdot EF)_{16} = (A \times 16^2) + (4 \times 16^1) + (F \times 16^0)$$
$$+ (E \times 16^{-1}) + (F \times 16^{-2})$$
$$= (10 \times 256) + (4 \times 16) + (15 \times 1)$$
$$+ (14 \times 0.0625) + (15 \times 0.004)$$
$$= 2560 + 64 + 15 + 0.875 + 0.060$$
$$= (2639.935)_{10}$$

$$(3FD \cdot AB)_{16} = (3 \times 16^2) + (F \times 16^1) + (D \times 16^0) + (A \times 16^{-1})$$
$$+ (B \times 16^{-2})$$

$$= (3 \times 256) + (15 \times 16) + (13 \times 1) + (10 \times 0.0625)$$
$$+ (11 \times 0.004)$$
$$= 768 + 240 + 13 + 0.625 + 0.044$$
$$= (1021.669)_{10}.$$

i.e. $(A4F \cdot EF)_{16} + (3FD \cdot AB)_{16} = (2639.935)_{10} + (1021.669)_{10}$
$$= (3661.604)_{10}.$$

Now,

$$(33661.604)_{10} = (\quad)_{16}$$

Base ÷ Decimal no. Remainder

16	33661	13	= D
16	2103	7	
16	131	3	
	8		

i.e. $(837D)_{16}.$

For fractional part,

Base × Decimal no. Integer of product

16	0.604	9
16	0.664	10 = A
16	0.624	9
	0.984	

Hence, $(0.604)_{10} = (0.9A9)_{16}.$

Hence, $(A4F \times EF)_{16} + (3FD \times AB)_{16} = (837D \times 9A9)_{16}.$ **(Ans.)**

(b) First, convert from base 9 to base 10.

$$(6488.43)_9 = (6 \times 9^3) + (4 \times 9^2) + (8 \times 9^1) + (8 \times 9^0)$$
$$+ (4 \times 9^{-1}) + (3 \times 9^{-2})$$
$$= (6 \times 729) + (4 \times 81) + (8 \times 9) + (8 \times 1)$$
$$+ (4 \times 0.11) + (3 \times 0.012)$$
$$= 4374 + 324 + 72 + 8 + 0.44 + 0.036$$
$$= (4778.476)_{10}.$$

$$(3837.78)_9 = (3 \times 9^3) + (8 \times 9^2) + (3 \times 9^1) + (7 \times 9^0)$$
$$+ (7 \times 9^{-1}) + (8 \times 9^{-2})$$
$$= (3 \times 729) + (8 + 81) + (3 \times 9) + (7 \times 1)$$
$$+ (7 \times 0.11) + (8 \times 0.012)$$
$$= 2187 + 648 + 27 + 7 + 0.77 + 0.096$$
$$= (2869.866)_{10}$$

i.e. $\quad (6488.43)_9 - (3837.78)_9 = (4778.476)_{10} - (2869.866)_{10}$
$$= (1908.61)_{10}$$

Now, $(1908.61)_{10} = (\quad)_9$

Base ÷ Decimal no. Remainder

9	1908	0
9	212	5
9	23	5
	2	

i.e. $\quad (1908)_{10} = (2550)_9.$

For fractional part,

Base × Fractional Integer

9	0.61	5
9	0.49	4
9	0.41	3
	0.69	

i.e. $\quad (0.61)_{10} = (0.543)_9.$

Hence,

$$(6488.43)_9 - (3837.78)_9 = (2550.543)_9. \qquad \textbf{(Ans.)}$$

(c) First, convert hexadecimal into decimal number

$$(FA12.35)_{16} = (F \times 16^3) + (A \times 16^2) + (1 \times 16^1) + (2 \times 16^0)$$
$$+ (3 \times 16^{-1}) + (5 \times 16^{-2})$$
$$= (15 \times 4096) + (10 \times 256) + (1 \times 16) + (2 \times 1)$$
$$+ (3 \times 0.0625) + (5 \times 0.004)$$
$$= 61440 + 2560 + 16 + 2 + 0.1875 + 0.020$$
$$= (64018.2075)_{10}.$$

$$(9BCD \times EC)_{16} = (9 \times 16^3) + (B \times 16^2) + (C \times 16^1) + (D \times 16^0)$$
$$+ (E \times 16^{-1}) + (C \times 16^{-2})$$
$$= (9 \times 4096) + (11 \times 256) + (12 \times 16) + (13 \times 1)$$
$$+ (14 \times 0.062) + (12 \times 0.004)$$
$$= 36864 + 2816 + 192 + 13 + 0.868 + 0.048$$
$$= (39885.916)_{10}$$

i.e. $\quad (FA12.35)_{16} - (9BCD \times EC)_{16} = (64018.2075)_{10} - (39885.916)_{10}$
$$= (24132.292)_{10}.$$

Now, convert decimal number into hexadecimal,

For Integer part,

Base ÷	Decimal number	Remainder
16	24132	4
16	1508	4
16	94	14 = E
	5	

i.e. $\qquad (24132)_{10} = (5E44)_{16}$.

For fractional part,

Base ×	fractional part	Integer of product
16	0.292	4
16	0.672	10 = A
16	0.752	12 = C

i.e. $\qquad (0.292)_{10} = (0.4AC)_{16}$

Hence, $(FA12.35)_{16} - (9BCD \times EC)_{16} = (5E44 \times 4AC)_{16}$. **(Ans.)**

Example 9.11 Convert the following:

(a) $(1010.101)_{10} \rightarrow (\ )_2 \rightarrow (\ )_4$

(b) $(C3A \times 47)_{16} \rightarrow (\ )_8 \rightarrow (\ )_2$ $\qquad$ **(UPTU 2007-08 (odd))**

Solution (a) First convert decimal number into binary, apply Rule-2.

For Integer part only,

Required base ÷	Decimal number	Remainder
2	1010	0
2	505	1
2	252	0
2	126	0
2	63	1
2	31	1
2	15	1
2	7	1
2	3	1
	1	

i.e. $\qquad (1010)_{10} = (1111110010)_2$.

For fractional part,

Required base × Fractional part Integer of product

2	0.101	0
2	0.202	0
2	0.404	0
2	0.808	1
	0.616	

i.e. $(0.101)_{10} = (0.0001)_2$

Hence, $(1010.101)_{10} = (1111110010.0001)_2$.

Now, convert binary number into base 4 number. Apply Rule-3. i.e., making a group of 2 from left to right.

$$= (\underbrace{11}\ \underbrace{11}\ \underbrace{11}\ \underbrace{11}\ \underbrace{00}\ \underbrace{10}\ .\ \underbrace{00}\ \underbrace{01})_2$$
$$= (333302.01)_4. \tag{Ans.}$$

(b) Convert hexadecimal to octal number, apply Rule-5.

First apply Rule-4.

Hexadecimal → C 3 A ·   4   7

Binary →   1100   0011   1010   ·   0100   0111

Now apply Rule-3 making a group of 3-bits

$$= (\underbrace{110}\ \underbrace{000}\ \underbrace{111}\ \underbrace{010}\ .\ \underbrace{010}\ \underbrace{001}\ \underbrace{110})_2$$
$$(6\ \ 0\ \ 7\ \ 2\ ·\ 2\ \ 1\ \ 6)_8$$
$$= (6\ 0\ 7\ 2\ ·\ 2\ 1\ 6)_8.$$

Now, convert octal to binary, apply Rule-3

Octal →   6   0   7   2   ·   2   ·   1   6

Binary → $(110\ \ 000\ \ 111\ \ 010\ ·\ 010\ ·\ 001\ \ 110)_2$

i.e. $(6072.216)_8 = (110\ 000\ 111\ 010\ ·\ 010\ ·\ 001\ 110)_2.$ **(Ans.)**

Example 9.12   Convert the following: **(UPTU 2007-08 (even))**

(a) $(2D6)_{16} = (\ )_2$

(b) $(011010110)_2 = (\ )_{16}$

Complete the following operation:

(c) $(8)_{16} + (F)_{16} = (\ )_{16}.$

(d) $00010100 + 0111001 = (\ )_2$

(e) $01001111 - 0000101 = (\ )_2$

Solution   (a) $(2D6)_{16} = (\ )_2$

Hexadecimal →   2   D   6

Binary →   0010   1101   0110

i.e. $(2D6)_{16} = (001011010110)_2.$ **(Ans.)**

(b) $(011010110)_2 = (\ )_{16}$

$$\left(\underbrace{0000}\ \ \underbrace{1101}\ \ \underbrace{0110}\right)_2 = (0D6)_{16}.$$

$$\Rightarrow\qquad 0\qquad D\qquad 6$$ (**Ans.**)

(c) $(8)_{16} + (F)_{16} = (17)_{16}.$ (**Ans.**)

(d) $0001\ 0100 + 0011\ 1001 = (\ )_2.$

$$\begin{array}{r} 0001\ 0100 \\ +\ 0011\ 1001 \\ \hline 0100\ 1101 \\ \hline \end{array}$$

i.e. $(0001\ 0100 + 0011\ 1001)_2 = (01001101)_2.$ (**Ans.**)

(e) $(0100\ 1111 - 0000\ 0101)_2 = (\ )_2$

$$\begin{array}{r} 0100\ 1111 \\ -\ 0000\ 0101 \\ \hline 0100\ 1010 \\ \hline \end{array}$$

i.e. $(01001111 - 00000101)_2 = (01001010)_2.$

Example 9.13 (a) Add and subtract without converting the following two hexadecimal numbers $A4FB$ and $3FDC$.

(b) Convert the following numbers

(i) $(6089 \cdot 25)_{10} = (\ )_8$

(ii) $(A6BF \cdot 5)_{16} = (\ )_2$

(iii) $(25 \cdot 26)_8 = (\ )_2$ **(UPTU 2008-09 (even))**

Solution   (a) Addition of two hexadecimal numbers.

$$\begin{array}{r} 1\ 1\ 1 \\ A\ 4\ F\ B \\ +\ 3\ F\ D\ C \\ \hline E\ 4\ D\ 7 \\ \hline \end{array}$$

Subtracting the hexadecimal number

$$\begin{array}{r} A\ 4\ F\ B \\ -\ 3\ \ \ F\ D\ C \\ \hline 6\ 5\ 1\ F \\ \hline \end{array}$$

(b) (i) $(6089.25)_{10} = (\ )_8$

For Integer part,

Base ÷	Decimal	Remainder
8	6089	1
8	761	1
8	95	7
8	11	3
	1	

i.e. $(6089)_{10} = (13711)_8$

For fractional part

Base ×	Fractional part	Product of integer
8	0.25	2
	0.00	

i.e. $(0.25)_{10} = (0.2)_8$

Hence, $(6089.25)_{10} = (13711.2)_8$.

(ii) $(A6BF \cdot 5)_{16} = (\ )_2$

Hexadecimal → $\quad A \quad\quad 6 \quad\quad B \quad\quad F \quad \cdot \quad 5$

Binary → $\quad 1010 \quad 0110 \quad 1011 \quad 1111 \cdot 0101$

i.e. $(A\,6\,B\,F \cdot 5)_{16} = (1010\ 0110\ 1011\ 1111 \cdot 0101)_2$. **(Ans.)**

(iii) $(25.26)_8 = (\ )_2$

Octal number → $\quad 2 \quad 5 \quad \cdot \quad 2 \quad 6$

Binary → $\quad 010\,101 \quad \cdot \quad 010 \quad 110$

i.e. $(25.26)_8 = (010101 \cdot 010110)_2$. **(Ans.)**

Example 9.14 (a) Convert $(725.25)_{10}$ to its equivalent in base 2, base 8, and base 16.

(b) Per for $M - N$ and $M + N$ if $M = 10101$ and $N = 1111$.

(UPTU 2008-09 (even))

Solution (a) $(725.25)_{10} = (\ )_2$

Base ÷	Decimal	Reminder
2	725	1
2	362	0
2	181	1
2	90	0
2	45	1
2	22	0
2	11	1
2	5	1
2	2	0

Base	× Fractional part	Integer of product
2	0.25	0
2	0.50	1
	0.00	

i.e. $(725.25)_{10} = (1011010101 \times 01)_2.$ **(Ans.)**

Decimal to octal conversion,

Base	÷ Decimal	Remainder
8	725	5
8	90	2
8	11	3
	1	

Base	× Fractional part	Integer of product
8	0.25	2
	0.00	

i.e. $(725.25)_{10} = (1325.2)_8.$ **(Ans.)**

Decimal to hexadecimal conversion,

Base	÷ Decimal	Remainder
16	725	5
16	45	D
	2	

Base	× Fractional part	Integer of product
16	0.25	4
	0.00	

i.e. $(725.25)_{10} = (2D5 \times 4)_{16}.$ **(Ans.)**

(b)

$$M - N = \begin{array}{r} 10101 \\ -\ 1111 \\ \hline 0110 \end{array}$$ **(Ans.)**

$$M + N = \begin{array}{r} 10101 \\ +\ 1111 \\ \hline 100100 \end{array}$$ **(Ans.)**

Example 9.15 Find the 10's complement of each of the following decimal number.

   (a) 25   (b) 462   (c) 2421

Solution   (a) First, we find 9's complement

$$
\begin{array}{r}
9\ 9 \\
-\ 2\ 5 \\
\hline
7\ 4
\end{array}
\quad \rightarrow \text{9's complement of 25}
$$

Now, add 1 in 9's complement

$$
\begin{array}{r}
7\ 4 \\
+\ 1 \\
\hline
7\ 5
\end{array}
\quad \rightarrow \text{10's complement}
$$

Hence, 10's complement of 25 is 75.   **(Ans.)**

(b)
$$
\begin{array}{r}
9\ 9\ 9 \\
-\ 4\ 6\ 2 \\
\hline
5\ 3\ 7
\end{array}
\quad \rightarrow \text{9's complement of 462}
$$

Now, add 1 in 9's complement

$$
\begin{array}{r}
537 \\
+\ 1 \\
\hline
538
\end{array}
\quad \rightarrow \text{10's complement}
$$

Hence, 10's complement of 462 is 538.   **(Ans.)**

(c)
$$
\begin{array}{r}
9\ 9\ 9\ 9 \\
-\ 2\ 4\ 2\ 1 \\
\hline
7\ 5\ 7\ 8
\end{array}
\quad \rightarrow \text{9's complement of 2421}
$$

Now, add 1 in 9's complement

$$
\begin{array}{r}
7\ 5\ 7\ 8 \\
+\quad\ 1 \\
\hline
7\ 5\ 7\ 9
\end{array}
\quad \rightarrow \text{10's complement}
$$

Hence, 10's complement of 2421 is 7579.   **(Ans).**

Example 9.16 Subtract 125 from 239 using 10's complement and verify the result by direct subtraction.

Solution   The subtrahend is 125.

   Now, we will find the 10's complement of subtrahend.

$$9\ 9\ 9$$
$$\underline{-\ 1\ 2\ 5}\quad \rightarrow \text{substrahend}$$
$$8\ 7\ 4\quad \rightarrow 9\text{'s complement}$$
$$\underline{+\ 1}$$
$$\underline{8\ 7\ 5}\quad \rightarrow 10\text{'s complement.}$$

The minuend is 239, now add minuend with 10's complement of subtrahend

$$2\ 3\ 9$$
$$\underline{+\ 8\ 7\ 5}$$
$$\text{neglecting carry} \rightarrow 1\ 1\ 1\ 4$$

Hence, result is 114.

Direct subtraction.

$$2\ 3\ 9$$
$$\underline{-\ 1\ 2\ 5}$$
$$\underline{1\ 1\ 4}$$

Hence, result obtained by 10's complement and direct subtraction is same, so the result is verified.

Example 9.17   Find 1's complement of binary number 1001011.

Solution   Binary No. $\rightarrow$ 1001011

1's complement $\rightarrow$ 0110100

i.e. 1's complement of binary number 1001011 is 0110100.   **(Ans.)**

Example 9.18   Subtract $(10011)_2$ from $(11001)_2$ using 1's complement and verify the result by direct subtraction.

Solution   The subtrahend is $(10011)_2$ i.e., smaller than minuend.

The 1's complement of subtrahend is

subtrahend $\rightarrow (10011)_2$

1's complement $\rightarrow (01100)_2$.

The minuend is $(11001)_2$ i.e., larger than subtrahend

i.e.
$$1\ 1\ 0\ 0\ 1\quad \rightarrow \text{minuend}$$
$$\underline{+\ 0\ 1\ 1\ 0\ 0}\quad \rightarrow 1\text{'s complement of subtrahend}$$
$$\text{carry} \leftarrow \textcircled{1}0\ 0\ 1\ 0\ 1$$
$$\underline{+\ 1}\quad \rightarrow \text{adding carry to result}$$
$$\underline{0\ 0\ 1\ 1\ 0}$$

Direct subtraction

$$\begin{array}{r} 1\,1\,0\,0\,1 \\ -\,1\,0\,0\,1\,1 \\ \hline 0\,0\,1\,1\,0 \end{array}$$

(Ans.)

Example 9.19 Subtract $(11001)_2$ from $(10011)_2$ using 1's complement and verify the result by direct subtraction.

Solution The subtrahend is $(11001)_2$ i.e., larger than minuend.

The 1's complement of subtrahend is

subtrahend $\rightarrow$ $(11001)_2$

1's complement $\rightarrow$ $(00110)_2$.

The minuend is $(10011)_2$ i.e., smaller than subtrahend.

i.e.

$$\begin{array}{ll} 1\,0\,0\,1\,1 & \rightarrow \quad \text{minuend} \\ +\,0\,0\,1\,1\,0 & \rightarrow \quad \text{1's complement of subtrahend} \\ \hline 1\,1\,0\,0\,1 & \rightarrow \quad \text{No carry means that the answer is} \\ & \qquad\quad \text{negative and in complement form.} \end{array}$$

Hence, 0 0 1 1 0.

(Ans.)

Direct subtraction

$$\begin{array}{r} 1\,0\,0\,1\,1 \\ -\,1\,1\,0\,0\,1 \\ \hline -\,0\,0\,1\,1\,0 \end{array}$$

(Ans.)

Hence, the result obtained by 1's complement and direct subtraction is same, so the result is verified.

Example 9.20 Find the 2's complement of 110001.

Solution First, we find 1's complement

Binary No. $\rightarrow$ 1   1   0   0   0   1

1's complement $\rightarrow$ 0   0   1   1   1   0

adding 1 is 1's complement. We find 2's complement

$$\begin{array}{r} 0\,0\,1\,1\,1\,0 \\ +\,1 \\ \hline 0\,0\,1\,1\,1\,1 \end{array}$$

Hence, 2's complement of 110001 is 1111.

(Ans.)

Example 9.21 Subtract $(1011)_2$ from $(1100)_2$ using 2's complement method. Also show direct subtraction for comparison.

Solution The subtrahend is $(1011)_2$

Now, we will find the 2's complement of subtrahend.

$$1\ 0\ 1\ 1\ \leftarrow \text{subtrahend}$$
$$0\ 1\ 0\ 0\ \leftarrow \text{1's complement of subtrahend}$$
$$+\ 1\ \leftarrow \text{adding 1}$$
$$0\ 1\ 0\ 1\ \leftarrow \text{2's complement of subtrahend.}$$

The minuend is $(1100)_2$. Now, adding 2's complement of subtrahend with minuend.

$$1\ 1\ 0\ 0$$
$$+\ 0\ 1\ 0\ 1$$
$$\text{neglecting carry} \leftarrow ①0\ 0\ 1$$

Hence, the result is 0001.

Direct subtraction

$$1\ 1\ 0\ 0$$
$$-\ 1\ 0\ 1\ 1$$
$$0\ 0\ 0\ 1$$

Hence, the result is verified.

Example 9.22   Simplify the expression
$$AB\overline{C} + AB\overline{C}D + A\overline{C}.$$

Solution $AB\overline{C} + AB\overline{C}D + A\overline{C}$

$$= AB\overline{C}\,(1 + D) + A\overline{C}$$
$$= AB\overline{C} \cdot 1 + A\overline{C} \quad \because \ (D + 1) = D$$
$$= AC\,(B + 1)$$
$$= A\overline{C} \qquad \because \ B + 1 = 1 \qquad \textbf{(Ans.)}$$

Example 9.23   Simplify the logical expression by Boolean algebra
$$(A + B) \cdot (\overline{A} + B) \cdot (A + \overline{B}).$$

Solution $(A + B) \cdot (\overline{A} + B) \cdot (A + \overline{B})$

$$= (A + B) \cdot (A\overline{A} + \overline{A}B + BA + B\overline{B})$$
$$\because \qquad A\overline{A} = 0 \quad \text{and} \quad B\overline{B} = 0$$
$$\therefore \qquad = (A + B)\,(0 + \overline{A}B + AB + 0)$$
$$= (A + B)\,(\overline{A}B + AB)$$
$$= A\overline{A}B + A \cdot A \cdot B + \overline{A}BB + A \cdot B \cdot B$$
$$\because \qquad A\overline{A} \times B\overline{B} = 0$$
$$\therefore \qquad = 0 \cdot \overline{B} + A \cdot A \cdot B + \overline{A} \cdot 0 + A \cdot B \cdot B$$

$$= A \cdot A \cdot B + A \cdot B \cdot B$$

$$\because \qquad A \cdot A = A \times B \cdot B = B$$

$$= AB + AB$$

$$= AB$$

$$\because \qquad A + A = A. \qquad \text{(Ans.)}$$

Example 9.24 Simplify the logical expression by Boolean algebra

$$Y = A\overline{B}\,\overline{C}D + \overline{A}\,\overline{B}D + BC\overline{D} + \overline{A}B + B\overline{C}.$$

$$\text{(UPTU 2006-07 (even))}$$

Solution $Y = A\overline{B}\,\overline{C}D + \overline{A}\,\overline{B}D + BC\overline{D} + \overline{A}B + B\overline{C}$

$$= \overline{B}D\,(A\overline{C} + \overline{A}) + B\,(C\overline{D} + \overline{C}) + \overline{A}B$$

$$= \overline{B}D\,(A + \overline{A})(\overline{C} + \overline{A}) + B\,(C + \overline{C})\,(\overline{D} + \overline{C}) + \overline{A}B$$

$$\because \qquad A + \overline{A} = 1 \quad \text{and} \quad C + \overline{C} = 1$$

$$= \overline{B}D\,(\overline{C} + \overline{A}) + B\,(\overline{D} + \overline{C}) + \overline{A}B$$

$$= \overline{B}\,\overline{C}D + \overline{A}\,\overline{B}D + B\overline{D} + B\overline{C} + \overline{A}B$$

$$= \overline{A}\,(\overline{B}D + B) + \overline{C}\,(B + \overline{B}D) + B\overline{D}$$

$$= \overline{A}\,(\overline{B} + B)\,(D + B) + \overline{C}\,(B + \overline{B})\,(B + D) + B\overline{D}$$

$$\because \qquad \overline{B} + B = 1$$

$$\therefore \qquad = \overline{A}\,(B + D) + \overline{C}\,(B + D) + B\overline{D}$$

$$= (B + D)\,(\overline{A} + \overline{C}) + B\overline{D}. \qquad \text{(Ans.)}$$

Example 9.25 Simplify the Boolean expression

$$Y = (A + \overline{B})\,C + ABC.$$

Solution $Y = (A + \overline{B})\,C + ABC$

$$= AC + \overline{B}C + ABC$$

$$= \overline{B}C + AC\,(1 + B)$$

$$\because \qquad 1 + B = 1$$

$$\therefore \qquad = \overline{B}C + AC$$

$$Y = (\overline{B} + A)\,C$$

$$Y = (A + \overline{B})\,C. \qquad \text{(Ans.)}$$

Example 9.26 Simplify the Boolean expression

$$Y = AB + A\,(B + C) + B\,(B + C).$$

$$\text{(UPTU 2006-07 (even))}$$

Solution $Y = AB + A(B + C) + B\,(B + C)$

$$= AB + AB + AC + B \cdot B + BC$$

$$\because \qquad A + A = 1 \quad \text{and} \quad B \cdot B = B$$

$$Y = AB + AC + B + BC$$

$$= AB + AC + B(1 + C)$$

$$= AB + AC + B \qquad \because \quad C + 1 = 1$$

$$= B (A + 1) + AC \qquad \because \quad A + 1 = 1$$

$$= B + AC. \qquad \textbf{(Ans.)}$$

Example 9.27   Simplify the logical expression

$$Y = \overline{\overline{(\overline{A} + B)} + \overline{(\overline{A} \cdot B)} + \overline{(A \cdot \overline{B})}}.$$

Solution $Y = \overline{\overline{(\overline{A} + B)} + \overline{(\overline{A} \cdot B)} + \overline{(A \cdot \overline{B})}}$

Applying De-Morgan's theorem,

$$Y = \overline{\overline{(\overline{A} + B)}} \cdot \overline{\overline{(\overline{A} \cdot B)}} \cdot \overline{\overline{(A \cdot \overline{B})}}$$

$$= (\overline{A} + B) \cdot (\overline{A} \cdot B) \cdot (A \cdot \overline{B}) \qquad \because \quad \overline{\overline{A}} = A$$

$$= (\overline{A} + B) \cdot (\overline{A}A \cdot B\overline{B}) \qquad \because \quad B\overline{B} = A\overline{A} = 0$$

$$= (\overline{A} + B) \cdot 0$$

$$= 0. \qquad \textbf{(Ans.)}$$

Example 9.28   Convert the following function into canonical forms

$$Y = AB + AC + AD + BCD.$$

(UPTU 2007-08 (even))

Solution   Given expression is in the sum of product form.

$$Y = AB + AC + AD + BCD$$

$$Y = AB (C + \overline{C}) (D + \overline{D}) + AC (B + \overline{B}) (D + \overline{D})$$

$$+ AD (B + \overline{B}) (C + \overline{C}) + BCD (A + \overline{A})$$

$$= AB (CD + C\overline{D} + \overline{C}D + \overline{C}\overline{D}) + AC (BD + B\overline{D} + \overline{B}D + \overline{B}\overline{D})$$

$$+ AD (BC + B\overline{C} + \overline{B}C + \overline{B}\overline{C}) + BCD (A + \overline{A})$$

$$= \underline{ABCD} + \underline{ABC\overline{D}} + \underline{AB\overline{C}D} + AB\overline{C}\overline{D} + \underline{ABCD} + \underline{ABC\overline{D}}$$

$$+ \underline{A\overline{B}CD} + \underline{A\overline{B}C\overline{D}} + \underline{ABCD} + \underline{AB\overline{C}D} + \underline{A\overline{B}CD} + A\overline{B}\,\overline{C}D$$

$$+ \underline{ABCD} + \overline{A}BCD$$

$$= ABCD + ABC\overline{D} + AB\overline{C}D + AB\overline{C}\overline{D} + A\overline{B}CD + A\overline{B}\,CD$$

$$+ A\overline{B}\,\overline{C}D + \overline{A}BCD. \qquad \textbf{(Ans.)}$$

Example 9.29   Simplify the logical expression

$$Y = C (B + C) (A + B + C) \qquad \textbf{(Pune University)}$$

$$\begin{aligned}
\textbf{Solution} \quad Y &= C\,(B + C)\,(A + B + C) \\
&= (CB + C \cdot C)\,(A + B + C) \\
&= (BC + C)\,(A + B + C) \qquad \because \quad C \cdot C = C \\
&= C\,(B + 1)\,(A + B + C) \qquad \because \quad B + 1 = 1 \\
&= C\,(A + B + C) \\
&= AC + BC + C \cdot C \qquad\qquad \because \quad C \cdot C = C \\
&= AC + BC + C \\
&= AC + C\,(B + 1) \\
&= AC + C \qquad\qquad\qquad \because \quad C + 1 = 1 \\
&= C\,(A + 1) \\
&= C. \qquad\qquad\qquad\qquad \because \quad A + 1 = 1 \quad \textbf{(Ans.)}
\end{aligned}$$

Example 9.30 Minimise the following function using Boolean algebra

$$Y = \overline{A}BCD + AB\overline{C}\,\overline{D} + AB\overline{C}D + ABCD + ABC\overline{D} + A\overline{B}\,\overline{C}D + \overline{A}\overline{B}CD + A\overline{B}CD.$$

(UPTU 2007-08 (even))

Solution

$$\begin{aligned}
Y &= \overline{A}BCD + ABCD + AB\overline{C}\,\overline{D} + AB\overline{C}D + ABC\overline{D} \\
&\quad + A\overline{B}\,\overline{C}D + A\overline{B}\,\overline{C}D + A\overline{B}CD \\
&= BCD\,(\overline{A} + A) + AB\overline{C}\,(\overline{D} + D) + AC\overline{D}\,(B + \overline{B}) \\
&\quad + A\overline{B}D\,(\overline{C} + C) \\
\because \quad A + \overline{A} &= B + \overline{B} = C + \overline{C} = D + \overline{D} = 1 \\
\therefore \qquad Y &= BCD + AB\overline{C} + AC\overline{D} + A\overline{B}D. \qquad \textbf{(Ans.)}
\end{aligned}$$

Example 9.31 Simplify the logical expression in,

$$Y = AB + AC + A\overline{D}$$

canonical form.

(UPTU 2008-09 (add))

$$\begin{aligned}
\textbf{Solution} \quad Y &= AB + AC + A\overline{D} \\
Y &= AB\,(C + \overline{C})\,(D + \overline{D}) + AC\,(B + \overline{B})\,(D + \overline{D}) \\
&\quad + A\overline{D}\,(B + \overline{B})\,(C + \overline{C}) \\
&= AB\,(CD + C\overline{D} + \overline{C}D + \overline{C}\,\overline{D}) + AC\,(BD + B\overline{D} + \overline{B}D + \overline{B}\,\overline{D}) \\
&\quad + A\overline{D}\,(BC + B\overline{C} + \overline{B}C + \overline{B}\,\overline{C}) \\
&= ABCD + ABC\overline{D} + AB\overline{C}D + AB\overline{C}\,\overline{D} + ABCD + ABC\overline{D} \\
&\quad + A\overline{B}CD + A\overline{B}C\overline{D} + ABC\overline{D} + AB\overline{C}\,\overline{D} + A\overline{B}C\overline{D} + A\overline{B}\,\overline{C}\,\overline{D} \\
&= ABCD + ABC\overline{D} + AB\overline{C}D + AB\overline{C}\,\overline{D} + A\overline{B}CD + A\overline{B}C\overline{D} \\
&\quad + A\overline{B}\,\overline{C}\,\overline{D}. \qquad\qquad\qquad\qquad\qquad \textbf{(Ans.)}
\end{aligned}$$

Example 9.32   Simplify the logical expression
$$Y = A + AB + A\overline{B}C.$$

Solution $Y = A + AB + A\overline{B}C$

$$= A(1 + B) + A\overline{B}C$$

$$= A + A\overline{B}C \qquad \because \quad B + 1 = 1$$

$$= (A + A) \cdot (A + \overline{B}C) \quad \text{Distributive law}$$

$$= A \cdot (A + \overline{B}C) \qquad \because \quad A + A = A$$

$$= A \cdot A + A\overline{B}C$$

$$= A + A\overline{B}C$$

$$= A(1 + \overline{B}C)$$

$$= A \cdot 1 \qquad\qquad \because \quad 1 + A = 1$$

$$= A. \qquad\qquad\qquad\qquad \textbf{(Ans.)}$$

Example 9.33   Simplify the logical expression
$$Y = \overline{(\overline{B} \cdot C \cdot D)} + \overline{(A \cdot \overline{C} D)}.$$

Solution $Y = \overline{(\overline{B} \cdot C \cdot D)} + \overline{(A \cdot \overline{C} D)}$

Applying De-Morgan's theorem,

$$Y = \overline{(\overline{B} \cdot C \cdot D)} \cdot \overline{(A \cdot \overline{C} D)}$$

$$= (\overline{\overline{B} \cdot C} + \overline{D}) \cdot \overline{(A \cdot \overline{C} D)}$$

$$= (\overline{B}C + \overline{D}) \cdot (A \cdot \overline{C} D)$$

$$= A\overline{B} \, C\overline{C} \, D + A \, D\overline{D} \, \overline{C}$$

$$\because \qquad C\overline{C} = 0 = D\overline{D}$$

$$= 0 + 0 = 0. \qquad\qquad \textbf{(Ans.)}$$

Example 9.34   Simplify the following logical expression by Boolean algebra
$$Y = (A + B + C) \cdot (A + B + C) \cdot (A + B + C).$$

Solution $Y = (A + B + C) \cdot (A + B + \overline{C}) \cdot (\overline{A} + B + C)$

$$= (A \cdot A + AB + A\overline{C} + BA + B \cdot B + B\overline{C} + AC + BC + C\overline{C})$$

$$\cdot (\overline{A} + B + C)$$

$$= (A + AB + A\overline{C} + AC + B + B\overline{C} + BC + 0) \cdot (\overline{A} + B + C)$$

$$= (A + AB + AC \, C + \overline{C}) + B (C + \overline{C}) + B) \cdot (\overline{A} + B + C)$$

$$= (A + B + AB + A + B) \cdot (\overline{A} + B + C)$$

$$\because \qquad\quad C + \overline{C} = 1$$

$$= (A + B + AB) \cdot (\overline{A} + B + C)$$

$$= (A + B (1 + A)) \cdot (\overline{A} + B + C)$$

$\because \qquad\quad A + 1 = 1$

$$= (A + B) (\overline{A} + B + C)$$

$$= A\overline{A} + AB + AC + \overline{A}B + B \cdot B + BC$$

$$= 0 + AB + AC + \overline{A}B + B + BC$$

$$= AB + \overline{A}B + AC + BC + B$$

$$= B (A + \overline{A}) + AC + B (C + 1)$$

$$= B \cdot 1 + AC + B \cdot 1$$

$$= AC + B. \qquad\qquad\qquad\text{(Ans.)}$$

Example 9.35 Simplify the following Boolean function

$$Y = A\overline{B}\,\overline{C} + \overline{A}\,\overline{B}\,C + \overline{A}B\overline{C} + \overline{A}\,\overline{B}\,C + \overline{A}BC.$$

Solution $\quad Y = \overline{A}BC + A\overline{B}\,\overline{C} + \overline{A}\,\overline{B}\,C + \overline{A}B\overline{C} + \overline{A}\,\overline{B}\,C$

$$= \overline{A}BC + A\overline{B}\,\overline{C} + \overline{A}\,\overline{B}\,C + \overline{A}B\overline{C} + \overline{A}\,\overline{B}\,C + \overline{A}BC + \overline{A}\,\overline{B}\,\overline{C}$$

$\because \qquad\qquad A + A + A = A.$

$$Y = \overline{A} (BC + \overline{B}\,\overline{C}) + \overline{B}\,\overline{C} (A + \overline{A}) + \overline{A}\,\overline{C} (B + \overline{B}) + \overline{A}\,\overline{B} (C + \overline{C})$$

$\because \qquad\quad A + \overline{A} = 1$

$\therefore \qquad\quad Y = \overline{A} (BC + \overline{B}\,\overline{C}) + \overline{B}\,\overline{C} \cdot 1 + \overline{A}\,\overline{C} \cdot 1 + \overline{A}\,\overline{B} \cdot 1$

$$= \overline{A}BC + \overline{B}\,\overline{C} + \overline{A}\,\overline{C} + \overline{A}\,\overline{B}$$

$$= \overline{A}BC + \overline{B}\,\overline{C} + \overline{A}\,\overline{C} + \overline{A}\,\overline{B}. \qquad\qquad\text{(Ans.)}$$

Example 9.36 Simplify the following expression

$$Y = A\overline{B} + \overline{A + C} + B.$$

Solution $\quad Y = \overline{AB} + \overline{A + C} + B$

$$Y = \overline{A} + \overline{B} + \overline{A} + \overline{C} + B \quad \text{By De-Morgan's theorem}$$

$$= \overline{A} + \overline{B} + \overline{A} \cdot \overline{C} + B$$

$$= \overline{A} + \overline{A} \cdot \overline{C} + (B + \overline{B})$$

$$= \overline{A} (1 + \overline{C}) + 1 \qquad \because \quad B + \overline{B} = 1$$

$$= \overline{A} \cdot 1 + 1 \qquad\qquad \because \quad 1 + \overline{C} = 1$$

$$= 1 + \overline{A} \qquad\qquad\quad \because \quad 1 + \overline{A} = 1$$

$$Y = 1. \qquad\qquad\qquad\qquad\qquad\text{(Ans.)}$$

Example 9.37 Simplify the following Boolean function

$$Y = \overline{A}\,\overline{B}\,\overline{C} + \overline{A}\,\overline{B}\,C + A\overline{B}\,\overline{C} + AB\overline{C}.$$

Solution $Y = \overline{A}\,\overline{B}\,\overline{C} + \overline{A}\,BC + A\overline{B}\,\overline{C} + AB\overline{C}$

$$= \overline{A}\,\overline{B}\,(C + \overline{C}) + A\overline{C}\,(B + \overline{B})$$

$$= \overline{A}\,\overline{B} \cdot 1 + A\overline{C} \cdot 1 \quad \because \quad C + \overline{C} = 1$$

$$Y = \overline{A}\,\overline{B} + A\overline{C}. \tag{Ans.}$$

Example 9.38 Prove that the expression

$$(A + B)\,(\overline{A}\,\overline{C} + C)\,(\overline{\overline{B} + AC}) = \overline{A}B.$$

Solution $= (A + B)\,(\overline{A}\,\overline{C} + C)\,(\overline{\overline{B}} \cdot \overline{AC})$. Applying DeMorgan theorem

$$= (A + B)\,(\overline{A}\,\overline{C} + C)\,(B \cdot \overline{AC})$$

$$= (A\overline{A}\,\overline{C} + AC + \overline{A}B\overline{C} + BC)\,[B \cdot (\overline{A} + \overline{C})] \text{ DeMorgan theorem}$$

$$= (0 + AC + \overline{A}B\overline{C} + BC)\,(B \cdot \overline{A} + B\overline{C}) \quad \because \quad A\overline{A} = 0$$

$$= ACB\overline{A} + ACB\overline{C} + \overline{A}B\overline{C}\,B\overline{A} + \overline{A}B\overline{C}\,B\overline{C} + BCB\overline{A} + BCB\overline{C}$$

$$= 0 + \overline{A}B\overline{C} + \overline{A}B\overline{C} + \overline{A}BC + 0 \quad \because \quad A\overline{A} = 0$$

$$= \overline{A}B\overline{C} + \overline{A}B\overline{C} + \overline{A}BC$$

$$= \overline{A}B\overline{C} + \overline{A}BC$$

$$= \overline{A}B\,(C + \overline{C})$$

$$= \overline{A}B \cdot 1 \quad\quad \because \quad C + \overline{C} = 1$$

$$= \overline{A}B \tag{Ans.}$$

Hence proved.

Example 9.39 Prove that the expression

$$AB + \overline{AC} + AB\overline{C}\,(AC + B) = AB + \overline{A} + \overline{C}.$$

Solution $AB + \overline{AC} + AB\overline{C}\,(AC + B)$

$$= AB + \overline{AC} + AB\overline{C} \cdot AC + AB\overline{C}B$$

$$= AB + \overline{AC} + 0 + AB\overline{C} \quad\quad \because \quad C\overline{C} = 0 \quad \text{and} \quad B \cdot B = B$$

$$= AB + AB\overline{C} + A\overline{C}$$

$$= AB\,(1 + \overline{C}) + A\overline{C}$$

$$= AB + A\overline{C} \quad\quad \because \quad \overline{C} + 1 = 1$$

$$= AB + \overline{A} + \overline{C}. \tag{Ans.}$$

Hence proved.

Example 9.40 Simplify the following Boolean function

(a) $f = B \cdot (A + B)$

(b) $f = \overline{A} + \overline{B} + A \cdot B \cdot \overline{C}$

(c) $f = \overline{A}\,\overline{B}\,\overline{C}D + \overline{A}\,\overline{B}\,CD$

(d) $f = \overline{A}C + \overline{A}B + A\overline{B}C + BC$ **(UP Teach Tutorial Question Bank)**

Solution   (a) $f = B \times (A + B)$

$$= B \cdot A + B \cdot B$$
$$= B \cdot A + B$$
$$= B (A + 1) \qquad \because \quad A + 1 = 1$$
$$= B \cdot 1$$
$$= B. \qquad\qquad \textbf{(Ans.)}$$

(b) $f = \overline{A} + \overline{B} + A \cdot B \cdot \overline{C}$

$$= \overline{A} \cdot (1 + B\overline{C}) + \overline{B} + AB\overline{C} \quad \because \quad 1 + B\overline{C} = 1$$
$$= \overline{A} + \overline{A}B\overline{C} + \overline{B} + AB\overline{C}$$
$$= B\overline{C} (A + \overline{A}) + \overline{A} + \overline{B}$$
$$= B\overline{C} + \overline{A} + \overline{B}. \qquad \because \quad A + \overline{A} = 1 \quad \textbf{(Ans.)}$$

(c) $f = \overline{A}\,\overline{B}\,\overline{C}D + \overline{A}\,\overline{B}\,\overline{C}\overline{D}$

$$= \overline{A}\,\overline{B}\,\overline{C} (D + \overline{D})$$
$$= \overline{A}\,\overline{B}\,\overline{C}. \qquad \because \quad D + D = 1 \; \textbf{(Ans.)}$$

(d) $f = \overline{A}C + \overline{A}B + A\overline{B}C + BC$

$$= \overline{A}B + \overline{A}C + A\overline{B}C + BC$$
$$= \overline{A}B + C (\overline{A} + A\overline{B} + B)$$
$$= \overline{A}B + C [\overline{A} + A\overline{B} + B (A + \overline{A})] \quad \because \quad A + \overline{A} = 1$$
$$= \overline{A}B + C [\overline{A} + A\overline{B} + AB + \overline{A}B]$$
$$= \overline{A}B + C [\overline{A} + A (B + \overline{B}) + \overline{A}B]$$
$$= \overline{A}B + C [\overline{A} + A \cdot 1 + \overline{A}B]$$
$$= \overline{A}B + C [1 + \overline{A}B] \qquad \because \quad 1 + \overline{A}B = 1$$
$$= \overline{A}B + C. \qquad\qquad \textbf{(Ans.)}$$

Example 9.41   Simplify the following expression by Boolean algebra and make a logic diagram with (a) Basic gates (b) NAND gate only

$$Y = A\overline{B}\,\overline{C} + \overline{A}\,\overline{B}\,\overline{C} + \overline{A}B\overline{C} + \overline{A}\,\overline{B}\,C$$

Solution $Y = A\overline{B}\,\overline{C} + \overline{A}\,\overline{B}\,\overline{C} + \overline{A}B\overline{C} + \overline{A}\,\overline{B}C$

$$= \overline{B}\,\overline{C} (A + \overline{A}) + \overline{A}B\overline{C} + \overline{A}\,\overline{B}C$$
$$= \overline{B}\,\overline{C} + \overline{A}B\overline{C} + \overline{A}\,\overline{B}C \qquad \because \quad A + \overline{A} = 1$$
$$= \overline{C} (\overline{B} + \overline{A}B) + \overline{A}\,\overline{B}C$$
$$= \overline{C} (\overline{B} + \overline{A}) + \overline{A}\,\overline{B}C \qquad \because \quad x + \overline{x}y = x + y$$
$$= \overline{B}\,\overline{C} + \overline{A}\,\overline{C} + \overline{A}\,\overline{B}C$$
$$= \overline{B} (\overline{C} + \overline{A}C) + \overline{A}\,\overline{C}$$

$$= \overline{B}\,(\overline{C} + \overline{A}) + \overline{A}\,\overline{C} \qquad \because \quad \overline{C} + \overline{A}C = \overline{C} + \overline{A}$$

$$= \overline{B}\,\overline{C} + \overline{A}\,\overline{B} + \overline{A}\,\overline{C}. \qquad\qquad \textbf{(Ans.)}$$

(a) Logic diagram with basic gates (AND, OR, NOT)

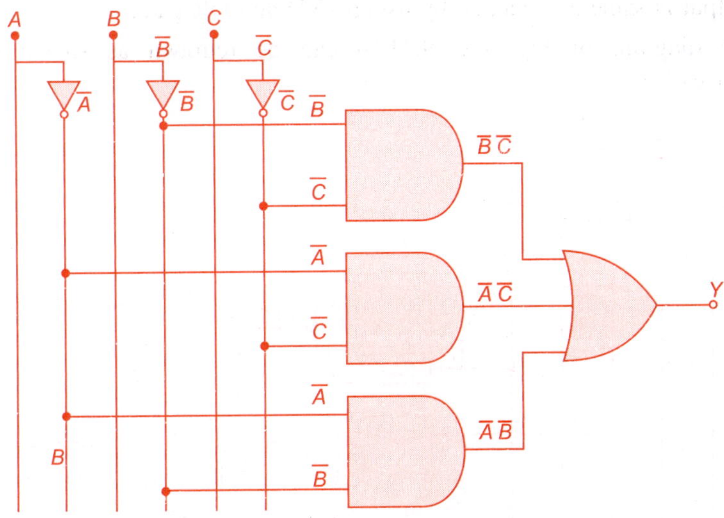

Fig. Ex. 9.41(a)   *AND-OR diagram of Ex. 9.41*

(b) Logic diagram with universal gate (NAND gate only) with the help of Fig. 9.13 (a), (b) and (c). The gates in Fig. Ex. 9.41(a) will be replaced by their equivalent NAND gate as shown in Fig. Ex. 9.41(b)

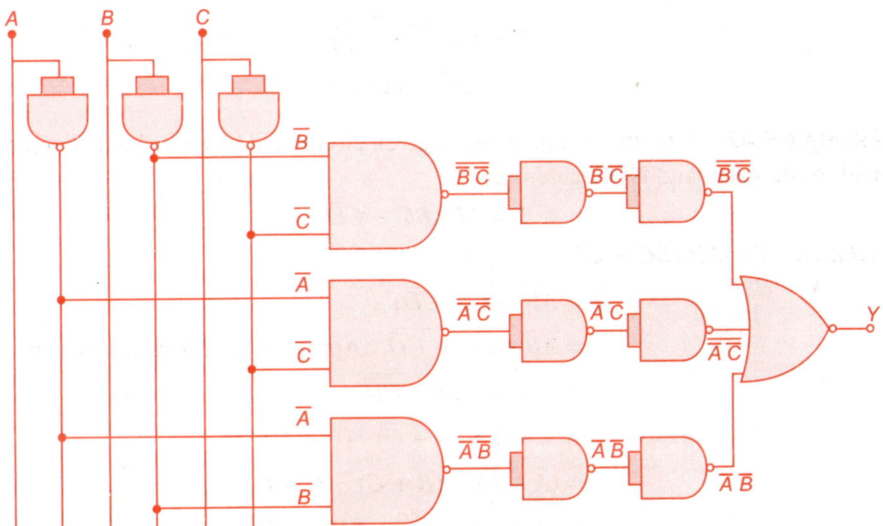

Fig. Ex. 9.41(b)   *NAND diagram of Ex. 9.41*

$$Y = (\overline{\overline{B}\,\overline{C}}) \cdot (\overline{\overline{A}\,\overline{C}}) \cdot (\overline{\overline{A}\,\overline{B}})$$

$$= \overline{\overline{\overline{B}\,\overline{C}}} + \overline{\overline{\overline{A}\,\overline{C}}} + \overline{\overline{\overline{A}\,\overline{B}}}$$

$$= \overline{B}\,\overline{C} + \overline{A}\,\overline{C} + \overline{A}\,\overline{B}$$

i.e., Output is same as obtained by using AND and OR gate.

The diagram of Fig. Ex. 9.41(b) can be redrawn as shown in Fig. Ex. 9.41(c)

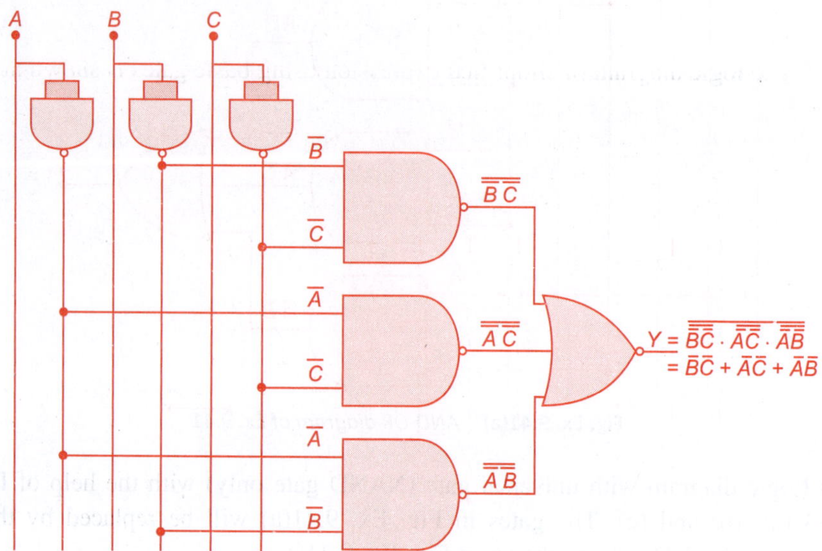

Fig. Ex. 9.41(c)   *Simplified diagram of Fig. Ex. 9.41(b)*

$$Y = \overline{\overline{\overline{B}\,\overline{C}} \cdot \overline{\overline{A}\,\overline{C}} \cdot \overline{\overline{A}\,\overline{B}}}$$

$$= \overline{B}\,\overline{C} + \overline{A}\,\overline{C} + \overline{A}\,\overline{B}.$$

Example 9.42   Find the complement of given circuit and make its logic diagram with basic gates and NOR gate only.

$$Y = AB\,(\overline{B}C + C\overline{D}).$$

Solution $Y = AB\,(\overline{B}C + C\overline{D})$

$$\overline{Y} = \overline{AB\,(\overline{B}C + C\overline{D})}$$

$$= \overline{AB} + \overline{\overline{B}C + C\overline{D}} \quad \text{Applying De-Morgan theorem}$$

$$= \overline{A} + \overline{B} + \overline{\overline{B}C} \cdot \overline{C\overline{D}}$$

$$= \overline{A} + \overline{B} + (\overline{\overline{B}} + \overline{C}) \cdot (\overline{C} + \overline{\overline{D}})$$

$$= (\overline{A} + \overline{B}) + (B + \overline{C}) \cdot (\overline{C} + D)$$

$$= (\overline{A} + \overline{B}) + (B\overline{C} + BD + \overline{C}\,\overline{C} + \overline{C}D)$$

$$= (\overline{A} + \overline{B}) + (B\overline{C} + BD + \overline{C} + \overline{C}D)$$

$$= (\overline{A} + \overline{B}) + (B\overline{C} + BD + \overline{C}(1 + D))$$

$$= (\overline{A} + \overline{B}) + (B\overline{C} + BD + \overline{C}) \quad \because \quad 1 + D = 1$$

$$= (\overline{A} + \overline{B}) + BD + B\overline{C} + \overline{C}$$

$$= \overline{A} + \overline{B} + BD + \overline{C}(B + 1)$$

$$= \overline{A} + \overline{B} + \overline{C} + BD \quad\quad\quad \because \quad B + 1 = 1$$

$$= \overline{A} + \overline{B} + BD + \overline{C}$$

$$= \overline{A} + \overline{B} + D + \overline{C}. \quad\quad \because \quad \overline{x} + xy = \overline{x} + y \quad (\textbf{Ans.})$$

The logic diagram of simplified expression using basic gates is shown in Fig. Ex. 9.42 (a) and logic diagram of simplified expression using universal gate (NOR gate only) is shown in Fig. Ex. 9.42(b) by using Figs. 9.15 (b) and (c).

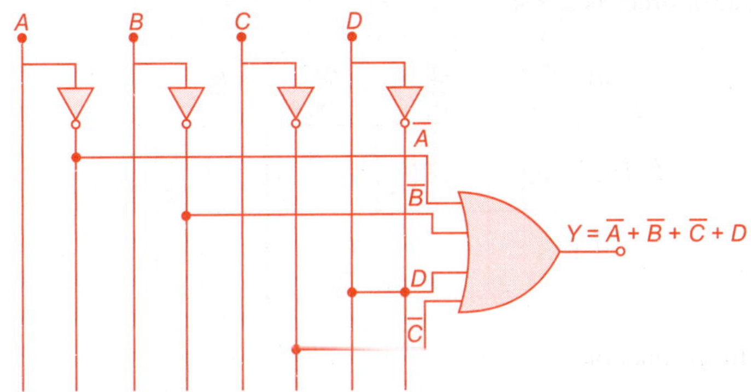

Fig. Ex. 9.42   (a) *AND-OR diagram of fig. Ex 9.42*

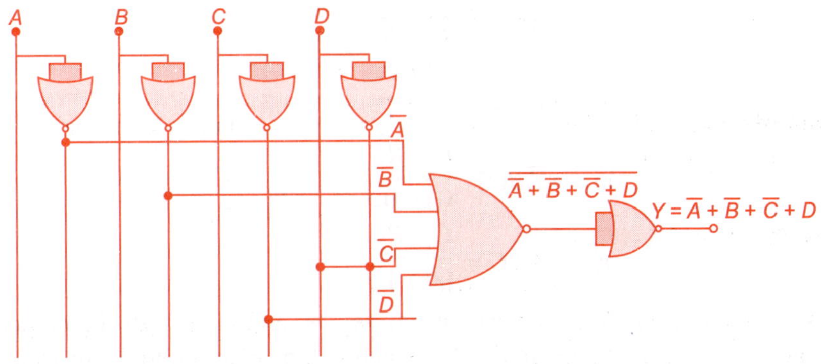

Fig. Ex. 9.42   (b) *NOR diagram of fig Ex 9.42*

Fig Ex 9.42. logic diagram of fig Ex. 9.42.

Example 9.43   Obtain the simplified expression in SOP for the following functions,

(a) $Y = \overline{A}\,\overline{B}\,\overline{C} + A\overline{B}\,\overline{C} + ABC + \overline{A}B\overline{C} + \overline{A}\,\overline{B}C$

(b) $Y = ABCD + \overline{A}\,\overline{B}\,\overline{C}D + \overline{A}BCD + A\overline{B}\overline{C}D + \overline{A}\,\overline{B}CD + ABC\overline{D}.$

Solution   (a) $Y = \overline{A}\,\overline{B}\,\overline{C} + A\overline{B}\,\overline{C} + ABC + \overline{A}B\overline{C} + \overline{A}\,\overline{B}C$

$$\downarrow \qquad \downarrow \qquad \downarrow \qquad \downarrow \qquad \downarrow$$
$$m_0 \qquad m_1 \qquad m_2 \qquad m_3 \qquad m_4$$

From the above function, there are three variables.

Let column variables $= A, B$ i.e., $n_c = 2$

row variable $= C$ i.e., $n_r = 1$.

The size of matrix $= 2^{n_c} \times 2^{n_r}$

$$= 2^2 \times 2^1$$
$$= 4 \times 2.$$

The matrix order is 2×4

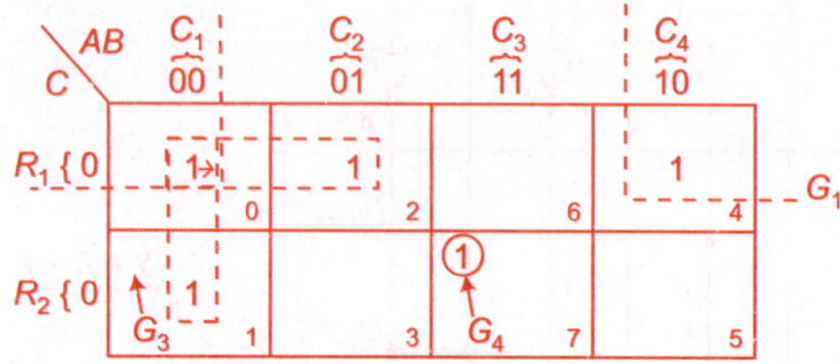

Now, from function

$$m_0 \rightarrow \overline{A}\,\overline{B}\,\overline{C} = \frac{C_1}{00} \frac{R_1}{0}$$

$$m_1 \rightarrow A\overline{B}\,\overline{C} = \frac{C_4}{10} \frac{R_1}{0}$$

$$m_2 \rightarrow ABC = \frac{C_3}{11} \frac{R_2}{1}$$

$$m_3 \rightarrow \overline{A}B\overline{C} = \frac{C_2}{01} \frac{R_1}{0}$$

$$m_4 \rightarrow \overline{A}\,\overline{B}C = \frac{C_1}{00} \frac{R_2}{1}$$

The function has five minterm m_0, m_1, m_2, m_3 and m_4 and all the minterms have all variables. So, there are five '1' which will be placed into the matrix, and making the graph,

$$G_1 = \overline{B}\,\overline{C}, \quad G_2 = \overline{A}\,\overline{C}, \quad G_3 = \overline{A}\,\overline{B}, \quad G_4 = ABC$$

i.e.
$$Y = G_1 + G_2 + G_3 + G_4$$
$$Y = \overline{B}\,\overline{C} + \overline{A}\,\overline{C} + \overline{A}\,\overline{B} + ABC \qquad \textbf{(Ans.)}$$

(b) $Y = ABCD + \overline{A}\,\overline{B}\,\overline{C}D + \overline{A}BCD + A\overline{B}CD + ABCD + ABC\overline{D}$

$$\begin{array}{cccccc} \downarrow & \downarrow & \downarrow & \downarrow & \downarrow & \downarrow \\ m_0 & m_1 & m_2 & m_3 & m_4 & m_5 \end{array}$$

From the above expression, there are four variables,

Let column variables = A, B

i.e. $\qquad\qquad n_c = 2$

and row variables = C, D

i.e. $\qquad\qquad n_r = 2$

The size of k-map matrix $= 2^{n_r} \times 2^{n_c}$

$$= 2^2 \times 2^2$$

$$= 4 \times 4$$

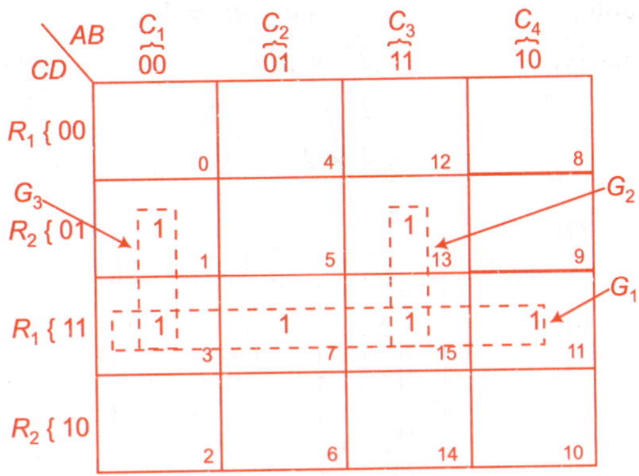

Now, from function,

$$m_0 \rightarrow ABCD = \overset{c_3}{\overbrace{11}}\ \overset{R_3}{\overbrace{11}}$$

$$m_1 \rightarrow \overline{A}\,\overline{B}\,\overline{C}D = \overset{c_1}{\overbrace{00}}\ \overset{R_2}{\overbrace{01}}$$

$$m_2 \rightarrow \overline{A}BCD = \overset{c_2}{\overbrace{01}}\ \overset{R_3}{\overbrace{11}}$$

$$m_3 \rightarrow A\overline{B}CD = \overset{c_4}{\overbrace{10}}\ \overset{R_3}{\overbrace{11}}$$

$$m_4 \rightarrow \overline{A}\,\overline{B}CD = \overset{c_1}{\overbrace{00}}\ \overset{R_3}{\overbrace{11}}$$

$$m_5 \rightarrow ABC\overline{D} = \overset{c_3}{\overbrace{11}}\ \overset{R_2}{\overbrace{01}}$$

The functions have six minterms $m_0, m_1 \ldots .. m_5$ and all the minterms have all the variables. So, there are six '1' which will be placed into the k-map matrix, making the graph.

$$G_1 = CD, \quad G_2 = ABD, \quad G_3 = \overline{A}\,\overline{B}D$$

i.e.

$$Y = G_1 + G_2 + G_3$$

$$Y = CD + ABD + \overline{A}\,\overline{B}D. \tag{Ans.}$$

Example 9.44   Simplify the Boolean function

$$Y(A, B, C) = \Sigma\,(2, 3, 4, 5).$$

Solution   The minterms are given in expression. It has three variable functions. Let the column variables $= A, B$

i.e.

$$n_c = 2$$

The row variable, $\qquad = C \quad$ i.e. $\quad n_r = 1$

The size of matrix, $\qquad = 2^{n_c} \times 2^{n_r}$

$$= 2^2 \times 2^1$$

$$= 4 \times 2.$$

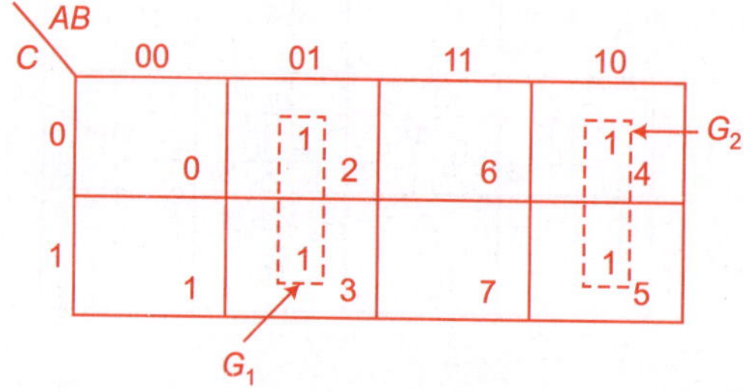

The matrix order is $= 2 \times 4$.

Firstly, 1 is marked in each minterm that represents the function. In mapping, there are two graphs G_1 and G_2

$$G_1 = \overline{A}B \quad \text{and} \quad G_2 = A\overline{B}$$

i.e.

$$Y = G_1 + G_2$$

$$Y = \overline{A}B + A\overline{B}. \tag{Ans.}$$

Example 9.45   Minimise the following using k-map,

$$f = \Sigma\,m\,(1, 2, 5, 7, 9, 15) + d\,\Sigma\,m\,(0, 3, 4, 6).$$

(UPTU 2006-07 (odd))

Solution   We observe that the minterms, is a four variable function in which minterms are given for function and (don't care) conditions minterms are also given.

Let column variables $= A, B$

i.e.

$$n_c = 2$$

The row variables, $\qquad = C, D$

i.e. $\qquad n_r = 2$

The size of matrix, $\qquad = 2^{n_r} \times 2^{n_c}$

$$= 2^2 \times 2^2$$
$$= 4 \times 4.$$

Firstly, place 1's in each minterms that represents the function place 'x' at each minterm of don't care expression. In mapping, there are three graphs G_1, G_2 and G_3.

$$G_1 = \overline{A}, \quad G_2 = BCD, \quad G_3 = \overline{B}\,\overline{C}D$$

i.e.
$$f = G_1 + G_2 + G_3$$
$$f = \overline{A} + BCD + \overline{B}\,\overline{C}D. \qquad \textbf{(Ans.)}$$

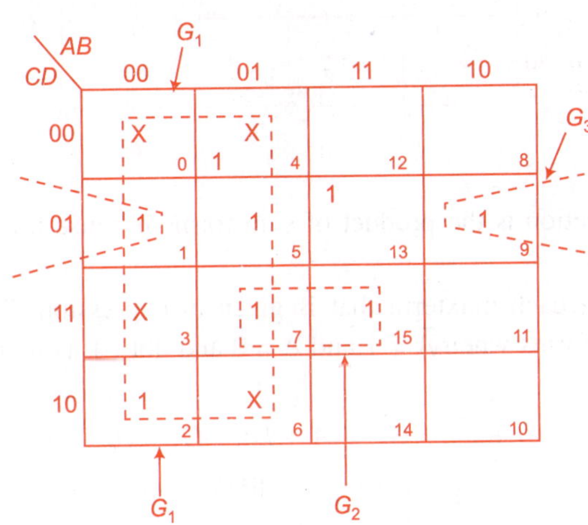

Example 9.46 Simplify the following expression using k-map and implement the output using fundamental gates

(a) $f(A, B, C, D) = \Sigma m\,(1, 3, 4, 6, 8, 9, 11, 13, 15) + d\,\Sigma\,2n\,(0, 2, 14)$

(b) $f(A, B, C, D) = \Pi m\,(0, 1, 3, 6, 7, 8, 9, 11, 13, 14, 15)$

(UPTU 2006-07 (even))

Solution (a) Column variables $= A, B$

i.e. $\qquad n_c = 2$

The row variables $= C, D$

i.e. $\qquad n_r = 2$

The size of matrix, $\qquad = 2^{n_c} \times 2^{n_r}$

$$= 2^2 \times 2^2$$
$$= 4 \times 4.$$

Placing 1's at each minterm that is given in expression. And placing 'X' at don't care condition minterms.

$$G_1 = \overline{B}\,\overline{C}, \quad G_2 = AD, \quad G_3 = \overline{B}D, \quad G_4 = \overline{A}\,\overline{D}$$

i.e.
$$f = G_1 + G_2 + G_3 + G_4 \quad \text{(in SOP form)}$$
$$f = \overline{B}\,\overline{C} + AD + \overline{B}D + \overline{A}\,\overline{D}. \tag{Ans.}$$

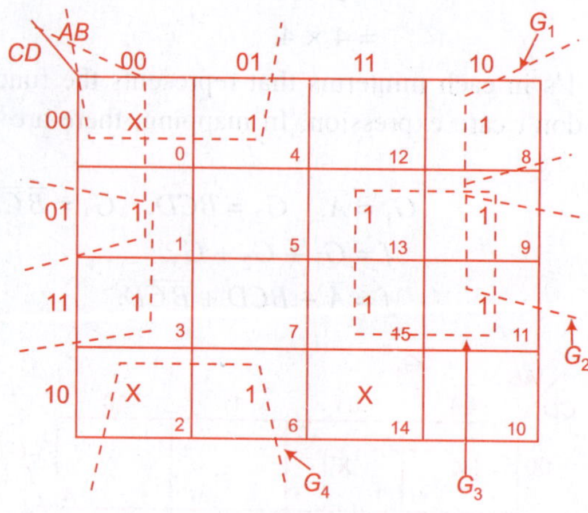

(b) Given function is the product of sum forms because function is given in maxterm.

   Placing 0's at each maxterm that is given in expression. POS form is just opposite to POS forms when $\overline{A} = 1$ and $A = 0$ and dot $(\cdot)$ is replaced by $(+)$ and vice versa.

$$G_1 = (\overline{B} + \overline{C}), \quad G_2 = (\overline{A} + \overline{D}), \quad G_3 = (B + \overline{D}), \quad G_4 = (B + C)$$

i.e.
$$f = G_1 \cdot G_2 \cdot G_3 \cdot G_4 \quad \text{(in POS form)}$$
$$f = (\overline{B} + \overline{C}) \cdot (\overline{A} + \overline{D}) \cdot (B + \overline{D}) \cdot (B + C). \tag{Ans.}$$

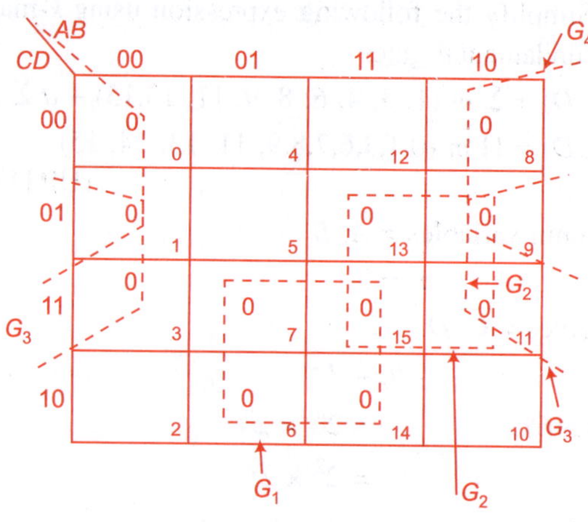

Example 9.47 Minimise the following using k-map
$$f(A, B, C, D) = \Sigma\, m\,(0, 1, 3, 4, 7, 9, 10, 14, 15)$$

(UPTU 2007-08 (odd))

Solution The size of matrix $= 2^{n_c} \times 2^{n_r}$

$$= 4 \times 4.$$

Firstly, placing 1's at each minterm of given expression and grouping these

$$G_1 = AC\overline{D}, \quad G_2 = \overline{B}\,\overline{C}D, \quad G_3 = \overline{A}CD,$$

$$G_4 = \overline{A}\,\overline{C}\,\overline{D} \quad \text{and} \quad G_5 = ABC$$

i.e. $$f(A, B, C, D) = G_1 + G_2 + G_3 + G_4 + G_5$$

$$f = AC\overline{D} + \overline{B}\,\overline{C}D + \overline{A}CD + \overline{A}\,\overline{C}\,\overline{D} + ABC. \qquad \textbf{(Ans.)}$$

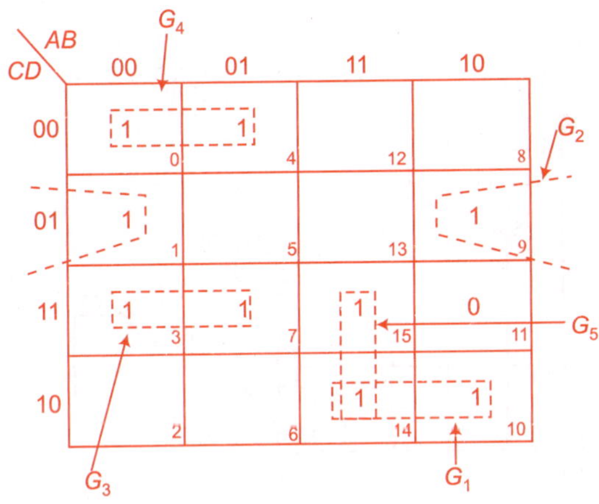

Example 9.48 Minimise the given Boolean function using k-map and implement the simplified function using NOR gate only

$$F(w, x, y, z) = \Sigma m(0, 1, 2, 9, 11, 15) + d(8, 10, 14).$$

(UPTU 2008-09 (odd))

Solution The size of matrix is $= 2^{n_c} \times 2^{n_r}$

$$= 2^2 \times 2^2$$

$$= 4 \times 4.$$

Now, placing 1's at each minterm into k-map and placing 'X' at don't care condition.

$$G_1 = wy, \quad G_2 = \overline{x}\,\overline{y}, \quad G_3 = \overline{x}\,\overline{z}$$

i.e. $$f = G_1 + G_2 + G_3$$

$$f = wy + \overline{x}\,\overline{y} + \overline{x}\,\overline{z}.$$

The implementation of above function which is shown below in Fig. Ex. 9.48

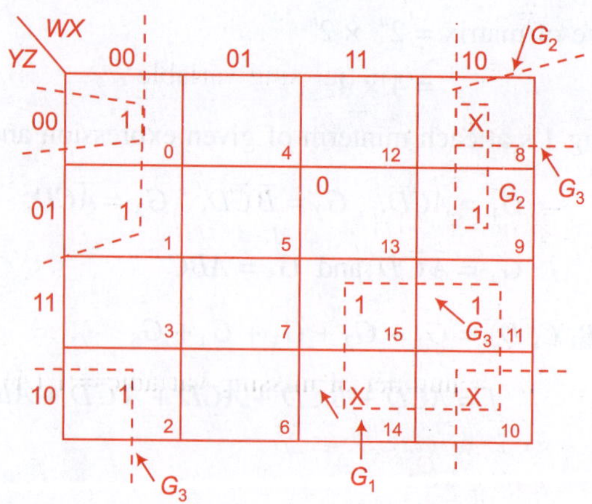

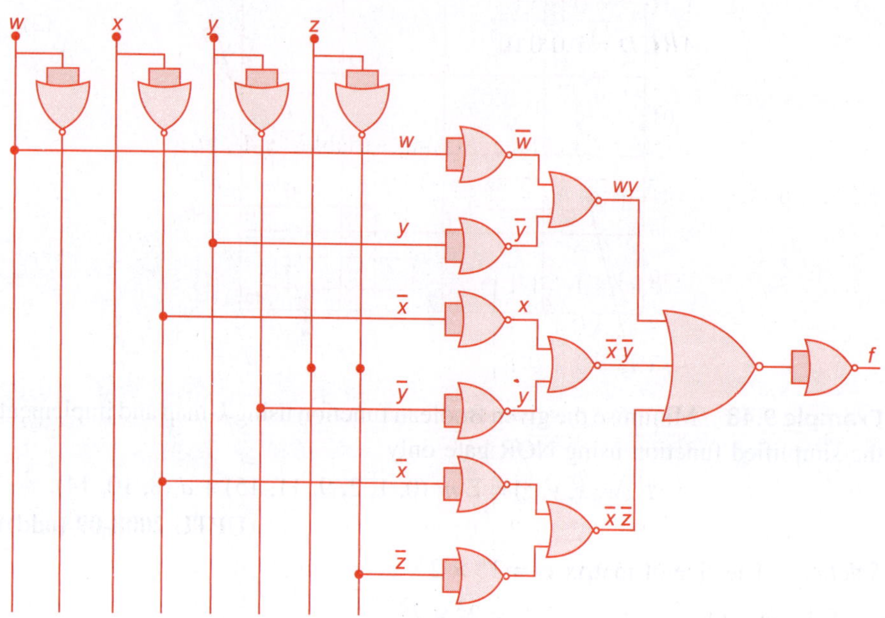

Fig. Ex. 9.48

Example 9.49 Simplify the following Boolean expression using four variable maps

$$f = A\overline{B}C + \overline{B}\,\overline{C}\,\overline{D} + BCD + AC\overline{D} + ABCD + \overline{A}\,\overline{B}C.$$

Solution: Firstly, we will find the minterms for given expression,

$$f = A\overline{B}C + \overline{B}\,\overline{C}\,\overline{D} + B + AC$$

$$\begin{array}{cccc} \downarrow & \downarrow & \downarrow & \downarrow \\ m_0 & m_1 & m_2 & m_3 \end{array}$$

For minterm m_0 $(A\bar{B}C)$

$$n_{v_m} = 1 \quad \text{(one missing variable } D)$$

The number of 1's due to $m_0 = 2n_{v_m}$

$$= 2^1 = 2$$

$$\left.\begin{array}{l} A\bar{B}CD = 1\ 0\ 1\ 1 \\ A\bar{B}C\bar{D} = 1\ 0\ 1\ 0 \end{array}\right\} \tag{2}$$

For minterm $m_1(\bar{B}\,\bar{C}\,\bar{D})$

$$n_{v_m} = \text{number of missing variable} = 1\ (A)$$

The number of 1's due to $m_1 = 2^{n_{v_m}} = 2^1$

$$= 2$$

$$\left.\begin{array}{l} \bar{A}\,\bar{B}\,\bar{C}\,\bar{D} = 0\ 0\ 0\ 0 \\ A\bar{B}\,\bar{C}\,\bar{D} = 1\ 0\ 0\ 0 \end{array}\right\} \tag{2}$$

For minterm m_2 (B)

$$n_{v_m} = 3 \text{ (Three missing variables } A,\ C,\ D)$$

The number of 1's due to $m_2 = 2^{n_{v_m}} = 2^3$

$$= 8$$

$$\left.\begin{array}{l} ABCD = 1\ 1\ 1\ 1 \\ \bar{A}BCD = 0\ 1\ 1\ 1 \\ AB\bar{C}D = 1\ 1\ 0\ 1 \\ ABC\bar{D} = 1\ 1\ 1\ 0 \\ AB\bar{C}\bar{D} = 1\ 1\ 0\ 0 \\ \bar{A}B\bar{C}\bar{D} = 0\ 1\ 0\ 0 \\ \bar{A}B\bar{C}D = 0\ 1\ 0\ 1 \\ \bar{A}BC\bar{D} = 0\ 1\ 1\ 0 \end{array}\right\} \tag{8}$$

For minterm m_3 (AC)

$$n_{v_m} = 2 \text{ (Two variables missing } B,\ D).$$

The number 1's due to $m_3 = 2\ n_{v_m} = 2^2 = 4$

$$\left.\begin{array}{l} ABCD = 1\ 1\ 1\ 1 \\ A\bar{B}CD = 1\ 0\ 1\ 1 \\ ABC\bar{D} = 1\ 1\ 1\ 0 \\ A\bar{B}C\bar{D} = 1\ 0\ 1\ 0 \end{array}\right\} \tag{4}$$

The Minterm Table 9.13 is shown below:

Table 9.13

	Variables			Output	
A	B	C	D	Y	Minterm
0	0	0	0	1	0
0	0	0	1	0	1
0	0	1	0	0	2
0	0	1	1	0	3
0	1	0	0	1	4
0	1	0	1	1	5
0	1	1	0	1	6
0	1	1	1	1	7
1	0	0	0	1	8
1	0	0	1	0	9
1	0	1	0	1	10
1	0	1	1	1	11
1	1	0	0	1	12
1	1	0	1	1	13
1	1	1	0	1	14
1	1	1	1	1	15

From above table the expression in minterms is

$$f(A, B, C, D) = \Sigma\, m\,(0, 4.\ 5, 6, 7, 8, 10, 11, 12, 13, 14, 15)$$

The size of matrix is $= 4 \times 4$.

Now, placing the 1's at each minterm in k-map.

$$G_1 = B, \quad G_2 = AC, \quad G_3 = \overline{C}\overline{D}$$

i.e.
$$f = G_1 + G_2 + G_3$$
$$f = B + AC + \overline{C}\overline{D}.$$

(**Ans.**)

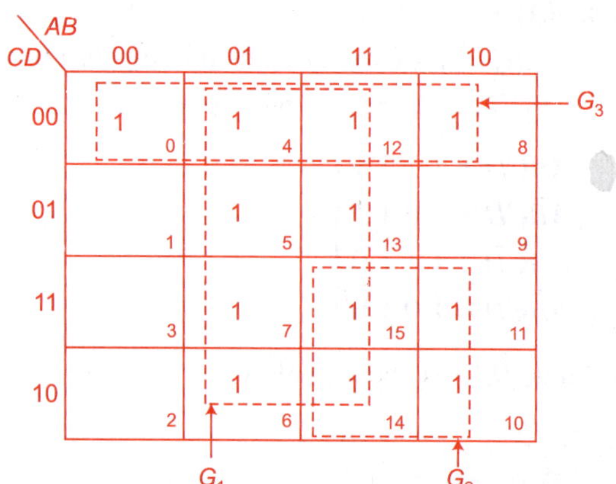

Example 9.50   Simplify the given expression in POS form by k-map

$$Y = (A + \bar{B} + C)(A + \bar{B} + \bar{C})(A + B + C)(\bar{A} + \bar{B} + \bar{C}).$$

Solution   Given expression is in product of sum form in which $\bar{A} = 1$ and $A = 0$.

$$Y = (A + \bar{B} + C) \quad (A + \bar{B} + \bar{C}) \quad (A + B + C) \quad (\bar{A} + \bar{B} + \bar{C})$$

$$\qquad\quad \downarrow \qquad\qquad \downarrow \qquad\qquad \downarrow \qquad\qquad \downarrow$$

$$\qquad\quad m_0 \qquad\qquad m_1 \qquad\qquad m_2 \qquad\qquad m_3$$

The Maxterm Table 9.14 is shown below in which 1's are placed those maxterms exists in expression.

Table 9.14

A	B	C	Y	Maxterm M
0	0	0	0	0
0	0	1	1	1
0	1	0	0	2
0	1	1	0	3
1	0	0	1	4
1	0	1	1	5
1	1	0	1	6
1	1	1	0	7

$$M_0 \rightarrow A + \bar{B} + C = 0\ 1\ 0$$
$$M_1 \rightarrow A + \bar{B} + \bar{C} = 0\ 1\ 1$$
$$M_2 \rightarrow A + B + C = 0\ 0\ 0$$
$$M_3 \rightarrow \bar{A} + \bar{B} + \bar{C} = 1\ 1\ 1.$$

As we know in POS form 0 is placed in maxterm table where the term is exists in expression at remaining maxterm 1's are placed.

The size of matrix $= 2^{n_r}_1 \times 2^{n_c}$

$$= 2^1 \times 2^2$$
$$= 2 \times 4.$$

$$\therefore \qquad n_c = A, \quad B = 2$$
$$n_r = C = 1$$

expression for Group 1 i.e., G_1

$$G_1 = (\bar{B} + \bar{C})$$

expression for Group 2 i.e., G_2

$$G_2 = (A + C).$$

The simplified expression in POS form is

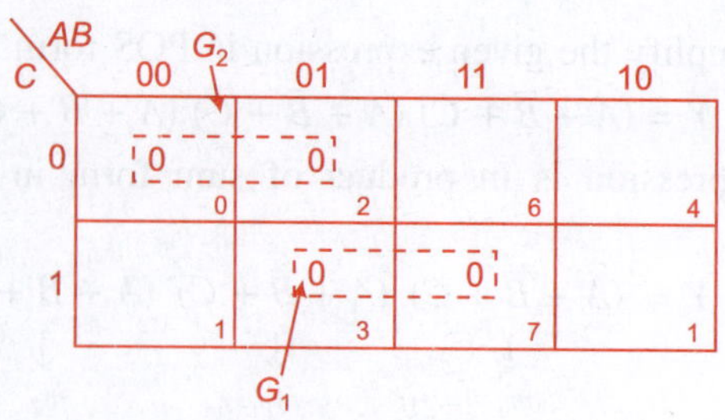

$$Y = G_1 \cdot G_2$$

$$Y = (\overline{B} + \overline{C}) \cdot (A + C). \tag{Ans.}$$

Example 9.51   Simplify the Boolean function F in sum of products using don't care conditions d (using k-map)

(a) $F = \overline{Y} + \overline{X}\overline{Z}$ and $d = YZ + XY$

(b) $F = \overline{B}\,\overline{C}\overline{D} + BC\overline{D} + ABC\overline{D}$ and $d = \overline{B}C\overline{D} + \overline{A}BC\overline{D}$

(UPTU 2008-09 (even))

Solution   (a) $F = \overline{Y} + \overline{X}\overline{Z}$ and $d = YZ + XY$

$$\begin{array}{cccc} \downarrow & \downarrow & & \downarrow & \downarrow \\ m_0 & m_1 & & dm_0 & dm_0 \end{array}$$

Considering the minterm m_0, m_1, dm_1 and dm_2 for given expression. After observing the expression, we obtain that it has three variables X, Y and Z.

The size of matrix $= 2^{n_r} \times 2^{n_c}$

$$= 2^1 \times 2^2$$

$$= 2 \times 4.$$

For minterm m_0 $(\overline{Y})$.

The number of 1's due to minterm $m_0 = 2^{n_{v_m}}$

$$= 2^2 = 4$$

$\because \qquad\qquad n_{v_m} = 2$ (two missing variables X, Z)

Hence,

$$\left.\begin{array}{l} \overline{X}\,\overline{Y}\overline{Z} = 0\ 0\ 0 \\ \overline{X}\,\overline{Y}Z = 0\ 0\ 1 \\ X\overline{Y}\,\overline{Z} = 1\ 0\ 0 \\ X\overline{Y}Z = 1\ 0\ 1 \end{array}\right\} \tag{4}$$

For minterm m_1 $(\overline{X}\,\overline{Z})$

$$n_{v_m} = 1 \text{ (one missing variable } Y)$$

So, the number of 1's due to minterm $m_1 = 2^{n_{v_m}}$

$$= 2^1 = 2.$$

Hence,

$$\left.\begin{array}{l}\overline{X}\,\overline{Y}\,\overline{Z} = 0\ 0\ 0 \\ \overline{X}Y\overline{Z} = 0\ 1\ 0\end{array}\right\} \tag{2}$$

For minterm dm_1 (YZ) don't care minterm.

$$n_{v_m} = 1 \text{ (one missing variable } X)$$

So, the number of 'Xs' due to minterm $m_2 = 2^{n_{v_m}}$

$$= 2^1 = 2.$$

Hence,

$$\left.\begin{array}{l}\overline{X}YZ = 0\ 1\ 1 \\ XYZ = 1\ 1\ 1\end{array}\right\} \tag{2}$$

For minterm dm_2 (XY) don't care minterm

$$n_{v_m} = 1 \text{ (one missing variable } Z)$$

So, the number of 'Xs' due to minterm $m_3 = 2^{n_{v_m}}$

$$= 2^1 = 2.$$

Hence,

$$\left.\begin{array}{l}XY\overline{Z} = 1\ 1\ 0 \\ XYZ = 1\ 1\ 1\end{array}\right\} \tag{2}$$

The Minterm Table 9.15 for given expression is shown below,

Table 9.15

X	Y	Z	F	Minterm
0	0	0	1	0
0	0	1	1	1
0	1	0	1	2
0	1	1	X	3
1	0	0	1	4
1	0	1	1	5
1	1	0	X	6
1	1	1	X	7

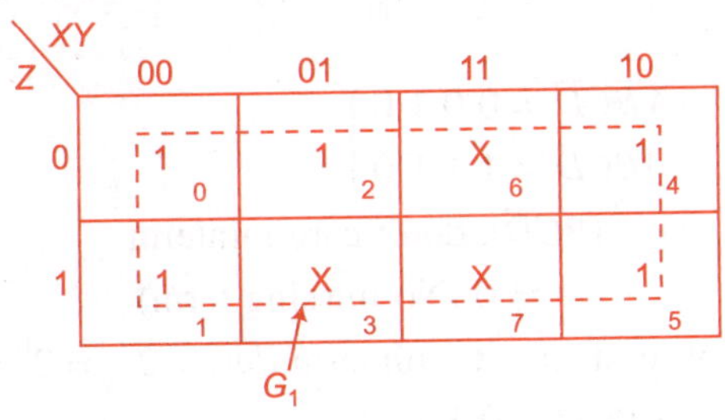

From the above table, the expression in minterms is,

$$F = (X, Y, Z) = \Sigma m\,(0, 1, 2, 4, 5) + d(3, 6, 7)$$

From k-map

$$G_1 = 1$$

The simplified expression SOP form is,

$$F\,(X, Y, Z) = 1. \qquad \textbf{(Ans.)}$$

(b) $\quad F = \overline{B}\,\overline{C}\overline{D} + BC\overline{D} + ABC\overline{D} \quad$ and $\quad d = \overline{B}C\overline{D} + \overline{A}\,B\overline{C}D$

$$\downarrow \qquad\quad \downarrow \qquad\quad \downarrow \qquad\qquad\qquad \downarrow \qquad\quad \downarrow$$
$$m_0 \qquad\quad m_1 \qquad\quad m_2 \qquad\qquad\qquad dm_1 \qquad\quad dm_2$$

For minterm m_0 ($\overline{B}\,\overline{C}\overline{D}$)

$$n_{v_m} = 1 \text{ (one missing variable } A)$$

The number of 1's due to minterm $m_0 = 2^{n_{v_m}} = 2^1 = 2$.
Hence,

$$\left.\begin{aligned}\overline{A}\,\overline{B}\,\overline{C}\overline{D} &= 0\ 0\ 0\ 0 \\ A\overline{B}\,\overline{C}\overline{D} &= 1\ 0\ 0\ 0\end{aligned}\right\} \qquad\qquad (2)$$

For minterm m_1 ($BC\overline{D}$)

$$n_{v_m} = 1 \text{ (one missing variable } A\,)$$

The number of 1's due to minterm $m_1 = 2^{n_{v_m}} = 2^1 = 2$
Hence,

$$\left.\begin{aligned}\overline{A}BC\overline{D} &= 0\ 1\ 1\ 0 \\ ABC\overline{D} &= 1\ 1\ 1\ 0\end{aligned}\right\} \qquad\qquad (2)$$

For minterm m_2 ($ABC\overline{D}$)

$$n_{v_m} = 0 \text{ (No missing term)}$$

The number of 1's due to minterm $m_2 = 2^{n_{v_m}} = 2^0 = 1$
Hence,

$$ABC\overline{D} = 1\ 1\ 1\ 0\ \} \qquad\qquad (1)$$

For minterm dm_1 ($\overline{B}C\overline{D}$) don't care minterm

$$n_{v_m} = 1 \text{ (one missing variable } A\,)$$

The number of 1's due to minterm $dm_1 = 2^{n_{v_m}} = 2^1 = 2$.
Hence,

$$\left.\begin{aligned}\overline{A}\,\overline{B}C\overline{D} &= 0\ 0\ 1\ 0 \\ A\overline{B}C\overline{D} &= 1\ 1\ 1\ 0\end{aligned}\right\} \qquad\qquad (2)$$

For minterm dm_2 ($\overline{A}\,B\overline{C}D$), don't care minterm

$$n_{v_m} = 0 \text{ (No missing term)}$$

The number of terms due to minterm $dm_2 = 2^{n_{v_m}} = 2^0 = 1$

$$\overline{A}B\overline{C}D = 0\ 1\ 1\ 0\} \qquad\qquad (1)$$

The Minterm Table 9.16 for the given expression is shown below,

Table 9.16

A	B	C	D	F	Minterm
0	0	0	0	1	0
0	0	0	1	0	1
0	0	1	0	X	2
0	0	1	1	0	3
0	1	0	0	0	4
0	1	0	1	0	5
0	1	1	0	1	6
0	1	1	1	0	7
1	0	0	0	1	8
1	0	0	1	0	9
1	0	1	0	X	10
1	0	1	1	0	11
1	1	0	0	0	12
1	1	0	1	0	13
1	1	1	0	1	14
1	1	1	1	0	15

From the above table, the expression in minterm form is,

$$F = (A, B, C, D) = \Sigma\, m\,(0, 6, 8, 14) + d(2, 10)$$

From *k*-map

$$G_1 = C\overline{D}$$
$$G_2 = \overline{B}\,\overline{D}$$

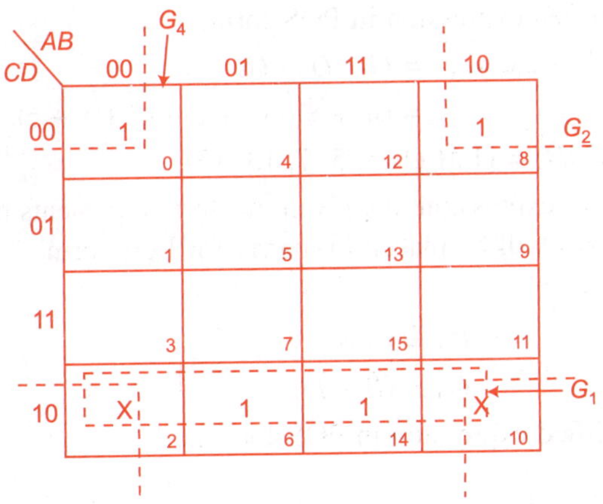

The simplified expression in SOP form is,

$$F = (A, B, C, D) = G_1 + G_2$$
$$= C\overline{D} + \overline{B}\,\overline{D}. \qquad \textbf{(Ans.)}$$

Example 9.52 Simplify the following Boolean functions in product of sum form, (a) $F(w, x, y, z) = \Sigma\, m\,(0, 2, 5, 6, 7, 8, 10)$

(b) $F(A, B, C, D) = \Pi\, M\,(1, 3, 5, 7, 13, 15)$.

Solution (a) $F(w, x, y, z) = \Sigma\, m\,(0, 2, 5, 6, 7, 8, 10)$.

In the above expression, the given minterm represents the 1's at minterm table i.e., the remaining minterm has 0's in table. Hence 0's will be placed in matrix for POS form.

From *k*-map

Group 1
$$G_1 = (\overline{w} + \overline{x})$$
$$G_2 = (x + \overline{z})$$
$$G_3 = (\overline{x} + y + z)$$

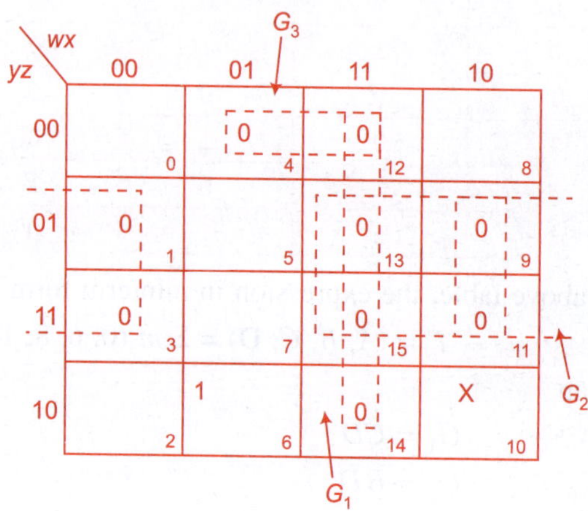

The simplified expression in POS form,

$$F(w, x, y, z) = G_1 \cdot G_2 \cdot G_3$$
$$= (\overline{w} + \overline{x}) \cdot (x + \overline{z}) \cdot (\overline{x} + y + z). \qquad \textbf{(Ans.)}$$

(b) $F(A, B, C, D) = \Pi\, M\,(1, 3, 5, 7, 13, 15)$.

In the above expression, the given maxterm represents the 0's at minterm table. the 0's will be placed in matrix for POS form.

From *k*-map

$$G_1 = (A + \overline{D})$$
$$G_2 = (\overline{B} + \overline{D})$$

The simplified expression in POS form is,

$$F\,(A, B, C, D) = G_1 \cdot G_2$$
$$F\,(A, B, C, D) = (A + \overline{D}) \cdot (\overline{B} + \overline{D}).$$ **(Ans.)**

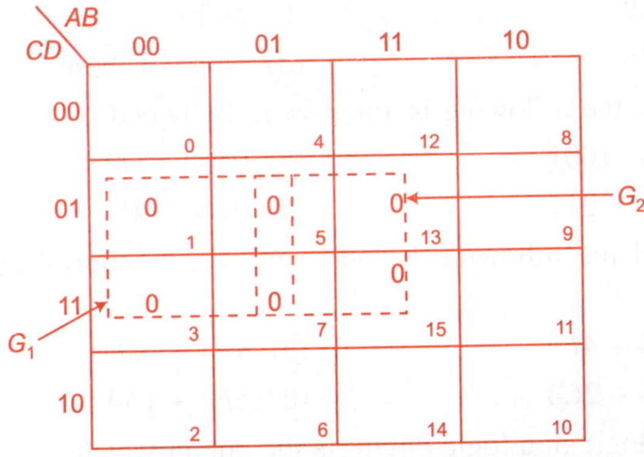

Example 9.53 Simplify the following Boolean function in product of sum form,

$$F\,(A, B, C, D) = \Pi\, M\,(0, 1, 2, 3, 4, 10, 11).$$

Solution The expression is $F\,(A, B, C, D) = \Pi\, M\,(0, 1, 2, 3, 4, 10, 11)$

From k-map

$$G_1 = (B + \overline{C})$$
$$G_2 = (A + B)$$
$$G_3 = (\,A + \overline{C} + D)$$

The simplified expression in POS form is,

$$F\,(A, B, C, D) = G_1 \cdot G_2 \cdot G_3$$
$$= (B + \overline{C}) \cdot (A + B) \cdot (A + C + D).$$ **(Ans)**

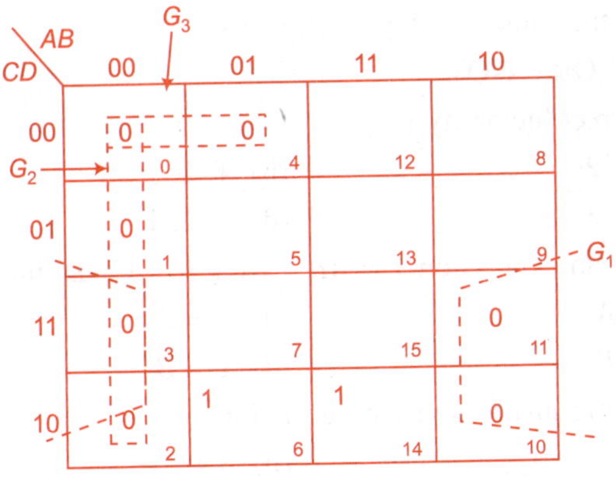

Objective Type Questions

1. The decimal equivalent of the hexadecimal number $(BAD)_{16}$ is
   - (a) 2989
   - (b) 5930
   - (c) 3425
   - (d) None of these

2. Which of the following is a non-valid BCD code?
   - (a) 0111 1001
   - (b) 0100 1000
   - (c) 0101 1011
   - (d) 0100 1001

3. Which of the following hexadecimal sum is equivalent to hexadecimal $B88$?
   - (a) $2C4 + 4F4$
   - (b) $4B4 + 6D4$
   - (c) $5A4 + 2C4$
   - (d) $5E4 + 1A4$

4. Positive logic in a logic circuit is the one in which
   - (a) Logic 0 and 1 are represented by the negative and positive voltages respectively
   - (b) Logic 0 and 1 are represented by 0 and positive voltages respectively
   - (c) Logic 1 voltage level is higher than logic 0 voltage level
   - (d) Logic 1 voltage level is lower than logic 0 voltage level

5. An x-OR gate produces output only when two inputs are
   - (a) High
   - (b) Low
   - (c) Different
   - (d) Equal

6. The logic function $f = \overline{(\bar{x} \cdot y) + (x \cdot \bar{y})}$ is the same as
   - (a) $f = (x + \bar{y}) \cdot (\bar{x} + y)$
   - (b) $f = (\bar{x} + y) \cdot (x + \bar{y})$
   - (c) $f = (x \times \bar{y}) \cdot (y + \bar{x})$
   - (d) $f = (\bar{x} + y) \cdot (x + \bar{y})$

7. Consider the following logic operators

   1. AND 2 OR 3 NOT

   Then, correct hierarchy is
   - (a) 3, 1, 2
   - (b) 1, 2, 3
   - (c) 2, 1, 3
   - (d) 3, 2, 1

8. The Boolean expression $Y = AB + (A + B)(\bar{A} + B)$ may be simplified as
   - (a) $Y = A$
   - (b) $Y = \bar{A}$
   - (c) $Y = B$
   - (d) $Y = \bar{B}$

9. Octal coding involves grouping the bits in
   - (a) 4's
   - (b) 3's
   - (c) 5's
   - (d) 7's

10. The number $(17)_8$ is equivalent to binary
    (a) 1111   (b) 1110
    (c) 111   (d) 10000
11. The output of a 2-input OR gate is zero only when its
    (a) Either input is 0   (b) Either input is 1
    (c) Both inputs are 0   (d) Both inputs are 1
12. The output of a 2-input Ex-OR gate is zero when its
    (a) Both inputs are different   (b) Both inputs are same
    (c) Any one input is zero   (d) Any one input is 1
13. Which of the following 4-bit combination is invalid in the BCD code?
    (a) 0010   (b) 0101
    (c) 1000   (d) 1010
14. The number $(100101)_2$ is equal to octal
    (a) 45   (b) 36
    (c) 25   (d) 52
15. Which of the following code is an unweighted code 3?
    (a) 8421   (b) Excess 3 code
    (c) 2421   (d) 63210
16. The logic performed by the circuit shown in Fig. OT 9.1 is
    (a) AND   (b) NAND
    (c) Ex-OR   (d) None of these

Fig. OT 9.1

17. The logic performed by the circuit shown in Fig. OT 9.2 is
    (a) AND   (b) OR
    (c) Ex-OR   (d) None of these

Fig. OT 9.2

18. The logic performed by the circuit shown in Fig. OT 9.3 is
    (a) AND
    (b) OR
    (c) NOT
    (d) Ex-OR

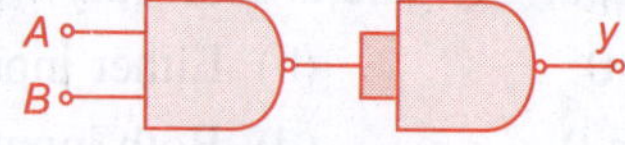

Fig. OT 9.3

19. The logic performed by the circuit shown in Fig. OT 9.4 is
    (a) AND
    (b) OR
    (c) NOT
    (d) Ex-OR

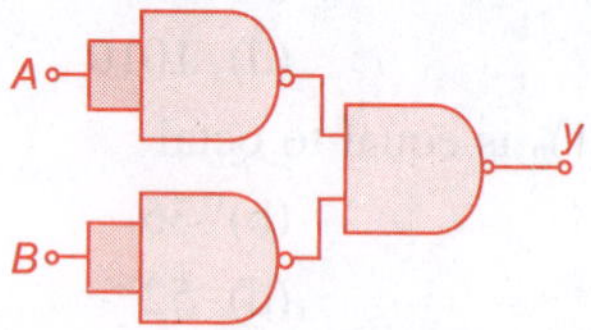

Fig. OT 9.4

20. The output A for the circuit shown in Fig. OT 9.5 is
    (a) $a + bc$
    (b) $\bar{a} + \bar{b}\bar{c}$
    (c) $\bar{a}\bar{b} + \bar{a}\bar{c}$
    (d) $a + b\bar{c}$

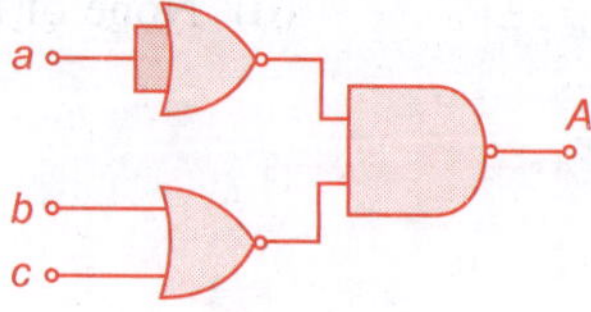

Fig. OT 9.5

21. NAND operation is
    (a) $\overline{(x + y)}$
    (b) $\bar{x} + \bar{y}$
    (c) $\bar{x} \cdot \bar{y}$
    (d) $x + \bar{x}$

22. NOR operation is
    (a) $\overline{(x + y)}$
    (b) $\bar{x}\bar{y}$
    (c) $(\bar{x} + \bar{y})(x + y)$
    (d) xy

23. The weight of the number system can be expressed as
    (a) $(n)^{base}$
    (b) $n \times base$
    (c) $(base)^n$
    (d) None of these

24. The number of symbols which are in decimal number system is
    (a) 8
    (b) 2
    (c) 10
    (d) 16
25. The number of symbols which are used in binary number system is
    (a) 2
    (b) 8
    (c) 10
    (d) None of these
26. The number .... is in octal form
    (a) $(18)_8$
    (b) $(29)_8$
    (c) $(37)_8$
    (d) None of these
27. The number $(17)_{10}$ is .... in octal number system
    (a) 20
    (b) 21
    (c) 15
    (d) None of these
28. The base of binary number system is
    (a) 2
    (b) 5
    (c) 8
    (d) 10
29. The base of octal number system is
    (a) 2
    (b) 8
    (c) 16
    (d) None of these
30. The base of hexadecimal number system is
    (a) 16
    (b) 8
    (c) 2
    (d) None of these
31. The base of any number system represents the .... available to represent that number system
    (a) Toggle
    (b) Symbols
    (c) Bit
    (d) None of these
32. When bit changes from 0 to 1 or 1 to 0 at any count, it is called as
    (a) Conversion
    (b) Toggle
    (c) Complement
    (d) None of these
33. Hexadecimal coding involves grouping the bits in
    (a) 3's
    (b) 2's
    (c) 4's
    (d) None of these
34. The addition of $1 + 1 + 1$ is equal to
    (a) 0 and carry 1
    (b) 1 and carry 0
    (c) 1 and carry 1
    (d) None of these
35. The addition of $1 + 1$ is equal to
    (a) 0 and carry 1
    (b) 1 and carry 0
    (c) 1 and carry 1
    (d) None of these

36. The complement of number system is to represent the ....
    (a) Positive Number
    (b) Negative Number
    (c) Both (a) and (b)
    (d) None of these
37. 9's complement is used in
    (a) Binary number system
    (b) Decimal number system
    (c) Octal number system
    (d) None of these
38. 1's complement is used in .... number system
    (a) Binary
    (b) Octal
    (c) Decimal
    (d) None of these
39. 10's complement is
    (a) 9's complement + 1
    (b) 2's complement + 1
    (c) 9's complement – 1
    (d) None of these
40. 2's complement is
    (a) 2's complement – 1
    (b) 1's complement + 1
    (c) 9's complement + 1
    (d) None of these
41. The 1's complement of binary number 01 is
    (a) 10
    (b) 11
    (c) 00
    (d) None of these
42. The 2's complement of binary number 1010 is
    (a) 1011
    (b) 0110
    (c) 1110
    (d) 1111
43. The 9's complement of 785 is
    (a) 214
    (b) 215
    (c) 213
    (d) None of these
44. In 8421 code, each number represents the
    (a) Positional bit
    (b) Positional weight of binary
    (c) Base
    (d) None of these
45. The BCD code of decimal number 256 is
    (a) 0010 0101 0110
    (b) 010 101 110
    (c) 0100 0110 0101
    (d) None of these
46. If A, B, C are certain objects, then,
    (a) $A + (B \cdot C) = (A + B) \cdot (A + C)$
    (b) $A + (B \cdot C) = (A \cdot B) + (A \cdot C)$
    (c) $A + (B \cdot C) = ABC$
    (d) None of these

47. If A is any object and $\overline{A}$ is a complement of A, then
    (a) $A + \overline{A} = 1$ and $A \cdot \overline{A} = 1$
    (b) $A + \overline{A} = 0$ and $A \cdot \overline{A} = 0$
    (c) $A + \overline{A} = 1$ and $A \cdot \overline{A} = 0$
    (d) None of these

48. $A + (\overline{A} \cdot B) = A + B$ is also called as
    (a) Distributive law
    (b) Commutative law
    (c) Involution law
    (d) None of these

49. $A \cdot (B + C) = A \cdot B + A \cdot C$ and $A + (B + C) = (A + B) + C$ is called as
    .... respectively.
    (a) Associative, Distributive
    (b) Distributive, Associative
    (c) Associative, Associative
    (d) Distributive, Distributive

50. DeMorgan theorem represents
    (a) $\overline{A + B} = \overline{A} \cdot \overline{B}$
    (b) $\overline{A \cdot B} = \overline{A} + \overline{B}$
    (c) Both (a) and (b)
    (d) None of these

51. When any input of OR gate is 1, then the output is
    (a) 0
    (b) 1
    (c) Both (a) and (b)
    (d) None of these

52. When any input of AND gate is 0, then the output is
    (a) 0
    (b) 1
    (c) Both (a) and (b)
    (d) None of these

53. If the expression has four variables, then the size of k-map matrix is
    (a) 2×2
    (b) 2×4
    (c) 4×4
    (d) None of these

54. The simplified form of the Boolean expression
    $F(x, y, z) = \Sigma (0, 1, 5, 7)$ is
    (a) $\overline{x}\,\overline{y} + xy$
    (b) $\overline{x}y + \overline{y}x$
    (c) $xy + \overline{x}y$
    (d) $\overline{x}\,\overline{y} + \overline{x}y$

55. Simplified expression of the Boolean function,
    $F(x, y, z) = \Sigma m (2, 3, 4, 5)$ is,
    (a) $\overline{x}y + \overline{x}\,\overline{y}$
    (b) $\overline{x}y + x\overline{y}$
    (c) $\overline{x}\,\overline{y} + xy$
    (d) None of these

56. The simplified expression of Boolean function
    $F(x, y) = \overline{x}y + x\overline{y} + xy$ is
    (a) $x + y$
    (b) $x - y$
    (c) xy
    (d) None of these

57. The simplified expression of Boolean function,

    $F(x, y, z) = \Pi(0, 2, 5, 7)$ is

    (a) $(\bar{x} + \bar{z})(x + z)$

    (b) $(\bar{x} + z)(x + \bar{z})$

    (c) $(\bar{x} + \bar{z})(x + \bar{z})$

    (d) None of these

ANSWER

1 (a)	2 (c)	3 (b)	4 (c)	5 (c)	6 (a)	7 (a)
8 (c)	9 (b)	10 (a)	11 (c)	12 (b)	13 (d)	14 (a)
15 (b)	16 (a)	17 (b)	18 (a)	19 (b)	20 (b)	21 (b)
22 (b)	23 (c)	24 (c)	25 (a)	26 (c)	27 (b)	28 (a)
29 (b)	30 (a)	31 (b)	32 (b)	3 (c)	34 (c)	35 (a)
36 (b)	37 (b)	38 (a)	39 (a)	40 (b)	41 (a)	42 (c)
43 (a)	44 (b)	45 (a)	46 (a)	47 (c)	48 (a)	49 (b)
50 (c)	51 (b)	52 (a)	53 (c)	54 (a)	55 (b)	56 (a)
57 (a)						

Exercise

9.1. What are basic logic gates and what are universal logic gates?

(UPTU-TQ-Bank)

9.2. What is k-map? Explain.

9.3. Simplify the Boolean function using k-map.

$F(A, B, C, D) = \Sigma m(0, 1, 2, 4, 5, 6, 8, 9, 12, 13, 14)$

[Ans. $F = \bar{C} + A\bar{D} + B\bar{D}$**]**

9.4. Simplify the Boolean function using k-map.

$F(A, B, C, D) = \overline{AB}\bar{C} + \overline{BC}\bar{D} + \overline{A}BC\bar{D} + A\bar{B}\bar{C}$

[Ans. $F = \bar{B}\bar{D} + \bar{B}\bar{C} + \bar{A}C\bar{D}$**]**

9.5. Simplify the following Boolean function:

    (a) SOP   and   (b) POS

$F(A, B, C, D) = \Sigma m(0, 1, 2, 4, 5, 8, 9, 10)$

[Ans. (a) $F = AB + CD + B\bar{D}$

(b) $F = (\bar{A'} + \bar{B})(\bar{B} + D)(\bar{C} + \bar{D})$**]**

9.6. What are the universal gates? Write their advantages.

9.7. What is the difference between canonical form and standard form? Also write the De-Morgan theorem.   **(UPTU 2002-03 (odd))**

9.8. Why is two input NAND gate called universal gate?

(UPTU 2000-01 (even))

9.9. State and prove De-Morgan theorem.   **(UPTU 2000-01 (even))**

9.10. What is the canonical form of a logic expression? Define minterms and maxterms.

9.11. What are Boolean postulate? State them.   **(UPTU 2000-01 (odd))**

9.12. Construct AND gate, NAND gate and OR gate with the help of NOR gate.   **(UPTU 2002-03 (odd))**

9.13. What are universal gates? Why are they required?   **(UPTU 2006-07 (odd))**

9.14. Write short notes on:

    (a) Minterm and Maxterm

    (b) BCD code and excess-3 code   **(UPTU 2007-08 (odd))**

9.15. What is/are universal gate (s) implement two input XOR gates using only 4 NAND gates?   **(UPTU 2008-09 (odd))**

9.16. Discuss the postulates of Boolean algebra. How, is it different from ordinary algebra?   **(UPTU 2008-09 (even))**

9.17. Explain the steps to convert other number system into decimal number.

9.18. Explain the steps to convert decimal number into other number system.

9.19. What is 2's complement? Explain 1's complement.

9.20. Explain the following

    (a) 9's complement   (b) 10's complement

9.21. Convert the following hexadecimal number into decimals:

    (a) A13B   (b) 7CA3   (c) 7FD6   **(UPTU TQ-Bank)**

9.22. Convert the following binary numbers into decimals:

    (a) 101.01   (b) 10101.0101   **(UPTU TQ-Bank)**

9.23. Simplify the following using k-map

$$F = \overline{A}\,\overline{B}\,\overline{C} + A\overline{C}\,\overline{D} + A\overline{B} + ABC\overline{D} + \overline{A}\,\overline{B}C \qquad \textbf{(UPTU TQ-Bank)}$$

9.24. Add the following hexadecimal numbers:

    (a) 93 + DE   (b) ABCD + EF12   **(UPTU 2001-02 (even))**

9.25. Simplify the following using NAND gates only without minimisation.

    (a) $A\overline{B}C + \overline{A}BC + ABC$ (b) $(1 + B)\,(ABC)$ **(UPTU 2001-02 (even))**

9.26. Add and multiply the following numbers without converting to decimal.

    (a) $(231)_4$ and $(13)_4$ (b) $(14.5)_6$ and $(43.0)_6$ **(UPTU 2002-03 (odd))**

9.27. Minimise the four variable logic function using k-map.

$$f\,(A, B, C, D) = \Sigma\, m\,(0, 1, 2, 3, 5, 7, 8, 9, 11, 14).$$

    (UPTU 2002-03 (odd))

9.28. Simplify the following Boolean function in POS form.

$$F\,(A, B, C, D) = \Sigma\,(0, 1, 2, 5, 8, 9, 10) \qquad \textbf{(UPTU 2002-03 (odd))}$$

9.29. Obtain a reduced expression for the following multiple output system, given as,

$$F_1 (A, B, C, D) = \Sigma m (0, 1, 2, 5, 7, 8, 9, 10, 13, 15)$$
$$F_2 (A, B, C, D) = \Sigma m (0, 1, 8, 10, 11, 14, 15)$$

and realise the minimised function using NOR gates only.

(UPTU 2002-03 (even))

9.30. Convert the following logic expression into canonical form.

(a) $AB + \overline{A}BC + BC\overline{D}$ (b) $A + \overline{A}B$

9.31. Convert the following binary numbers to octal.

(a) 11001   (b) 1001 1100

9.32. Determine the binary numbers represented by the following decimal numbers.

(a) 285   (b) 265.325

9.33. Convert binary number 110011 into decimal number.

9.34. Convert following number (42511) into hexadecimal.

9.35. Perform the following subtractions:

(a) 1101 − 1010   (b) 10011011 − 1100011

9.36. Perform the following additions:

(a) 1101 + 1001   (b) 110001 + 101110

9.37. Find 1's and 2's complement of the following binary number.

(a) 1011   (b) 11100

9.38. Simplify the following Boolean functions:

(a) $(\overline{A} + \overline{B} + C) AB\overline{C}$ (b) $A \cdot B (B + C + A) + \overline{A}B$

9.39. Prove the following:

(a) $A + AB = A$ (b) $A + \overline{A}B = A + B$

9.40. Simplify the following function in (a) POS and (b) SOP form.

(i) $F (A, B, C, D) = \Sigma m(0, 2, 4, 5, 6, 7, 8, 10, 13, 15)$

(ii) $F (w, x, y, z) = \Pi m(0, 1, 2, 3, 5, 6, 11, 13, 14, 15)$

Introduction to Electronic Instruments

10

10.1 INTRODUCTION

Measurement of any quantity plays a vital role in science and engineering, medicine and in our day-to-day life. The measurement in very important in advancement of science because whenever we are designing some new system, discovering the new phenomenon, new relationship, the laws of nature etc., then we need to compare this unknown quantity with a predefined standard. In science and engineering, the new systems and equipments are designed. The measurement is very important, whether they are giving required output or need to redesign. In other words, we can say that for the validation of the designed system, measurement is very important. After the design of systems or equipments, the operation, control and the maintenance of such equipments and the process is also one of the important functions of science and engineering. All these functions are based on proper measurement.

In other words, we can state that measurement is a process by means of which we monitor a operation, using an instrument to express the parameter, quantity or a variable in terms of meaningful numbers. Such measurement helps us in further modifications, if required.

An electronic instrument is the instrument which is based on electronic principles for its measurement. The measurement of any electronic or electrical quantity or variables is termed as an electronic measurement.

10.2 FUNCTIONAL BLOCK DIAGRAM OF AN INSTRUMENT

Any instrument or a measuring system can be represented in terms of their block diagram. The study of any instrument in the form of functional block diagram, makes it easy. The block diagram indicates the necessary elements and their functions in a general measuring system. Functional block diagram of an instrument is shown in Fig. 10.1. It contains three major blocks as primary sensing element, data conditioning element and data presentation element. The function of each element is as follows:

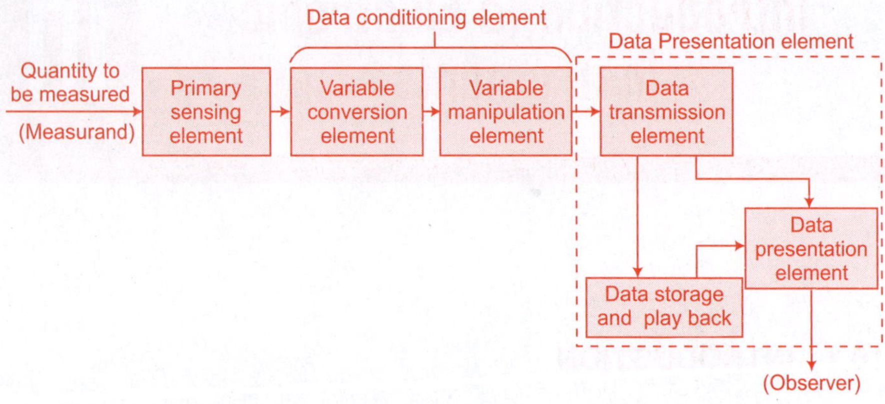

Fig. 10.1 *Functional block diagram of an instrument*

10.2.1 Primary Sensing Element

Primary sensing element is a transducer which converts the measurand into a corresponding electrical signal. So, we can say that the first detection of the quantity to be measured is done by the primary sensing element. For example, in ammeter, coil is a primary sensing element which carries the current to be measured.

10.2.2 Variable Conversion Element

The output obtained from primary sensing element is in the electrical form such as voltage, current, frequency or any other electrical parameter. Such an output may not be suitable for the actual measurement system. So, the variable conversion element converts the output into the acceptable form of measurement system. For example, if the measurement system is digital, then the analog to digital converter is used as a variable conversion element which converts an analog output obtained from primary sensing element, into the digital form.

10.2.3 Variable Manipulation Element

Variable manipulation element is used when the level of output, obtained from previous stage (variable conversion element) is not enough to drive the next stage. For example, an amplifier is used as variable manipulation element which amplifies the magnitude of its input. It is not secondary that the variable manipulation element is used after the variable conversion element. It may be used before the variable conversion element, if required.

10.2.4 Data Transmission Element

When the Primary sensing element and data conditioning element are physically separated from data presentation element, then it is necessary to transmit the data from one stage to other. This is done by data transmission element.

10.2.5　Data Presentation Element

The data presentation element is used for monitoring the data which is further used for controlling the analysis purposes. If the data is to be monitored, then visual display devices are used as data presentation element. If the signal is to be recorded for analysis purpose, then recorders, cameras, magnetic tapes are used as data presentation element, the micro controllers, microprocessor and computer may be used as data presentation element.

10.3　DIGITAL VOLTMETER

Digital voltmeter is an electronic instrument which is used to measure the voltage. In other words, we can say that it converts the analog signal into digital signal and displays the voltages to be measured as discrete numerical inplace of pointer deflection, on the digital displays. It is also called as DVM.

Basic block diagram of digital voltmeter is shown in Fig. 10.2. In the Fig., we can see that the first block is Analog to Digital Converter (ADC) which converts analog signal into digital form. Every ADC requires a reference signal. This reference signal is generated internally by the reference generator. The circuit of reference generator depends on the type of ADC techniques used.

After the ADC, the signal processing block is used i.e., the output of ADC is feed to the input at signal processing stage, in which the signal is processed or decoded. Such a decoding is necessary to drive the seven segments display. The data from decoder is then transmitted to the display.

The data transmission element is used to transmit the data from decoding stage to the display unit. The data transmission element may be batches, counters etc., according to the requirement.

Finally, a seven segment display is used to display the result of the measurement.

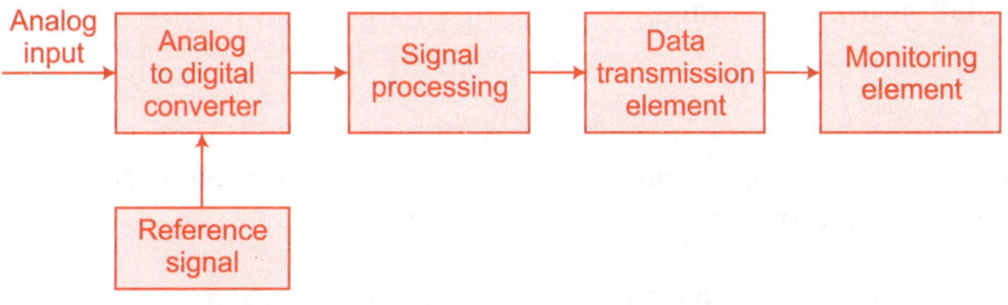

Fig. 10.2　*Basic block diagram of digital voltmeter*

10.3.1　Resolution and Sensitivity

The resolution of digital voltmeter is given by,

$$R = \frac{1}{10^n} \tag{10.1}$$

where, R = Resolution

n = Number of full digits.

Thus for 4-digit display,

$$n = 4$$

$$\therefore \quad R = \frac{1}{10^4} = 10^{-4} = 0.0001$$

$$\boxed{R = 0.01\%.}$$

The smallest change in the input which a digital meter should be able to detect, is called Sensitivity. It is defined by the full scale value of the lowest range multiplied by the resolution of the meter.

Hence, $\boxed{S = (f_s)_{\min} \times R.}$ (10.2)

where, S = Sensitivity

$(f_s)_{\min}$ = full scale value on minimum range

R = Resolution.

Example 10.3.1 What is the resolution of a $4 - \frac{1}{2}$ digit display on 1 V and 50 V ranges?

Solution Given that,

$$n = 4 \text{ (number of full digits)}$$

$$\therefore \quad \text{Resolution,} \qquad R = \frac{1}{10^4} = 0.0001.$$

Thus, meter cannot distinguish between the values that differs from each other by less than 0.0001 of full scale.

Resolution for 1 V range,

$$= 1 \times 0.0001$$

$$= 0.0001 \text{ V}$$

Resolution for 50 V range,

$$= 50 \times 0.0001$$

$$= 0.005 \text{ volt.}$$

Thus, on 50 V range, the meter cannot distinguish between the readings that differ by less than 0.005 V.

Example 10.3.2 A voltmeter uses $3 - \frac{1}{2}$ digit display, then,

  (i) Find its resolution.

  (ii) How would 12.53 V be displayed on a 10 V range?

  (iii) How would 0.443 be displayed on 1 V and 10 V ranges?

Solution Given that,

$$n = 3 \text{ (number of full digits)}$$

(i) $R = \dfrac{1}{10^3} = 0.001$

$\boxed{R = 0.001.}$ **(Ans.)**

(ii) There are 4-digit places in $3\,\dfrac{1}{2}$ digits. Hence, 12.53 would be displayed as 12.53.

(iii) (a) Resolution on 1 V range is $1\,V \times 0.001 = 0.001\,V$. Hence, any reading up to 3rd decimal can be displayed.

Hence, 0.443 will be displayed as 0.443.

(b) But resolution on 10 V range is $10\,V \times 0.001 = 0.01\,V$.

Hence, decimal up to 2nd place can be displayed. Therefore, on 10 V range, the reading will be displayed as 0.44 rather than 0.443.

10.3.2   Classification of Digital Voltmeter

The digital voltmeters are classified mainly based on the technique used for the ADC. Based on same the classification of DVM is shown in Fig. 10.3.

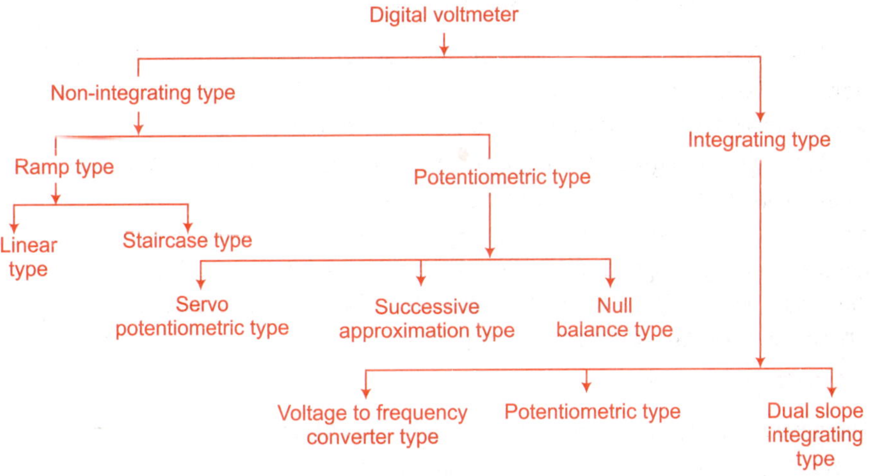

Fig. 10.3   *Classification of DVM*

10.3.3   Some Important Digital Voltmeter Techniques

We have seen the classification of digital voltmeter in the previous section. In this section, we will study the working of block diagram of some important digital voltmeters.

(i) Successive approximation type digital voltmeter

The block diagram of successive approximation type digital voltmeter is shown in Fig. 10.4. In successive approximation type DVM, the comparator circuit compares the output of digital to analog converts with unknown voltage V_{in}.

According to the comparison, the comparator provides the output as logic high or low. The output of comparator is given to the logic control and sequencer.

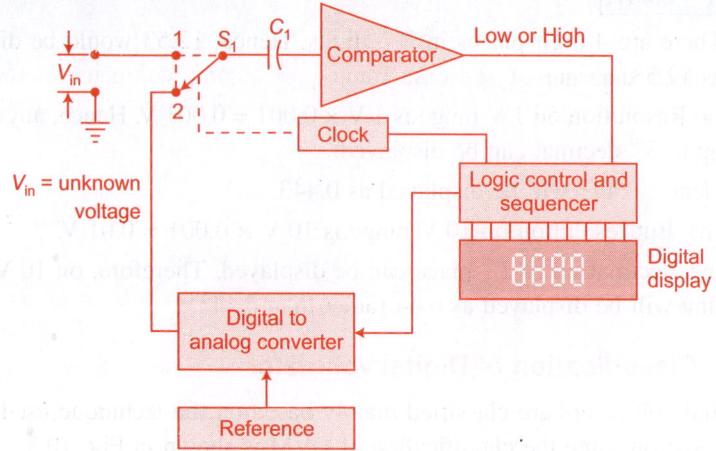

Fig. 10.4 *Successive approximation type digital voltmeter*

The digital to analog converter successively generates the set pattern of signals. The generation in of pattern continues till the output of the digital to analog converter becomes equal to the unknown voltage V_{in}.

The logic control and sequencer unit generates the sequence of code which gives input to the digital to analog converter. The switch S_1 moves between position 1 and 2. When the switch S_1 is at position 1, then it will receive the unknown voltage V_{in} and when the switch S_1 is at position 2, then it will receive the output from digital to analog converter. Clock is also divided by the logic control to move the switch between position 1 and 2 as per the requirement.

Advantages

The advantages of successive approximation DVM are:

    (a) The speed of successive approximation type DVM is very high (approximately 100 readings/second).

    (b) The resolution up to 5-digit is possible.

    (c) The accuracy is high.

Disadvantages

The disadvantages of successive approximation DVM are:

    (a) The circutary is complex.

    (b) The digital to analog converter is also required.

    (c) The decision made by comparator may be incorrect. This noise can cause error.

(ii) *Ramp type digital voltmeter*

Ramp type digital voltmeter uses a linear ramp technique or staircase ramp technique. The staircase ramp technique is simpler than the linear ramp technique. Therefore, we will discuss the staircase ramp technique.

The block diagram of staircase ramp type digital voltmeter is shown in Fig. 10.5. In this type of voltmeter, the staircase ramp signal is used. Staircase ramp signal is generated by the digital to analog converter. The input voltage is properly attenuated by the attenuator and applied to Null detector. The second input of Null detector is staircase signal which is continuously compared with the input signal.

The output of Null detector is given input to the logic control circuit. Logic control circuit gives the reset signal to the counter which is used to reset the counter as well as digital to analog converter. It also provides the starting pulse to the gate.

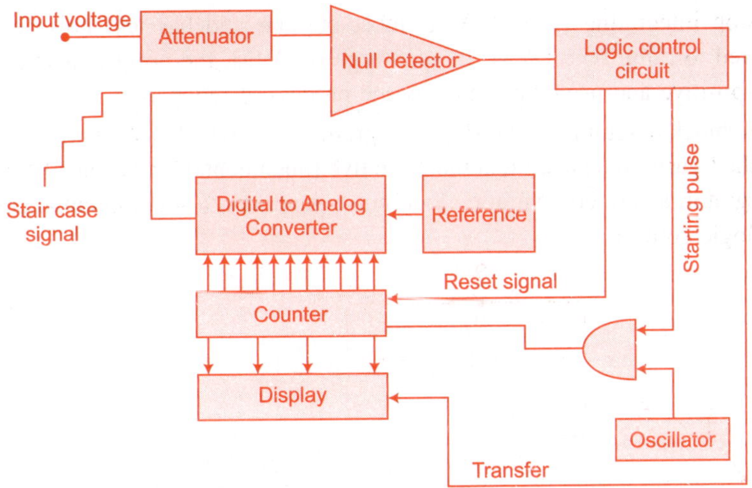

Fig. 10.5 *Staircase ramp type digital voltmeter*

At the starting of measurement, the logic control circuit sends a starting pulse to the gate. Another pulse to the gate is given by a local oscillator and the output of gate is given to the counter which counts the pulses generated by the local oscillator. The output of the counter is given to digital to analog converter which generates the staircase ramp signal. At every count, there is an incremental change in the ramp generated. Due to this, the stair case ramp is generated at the output of the digital to analog converter. The increase in ramp continues till it achieves the voltage equal to input voltage.

When two inputs of Null detector are equal, they generate a signal which initiates the logic control circuit. In other words, we can say that the logic control circuit sends a stop pulse, which closes the gate and the counter stops counting.

At the same time, the logic control circuit generates a transfer signal due to which the counter information is transferred to the display. The display shows the digital result of the count.

Advantages

(a) Accuracy is greater than linear ramp type.

(b) The design of circuitry is simple and more economical.

(c) The input impedance of DAC is high, when the compensation is reached.

Disadvantages

(a) The accuracy of meter is dependent upon the accuracy of digital to analog converter and its reference.

(b) The speed is low (approximatly 10 readings/second).

(iii) Dual slop integrating type digital voltmeter

Dual slop integrating type DVM is most popular and has less noise error as compared to ramp type DVM. In such type of DVM, the noise is averaged out by the positive and negative ramps by the process of integration.

The block diagram of dual slope integrating type DVM is shown in Fig. 10.6. From the figure, we can see that there are five fundamental blocks, an OP-Amp as an integrator, a zero comparator, clock pulse generator, a set of decimal counters and a logic control.

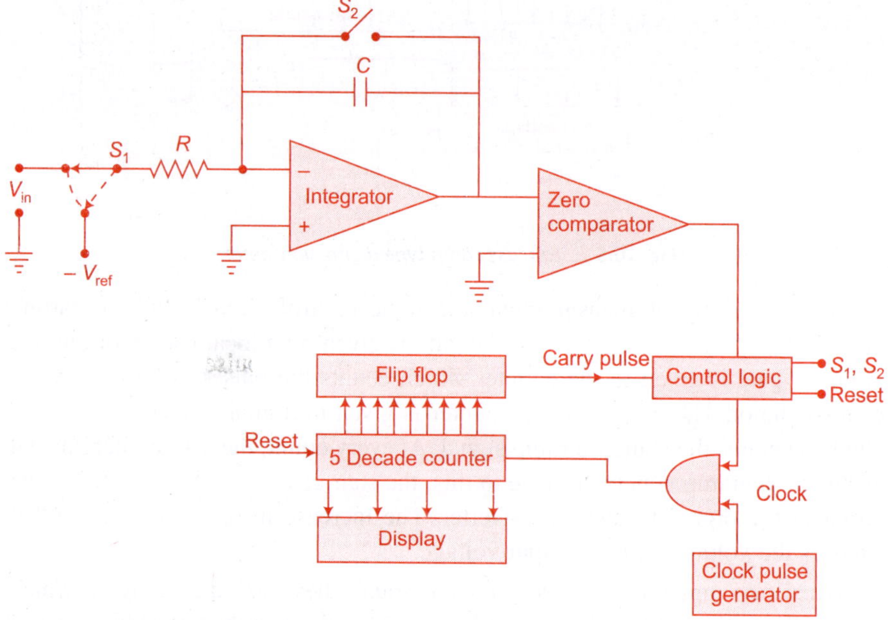

Fig. 10.6 *Dual slope integrating type DVM*

When the measurement starts, the counter is reset to zero. Due to this, the output of flip flop is also zero. The output of flip-flop is given to the control logic. This control sends a signal to close switch S_1 to position 1 and integration of input voltage V_{in} starts. Integration continues till the time period t_1.

As the output of integrator changes from its zero value, due to the output of integrator the state of output of zero comparator changes. The output of zero comparator given to the control logic which opens the gate and the counting of the clock pulse starts.

When the counter reaches to 9999, it generates a carry pulse and all digits become zero. The output of flip-flop is activated to logic 1 and it activates the control logic. The control logic sends a signal which changes the position of switch S_1 from 1 to 2. Thus, $-V_{ref}$ gets connected to the OP-Amp.

Capacitor C will discharge due to the opposite polarity of V_{ref}. The integrator output will have constant negative slope as shown in Fig. 10.7. The output of integrator decreases linearly and after time t_2, it attains zero value, when the capacitor C gets fully discharged.

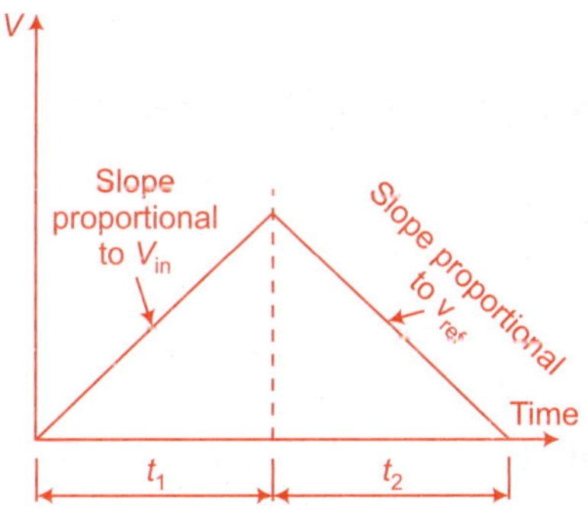

Fig. 10.7   *Dual slope method integrating output*

At the same time, the zero comparator output changes its state and sends a signal to the control logic. It provides the signal to the gate and it will be closed. The gate remains open for the period of $t_1 + t_2$. The counting stops at this instant. The counts are then transferred to the display.

Thus, the pulses counted by the counter have a direct relation with the input voltage as follows:

$$V_{in} = V_{ref} \frac{t_2}{t_1} \tag{10.3}$$

Let us assume that time period of clock oscillator be T and counter has counted the counts n_1 and n_2 during the period t_1 and t_2 respectively.

$$\therefore \qquad V_{in} = V_{ref} \frac{n_2\,T}{n_1\,T}$$

$$V_{in} = V_{ref} \frac{n_2}{n_1} \tag{10.4}$$

From Eqn. (10.4), we can state that the voltage measurement is not dependent upon the clock frequency, but dependent upon the counts measured by the counter.

Advantages

(a) During the process of integration, the noise is averaged out.

(b) In the input sample and hold circuit is not necessary because the integrator responds to the average value.

(c) The accuracy is high.

(d) *RC* time constant does not affect the measurement.

(e) The capacitor is connected by an electronic switch. So, it avoids the effects of offset voltage.

Disadvantages

Dual slope integrator type DVM has only one disadvantage that its speed is slow.

10.3.4 Specifications of Digital Voltmeter

The digital voltmeter specifications are as follows:

Display: $3 - \frac{1}{2}$ digits, LCD

Unit measured: mV, V, mA, Ω, kΩ, MΩ LOW BAT, diode test

Maximum indication: 1999 or $-$ 1999

Over range indication: 1 or ($-$ 1) displayed at MSB

Functions: DC Volts, AC Volts, DC Amps, AC Amps, Ohms, Continuity test, Diode test

Sampling rate: 2 samples/second

Ranging: Automatic or manual

Polarity: Auto negative polarity indication

Zero adjustment: Automatic

Accuracy: $\pm$ 0.5% to 0.7%

Temperature: 0 to 40°C

Input impedance: 100 MΩ to 1000 MΩ

Power: 2 AA size 1.5 V battery or other type.

10.4 DIGITAL MULTIMETER

Multimeter is an electronic instrument which is used to measure electrical quantities such as voltages, currents and resistances. It can be used for measuring

AC as well as DC. When the output of multimeter is digital, then it is called as digital multimeter where as in analog multimeter the value (output) of a quantity measured, is displayed with the help of an electromechanical movement of a pointer over a calibrated scale. The output of digital multimeter is shown on the digital display such as seven segment display, alphanumeric display or Liquid Crystal Displays (LCDs).

The basic circuit of a digital multimeter is always a DC voltmeter as shown in Fig. 10.8. The block diagram of digital multimeter is shown in Fig. 10.8. As we have discussed earlier that a digital multimeter is a very important laboratory instrument. It is used to measure AC/DC voltage, AC/DC current and resistance. Since it gives digital display, it is very accurate. The accuracy is sometimes called as resolution of digital multimeter. Resolution is also related with sensitivity of multimeter. Greater the sensitivity, higher is the resolution.

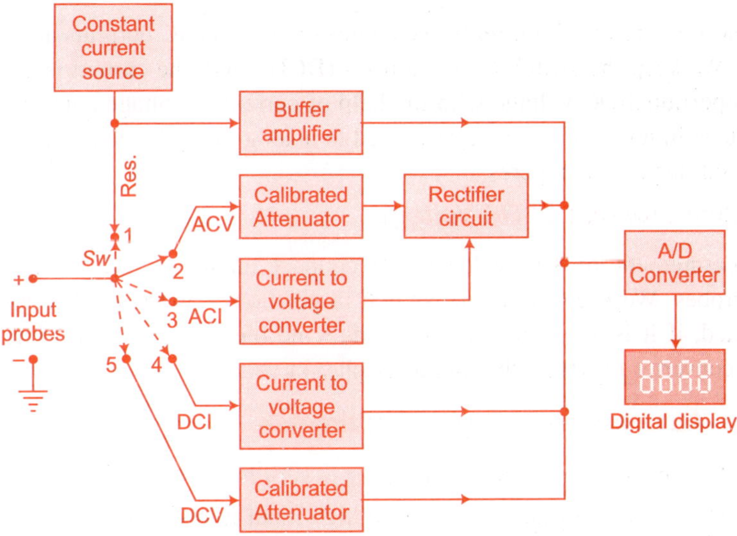

Fig. 10.8   *The block diagram of digital multimeter*

(i) *Measurement of resistance*

For measurement of resistance, connect unknown resistance across its input probes and select a rotary switch at position-1 (Res.) Due to the constant current source, some current flows through the resistor. The voltage across the resistor will be produced due to Ohm's Law. This voltage is directly proportional to the resistance, and this voltage is buffered by the buffer amplifier and then feed to Analog to Digital convecter (A/D convecter). Finally, it will display on digital display in Ohms.

(ii) *Measurement of AC Voltage*

For measurement of AC voltage, we connect unknown voltage across its input probes. And we select rotary switch at position-2 (ACV). If the voltage is above the selected range, then the voltage is attenuated and feed to the rectifier to convert it into proportional DC voltage. After that the output of rectifier is feed to analog to digital converter to obtain the digital display in volts.

(iii) Measurement of AC current

For measurement of AC current, we connect unknown AC current across input probes. This circuit (Fig. 10.8) measures the current indirectly. Since the circuit can measure only voltage, therfore, the current is converted into proportional voltage first and then measured. We keep the rotary switch at position - 3 (ACI). The current is converted proportionally into voltage with the help of current to voltage converter and then feed to rectifier. Now, the voltage interms of AC currents is feed to analog to digital converter to obtain digital display in amperes.

(iv) Measurement of DC Current

For measurement of DC currents, we connect unknown DC currents across input probes. We keep the switch at position - 4 (DCI). First, the current is connected into proportional DC voltage with the help of current to voltage converter. Now, the voltage in terms of DC current is feed to analog to digital converter to obtain the digital display is amperes.

(v) Measurement of DC Voltage

For measurement of DC voltage, we connect unknown DC voltage across input probes. We keep the rotary switch in position – 5 (DCV). The voltage is attenuated, if it is above the selected range and then directly feed to analog to digital converter to obtain the digital display in volts.

10.4.1 Advantages of Digital Multimeter

  (i) The accuracy of digital multimeter is very high.

  (ii) There is no loading effect due to high input impedance.

  (iii) The output is in electrical form, so it can be used for interfacing with external equipment.

  (iv) Cost is low, due to the improvement in integrated technology.

  (v) Size is small.

  (vi) At larger distance, the reading is unambiguous.

10.4.2 Specifications of digital multimeter

Specifications of digital multimeter is as follows:

(i) Resolution: Resolution is the measure of the smallest increment in the value which is clearly distinguished. It is denoted by R and is defined as,

$$R = \frac{1}{10^n} \tag{10.5}$$

where,

$$n = \text{Number of full digits}$$

For example, 4-digit display, $n = 4$, so the resolution R is,

$$R = \frac{1}{10^4} = 0.0001$$

$$\boxed{R = 0.01\%.} \tag{10.6}$$

From Eqn. (10.6), we can say that a display with 4-digit cannot distinguish between the values that differ from each other by less than 0.0001 of full scale.

(ii) Sensitivity: Sensitivity is the ability of the multimeter to detect the smallest change in the input signal of the meter. It depends upon the resolution and lowest measurement range of the meter. It is denoted by S.

$$S = (fs)_{\text{min}} \times R \tag{10.7}$$

where,

$$(fs)_{\text{min}} = \text{Full scale value of minimum range}$$
$$R = \text{Resolution.}$$

(iii) Accuracy: Accuracy is the degree of measurement in which the reading of meter approaches to the actual value of the quantity to be measured. In other words, we can say that there is very little difference between measure value and actual value if accuracy is better.

(iv) Display Digits: The resolution of the meter depends upon the number of digits used in the display. $4 - \frac{1}{2}$ digit display is shown in Fig. 10.9. In the Fig., one half digit is shown which may be 0 or 1 and 4 full digits those values varies from 0 to 9999 mv with the smallest increment of 1 mv.

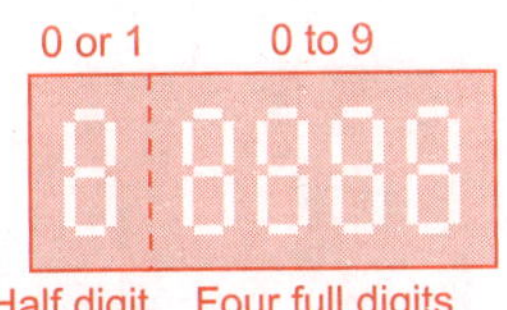

Fig. 10.9 $4 - \frac{1}{2}$ *digit display*

(v) Uncertainty: The uncertainty of meter cannot be separated from accuracy because these two are defined simultaneously as follows:

$$\text{Uncertainty} = 100\% - \% \text{ Accuracy} \tag{10.8}$$

For example, a meter has 97% accuracy as the measurement, then it has

$$\text{Uncertainty} = 100\% - 97\%$$
$$= 3\%$$

i.e. 3% of uncertainty in the measurement is nothing but accuracy of 97%.

(vi) Repeatability Repeatability is defined as the degree of closeness with which the given value may be measured repeatedly.

10.5 CATHODE RAY OSCILLOSCOPE (CRO)

The Cathode Ray Oscilloscope is an extremely useful and versatile laboratory instrument used for studying wave shapes of AC currents and voltages for the measurement of voltage, current, power and frequency, in all the quantities that involves amplitude and waveform, while studying various electronic, electrical network and systems in which the signals are the functions of time, often occurs. Such signal may be periodic or non-periodic in nature. The device which allows the amplitude of such signals, to be displayed primarily as a function of time is called Cathode Ray Oscilloscope, commonly by known as CRO. It is a test instrument which allows you to look at the shape of electrical signals by displaying a graph of voltage versus time on its screen. We observe how the voltage varies with time on its screen.

The block diagram of general purpose CRO is shown in Fig. 10.10. It is widely used for trouble shooting radio and TV receivers as well as laboratory work involving research and design. The block diagram of CRO is shown in Fig. 10.10 in which the Cathode Ray Tube (CRT) is the heart of the oscilloscope.

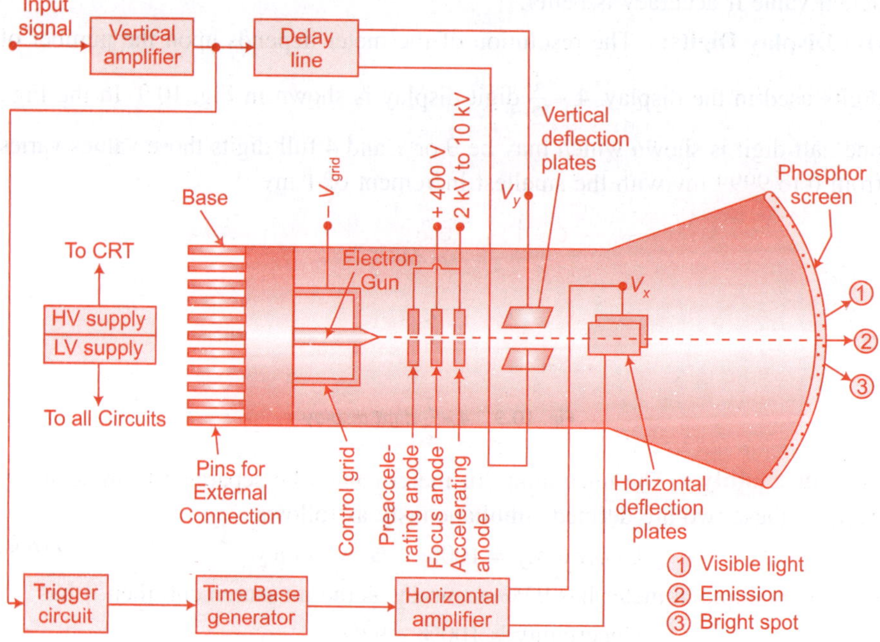

Fig. 10.10 *Block diagram of a general purpose CRO*

(i) Electron gun or cathode: It generates the electron beam.

(ii) Control Grid: Control grid is used to control the electron beam i.e., controls the number of electrons in the beam. Grid is given negative potential with respect to cathode.

(iii) Anodes: There are three anodes, two are accelerating anodes and the third one is focusing anode which accelerates the beam to a high velocity.

(iv) Deflection plates: There are two pairs of deflection plates, in which one pair is vertical deflection plate and the other pair is horizontal deflection plate. One of the plates in each set is connected to ground.

As shown in Fig. 10.10, the electron beam passes through these plates. A positive voltage applied to y input terminal (V_y) causes the beam to deflect vertically upwards due to the attraction forces, while a negative voltage applied to the y-input terminal will cause the electron beam to deflect vertically downward, due to the repulsion forces.

Similarly, a positive voltage applied to x-input terminal (V_x) will cause the electron beam to deflect horizontally towards the right; while a negative voltage applied to x-input terminal will cause the electron beam to deflect horizontally towards the left of the screen. The amount of vertical or horizontal deflection is directly proportional to the correspondingly applied voltage. When the voltages are applied simultaneously to vertical and horizontal deflecting plates, the electron beam is deflected due to the resultant of these two voltages. Deflected beam creates the image, and contains a phosphor screen where the electron beam eventually becomes visible.

(v) Power Supply: Low voltage (LV) supply is required for the heater of the electron gun for generation of electron beam and high voltage (HV) of the order of few thousands volts is required for cathode ray tube to accelerate the beam. Normal voltage supply, say a few hundred volts, is required for other control circuits of the oscilloscope.

(vi) Phosphor screen: Electron beam strikes the screen and creates a visible spot. This spot is deflected on the screen in horizontal direction (x-axis) with constant time dependent rate.

(vii) Time base circuit: The deflection in horizontal direction with constant time dependent rate is accomplished by a time base circuit.

(viii) Vertical Amplifier: The signal to be viewed is supplied to the vertical deflection plates through the vertical amplifier, which raises the potential of the input signal to a level that provides usable deflection of the electron beam.

(ix) Trigging circuit: Trigging circuit is provided for synchronizing two types of deflections so that horizontal deflection starts at the same point of the input vertical signal each time it sweeps.

10.5.1   Measurement of Voltage Using CRO

The voltage of an input signal can be measured using CRO. Before any measurement from CRO, there is a need to set up an oscilloscope. It requires same case to set up. If we do not set the controls of CRO, then it may give wrong information which we require. So, the following instructions are required to be adapted for the instrument.

1. Switch on the CRO and wait for one or two minutes to warm up.
2. Do not connect input signal to CRO at this stage.
3. Set the AC/GND/DC switch to DC.
4. Set the SWP/*x-y* switch to SWP (sweep).
5. Set trigger level to Auto.
6. Set trigger source to INT (internal).
7. Set the *y*-amplifier to moderate value (approximately 5 V/division).
8. Set the Time base to a moderate speed (approximately 10 ms/division).
9. Turn the time base variable control to 1 or cal.
10. Adjust *y*-shift (up/down) and *x*-shift (left/right) to set a trace at the middle of the screen as shown in Fig. 10.11.
11. Adjust intensity and focus to give a bright and sharp trace.

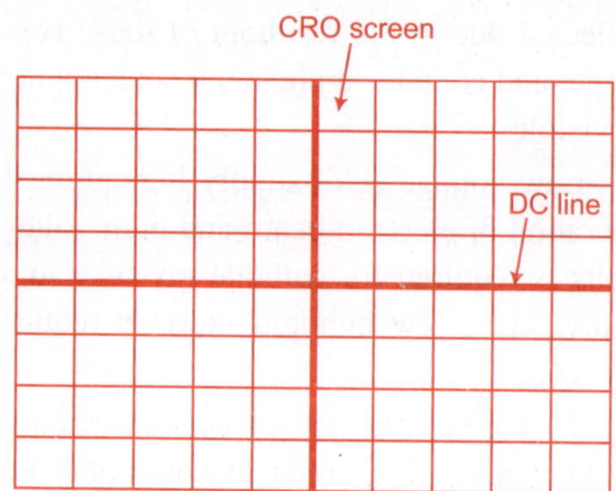

Fig. 10.11   *DC line at centre*

The trace on a CRO is a graph of voltage against time as show in Fig. 10.12. The shape of this graph is dependent upon the nature of the input signal.

The diagram shown in Fig. 10.12 is a sine wave but the following properties apply to any signal with a constant shape.

(i) Amplitude (A):   It is the maximum voltage reached by the signal. It is measured in volts, V. It is also called as peak voltage.

(ii) Peak to-peak voltage (V_{pp}):   Peak-to-peak voltage is twice the peak voltage or amplitude. When we read it on CRO trace, it is easy to measure peak-to-peak voltage.

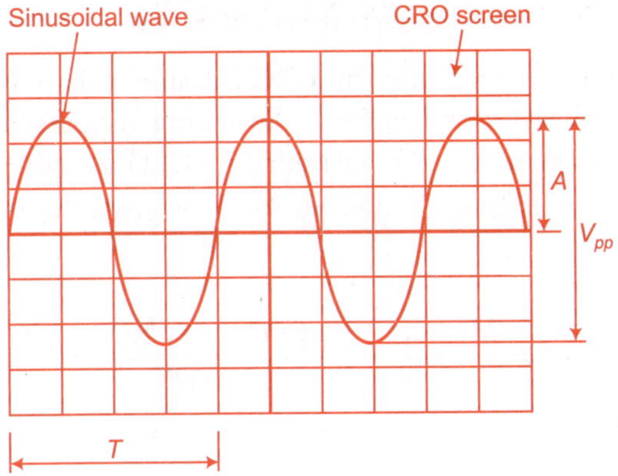

Fig. 10.12   *Measurement of voltage*

(iii) Time Period (T):   It is taken twice for a single to complete the cycle. It is measured in second (s) or millisecond (ms).

(iv) Frequency (f):   Frequency is nothing but the number of cycles per second. It is measured in Hz (Hertz). It can also be determined from time period by the following relation:

$$\text{frequency } (f) = \frac{1}{\text{Time period } (T)}. \tag{10.9}$$

For the measurement of voltage or amplitude, use the following steps.

1. Note down the volts/division by y-amplifier let us assume 2 V/division. Usually peak-to-peak voltage is measured because it can be read correctly even if the position of 0 Volt is not known.

2. Note down peak-to-peak value interms of the number of division on screen. From Fig. 10.12 it is observed as 4.4.

3. Use the following relation to obtain peak-to-peak value in volts.

$$V_{pp} = (\text{number of division on } y\text{-axis}) \times (\text{volts/division}) \tag{10.10}$$

Hence,

$$V_{pp} = (4.4 \text{ division}) \times (2 \text{ V/division})$$

$$\boxed{V_{pp} = 8.8 \text{ V.}}$$

4. The amplitude can be calculated as,

$$A_m = \frac{V_{pp}}{2} \tag{10.11}$$

5. The r.m.s. value of Sinusoidal signal can be obtained as,

$$V_{RMS} = \frac{A_m}{\sqrt{2}} = \frac{V_{pp}}{2\sqrt{2}} \text{ (only for sine wave).} \tag{10.12}$$

10.5.2 Measurement of Current Using CRO

Basically, the voltage is measured by CRO because it is a voltage indicating device. Hence, to measure the current, the current is passed through known resistor (R) and voltage across R is measured by CRO as shown in Fig. 10.13.

This measured voltage is divided by R and it gives the value of unknown current.

Hence,

$$I = \frac{\text{Measure Voltage } (V) \text{ on CRO}}{\text{Known Resistance } (R)} \tag{10.13}$$

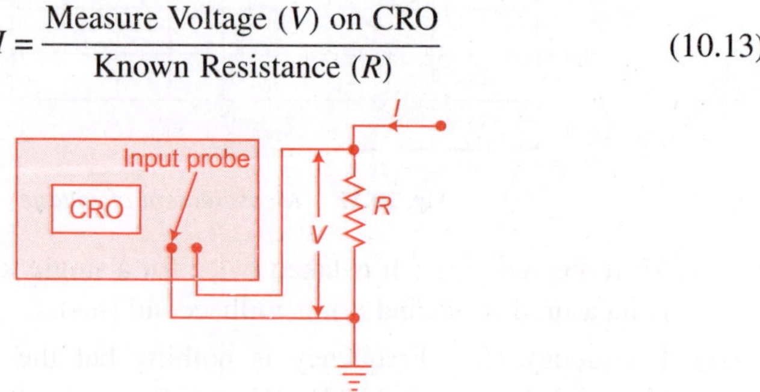

Fig. 10.13   *Measurement of current*

10.5.3 Measurement of Time Period and Frequency Using CRO

1. As shown in Fig. 10.12, the time is shown on the horizontal axis (x-axis) and scale is determined by the time box control. Let it 5ms/division.
2. Note down the value of time in terms of number of division occupied by one cycle on screen. From Fig. 10.12, it is observed 4.
3. Use the following relation to obtain the time periods,

$$\text{Time period } (T) = \left(\begin{array}{c} \text{number of division} \\ \text{occupied by 1 cycle} \end{array} \right) \times (\text{Time/division}) \tag{10.14}$$

Hence, $T = (4 \text{ division}) \times (5\text{ms/division})$

$$\boxed{T = 20 \text{ ms.}}$$

The frequency is the reciprocal of time period. Hence, from Eqn. (10.9), we know that,

$$f = \frac{1}{T}$$

so,

$$f = \frac{1}{20\text{ms}} = 50 \text{ Hz}$$

$$\boxed{f = 50 \text{ Hz.}}$$

10.5.4 Measurement of Phase Difference

The phase difference can be measured between two waveform whose frequency and amplitude is same. The phase difference between the resultant voltage and the current through RC circuit can also be measured.

Before calculating the phase difference you should know the following terms used in determination of phase difference.

Phase: The phase of alternating quantity is the fraction of the time period of the alternating voltage which has elapsed since the voltage last passes through the zero position of reference.

Phase difference: Consider two alternating quantity of same frequency reaching their peak/zero value at different instant of time. This gives the phase difference between two waves as shown in Fig. 10.14.

Lissajous pattern: A lissajous pattern is produced on the screen when two sine wave voltage are applied simultaneously to both pairs of deflection plates of a CRO.

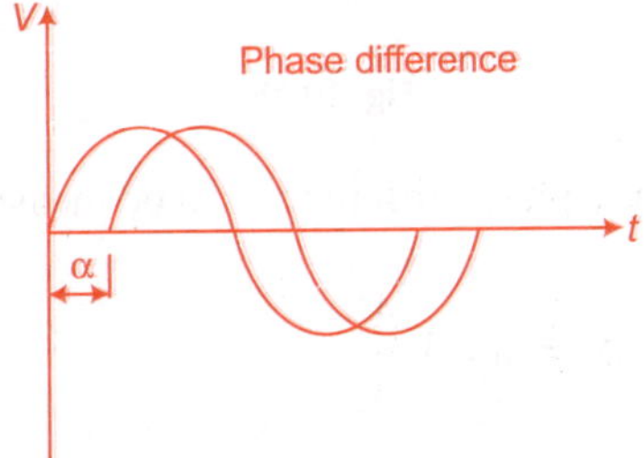

Fig. 10.14   *Phase difference between two waves*

(i) *Measurement of phase difference between the resultant voltage and the current through RC Circuit*

When an AC current is passed through RC circuit, the current direction is same in both the elements, R and C. But the voltage direction is different. The voltage across the resistor is in the direction of current and the voltage across the capacitor, lags behind the current by 90°. Due to this, the resultant voltage also lags behind the current by some angle (ϕ) called phase difference. Since the current cannot be measured directly by CRO, the voltage across the resistor is given to CRO as explained in section 10.5.3, which represents the current directions. Hence, the phase difference is the angle between the voltage across the resistor and the resultant voltage and is represented by equation below,

$$\text{The phase difference, } \phi = \tan^{-1} \frac{1}{\text{WCR}} \tag{10.15}$$

where,

$\quad\quad C$ = Capacitance of the capacitor (µF)

$\quad\quad R$ = Resistance of the resistor (Ω)

$\quad\quad w$ = $2\pi f$ Rad/sec

$\quad\quad f$ = Applied frequency (Hz)

The measurement of phase difference using CRO is shown in Fig. 10.15 (a). The pattern obtained is shown in Fig. 10.15 (b).

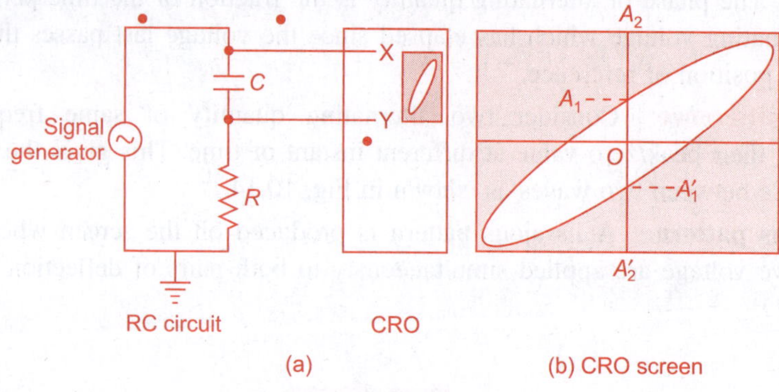

Fig. 10.15

From Fig. 10.15 (b), the phase difference can be measured by CRO practically as follows:

$$\phi = \sin{-1} \frac{A'_1}{A_2} \qquad (10.16)$$

where,

$$A_1 = \text{Vertical deflection, at } t = 0$$
$$A_2 = \text{Maximum vertical deflections}$$

From Fig. 10.14(b), we know that,

$$A_1 = A_1 \frac{A'_1}{2}$$

$$A_2 = A_2 \frac{A'_2}{2}$$

Hence,

$$\phi = \sin^{-1} \frac{A_1 A'_1}{A_2 A'_2}. \qquad (10.17)$$

(ii) Measurement of phase difference between two waves of the same frequency and amplitude

The two voltage signal of same frequency and amplitude is given to the CRO, as shown in Fig. 10.16. Consider the lissajous figure obtained on CRO with an unknown phase difference ϕ as shown in Fig. 10.17 (a).

The parameters x_1, x_2 or y_1, y_2 can be measured in the Fig. 10.17 (a). Then, the phase angle can be obtained as,

$$\phi = \sin^{-1} \frac{y_1}{y_2} = \sin^{-1} \frac{x_1}{x_2}. \qquad (10.18)$$

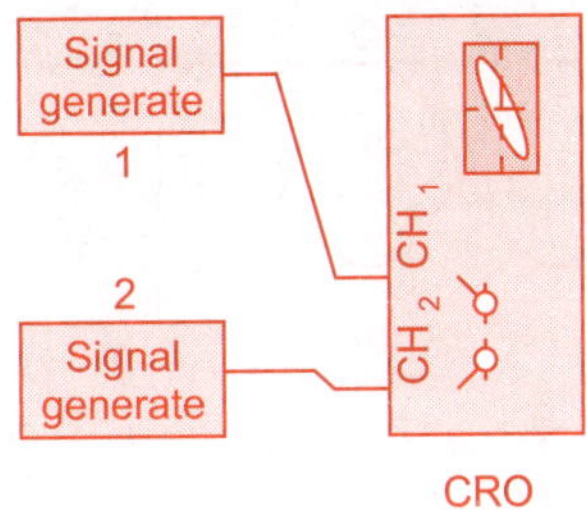

Fig. 10.16   *Measurement of phase*

If the pattern obtained is same as shown in Fig. 10.17 (b), then the phase ϕ is given by,

$$\phi = 180° - \sin^{-1}\frac{y_1}{y_2} \tag{10.19}$$

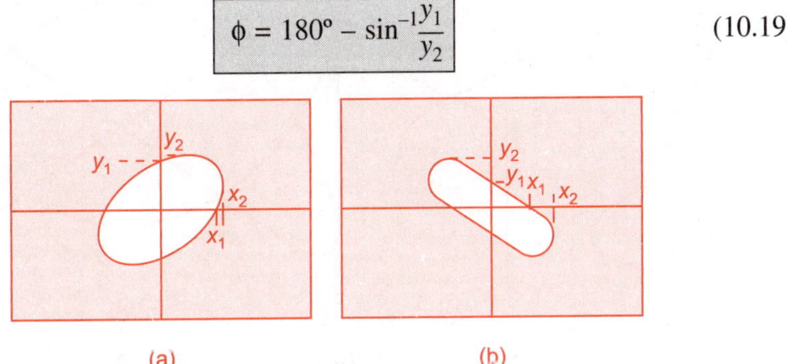

(a)   (b)

Fig. 10.17

(iii) Measurement of frequency

A known frequency f_H is applied to the horizontal input and an unknown frequency f_v is applied to the vertical input. Then a lassijous pattern with loops is obtained. The unknown frequency f_v can be measured by the relation;

$$F_v = \frac{\text{No of loops cut by horizontal line (NLM)}}{\text{No of loops cut by vertical line (NLV)}} f_H \tag{10.20}$$

The pattern depends on the ratio of the frequencies f_H and f_v as shown in Eqn. (10.20). The ratio of f_H and f_v for different tangents are shown in Fig. 10.18.

$$f_v = \frac{3}{2} f_H$$

$$\boxed{f_v = 1.5\, f_H.}$$

From Fig. 10.18 (a),

$$NL_H = 2 \text{ and } NL_V = 1$$

Hence,
$$f_v = \frac{2}{1} f_H$$

$\Rightarrow$
$$\boxed{f_v = 2 f_H}$$
From Eqn. (10.20)

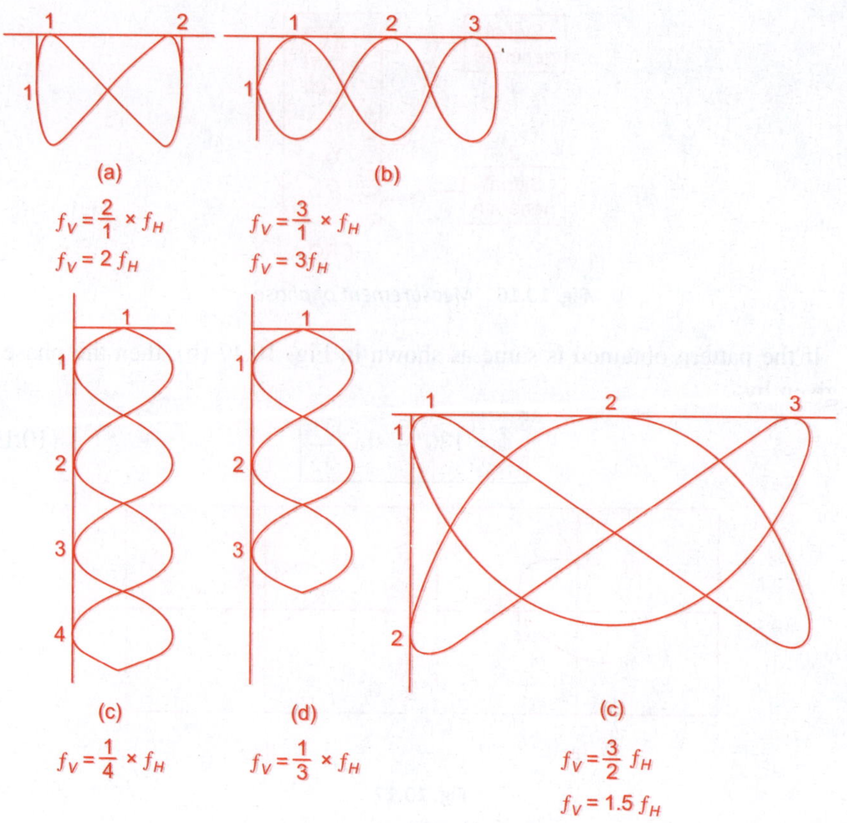

Fig. 10.18

From Fig. 10.18 (b),

$$NL_H = 3, \quad NL_V = 1$$

Hence,

$$f_v = \frac{3}{1} f_H$$

$\Rightarrow$ $\boxed{f_v = 3f_H}$ From Eqn. (10.20)

From Fig. 10.18 (c),

$$NL_H = 1, \quad NL_V = 4$$

Hence,

$$f_v = \frac{1}{4} f_H$$ From Eqn. (10.20)

From Fig. 10.18 (d),

$$NL_H = 1, \quad NL_V = 3$$

Hence, $\boxed{f_v = \frac{3}{2} f_H}$ From Eqn. (10.20)

From Fig. 10.18 (e),

$$NL_H = 3, \ NL_V = 2$$

Hence,

$$f_v = \frac{3}{2} f_H$$

$$\boxed{f_v = 1.5 \, f_H.}$$

From Eqn. (10.20)

Objective Type question

1. The meter used for measuring electrical quantities are:
   - (a) Tachometer
   - (b) Micrometer
   - (c) Measuring Instruments
   - (d) Speedometer

2. The meter used for measuring potential difference of a circuit is called
   - (a) Ammeter
   - (b) Voltmeter
   - (c) Energy meter
   - (d) Ohm meter

3. The meter used for measuring current of an electrical circuit is called
   - (a) Voltmeter
   - (b) Ammeter
   - (c) Potentiometer
   - (d) Multimeter

4. The cost of ammeter as compared to voltmeter is
   - (a) Same
   - (b) Higher
   - (c) Lower
   - (d) Very high

5. Millimeter can measure
   - (a) Current
   - (b) Voltage
   - (c) Resistance
   - (d) All of the above

6. Millimeter can be used for measuring
   - (a) AC quantities
   - (b) DC quantities
   - (c) AC × DC both
   - (d) None of the above

7. The meter of millimeter will work on
   - (a) DC supply
   - (b) AC supply
   - (c) AC × DC supply both
   - (d) H.W. rectified AC

8. An operating voltage of a particular millimeter is
   - (a) 230V, AC
   - (b) 230V, DC
   - (c) 9V, DC
   - (d) 9V, AC

9. Higher value of resistance can be measured by keeping the selection switch position on
   - (a) $R \times 1$
   - (b) $R \times 100$
   - (c) $R \times 100$
   - (d) $R \times 1000$

10. If the battery of the millimeter is weak, it will give
    - (a) Accurate reading
    - (b) More reading
    - (c) Less reading
    - (d) No reading

11. The reliability of an instrument means
    - (a) The life of instrument
    - (b) The extent in which the characteristics remain linear
    - (c) Degree to which repeatability continues to remain with specific limits
    - (d) All of these

12. Sensitivity of a voltage is expressed as
    - (a) Volt/Ohms
    - (b) Ohms/volts
    - (c) Ohms volt
    - (d) 1/Ohms volt

13. A CRO uses
    - (a) Electromagnetic focussing
    - (b) Electrostatic focussing
    - (c) Both focussing techniques
    - (d) No focussing technique

14. A CRO can display
    - (a) AC signal
    - (b) DC signal
    - (c) Both (a) × (b)
    - (d) Time invariant signal

15. Which meter is suitable for the measurement of 10mV at 50 MHz?
    - (a) MI voltmeter
    - (b) VTVM
    - (c) Electrostatic voltmeter
    - (d) CRO

16. An oscilloscope indicates
    - (a) Peak-to-Peak values of voltage
    - (b) DC value of voltage
    - (c) R.M.S. value
    - (d) Average values

17. A 10 MHz, CRO has
    - (a) 5 MHz sweep
    - (b) 10 MHz vertical oscillator
    - (c) 10 MHz horizontal oscillator
    - (d) 10 MHz supply frequency

18. What will happen if a voltmeter is connected like an ammeter in series to the load?
    - (a) The meter will burn out
    - (b) The measurement will be too high

(c) An inadmissably high current will flow

(d) There will be almost no current in the circuit

19. Which of the following statement is not a disadvantage of a digital voltmeter?

    (a) It cannot measure non-linear parameter

    (b) It is comparatively high priced

    (c) Its circuitry is more complex

    (d) It's accuracy change with the reading taken

20. A digital voltmeter has $4\frac{1}{2}$ digit display. The 1V range can read upto

    (a) 9999   (b) 9.999

    (c) 1.9999   (d) 19999

21. The two sine wave of same amplitude and phase are applied to the vertical and horizontal inputs of a oscilloscope. The pattern on the screen will be

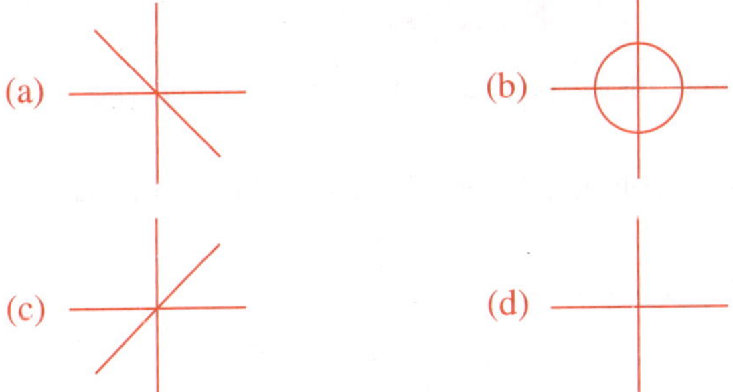

22. A Lissajous pattern is used to measure

    (a) Voltage and frequency

    (b) Frequency and phase shift

    (c) Frequency and amplitude distortion

    (d) Amplitude and flux

23. Two sinusoidal voltage of unequal frequencies are simultaneously applied to two pairs of the deflection plates in a CRT. The figure traced out on the screen is shown in Fig. OT 10.1. The ratio of the frequencies of the two voltages will be

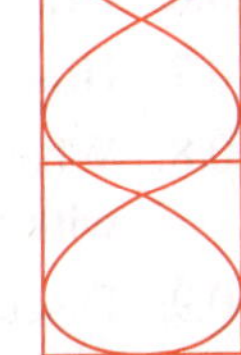

Fig. OT 10.1

    (a) 4:3   (b) 5:2

    (c) 5:4   (d) 7:4

24. The sinusoidal voltage of equal frequencies and difference of 120° are simultaneously applied to the pair of the deflection plates of a CRT. The lissajous figure traced on the screen will be

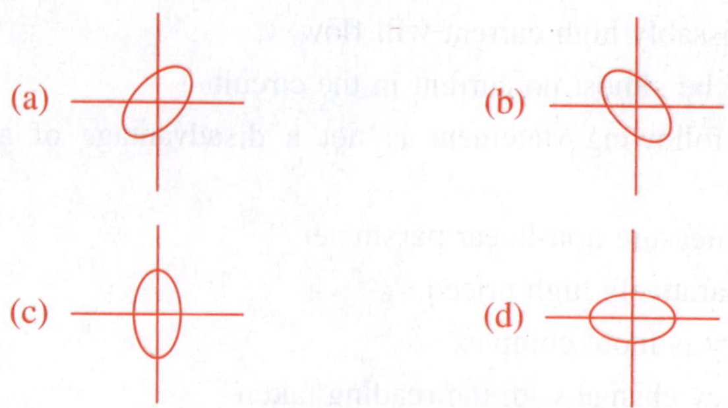

ANSWERS

1 (c)	2 (b)	3 (b)	4 (c)	5 (d)	6 (c)	7 (a)
8 (c)	9 (d)	10 (c)	11 (c)	12 (b)	13 (b)	14 (c)
15 (d)	16 (a)	17 (c)	18 (d)	19 (d)	20 (c)	21 (c)
22 (b)	23 (b)	24 (b)				

Exercise

10.1. Draw a functional block diagram of an instrument and explain each block.

10.2. What is a digital voltmeter? Draw a basic block diagram of digital voltmeter.

10.3. Explain the different terms

    (a) Resolution   (b) Sensitivity

    (c) Accuracy   (d) Uncertainty

10.4. Explain the following types digital voltmeter

    (a) Successive approximation type

    (b) Ramp type

    (c) Dual slope integrating type

10.5. Explain digital multimeter and draw its block diagram.

10.6. Draw a functional block diagram of CRO and explain each block.

10.7. How voltage, current, phase and frequency is measured using CRO?

10.8. What is a digital voltmeter? What are its advantages? Explain its working with the help of diagram.

10.9. Discuss various characteristics of digital meters.

10.10. How would you measure the phase difference with the help of CRO? Illustrate.

10.11. Illustrate the way the frequency of a voltage signal, is measured.

10.12. Explain briefly the working principal of a digital millimeter with the help of a block diagram. What are the characteristics of digital voltmeter used in a typical digital multimeter? **(UPTU 2007-08)**

10.13. State the main application of a CRO. Briefly explain each of them. Explain how will you quickly measure the frequency of waveform displayed on the CRO? **(UPTU 2007-08)**

10.14. (a) Explain the working of a digital multimeter. What are its applications?

(b) Discuss CRO in detail. How it is used for measurement of frequency? **(UPTU 2009-10)**

10.15. Explain briefly the working principle of a digital voltmeter. What are the advantages obtained by numeric read out? **(UPTU 2009-10)**

10.16. Draw the lissajous pattern you expect when the ratio of the frequency of the vertical input to that of the horizontal input is 1:2. Explain with the help of a neat diagram.Why you get this pattern? **(UPTU 2009-10)**

Index